Cell Growth

CONTROL OF CELL SIZE

COLD SPRING HARBOR MONOGRAPH SERIES

The Lactose Operon
The Bacteriophage Lambda
The Molecular Biology of Tumour Viruses
Ribosomes
RNA Phages
RNA Polymerase
The Operon
The Single-Stranded DNA Phages
Transfer RNA:
    Structure, Properties, and Recognition
    Biological Aspects
Molecular Biology of Tumor Viruses, Second Edition:
    DNA Tumor Viruses
    RNA Tumor Viruses
The Molecular Biology of the Yeast *Saccharomyces:*
    Life Cycle and Inheritance
    Metabolism and Gene Expression
Mitochondrial Genes
Lambda II
Nucleases
Gene Function in Prokaryotes
Microbial Development
The Nematode *Caenorhabditis elegans*
Oncogenes and the Molecular Origins of Cancer
Stress Proteins in Biology and Medicine
DNA Topology and Its Biological Effects
The Molecular and Cellular Biology of the Yeast *Saccharomyces:*
    Genome Dynamics, Protein Synthesis, and Energetics
    Gene Expression
    Cell Cycle and Cell Biology
Transcriptional Regulation
Reverse Transcriptase
The RNA World
Nucleases, Second Edition
The Biology of Heat Shock Proteins and Molecular Chaperones
Arabidopsis
Cellular Receptors for Animal Viruses
Telomeres
Translational Control
DNA Replication in Eukaryotic Cells
Epigenetic Mechanisms of Gene Regulation
C. elegans II
Oxidative Stress and the Molecular Biology of Antioxidant Defenses
RNA Structure and Function
The Development of Human Gene Therapy
The RNA World, Second Edition
Prion Biology and Diseases
Translational Control of Gene Expression
Stem Cell Biology
Prion Biology and Diseases, Second Edition
Cell Growth: Control of Cell Size

Cell Growth

CONTROL OF CELL SIZE

EDITED BY

Michael N. Hall
Biozentrum University of Basel, Switzerland

Martin Raff
University College, London

George Thomas
Friedrich Miescher Institute, Basel, Switzerland

CSHL PRESS

COLD SPRING HARBOR LABORATORY PRESS
Cold Spring Harbor, New York

Cell Growth

CONTROL OF CELL SIZE

Monograph 42
©2004 by Cold Spring Harbor Laboratory Press
All rights reserved
Printed in China

Publisher	John Inglis
Acquisitions Editor	David Crotty
Production Manager	Denise Weiss
Project Coordinator	Joan Ebert
Production Editors	Melissa Frey and Patricia Barker
Desktop Editor	Danny deBruin
Interior Book Designer	Denise Weiss and Emily Harste
Cover Designer	Ed Atkeson and Denise Weiss

Cover: Insulin receptors promote the growth of *Drosophila* fat body cells. Insulin receptors were over-expressed clonally, just in those cells with green membranes. Cells with red membranes are normal. Insulin-driven growth also increases the nuclear DNA (blue) in these highly polyploid cells. (Photo courtesy of Ling Li; for details, see J.S. Britton et al. [2002] *Dev. Cell* **2:** 239–249.)

Library of Congress Cataloging-in-Publication Data

Cell growth : control of cell size / edited by Michael N. Hall, Martin Raff,
George Thomas.
    p. cm. -- (Monograph ; 42)
  Includes bibliographical references and index.
  ISBN 0-87969-672-9 (hardcover : alk. paper)
  1. Cells--Growth. 2. Cellular control mechanisms. 3. Cell cycle. 4.
Cell death. I. Hall, Michael N. II. Raff, Martin C. III. Thomas, G.
(George), 1947- IV. Cold Spring Harbor monograph series
  QH604.7.C446 2004
  571.8'49--dc22

                                                                    2004001496

10 9 8 7 6 5 4 3 2 1

All Cold Spring Harbor Laboratory Press publications may be ordered directly from Cold Spring Harbor Laboratory Press, 500 Sunnyside Boulevard, Woodbury, New York 11797-2924. Phone: 1-800-843-4388 in Continental U.S. and Canada. All other locations: (516) 422-4100. FAX: (516) 422-4097. E-mail: cshpress@cshl.edu. For a complete catalog of Cold Spring Harbor Laboratory Press publications, visit our World Wide Web Site http://www.cshlpress.com/

Contents

v

Preface

CELL GROWTH (INCREASE IN CELL MASS OR SIZE) is a highly regulated process, being subject to both temporal and spatial controls. It is usually coupled with cell division (increase in cell number) to give rise to an organ or organism of a characteristic size. In other cases, cell growth and cell division are unlinked; examples include oogenesis, muscle hypertrophy in response to an increased workload, and synapse strengthening during memory storage. Many of the molecules and mechanisms that control cell growth have been conserved in evolution from yeast to humans. Indeed, cell growth, along with cell division and cell death, is one of the most basic aspects of cell behavior. Furthermore, dysfunction of the signaling pathways controlling cell growth results in cells of altered size and can lead to developmental errors and contribute to a wide variety of pathological conditions, including cancer, diabetes, and inflammation. Considering its fundamental importance and clinical relevance, cell growth has not received as much attention as it deserves. We hope that this book will help redress this situation.

Creating this volume was an ambitious task, as cell growth is a broad and multifaceted subject. In 1957 (*Cancer Research* **17:** 727–757), M.M. Swann defined cell growth as "a not too precise shorthand word for those synthetic processes that do not appear, as yet, to be immediately connected with division and which provide the bulk of new protoplasm." To provide boundaries, while at the same time leaving some leeway, we asked the authors to provide their unique perspective on Swann's "processes." The first four chapters discuss cell growth in the context of development and/or cell division; Chapters 5 through 13 focus on individual molecules and mechanisms that control cell growth, and Chapters 14 through 20 describe cell growth in specific tissues.

We are extremely grateful to the authors for their contributions and enthusiasm, to Paul Nurse for writing a foreword while dealing with a trans-Atlantic move, and to our colleagues at Cold Spring Harbor

Laboratory Press (John Inglis, David Crotty, Patricia Barker, Melissa Frey, and, in particular, Joan Ebert), with whom collaboration was a pleasure.

MICHAEL N. HALL
MARTIN RAFF
GEORGE THOMAS

Foreword

IN ADDITION TO BEING THE UNIT OF LIFE, the cell is also the unit of growth. Understanding what regulates the overall growth of a cell and how cell growth is coordinated with progression through the cell cycle are important problems in cell biology. In steady-state conditions, proliferating cells usually maintain a particular size, suggesting a strong link between cell growth and division, but how this is brought about remains unknown for most cells. Thirty years ago, these problems were at the forefront of cell-cycle research (see Murdoch Mitchison's classic book *The Biology of the Cell Cycle,* Cambridge University Press 1971) but, more recently, they have been relatively neglected, which is why the appearance of this volume on cell growth is so timely.

For steady state proliferating cells to maintain their size, the amount of an individual component and the size or mass of a cell must double in amount every cell cycle. Different patterns of growth during the cycle can achieve this. The simplest pattern is a step accumulation of the component or cell mass, with a doubling occurring over a restricted period of the cell cycle. A good example is the replication of a gene, which occurs during a limited part of S-phase. Such a pattern can apply to individual components but is unlikely to apply to the overall growth of a cell, as it would mean that all cellular components would be synthesized within the same restricted time period. In fact, a step pattern is deceptively simple, because a regulatory mechanism must be in place to ensure that synthesis stops once the amount of the component has doubled. When synthesis is determined by a template, as in the case of semi-conservative replication of a gene, monitoring such a doubling is straightforward. For other components, however, the regulatory mechanism must be able to monitor the absolute amount or concentration of the component during the period of synthesis so that this can be switched off once a doubling in amount has occurred.

Another pattern is where growth is linear with a rate doubling during a restricted period of the cell cycle. This occurs when the synthesis of a component or the overall growth of the cell is constant, but then doubles in a stepwise fashion during a limited period of the cycle. This is an interesting pattern because it suggests that some mechanism is limiting accumulation of the component or overall mass, and that this limitation is relaxed during the stepwise increase. An example of this would be when a gene is limiting for transcription; once the gene is replicated during S-phase, the presence of two templates doubles the rate of transcription. A linear pattern of overall cell growth implies that a single component, or a very restricted set of components, is limiting for growth; these limitations become relaxed at the rate-doubling point. In this situation, an increase in the levels of these limiting components leads to an immediate increase in growth rate. For free-living single-celled organisms such as yeasts, which are likely to have been under selective pressure to grow at a maximal growth rate, a limitation like this would probably be evolutionarily unstable. If only a single component is limiting, then it should be straightforward for the cell to evolve and make that limiting component more efficient, allowing an increase in growth rate. In multicellular organisms such as metazoans, selection for rapid growth may be unimportant, in which case certain components might well be limiting for overall cell growth. If this is the case, it should be possible to identify gene functions which, when more active, could drive growth more rapidly. The identification of such genes should indicate what limits overall growth rate in such cells. Given that cells in multicellular organisms are "social," communicating with other cells in the body, growth control may be regulated more by extracellular signaling mechanisms than by intracellular components. Another interesting aspect of linear growth followed by rate doubling is how the increase in rate of synthesis is controlled, because, again, some regulatory mechanism must operate to ensure that there is a step doubling in synthesis at the change point.

A third deceptively simple pattern is exponential growth, often favored by theoretical and more mathematically inclined biologists. An exponential pattern of growth means that the rate of increase of a component or overall cell mass is determined by the absolute amount of the component or cell mass at that time. As a cell gets gradually larger during the cell cycle, the rate of increase gets gradually larger, too. Such a pattern requires specific regulatory mechanisms to ensure an autocatalytic-like pattern of accumulation. It can be very difficult to distinguish experimentally between a linear with rate-doubling pattern and an exponential pattern, even though the explanations for these patterns are very different.

In exponential growth, the patterns of accumulation of different components must be coordinated so that they are all similar to the overall growth of the cell.

A second problem in cell growth control is how cell growth and division are coordinated to maintain cell-size homeostasis. The simplest explanation for cell-size homeostasis is that progression through the cell cycle is restrained by the need for the cell to attain a critical cell mass before certain cell-cycle events can be completed. Such controls have been demonstrated in yeast, where they have been shown to operate before the onset of S-phase and/or mitosis. If a cell is too small, it needs to grow more before carrying out these events. This is achieved by lengthening the cell cycle, thus restoring a normal cell mass. In contrast, if a cell is too large, it grows less, shortening the cell cycle, once again restoring a normal cell mass. The best evidence for critical-mass cell-cycle controls in eukaryotic cells are experiments in both budding and fission yeasts, which have shown that undersized and oversized cells largely return to a normal size within one or two cell cycles. However, evidence for such controls in metazoan cells is much less compelling (see Chapter 3).

Despite the fact that experiments establishing the existence of such critical-mass controls in regulating cell-cycle progression have been in place for many years, the nature of the molecular mechanisms that monitor cell mass remain unknown. Many models have been proposed, but working out which operate has been very difficult. One interesting feature of a major class of these models is the requirement to count absolute numbers of a component in the mass-monitoring network. This requirement comes about because there is often a concentration term in the equations used, with an absolute amount of the component divided by cell volume (where volume is assumed to be proportional to cell mass). In these models, measurements of a critical mass or volume require monitoring or generating a constant amount of a component. One component present in a constant amount per cell is the genome and molecules associated with it. Thus, a general class of cell-mass-monitoring mechanism could be imagined that involves titrating out a fixed number of sites associated with the genome. The genome is relevant here because the major events of the cell cycle controlled by the cell-mass controls are S-phase and mitosis, both of which involve the genome. This role of the genome is also consistent with the fact that bacterial, yeast, fungal, plant, and animal cells of increased ploidy have a proportionally increased cell mass, which could be a consequence of having more sites associated with the genome.

Another way to bring about cell-size homeostasis is for cell-cycle progression to be regulated by a timer or oscillator that measures an absolute period of time from cell division to cell division, combined with a linear pattern of cell-mass increase. Time could be measured between the same events in succeeding cycles, such as S-phase or mitosis.

If the pattern of cell-mass accumulation is linear rather than exponential, cell-size homeostasis can be achieved without critical-mass controls, although it is inefficient. This comes about because, if growth is linear, with a rate that does not increase with cell size, then a small cell will accumulate mass faster per unit mass than a large cell. Thus, in a constant period of time between successive cycles, an initially small cell will grow proportionately more than an initially large cell. Over successive divisions, this will gradually shift smaller and larger cells back to the mean size of the population. This mechanism can work, but it is slow to achieve cell-size homeostasis, although to date there is little experimental evidence of support for this view (but see Conlon et al., Chapter 3).

This foreword has briefly touched upon some of the problems associated with cell growth and the cell cycle that were being discussed 30 years ago. Clearly, over this period partial answers have been provided to some of the issues described here. For example, we now know that cell growth is generally steady and continuous throughout most of the cell cycle. However, it is to be hoped that the renewed interest in these areas, as evidenced by the production of this volume, will allow some of these interesting issues to be resolved and extended. Not only does the mechanism by which cell growth is coupled to cell division need to be resolved, but also the regulation of overall cell growth in response to nutrients, growth factors, and other stimuli remains to be elucidated. Although the conversations occurred mostly long ago, it is a pleasure to acknowledge many fruitful discussions about these areas, particularly with Peter Fantes, Murdoch Mitchison, John Pringle, John Tyson, and Robert Brooks.

PAUL NURSE

1

How Metazoans Reach Their Full Size: The Natural History of Bigness

Patrick H. O'Farrell

Department of Biochemistry and Biophysics,
University of California, San Francisco,
California 94143-2200

LIFE FORMS RANGE ENORMOUSLY IN SIZE, but as yet, we have little under-standing of the mechanisms that determine the size of a cell or the size of an organism. Not only do we have little understanding of the mecha-nisms, but there is also a lack of appreciation of what is represented by growth control. Until recently, growth control was equated with the con-trol of proliferation (Raff 1996; Su and O'Farrell 1998). However, the sim-ple observations that cells can grow to different sizes, and that cells can divide without growth to produce larger numbers of smaller cells, suggest that the processes ought to be considered separately. It is recognized that growth should be considered in terms of increase in mass rather than increase in cell number. To many investigators, this appears to be a for-malism, because they are so familiar with growth situations in which the two go hand in hand. The formalism becomes much more concrete when examining growth and cell proliferation in metazoans. It turns out that there is an extraordinary and fundamental segregation of organism growth and cell proliferation in the life histories of most metazoans.

There has been a recent surge of interest in growth control and wider recognition of the distinctions between the control of mass increase and the control of cell proliferation. Despite newly emphasized distinctions between growth and cell proliferation, in most of the systems that are cur-rently being investigated, growth is largely exponential and cells are actively proliferating in close parallel. This includes proliferating yeast (Crespo and Hall 2002), expanding tissue culture populations, and the multiplying cells of the *Drosophila* imaginal disc (Neufeld et al. 1998). These systems provide experimental access to numerous questions about the mechanisms that couple cell-cycle progression to increase in cell mass and the interactions that coordinate growth with nutritional conditions. Furthermore, since proliferative growth marks the lifestyles of the unicellular predecessors of metazoans, this mode of growth is presumably primordial and fundamental. However, if one looks beyond these systems and examines how growth is integrated into the life cycles of different metazoan organisms, one finds that growth often does not parallel cell proliferation. Indeed, they are often completely out of synchrony, such that growth and cell proliferation occur in different parts of the life cycle. I describe the major metazoan growth programs and the integration of growth into the life cycles of different organisms. It will be seen that the separation of growth and cell proliferation are early adaptations in metazoan evolution. To understand the origin and purpose of these features of growth regulation, I have built a discussion based on the fact that the evolutionary history of metazoans influences present-day regulatory programs.

The considerations introduced in this chapter also promote a broader recognition of the importance of growth control in the natural history of organisms. Growth control goes beyond controlling the size of cells or the size of an organism. It is growth control that properly proportions different body parts and consequently shapes the body. My interest in growth (size) control was first piqued by its astonishing precision—a precision that is illustrated by near perfection of bilateral symmetry. This interest was fueled and sustained by recognition of the huge contributions that growth control makes to development and evolution. Each and every part of a body is sized appropriately, and evolutionary adaptations—the specialized finger of the aye-aye or the nose/trunk of an elephant—involve extraordinary modifications in the relative sizes of different body parts. To appreciate all these issues, it is important to examine how metazoans grow.

I begin this chapter with an outline of what I call biological constraints—factors that affect every metazoan to create "problems" that must be solved or evaded in the development and growth of every organism. It appears that a few of these constraints limited, and hence directed, the evolutionary history of current growth paradigms. Among other things, these constraints favor a separation in which growth and cell proliferation occur at different stages of an organism's life cycle. Indeed, consideration of when and where growth occurs and when it is limited will lead us to identify several distinct programs of growth control.

CONSTRAINTS ON BIOLOGY AND GROWTH

Conservation of Mass

Even evolution cannot evade the dictates of the first law of thermodynamics. Consequently, an organism without a mouth or alternative means of nutrient uptake cannot grow (increase in mass). Although this may sound trite, it has an enormous impact. Eggs generally come with protective shells to isolate them from the harsh environments in which they are laid. Without input of new mass, the hatchling can be no larger than the egg that was laid. Indeed, since the vast majority of metazoan embryos have no means of nutrient uptake, they must develop at least until they are competent to feed before growth can commence. Feeding usually requires at least a mouth and an alimentary canal, structures that can only be produced following significant development. Thus, this constraint predicts that extensive cell proliferation and tremendous steps in development precede growth. Indeed, as discussed below, in the life cycles of many organisms, cell proliferation precedes growth, and it is likely that this separation was a feature of primitive programs for metazoan growth.

Mitosis Is Incompatible with Many Differentiated Structures and with Many Steps in Morphogenesis

Mitosis is remarkable, not only because of what it accomplishes, but also because of the way in which it appropriates or disrupts much of the cellular machinery in order to achieve its ends. The microtubular cytoskeleton is remodeled from its interphase arrangement to assemble the mitotic spindle. The actin cytoskeleton is rearranged to create a cytokinesis furrow and to drive its invagination through the cell. The Golgi is disassembled, transcription arrested, translation suppressed, and cellular processes withdrawn. These demands of mitosis would wreak havoc with the structural specializations of many differentiated cells. It is obvious that syncytial muscle cells cannot replicate mitotically. Similarly, the specializations of neurons, or the crystal cells composing the lens of the eye, or the Schwann cells that encase the axon in myelin are not compatible with mitotic proliferation. Although mitosis may successfully duplicate less dramatically differentiated cells, it remains a disruptive process. For example, even though mitotic epithelial cells retain connections to the rest of the epithelium, mitosis nonetheless locally compromises the ability of the endothelium to prevent leakage of large molecules from the vasculature (Lin et al. 1988; Baker and Garrod 1993). Consequently, mitosis tends to interfere with the function of terminally differentiated tissues. In the embryos of many organisms, this difficulty does not arise because proliferation is largely limited to the early stages of development prior to differentiation.

Distance-dependent Mechanisms

Some phenomena, such as the transport of oxygen by diffusion, are distance dependent. It is easy to recognize that large body sizes are associated with structures and mechanisms designed to transport oxygen and eliminate waste. Unfortunately, it is perhaps also easy to overlook the fact that, during evolution, body size was constrained by the ability to transport oxygen and eliminate waste.

This constraint is not limited to the difficulty of getting things in and out of a large organism. Some biological mechanisms, most notably those that pattern an embryo, are founded on distance-dependent phenomena. One example is the conserved mechanism involving a secreted *dpp*/ TGF-β signal from one side of an embryo interacting with an inhibitory signal (*short gastrulation*/Chordin) from the other side to create a morphogenetic gradient that patterns the dorsal/ventral axis (Ferguson 1996). Hence, patterning works best at particular scales, and the evolutionary conservation of the mechanisms that pattern the body axis might demand conservation of size at crucial stages in the patterning process.

PROGRAMS OF GROWTH IN METAZOANS

The Big Egg Paradigm

Although there is a natural tendency to focus on our own biology, uterine support of embryonic growth is extraordinarily rare (about 4,000 eutherian species out of about 1,000,000 total metazoan species). In most extant species, including the evolutionary predecessors of uterine mammals, a mother must pack into the egg sufficient reserves to carry development far enough to produce a feeding animal, and no growth can occur until feeding starts. To achieve this end, eggs are big—sometimes very big. Frog eggs are about 10^5 times larger than a conventional somatic cell. Let us consider the significance of the production of a large egg in terms of the total growth of an organism during its life cycle.

Caenorhabditis elegans provides a simple example of metazoan growth issues. A germ-line stem "cell" grows extremely rapidly in a syncytial gonad to about 1000 times its starting size (Fig. 1). Cell proliferation is largely confined to the early part of embryogenesis and occurs within a closed egg shell without any growth. Morphogenesis then ensues, again without growth, to produce (at hatching) a small L1 juvenile. The worm then feeds and grows about 100-fold to produce the adult worm. This growth is almost exclusively by cell enlargement, as there is little postembryonic cell division other than cell proliferation associated with the development of the reproductive organs.

Several features of this description are notable. First, at least in terms of exponential growth or fold increase, the growth that produces the oocyte makes a contribution that is comparable to or larger than the growth during zygotic life. Second, there is an almost complete separation of growth, cell proliferation, and differentiation/morphogenesis. Third, growth can be divided into two phases, one prior to proliferation (and prior to fertilization), in which the growth of the oocyte is supported by the mother, and one after the period of proliferation, which is supported by the feeding of the hatchling.

The *C. elegans* program, with its sequence of growth, proliferation, and morphogenesis, followed by a second period of growth, satisfies the constraints that I described above. The egg is large enough to produce the 550 cells of the hatchling worm without growth, and, by limiting the vast majority of the cell division to the early stages of embryogenesis, the program largely avoids the difficulty of dividing differentiated cells. The early developmental paradigm of production of a large egg, followed by rapid division and subsequent differentiation, has been largely conserved, even if its features are masked in eutherian mammals.

C. elegans is very small. How are larger organisms produced? It appears that larger size is achieved by adding later stages of growth. Below, I describe a stage of development that is conserved between diverse organisms. This so-called "phylotypic stage" will serve as a useful reference point for comparison of the growth programs of various organisms, small and large, and for identification of the stage of development at which evolution has introduced variations in size.

A Universal Embryonic Size

As noted by von Baer and emphasized by Haeckel during the early years of the incipient field of developmental biology, embryos of very different species look remarkably similar at the stage following gastrulation, just after the body axis is morphologically established. The similarity applies to size as well as to morphology. Whether an embryo will produce a mouse or a whale, a minnow or a tuna, the postgastrulation embryos are of very similar size. Indeed, embryos of organisms as diverse as leech,

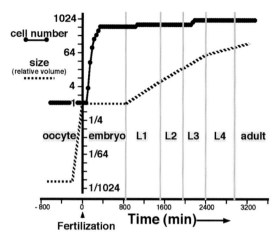

Figure 1. Growth and cell proliferation occur at different and largely nonoverlapping stages in the life cycle of the nematode *C. elegans*. Time relative to fertilization is indicated along the bottom, and the vertical axis, a logarithmic scale divided in intervals of a factor of two, indicates both cell number and size relative to the size of an egg. Expansion of somatic cell number (*solid line with filled circles*) by cell proliferation occurs almost exclusively during the first third of embryogenesis. This proliferative phase precedes much of the morphogenesis and differentiation, which occur during the later two-thirds of embryogenesis. Because differentiation and morphogenesis are segregated to a later stage, they avoid the potentially disruptive influence of mitosis (see text). Growth (*segmented line*) is also dramatically separated from cell proliferation. There is no growth during embryogenesis, which occurs within an enclosed environment inside the egg shell. Growth awaits hatching and the onset of feeding of the juvenile worm. Because this growth phase occurs after the major cell proliferative phase, growth depends largely on expansion of cell volume. In addition to this growth, the mother sponsors embryonic development via a specialized form of growth in which the oocyte grows tremendously. The germ-line portion of the ovary is syncytial, so that numerous nuclei make a communal contribution to the production of additional syncytial cytoplasm (growth) that supports the production of oocytes. It is difficult to plot this growth. Instead, I have plotted the extraordinarily rapid expansion of incipient oocytes in the distal part of the linearly arranged ovary. Beyond mitotic and meiotic zones, the syncytial cytoplasm is divided into modestly sized "cells," as membrane surrounds a single nucleus and an allotment of syncytial cytoplasm. The newly formed "cell" is only incompletely cellularized, as it maintains a cytoplasmic connection to the syncytium. A flow of cytoplasm to the incipient oocyte drives its very rapid growth, as indicated in the graph.

Drosophila, fish, frog, and mouse are all about 1 mm long when they first elongate along the anterior–posterior body axis. Although size constancy is not precise, and the *C. elegans* embryo is substantially shorter (~100 microns long after elongation [Priess and Hirsh 1986]), the differences are tiny in comparison to the 10^{10}-fold differences in size (mass or volume) separating the adults of large and small organisms.

Although it is not entirely certain why such diverse embryos should be so similar in size, I suggest that it is because of the final constraint introduced above—that some mechanisms of pattern formation are inherently size dependent. Studies of embryonic patterning have confirmed that distance-dependent signals contribute importantly to defining the body plan. Hence, the patterning mechanisms ought to work only in a range of scales. Since there is considerable conservation of the genetic programs that pattern the body axis during early development, the conservation of developmental mechanisms would be expected to enforce conservation of size at this crucial stage in the patterning process. Even if we cannot be sure of the correctness of this explanation for why the sizes are so similar, it is clear that evolution has been constrained to maintain the size of the early embryo. Indeed, we can roughly estimate the size of a "universal" postgastrulation embryo as about 10 µg (wet weight of the embryo proper without large yolk deposits).

The common size of metazoan embryos provides a reference point for comparison between species and allows us to divide the problem of growth control into two issues. How do embryos grow to reach this universal embryonic size, and how do they grow to obtain their final size? Here, we emphasize the late growth, and, in a separate report, will discuss the remarkable conservation of the early embryonic schedule of rapid cell proliferation followed by gastrulation (J. Stumpff et al., in prep.).

GETTING BIGGER

Large organisms must grow hugely after establishment of the body plan, whereas small organisms grow more modestly. From the postgastrulation stage, a blue whale will grow about 10^{13}-fold to reach a final weight of 150 tons, whereas a mouse will grow about 10^{6}-fold to a weight of about 15 g—a 10^{7}-fold difference. The nematodes provide another, perhaps less familiar, example. *Ascaris lumbricoides*, one of the largest nematodes, has an egg the same size as *C. elegans*, but it grows to produce a worm that is about 40 cm long, compared to the 1-mm-long *C. elegans* adult (about 10^{7} times larger volume). The late growth that produces these large differences in size is very different in character in nematodes and vertebrates. Furthermore, examination of the growth and cell proliferation characteristics within an organism suggests the existence of diverse types of controls and indicates that evolution has selected among several options in designing the late-growth programs of different organisms, and even of different tissues.

The constraints that we have discussed suggest that late growth programs have to contend with the difficulty of dividing cells that are terminally differentiated. Two candidate solutions introduce their own difficulties. One of these candidate solutions is to defer differentiation until late in the life cycle to allow an extended early period of cell proliferation. The problem that accompanies this solution is connected to the issue of conservation of mass. In order to grow, a developing organism must be nourished, and this is not easily done without at least some differentiation to produce a feeding apparatus. Hence, if differentiation is to be delayed to allow cell proliferation, the early period of proliferation would have to occur without growth, unless the delay in differentiation is selective so that those structures required for feeding/nourishment develop early. Another candidate solution is to simply allow cells to grow bigger without division. This later solution is inherently limited because the nucleus appears to be overwhelmed by the task of directing the growth of extremely large cells, and because of the difficulties of constructing elaborate body plans with few cells. Nonetheless, these solutions are among the ones that are used, sometimes with additional adaptations to deal with the specific problems that these schemes introduce.

GROWTH OF NEMATODES

C. elegans

The thoroughness of our knowledge of the cellular anatomy and development of *C. elegans* make it particularly instructive. Although its growth is modest in comparison to many larger animals, it nonetheless increases its body size about 100-fold after hatching. Most of this growth occurs without cell division. However, even though they do not divide, some cells, such as the gut cells, carry out an abbreviated cell cycle in conjunction with growth (Hedgecock and White 1985). In the first molt, the anterior gut cells undergo mitosis without cytokinesis to produce binucleate cells (an endomitotic cycle). Otherwise, at each molt the DNA of the gut cells replicates without intervening mitosis (endoreplication cycles) to create polyploid nuclei that ultimately reach 32 N. Other cells simply enlarge to an enormous degree without amplifying their DNA. This growth without division is consistent with the predicted conflict between terminal differentiation and division, and the amplification of DNA without division might be an adaptation that helps support the growth of especially large cells. Indeed, we see similar adaptations in the growth program of

Drosophila larvae, wherein there is experimental support for the importance of genome amplification for growth of some large cells.

Although increase in cell size is the predominant mode of growth (Fig. 1), there is some proliferation, and this proliferation appears to fall into two categories. Some cells, such as the Z1 and Z4 cells that act as the progenitors of the somatic gonad, remain relatively undifferentiated during embryogenesis. They appear to represent a case of deferred development. They initiate a complex and stereotyped lineage during the larval stages. The proliferation of these cells to produce the gonadal primordium is associated with growth, and in this way differs from the proliferation of undifferentiated cells of the embryonic stage. However, as we saw in the embryo, significant differentiation occurs only late in the larval stages, after proliferation. In contrast, some other cells (e.g., hypodermal cells) differentiate but retain the ability to divide. This latter category emphasizes that the conflict between differentiation and proliferation is not absolute; indeed, it is more appropriate to think that mitosis and cytokinesis are inconvenient for differentiated cells, but often not impossible.

In summary, *C. elegans* exhibits at least three types of "solutions" to allow growth after differentiation of many of its tissues. First, the vast majority of the growth is due to an increase in cell size without division—a process that avoids the potential conflict between differentiation and division (Fig. 1). Second, the continued presence of undifferentiated cells represents a selective deferral of differentiation that maintains a separation between proliferation and differentiation. Third, some types of cells remain competent to divide despite their differentiation. The different solutions to growth seen in *C. elegans* are also seen in different forms and abundance in different organisms.

Ascaris Carries the *C. elegans* Paradigm to Extremes

The different species of nematodes have a wide range of lifestyles and habitats but share a common body plan. Some of the parasitic nematodes reach large sizes, apparently just by exaggerating the growth programs seen in *C. elegans* (Chitwood and Chitwood 1974). *Ascaris lumbricoides* is close to the same size as *C. elegans* at hatching, but it subsequently grows nearly 10^9-fold. As in *C. elegans*, there is very little cell division during this growth, which is largely due to an increase in cell size. In contrast to *C. elegans*, which has mostly rather small cells, the cells of large parasitic nematodes become extremely large. Chitwood and Chitwood (1974) give the size of a single intestinal cell in the large parasitic nematode (*Strongylus equinus*) as 4 mm long by 500 microns wide, or several times the size of an entire adult *C. elegans* worm. The nervous system provides a very clear example of the character of the difference between the large and small nematodes. The nervous systems of *C. elegans* and *Ascaris* are remarkably similar, with the larger nematode having virtually the same number of neurons and similar wiring; however, the neurons are very much larger (Angstadt et al. 1989). Thus, *Ascaris* represents a case in which size increases immensely without a substantial increase in cell proliferation. As in *C. elegans*, cells can have multiple nuclei, and these can be polyploid. Indeed, such increases in DNA content are more dramatic in *Ascaris*, consistent with a contribution to the greatly increased size of the cells.

GETTING BIGGER DURING A LARVAL STAGE

Many organisms have an indirect program of development in which a larval stage is inserted between embryonic development and formation of the adult. The larval stages are generally periods of substantial growth. The holometabolous insects such as *Drosophila* provide a well-characterized example of such programs and illustrate how this type of developmental program adds developmental complexity and size while neatly avoiding the constraints described above. Indeed, one can rationalize the evolution of the two-step larval plan of development as an adaptation to solve constraints associated with growth.

The *C. elegans* Growth Paradigm Is Used to Make a *Drosophila* Larva

Drosophila uses an elaborate scheme during oogenesis to produce a large egg that is about 10^5 times the volume of an average somatic cell. Embryonic development produces a larva of about 50,000 cells without growth. This production of the larva begins with an extraordinary proliferative phase in which 13 cycles rapidly subdivide the egg into about 6,000 cells in 2 hours. This is immediately followed by a short phase in which three slower mitotic cycles are dovetailed with the initial stages of morphogenesis, and then by a longer phase of morphogenesis and differentiation in which there are few divisions. As in *C. elegans*, the rapid embryonic cell proliferation and subsequent differentiation are separated in time and occur without growth (Fig. 2).

Once hatched, *Drosophila* larvae feed voraciously and grow nearly 1,000-fold in volume during three larval stages, or instars, that are separated by molts (Ashburner 1989). This growth is independent of cell prolifer-

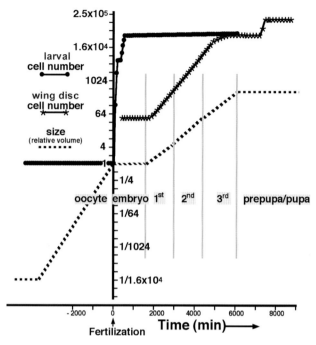

Figure 2. Growth and cell proliferation in *Drosophila*. The axes are as in Fig. 1. After fertilization (time=0), increase in the number of somatic cells contributing to the larva proper (*dots*) occurs almost exclusively during the first quarter of embryogenesis. This proliferative phase precedes much of the morphogenesis and the differentiation, which occur during the later two-thirds of embryogenesis. Growth (*segmented lines*) is also dramatically separated from proliferation. There is dramatic growth of the oocyte, which occurs in a specialized growth process supported by nurse cells in the egg chamber. After fertilization, there is no growth during embryogenesis, which occurs within an enclosed environment inside the egg shell. Growth awaits hatching and the onset of feeding by the larva. This larval growth phase, which can occur in the complete absence of cell proliferation, is the result of expansion of cell volume in conjunction with endoreplicative cell cycles, which increase ploidy without cell division. Imaginal cells are special cells that make no essential contribution to the larva but will differentiate into the adult structures at metamorphosis. Unlike the larval cells, most of these cells proliferate during the larval stages (see text for description of diversity of the imaginal growth programs). The number of wing disc cells is shown as an example of the program of proliferation of the imaginal cells (*asterisks*). These imaginal cells grow as they proliferate. Each group of imaginal cells follows a substantially distinct program to develop into different structures of the adult. Notably, at the time of formation of the pupa, the bulk of the mass of the organism is contributed by the larval cells. During morphogenesis, there is transfer of mass from the larval tissues, which largely degenerate, to the imaginal tissues. In this way, the feeding of the larva and its growth fund the growth of the imaginal tissues.

ation, as demonstrated by successful growth of larvae that are mutant in genes required for continued proliferation (Gatti and Baker 1989). The growth is accomplished by tremendous increases in cell volume. Most of the larval cells undergo endoreplication cycles, in which regulated rounds of DNA replication without intervening mitosis amplify the DNA. Different numbers of endoreplication cycles occur in different tissues, and the most extensive amplification occurs in tissues with particularly large cells, such as the salivary glands. Experiments in which the endoreplication cycles are prevented in a particular tissue result in a retardation of growth in the targeted tissue, indicating that the amplification of the genome contributes to growth (Hayashi 1996; Follette et al. 1998; Weiss et al. 1998). Endoreplication cycles occur in fully differentiated cells. The amplification of DNA in the individual nuclei of multinucleated skeletal muscle cells indicates that, by avoiding mitosis, this abbreviated cell cycle avoids incompatibilities between terminal differentiation and proliferation.

In summary, like the paradigm followed by *C. elegans*, growth through the larval stage in *Drosophila* is segregated from cell proliferation. Again, we see two dramatic phases of growth: one preceding fertilization, in which the mother supports a tremendous increase in the mass of the oocyte, and a second that occurs after hatching, in which increases in the mass of larval somatic cells is supported by larval feeding and metabolism (Fig. 2). Proliferation occurs during embryogenesis between these growth stages and prior to overt differentiation.

Additional Programs of Growth Produce the Fly

The *Drosophila* larva is only an intermediate in the production of a fly, and one must consider a second program of late growth that overlaps and extends beyond the expansion of larval cells. In addition to the cells that make up the feeding larva, there is a second population of cells, imaginal cells. These are undifferentiated cells that are carried within the larvae for the purpose of making the adult during the pupal stages. Several diverse groups of cells comprise the imaginal population, and each group of imaginal cells acts as a primordium for one adult organ, or structure. Each of these diverse groups can be considered as analogous to the primordium of the gonad in *C. elegans*. Just as the Z1 and Z4 cells of *C. elegans* defer differentiation until after embryogenesis, *Drosophila* imaginal cells, which are set aside in embryonic development, defer their differentiation. Similarly, just as the Z1 and Z4 cells proliferate in the larva to make a multicellular primordium that subsequently differentiates into a reproductive tract, each group of *Drosophila* imaginal cells proliferates as undifferentiated cells to create a primordium that later (most frequently in pupae) undergoes morphogenesis and differentiation, each according to its unique program, to produce particular adult structures.

The growth of the cells composing the wing imaginal disc illustrates the growth program of imaginal cells. Toward the end of the embryonic proliferative divisions, the fates of cells begin to be defined, so that an embryonic group of about 40 cells becomes committed to development into a wing disc. These cells lie quiescent for the last two-thirds of embryogenesis and the early part of larval life. Roughly two-thirds of the way through the first larval instar, but only after feeding (Britton and Edgar 1998), the wing disc cells begin proliferative growth, in which they double their number and total cell mass in fairly close concordance every 8.5 hours, until most of the cells begin to slow to a G_2 arrest in the third instar (Graves and Schubiger 1982; Johnston and Edgar 1998). These cells remain together as a coherent group that forms a sac-like epithelial structure, the disc, in which cells exhibit only modest structural specializations. Investigators have posed many interesting questions about this growth program. What triggers the onset of the proliferation (Britton et al. 2002)? What mechanisms coordinate the proliferation and growth of the cells (Neufeld et al. 1998)? Is disc size defined by cell number or regulated by additional and distinct mechanisms (Weigmann et al. 1997)? What determines final disc size? How is growth coupled to the patterning genes that influence the growth of the disc and ultimately the shape, form, and structure of the adult wing (Serrano and O'Farrell 1997)? Although these are important questions in current studies of growth control in metazoans, the exponential expansion of the wing discs during larval stages is but one step in the overall program of growth. Here, rather than focusing on this step, we hope to reveal more of its context in the life history of growth.

Diversity is a little-emphasized but important feature of imaginal programs of cell proliferation and growth. The time at which each imaginal tissue breaks the quiescence that began in the embryo and initiates proliferation is different. Most of the discs begin proliferation during the first larval instar, but at distinct times. Other imaginal tissues exhibit more diverse programs. The dispersed imaginal cells of the gut remain quiescent until the third instar, and the clusters of imaginal cells in the abdomen (the histoblasts) remain quiescent until pupariation, whereupon they proliferate with an exceptionally short cell cycle (about 2.8 hours) (Milan et al. 1996). The different imaginal tissues also make the transition from proliferation to differentiation at different times and in different ways. Whereas the cells of the wing disc divide two more times about 18 hours after pupariation (Milan et al. 1996), cells of the eye disc begin to exit the cell cycle and differentiate midway through the third instar. Furthermore, the character of larval proliferation can differ. Thus, disc cells proliferate exponentially and with few lineage restrictions, while larval neuroblasts proliferate in a stem cell pattern, in which only one of the daughter cells continues to proliferate, and the other forms two neurons after one more division. Since the neuroblast proliferation pattern is not exponential, it takes many divisions to produce all the cells of the adult nervous system. Normally, the daughter of a stem cell division that is committed to a neuronal fate would differentiate almost immediately. This would give rise to an incongruous situation in which the very long larval program of neuroblast proliferation would begin to generate adult neurons in the young larva, long before the differentiation of target structures. However, this does not occur, because the blast cell lineages are specialized, in that the larval-generated neurons arrest and defer their differentiation until pupal stages, when differentiation would be more timely (Truman and Bate 1988). Even though all the different imaginal cells share the same endocrine and nutritional environment as the larva matures, they exhibit different programs of growth, proliferation, and differentiation that are largely autonomous to the individual group of imaginal cells.

Although pupariation ends the major period of feeding and overall body growth, the situation in pupae is far from static. Larval tissues are largely broken down, and there is an effective conversion of mass from the larval tissues to the imaginal tissues. Thus, most of the larval mass will be used to support growth of the imaginal cells during the pupal stages. In this way, the feeding and increase in mass of the larva contributes to the growth of the adult. The wing disc, for example, will expand to produce adult structures that appear many times the size of the larval disc. At present, little can be said about this pupal growth, because we do not know, even at a descriptive level, how the mass that was accumulated in larval tissues during larval growth is transferred to the different imaginal tissues upon breakdown.

In summary, the imaginal tissues present dramatic examples of deferred differentiation. The larval tissues differentiate immediately to produce an efficient feeding machine that supports the growth of these imaginal tissues. This growth includes a proliferative phase, but the timing, rate, and type of cell proliferation (i.e., exponential proliferation versus a stem-cell pattern of proliferation) vary from one imaginal tissue to another. During pupal development, the mass of the larva is sacrificed to feed the growth of the imaginal tissues within the closed environment of the pupal case. Thus, in terms of the constraints discussed above, cell enlargement without proliferation is the predominant solution used by larval tissues, and deferred differentiation is the predominant solution used by imaginal/adult tissues. It is notable that the extended late growth phase of all the imaginal tissues involves cell proliferation, followed by deferred development. However, within this common arrangement, there is great divergence of the individual programs.

HOW VERTEBRATES GET SO BIG

It may seem that much of the above is irrelevant to growth in vertebrates. After all, vertebrates do not have larvae with giant polytene cells, and unlike the difference between *C. elegans* and *Ascaris*, the difference between the sizes of a whale and a mouse is due to differences in the number of cells, not in cell size. The constraints on growth still apply, however, and it is of interest to consider whether vertebrates have evolved totally distinct solutions or whether they have simply selected among alternatives differently.

Yolk-supported Growth in Chordates

In considering overall growth programs of vertebrates, it is helpful to consider their evolutionary history. First, consider egg sizes among the fish. Anyone who has fished using salmon eggs as bait or is an aficionado of sushi knows that salmon eggs are big, about 6 mm in diameter. This is among the larger eggs of the bony fish (teleost fish), but if we look back at more primitive species of fish, we find extremely large eggs. The modern jawless fish (lampreys and hagfish) are thought to represent the most primitive of vertebrates. The oblong egg of a hagfish is more than 2 cm long and about 7 mm across (Morisawa 1999). Even larger eggs are found among the early jawed fish, which are represented by the cartilaginous fish (Chondrichthyes) such as rays and sharks. For example, the shark *Hemiscyllium ocellatum* has 2.5-cm eggs (Heupel et al. 1999), and the average size of hatchlings of various elasmobranch species is more than ten times larger (in length) than the size of the hatchlings of teleost fish species (Goodwin et al. 2002). This suggests that early chordate evolution had developed the big egg paradigm to an extreme not seen in prechordate evolution. Indeed, the very large size of these eggs (the shark egg being 10^{10} times the size of a normal cell) and of the resulting embryos forces recognition of a simplification that I have made in the presentation of the big egg paradigm.

Earlier in this chapter, I pointed out that the conservation of mass precludes real growth during embryonic development within an enclosed egg shell and that mass increase has to await feeding. I also indicated that embryos of all species have a very similar size at about the time of formation of the body axis. You might then ask, if embryos do not change in size during embryogenesis, and there is one stage at which embryos of all species are a similar size, should they not be the same size throughout embryogenesis? But, by the time of hatching, a shark is 10^6 times and an ostrich is 10^9 times the size of a *Drosophila* embryo. There appears to be a contradiction! Clearly, the missing element is that eggs include nutrient stores in the form of yolk, and embryogenesis is accompanied by a conversion of mass from the yolk stores to the embryo proper—although the mass of the egg as a whole does not increase. Yolk makes important contributions to metabolism of embryos from *C. elegans* to nonmammalian vertebrates, but the degree to which it contributes to growth varies enormously. Some insect eggs are considerably larger than *Drosophila* eggs and include yolk stores that support substantial embryonic growth. Nonetheless, nowhere else is yolk-supported growth used to the extent that is seen in chordates from sharks to dinosaurs.

Whereas the constraints on the size of the postgastrulation embryo appear to limit the cytoplasmic volume that a mother can contribute to an embryo, the stores provided as yolk appear unlimited. Thus, growth beyond the postgastrulation stage (which I have called late growth) can be supported by yolk, and in chordates this stage is extended and leads to very large increases in size.

Placental Nutrition Substitutes for Yolk in Mammals

Uterine mammals have developed a different strategy for late growth. Although the embryos still have a yolk sac, it is the placenta that provides nutrients during the comparable stages. Primitive forms of a placenta-supported growth that occur in some of the species of shark that give live birth, and in some viviparous reptiles (skinks), suggest that evolution of the placenta extended the phase of growth supported by yolk. For example, the shark *Hemiscyllium ocellatum* retains its eggs in a uterus, where they initially grow using yolk supplies. But when this yolk begins to be depleted, the epithelium of the yolk sac develops extensive interactions with the uterine walls. Both the yolk sac epithelium and the uterine wall epithelium develop specializations to produce a placental structure that feeds the developing shark through an umbilical cord (Heupel et al. 1999). Because a single epithelium serves a role first in retrieving nutrients from the yolk and then in retrieving nutrients from the mother, it is clear that the processes are related.

The mammalian placenta does not arise directly from the yolk sac as it does in the shark. The mammalian placenta made its appearance only after a major, but poorly appreciated, transition in chordate evolution. Fully terrestrial vertebrates have an amniote egg that differs in major ways from the eggs of chordates that lay their eggs in water. The evolution of the amniote egg, with its extraembryonic membranes, influenced many features of early embryogenesis and provided the setting in which the mammalian placenta evolved. However, in the context of late growth of the embryo, the complexities and uncertainties regarding evolution of the mammalian chorioallantoic placenta do not call into question the overall role of the placenta in nourishing the embryo during a prolonged period of embryonic/fetal growth (Blackburn and Vitt 2002; Thompson et al. 2002). Importantly, placental support of growth provides a means of further extending a growth period that was already a major part of the growth of chordates with yolk-rich eggs.

To appreciate the magnitude of uterine growth and its importance in evolution, it is of interest to compare the growth histories of mouse, humans, and whales. From time of birth to full-size adult, a mouse pup, human infant, and whale calf will each grow about 20-fold in mass. The big differences in size of the adult are fully anticipated by big differences in size of the newborn progeny. The enormous difference between a one-gram mouse pup and a seven-ton blue whale calf came about because of extraordinary differences in the amount of uterine growth from early embryos of similar size. Even the mouse pup represents an impressive amount of growth, about 10^5-fold, and the nearly 10^{12}-fold increase in mass of the embryonic/fetal whale shows the extraordinary capacity of this growth phase.

Analogies among Programs of Late Growth

Its remarkable achievements aside, fetal growth in a uterus has phylogenetic connections with other types of late growth. I have pointed out that placenta-supported growth represents an evolutionary modification of yolk-supported growth. I argue that it is also analogous to growth in a larval stage. A relationship between yolk-supported growth and larval-type growth is suggested by comparison of direct and indirect developing animals. Among vertebrates, the frogs are the best known of the indirect developers. A frog such as *Xenopus* has an egg that is large (1 mm in diameter) by some standards, with substantial yolk stores. However, this egg does not match the size and yolk content of many direct-developing, egg-laying chordates, such as birds and sharks. Much like the program described for *Drosophila*, *Xenopus* embryonic development includes successive stages of rapid cell proliferation, followed by morphogenesis and differentiation, to produce a tadpole the same size as the starting egg. In both organisms, there is some mass conversion from yolk to embryo proper, and in *Xenopus* this is somewhat more significant. Nonetheless, the tadpole is small, and it is the tadpole that shows extensive growth (10^5-fold in the larger frog species), which is supported by feeding. In contrast, a direct developing species of frog has a considerably larger egg (about 30-fold larger in volume), with a large yolk supply that promotes late embryonic growth (Elinson and Fang 1998). A similar relationship between direct and indirect developing organisms in other groups suggests that a larval stage is used as an alternative to yolk-supported embryonic growth.

Nutritional Strategies Compatible with Deferral of Differentiation

As discussed above, mitotic division of cells is not easily compatible with many types of cell differentiation because it requires an extraordinary level of remodeling of cellular structure. In *Drosophila* and nematodes, cell proliferation and differentiation are largely separated in time. Indeed, as a result of the lack of late proliferation in *Ascaris*, growth occurs via an extraordinary expansion of cell mass. Vertebrates have selected a different program of late growth, in which cells proliferate in proportion to growth. What is the nature of this program, and how is it organized to avoid conflicts between cell division and differentiation? We are far from knowing an answer to this question, but analogies to invertebrate systems suggest features that we might expect and possible models for their analysis.

In addition to cell expansion, both *C. elegans* and *Drosophila* use a second strategy to maintain separation of proliferation and differentiation. If differentiation is deferred, as in the stem cells of *C. elegans* gonad and in the larval discs of *Drosophila*, cells can grow and proliferate without impediments that might be imposed by specialized cell structures. This strategy is well suited to the vertebrate paradigm of growth. Since extensive growth is supported by maternal resources, supplied either as a nutritive yolk or more directly through a placenta, the embryo can defer the differentiation of structures required for free-living existence and feeding. This deferral is not complete, as the large size of the embryo imposes some requirements. A circulatory system is needed to transport and distribute nutrients, oxygen, and wastes, and specializations in the yolk sac are required to retrieve nutrient stores. Whereas these systems develop very early, morphogenesis and differentiation of many other structures occur later and progressively, each with sophisticated and distinctive developmental programs and specialized arrangements to allow growth.

Diversity of Late Growth Programs

Even in *C. elegans*, there is some diversity in the programs for late growth. In addition to cell enlargement, the gonadal precursors undergo a concerted program of cell growth and proliferation followed by differentiation, whereas hypodermal cells differentiate and then divide. Similarly, within the context of the exaggerated separation of growth and final differentiation that characterizes development of adult *Drosophila*, there is great diversity in the programs of development of different tissues. In vertebrate late growth, growth and proliferation often occur together and in patterns that are closely integrated with the developmental programs, but, nonetheless, many of these developmental programs include elaborate tissue-specific arrangements that maintain substantial separation of proliferation and differentiation.

Within the late embryonic growth phase of vertebrates, we can recognize different types of growth (and associated proliferation): (1) Growth during deferred differentiation occurs when relatively undifferentiated cells, such as mesenchymal cells, grow and proliferate prior to a morphogenetic and differentiative process; (2) developmentally programmed proliferation occurs when the development of a structure (such as a limb) occurs via a process (limb bud outgrowth) that incorporates growth and proliferation in the process; (3) growth and proliferation of differentiated cells occur when a particular cell type such as the cells of the endothelial lining of the blood vessels maintain a capacity to proliferate; (4) stem-cell-dependent growth occurs in a wide variety of vertebrate tissues that often have specialized but undifferentiated cells that can grow and proliferate to allow continued growth of a wide variety of differentiated structures, extending from muscle to brain. Although vertebrate growth uses all these processes, its reliance on stem cell lineages is particularly marked.

We are remarkably ignorant of the details that control the extensive late embryonic growth in vertebrates. There are specializations in the stem-cell-supported growth programs of different tissues such as bone, skin, and muscle. These distinctions suggest that there might be equally marked distinctions in the regulation of growth in each of these situations. To what extent has evolution operated on the processes controlling growth of each tissue independently? Ed Lewis and other investigators studying *Drosophila* introduced the idea that once evolution produced a segmental body plan, development of each segment could be independently guided during development, and the programs that accomplished this could be independently molded by evolution. Hence, the morphology of each segment could be designed with specializations useful for the animal as a whole (Lewis 1978). This idea has received strong support from studies of the ties between development and evolution (Carroll 1998). If a similar independence also applies to the evolution of the developmental programs that shape each tissue, we may be faced with great diversity in the mechanisms that control growth and proliferation during the late growth phase. Indeed, the tremendous plasticity of body form in vertebrates begs an explanation and might be taken as an argument that this is what we will find. Although this might leave us with many specific problems to solve, uncovering the general schema of growth control might give us insight into the evolutionary transformations that produced the trunk of an elephant and the finger of an aye-aye.

SUMMARY

A broad overview of growth in metazoans shows that there is remarkable separation in the timing of growth and cell proliferation in most complex organisms. An underlying early embryonic program is clearly evident in metazoans, from *C. elegans* to the immediate evolutionary precursors of eutherian vertebrates. In this program, the mother supports tremendous growth of an oocyte, and fertilization is followed first by rapid cell proliferation (cleavage) without growth and, later, by morphogenesis to establish the body plan, still without growth. At the time of establishment of the body plan, embryos of almost all species are roughly the same size. The small embryo subsequently grows to very different degrees and in very different ways, all of which seem largely to avoid cell division of fully differentiated cells. Examination of different species reveals some of the diversity of this late growth, some of which is coupled to the variety of mechanisms used to provide the nourishment essential to support this growth. Larval feeding, yolk-supported growth, and nourishment via a placenta appear to satisfy similar needs, and the phases of growth supported by these varied modes of nourishment appear to be evolutionarily related. The late growth processes themselves are extremely diverse, and evolutionary history suggests the possibility that the control of growth and proliferation of each tissue has evolved separately as "a separate experiment in evolution."

ACKNOWLEDGMENTS

I thank Mikiya Nakanishi, Pascale Dijkers, and Jean-Karim Hériché for comments on the manuscript, and The National Institutes of Health and the National Science Foundation for support.

REFERENCES

Angstadt J.D., Donmoyer J.E., and Stretton A.O. 1989. Retrovesicular ganglion of the nematode *Ascaris*. *J. Comp. Neurol.* **284:** 374–388.

Ashburner M. 1989. Drosophila: *A laboratory manual*. Cold Spring Harbor Laboratory Press, Cold Spring Harbor, New York.

Baker J. and Garrod D. 1993. Epithelial cells retain junctions during mitosis. *J. Cell Sci.* **104:** 415–425.

Blackburn D.G. and Vitt L.J. 2002. Specializations of the chorioallantoic placenta in the Brazilian scincid lizard, *Mabuya heathi:* A new placental morphotype for reptiles. *J. Morphol.* **254:** 121–131.

Britton J.S. and Edgar B.A. 1998. Environmental control of the cell cycle in *Drosophila*: Nutrition activates mitotic and endoreplicative cells by distinct mechanisms. *Development* **125:** 2149–2158.

Britton J.S., Lockwood W.K., Li L., Cohen S.M., and Edgar B.A. 2002. *Drosophila's* insulin/PI3-kinase pathway coordinates cellular metabolism with nutritional conditions. *Dev. Cell* **2:** 239–249.

Carroll S.B. 1998. From pattern to gene, from gene to pattern. *Int. J. Dev. Biol.* **42:** 305–309.

Chitwood B.G. and Chitwood M.B. 1974. *Introduction to nematology*. University Park Press, Baltimore, Maryland.

Crespo J.L. and Hall M.N. 2002. Elucidating TOR signaling and rapamycin action: Lessons from *Saccharomyces cerevisiae*. *Microbiol. Mol. Biol. Rev.* **66:** 579–591.

Elinson R.P. and Fang H. 1998. Secondary coverage of the yolk by the body wall in the direct developing frog, *Eleutherodactylus coqui:* An unusual process for amphibian embryos. *Dev. Genes Evol.* **208:** 457–466.

Ferguson E.L. 1996. Conservation of dorsal-ventral patterning in arthropods and chordates. *Curr. Opin. Genet. Dev.* **6:** 424–431.

Follette P.J., Duronio R.J., and O'Farrell P.H. 1998. Fluctuations in cyclin E levels are required for multiple rounds of endocycle S phase in *Drosophila*. *Curr. Biol.* **8:** 235–238.

Gatti M. and Baker B.S. 1989. Genes controlling essential cell-cycle functions in *Drosophila melanogaster*. *Genes Dev.* **3:** 438–453.

Goodwin N.B., Dulvy N.K., and Reynolds J.D. 2002. Life-history correlates of the evolution of live bearing in fishes. *Philos. Trans. R. Soc. Lond. B Biol. Sci.* **357:** 259–267.

Graves B.J. and Schubiger G. 1982. Cell cycle changes during growth and differentiation of imaginal leg discs in *Drosophila melanogaster*. *Dev. Biol.* **93:** 104–110.

Hayashi S. 1996. A Cdc2 dependent checkpoint maintains diploidy in *Drosophila*. *Development* **122:** 1051–1058.

Hedgecock E.M. and White J.G. 1985. Polyploid tissues in the nematode *Caenorhabditis elegans*. *Dev. Biol.* **107:** 128–133.

Heupel M.R., Whittier J.M., and Bennett M.B. 1999. Plasma steroid hormone profiles and reproductive biology of the epaulette shark, *Hemiscyllium ocellatum*. *J. Exp. Zool.* **284:** 586–594.

Johnston L.A. and Edgar B.A. 1998. Wingless and Notch regulate cell-cycle arrest in the developing *Drosophila* wing. *Nature* **394:** 82–84.

Lewis E.B. 1978. A gene complex controlling segmentation in *Drosophila*. *Nature* **276:** 565–570.

Lin S.J., Jan K.M., Schuessler G., Weinbaum S., and Chien S. 1988. Enhanced macromolecular permeability of aortic endothelial cells in association with mitosis. *Atherosclerosis* **73:** 223–232.

Milan M., Campuzano S., and Garcia-Bellido A. 1996. Cell cycling and patterned cell proliferation in the *Drosophila* wing during metamorphosis. *Proc. Natl. Acad. Sci.* **93:** 11687–11692.

Morisawa S. 1999. Fine structure of micropylar region during late oogenesis in eggs of the hagfish *Eptatretus burgeri* (Agnatha). *Dev. Growth Differ.* **41:** 611–618.

Neufeld T.P., de la Cruz A.F., Johnston L.A., and Edgar B.A. 1998. Coordination of growth and cell division in the *Drosophila* wing. *Cell* **93:** 1183–1193.

Priess J.R. and Hirsh D.I. 1986. *Caenorhabditis elegans* morphogenesis: The role of the cytoskeleton in elongation of the embryo. *Dev. Biol.* **117:** 156–173.

Raff M.C. 1996. Size control: The regulation of cell numbers in animal development. *Cell* **86:** 173–175.

Serrano N. and O'Farrell P.H. 1997. Limb morphogenesis: Connections between patterning and growth. *Curr. Biol.* **7:** R186–R195.

Su T.T. and O'Farrell P.H. 1998. Size control: Cell proliferation does not equal growth. *Curr. Biol.* **8:** R687–R689.

Thompson M.B., Stewart J.R., Speake B.K., Hosie M.J., and Murphy C.R. 2002. Evolution of viviparity: What can Australian lizards tell us? *Comp. Biochem. Physiol. B Biochem. Mol. Biol.* **131:** 631–643.

Truman J.W. and Bate M. 1988. Spatial and temporal patterns of neurogenesis in the central nervous system of *Drosophila melanogaster*. *Dev. Biol.* **125:** 145–157.

Weigmann K., Cohen S.M., and Lehner C.F. 1997. Cell cycle progression, growth and patterning in imaginal discs despite inhibition of cell division after inactivation of *Drosophila* Cdc2 kinase. *Development* **124:** 3555–3563.

Weiss A., Herzig A., Jacobs H., and Lehner C.F. 1998. Continuous cyclin E expression inhibits progression through endoreduplication cycles in *Drosophila*. *Curr. Biol.* **8:** 239–242.

2

Growth and Cell Cycle Control in *Drosophila*

Bruce A. Edgar
Division of Basic Sciences
Fred Hutchinson Cancer Research Center
Seattle, Washington 98109

H. Frederik Nijhout
Department of Biology
Duke University
Durham, North Carolina 27708

UNICELLULAR ORGANISMS TEND TO PROLIFERATE whenever environmental conditions—nutrients, oxygen, temperature, etc.—permit, and they cease growth when one or another of these conditions throttles their metabolism. In such cells, the progress of the division cycle is generally tightly coupled to cell growth. In contrast, animal cells exist in physiologically controlled environments that are nearly always nutrient-rich, yet they proliferate selectively. Specific autocrine, paracrine, and endocrine signals stimulate or limit cell growth and proliferation according to rules that benefit the organism as a whole. Moreover, the division cycle in multicellular systems is not always growth-coupled. Some cell types, like neurons and oocytes, grow without dividing, whereas others, such as spermatocytes, early blastomeres, and neuroblasts, can divide many times with little or no growth. Thus, the controls for cell growth and cell-cycle progression, although overlapping, are not the same in animal cells. In this chapter, we discuss the various modes of cell cycling and cell growth that occur during *Drosophila* development, and we examine the relationship between these two processes.

OUTLINE OF CELL GROWTH AND PROLIFERATION DURING *DROSOPHILA* DEVELOPMENT

Growth-independent Cell Cycles in Embryogenesis

Careful descriptive studies of *Drosophila* from as long ago as the 1920s provide a rich literature on cell-cycle dynamics during development (Lee and Orr-Weaver 2003). The first cell cycles of embryogenesis are the female meiotic divisions, which are triggered by ovulation (Huettner 1924). Following meiosis and fertilization, the embryonic nuclei undergo 13 rapid mitotic divisions, which are driven by maternal proteins and mRNAs (Rabinowitz 1941; Foe and Alberts 1983). Interphases begin to lengthen after the 10th mitosis due to increases in the nucleo-cytoplasmic ratio, and after the 13th mitosis, the embryo arrests in an extended G_2 phase due to the degradation of maternal stores of a protein phosphatase encoded by the *Cdc25* homolog *string (stg)* (Edgar and Datar 1996), which is required to activate mitotic Cyclin/Cdk1 complexes (Edgar et al. 1994b). A discrete switch to zygotic cell-cycle control occurs during interphase 14, and a series of slower cell cycles (typically three) follow (Hartenstein and Campos-Ortega 1985; Foe and Odell 1989). These cell cycles are regulated at G_2/M transitions (Edgar and O'Farrell 1990), have durations of 1–3 hours, and occur in invariant spatiotemporal patterns that are controlled by patterned transcription of the *stg* gene (Edgar and O'Farrell 1989; Edgar et al. 1994a). Interestingly, these patterns are determined by the same gene network that controls cell identities in the embryo (Arora and Nüsslein-Volhard 1992; Lehman et al. 1999). In the seventh hour of embryogenesis, cells begin to exit the cell cycle and arrest in their first G_1 phase (Edgar and O'Farrell 1990). This is due to the transcriptional down-regulation of *cyclin E* (Knoblich et al. 1994), and induction of the Cyclin E/Cdk2 inhibitor *dacapo* (de Nooij et al. 1996; Lane et al. 1996). Both of these processes are also controlled according to cell type, by developmentally programmed transcriptional control.

Few of the embryonic cell cycles are expected to be dependent on, or regulated by, cell growth. This is because most of the embryonic cells decrease in size with each division, rather than doubling their mass, and because their cell cycles completely lack the G_1 phase, during which cells are thought to monitor growth. Observations of wild-type and cell-cycle-arrested *stg* embryos do suggest, however, that cells in the embryo grow by converting yolk to cytoplasm (Hartenstein and Posakony 1990). Although much of this growth appears to occur in the postproliferative phase, it remains possible that cell growth is an important regulatory parameter for the cycling of select subsets of embryonic cells. The most likely candidates are embryonic neuroblasts in the CNS, which undergo 8–10 rather than 3 postblastoderm divisions (Campos-Ortega and Hartenstein 1985); cells in the gut, which initiate endoreplication cycles late in embryogenesis (Smith and Orr-Weaver 1991); and a patch of thoracic epidermal cells that transiently arrest in G_1 of cycle 17, but then undergo another S phase and mitosis (Jones et al. 2000). Studies of zygotic-effect growth control mutants (see below) are generally uninformative with respect to whether these cycles actually require cell growth, since the embryo inherits a surplus of most of the gene-regulatory gene products maternally. In this respect, it is interesting, however, that null mutations in the insulin receptor gene, which are important for cell growth, cause embryonic lethality with neural defects (Fernandez et al. 1995).

Although studies of *Drosophila* embryogenesis do not inform us about specific modes of growth control, they do provide clear examples of how cell proliferation can be regulated independently of cell growth. Of particular note is the direct transcriptional control, by the patterning system, of three limiting cell-cycle regulators, *stg, cyclin E,* and *dacapo.* Studies of the *cis*-regulatory regions of each of these genes has revealed extended regulatory DNA sequences of up to 50 kb, composed of modular elements that function in different types of cells and presumably respond to different combinations of transcription factors (Edgar et al. 1994a; Lehman et al. 1999; Jones et al. 2000; Liu et al. 2002; Meyer et al. 2002). How the transcription of these genes is controlled in cell types that utilize a growth-regulated or growth-dependent cell cycle (see below) remains an interesting and unresolved question.

Neuroblast Growth and Proliferation

The program of cell proliferation for neurogenesis in *Drosophila* is complex and displays characteristics and regulatory mechanisms distinct from any other cell type. Neuroblasts segregate from the ectoderm in waves following the blastoderm stage in the early embryo. After segregating, the neuroblasts enlarge considerably, indicating increased cell growth relative to other cell types. They then divide 8–10 times, giving rise at each division to a smaller ganglion mother cell that is thought to divide only once (Campos-Ortega and Hartenstein 1985; Hartenstein et al. 1987). The neuroblast cell cycle is very rapid (~55 minutes) and thus probably lacks a G_1 phase, as do other embryonic cell cycles. At the end of their embryonic proliferation program, neuroblasts arrest in G_1 and have reduced their cell size so much that they are indistinguishable from surrounding epithelial cells. Thus, although neuroblasts clearly grow, their cell cycle is not obviously growth-coupled, at least on a cycle-by-cycle basis. In the early larva, most neuroblasts remain G_1-arrested until the influx of nutrients from feeding allows the cells to enlarge again. After more than 12 hours of growth, they initiate S phases according to a stereotyped spatiotemporal pattern (Truman and Bate 1988). This first G_1/S transition, which appears to be growth-dependent, is then followed by a remarkably rapid series of cell cycles that average 1–2 hours. Each neuroblast produces ~100 progeny cells during this larval proliferation phase. Analysis of cell-cycle gene expression patterns in the larval CNS showed that proliferating neuroblasts constitutively express *Cyclin E* but express *Cdc25/stg* periodically, suggesting that their cell cycles may be G_2- but not G_1-regulated, as in the embryo (Britton 2000). Experiments in which nutritional protein was withdrawn from larvae at different stages showed that neuroblasts which had initiated their proliferative program would sustain the entire program even after the larva was starved, whereas G_1-arrested neuroblasts, which had not yet initiated proliferation, remained arrested (Britton and Edgar 1998). Thus, in contrast to the growth- and nutrition-regulated cell cycles used by other larval tissues (see below), neuroblasts appear to have a cell-autonomous cell proliferation program in which only the decision to activate is overtly growth-dependent.

Growth-dependent Cycles in Larval Tissues

Drosophila larvae increase in mass ~ 200-fold during their 5 days of development (Church 1965; Ashburner 1989). Most of this growth occurs in the differentiated, polyploid cells that constitute virtually all of the animal's functional organs and tissues. Cells in these "larval-specific" tissues do not divide, but undergo many cycles of DNA endoreplication, reaching ploidies ranging from 16C to 2048C, depending on cell type (Rudkin 1972; Lamb 1982). Virtually all of these cells are quiescent when the larva first hatches, and they initiate (or reinitiate in some cases) endoreplication only after larval feeding has begun and cell growth has caused some cell enlargement. Cell size in the larval endoreplicating tissues (ERTs) is roughly proportional to ploidy and is highly cell-type-specific. The mechanisms that determine the characteristic sizes and ploidies achieved by the different larval cell types are not well-understood. Although the specific mechanisms of endocycle control are also obscure, sufficient data are available to speculate as to the mechanisms at play (see Edgar and Orr-Weaver 2001; and see below). These cycles employ the conserved S-phase regulatory kinase Cyclin E/Cdk2, the CDK inhibitor Dacapo, and the cell-cycle regulatory transcription factor E2F1. Mitotic control genes such as *cdk1, cyclin B*, and *cdc25/Stg* are not expressed (Klebes et al. 2002). As described below, the larval endocycles are nutrition-dependent (Britton and Edgar 1998), and genetic manipulations that specifically promote cell growth (i.e., accumulation of mass) can drive DNA replication and result in hyperpolyploidy in the ERTs (Figs. 1 and 2). This suggests that the cell-cycle control mechanism in the ERT cells is growth-regulated (Fig. 3A) (Britton and Edgar 1998; Datar 2000; Edgar and Orr-Weaver 2001; Britton et al. 2002).

Imaginal Disc Growth and Development

Clusters of progenitor cells in most larval tissues refrain from differentiating, remain diploid, and proliferate mitotically in preparation for metamorphosis to the adult stage. Such cells make up the imaginal discs that give rise to wings, legs, eyes, and most other ectodermal structures in the adult. These "imaginal" cells are also found in clusters in the gut, salivary glands, and abdominal epidermis, which are used in remodeling these structures during metamorphosis. Imaginal disc cells are G_1-arrested when the embryo hatches, and enter S phase only after sustaining a substantial amount of cell growth, which increases their mass as much as six-fold (Madhavan and Schneiderman 1977). They then proliferate exponentially, with cell size gradually decreasing until the cells differentiate 5 days later (Fain and Stevens 1982; Neufeld et al. 1998). The wing disc, the largest and most thoroughly characterized, contains about 40 cells when the larva hatches, and these proliferate to >50,000 before differentiating to form the adult wing 4 days later (García-Bellido and Merriam 1971; Madhavan and Schneiderman 1977). Cell proliferation occurs nearly ubiquitously throughout the wing disc during the first 6–8 cycles (Madhavan and Schneiderman 1977; Fain and Stevens 1982; Graves and Schubiger 1982; Adler and MacQueen 1984; Neufeld et al. 1998). Although there are few overt patterns of cell division until the onset of cell differentiation, clonal analysis has revealed regional differences in cell proliferation rates (González-Gaitán et al. 1994; Resino et al. 2002). Thus, differential patterning of cell proliferation (zonal growth) is a plausible determinant of wing morphogenesis. Proliferative control in the eye discs has also been intensively studied and is reviewed by Baker (Baker 2001; Baker and Yu 2001; Yang and Baker 2003). Cell-cycle controls operating in the imaginal discs are similar in many respects to those seen in vertebrate

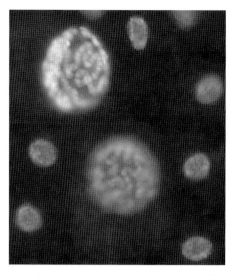

Figure 1. dMyc promotes hyperpolyteny in larval fat body. The two large cells over-express dMyc and are >4096C. The other cells are normal and are ~256C. DNA stained with DAPI.

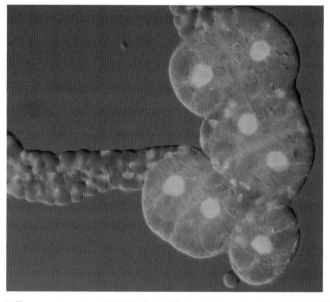

Figure 2. Cells overexpressing PI3K in larval fat body (marked green with GFP), grow under starvation conditions, whereas normal cells (clear with blue nuclei) arrest.

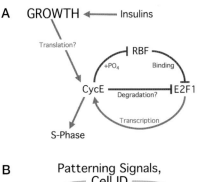

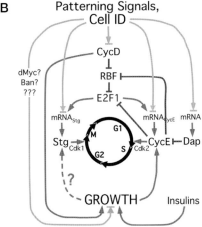

Figure 3. (*A*) Model for endocycle control in the larval-specific tissues. An autonomous CycE/E2F1-based oscillator responds to insulin-stimulated cell growth. (*B*) Model for control of the mitotic cell cycle in imaginal discs. Growth affects cell-cycle progression primarily at G_1/S transitions via regulating CycE protein levels, and patterning signals and cell identities (Cell ID) affect cell cycle and growth regulators transcriptionally. Arrows with blocked heads denote effects, primarily transcriptional, that can be either positive or negative, depending on the upstream signal and the developmental context.

cells. Transcription of the G_1/S regulator *cyclin E* (*cycE*) and the G_2/M regulator *Cdc25/stg* limit progression of the wing and eye disc cell cycles (Milán et al. 1996a,b; Johnston and Edgar 1998; Neufeld et al. 1998). E2F1/DP (Dyson 1998) is normally required for disc cell proliferation and can activate both *cyclin E* and *stg* transcription when overexpressed (Asano et al. 1996; A. Brook et al. 1996; Royzman et al. 1997; Dyson 1998; Neufeld et al. 1998). Transcription of both *stg* and *cycE*, however, is also subject to many other regulatory inputs that act through extensive *cis*-acting regulatory regions in their promoters (Lehman et al. 1999; Jones et al. 2000).

Cell Growth in Adults

In adult flies, the majority of cell growth is confined to blood cells (hemocytes), male and female germ-line cells, and somatic follicle cells in the ovary. Hemocyte proliferation has not been studied in depth, but it is known that these circulating cells proliferate for the life of the fly (for review, see Dearolf 1998) and that their proliferation can be increased dramatically by hyperactivation of Ras or JAK/STAT signaling (Dearolf 1998; Asha et al. 2003). Male germ stem cells and a supporting cell type, cyst progenitor cells, divide nearly as long as the male lives. Differentiating male germ cells form cysts of 16 primary spermatocytes by mitotic division. During the extended G_2 phase that constitutes meiotic prophase, these primary spermatocytes increase in volume 25-fold in 3.5 days. This remarkable growth period is followed by meiosis and spermatid differentiation, which occur with little or no transcription (for review, see Fuller 1998). Large amounts of cell growth also occur during oogenesis in both the nurse cells, which derive from the germ line and become highly polyploid, and the somatic follicle cells, which also support endocycles. Although cell-cycle controls in these cell types have been well-studied (Spradling 1993; Lilly and Spradling 1996; Deng et al. 2001; Edgar and Orr-Weaver 2001), the importance of cell growth as a regulatory mechanism has not been studied in detail. Oogenesis is known to be nutrition-dependent, and it is also supersensitive to mutations in genes that are important for cell growth in the developing larva (see below). For instance, null mutations in *cyclin dependent kinase 4 (cdk4)* or *cyclin D (cycD)* have minor effects on larval development yet suppress fertility profoundly in both males and females (Meyer et al. 2000), and hypomorphic alleles of *dmyc* and the insulin receptor (*dInr*), which have minor effects on larval development, also cause female sterility. Experiments in which the abdomens of living adult females were used as vessels for organ culture show that females must be fed a proteinaceous diet to support the growth of implanted tissues (Schubiger and Schneiderman 1971). This indicates that nutrition-dependent circulating growth factors, probably insulins (dILPs), imaginal disc growth factors (IDGFs), and/or ecdysone, are required to support cell growth in the female abdomen. Since oogenesis is also nutrient-dependent, it is presumed that the ovary is responsive to the same nutrition-dependent factors.

GENES AND PATHWAYS KNOWN TO REGULATE
CELL GROWTH IN THE FLY

In this section, we outline some of what is known about the genes and systems that are dedicated to regulating cell growth in the fly. Genes that affect tissue growth via effects on cell viability, cell-fate specification, and cell adhesion, despite their developmental importance, are not considered.

dMyc

The fly homolog of the cMyc transcription factor, dMyc (Gallant et al. 1996; Schreiber-Agus et al. 1997; Johnston et al. 1999; Prober and Edgar 2000, 2002) is a potent dose-dependent regulator of cell growth. Animals with reduced dMyc function are sterile and small (Johnston et al. 1999), and amorphic alleles exhibit lethality with defective larval growth (Bourbon et al. 2002; S. Pierce and R.N. Eisenman, in prep.). Overexpressed dMyc accelerates growth in both mitotic and endoreplicative cells in vivo. In polytene larval cells, ectopic dMyc accelerates the endocycle, causing hyperpolyteny (Fig. 1). In mitotic cells in the wing disc, dMyc-induced growth accelerates G_1/S progression but fails to alter overall rates of cell division (Johnston et al. 1999). Consequently, wing cells that overexpress dMyc become very large (see Fig. 5, below).

dMyc causes dramatic increases in nucleolar mass, suggesting that it may mediate its growth effects by increasing protein synthetic capacity. This notion has been corroborated by mRNA expression profiling studies using microarrays, which indicate that dMyc primarily activates genes involved in ribosome biogenesis, mRNA processing, and translation (Zaffran et al. 1998; Orian et al. 2003). *mRNAs* that are up-regulated include translation elongation and initiation factors, tRNA synthetases, ribosomal proteins, and many other factors used in ribosome production. Consistent with a role in protein synthetic control, ectopic dMyc also rapidly increases rates of larval rRNA synthesis in vivo in *Drosophila* larvae (S. Grewal and B. Edgar, unpubl.). A recent study aimed at defining the chromatin-binding sites of dMyc in cultured *Drosophila* cells suggests that many of the dMyc-responsive genes defined by expression profiling may be direct targets (Orian et al. 2003). Functional and target identification studies of cMyc in mammals concur with the work in flies in supporting a role for cMyc in regulating cellular growth via stimulating protein synthesis (Schmidt 1999; Schuhmacher et al. 1999; Coller et al. 2000; Guo et al. 2000; Kim et al. 2000; Boon et al. 2001; de Alboran et al. 2001; Douglas et al. 2001; Neiman et al. 2001; Watson et al. 2002). These reports have also indicated that some cell-cycle genes are cMyc targets, and the question of whether cMyc regulates cell-cycle progression directly via these targets or indirectly via effects on cell growth has generated ample discussion (see Trumpp et al. 2001; Iritani et al. 2002). dMyc, however, has not been observed to up-regulate critical cell-cycle control genes such as *e2f1, cyclin E, stg, pcna, rnr2,* or *dna pol-α* in *Drosophila* larvae (Orian et al. 2003), implying that its effects on cell-cycle progression are indirect consequences of potentiating general biosynthesis.

The finding that experimentally manipulated dMyc behaves as a dose-dependent regulator of cell growth in vivo suggests that differential expression of dMyc could be used to orchestrate patterned, differential growth during morphogenesis. Is this a mechanism that is used during fly development? Although altered levels of dMyc can change the ploidy and size of the larval cells, overexpressed dMyc does not drive normal growth in most of these cell types. dMyc-expressing cells develop disproportionately large nuclei containing excessive nucleolar material, and their cytoplasmic volume is proportionally reduced. This, and studies of other growth-regulatory pathways, indicate that dMyc levels are probably not the sole normal determinant of cell growth and size in ERT cells. Nevertheless, several observations suggest that dMyc may serve as an intermediary between patterning signals, which determine cell-type-specific growth characteristics, and the metabolism of growth. In the developing wing, for instance, dMyc is expressed in a stereotyped spatial pattern—high in the wing blade and notum, low in the presumptive hinge—that aligns with fate specification. Experimental manipulation of Wg or Dpp, two graded morphogens that pattern cell fates in the wing, alters the spatial expression of dMyc (Johnston et al. 1999; C. Martin-Castellanos and B. Edgar, unpubl.) Another spatially controlled signaling system, the EGFR/Ras pathway, also affects levels of endogenous dMyc in the wing (Prober and Edgar 2000, 2002). In addition, dMyc levels in the larval salivary gland vary along the gland's length and show a positive correlation with the final ploidy of these cells. These observations suggest that dMyc could act as an important mode of growth control utilized by diverse patterning signals. Several of the effects listed above have been observed at the transcriptional level, suggesting that studies of *dMyc*'s *cis*-acting transcription control elements will be interesting.

Other Growth Regulators Related to Protein Synthesis

In addition to dMyc, numerous factors believed to be involved in ribosome biogenesis and protein synthesis have been genetically characterized and shown to be required for normal growth in *Drosophila*. These include the *Minutes,* most of which encode ribosomal proteins (Morata and Ripoll 1975; Lambertsson 1998); translation initiation and elongation factors (Galloni and Edgar 1999); and factors involved in rRNA processing and ribosome assembly (Zaffran et al. 1998). Many of these show early larval-growth-arrest phenotypes when mutated, an expected consequence of the depletion of maternal gene products. With the exception of S6-kinase, however (Montagne et al. 1999; see Chapter 9), none of these factors has been reported to increase cell growth when overexpressed, suggesting that they do not function as dose-dependent regulators of growth. One interesting exception, however, is the *brat* gene (Frank et al. 2002), which encodes a novel cytoplasmic protein thought to be a translational repressor (Arama et al. 2000; Sonoda and Wharton 2001). Loss of *brat*, or of its *C. elegans* ortholog *ncl-1* (Frank and Roth 1998), causes the enlargement of nucleoli, suggesting that these gene products act to suppress ribosome biogenesis. Loss of *brat* in *Drosophila* wing discs also led to increases in rRNA, cell size, and clonal growth, whereas overexpressed *brat* suppressed cell and tissue growth in several assays. These observations suggest that *brat* acts to suppress cell growth via negative effects on protein synthesis. Whether *brat* is used to regulate cell growth differentially during development remains unclear, but the tumorous, metastatic overgrowth of neural and imaginal tissues observed in *brat* mutants (Kurzik-Dumke et al. 1992; Woodhouse et al. 1998) suggests an important function.

Cyclin D

Molecular studies in vertebrate cells have focused on the Cyclin D (CycD)-dependent kinases as specific promoters of the G_1/S transition of the cell cycle (Dyson 1998). The prevailing dogma holds that CycD/Cdk4 complexes promote G_1 progression by phosphorylating and thereby neutralizing "pocket proteins" of the *retinoblastoma* family. In the hypophosphorylated state, these pocket proteins bind E2F family members, inhibiting their function and in some cases converting them into repressors of transcription. Upon phosphorylation by CycD/Cdk4, the pocket proteins release E2F, thereby allowing it to *trans*-activate a large set of genes that promote cell-cycle progression, including those encoding enzymes for nucleotide synthesis and G_1 cyclins such as CycE. Suppression of CycD/Cdk activity in many vertebrate cells slows or arrests the cell cycle (see, e.g., Fingar et al. 2002), and deletion mutations of each of the three D Cyclins or their conjugate Cdks in mice cause tissue-specific defects in cell proliferation (Fantl et al. 1995; Sicinski et al. 1995, 1996; Rane et al.

1999; Tsutsui et al. 1999). Although few studies of the mammalian E2F/Rb/CycD network in vertebrates have directly tested the possibility that the network functions in cell growth, as opposed to cell-cycle progression, this possibility is at least consistent with much of the extant data from vertebrate systems. The mouse knockouts show decreased growth in select tissues, but it remains unclear whether these defects are due to defects in cell growth or to defects in cell proliferation, as suggested by Geng et al. (1999) and others. In this respect, it is interesting that the pocket proteins, the best-characterized targets of CycD/CDK complexes, have been implicated as negative regulators of RNA polymerase I- and polymerase III-mediated transcription (Brown et al. 2000; Ciarmatori et al. 2001; see Chapter 12, this volume). Given that pol I and pol III are essential for maintaining the cell's protein synthetic capacity, this function is expected to be important in cell growth regulation, but it has not yet been tested in vivo.

Genetic studies of *Drosophila* E2F1, the retinoblastoma homolog RBF, and their targets have validated the accepted paradigm of cell-cycle regulation to a remarkable extent (Dyson 1998). These studies have also indicated, however, that one component of this network, CycD/Cdk4, has the capability to promote cell growth, in addition to its expected functions in promoting cell-cycle progression (Datar 2000; Meyer et al. 2000). The *Drosophila* CycD/Cdk4 complex's functional characteristics differ markedly from those of other cell-cycle regulators, including E2F1, RBF, and CycE (Datar 2000; Meyer et al. 2000). When overexpressed in wing cell clones, CycD/Cdk4 forced cells to divide faster than controls without altering cell size, and so yielded larger cell clones than controls. In the postmitotic, differentiating eye, ectopic CycD/Cdk4 caused cell enlargement (hypertrophy) in multiple differentiated cell types (pigment, cone, and photoreceptor), enlargement of interommatidial bristles, ommatidia (eye facets), and the entire eye. Ectopic CycD/Cdk4 also promoted extensive cell proliferation in differentiating cells in the pupal eye, which are normally quiescent, but did not overtly block differentiation as reported by several studies in vertebrates (Skapek et al. 1995; Lazaro et al. 2002). In addition, CycD/Cdk4 complexes promote cell growth and concomitant increases in ploidy in the polyploid tissues of the larva. Complementary studies of the *cdk4* loss-of-function mutants show that Cdk4 is not essential for viability or cell-cycle progression in *Drosophila* but is in fact required for normal rates of cell growth (Meyer et al. 2000). These results are not dissimilar to those reported for the CycD and Cdk4 mutants of mice (Fantl et al. 1995; Sicinski et al. 1995, 1996; Rane et al. 1999; Tsutsui et al. 1999). In summary, these studies indicated that in *Drosophila*, CycD/Cdk4 positively regulates both cell growth and cell-cycle progression.

The data from *Drosophila* are consistent with RBF and E2F1 being the critical targets of CycD/Cdk4 for stimulating cell-cycle progression, but the growth-regulatory targets of the complex remain uncertain. Gain of E2F1 function or loss of RBF function does not appear to augment growth in this system, suggesting that other targets must be used. Although comprehensive screens to identify these targets have not yet been reported, one recent paper has implicated the Stat92E transcription factor as a target of CycD/Cdk4 that might mediate its cell-growth effects (Chen et al. 2003). Stat92E is a downstream effector in *Drosophila's* JAK/STAT pathway, which is functionally conserved with that found in vertebrates (see below). This paper shows that CycD/Cdk4 can stabilize Stat92E, increase its activity as a transcription factor, and also synergize with overproduced STAT in driving overgrowth of the eye.

How are the levels and activity of CycD/Cdk4 regulated in the fly? Might CycD be used to control differential cell growth during organogenesis? Studies of mutants for the three D-type cyclins in mice have highlighted tissue-specific expression and function (Fantl et al. 1995; Sicinski et al. 1995,1996; Rane et al. 1999; Tsutsui et al. 1999) and thus support this possibility, although the emphasis has been on cell cycling as opposed to cell growth. In the fly, some insights come from a recent study showing that signaling through the *hedgehog (hh)* pathway is essential for patterning CycD protein expression in the developing eye (Duman-Scheel et al. 2002). CycD expression exhibits a pattern in the eye that corresponds to the wave of photoreceptor differentiation and the final round of cell division in undifferentiated neuronal precursors, termed the "second mitotic wave." This last wave of mitosis requires signaling from the Hh ligand via its receptors, Patched (Ptc) and Smoothened (Smo), to a transcription factor, Cubitus interruptus (Ci). Using a variety of functional genetic tests, Duman-Scheel et al. (2002) showed that Ci is required for, and can promote, the expression of both CycE and CycD in this region of the eye. CycE is required for the S phase that precedes the second mitotic wave, whereas CycD may promote the growth that accompanies division of these cells, although this was not demonstrated. This study also showed that Ptc and Ci can affect the growth rates of cell clones in the wing, and that this required the presence of Cdk4. Thus, it was proposed that the growth effects of Hh signaling are mediated, at least in part, by regulating levels of CycD/Cdk4 activity, probably via affecting *cycD* transcription. No correspondence has been reported between the normal expression patterns of Hh and CycD in the wing, however, suggesting that *cycD* transcription either has multiple inputs, or that the effects of Hh signaling on CycD/Cdk4 activity in the wing may be posttranscriptional.

JAK/STAT and Csk/Src Signaling

Several recent papers, again focusing mostly on the eye, show that the fly orthologs of these factors have profound effects on cell proliferation and growth. Although these genes affect other cellular processes in some contexts (e.g., cell patterning and fate specification), the effects observed in the eye are most easily interpreted as specific effects on cell growth. Loss of function of Stat92E, the downstream transcription factor in this system, or of its upstream regulators *hopscotch* (*hop*, a JAK ortholog) and *unpaired* (*unp*, a ligand), reduces eye growth, whereas gain of function of *hop* or *unp* increases cell proliferation in the eye (Chen et al. 2002; E. Bach et al., in prep.) and other tissues such as hemocytes (Dearolf 1998). Massive eye overgrowth was observed when the *Drosophila* C-Src kinase (*dCsk*), a negative regulator of *Drosophila Src*, was specifically deleted in the developing eye (R. Read and R. Cagan, unpubl.). In this case, larval size was also increased in *dCsk* mutants relative to age-matched controls, suggesting general growth effects in the differentiated larval cells. Consistent with studies in vertebrate cells, epistasis tests performed in the eye indicated that *Drosophila's* two *Src* genes, and two downstream kinases *Btk* and *bsk* (the *Drosophila* JNK ortholog), were required for the overproliferation of *dCsk* mutant cells. Epistasis tests also indicated that the JAK/STAT pathway was required for this overgrowth (R. Read and R. Cagan, unpubl.). The eye overgrowth of dCsk mutant tissue was attributed to increased cell proliferation and increased ommatidia (eye facet) number, without detectable effects on patterning or cell size at the adult stage. This suggests that this pathway regulates cell growth in a different manner than does the insulin/Tor signaling pathway, which has mild effects on ommatidia number but profoundly increases cell size. As noted above, a functional connection between Stat92E and CycD has been observed in *Drosophila* (Chen et al. 2003), suggesting another possible mode of growth control.

Bantam

This interesting locus was discovered by several groups in screens to identify genes that increase wing or eye size when overexpressed in those organs. The initial report (Hipfner et al. 2002) described not only gain-of-function overgrowth of the eye and wing, but also a loss-of-function mutant with a larval growth defect. This established the *bantam (ban)* locus as a dose-dependent regulator of cell and organ growth. A clever bit of detective work from Brennecke et al. (2003) identified *bantam* as a micro-RNA (Ambros 2001), thus defining an entirely new mechanism of growth control. Although the growth-regulatory targets of *ban* have not yet been defined, studies of other miRNAs and *hid*, a validated *ban* target involved in cell death, indicate that *ban* probably stimulates cell growth by suppressing the translation of mRNAs that encode growth inhibitory factors. The *ban* miRNA is conserved in *diptera* and has relatives in *C. elegans*, suggesting that this mechanism could be widespread. Using an RNAi-based reporter system, it was also shown that *bantam* activity is patterned in the developing fly wing, raising the exciting possibility that *ban*, like *dmyc*, may integrate patterning information and regulate patterned cell growth accordingly. In this case, *ban* activity was shown to be responsive to patterned *wingless* signaling and also correlated with the pattern of *hedgehog* signaling.

Insulin Signaling

One important regulator of cell growth in imaginal disc cells and the ERTs is insulin signaling (see Chapter 6). *Drosophila* produce seven *insulin-like peptides* (dILPs), which are expressed by the gut, imaginal discs, and neurosecretory cells in the brain (Brogiolo et al. 2001; Ikeya et al. 2002; Rulifson et al. 2002). dILPs are presumed to be secreted into the hemolymph and to act on all the cells in the animal in an endocrine fashion. A single insulin receptor (InR) acts via a canonical phosphatidylinositide 3-kinase (PI3K)/protein kinase B (PKB/AKT) pathway, which has been well characterized in both *Drosophila* and mammals. How insulin signaling stimulates cell growth is not fully understood. Genetic analysis indicates that the nutrient-sensing protein kinase, target of rapamycin (Tor), is required for insulin-stimulated growth. Tor is thought to regulate protein synthesis via the S6 kinase (S6K) and 4EBP, a suppressor of the translation initiation factor eiF4E, as well as other mechanisms (Schmelzle and Hall 2000). Tor appears to be linked to insulin signaling via the Tsc1/2 complex, which is growth-suppressive (Gao et al. 2002; Potter et al. 2002), and Rheb, a Ras-related G-protein that has positive growth effects and is a direct target of the GTPase-activating domain of Tsc2 (Patel et al. 2003; Saucedo et al. 2003; Stocker et al. 2003; Zhang et al. 2003). Tor is also thought to respond to nutrient availability directly, in an insulin-independent fashion. Cellular levels of ATP, amino acids, or amino-acyl-tRNAs have been suggested as inputs, but it remains unclear which metabolites might be sensed by Tor. Although protein synthetic control is an attractive mechanism of growth control, the in vivo data supporting S6K and 4EBP as growth regulators in the fly are rather weak

(Bernal and Kimbrell 2000; Montagne 2000; Miron et al. 2001). Although many components of this network are essential, S6K is not (Montagne 2000), and 4EBP has not been clearly implicated in growth control (Miron et al. 2001). One plausible idea is that insulin signaling promotes growth primarily by regulating cellular import and storage of nutrients (sugars and/or amino acids). Studies in mammalian cells show that nutrient import and storage are insulin-dependent, and implicate molecules such as Glut4 (a glucose transporter), glyocgen synthase (stores glucose as glycogen), and hexokinase (used in ATP generation) as mediators of these effects (Saltiel and Kahn 2001; Edinger and Thompson 2002). Nutrient import is one plausible connection between insulin signaling and Tor, which would presumably sense any nutrient influx stimulated by insulins.

Several studies indicate that *Drosophila* use the insulin signaling system to coordinate cell metabolism and growth with the availability of nutrients from the diet (Britton et al. 2002; Ikeya et al. 2002). This was first suspected because mutations leading to loss of InR/PI3K activity phenocopy starvation, both at the cell and the organismal level. In addition, experimental activation of the insulin/Tor network has the ability to drive cell growth and DNA replication in starved, quiescent larvae, which normally display no cell growth whatsoever (Fig. 2) (Britton et al. 2002; Saucedo et al. 2003). This property is unique to InR/PI3K/Tor pathway components and is not shared by other growth regulators such as dMyc or CycD. The idea that InR/PI3K activity is nutritionally regulated was tested using a pleckstrin homology (PH)-GFP fusion protein (GPH), which functions as a cellular indicator for PI3K activity by virtue of associating with the PI3K second messenger, PIP_3, at cell membranes. Studies using this reporter indicated that PI3K activity is suppressed in diverse cell types when a larva is starved for protein (Britton et al. 2002). Moreover, bovine insulin, when added to tissues from starved animals in culture, restored membrane association of the PH-GPH reporter within 15 minutes (D. Chiarelli and B. Edgar, unpubl.). This indicates that the dILPs (insulins) probably become limiting for PIP_3 production during starvation. Consistent with this observation, it has been shown that expression of the dILP3 and dILP5 mRNAs is repressed when larvae are starved (Ikeya et al. 2002). Thus, *Drosophila* appear to use at least some of their seven dILPs to coordinate cell growth and metabolism with diet (Britton et al. 2002). Columbani et al. (2003) contributed another twist to this physiological paradigm by showing that suppression of amino acid import specifically in the fat body could inhibit PI3K activity remotely, in peripheral tissues such as epidermis, and that this also caused non-autonomous suppression of growth of the organism. Furthermore, this study showed that the fat body in fed, but not starved, larvae produces dALS (acid labile subunit), a secreted glycoprotein that probably binds the dILPs in the hemolymph and potentiates their action. Thus, although the fat body has not been reported to be a site of dILP production, it does seem to serve an important role as a tissue that senses hemolymph amino acid levels, and then modulates insulin signaling activity and cell growth rates throughout the animal accordingly.

Is the insulin signaling system used to control differential cell growth during organogenesis in *Drosophila*? So far, most of the data pertaining to this interesting question are negative. Consistent with the idea that PI3K activity is regulated by circulating dILPs rather than by local patterning cues like Dpp, Wg, or the EGFR ligand Vein, the membrane association of the PH-GFP reporter introduced above was found to be spatially uniform in wing imaginal discs (Prober and Edgar 2002). Moreover, ectopic activation of EGFR or Dpp signaling had no effect on PH-GFP localization (Martin-Castellanos and Edgar 2002; Prober and Edgar 2002). These observations indicate that PIP_3 is not used as a second messenger by diverse growth-regulatory pathways and is a specific intermediary in insulin signaling. It is quite possible, however, that some of the patterning signals that regulate cell growth do so by feeding into the insulin/PI3K/Tor network downstream of PIP_3.

Other Endocrine Factors

In addition to insulin signaling, there is circumstantial evidence that several other insect hormones are involved in cell and tissue growth in *Drosophila*. For instance, the slow growth of imaginal discs in *giant (gt)* mutant larvae is attributable to a lowered ecdysone level in these larvae. Normal disc growth can be restored by feeding larvae an ecdysone-fortified diet (Schwartz et al. 1984), indicating that ecdysone signaling may be important for normal disc growth. Low levels of ecdysone have also been clearly demonstrated to be required for cell proliferation in the CNS of another insect, *Manduca sexta* (Champlin and Truman 2000). Early attempts to culture intact larval imaginal discs of *Drosophila* in vitro (Davis and Shearn 1977) found that supplementation of the culture medium with juvenile hormone (JH) was also required to obtain growth. Interestingly, a decline in the rate of wing disc growth in the middle of the third larval instar (Bryant and Levinson 1985) coincides with a decline in the juvenile hormone titer (Riddiford 1993a). In addition to ecdysone and JH, studies of cloned wing imaginal disc cells in culture indicate that a family of secreted chitinase-related proteins called imaginal disc growth factors (IDGFs; Kawamura et al. 1999) and another family of adenosine deaminase-related growth factors (ADGFs; Zurovec et al. 2002) are required for cell growth and proliferation. The ADGFs are particularly interesting since they require enzymatic activity for function and appear to stimulate cell growth and proliferation by depleting extracellular pools of adenosine, which is growth-suppressive.

Tumor Suppressor Genes

Another class of genes that have been characterized as regulators of cell growth in flies are the so-called "tumor suppressor" genes studied by Gateff, Bryant, and others. These include *fat, lethal giant discs (lgd), warts (wts), lethal discs large (dlg), scribble (scrib),* and as many as 40 other genes (for reviews, see Gateff 1994; Watson et al. 1994). These genes cause over-growth of the imaginal discs, brain, or blood cells in mutant animals. In a few cases (e.g., *wts, scrib, salvador/shar pei, hippo*), clones of mutant cells induced in wild-type imaginal discs will also overgrow (Xu and Rubin 1993; Justice et al. 1995; Kango-Singh et al. 2002; Tapon et al. 2002; Bilder et al. 2003; Harvey et al. 2003; Pantalacci et al. 2003; Udan et al. 2003; Wu et al. 2003), but in the other cases that have been tested, mutant cells in a mosaic are out-competed and eliminated by wild-type cells. Several of these tumor suppressor gene products (*fat, dlg, lgl, scrib*) localize to cell–cell junctional complexes and are required to maintain epithelial integrity and thus appear to function in cell adhesion and polarity (Bryant 1997; Bilder et al. 2003). Accordingly, it has been suggested that the over-proliferation phenotypes seen in these mutants stem from aberrant cell adhesion or communication, rather than from direct effects on cell growth. Another group of mutations, composed of *warts, salvador,* and *hippo,* encode cytoplasmic proteins that form a protein kinase complex (Xu and Rubin 1993; Justice et al. 1995; Kango-Singh et al. 2002; Tapon et al. 2002; Harvey et al. 2003; Pantalacci et al. 2003; Udan et al. 2003; Wu et al. 2003). Deletion of any of these genes in cell clones in the eye allows cells to continue proliferating, and also suppresses apoptosis, in regions of the eye disc where cells would normally undergo G_1 arrest in conjunction with the onset of differentiation. Such mutant cells show increased levels of DIAP1 and Cyclin E mRNA and protein, which respectively inhibit apoptosis and promote $G_1 \rightarrow S$ progression. These targets seem insufficient to explain the dramatic overgrowth phenotypes caused by deletion of *hippo, warts,* or *salvador,* however, and thus this interesting complex is expected to target other growth-regulatory factors.

COORDINATING GROWTH WITH CELL-CYCLE PROGRESSION

Studies in fission and budding yeast, as well as in human cells, have suggested at least one attractive molecular model to explain how cell-cycle progression is coordinated with cell growth (see Chapter 4). This posits that the level of a critical, unstable, limiting regulator of cell-cycle progression is sensitive to the translational capacity of the cell. Since levels of this cell-cycle regulator would be limiting for cycle progression, this relationship would result in the rate of cell-cycle progression also being coupled to translational efficiency, and hence to cell growth. Such a mechanism has been supported by detailed studies of the G_1 cyclin CLN3 in budding yeast (Polymenis and Schmidt 1997) and of the G_2/M regulator Cdc25 in fission yeast (Daga and Jimenez 1999). In both cases, the coupling mechanism appears to involve translational repression of the cell-cycle regulatory mRNA by upstream open reading frames (uORFs), which prevent the messages from being productively translated when growth conditions are poor and, presumably, ribosome concentrations and rates of translational initiation are low. Such a mechanism has been proposed to apply in vertebrate cells (Rosenwald et al. 1993) and, as discussed below, is consistent with work in *Drosophila* (Prober and Edgar 2000, 2002; Kango-Singh et al. 2002; Tapon et al. 2002).

Endocycles are Growth-coupled

In *Drosophila's* larval endoreplicating tissues (ERTs), growth and cell-cycle progression proceed in concert. These tissues are fully differentiated and lack mitotic control, and so provide a simple context for understanding how cell growth might affect G_1/S progression. Numerous observations suggest that the larval endocycles are in fact controlled rather directly by rates of cell growth. First, endocycle progression is dependent on an ongoing influx of dietary protein from feeding. These cycles can be stopped, restarted, slowed, or accelerated by altering levels of amino acids in the diet (Britton and Edgar 1998). Altering nutritional conditions, temperature, or the hormonal system that defines the feeding period also changes the final ploidy of the larval tissues. Treatments that delay or prevent pupariation and metamorphosis, for instance, extend the feeding period and allow the larva to achieve higher than normal levels of ploidy. Second, since the endocycles require cell growth for their progression, mutations in genes required for protein synthesis and cell growth generally block endocycling (Zhang et al. 2000). Third, the endocycles can be "jump-started" in protein-starved, quiescent animals by genetic manipulations that autonomously promote cell growth. These include overexpression of the *Drosophila* insulin receptor (InR), PI3K (Britton et al. 2002), or Rheb (Saucedo et al. 2003), or loss of the Tsc1/2 complex. Growth-promoting manipulations such as the overexpression of InR, PI3K, dMyc, activated EGFR, Ras, and CycD/Cdk4 can also promote hyperpolyploidy of endocycling tissues in fed animals. For instance, the larval fat body, which normally endoreplicates to 256C, can be driven to >2000C by overexpression of dMyc (Fig. 1) (Edgar and Orr-Weaver 2001).

How do these growth-promoting maneuvers drive the DNA endoreplication cycle? By analogy with the examples from yeast cited above (Polymenis and Schmidt 1997; Daga and Jimenez 1999), one attractive mechanism is that an unstable, translationally regulated protein limits its G_1/S transitions. Consistent with this idea, genes that promote cell growth in *Drosophila* when overexpressed (Ras^{V12}, dMyc, InR), or lost (Tsc2, *salvador*, *hippo*), increase levels of CycE protein, which is the limiting factor for G_1/S progression (Prober and Edgar 2000). In the case of Ras^{V12} and dMyc, this effect has been shown to be posttranscriptional (Prober and Edgar 2000, 2002). Although these observations were made mostly in mitotic imaginal discs rather than ERTs, they are provocative. Both of the *Drosophila* CycE mRNAs contain several upstream open reading frames (uORFs) in their 5′UTRs, which could potentially suppress translational initiation under suboptimal growth conditions (Polymenis and Schmidt 1997; Daga and Jimenez 1999; Geballe and Sachs 2000). This suggests that the growth-dependent up-regulation of CycE could be affected translationally, although stabilization of the protein under optimal growth conditions has not been ruled out.

When considered in the context of other studies of the endocycles, growth-sensitive expression of CycE suggests a simple, albeit speculative, model for endocycle control (Fig. 3A). This model posits that alternating pulses of E2F1 and CycE expression are generated by positive and negative feedback between CycE/Cdk2 and the transcriptional regulators, E2F1/Dp and Rbf. According to this model, E2F1 stimulates CycE transcription during gap (G) phases, initiating positive feedback in which rising CycE/Cdk2 activity augments E2F1 activity by inhibiting the E2F corepressor, RBF. The resulting high levels of CycE/Cdk2 activity eventually trigger the G/S transition, but also initiate negative feedback in which E2F1 is down-regulated, most probably by CycE/Cdk2-dependent degradation (Heriche et al. 2003; T. Reis and B. Edgar, in prep.). The subsequent loss of E2F1 then reduces *CycE* transcription and lowers levels of Cdk2 activity, allowing the reassembly of pre-replication complexes on the DNA during the next G phase (see Diffley and Labib 2002). Then the cycle resumes. By analogy with findings in vertebrate cells, autophosphorylation of CycE might also contribute to its periodic degradation (Strohmaier et al. 2001). This model is supported by much experimental data (for review, see Edgar and Orr-Weaver 2001). Notably: (1) *cycE* mRNA and protein oscillate in the ERTs (Duronio and O'Farrell 1995; Lilly and Spradling 1996); (2) CycE down-regulates its own transcription (Sauer et al. 1995; Lilly and Spradling 1996); (3) constitutive expression of *cycE* mRNA blocks endocycle progression (Follette et al. 1998; Weiss et al. 1998); (4) E2F1 protein cycles with a cell-cycle periodicity (Heriche et al 2003; T. Reis and B. Edgar, in prep.); and (5) overexpressed E2F1 can accelerate endocycle progression, leading to hyperpolyploidy (J. Evans and B. Edgar, unpubl.). In this model, the feedback relationships between CycE/Cdk2 and E2F1/Dp/Rbf constitute an autonomous oscillator, the speed of which is determined by rates of CycE production and turnover. Growth-dependent modulation of these parameters could, in principle, serve to couple rates of endocycle progression to rates of cell growth.

Growth and Cell-cycle Control in Mitotic Cells: A More Complex Story

How is cell growth coordinated with cell-cycle progression in mitotically proliferating cells? Studies of the *cdc* mutants in *Saccharomyces cerevisiae* and *Schizosaccharomyces pombe* performed in the 1970s and 1980s showed that specifically slowing or arresting cell-cycle progression does not immediately block cell growth and so can result in very large cells (Hartwell 1971; Nurse et al. 1976). More recently, similar experiments have been performed in plant cells using *Cdc2* mutations (Hemerly et al. 1995), in mammalian cells by overexpressing CDK inhibitors (Sheikh et al. 1995; Fingar et al. 2002), and in the wing disc of *Drosophila* using both approaches (Weigmann et al. 1997; Neufeld et al. 1998). The results are analogous to what was found in yeast: Cells forced to divide more slowly became much larger than controls (Fig. 4). Thus, in multicellular systems like yeast, cell growth is not immediately dependent on cell-cycle progression, and can be uncoupled from it. In contrast, treatments that slow growth generally do slow rates of cell division in both yeasts (Johnston et al. 1977) and *Drosophila* (Galloni and Edgar 1999), indicating that cell-cycle progression is often growth-dependent.

Recent advances have also made it possible to artificially accelerate cell-cycle progression in vivo, and thus to ask whether the cell cycle can potentiate or "drive" cell and tissue growth. This was accomplished in the fly wing by overexpressing either the combination of CycE and Cdc25/Stg,

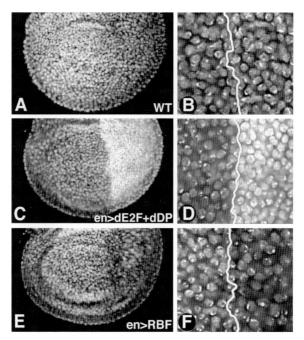

Figure 4. Altering rates of cell-cycle progression in posterior compartments of the wing alters cell size and number but does not affect compartment mass. Wing imaginal discs stained for DNA with DAPI are shown at low and high magnifications, with posterior compartments to the right. The AP borders are highlighted in the high-magnification photos (*B, D, F*). (*A, B*): Wild-type control. (*C, D*) A disc in which the E2F1/DP transcription factor was overexpressed using the engrailed-Gal4 driver in the posterior compartment. This decreases the cell doubling time from ~12 hours to ~9 hours. (*E, F*) A disc in which the *retinoblastoma* ortholog, RBF, was overexpressed using en-Gal4 in the posterior compartment. This increases the cell dividing time to ~18 hours. (Adapted, with permission, from Neufeld et al. 1998 [copyright Elsevier].)

which, respectively, limit G_1/S and G_2/M progression, or the E2F1/Dp transcription factor complex, which increases transcription of *cycE*, *stg*, and other cell-cycle genes (Neufeld et al. 1998). Either treatment accelerated the rate of wing cell division but, interestingly, failed to accelerate growth, in terms of mass increase, of cell clones or wing compartments. Consequently, cell size and viability decreased, presumably because cells cycled faster than they were able to double their mass (Fig. 4). This decrease in cell size is not unlike what is seen in *S. cerevisiae* cells that overexpress Cln3, the rate-limiting factor for cell-cycle progression at START, or *S. pombe* cells mutant for *wee1*, a limiting negative regulator of

Cdc2 and G_2/M progression. In the yeasts, however, cell doubling times have been reported to be normal, and the size decrease is thought to occur only during the first cycle after the mutations take effect. This is not the case when E2F1 is overexpressed in the fly wing, in which case rates of cell division are increased for many cycles. Perhaps as a consequence, E2F1 overexpression produces a continuous flow of very small cells that are eliminated by apoptosis. An important conclusion from these studies is that Cdc25/Stg, Cdk2/CycE, E2F1/DP, RBF, and their targets all function as cell-cycle-specific regulators and have no positive regulatory function in controling cell growth rates. A noteworthy corollary is that specific increases in cell-cycle progression are not sufficient to promote cell or tissue growth in the context of the wing. Yet another implication is that the developmental program that "measures" the growing wing does not use cell number as the critical metric, but relies on some facet of tissue mass.

Notwithstanding these conclusions, it should be emphasized that experimental treatments that inhibit rates of DNA replication or cell division can effectively throttle tissue growth (Du et al. 1996; Weigmann et al. 1997; Neufeld et al. 1998). This can be attributed to the essential requirement of genomic DNA for all metabolic processes. Thus, whereas specific cell-cycle regulators may not be used to promote growth, they could, in principle, be used to limit tissue growth as part of a developmental program. Although this mechanism has not yet been reported in flies, studies of CDK inhibitor mutants in mice (Fero et al. 1996; Kiyokawa et al. 1996; Nakayama et al. 1996) and *C. elegans* (Kipreos et al. 1996) provide in vivo examples of this mechanism of growth control in action.

How does cell growth affect the mitotic cell cycle in *Drosophila*? In wing discs, mutations that slow growth, such as *Minutes*, *dmyc*, or lesions in signaling pathways required for growth do, as a rule, slow the cell cycle. Reductions in insulin signaling components or *dmyc* also decrease cell size, but other growth-defective mutations, like the *Minutes* or *cdk4*, do not (Meyer et al. 2000; Martin-Castellanos and Edgar 2002). In addition, high rates of cell death occur when the machinery of cell growth is compromised, even in non-mosaic situations that do not involve competition with wild-type cells. Overall, experiments using loss-of-function mutations in growth genes indicate a tight dependency of cell-cycle progression on growth in the wing. Although it has not been directly tested, the same dependency would not be expected in embryonic cells, or perhaps even larval neuroblasts, which have fast cell cycles that progressively reduce the amount of cytoplasm per cell.

Studies in which cell growth was accelerated in wing and eye disc cells add some complexity to the conception of growth-regulated cell cycles.

Although growth stimulators such as dMyc, activated Ras, and components of the insulin/PI3K/Tor network do in fact accelerate cell growth in the wing when overexpressed, these factors appear to be incapable of actually speeding up rates of cell-cycle progression in this context (Johnston et al. 1999; Verdu et al. 1999; Weinkove et al. 1999; Prober and Edgar 2000, 2002; Saucedo et al. 2003). Consequently, the increased growth brought by these factors causes marked increases in cell size. FACS analysis indicates that all of these growth regulators accelerate G_1/S transitions, and thus truncate the G_1 phase, but that rates of G_2/M progression are unaffected (Fig. 5). The acceleration of G_1/S progression has been correlated in some cases with increased levels of CycE (Prober and Edgar 2000, 2002), suggesting that the same growth-sensitive mechanism proposed for the endoreplicating cells (Fig. 3A) also operates in the mitotic wing (Fig. 3B). The finding that increased cell growth does not necessarily accelerate rates of cell division in the wing suggests that the limiting factor for G_2/M progression, Cdc25/Stg, may not act as a "growth sensor" like CycE does. Nevertheless, it is interesting to note that the Cdc25/Stg protein is unstable, being periodically degraded at mitosis (Edgar and Datar 1996), and that the *stg* mRNA's 5′UTR has three upstream ORFs. This suggests that growth-dependent regulation of Cdc25/Stg activity could occur, even if it is not a major mode of control.

Studies in both embryos and wing discs indicate that the principal mode of control over *stg* activity is transcriptional (Edgar and O'Farrell 1989; Edgar et al. 1994a; Milán et al. 1996a,b; Lehman et al. 1999). In late-stage wing discs, *stg* is subject to dominant transcriptional control by patterning genes, much as described above for the embryonic cell cycles. *stg* mRNA expression is repressed at the dorsoventral compartment boundary in response to *wingless (wg)* at the L3 stage (Johnston and Edgar 1998), and *stg* mRNA displays vein/intervein patterning in the pupal wing, presumably in response to *EGFR, Notch*, or *decapentaplegic (dpp)* signaling. These patterns affect mitotic patterning in the late-stage wing disc (Schubiger and Palka 1987; Milán et al. 1996a,b). Interestingly, *stg*'s pattern sensing capability appears not to affect the cell cycle significantly during early disc growth. Cells in early discs cycle very rapidly, express *stg* RNA constitutively, and have a very short G_2 period relative to more mature discs (Fain and Stevens 1982; Neufeld et al. 1998). This suggests that the cell cycle in very early wing discs is primarily regulated at G_1/S transitions and that, as in the ERTs, cell growth could be the limiting parameter. As development progresses, however, distinct on/off patterns of *stg* expression resolve, and the disc cells slow their cycle by accumulating in G_2. In this way, patterned control of G_2/M by *stg* appears to act as a brake

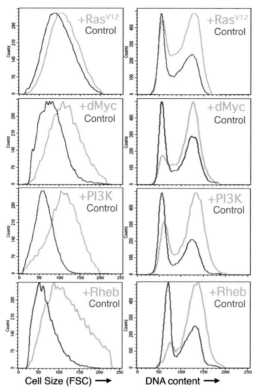

Figure 5. Genes that promote growth increase cell size and accelerate G$_1$/S progression in the wing. FACS analysis shows cell size as assayed by forward light scatter (FSC; *left* panels) and cellular DNA content (*right* panels). The indicated genes were overexpressed for 48 hours using the Flp/Gal4 technique in wing imaginal discs. Results with experimental cells are shown with pink lines; and results with green, internal control cells are shown with green lines.

to cell proliferation (and hence growth), even when growth promoters like insulin signaling or dMyc are inappropriately activated (Figs. 3B, 5). The contrast to the ERT cell cycle, which lacks this fail-safe G$_2$/M checkpoint and runs faster in response to increased growth, is intriguing.

It is also interesting to note that promoters of cell growth including dMyc and components of the insulin-signaling network (PI3K, InR, Rheb) are incapable of prolonging the proliferative phase of development in imaginal wing and eye discs. Expression of these factors using Gal4 drivers that are active in postmitotic cells, such as GMR-Gal4 in the eye or en-Gal4 in the wing, does promote postmitotic cell growth but fails to drive cell-cycle progression in the differentiating cells, which remain arrested in G_1. Such observations are consistent with the notion that the developmental program that shuts off transcription of the cell-cycle control genes at cell differentiation runs independently of cell growth. Given that some cell types (e.g., eye photoreceptors) normally sustain substantial amounts of growth after they cease cycling and begin to differentiate, this is perhaps not surprising.

ORGAN SIZE CONTROL: THE *DROSOPHILA* WING

Patterned Growth during Organogenesis: Wing Development

One of the great unsolved mysteries of development concerns how organ size, shape, and form are determined and realized. Although our understanding of the molecular/genetic basis of pattern formation in terms of signaling and transcriptional control has advanced immeasurably in recent years, the connections of patterning systems to cell growth and proliferation remain surprisingly obscure. The wings and eyes of *Drosophila* are among the most intensively studied organs in terms of pattern formation and growth. The wing, being a relatively simple organ consisting of a folded epithelial sheet and containing only a few cell types, would seem more likely than many organs to yield insight into patterned growth control. Here, we outline what is known about patterning and growth in the wing and discuss the models that have been proposed to connect the one to the other. A detailed discussion of the eye is not included, but this can be found in Baker (2001).

As outlined above, the embryo specifies a cluster of 35–40 epidermal cells to become each wing. These remain diploid and undifferentiated during the latter stages of embryogenesis. During the first 12 hours of larval development, these cells increase their mass about sixfold without dividing (Madhaven and Schneiderman 1977) and then proliferate exponentially throughout the three larval instars. The wing disc cell cycle is rapid at first (6–8 hours) but slows progressively, such that at the end of the third instar it is >15 hours long. Cell size shows a small but progressive decrease during the progression from faster to slower cycles (Fain and Stevens 1982; Neufeld et al. 1998), and thus the disc cells divide a bit more rapidly than they double their mass. Two mitoses with an intervening S phase occur during early pupal development (Schubiger and Palka 1987; Milán 1996b), and then the cells arrest and differentiate to form the adult wing. Studies using BrdU incorporation (Milán et al. 1996a,b) and clonal analysis (García-Bellido and Merriam 1971; González-Gaitán et al. 1994) show that virtually all cells in the wing disc proliferate during most of its growth phase. These mosaic analyses also showed that growth zones, in which cells proliferate faster or slower, can be mapped and that cell lineages in the discs are highly variable and quite plastic.

Mosaics composed of mixed fast- and slow-growing cells have been extremely informative. These were created using mitotic recombination and *Minute* mutations, which encode ribosomal proteins (Morata and Ripoll 1975; Lambertsson 1998) and have dominant suppressive effects on cell growth. Studies using *Minute* mutations defined the phenomenon of "cell competition," in which a few fast-growing progenitor cells can populate very large parts of the wing, effectively eliminating slow-growing cells, without altering the size or form of the final organ (Fig. 6) (Morata and Ripoll 1975; Simpson and Morata 1981). Furthermore, these experiments defined for the first time the existence of lineage-restricted cellular compartments within the discs that function as units of growth control (see Dahmann and Basler 1999; Milán and Cohen 2000). The wing primordium has just two compartments when it is established, anterior (A) and posterior (P), and cells within a given lineage never cross the compartment boundary. Later in wing development, compartment boundaries arise to restrict lineages to the dorsal and ventral halves, and, later still, cell lineage boundaries appear along most of the wing veins (González-Gaitán et al. 1994). Compartment boundaries appear to be boundaries of cell adhesion (Milán et al. 2002), and interactions across the boundaries are essential to set up the signaling centers that direct patterned growth (see below).

Experiments in which the imaginal discs were surgically manipulated or transplanted define other relevant characteristics of their growth. When transplanted from a larva into an adult host, imaginal discs will continue to grow until reaching their normal final size and shape and then cease growth (Bryant and Levinson 1985). This indicates that final disc size and shape are controlled by signals intrinsic to the organ, rather than by external factors. When a fragment of an imaginal disc is excised and then the disc is allowed to heal in in vivo culture, the disc will regenerate the missing quadrant, or, if a large portion was excised, duplicate (French et al. 1976). This finding, together with experiments from other insects, gave rise to the idea that the discs contain an intrinsic "map" of positional information that can be restored during intercalar, regenerative growth (see below) (French et al. 1976; Lawrence and Struhl 1996; Day and Lawrence 2000).

Like the ERTs, imaginal cells require insulin signaling for sustained growth and cell-cycle progression (see Chapter 6). Discs appear to make

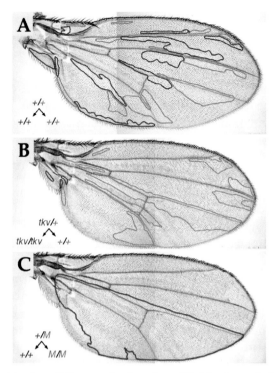

Figure 6. Plasticity in cell lineages in the wing. (*A*) Mosaic analysis using mitotic recombination shows that marked cell clones (*red* and *blue* outlined areas) take on random, irregular patterns but respect compartment boundaries at the wing margin (the DV compartment border), the AP boundary, and along veins. Sister clones deriving from a single mother cell are marked red or blue and are often found as adjacent pairs. Each small hair (*dark dots*) demarks one cell. (*B*) A mosaic in which cells mutant for the Dpp receptor *tkv* were generated. The *tkv* mutant clones (*red*) were eliminated by cell competition, leaving only the wild-type sister clones (*blue*), but the wing is normal. (*C*) A mosaic in which a clone of fast-growing +/+ cells was generated in a slow-growing *Minute* (+/M) background. A single +/+ cell has proliferated to fill nearly the entire posterior compartment of the wing (*red* outline). M/M cells (*blue*), which cannot make functional ribosomes, have all died. (*A, B,* Adapted, with permission, from Burke and Basler 1996 [copyright Company of Biologists]; *C,* adapted, with permission, from Simpson and Morata 1981 [copyright Elsevier].)

their own insulin-like peptide (Brogiolo et al. 2001), which might explain their ability to complete their growth even in larvae that have been starved early in the third instar, at a size less than half of that normally achieved by pupariation (Beadle et al. 1938).

In addition to endocrine factors such as the dILPs and IDGFs (Kawamura et al. 1999), locally secreted signals direct disc growth in time and space to allow the morphogenesis of complex structures. In the wing disc, these signals include decapentaplegic (Dpp), wingless (Wg), hedgehog (Hh), Notch (N), and the epidermal growth factor/neuregulin homolog Vein (Vn). Mutant phenotypes indicate that each of these factors is required for normal wing growth. Unlike the insulins, these signals also pattern cell fates, a process that occurs coincidentally with disc growth. Each of these signals is expressed and secreted in a stereotypic pattern, and these patterns form overlapping morphogen gradients, such that cells in different positions in the wing receive different combinations of signals (Fig. 7). These patterns direct differential growth within the tissue and are also involved in setting up the compartment boundaries that partition the organ (see Neumann and Cohen 1997; Dahmann and Basler 1999; Milán and Cohen 2000). As the disc grows and changes shape, the expression patterns of these signals also change (W.J. Brook et al. 1996). Orthologs of these factors are used to control differential growth during vertebrate development, so the problem of growth control by such signals is a very general one (St-Jacques et al. 1999; Minina et al. 2001; Gritli-Linde et al. 2002; Kioussi et al. 2002; van de Wetering et al. 2002; Yu et al. 2002).

Dpp, a TGFβ superfamily member orthologous to BMP4, is secreted in a stripe along the anterior edge of the anteroposterior (AP) compartment border of the wing (Fig. 7) (Basler and Struhl 1994; Tabata et al. 1995; Zecca et al. 1995; Teleman and Cohen 2000). This localized expression is a response of anterior cells to Hh, which is secreted from the posterior compartment and acts only at short range in anterior cells. Dpp provides a long-range morphogen gradient that orchestrates pattern formation and growth along the AP axis (Grieder et al. 1995; Lecuit et al. 1996; Nellen et al. 1996; Wiersdorff et al. 1996). Its importance in controlling wing growth is illustrated by experiments showing that loss of Dpp results in tiny wings (Fig. 8) (Zecca et al. 1995), and that cells lacking the Dpp receptor, Tkv, grow very slowly and are eliminated by cell competition (Fig. 6) (Simpson and Morata 1981; Zecca et al. 1995; Burke and Basler 1996; Kim et al. 1996; Lecuit et al. 1996; Martin-Castellanos and Edgar 2002). Conversely, ectopic Dpp causes wing duplications involving massive overproliferation (Fig. 8). Clones of cells expressing an activated Dpp receptor Tkv^{Q235D} also overgrow (Burke and Basler 1996; Martin-Castellanos and Edgar 2002). Interestingly, increasing Dpp expression in its own domain, or expressing Dpp or activated Tkv ubiquitously throughout the wing disc, causes a dramatic expansion of the wing, but only in the A← →P dimension (Basler and Struhl 1994; Capdevila and Guerrero 1994; Burke and Basler 1996; Lecuit et al. 1996; Nellen et al. 1996).

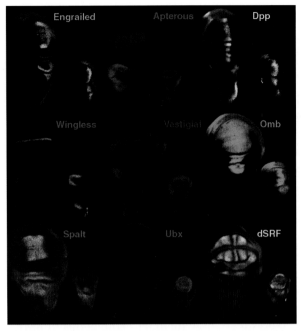

Figure 7. Patterns of signaling and transcription factors in wing and haltere imaginal discs. Pairs of wing discs (larger, *left*) and haltere discs (smaller, *right*) illustrate the expression patterns of the various factors discussed in the text. Ventral is up; posterior is right. Wingless and Dpp are secreted ligands that form diffusion gradients from their sources at the DV and AP compartment boundaries. The other gene products displayed are transcription factors and/or DNA-binding proteins that are, in some cases, targets of Wg (i.e., vestigial, dSRF) or Dpp (i.e., Omb, Spalt) signaling. Engrailed acts upstream of Dpp, whereas Apterous acts upstream of Wg. Ubx, which is expressed only in the haltere, modifies the effects of these signals, resulting in less growth (for details, see Weatherbee et al. 1998). (Courtesy of S. Carrol and S. Weatherbee.)

Wg is secreted in several regions of the wing disc in a dynamic pattern. The most thoroughly characterized sub-pattern is a stripe that bisects the wing primordium, orthogonal to Dpp and the AP boundary, and defines the dorsoventral (DV) compartment boundary. Notch (N) signaling is also involved in specifying the DV boundary, whereas Vn is expressed in a dynamic pattern that is essential for growth early in development and for determining vein/intervein boundaries at later stages. Studies in which Hh, N, Wg, and Vn signaling have been experimentally manipulated yield results analogous to those seen with Dpp: Ectopic expression of these ligands can cause wing duplications involving extra

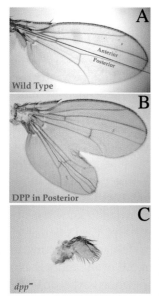

Figure 8. Dpp signaling affects wing size and pattern. In developing wild-type wings (*A*) Dpp is expressed in a narrow stripe along the AP border (*red* line). When Dpp is ectopically expressed in a posterior clone of cells (*red* outline), a duplication arises that involves massive overproliferation (*B*). When *dpp* function is reduced in the wing, cell proliferation is severely curtailed (*C*). (Adapted, with permission, from Zecca et al. 1995 [copyright Company of Biologists].)

growth, whereas their loss blocks growth in specific wing regions (Struhl and Basler 1993; Basler and Struhl 1994; Johnston and Schubiger 1996; Go et al. 1998; Baonza and García-Bellido 2000). As with the Dpp pathway, loss of intracellular signaling components in any of these pathways has deleterious, cell-autonomous effects on cell proliferation and viability. Recent studies using activated receptors and other cell-autonomous components to activate each pathway also provide clear evidence that these signals are cell-autonomously required for, and capable of, promoting cell growth and proliferation in the wing (Prober and Edgar 2000, 2002; Martin-Castellanos and Edgar 2002). Since the different signaling systems have unique regional patterns of activity, the specific regions of the wing affected after genetic manipulations vary according to the signal affected. The observation that expression of activated receptors in clones can accelerate cell growth and proliferation by these factors argues against models in which morphogen gradients (i.e., signaling differentials between cells) are essential for disc growth (Day and Lawrence 2000; see below).

Although most studies of disc growth and patterning work from a paradigm in which the signals that orchestrate growth travel laterally through the disc epithelium, recent papers suggest that signaling between the disc's two cell layers, the peripodial membrane and the columnar epithelium, is also important (Cho et al. 2000; Gibson and Schubiger 2000; Gibson et al. 2002).

Models of Patterned Growth Control in the Wing

Several models have been proposed to explain patterned growth in the wing (Serrano and O'Farrell 1997; García-Bellido and García-Bellido 1998; Day and Lawrence 2000). These are descended from early, abstract proposals addressing how positional information is generated by morphogen gradients (Wolpert 1971; French et al. 1976). They attempt to explain not only normal growth, but also the intercalar, regenerative growth that occurs when fragments of an organ are surgically removed or, as in amphibian and avian limbs, when a fragment is grafted on (for reviews, see Bryant and Simpson 1984; Held 2002). In general, these models remain abstract in that the cell–cell interactions envisioned as determining patterned growth have not been ascribed in detail to known genes and factors and that regulators of cell growth (e.g., dMyc, CycD, Ban, Tor) have not been incorporated. Nevertheless, the models are interesting to consider because they serve to define the problem. All of the models aim to address a few key characteristics of wing growth defined experimentally. One key observation is that, although the morphogens that orchestrate patterned growth (Dpp, Wg, etc.) display gradients of expression, corresponding gradients of cell proliferation are not observed (González-Gaitán et al. 1994; Milán et al. 1996a). A second property to be explained is that discontinuities in positional information, induced by surgical or genetic intervention, stimulate local cell proliferation. This proliferation eventually eliminates the discontinuity and, when this happens, proliferation ceases. A third aspect of wing development that must be considered is the role of compartment boundaries, which organize and shape groups of cells that otherwise display a striking degree of plasticity in cell lineages.

Model 1: Patterned Growth Factors

The extant data are superficially consistent with the notion that patterning signals shape the wing by acting as locally secreted factors that stimulate cell growth. One can imagine that the combined gradients of Dpp, Wg, Hh, and Vn, deriving from overlapping signaling centers, deliver a unique mix of growth factors to cells in different zones of the wing. The combinations of these factors might alter the metabolic activities of cells and thus affect the patterns of zonal cell growth and proliferation observed (Fig. 7). In principle, such a mechanism could determine the size and proportioning of the wing. The progressive sharpening of the spatial patterns of signaling, and coincident increases in signaling intensities, might eventually generate a signaling pattern that was incompatible with further growth and also trigger the onset of differentiation (for an example in the eye, see Yang and Baker 2003).

Although a patterned growth factor model is appealing, there is little evidence to support this simple view. Because the spatiotemporal dynamics of both the morphogen gradients and cell proliferation rates in the wing are complex, it has not been feasible to accurately correlate the one with the other. Real-time analysis could facilitate this but would require culture conditions sufficient to sustain disc growth, which have not yet been established. Numerous other considerations also indicate that this model is too simple to be correct. For instance, it is clear that differential cell adhesion along compartment boundaries, and perhaps within them, and the nonrandom alignments assumed by cells after mitosis (Resino et al. 2002) have a profound influence on the final shape, if not size, of the wing. Since the adhesive interactions between cells are regulated according to cell identity by the same patterned signaling factors (Dpp, Hh, Wg, N) that orchestrate differential growth, it has proven difficult to experimentally separate the effects of differential adhesion from differential cell growth and proliferation. Furthermore, it is difficult to reconcile this type of model with the observation that growth promotion by signals such as Wg, Dpp, Notch, and Vein is highly context-dependent. Each of these signals only promotes cell growth and proliferation in certain regions of a disc and at certain developmental stages (Neumann and Cohen 1996; Johnston 1998; Johnston and Edgar 1998; Martin-Castellanos and Edgar 2002). These signals may be neutral or even negative toward cell growth and proliferation in other contexts.

A striking example of context-dependent growth effects is Wg signaling, which is required for wing growth and presumed to drive it specifically in the notum (dorsal thorax) and wing hinge in early development (Neumann and Cohen 1996). Wg actually suppresses growth at the dorsoventral wing boundary during later development (Johnston 1998) and thus can be either a growth stimulator or growth suppressor, depending on context. One view of this type of context-dependence is that it results from inherent cell responsiveness to a given signal, based on the different combinations of the transcription factors a cell expresses and uses to respond to the signal. A pertinent example is the haltere, a rudimentary dorsal appendage that is homologous to the wing, but which is >5x smaller than the wing at the L3 disc stage (Fig. 7) and perhaps 20x smaller in the adult. During its development, the haltere expresses Hh, Dpp, Wg, etc. in patterns similar to those seen in the wing (Fig. 7), but it achieves a vastly different size. The size differential between wing and haltere is due solely to the homeobox transcription factor *Ubx*, which is expressed in the haltere but not in the wing (Fig. 7). Thus Ubx is, formally, a growth suppressor that presumably acts by dampening the growth response to patterned signals like Dpp (Weatherbee et al. 1998).

A common interpretation of the context-dependent growth observed with patterning signals is that these actually represent alterations in cell specification, rather than direct effects on cell growth and proliferation. In this view, the loss of growth observed in the *wg¹* mutant, for instance, is attributed to the transformation of wing blade tissue to notum (dorsal thoracic body), and the loss of growth in *dpp* mutant wings is viewed as the loss of distal (wing blade) cell fates. Likewise, the wing duplications resulting from ectopic Wg, Dpp, or Hh (e.g., Fig. 8) can be viewed as an indirect consequence of generating ectopic AP or DV boundary organizers. Such interpretations are consistent with a large body of data but do not necessarily exclude the possibility that these signals act as growth factors. To address this issue, it would be desirable to experimentally separate the growth and pattern-specification functions of each patterning signal, but this has proven to be difficult. There are only a few clues as to how the patterning signals might regulate the metabolic events that affect cell growth and proliferation (Duman-Scheel et al. 2002; Martin-Castellanos and Edgar 2002; Prober and Edgar 2002; Brennecke et al. 2003), and these do not yet clearly answer the question. Given that the patterning signals regulate cell growth and proliferation in some instances and not others, the expectation is that the specific connections made to metabolic control will be variable and indirect.

Model 2: Autocrine/Paracrine Signaling

Taking account of the fact that many tissues appear to control their growth via autocrine/paracrine signaling loops, Nijhout (2003) proposed a simple model for tissue size regulation based on auto-activation and inhibition. This model assumes that a tissue produces its own growth activator and also produces an inhibitor. This inhibitor could act either by sequestering the activator (as a receptor or a binding protein might) or by breaking down the activator (as a catabolic enzyme might). It is simple to find conditions under which the tissue will grow exponentially up to a well-defined size and then slow and eventually stop growing (Fig. 9). The slowdown in growth occurs because the rising concentration of the inhibitor eventually removes the activator faster than it is being synthesized. This model only requires that cells be sensitive to the concentrations of these molecules, not to the gradient across their diameter. If the activator is secreted as a paracrine signal that can also enter the circulatory system, this mechanism provides a way for discs to communicate and affect each other's growth (Emlen and Nijhout 2000). The disappearance of the growth stimulator from the circulatory system not only stops growth, but can also be used as a systemic signal to indicate that growth is completed. Such a systemic signal is believed to be required for the regulation of body size (see below). Likely candidates for the activator are insulin and IDGFs. Like some of the other models, this one accounts for the regulation of the total mass of a tissue but does not address the problem of pattern and shape.

Model 3: A Gradient of Responsiveness

Another model that utilizes auto-activation and inhibition, but which does address the problem of shape, was discussed by Serrano and O'Farrell (1997). They proposed a "gradient of responsiveness" that opposes the growth-stimulatory effects of Dpp and Wg. The Dpp and Wg gradients lie orthogonally to each other, Dpp along the anterior–posterior compartment boundary and Wg along the dorsoventral compartment boundary (Fig. 7). To flatten out the positive effects of the Dpp and Wg on cell growth and to produce uniform proliferation, this model proposes graded expression of an inhibitor that opposes the positive effects of Dpp and Wg, which are presumed to act synergistically to stimulate cell growth and proliferation (Fig. 10). Potential candidates for such an inhibitor include nitric oxide (Kuzin et al. 1996); *Dad*, a negative effector of Dpp signaling, the expression of which is Dpp-responsive (Tsuneizumi et al. 1997); *Brinker*, a negative effector of Dpp signaling, the expression of which is inhibited by Dpp (Campbell and Tomlinson 1999; Jazwinska et al. 1999; Minami et al. 1999); and *"wingful"* or *"notum,"* an inhibitor of Wg signaling, the expression of which is Wg-responsive (Gerlitz and Basler 2002; Giraldez et al. 2002). In this model, wing disc growth is complete, and therefore ceases, when the gradients of growth factors (Dpp and Wg) exactly counterbalance the gradients of the inhibitors (e.g., *Dad* and/or *wingful*).

Although this model was formulated before these inhibitors were discovered and is simplistic in that a single inhibitor is proposed, it has some compelling features. First, it explains how graded growth factors might elicit uniform growth. Second, subtle changes in the shapes of the gradients could, in principle, account for wing proportioning. Third, it appears to provide satisfactory explanations for intercalar growth during regeneration (Fig. 10). In the latter case, the proposal is that cells experiencing

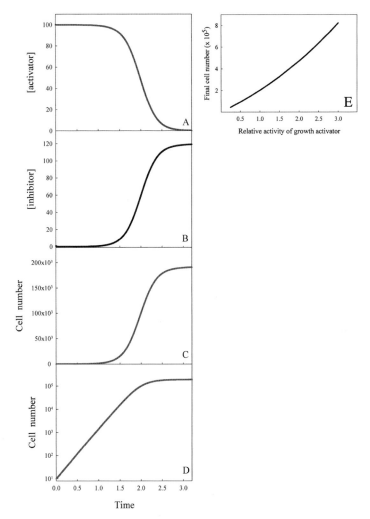

Figure 9. Wing growth (Model 2): Autoregulatory control of growth and size. Discs are assumed to synthesize their own growth activator and growth inhibitor, each at a disc-size-dependent rate (for mathematical details of the model, see Nijhout 2003). Panels *A* and *B* illustrate the time-dependent profiles of activator and inhibitor concentrations, and panels *C* and *D* illustrate the growth trajectory of the disc. Panel *D* shows the same data as panel *C*, except plotted on semi-logarithmic axes to show that growth is exponential during most of the growth phase. (Concentration and time are in arbitrary units and could be interpreted as: concentration = nM, time = days.) (Panel *E*) Dependence of the final size of an imaginal disc on the level of signaling by the growth activator. Different levels of activator signaling were simulated by different rates of activator synthesis. At higher levels of activator signaling, it takes longer for the activator to be eliminated, thus the discs grow to a larger size.

high levels of a growth-stimulatory factor (such as Dpp or Wg) would normally have a low level of responsiveness, but that if such cells were juxtaposed to cells with a high level of responsiveness, the latter would respond to the high local levels of Dpp by increasing their growth. Some of the model's specific predictions, such as asymmetric growth following certain types of surgical excision, are difficult to relate to real observations. Nevertheless, the discovery of *Dad* and *wingful/notum*, which could provide the postulated gradients of responsiveness, is consistent with one of the model's central premises. As the model predicts, loss of *wingful/notum* causes overgrowth of the wing disc, whereas overexpression of this inhibitor virtually eliminates wing blade growth by transforming wing to notum (Gerlitz and Basler 2002; Giraldez et al. 2002). Nevertheless, patterns of known Dpp and Wg targets do seem to be graded (Fig. 7), and so the expectation is that the expression of growth effectors targeted by these signals will also be graded, rather than uniform.

Model 4: Entelechia

Another interesting model, named "Entelechia" (completion), was proposed by García-Bellido and García-Bellido (1998) and draws from the latter's lifelong study of the wing. This model appears, after a computer simulation, to retain many of the properties of wing growth, such as (1) the wing's ability to attain a relatively constant number of cells regardless of culture conditions; (2) the restriction of cell clones by compartment boundaries; (3) the ability of dissociated or wounded discs to reconstruct normal patterns; and (4) the context-dependent effects on growth observed in wing discs that are genetically mosaic for patterning mutations. A unique aspect of the Entelechia model is that it relies solely on local, short-range interactions between cells. A molecular switching mechanism is proposed, using receptors, ligands, and transcription factors, that causes the division of any cell whose "martial value"—the activity of a special transcription factor—is lower than that of neighboring cells (Fig. 11). The activation of "martial gene" expression is proposed to arise from interactions between cells across compartment borders. The martial genes stimulate their own expression and are thus subject to cell-autonomous positive feedback. They also stimulate the expression of secreted ligands that suppress their own expression and are thus subject to non-cell-autonomous negative feedback. The net result of these interactions, when simulated in a growing population of cells surrounding a compartment boundary, is that martial gene expression and ligand expression form graded profiles extending away from the compartment boundaries. Cell proliferation, however, is relatively uniform, occurring in cascades that originate from compartment boundaries but propagate throughout the organ. In simulations, this model exhibits many of the growth properties exhibited in living discs that have been surgically or genetically manipulated, such as plasticity in cell clone shapes and sizes within compartments and the ability of the disc to stop growing when reaching a final, genetically encoded size and shape.

The entelechia model remains abstract, however, in that it includes a number of ad hoc proposals that are not yet supported by experimental data. Several of its components—most importantly the "martial genes"—have no known correlates among the many patterning genes identified to date. Furthermore, this model does not use the Dpp and Wg gradients as growth effectors, even though these gradients arise from interactions at compartment boundaries, a central property of the model, and are known

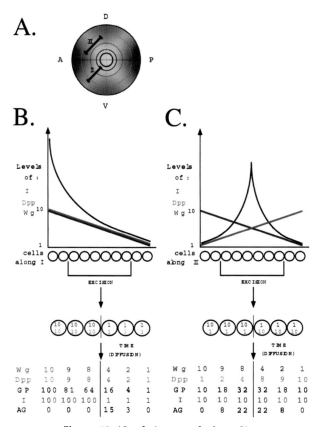

Figure 10. (*See facing page for legend.*)

to affect growth. Instead, the EGFR/Vein and Notch/Delta pathways are suggested as playing the roles of graded receptors/ligands. Finally, this model posits that wing size is measured by counting cell numbers rather than areas or distances, an assumption that seems to be at odds with observations in many organs, including the wing (Fig. 4), indicating that cell numbers are not the crucial metric in size determination (Weigmann et al. 1997; Neufeld et al. 1998; Day and Lawrence 2000). Nevertheless, this is a thought-provoking proposal that undoubtedly mirrors some essential facets of growth control in the imaginal discs.

Model 5: Measuring the Morphogen Gradient

A fifth proposal, articulated nicely by Lawrence and Struhl (1996) and Day and Lawrence (2000), incorporates notions that arose when the polar coordinate model was elaborated about 30 years ago (Wolpert 1971; French et al. 1976; Bryant and Simpson 1984; Held 2002). In this case, the central idea is that cells in the wing measure the steepness of the Dpp and

---

Figure 10. Wing growth (Model 3): A gradient of responsiveness. (*A*) An idealized diagram of the graded distribution of Wg and Dpp within the wing disc. The distributions of Wg (*red*) and Dpp (*green*) expression within the disc are shown by the gradients of color intensity; yellow represents regions where Wg and Dpp expression overlap. The intersecting orthogonal lines represent the AP (*vertical line*) and DV (*horizontal line*) boundaries. Pale blue concentric circles represent proximal-distal distinctions that might result from the joint action of high levels of Dpp and Wg at successive times during the growth of the disc. Lines I and II represent rows of cells: The cells along I maintain the same ratio of Dpp and Wg levels (see *B*) but the absolute levels drop with distance from the sources of morphogen; cells along II differ dramatically in the ratio of Dpp and Wg sources (see *C*) depending on the relative distance from the Wg and Dpp sources. (*B*, *C*) The graphs show how the concentration levels, in arbitrary numbers, of Wg (*red*), Dpp (*green*), and a hypothetical growth inhibitor (*blue*) vary along lines I (*B*) and II (*C*); linear gradients of morphogen concentration are used for simplicity. Below the graphs, we illustrate the consequence of an excision of some of the cells, which juxtaposes cells from distant positions giving discontinuities in the gradients of Wg (*red*) and Dpp (*green*). The table of numbers at the bottom gives the expected levels of the growth factors Wg and Dpp after some diffusion across the wound. Using the model, the table also indicates the growth potential (GP) based on synergy between Dpp and Wg (GP=WgxDpp), the presumed level of inhibitor (I), and the net or actual growth (AG). The actual growth is determined by the formula: AG=(WgxDpp)-I. The numbers, of course, are arbitrary and are used only to illustrate the principle. (Reprinted, with permission, from Serrano and O'Farrell 1997 [copyright Elsevier].)

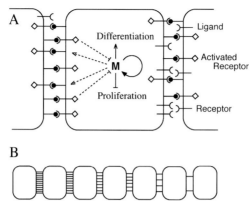

Figure 11. Wing growth (Model 4): Entelechia. (*A*) Components of the Entelechia model, showing how cell communication might control cell division. The activated form of the receptors may be associated with specific adhesion molecules, not shown in the figure. The cell will differentiate, or continue to proliferate, depending on the balance between M (martial) gene expression and M gene titration by the activated receptors. M gene expression promotes ligand production, and is also autoregulatory. Receptors (*open diamonds*) that are activated by ligands (*filled circles*) repress M gene expression. (*B*) A row of cells with different amounts of activated receptors, illustrating how the distribution of activated receptors polarizes each cell. (Reprinted, with permission, from García-Bellido and García-Bellido 1998 [copyright University of the Basque Country Press].)

Wg gradients and continue to proliferate as long as a sufficient concentration gradient can be detected (Fig. 12A). If the high and low endpoints of the gradient remain fixed at set concentrations, the steepness of the morphogen gradient measured across any individual cell decreases as the number of cells in the field increases. Accordingly, as proliferation expands the field of cells within the gradient, the slope of the concentration gradient across individual cells decreases progressively. When cells can no longer sense the gradient, cell growth and proliferation stop. This is perhaps the most widely discussed of the models described here, and it is attractive in many respects. It is of course consistent with much of the experimental data from which it derives, such as intercalary growth following the juxtaposition of cells from different disc regions and the cessation of disc growth at a genetically determined, organ-autonomous, size. It might also explain why treatments that enlarge the disc cells, such as increased PI3K activity, allow wings or wing compartments to grow to larger sizes (Teleman and Cohen 2000), since larger cells could sense a shallower gradient. This model does not integrate the organizing properties of compartment boundaries, however, and it has not yet been put to test in rigorous simulations. Perhaps more worrisome, it also seems at

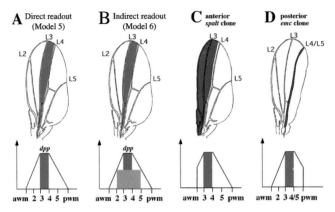

Figure 12. Long-range vs. local growth control. (*A*) Model 5, depicting direct long-range influence of Dpp on growth and patterning along the AP axis of the wing. The slope and amplitude of the gradient would directly control the size of A and P compartments. (*B*) Model 6, illustrating an indirect role for Dpp mediated through secondary subdivision of the axis. Different thresholds of Dpp induce the activation of different target genes, such as spalt (*green*) which has been implicated in specifying vein positions. (*C*) Depiction of a *spalt* mutant clone (*red*) filling the A compartment. Such clones cause loss of vein L2. The A compartment is reduced to two intervein regions of approximately normal size. According to the long-range gradient hypothesis (Model 5), the first vein (at the anterior margin) would be specified at a higher level in the Dpp gradient than in normal discs (*B*). Similar effects are seen in the posterior compartment where the clones lead to loss of vein 5 (not shown). (*D*) Depiction of an *extramacrochaetae (emc)* clone (*red*) causing loss of the region between veins 4 and 5 and fusion of these veins. The size of the region posterior to vein L5 is relatively normal, suggesting that posterior compartment size is not directly determined by the slope or amplitude of the Dpp gradient. (Reprinted, with permission, from Milán and Cohen 2000 [copyright Elsevier].)

odds with the findings that obliterating the Wg or Dpp gradients by ubiquitous expression of the ligands or activated intracellular transducers does not suppress the growth of the wing cells, but actually stimulates it (Nellen et al. 1996; Nagaraj et al. 1999; Martin et al. 2000). According to this model, any manipulation that effectively flattened out a morphogen gradient should arrest growth, rather than promote it. Model 3 seems more apt to concur with this particular observation.

Model 6: Compartments as Units of Growth Control

Milán and Cohen (2000) proposed a model in which the Dpp gradient is not used to control cell growth and proliferation directly but serves to subdivide the wing into smaller compartments, demarcated by veins, that autonomously regulate their sizes (Fig. 12B). In this view, Dpp would not act as a growth factor; it would simply subdivide the wing into subregions (intervein compartments), which would then use other signals to autonomously regulate their size. Work on patterning in the wing shows that Dpp levels do indeed determine the locations of the wing veins (e.g., vein L2 forms at the anterior edge of the Spalt-expression domain, and Spalt is controlled by Dpp; Figs. 7 and 12), and that the veins do function as compartment borders to suppress the mixing of adjacent cell populations (Fig. 6). An important observation relating to this model is that the loss of a single intervein compartment does not necessarily affect the sizes of neighboring compartments (Fig. 12D) (de Celis et al. 1995), as would be expected in models 3, 4, and 5 above. This shows that the control of intervein compartment sizes is probably compartment-autonomous. In addition, mutant wings that lack a vein lose the equivalent of a single compartment, just in the region of the missing vein, without affecting the sizes of the other compartments, and mutants that fail to form wing veins at all develop significantly smaller wings (García-Bellido et al. 1994). These observations suggest that the veins act as signaling centers to promote intervein growth. How vein-demarcated compartments would "measure" their size is not clear, but this could involve either the patterns of Vein/EGFR signaling that set up vein/intervein borders or the patterns of Notch/Delta and Dpp signaling that refine and maintain these borders (Bier 2000). This model is perhaps the most consistent with our current understanding of wing patterning but, like the other models, it does not address how metabolic processes impinging on cell growth, proliferation, or other cell behaviors such as adhesion might play into patterned growth.

In summary, the great diversity among the proposed models for growth control in the wing illustrate how superficial our understanding of organ size control really is. All of the models discussed remain abstract in that they cannot yet attribute all, or even many, of their essential characteristics to known gene functions. Of particular note here is that these models still reside on the level of pattern formation and signaling and have not yet made substantial connections to growth-effector genes of the type discussed above (p. 30). Moreover, the importance of differential cell growth and proliferation in shaping the wing, as opposed to other cell behaviors such as differential cell adhesion, is not entirely clear. Thus, despite copious experimental data, the problem of patterned growth in the wing, or any other organ, remains a significant challenge.

THE REGULATION OF BODY SIZE

As in other insects, adult body size of *Drosophila* is determined by the size of the larva at the time of metamorphosis. Thus, body size is effectively regulated by the mechanism that controls the timing of the onset of meta-morphosis during the third larval instar. Larvae that are large at this time will metamorphose into larger adults than larvae that are small. To under-stand the control of adult body size, it is therefore essential to understand how the size of a larva is related to the timing of its metamorphosis.

The endocrine control of metamorphosis in *Drosophila* is similar to that found in other holometabolous insects (Riddiford and Ashburner 1991; Riddiford 1993a, Nijhout 1994) and involves an interaction between three hormones: the prothotacicotropic hormone (PTTH), ecdysone, and juvenile hormone (JH). During the first half of the third larval instar, high JH levels inhibit the secretion of PTTH (a brain neurosecretory hormone) and ecdysone (a steroid hormone secreted by the ring gland). When JH declines in the latter half of the third instar, this inhibition is relieved. The positive stimulus for the secretion of PTTH is unknown in *Drosophila*, but it is known in several other insects. In *Manduca sexta*, PTTH secretion is regulated by a photoperiodic clock and occurs during the first photoperi-odic gate that follows the disappearance of JH (Truman and Riddiford 1974; Riddiford 1985; Kremen and Nijhout 1989; Nijhout 1994). In *Rhodnius prolixus, Dipetalogaster maximus,* and *Oncopeltus fasciatus,* PTTH secretion is stimulated by abdominal stretch receptors that are acti-vated when a particular critical distension of the abdomen is reached (Nijhout 1979, 1981, 1984; Chiang and Davey 1988). In *Onthophagus tau-rus,* exhaustion of the food supply, or simply removing a larva from food, triggers PTTH and ecdysone secretion (Shafiei et al. 2001). Thus, the trig-gers for PTTH secretion are patently diverse and, in general, appear to be adaptations to a particular life history (Nijhout 2003). The only known function for PTTH is to stimulate secretion of ecdysone by the ring gland (Riddiford 1993b; Nijhout 1994). In *Drosophila*, at the end of the third instar (about 100–106 hours after hatching), there is a brief period of ecdysone secretion that causes the larva to stop feeding and enter the wan-dering stage. This is followed about 12 hours later by a larger peak of ecdysone secretion that stimulates pupariation (Borst et al. 1974; Klose et al. 1980; Schwartz et al. 1984; Riddiford 1993b). The mechanism that con-trols the timing of the first pulse of ecdysone secretion terminates the growth of the larva and thus effectively determines the size of the adult.

Experimental Alteration of Adult Body Size: Environmental Effects

If *Drosophila* larvae are starved or underfed, their metamorphosis is delayed, but many will eventually metamorphose into miniature, but nor-mally proportioned, adults (Robertson 1963; Bryant and Simpson 1984).

The normal size-regulating mechanism can thus be overridden by variation in nutrition.

Rearing larvae at low temperatures increases the size at metamorphosis and results in adults of larger body size (Powell 1974; Partridge et al. 1994; de Moed et al. 1997). The dependence of body size on rearing temperature is a common finding in insects in general (Atkinson 1994), but the physiological or developmental basis of this response is not well understood. Interestingly, after many years of rearing *Drosophila* at a low temperature, the increases in body size can become genetically fixed, as shown by the fact that the large body size is maintained when larvae are reared at elevated temperatures (Partridge et al. 1994). Evidently, the increase in body size at lowered temperatures increases fitness sufficiently to favor the fixation of alleles that enhance body size. Both the environmental and the genetic response to temperature are due to an increase in cell size, not in cell number (Partridge et al. 1994; Azevedo et al. 2002).

Experimental Alteration of Adult Body Size: Genetic Effects

Selection on body size can readily increase or decrease the adult size of *Drosophila* (Robertson 1960). The heritability (the ratio of additive genetic variance to phenotypic variance; Falconer and Mackay [1996]) of body size is ~0.35 (Robertson 1960), indicating that there is substantial genetic variation for body size in natural populations of *Drosophila*. Unfortunately, it is impossible to tell exactly what kinds of genes are being selected for (or against) in such selection experiments, so they cannot be used as a tool for understanding the developmental regulation of body size.

Body size can be altered dramatically by interfering with the insulin-signaling pathway (see Chapter 6). Overexpression of *dilp-2* (one of the *Drosophila* insulin-like proteins) increases body size by increasing both cell size and cell number (Brogiolo et al. 2001). Conversely, ablation of DILP-producing neurosecretory cells results in miniature but normally proportioned adults, and this effect can be reversed by induced expression of a *dilp-2* transgene (Rulifson et al. 2002). Mutation in *DInr* (the *Drosophila* insulin receptor) can result in the development of dwarf adults (Tatar et al. 2001), as does the *chico*[1] mutation (Clancy et al. 2001). In contrast, overexpression of *DInr* causes an increase in adult body size. The expression level of *dilp-2* is limiting for the *DInr* overexpression phenotype, suggesting that it is the insulin level, rather than the level of receptor expression, that normally regulates growth and size (Brogiolo et al. 2001).

Perhaps the most interesting mechanism for the regulation of body size was suggested by studies on imaginal disc regeneration. Simpson and

Schneiderman (1975) and Simpson et al. (1980) studied the effects of a sex-linked, recessive, heat-sensitive, cell-lethal mutation in *Drosophila*. Gynandromorphs exhibited heat-sensitive clones of different sizes, and heat-treatment of gynandromorphic larvae caused a variable amount of cell death in the imaginal discs and delayed pupariation of the treated larvae. The delay in pupariation was proportional to the size of the killed clone and the degree of subsequent regeneration. Interestingly, elimination of the entire disc eliminated the possibility of regeneration and caused no delay of pupariation. Evidently, regenerating discs were able to delay pupariation until regeneration was complete. These results parallel similar findings on the metamorphosis-delaying effects of wing disc regeneration in the Lepidoptera (Pohley 1965; Madhavan and Schneiderman 1969; Rahn 1972).

Larvae with the disc overgrowth mutations *dco* and *dlg* grow at the normal rate but do not cease growing at the normal time. Instead, they continue to grow and pupariate at about double the weight of a wild-type larva (Sehnal and Bryant 1993). The delay in pupariation is due to the fact that in these mutants the peaks of ecdysone secretion that initiate the wandering stage and pupariation fail to occur (Sehnal and Bryant 1993). *dlg* mutants show an accumulation of PTTH in the larval brains, which indicates that the failure to secrete ecdysone and pupariate is due to a failure to release PTTH (Zitnan et al. 1993). The same high accumulation of PTTH is true for the *ecd-1* mutant, which is deficient in ecdysone production (Henrich et al. 1987). The *dlg* mutant also has a very high level of allatotropic hormone activity in its cerebral neurosecretory cells, suggesting strong stimulation of JH (Zitnan et al. 1993). Since JH inhibits PTTH and ecdysone secretion (Nijhout and Williams 1974; Rountree and Bollenbacher 1986), this may provide the mechanism that inhibits PTTH and ecdysone secretion in disc overgrowth mutants.

Mutant imaginal discs also overgrow when cultured in female abdomens under conditions where wild-type discs stop growing when they reach their normal size. Therefore, these overgrowth mutations appear to be disc-autonomous and are not a secondary effect of prolonged larval life or of systemic changes in the levels of hormones or growth factors. Interestingly, the ring glands of *dco* and *dlg* mutants secrete ecdysone at about 1/6 the rate of wild-type glands when cultured in vitro (Sehnal and Bryant 1993), so unless the inhibitory effects of the imaginal disc persist after the glands are isolated, the ability of the ring glands to produce ecdysone is also autonomously compromised in these mutants. The *giant* (*gt*) mutant has diminished ecdysone titers, slow-growing imaginal discs, and delayed metamorphosis. Hence, the larvae grow to a giant size before metamorphosis because it takes longer for discs to achieve their final size. The rate of disc growth can be increased by feeding larvae on exogenous ecdysone and results in normal-sized adults (Schwartz et al. 1984). Evidently, the growth rate of intact imaginal discs depends on the ecdysone level, as is the case in Lepidoptera (Nijhout and Grunert 2002).

The Role of Imaginal Discs in Body Size Regulation

Several lines of evidence show that growing and regenerating imaginal discs delay metamorphosis by interfering with the secretion of PTTH and ecdysone. The implication of these findings is that in normal development growing discs may also inhibit metamorphosis, presumably by producing (or depleting) a regulatory factor for PTTH and ecdysone secretion while they are growing. Size regulation of imaginal discs is disc-autonomous (Bryant and Simpson 1984; Bryant and Levinson 1985), so when discs reach their programmed final size and stop growing, their inhibitory effect on PTTH and ecdysone secretion presumably ceases and metamorphosis is initiated (Poodry and Woods 1990; Sehnal and Bryant 1993). It is important to note that, based on studies with an imaginal-disc-derived cell line, disc growth requires both insulin and IDGF signaling (Kawamura et al. 1999; Bryant 2001), and we do not yet know whether discs stop growing because either insulin or IDGFs disappear, or because one of the receptor-response pathways for these growth factors is interrupted.

If the natural mechanism that controls the onset of metamorphosis in *Drosophila* is based entirely on disc-derived signals, this is likely to be a derived feature, possibly a specialization of the higher Diptera, which, uniquely among insects, develop almost their entire adult body from specialized pre-imaginal structures (such as imaginal discs and histoblasts). In flies, the regulation of adult body size may thus be nothing more than the sum of the independent regulation of the sizes of its imaginal parts. In other words, overall body size regulation may be a secondary consequence of imaginal disc size regulation.

This neat and simple view, however, needs to be reconciled with the fact that it is possible to produce miniature flies and giant flies by environmental variation, such as temperature and nutrition, as well as by artificial changes in insulin signaling. For instance, when a larva is starved, growth of all its tissues, including the imaginal discs, ceases, but metamorphosis begins only after a very long delay, if at all. Therefore, a simple cessation of disc growth cannot be a sufficient condition for the initiation of pupariation. It may be that only the cessation of growth that is due to the achievement of the preprogrammed final size gives the right signal to the endocrine system, which would imply the existence of a specific signal for this event. This final size, evidently, is not entirely genetically determined but is strongly influenced by temperature and nutrition.

Finally, changes in insulin and phosphoinositol signaling can alter adult size dramatically (Oldham et al. 2000, 2002; Brogiolo et al. 2001); therefore, in addition to controlling growth, insulin signaling must somehow be involved in the size-regulating mechanism and the cessation of growth. The paracrine mechanism for disc size regulation proposed by Nijhout (Nijhout 2003) suggests one possibility for the effect of insulin signaling in both growth and size regulation. If the disc-growth stimulator in this model is insulin, the cessation of growth is signaled by the disappearance of insulin from the circulatory system. If the central nervous system can detect the circulating insulin level, its disappearance would serve to trigger PTTH and ecdysone secretion, which would terminate larval growth. The model (Fig. 9) shows that excess insulin will effectively revise the final sizes of the discs upward, which would be expected to be accompanied by a proportional increase in overall body size in the adult.

ACKNOWLEDGMENTS

The authors thank Savraj Grewal for his comments on the manuscript, Michel Gregory for his editorial assistance, and all the authors who contributed figures and unpublished communications. B.A.E. acknowledges National Institutes of Health grants GM-51186 and GM-61805 for support.

REFERENCES

Adler P.N. and MacQueen M. 1984. Cell proliferation and DNA replication in the imaginal wing disc of *Drosophila melanogaster*. *Dev. Biol.* **103**: 28–37.

Ambros V. 2001. microRNAs: Tiny regulators with great potential. *Cell* **107**: 823–826.

Arama E., Dickman D., Kimchie Z., Shearn A., and Lev Z. 2000. Mutations in the beta-propeller domain of the *Drosophila* brain tumor (brat) protein induce neoplasm in the larval brain. *Oncogene* **19**: 3706–3716.

Arora K. and Nüsslein-Volhard C. 1992. Altered mitotic domains reveal fate map changes in *Drosophila* embryos mutant for zygotic dorsoventral patterning genes. *Development* **114**: 1003–1024.

Asano M., Nevins J.R., and Wharton R.P. 1996. Ectopic E2F expression induces S-phase and apoptosis in *Drosophila* imaginal discs. *Genes Dev.* **10**: 1422–1432.

Asha H., Nagy I., Kovacs G., Stetson D., Ando I., and Dearolf C.R. 2003. Analysis of ras-induced overproliferation in *Drosophila* hemocytes. *Genetics* **163**: 203–215.

Ashburner M. 1989. *Drosophila: A laboratory handbook.* Cold Spring Harbor Laboratory Press, Cold Spring Harbor, New York.

Atkinson D. 1994. Temperature and organism size: A biological law for ectotherms? *Adv. Ecol. Res.* **25**: 1–58.

Azevedo R.B.R., French V., and Partridge L. 2002. Temperature modulates epidermal cell size in *Drosophila melanogaster*. *J. Insect Physiol.* **48**: 231–237.

Baker N.E. 2001. Cell proliferation, survival, and death in the *Drosophila* eye. *Semin. Cell Dev. Biol.* **12**: 499–507.

Baker N.E. and Yu S.Y. 2001. The EGF receptor defines domains of cell-cycle progression and survival to regulate cell number in the developing *Drosophila* eye. *Cell* **104**: 699–708.

Baonza A. and García-Bellido A. 2000. Notch signaling directly controls cell proliferation in the *Drosophila* wing disc. *Proc. Natl. Acad. Sci.* **97**: 2609–2614.

Basler K. and Struhl G. 1994. Compartment boundaries and the control of *Drosophila* limb pattern by *hedgehog* protein. *Nature* **368**: 208–214.

Beadle G.W., Tatum E.L., and Clancy C.W. 1938. Food level in relation to rate of development and eye pigmentation in *Drosophila melanogaster*. *Biol. Bull.* **75**: 447–462.

Bernal A. and Kimbrell D.A. 2000. *Drosophila* Thor participates in host immune defense and connects a translational regulator with innate immunity. *Proc. Natl. Acad. Sci.* **97**: 6019–6024.

Bier E. 2000. Drawing lines in the *Drosophila* wing: Initiation of wing vein development. *Curr. Opin. Genet. Dev.* **10**: 393–398.

Bilder D., Schober M., and Perrimon N. 2003. Integrated activity of PDZ protein complexes regulates epithelial polarity (comment). *Nat. Cell Biol.* **5**: 53–58.

Boon K., Caron H.N., van Asperen R., Valentijn L., Hermus M.C., van Sluis P., Roobeek I., Weis I., Voute P.A., Schwab M., and Versteeg R. 2001. N-myc enhances the expression of a large set of genes functioning in ribosome biogenesis and protein synthesis. *EMBO J.* **20**: 1383–1393.

Borst D.W., Bollenbacher W.E., O'Connor J.D., King D.S., and Fristrom J.W. 1974. Ecdysone levels during metamorphosis of *Drosophila melanogaster*. *Dev. Biol.* **39**: 308–316.

Bourbon H.M., Gonzy-Treboul G., Peronnet F., Alin M.F., Ardourel C., Benassayag C., Cribbs D., Deutsch J., Ferrer P., Haenlin M., Lepesant J.A., Noselli S., and Vincent A. 2002. A P-insertion screen identifying novel X-linked essential genes in *Drosophila*. *Mech. Dev.* **110**: 71–83.

Brennecke J., Hipfner D.R., Stark A., Russell R.B., and Cohen S.M. 2003. *bantam* encodes a developmentally regulated microRNA that controls cell proliferation and regulates the proapoptotic gene *hid* in *Drosophila*. *Cell* **113**: 25–36.

Britton J.S. 2000. "Genetic and environmental control of growth and the cell cycle during larval development of *Drosophila melanogaster*." Ph.D. thesis, University of Washington, Seattle.

Britton J.S. and Edgar B.A. 1998. Environmental control of the cell cycle in *Drosophila*: Nutrition activates mitotic and endoreplicative cells by distinct mechanisms. *Development* **125**: 2149–2158.

Britton J.S., Lockwood W.K., Li L., Cohen S.M., and Edgar B.A. 2002. *Drosophila's* insulin/PI3-kinase pathway coordinates cellular metabolism with nutritional conditions. *Dev. Cell* **2**: 239–249.

Brogiolo W., Stocker H., Ikeya T., Rintelen F., Fernandez R., and Hafen E. 2001. An evolutionarily conserved function of the *Drosophila* insulin receptor and insulin-like peptides in growth control. *Curr. Biol.* **11:** 213–221.

Brook A., Xie J.-E., Du W., and Dyson N. 1996. Requirements for dE2F function in proliferating cells and in post-mitotic differentiating cells. *EMBO J.* **15:** 3676–3683.

Brook W.J., Diaz-Benjumea F.J., and Cohen S.M. 1996. Organizing spatial pattern in limb development. *Annu. Rev. Cell. Dev. Biol.* **12:** 161–180.

Brown T.R., Scott P.H., Stein T., Winter A.G., and White R.J. 2000. RNA polymerase III transcription: Its control by tumor suppressors and its deregulation by transforming agents. *Gene Expr.* **9:** 15–28.

Bryant P.J. 1997. Junction genetics. *Dev. Genet.* **20:** 75–90.

———. 2001. Growth factors controlling imaginal disc growth in *Drosophila* (also see discussion). *Novartis Found. Symp.* **237:** 182–202.

Bryant P.J. and Levinson P. 1985. Intrinsic growth control in the imaginal primordia of *Drosophila,* and the autonomous action of a lethal mutation causing overgrowth. *Dev. Biol.* **107:** 355–363.

Bryant P.J. and Simpson P. 1984. Intrinsic and extrinsic control of growth in developing organs. *Q. Rev. Biol.* **59:** 387–415.

Burke R. and Basler K. 1996. Dpp receptors are autonomously required for cell proliferation in the entire developing *Drosophila* wing. *Development* **122:** 2261–2269.

Campbell G. and Tomlinson A. 1999. Transducing the Dpp morphogen gradient in the wing of *Drosophila:* Regulation of Dpp targets by brinker. *Cell* **96:** 553–562.

Campos-Ortega J.A. and Hartenstein V. 1985. *The embryonic development of* Drosophila melanogaster. Springer-Verlag, Berlin.

Capdevila J. and Guerrero I. 1994. Targeted expression of the signaling molecule decapentaplegic induces pattern duplications and growth alterations in *Drosophila* wings. *EMBO J.* **13:** 4459–4468.

Champlin D.T. and Truman J.W. 2000. Ecdysteroid coordinates optic lobe neurogenesis via a nitric oxide signaling pathway. *Development* **127:** 3543–3551.

Chen H.W., Chen X., Oh S.W., Marinissen M.J., Gutkind J.S., and Hou S.X. 2002. mom identifies a receptor for the *Drosophila* JAK/STAT signal transduction pathway and encodes a protein distantly related to the mammalian cytokine receptor family. *Genes Dev.* **16:** 388–398.

Chen X., Oh S.W., Zheng Z., Chen H.W., Shin H.H., and Hou S.X. 2003. Cyclin D-Cdk4 and cyclin E-Cdk2 regulate the Jak/STAT signal transduction pathway in *Drosophila* (comment). *Dev. Cell* **4:** 179–190.

Chiang G.R. and Davey K.G. 1988. A novel receptor capable of monitoring applied pressure in the abdomen of an insect. *Science* **241:** 1665–1667.

Cho K.O., Chern J., Izaddoost S., and Choi K.W. 2000. Novel signaling from the peripodial membrane is essential for eye disc patterning in *Drosophila* (comment). *Cell* **103:** 331–342.

Church R.B. 1965. "Genetic and biochemical studies of growth in *Drosophila*." Ph. D. thesis. University of Edinburgh, Edinburgh, Scotland, United Kingdom.

Ciarmatori S., Scott P.H., Sutcliffe J.E., McLees A., Alzuherri H.M., Dannenberg J.H., te Riele H., Grummt I., Voit R., and White R.J. 2001. Overlapping functions of the pRb family in the regulation of rRNA synthesis. *Mol. Cell. Biol.* **21:** 5806–5814.

Clancy D.J., Gems D., Harshman L.G., Oldham S., Stocker H., Hafen E., Leevers S.J., and Partridge L. 2001. Extension of life-span by loss of CHICO, a *Drosophila* insulin receptor substrate protein (see comments). *Science* **292:** 104–106.

Coller H.A., Grandori C., Tamayo P., Colbert T., Lander E.S., Eisenman R.N., and Golub T.R. 2000. Expression analysis with oligonucleotide microarrays reveals that MYC regulates genes involved in growth, cell cycle, signaling, and adhesion. *Proc. Natl. Acad. Sci.* **97:** 3260–3265.

Colombani J., Raisin S., Pantalacci S., Radimerski T., Montagne J., and Leopold P. 2003. A nutrient sensor mechanism controls *Drosophila* growth. *Cell* **114:** 739–749.

Daga R.R. and Jimenez J. 1999. Translational control of the cdc25 cell cycle phosphatase: A molecular mechanism coupling mitosis to cell growth. *J. Cell Sci.* **112:** 3137–3146.

Dahmann C. and Basler K. 1999. Compartment boundaries: At the edge of development. *Trends Genet.* **15:** 320–326.

Datar S.A. 2000. "Developmental regulation of growth and cell-cycle progression in *Drosophila melanogaster*: A larval growth arrest screen, and molecular and genetic analysis of the Cyclin D/Cdk4 complex." Ph. D. thesis. University of Washington, Seattle.

Davis K.T. and Shearn A. 1977. In vitro growth of imaginal disks from *Drosophila melanogaster*. *Science* **196:** 438–440.

Day S.J. and Lawrence P.A. 2000. Measuring dimensions: The regulation of size and shape. *Development* **127:** 2977–2987.

de Alboran I.M., O'Hagan R.C., Gartner F., Malynn B., Davidson L., Rickert R., Rajewsky K., DePinho R.A., and Alt F.W. 2001. Analysis of C-MYC function in normal cells via conditional gene-targeted mutation. *Immunity* **14:** 45–55.

Dearolf C.R. 1998. Fruit fly "leukemia." *Biochim. Biophys. Acta* **1377:** M13-M23.

de Celis J.F., Baonza A., and García-Bellido A. 1995. Behavior of extramacrochaetae mutant cells in the morphogenesis of the *Drosophila* wing. *Mech. Dev.* **53:** 209–221.

de Moed G.H., De Jong G., and Scharloo W. 1997. Environmental effects on body size variation in *Drosophila melanogaster* and its cellular basis. *Genet. Res.* **70:** 35–43.

Deng W.M., Althauser C., and Ruohola-Baker H. 2001. Notch-Delta signaling induces a transition from mitotic cell cycle to endocycle in *Drosophila* follicle cells. *Development* **128:** 4737–4746.

de Nooij J.C., Letendre M.A., and Hariharan I.K. 1996. A cyclin-dependent kinase inhibitor, *dacapo*, is necessary for timely exit from the cell cycle during *Drosophila* embryogenesis. *Cell* **87:** 1237–1247.

Diffley J.F. and Labib K. 2002. The chromosome replication cycle. *J. Cell Sci.* **115:** 869–872.

Douglas N.C., Jacobs H., Bothwell A.L., and Hayday A.C. 2001. Defining the specific physiological requirements for c-Myc in T cell development. *Nat. Immunol.* **2:** 307–315.

Du W., Vidal M., Xie J.-E., and Dyson N. 1996. *RBF*, a novel RB-related gene that regulates E2F activity and interacts with *cyclin E* in *Drosophila*. *Genes Dev.* **10:** 1206–1218.

Duman-Scheel M., Weng L., Xin S., and Du W. 2002. Hedgehog regulates cell growth and proliferation by inducing Cyclin D and Cyclin E. *Nature* **417:** 299–304.

Duronio R.J. and O'Farrell P.H. 1995. Developmental control of the G1 to S transition in *Drosophila*: Cyclin E is a limiting downstream target of E2F. *Genes Dev.* **9:** 1456–1468.

Dyson N. 1998. The regulation of E2F by pRB-family proteins. *Genes Dev.* **12:** 2245–2262.

Edgar B.A. and Datar S.A. 1996. Zygotic degradation of two maternal Cdc25 mRNAs terminates *Drosophila*'s early cell cycle program. *Genes Dev.* **10:** 1966–1977.

Edgar B.A. and O'Farrell P.H. 1989. Genetic control of cell division patterns in the *Drosophila* embryo. *Cell* **57:** 177–187.

———. 1990. The three postblastoderm cell cycles of *Drosophila* embryogenesis are regulated in G2 by string. *Cell* **62:** 469–480.

Edgar B.A. and Orr-Weaver T.L. 2001. Endoreplication cell cycles: More for less. *Cell* **105:** 297–306.

Edgar B.A., Lehman D., and O'Farrell P.H. 1994a. Transcriptional regulation of *string (cdc25):* A link between developmental programming and the cell cycle. *Development* **120:** 3131–3143.

Edgar B.A., Sprenger F., Duronio R.J., Leopold P., and O'Farrell P.H. 1994b. Distinct molecular mechanisms regulate cell cycle timing at successive stages of *Drosophila* embryogenesis. *Genes Dev.* **8:** 440–452.

Edinger A.L. and Thompson C.B. 2002. Akt maintains cell size and survival by increasing mTOR-dependent nutrient uptake. *Mol. Biol. Cell* **13:** 2276–2288.

Emlen D.J. and Nijhout H.F. 2000. The development and evolution of exaggerated morphologies in insects. *Annu. Rev. Entomol.* **45:** 661–708.

Fain M.J. and Stevens B. 1982. Alterations in the cell cycle of *Drosophila* imaginal disc cells precede metamorphosis. *Dev. Biol.* **92:** 247–258.

Falconer D.S. and Mackay T.F.C. 1996. *An introduction to quantitative genetics.* Addison Wesley Longman, Essex, United Kingdom.

Fantl V., Stamp G., Andrews A., Rosewell I., and Dickson C. 1995. Mice lacking cyclin D1 are small and show defects in eye and mammary gland development. *Genes Dev.* **9:** 2364–2372.

Fernandez R., Tabarini D., Azpiazu N., Frasch M., and Schlessinger J. 1995. The *Drosophila* insulin receptor homolog: A gene essential for embryonic development encodes two receptor isoforms with different signaling potential. *EMBO J.* **14:** 3373–3384.

Fero M.L., Rivkin M., Tasch M., Porter P., Carow C.E., Firpo E., Perlmutter R.M., Kaushansky K., and Roberts J.M. 1996. A syndrome of multiorgan hyperplasia with features of gigantism, tumorigenesis, and female sterility in p27*Kip1*-deficient mice. *Cell* **85:** 733–744.

Fingar D.C., Salama S., Tsou C., Harlow E., and Blenis J. 2002. Mammalian cell size is controlled by mTOR and its downstream targets S6K1 and 4EBP1/eIF4E. *Genes Dev.* **16:** 1472–1487.

Foe V.E. and Alberts B.M. 1983. Studies of nuclear and cytoplasmic behaviour during the five mitotic cycles that precede gastrulation in *Drosophila* embryogenesis. *J. Cell Sci.* **61:** 31–70.

Foe V.E. and Odell G.M. 1989. Mitotic domains partition fly embryos, reflecting early biological consequences of determination in progress. *Am. J. Zool.* **29:** 617–632.

Follette P.J., Duronio R.J., and O'Farrell P.H. 1998. Fluctuations in cyclin E levels are required for multiple rounds of endocycle S phase in *Drosophila*. *Curr. Biol.* **8:** 235–238.

Frank D.J. and Roth M.B. 1998. ncl-1 is required for the regulation of cell size and ribosomal RNA synthesis in *Caenorhabditis elegans*. *J. Cell Biol.* **140:** 1321–1329.

Frank D.J., Edgar B.A., and Roth M.B. 2002. The *Drosophila melanogaster* gene brain tumor negatively regulates cell growth and ribosomal RNA synthesis. *Dev. Suppl.* **129:** 399–407.

French V., Bryant P.J., and Bryant S.V. 1976. Pattern regulation in epimorphic fields. *Science* **193:** 969–981.

Fuller M.T. 1998. Genetic control of cell proliferation and differentiation in *Drosophila* spermatogenesis (comment). *Semin. Cell Dev. Biol.* **9:** 433–444.

Gallant P., Shiio Y., Cheng P.F., Parkhurst S.M., and Eisenman R.N. 1996. Myc and Max homologs in *Drosophila*. *Science* **274:** 1523–1527.

Galloni M. and Edgar B.A. 1999. Cell-autonomous and non-autonomous growth-defective mutants of *Drosophila melanogaster*. *Development* **126:** 2365–2375.

Gao X., Zhang Y., Arrazola P., Hino O., Kobayashi T., Yeung R.S., Ru B., and Pan D. 2002. Tsc tumour suppressor proteins antagonize amino-acid-TOR signalling (see comments). *Nat. Cell Biol.* **4:** 699–704.

García-Bellido A. and Merriam J.R. 1971. Parameters of the wing imaginal disc development in *Drosophila melanogaster*. *Dev. Biol.* **26:** 61–87.

García-Bellido A.C. and García-Bellido A. 1998. Cell proliferation in the attainment of constant sizes and shapes: The Entelechia model. *Int. J. Dev. Biol.* **42:** 353–362.

García-Bellido A., Cortes F., and Milán M. 1994. Cell interactions in the control of size in *Drosophila* wings. *Proc. Natl. Acad. Sci.* **91:** 10222–10226.

Gateff E. 1994. Tumor suppressor and overgrowth suppressor genes of *Drosophila melanogaster:* Developmental aspects. *Int. J. Dev. Biol.* **38:** 565–590.

Geballe A.P. and Sachs M.S. 2000. Translational control by upstream open reading frames. In *Translational control of gene expression* (ed. N. Sonenberg et al.), pp. 595–614. Cold Spring Harbor Laboratory Press, Cold Spring Harbor, New York.

Geng Y., Whoriskey W., Park M.Y., Bronson R.T., Medema R.H., Li T., Weinberg R.A., and Sicinski P. 1999. Rescue of cyclin D1 deficiency by knockin cyclin E. *Cell* **97:** 767–777.

Gerlitz O. and Basler K. 2002. Wingful, an extracellular feedback inhibitor of Wingless. *Genes Dev.* **16:** 1055–1059.

Gibson M.C. and Schubiger G. 2000. Peripodial cells regulate proliferation and patterning of *Drosophila* imaginal discs (comment). *Cell* **103:** 343–350.

Gibson M.C., Lehman D.A., and Schubiger G. 2002. Lumenal transmission of decapentaplegic in *Drosophila* imaginal discs. *Dev. Cell* **3:** 451–460.

Giraldez A.J., Copley R.R., and Cohen S.M. 2002. HSPG modification by the secreted enzyme Notum shapes the Wingless morphogen gradient. *Dev. Cell* **2:** 667–676.

Go M.J., Eastman D.S., and Artavanis-Tsakonas S. 1998. Cell proliferation control by Notch signaling in *Drosophila* development. *Development* **125:** 2031–2040.

González-Gaitán M., Capdevila M.P., and García-Bellido A. 1994. Cell proliferation patterns in the wing imaginal disc of *Drosophila*. *Mech. Dev.* **40:** 183–200.

Graves B.J. and Schubiger G. 1982. Cell cycle changes during growth and differentiation of imaginal leg discs in *Drosophila melanogaster*. *Dev. Biol.* **93:** 104–110.

Grieder N.C., Nellen D., Burke R., Basler K., and Affolter M. 1995. *Schnurri* is required for *Drosophila* dpp signaling and encodes a zinc-finger protein similar to the mammalian transcription factor prdii-bf1. *Cell* **81:** 791–800.

Gritli-Linde A., Bei M., Maas R., Zhang X.M., Linde A., and McMahon A.P. 2002. Shh signaling within the dental epithelium is necessary for cell proliferation, growth and polarization. *Development* **129:** 5323–5337.

Guo Q.M., Malek R.L., Kim S., Chiao C., He M., Ruffy M., Sanka K., Lee N.H., Dang C.V., and Liu E.T. 2000. Identification of c-myc responsive genes using rat cDNA microarray. *Cancer Res.* **60:** 5922–5928.

Hartenstein V. and Campos-Ortega J.A. 1985. Fate-mapping in wild-type *Drosophila melanogaster*. I. The spatio-temporal pattern of embryonic cell divisions. *Wilhelm Roux's Arch. Dev. Biol.* **194:** 181–195.

Hartenstein V. and Posakony J.W. 1990. Sensillum development in the absence of cell division: The sensillum phenotype of the *Drosophila* mutant string. *Dev. Biol.* **138:** 147–158.

Hartenstein V., Rudloff E., and Campos-Ortega J.A. 1987. The pattern of proliferation of the neuroblasts in the wild-type embryo of *Drosophila melanogaster*. *Wilhelm Roux's Arch. Dev. Biol.* **196:** 473–485.

Hartwell L.H. 1971. Genetic control of the cell division cycle in yeast. II. Genes controlling DNA replication and its initiation. *J. Mol. Biol.* **59**: 183–194.

Harvey K.F., Pfleger C.M., and Hariharan I.K. 2003. The *Drosophila* Mst ortholog, hippo, restricts growth and cell proliferation and promotes apoptosis. *Cell* **114**: 457–467.

Held L.I. 2002. *Imaginal discs: The genetic and cellular logic of pattern formation.* Cambridge University Press, Cambridge, United Kingdom.

Hemerly A., de Almeida Engler J., Bergounioux C., Van Montagu M., Engler G., Inzé D., and Ferreira P. 1995. Dominant negative mutants of the Cdc2 kinase uncouple cell division from iterative plant development. *EMBO J.* **14**: 3925–3936.

Henrich V.C., Tucker R.L., Maroni G., and Gilbert L.I. 1987. The ecdysoneless (ecd1ts) mutation disrupts ecdysteroid synthesis autonomously in the ring gland of *Drosophila melanogaster. Dev. Biol.* **120**: 50–55.

Heriche J.K., Ang D., Bier E., and O'Farrell P.H. 2003. Involvement of an SCFSlmb complex in timely elimination of E2F upon initiation of DNA replication in *Drosophila. BMC Genet.* **4**: 9.

Hipfner D.R., Weigmann K., and Cohen S.M. 2002. The bantam gene regulates *Drosophila* growth. *Genetics* **161**: 1527–1537.

Huettner A.F. 1924. Maturation and fertilization in *Drosophila melanogaster. J. Morphol.* **124**: 143–166.

Ikeya T., Galic M., Belawat P., Nairz K., and Hafen E. 2002. Nutrient-dependent expression of insulin-like peptides from neuroendocrine cells in the CNS contributes to growth regulation in *Drosophila. Curr. Biol.* **12**: 1293–1300.

Iritani B.M., Delrow J., Grandori C., Gomez I., Klacking M., Carlos L.S., and Eisenman R.N. 2002. Modulation of T-lymphocyte development, growth and cell size by the Myc antagonist and transcriptional repressor Mad1. *EMBO J.* **21**: 4820–4830.

Jazwinska A., Kirov N., Wieschaus E., Roth S., and Rushlow C. 1999. The *Drosophila* gene brinker reveals a novel mechanism of Dpp target gene regulation. *Cell* **96**: 563–573.

Johnston G.C., Pringle J.R., and Hartwell L.H. 1977. Coordination of growth with cell division in the yeast *Saccharomyces cerevisiae. Exp. Cell Res.* **105**: 79–98.

Johnston L.A. 1998. Uncoupling growth from the cell cycle. *Bioessays* **20**: 283–286.

Johnston L.A. and Edgar B.A. 1998. Wingless and Notch regulate cell-cycle arrest in the developing *Drosophila* wing. *Nature* **394**: 82–84.

Johnston L.A. and Schubiger G. 1996. Ectopic expression of *wingless* in imaginal discs interferes with *decapentaplegic* expression and alters cell determination. *Development* **122**: 3519–3529.

Johnston L.A., Prober D.A., Edgar B.A., Eisenman R.N., and Gallant P. 1999. *Drosophila myc* regulates cellular growth during development. *Cell* **98**: 779–790.

Jones L., Richardson H., and Saint R. 2000. Tissue-specific regulation of cyclin E transcription during *Drosophila melanogaster* embryogenesis. *Development* **127**: 4619–4630.

Justice R.W., Zilian O., Woods D.F., Noll M., and Bryant P.J. 1995. The *Drosophila* tumor-suppressor gene warts encodes a homolog of human myotonic-dystrophy kinase and is required for the control of cell-shape and proliferation. *Genes Dev.* **9**: 534–546.

Kango-Singh M., Nolo R., Tao C., Verstreken P., Hiesinger P.R., Bellen H.J., and Halder G. 2002. Shar-pei mediates cell proliferation arrest during imaginal disc growth in *Drosophila. Development* **129**: 5719–5730.

Kawamura K., Shibata T., Saget O., Peel D., and Bryant P.J. 1999. A new family of growth factors produced by the fat body and active on *Drosophila* imaginal disc cells.

Development **126:** 211–219.

Kim J., Sebring A., Esch J.J., Kraus M.E., Vorwerk K., Magee J., and Carroll S.B. 1996. Integration of positional signals and regulation of wing formation and identity by *Drosophila vestigial* gene. *Nature* **382:** 133–138.

Kim S., Li Q., Dang C.V., and Lee L.A. 2000. Induction of ribosomal genes and hepatocyte hypertrophy by adenovirus-mediated expression of c-Myc in vivo. *Proc. Natl. Acad. Sci.* **97:** 11198–11202.

Kioussi C., Briata P., Baek S.H., Rose D.W., Hamblet N.S., Herman T., Ohgi K.A., Lin C., Gleiberman A., Wang J., Brault V., Ruiz-Lozano P., Nguyen H.D., Kemler R., Glass C.K., Wynshaw-Boris A., and Rosenfeld M.G. 2002. Identification of a Wnt/Dvl/beta-Catenin → Pitx2 pathway mediating cell-type-specific proliferation during development. *Cell* **111:** 673–685.

Kipreos E.T., Lander L.E., Wing J.P., He W.W., and Hedgecock E.M. 1996. Cul-1 is required for cell-cycle exit in *C. elegans* and identifies a novel gene family. *Cell* **85:** 829–839.

Kiyokawa H., Kineman R.A., Manova-Todorova K.O., Soares V.C., Hoffman E.S., Ono M., Khanam D., Hayday A.C., Frohman L.A., and Koff A. 1996. Enhanced growth of mice lacking the cyclin-dependent kinase inhibitor function of p27Kip1. *Cell* **85:** 721–732.

Klebes A., Biehs B., Cifuentes F., and Kornberg T.B. 2002. Expression profiling of *Drosophila* imaginal discs. *Genome Biol.* **3:** RESEARCH0038.

Klose W., Gateff E., Emmerich H., and Beikrich H. 1980. Developmental studies on two ecdysone deficient mutants in *Drosophila melanogaster*. *Wilhem Roux' Arch.* **189:** 57–67.

Knoblich J.A., Sauer K., Jones L., Richardson H., Saint R., and Lehner C.F. 1994. Cyclin E controls S-phase progression and its down-regulation during *Drosophila* embryogenesis is required for the arrest of cell proliferation. *Cell* **77:** 107–120.

Kremen C. and Nijhout H.F. 1989. Juvenile hormone controls the onset of pupal commitment in the imaginal discs and epidermis of *Precis coenia* (Lepidoptera: Nymphalidae). *J. Insect Physiol.* **35:** 603–612.

Kurzik-Dumke U., Phannavong B., Gundacker D., and Gateff E. 1992. Genetic, cytogenetic and developmental analysis of the *Drosophila melanogaster* tumor suppressor gene lethal(2)tumorous imaginal discs (1(2)tid). *Differentiation* **51:** 91–104.

Kuzin B., Roberts I., Peunova N., and Enikolopov G. 1996. Nitric oxide regulates cell proliferation during *Drosophila* development. *Cell* **87:** 639–649.

Lamb M.J. 1982. The DNA content of polytene nuclei in midgut and malpighian tubule cells of adult *Drosophila melanogaster*. *Wilhelm Roux's Arch.* **191:** 381–384.

Lambertsson A. 1998. The minute genes in *Drosophila* and their molecular functions. *Adv. Genet.* **38:** 69–134.

Lane M.E., Sauer K., Wallace K., Jan Y.N., Lehner C.F., and Vaessin H. 1996. Dacapo, a cyclin-dependent kinase inhibitor, stops cell proliferation during *Drosophila* development. *Cell* **87:** 1225–1235.

Lawrence P.A. and Struhl G. 1996. Morphogens, compartments, and pattern: Lessons from *Drosophila*? *Cell* **85:** 951–961.

Lazaro J.B., Bailey P.J., and Lassar A.B. 2002. Cyclin D-cdk4 activity modulates the subnuclear localization and interaction of MEF2 with SRC-family coactivators during skeletal muscle differentiation. *Genes Dev.* **16:** 1792–1805.

Lecuit T., Brook W.J., Ng M., Calleja M., Sun H., and Cohen S.M. 1996. Two distinct mechanisms for long-range patterning by Decapentaplegic in the *Drosophila* wing. *Nature* **381:** 387–393.

Lee L.A. amd Orr-Weaver T.L. 2003. Regulation of cell cycles in *Drosophila* development: Intronsic and extrinsic cues. *Annu. Rev. Genet.* **37:** 545–578.

Lehman D.A., Patterson B., Johnston L.A., Balzer T., Britton J.S., Saint R., and Edgar B.A. 1999. *Cis*-regulatory elements of the mitotic regulator, *string/Cdc25. Development* **126:** 1793–1803.

Lilly M.A. and Spradling A.C. 1996. The *Drosophila* endocycle is controlled by Cyclin E and lacks a checkpoint ensuring S-phase completion. *Genes Dev.* **10:** 2514–2526.

Liu T.H., Li L., and Vaessin H. 2002. Transcription of the *Drosophila* CKI gene dacapo is regulated by a modular array of cis-regulatory sequences. *Mech. Dev.* **112:** 25–36.

Madhavan K. and Schneiderman H.A. 1969. Hormonal control of imaginal disc regeneration in *Galleria mellonella* (Lepidoptera). *Biol. Bull.* **137:** 321–331.

———. 1977. Histological analysis of the dynamics of growth of imaginal discs and histoblast nests during the larval development of *Drosophila melanogaster. Wilhelm Roux's Arch.* **183:** 269–305.

Martin J.F., Hersperger E., Simcox A., and Shearn A. 2000. minidiscs encodes a putative amino acid transporter subunit required non-autonomously for imaginal cell proliferation. *Mech. Dev.* **92:** 155–167.

Martin-Castellanos C. and Edgar B.A. 2002. A characterization of the effects of Dpp signaling on cell growth and proliferation in the *Drosophila* wing. *Dev. Suppl.* **129:** 1003–1013.

Meyer C.A., Jacobs H.W., Datar S.A., Du W., Edgar B.A., and Lehner C.F. 2000. *Drosophila Cdk4* stimulates growth and is dispensable for cell-cycle progression. *EMBO J.* **19:** 4533–4542.

Meyer C.A., Kramer I., Dittrich R., Marzodko S., Emmerich J., and Lehner C.F. 2002. *Drosophila* p27Dacapo expression during embryogenesis is controlled by a complex regulatory region independent of cell-cycle progression. *Development* **129:** 319–328.

Milán M. and Cohen S.M. 2000. Subdividing cell populations in the developing limbs of *Drosophila:* Do wing veins and leg segments define units of growth control? *Dev. Biol.* **217:** 1–9.

Milán M., Campuzano S., and García-Bellido A. 1996a. Cell cycling and patterned cell proliferation in the *Drosophila* wing during metamorphosis. *Proc. Natl. Acad. Sci.* **93:** 11687–11692.

———. 1996b. Cell cycling and patterned cell proliferation in the wing primordium of *Drosophila. Proc. Natl. Acad. Sci.* **93:** 640–645.

Milán M., Perez L., and Cohen S.M. 2002. Short-range cell interactions and cell survival in the *Drosophila* wing. *Dev. Cell* **2:** 797–805.

Minami M., Kinoshita N., Kamoshida Y., Tanimoto H., and Tabata T. 1999. brinker is a target of Dpp in *Drosophila* that negatively regulates Dpp-dependent genes. *Nature* **398:** 242–246.

Minina E., Wenzel H.M., Kreschel C., Karp S., Gaffield W., McMahon A.P., and Vortkamp A. 2001. BMP and Ihh/PTHrP signaling interact to coordinate chondrocyte proliferation and differentiation. *Development* **128:** 4523–4534.

Miron M., Verdu J., Lachance P.E., Birnbaum M.J., Lasko P.F., and Sonenberg N. 2001. The translational inhibitor 4E-BP is an effector of PI(3)K/Akt signalling and cell growth in *Drosophila. Nat. Cell Biol.* **3:** 596–601.

Montagne J. 2000. Genetic and molecular mechanisms of cell size control. *Mol. Cell. Biol. Res. Commun.* **4:** 195–202.

Montagne J., Stewart M.J., Stocker H., Hafen E., Kozma S.C., and Thomas G. 1999.

Drosophila S6 kinase: A regulator of cell size. *Science* **285:** 2126–2129.

Morata G. and Ripoll P. 1975. Minutes: Mutants of *Drosophila* autonomously affecting cell division rate. *Dev. Biol.* **42:** 211–221.

Nagaraj R., Pickup A.T., Howes R., Moses K., Freeman M., and Banerjee U. 1999. Role of the EGF receptor pathway in growth and patterning of the *Drosophila* wing through the regulation of vestigial. *Development* **126:** 975–985.

Nakayama K., Ishida N., Shirane M., Inomata A., Inoue T., Shishido N., Horii I., Loh D.Y., and Nakayama K. 1996. Mice lacking p27Kip1 display increased body size, multiple organ hyperplasia, retinal dysplasia, and pituitary tumors. *Cell* **85:** 707–720.

Neiman P.E., Ruddell A., Jasoni C., Loring G., Thomas S.J., Brandvold K.A., Lee R., Burnside J., and Delrow J. 2001. Analysis of gene expression during myc oncogene-induced lymphomagenesis in the bursa of Fabricius. *Proc. Natl. Acad. Sci.* **98:** 6378–6383.

Nellen D., Burke R., Struhl G., and Basler K. 1996. Direct and long-range action of a DPP morphogen gradient. *Cell* **85:** 357–368.

Neufeld T.P., de la Cruz A.F.A., Johnston L.A., and Edgar B.A. 1998. Coordination of growth and cell division in the *Drosophila* wing. *Cell* **93:** 1183–1193.

Neumann C.J. and Cohen S.M. 1996. Distinct mitogenic and cell fate specification functions of *wingless* in different regions of the wing. *Development* **122:** 1781–1789.

———. 1997. Morphogens and pattern formation. *Bioessays* **19:** 721–729.

Nijhout H.F. 1979. Stretch-induced moulting in *Oncopeltus fasciatus*. *J. Insect Physiol.* **25:** 277–281.

———. 1981. Physiological control of molting in insects. *Am. Zool.* **21:** 631–640.

———. 1984. Abdominal stretch reception in *Dipetalogaster maximus* (Hemiptera: Reduviidae). *J. Insect Physiol.* **30:** 629–633.

———. 1994. *Insect hormones.* Princeton University Press, Princeton, New Jersey.

———. 2003. The control of body size in insects. *Dev. Biol.* **261:** 1–9.

Nijhout H.F. and Grunert L.W. 2002. Bombyxin is a growth factor for wing imaginal disks in Lepidoptera. *Proc. Natl. Acad. Sci.* **99:** 15446–15450.

Nijhout H.F. and Williams C.M. 1974. Control of moulting and metamorphosis in the tobacco hornworm, *Manduca sexta* (L.): Cessation of juvenile hormone secretion as a trigger for pupation. *J. Exp. Biol.* **61:** 493–501.

Nurse P., Thuriaux P., and Nasmyth K. 1976. Genetic control of the cell division cycle in the fission yeast *Schizosaccharomyces pombe*. *Mol. Gen. Genet.* **146:** 167–178.

Oldham S., Bohni R., Stocker H., Brogiolo W., and Hafen E. 2000. Genetic control of size in *Drosophila*. *Philos. Trans. R. Soc. Lond. B Biol. Sci.* **355:** 945–952.

Oldham S., Stocker H., Laffargue M., Wittwer F., Wymann M., and Hafen E. 2002. The *Drosophila* insulin/IGF receptor controls growth and size by modulating PtdInsP(3) levels. *Development* **129:** 4103–4109.

Orian A., van Steensel B., Delrow J., Bussemaker H.J., Li L, Sawado T., Williams E., Loo L.W., Cowley S.M., Yost C., Pierce S., Edgar B.A., Parkhurst S.M., and Eisenman R.N. 2003. Genomic binding by the *Drosophila* Myc, Max, Mad/Mnt transcription factor network. *Genes Dev.* **17:** 1101–1114.

Pantalacci S., Tapon N., and Leopold P. 2003. The Salvador partner Hippo promotes apoptosis and cell-cycle exit in *Drosophila*. *Nat. Cell Biol.* **5:** 921–927.

Partridge L., Barrie B., Fowler K., and French V. 1994. Evolution and development of body size and cell size in *Drosophila melanogaster* in response to temperature. *Evolution* **48:** 1269–1276.

Patel P.H., Thapar N., Guo L., Martinez, M., Maris J., Gau C.-J., Lengyel J.A., and Tamanoi F. 2003. *Drosophila* Rheb GTPase is required for cell cycle progression and cell growth. *J. Cell Sci.* **116:** 3601–3610.

Pohley H.-J. 1965. Regeneration and the moulting cycle in *Ephestia kühniella*. In *Regeneration in animals and related problems* (ed. V. Kiortsis and H.A.L. Trampusch), pp. 324-330. North Holland, Amsterdam.

Polymenis M. and Schmidt E.V. 1997. Coupling of cell division to cell growth by translational control of the G_1 cyclin CLN3 in yeast. *Genes Dev.* **11:** 2522–2531.

Poodry C.A. and Woods D.A. 1990. Control of the developmental, timer for *Drosophila* pupariation. *Roux's Arch. Dev. Biol.* **199:** 219–227.

Potter C.J., Pedraza L.G., and Xu T. 2002. Akt regulates growth by directly phosphorylating Tsc2 (see comments). *Nat. Cell Biol.* **4:** 658–665.

Powell J.R. 1974. Temperature related genetic divergence in *Drosophila* body size. *J. Hered.* **65:** 257–258.

Prober D.A. and Edgar B.A. 2000. Ras promotes cellular growth in the *Drosophila* wing. *Cell* **100:** 435–446.

———. 2002. Interactions between Ras1, dMyc, and dPI3K signaling in the developing *Drosophila* wing. *Genes Dev.* **16:** 2286–2299.

Rabinowitz M. 1941. Studies on the cytology and early embryology of the egg in *Drosophila melanogaster. J. Morphol.* **69:** 1–49.

Rahn P. 1972. Untersuchungen zur Entwicklung von Ganz- unt Teilimplanten der Flügelimaginalscheibe von *Ephestia kühniella* Z. *Wilhelm Roux' Arch.* **170:** 48–82.

Rane S.G., Dubus P., Mettus R.V., Galbreath E.J., Boden G., Reddy E.P., and Barbacid M. 1999. Loss of Cdk4 expression causes insulin-deficient diabetes and Cdk4 activation results in beta-islet cell hyperplasia. *Nat. Genet.* **22:** 44–52.

Resino J., Salama-Cohen P., and García-Bellido A. 2002. Determining the role of patterned cell proliferation in the shape and size of the *Drosophila* wing. *Proc. Natl. Acad. Sci.* **99:** 7502–7507.

Riddiford L.M. 1985. Hormone action at the cellular level. In *Comprehensive insect physiology, biochemistry and pharmacology* (ed. G.A. Kerkut and L.I. Gilbert), pp. 37–84. Pergamon Press, Oxford, United Kingdom.

———. 1993a. Hormone receptors and the regulation of insect metamorphosis. *Receptor* **3:** 203–209.

———. 1993b. Hormones and *Drosophila* development. In *The development of* Drosophila melanogaster (ed. M. Bate and A. Martinez Arias), pp. 899–939. Cold Spring Harbor Laboratory Press, Cold Spring Harbor, New York.

Riddiford L.M. and Ashburner M. 1991. Effects of juvenile hormone mimics on larval development and metamorphosis of *Drosophila melanogaster. Gen. Comp. Endocrinol.* **82:** 172–183.

Robertson F.W. 1960. The ecological genetics of growth in *Drosophila*. 2. Selection for large body size on different diets. *Genet. Res.* **1:** 305–318.

———. 1963. The ecological genetics of growth in *Drosophila*. VI. The genetic correlation between the duration of the larval period and body size in relation to larval diet. *Genet. Res.* **4:** 74–92.

Rosenwald I.B., Lazaris-Karatzas A., Sonenberg N., and Schmidt E.V. 1993. Elevated levels of cyclin D1 protein in response to increased expression of eukaryotic initiation factor 4E. *Mol. Cell. Biol.* **13:** 7358–7363.

Rountree D.B. and Bollenbacher W.E. 1986. The release of the prothoracicotropic hormone in the tobacco hornworm, *Manduca sexta*, is controlled intrinsically by juvenile hormone. *J. Exp. Biol.* **120:** 41–58.

Royzman I., Whittaker A.J., and Orr-Weaver T.L. 1997. Mutations in *Drosophila DP* and *E2F* distinguish G_1-S progression from an associated transcriptional program. *Genes Dev.* **11:** 1999–2011.

Rudkin G.T. 1972. Replication in polytene chromosomes. *Results Probl. Cell Differ.* **4:** 59–85.

Rulifson E.J., Kim S.K., and Nusse R. 2002. Ablation of insulin-producing neurons in flies: Growth and diabetic phenotypes. *Science* **296:** 1118–1120.

Saltiel A.R. and Kahn C.R. 2001. Insulin signalling and the regulation of glucose and lipid metabolism. *Nature* **414:** 799–806.

Saucedo L.J., Gao X., Chiarelli D.A., Li L., Pan D., and Edgar B.A. 2003. Rheb promotes cell growth as a component of the insulin/TOR signalling network. *Nat. Cell Biol.* **5:** 566–571.

Sauer K., Knoblich J.A., Richardson H., and Lehner C.F. 1995. Distinct modes of cyclin E/cdc2c kinase regulation and S-phase control in mitotic and endoreduplication cycles of *Drosophila* embryogenesis. *Genes Dev.* **9:** 1327–1339.

Schmelzle T. and Hall M.N. 2000. TOR, a central controller of cell growth (see comments). *Cell* **103:** 253–262.

Schmidt E.V. 1999. The role of c-myc in cellular growth control. *Oncogene* **18:** 2988–2996.

Schreiber-Agus N., Stein D., Chen K., Goltz J.S., Stevens L., and DePinho R.A. 1997. *Drosophila* Myc is oncogenic in mammalian cells and plays a role in the diminutive phenotype. *Proc. Natl. Acad. Sci.* **94:** 1235–1240.

Schubiger M. and Palka J. 1987. Changing spatial patterns of DNA replication in the developing wing of *Drosophila*. *Dev. Biol.* **123:** 145–153.

Schubiger M. and Schneiderman H.A. 1971. Nuclear transplantation in *Drosophila melanogaster*. *Nature* **230:** 185–186.

Schuhmacher M., Staege M.S., Pajic A., Polack A., Weidle U.H., Bornkamm G.W., Eick D., and Kohlhuber F. 1999. Control of cell growth by c-Myc in the absence of cell division. *Curr. Biol.* **9:** 1255–1258.

Schwartz M.B., Imberski R.B., and Kelly T.J. 1984. Analysis of metamorphosis in *Drosophila melanogaster:* Characterization of giant, an ecdysteroid-deficient mutant. *Dev. Biol.* **103:** 85–95.

Sehnal F. and Bryant P.J. 1993. Delayed pupariation in *Drosophila* imaginal disc overgrowth mutants is associated with reduced ecdysteroid titer. *J. Insect Physiol.* **12:** 1051–1059.

Serrano N. and O'Farrell P.H. 1997. Limb morphogenesis: Connections between patterning and growth. *Curr. Biol.* **7:** R186–R195.

Shafiei M., Moczek A.P., and Nijhout H.F. 2001. Food availability controls the onset of metamorphosis in the dung beetle *Onthophagus taurus* (Coleoptera: Scarabeidae). *Physiol. Entomol.* **26:** 173–180.

Sheikh M.S., Rochefort H., and Garcia M. 1995. Overexpression of p21[WAF1/CIP1] induces growth arrest, giant cell formation and apoptosis in human breast carcinoma cell lines. *Oncogene* **11:** 1899–1905.

Sicinski P., Donaher J.L., Parker S.B., Li T., Fazeli A., Gardner H., Haslam S.Z., Bronson R.T., Elledge S.J., and Weinberg R.A. 1995. Cyclin D1 provides a link between development and oncogenesis in the retina and breast. *Cell* **82:** 621–630.

Sicinski P., Donaher J.L., Geng Y., Parker S.B., Gardner H., Park M.Y., Robker R.L., Richards J.S., McGinnis L.K., Biggers J.D., Eppig J.J., Bronson R.T., Elledge S.J., and

Weinberg R.A. 1996. Cyclin D2 is a FSH-responsive gene involved in gonadal cell proliferation and oncogenesis. *Nature* **384:** 470–474.

Simpson P. and Morata G. 1981. Differential mitotic rates and patterns of growth in compartments in the *Drosophila* wing. *Dev. Biol.* **85:** 299–308.

Simpson P. and Schneiderman H.A. 1975. Isolation of temperature sensitive mutations blocking clone development in *Drosophila melanogaster*, and the effects of a temperature sensitive cell lethal mutation on pattern formation in imaginal discs. *Wilhelm Roux' Arch. Dev. Biol.* **178:** 247–275.

Simpson P., Berreur P., and Berreur-Bonnenfant J. 1980. The initiation of pupariation in *Drosophila:* Dependence on growth of the imaginal discs. *J. Embryol. Exp. Morphol.* **57:** 155–165.

Skapek S.X., Rhee J., Spicer D.B., and Lassar A.B. 1995. Inhibition of myogenic differentiation in proliferating myoblasts by cyclin D1-dependent kinase. *Science* **267:** 1022–1024.

Smith A.V. and Orr-Weaver T.L. 1991. The regulation of the cell cycle during *Drosophila* embryogenesis—The transition to polyteny. *Development* **112:** 997–1008.

Sonoda J. and Wharton R.P. 2001. *Drosophila* Brain Tumor is a translational repressor. *Genes Dev.* **15:** 762–773.

Spradling A.C. 1993. Developmental genetics of oogenesis. In *The development of* Drosophila melanogaster (ed. M. Bate and A. Martinez Arias), pp. 1–70. Cold Spring Harbor Laboratory Press, Cold Spring Harbor, New York.

St-Jacques B., Hammerschmidt M., and McMahon A.P. 1999. Indian hedgehog signaling regulates proliferation and differentiation of chondrocytes and is essential for bone formation (erratum in *Genes Dev.* [1999] **13:** 2617). *Genes Dev.* **13:** 2072–2086.

Stocker H., Radimerski T., Schindelholz B., Wittwer F., Belawat P., Daram P., Breuer S., Thomas G., and Hafen E. 2003. Rheb is an essential regulator of S6K in controlling cell growth in *Drosophila*. *Nat. Cell Biol.* **5:** 559–565.

Strohmaier H., Spruck C.H., Kaiser P., Won K.A., Sangfelt O., and Reed S.I. 2001. Human F-box protein hCdc4 targets cyclin E for proteolysis and is mutated in a breast cancer cell line (see comments). *Nature* **413:** 316–322.

Struhl G. and Basler K. 1993. Organizing activity of wingless protein in *Drosophila*. *Cell* **72:** 527–540.

Tabata T., Schwartz C., Gustavson E., Ali Z., and Kornberg T.B. 1995. Creating a *Drosophila* wing de novo, the role of *engrailed*, and the compartment border hypothesis. *Development* **121:** 3359–3369.

Tapon N., Harvey K.F., Bell D.W., Wahrer D.C., Schiripo T.A., Haber D.A., and Hariharan I.K. 2002. *salvador* promotes both cell cycle exit and apoptosis in *Drosophila* and is mutated in human cancer cell lines (see comments). *Cell* **110:** 467–478.

Tatar M., Kopelman A., Epstein D., Tu M.P., Yin C.M., and Garofalo R.S. 2001. A mutant *Drosophila* insulin receptor homolog that extends life-span and impairs neuroendocrine function (comment). *Science* **292:** 107–110.

Teleman A.A. and Cohen S.M. 2000. Dpp gradient formation in the *Drosophila* wing imaginal disc. *Cell* **103:** 971–980.

Truman J.W. and Bate M. 1988. Spatial and temporal patterns of neurogenesis in the central nervous system of *Drosophila melanogaster*. *Dev. Biol.* **125:** 145–157.

Truman J.W. and Riddiford L.M. 1974. Physiology of insect rhythms. 3. The temporal organization of the endocrine events underlying pupation of the tobacco hornworm. *J. Exp. Biol.* **60:** 371–382.

Trumpp A., Refaeli Y., Oskarsson T., Gasser S., Murphy M., Martin G.R., and Bishop J.M.

2001. c-Myc regulates mammalian body size by controlling cell number but not cell size. *Nature* **414**: 768–773.

Tsuneizumi K., Nakayama T., Kamoshida Y., Kornberg T.B., Christian J.L., and Tabata T. 1997. Daughters against dpp modulates dpp organizing activity in *Drosophila* wing development (see comments). *Nature* **389**: 627–631.

Tsutsui T., Hesabi B., Moons D.S., Pandolfi P.P., Hansel K.S., Koff A., and Kiyokawa H. 1999. Targeted disruption of CDK4 delays cell cycle entry with enhanced p27 (Kip1) activity. *Mol. Cell. Biol.* **19**: 7011–7019.

Udan R.S., Kango-Singh M., Nolo R., Tao C., and Halder G. 2003. Hippo promotes proliferation arrest and apoptosis in the Salvador/Warts pathway. *Nat. Cell Biol.* **5**: 914–920.

van de Wetering M., Sancho E., Verweij C., de Lau W., Oving I., Hurlstone A., van der Horn K., Batlle E., Coudreuse D., Haramis A.P., Tjon-Pon-Fong M., Moerer P., van den Born M., Soete G., Pals S., Eilers M., Medema R., and Clevers H. 2002. The beta-catenin/TCF-4 complex imposes a crypt progenitor phenotype on colorectal cancer cells. *Cell* **111**: 241–250.

Verdu J., Burtovich M.A., Wilder E.L., and Birnbaum M.J. 1999. Cell autonomous regulation of cell and organ growth in *Drosophila* by AKT/PKB. *Nat. Cell Biol.* **1**: 500–513.

Watson J.D., Oster S.K., Shago M., Khosravi F., and Penn L.Z. 2002. Identifying genes regulated in a Myc-dependent manner. *J. Biol. Chem.* **277**: 36921–36930.

Watson K.L., Justice R.W., and Bryant P.J. 1994. *Drosophila* in cancer research: The first fifty tumor suppressor genes. *J. Cell Sci.* **18**: 19–33.

Weatherbee S.D., Halder G., Kim J., Hudson A., and Carroll S. 1998. Ultrabithorax regulates genes at several levels of the wing-patterning hierarchy to shape the development of the *Drosophila* haltere. *Genes Dev.* **12**: 1474–1482.

Weigmann K., Cohen S.M., and Lehner C.F. 1997. Cell-cycle progression, growth and patterning in imaginal discs despite inhibition of cell division after inactivation of *Drosophila* Cdc2 kinase. *Development* **124**: 3555–3563.

Weinkove D., Neufeld T.P., Twardzik T., Waterfield M.D., and Leevers S.J. 1999. Regulation of imaginal disc cell size, cell number and organ size by *Drosophila* class I_A phosphoinositide 3-kinase and its adaptor. *Curr. Biol.* **9**: 1019–1029.

Weiss A., Herzig A., Jacobs H., and Lehner C.F. 1998. Continuous Cyclin E expression inhibits progression through endoreduplication cycles in *Drosophila*. *Curr. Biol.* **8**: 239–242.

Wiersdorff V., Lecuit T., Cohen S.M., and Mlodzik M. 1996. *Mad* acts downstream of Dpp receptors, revealing a differential requirement for *dpp* signaling in initiation and propagation of morphogenesis in the *Drosophila* eye. *Development* **122**: 2153–2162.

Wolpert L. 1971. Positional information and pattern formation. *Curr. Top. Dev. Biol.* **6**: 183–224.

Woodhouse E., Hersperger E., and Shearn A. 1998. Growth, metastasis, and invasiveness of *Drosophila* tumors caused by mutations in specific tumor suppressor genes. *Development Genes Evol.* **207**: 542–550.

Wu S., Huang J., Dong J., and Pan D. 2003. hippo encodes a Ste-20 family protein kinase that restricts cell proliferation and promotes apoptosis in conjunction with salvador and warts. *Cell* **114**: 445–456.

Xu T. and Rubin G.M. 1993. Analysis of genetic mosaics in developing and adult *Drosophila* tissues. *Development* **117**: 1223–1237.

Yang L. and Baker N.E. 2003. Cell cycle withdrawal, progression, and cell survival regulation by EGFR and its effectors in the differentiating *Drosophila* eye. *Dev. Cell* **4**: 359–369.

Yu J., Carroll T.J., and McMahon A.P. 2002. Sonic hedgehog regulates proliferation and differentiation of mesenchymal cells in the mouse metanephric kidney. *Development* **129:** 5301–5312.

Zaffran S., Chartier A., Gallant P., Astier M., Arquier N., Doherty D., Gratecos D., and Semeriva M. 1998. A *Drosophila* RNA helicase gene, pitchoune, is required for cell growth and proliferation and is a potential target of d–Myc. *Development* **125:** 3571–3584.

Zecca M., Basler K., and Struhl G. 1995. Sequential organizing activities of engrailed, hedgehog and decapentaplegic in the *Drosophila* wing. *Development* **121:** 2265–2278.

Zhang H., Stallock J.P., Ng J.C., Reinhard C., and Neufeld T.P. 2000. Regulation of cellular growth by the *Drosophila* target of rapamycin dTOR. *Genes Dev.* **14:** 2712–2724.

Zhang Y., Gao X., Saucedo L.J., Ru B., Edgar B.A., and Pan D. 2003. Rheb is a direct target of the tuberous sclerosis tumour suppressor proteins. *Nat. Cell Biol.* **5:** 578–581.

Zitnan D., Sehnal F., and Bryant P.J. 1993. Neurons producing specific neuropeptides in the central nervous system of normal and pupariation-delayed *Drosophila*. *Dev. Biol.* **156:** 117–135.

Zurovec M., Dolezal T., Gazi M., Pavlova E., and Bryant P.J. 2002. Adenosine deaminase-related growth factors stimulate cell proliferation in *Drosophila* by depleting extracellular adenosine. *Proc. Natl. Acad. Sci.* **99:** 4403–4408.

3

Coordination of Cell Growth and Cell-cycle Progression in Proliferating Mammalian Cells

Ian Conlon,[1] Alison Lloyd, and Martin Raff
MRC Laboratory for Molecular Cell Biology and Cell Biology Unit,
University College London,
London WC1E 6BT, UK

THE SIZE OF AN ANIMAL DEPENDS MAINLY on the size and number of the cells it contains. Although animal growth depends on both cell growth and cell division, cell growth has, inexplicably, received much less attention than cell division. In this chapter, we consider one aspect of cell growth—how it and progression through the cell division cycle are coordinated in mammalian cells.

This coordination problem has mainly been studied in yeasts, in which cell growth appears to be rate-limiting for cell-cycle progression (Johnston et al. 1977). There is good evidence that yeasts have intracellular cell-size checkpoints, where the cell cycle can pause until cell size reaches a threshold value before progressing to the next phase of the cycle (Nurse et al. 1976; Fantes and Nurse 1977; Johnston et al. 1977; Nurse and Thuriaux 1977). In contrast, our studies on primary rat Schwann cells in culture suggest that the coordination of cell growth and cell-cycle progression in mammalian cells may be different from that in yeasts and may rely on extracellular signals rather than on intracellular controls (Conlon et al. 2001; Conlon and Raff 2003).

[1]Present address: Great Minster House, 77 Marsham Street, London SW1P 4DR, United Kingdom.

SCHWANN CELL GROWTH DEPENDS ON EXTRACELLULAR GROWTH FACTORS

We study Schwann cells purified from newborn rat sciatic nerve (Cheng et al. 1998). We assess cell size by removing the cells from the culture dish and measuring cell volume in a Coulter counter, and we analyze cell growth independently of cell-cycle progression by arresting the cell cycle in S phase with aphidicolin, which blocks DNA synthesis by inhibiting DNA polymerase α (Conlon et al. 2001). We find that cells in serum that are arrested with aphidicolin for 24 hours are much larger on average than cells that are allowed to continue to proliferate (Fig. 1a) (Conlon et al. 2001). Thus, as for yeast cells, Schwann cells do not need to progress through the cell cycle to grow. Similar observations have recently been published for several fibroblast cell lines (Fingar et al. 2002).

To assure ourselves that the Coulter counter measurement of cell volume reflects cell mass, we collaborated with Graham Dunn and used density interference microscopy to study dry cell mass. We obtain the same results as with the Coulter counter: The average dry mass of cells arrested for 24 hours in aphidicolin is much greater than that of proliferating cells (Conlon et al. 2001).

As expected, the growth of Schwann cells seems to depend on extracellular signals, as aphidicolin-arrested cells do not grow in the absence of serum and growth factors (Fig. 1b) (Conlon et al. 2001). They also do not grow in suspension, despite the presence of serum and growth factors

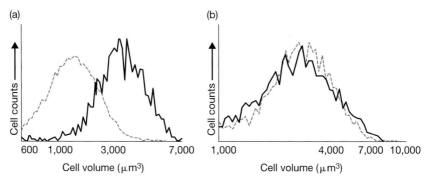

Figure 1. (*a*) Aphidicolin-arrested Schwann cells (*solid line*) are larger on average after 24 hours in 3% fetal calf serum (FCS) than untreated proliferating cells (*dashed line*). (*b*) In contrast, aphidicolin-arrested cells after 24 hours in the absence of FCS (*dashed line*) are the same size as aphidicolin-arrested cells after 1 hour in the absence of FCS (*solid line*). Cell volume was measured in a Coulter counter. (Reprinted, with permission, from Conlon et al. 2001 [copyright Nature Publishing Group].)

(I. Conlon, unpubl.). Thus, unlike yeast cells, which depend solely on nutrients and grow in suspension, it seems that Schwann cells (and probably many other adherent mammalian cells) require extracellular signals and attachment to a substratum both to grow and to proliferate.

IGF-1 IS A GROWTH FACTOR FOR SCHWANN CELLS; GGF-2 IS ONLY A MITOGEN

We have studied two "growth factors" that have been shown to promote Schwann cell proliferation—insulin-like growth factor 1 (IGF-1) and glial growth factor 2 (GGF-2) (Conlon et al. 2001). We find that, in the absence of serum, IGF-1 on its own stimulates aphidicolin-arrested cells to grow (Fig. 2a), whereas GGF-2 does not (Fig. 2b). Moreover, GGF-2 does not increase the growth stimulation seen with IGF-1 alone. Thus, IGF-1 is a true growth factor for Schwann cells, whereas GGF-2 is not.

GGF-2 is, however, a mitogen for Schwann cells. It induces Schwann cells cultured without serum or aphidicolin to incorporate BrdU into DNA; IGF-1, in contrast, has almost no ability to do this on its own, but it greatly increases the ability of GGF-2 to do so (Fig. 3) (Conlon et al. 2001). Thus, for Schwann cells, GGF-2 is a mitogen but not a growth factor, whereas IGF-1 is a growth factor and also promotes cell-cycle progression in the presence of GGF-2.

Similar observations have been reported for other mammalian cell types. EGF (Zetterberg et al. 1984) and thyroid-stimulating hormone

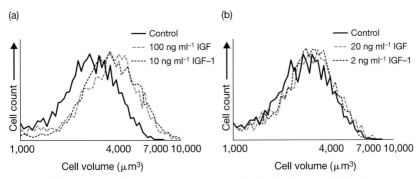

Figure 2. (*a*) IGF-1 at either 10 ng/ml or 100 ng/ml stimulates aphidicolin-arrested Schwann cells to grow in the absence of FCS, as assessed after 24 hours; control cells were cultured without FCS or growth factors. (*b*) In contrast, GGF-2 at either 2 ng/ml or 10 ng/ml does not stimulate the cells to grow. (Reprinted, with, permission from Conlon et al. 2001 [copyright Nature Publishing Group].)

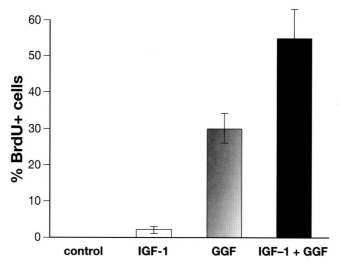

Figure 3. GGF-2 (20 ng/ml) is more potent than IGF-1 (100 ng/ml) at stimulating BrdU incorporation in Schwann cells that have been arrested for 48 hours in the absence of FCS and treated with growth factors and BrdU for a further 24 hours. IGF-1, however, synergizes with GGF-2 in stimulating BrdU incorporation. (Modified, with permission, from Conlon et al. 2001 [copyright Nature Publishing Group].)

(Deleu et al. 1999), for example, promote cell-cycle progression in Swiss 3T3 cells and dog thyrocytes, respectively, without promoting cell growth.

Why do IGF-1 and GGF-2 have such different effects on Schwann cells when they both act via receptor tyrosine kinases? One reason is that their receptors activate intracellular signaling pathways with different kinetics. Whereas IGF-1 stimulates a sustained activation of Akt and a transient activation of MAP kinases, GGF-2 does the opposite (Fig. 4) (I. Conlon and A. Lloyd, unpubl.). IGF-1 does not increase the activation of MAP kinases stimulated by GGF-2 (Fig. 4) (I. Conlon and A. Lloyd, unpubl. observations), suggesting that MAP kinases are not the point of convergence at which IGF-1 and GGF-2 synergize to stimulate cell-cycle progression.

CELL GROWTH IS NOT THE SOLE DRIVER OF CELL-CYCLE PROGRESSION IN SCHWANN CELLS

The finding that IGF-1 and GGF-2 have different effects on Schwann cells has enabled us to assess the role of cell growth in cell-cycle progression. We have cultured Schwann cells without serum or aphidicolin in variable

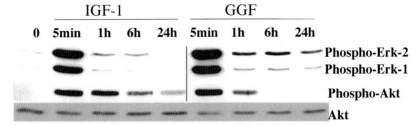

Figure 4. Western blots showing that IGF-1 and GGF-2 stimulate signaling through the PI3 kinase and MAP kinase pathways with different kinetics in Schwann cells. The cells were arrested in aphidicolin in 3% FCS for 24 hours and then in aphidicolin without FCS for a further 24 hours, before the growth factors were added for various time periods as indicated. Blotting was performed with antibodies specific for Akt or for phosphorylated (activated) Erk-1, Erk-2, or Akt.

concentrations of GGF-2, but in a constant concentration of IGF-1, thereby keeping cell growth constant. We find that the cells go through the cycle faster in the high concentration of GGF-2 than in the low concentration, even though the cells are growing at the same rate (Fig. 5a) (Conlon et al. 2001). This finding indicates that cell growth cannot be the only rate-limiting event that controls cell-cycle progression in Schwann cells. If cells in high GGF-2 cycle faster but grow at the same rate as cells in low GGF, the cells should divide at a smaller size in high GGF-2 than in low GGF-2, and this is the case: The cells in high GGF are smaller at all phases of the cell cycle (Fig. 5b–d) (Conlon et al. 2001).

DO MAMMALIAN CELLS HAVE CELL-SIZE CHECKPOINTS?

The above findings suggest that yeast cells and Schwann cells may use different strategies to coordinate cell growth and cell-cycle progression. They do not, however, exclude the possibility that Schwann cells have cell-size checkpoints, in which the size thresholds required to progress through the cell cycle are decreased by mitogens such as GGF-2.

Indeed, some observations on proliferating mammalian cells in culture have been interpreted to suggest the presence of cell-size checkpoints (for review, see Wells 2002). Fibroblasts that divide at a smaller size than the mean, for example, spend longer on average in the subsequent cell cycle and therefore grow more than the average cells before dividing again (Killander and Zetterberg 1965). Other findings with fibroblasts, however, argue against the operation of a cell-size checkpoint. Although the

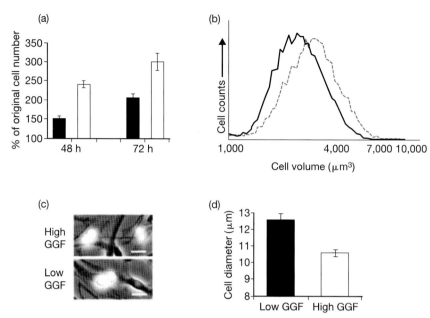

Figure 5. A higher concentration of GGF-2 shortens the cell-cycle time and decreases cell size. (*a*) At 20 ng/ml (*open columns*), cell number increases faster than at 2 ng/ml (*filled columns*), even though IGF-1 levels are constant (100 ng/ml) so that cell growth is the same at both GGF concentrations. (*b*) The average volume of cells proliferating in high GGF-2 (*solid line*) is smaller than that of cells proliferating in low GGF-2 (*dashed line*). (*c*, *d*) Mitotic cells (phase-bright rounded cells in *c*) in high GGF-2 are also smaller than mitotic cells in low GGF-2. (Reprinted, with permission, from Conlon et al. 2001 [copyright Nature Publishing Group].)

average cell-cycle time of smaller cells is longer, the variability in cycle times is much greater than for larger cells (Shields et al. 1978; Brooks 1981). Most importantly, the minimum cell-cycle time is the same for the large and small cells (Brooks 1981). If a cell-size checkpoint were operating, the variability in cell-cycle times should be similar for large and small cells, and the minimum cell-cycle time should be longer for small cells than large cells. Similarly, fibroblasts proliferating in low concentrations of serum grow and cycle more slowly than cells proliferating in high concentrations of serum, but the minimum cycle times are the same, again arguing against the operation of cell-size checkpoints (Brooks 1981). Moreover, although there are reports that a small mammalian daughter cell takes longer to go through its next G_1 phase than does its larger sister (Killander and Zetterberg 1965), there are other cases where no correlation is observed between cell size after division and the length of the next G_1 (Fox and Pardee 1969). Thus, the evidence for cell-size checkpoints in mammalian cells is controversial and not compelling.

YEAST CELLS MAY NEED A CELL-SIZE CHECKPOINT BECAUSE BIG YEAST CELLS GROW FASTER THAN SMALL CELLS

One line of evidence that yeast cells have cell-size checkpoints is that they maintain a constant size distribution over time as they proliferate, even though big yeast cells grow faster than smaller yeast cells, at least when blocked in S phase by a mutation (Nurse et al. 1976). If yeast cells did not have cell-size checkpoints, cells that are born larger than the mean birth size (through natural variation) would grow faster than those that are born smaller, and these larger cells would produce still larger daughters, which would then grow even faster. Thus, the spread of sizes in the population would increase over time, which does not happen, implying that yeast cells have cell-size checkpoints (for discussion, see Brooks 1981).

Cell-size checkpoints are required to maintain a constant size distribution over time, however, only if large cells grow faster than small cells at the same point in the cell cycle. This can be illustrated by the following example. Suppose a cell divides unevenly to produce one daughter cell of mass 10 (arbitrary units) and another daughter cell of mass 1. If big cells grow and go through the cell cycle at the same rate as small cells, the two daughter cells and their progeny would tend to converge to a mean size of 5.5 (Fig. 6) (Brooks 1981). The sizes converge because the larger cells do not double their cell mass in one cycle, and the smaller cells more than double their cell mass in one cycle. Thus, such cells will maintain a constant size distribution even without cell-size checkpoints.

DO BIG SCHWANN CELLS GROW FASTER THAN SMALL SCHWANN CELLS?

When we follow the serum-stimulated growth of Schwann cells that are arrested in S phase by the addition of aphidicolin, we find that cell growth is strikingly linear, with the cells adding the same amount of volume, and presumably cell mass, per day, independent of their size (Fig. 7) (Conlon and Raff 2003). Thus, unlike yeast cells blocked in S phase, where big cells grow faster than small cells, big Schwann cells blocked in S phase do not grow faster than small Schwann cells blocked in S phase. If the same growth rules apply to cycling cells as apply to cells blocked in S phase, then

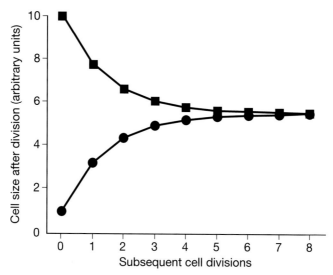

Figure 6. Hypothetical model (based on Brooks 1981) showing why the progeny of large and small daughter cells eventually return to the mean population size over time if large and small cells grow and progress through the cell cycle at the same rates. The points indicate the cell sizes at division of two daughter cells produced by an unequal cell division and their progeny cells produced by eight subsequent divisions. The initial division is unequal and produces one cell of mass 10 units and one cell of mass 1 unit; the subsequent eight divisions of the progeny cells are equal. Following the first division, each cell grows 5.5 mass units in each cycle. Thus, the initial small daughter cell grows to 6.5 units before it divides to produce two daughters of about 3.2 units each, whereas the initial large daughter cell grows to 15.5 units before it divides to produce two daughters of about 7.8 units.

Schwann cells, unlike yeast cells, should not need a cell-size checkpoint to maintain a constant distribution of sizes as they proliferate.

Although the growth rate of aphidicolin-blocked Schwann cells is independent of size, it is not fixed. It depends on the concentration of extracellular growth factors. As the concentration of serum is increased, even up to 50% (not shown), cell growth remains linear, but the slope increases (see Fig. 7) (Conlon and Raff 2003). It seems that the levels of extracellular growth factors, rather than anything inside the cell, normally limit Schwann cell growth in culture.

Is it possible that the apparent difference between yeast cell growth and Schwann cell growth is an artifact of the way the experiments are done, although in both cases cells blocked in S phase were studied? In the Schwann cell experiments, for example, we measure cell volume (Conlon

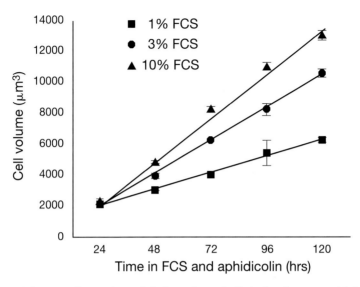

Figure 7. Schwann cell growth rate is independent of cell size but increases with increasing concentrations of FCS. The cells were arrested for various periods of time in aphidicolin and 1% FCS, 3% FCS, or 10% FCS. (Reprinted from Conlon and Raff 2003.)

and Raff 2003), whereas in the yeast experiments protein per cell was measured (Nurse et al. 1976). This is an unlikely explanation, as measuring protein per cell gives the same results in Schwann cells as measuring cell volume (Conlon and Raff 2003). Is it because aphidicolin is used to arrest Schwann cells in S phase (Conlon and Raff 2003), whereas a mutation arrested the yeast cell cycle in S phase (Nurse et al. 1976)? Experiments done by Hudson and Mortimer suggest that this is unlikely to be the explanation for the differences between yeast cell growth and Schwann cell growth (Hutson and Mortimore 1982). They starved mice, which causes the liver to shrink rapidly. All the shrinkage was due to hepatocyte shrinkage, rather than to cell death or a decrease in cell proliferation. When they re-fed the mice, the liver re-grew rapidly—entirely due to hepatocyte growth, which was clearly linear. Thus, the growth of hepatocytes in vivo, which are mainly in G_0, is apparently independent of cell size. Moreover, when quiescent 3T3 cells are separated by size and re-stimulated to enter the cell cycle, large and small cells are observed to grow at the same rate (Brooks and Shields 1985). It seems, therefore, that many, if not all, mammalian cells grow linearly, and thus do not require cell-size checkpoints to maintain a constant size distribution as they proliferate.

These experiments do not prove, however, that Schwann cells and other mammalian cells do not have cell-size checkpoints.

PROLIFERATING SCHWANN CELLS INCREASE THEIR SIZE ONLY SLOWLY WHEN SWITCHED FROM SERUM-FREE TO SERUM-CONTAINING CULTURE MEDIUM

Some of the strongest evidence that yeast cells have cell-size checkpoints is their rapid adjustment to a new size when nutrient conditions are altered (Fantes and Nurse 1977). Yeast cells divide at a size that is characteristic for a particular culture medium; generally, the more nutrient-rich the medium, the larger is the size at division. When switched to poorer nutrient conditions, yeast cells rapidly alter their size threshold within one cell cycle and divide at a smaller size that is characteristic for the new condition; conversely, when switched to richer nutrient conditions, they rapidly adjust and divide at the larger size characteristic of the new condition (Fantes and Nurse 1977). Thus, the cell-size threshold can change quickly in response to changes in extracellular conditions.

We have performed an equivalent experiment with purified rat Schwann cells (Conlon and Raff 2003). These cells can proliferate indefinitely on a laminin-coated culture dish, in either serum-containing or serum-free medium. They maintain a constant cell-size distribution if passaged regularly. The cells in serum grow faster and divide at an average size that is more than twice that for cells proliferating without serum (in GGF-2 + IGF-1); when cells proliferating without serum are switched into serum, they take about 6 divisions and 10 days before they adopt the size of the cells that had always been proliferating in serum (Fig. 8) (Conlon and Raff 2003). This behavior is very different from that of yeast cells. It closely resembles that predicted by the hypothetical model illustrated in Figure 6, for cells that grow linearly do not have cell-size checkpoints and divide at a size that depends on the levels of extracellular signals that stimulate cell growth, cell-cycle progression, or both.

WHY DO SCHWANN CELLS AND YEAST CELLS COORDINATE THEIR GROWTH AND CELL-CYCLE PROGRESSION SO DIFFERENTLY?

The lifestyles of yeast cells and animal cells are very different. Yeast cell proliferation is controlled mainly by nutrients, whereas mammalian cell proliferation is controlled mainly by signals from other cells. As a unicellular organism, each yeast cell grows and divides as fast as the nutrient supply allows and must quickly adapt to changing extracellular conditions. In the

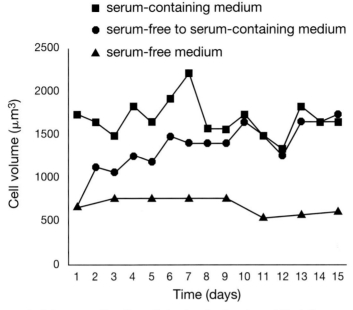

Figure 8. Schwann cells adjust their size slowly when shifted from serum-free to serum-containing medium. The mean volume of cells proliferating in serum-free medium is smaller than that of cells proliferating in serum-containing medium; in both cases, IGF -1 and GGF-2 are present at saturating concentrations. When the latter cells are shifted to serum-containing medium, the cells gradually attain the size of the former cells, but this takes about six divisions and 10 days. The cell-cycle time is not significantly different for the switched cells and those proliferating continuously in FCS (not shown). (Reprinted from Conlon and Raff 2003.)

experiments that show that big yeast cells grow faster than small yeast cells (Nurse et al. 1976), nutrients were presumably saturating, and cell growth was probably limited by intracellular factors such as ribosomes.

The growth and division of mammalian cells, in contrast, must be carefully controlled and coordinated for the good of the animal as a whole, and this control depends on intercellular signals. In our experiments, Schwann cell growth is linear, presumably because it is mainly limited by extracellular conditions, rather than by anything inside the cells (see Fig. 7) (Conlon and Raff 2003). The levels of extracellular signals in animals are usually limiting rather than saturating (Conlon and Raff 1999; van Heyningen et al. 2001), so that small changes in signal concentration can powerfully influence the rate of cell growth and cell-cycle progression. Despite their importance, little is known about how the levels of most extracellular signals are controlled in animals.

LINEAR CELL GROWTH IMPLIES TIGHT COUPLING BETWEEN PROTEIN SYNTHESIS AND PROTEIN DEGRADATION

It is remarkable that a mammalian cell can grow linearly, independent of its size, as many things must change as the cell gets bigger. Big cells have a much smaller surface-area-to-volume ratio compared to small cells, for example, which would be expected to affect the rates of transport across the plasma membrane. Mammalian cells must have special mechanisms to ensure that cell growth rate is independent of cell size.

Cells grow when macromolecular synthesis is greater than macro-molecular degradation (and secretion). Does linear growth imply that big cells and little cells make macromolecules at the same rate, independent of their size? This seems not to be the case. Schwann cells that have been arrested in aphidicolin for three days in serum are much larger than cells arrested in aphidicolin for only one day, and the amount of protein they make, as measured by [^{35}S]methionine and [^{35}S]-cysteine incorporation during a two-hour pulse, is greater (Fig. 9a) (Conlon and Raff 2003). Because the big cells grow at the same rate as the smaller cells, they must degrade (and/or secrete) proteins faster, which is indeed the case (Fig. 9b) (Conlon and Raff 2003). Thus, remarkably, even though protein synthesis and degradation rates can vary with cell size, the cell can keep the difference between synthesis and degradation constant over a large range of sizes. It is not known how cells achieve this.

Studies of sympathetic neurons suggest that the coupling between protein synthesis and degradation can depend on extracellular signals (Franklin and Johnson 1998). The size of these neurons in an adult rodent depends directly on the amount of nerve growth factor (NGF) secreted by the target cells they innervate: If the level of NGF is experimentally increased in the adult, the neurons get bigger; if the level is reduced (by treatment with anti-NGF antibody), the neurons get smaller (Purves et al. 1988). NGF somehow helps these cells to couple the rate of protein synthesis to the rate of protein degradation, at least in culture. When protein synthesis is blocked by cycloheximide treatment in the presence of NGF, the degradation rate of long-lived proteins also decreases, and the cells maintain their size; in the absence of NGF, however, the degradation rate does not decrease when the cells are treated with cycloheximide, and the neurons rapidly atrophy (Franklin and Johnson 1998). It is not known how NGF promotes this remarkable coupling between protein synthesis and degradation. This is a fundamental problem in cell biology, and, like cell growth itself, merits further study.

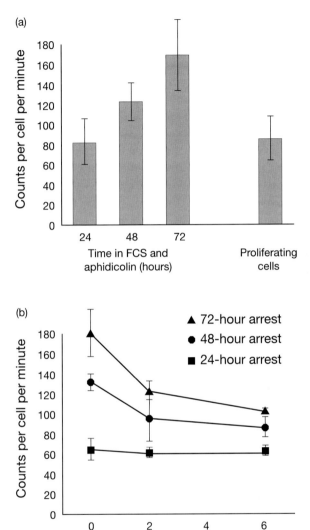

Figure 9. Large Schwann cells synthesize (*a*) and degrade (*b*) protein faster than smaller cells. (*a*) The amount of radioactivity incorporated into protein after a 2-hour pulse of [³⁵S]methionine and [³⁵S]cysteine was measured in cells arrested in aphidicolin in 3% FCS for various times. The amount of radioactivity incorporated into protein in proliferating cells is shown for comparison. (*b*) Cells were treated as in *a* for 24, 48, or 72 hours. They were then either harvested immediately (0 hours after pulse) to assess the rate of total protein synthesis or washed and "chased" with medium containing nonradioactive methionine and cysteine for 2 or 6 hours before harvesting to assess the rate of protein degradation (and/or secretion), which is indicated by the slope of the line. (Reprinted from Conlon and Raff 2003.)

SUMMARY

It seems that yeast cells and at least some mammalian cells, use very different strategies to coordinate cell growth and cell-cycle progression. Whereas yeast cells rely mainly on intracellular mechanisms, mammalian cells rely mainly on extracellular signals that regulate cell growth, cell-cycle progression, or both (Conlon et al. 2001). There is little doubt that the intracellular signaling pathways activated by these extracellular signals contribute to the coordination; PI3 kinase, for example, can activate downstream pathways that can influence both cell growth and cell-cycle progression, at least in some cell types (Katso et al. 2001). It is a major challenge to understand how the levels of these extracellular signals are controlled and how the intracellular signals they activate communicate to the cell-cycle control system, the protein-synthesis apparatus, and the protein-degradation machinery to regulate the rates of cell-cycle progression and cell growth.

ACKNOWLEDGEMENTS

I.C. and M.R. were supported by the Medical Research Council, and A.L. is supported by Cancer Research UK. We are grateful to Graham Dunn and Anne Mudge for help with some experiments and to Sir Paul Nurse and Robert Brooks for helpful discussions.

REFERENCES

Brooks R.F. 1981. Variability in the cell cycle and the control of proliferation. In *The cell cycle* (ed. P.C.L. John, pp. 35–61). Cambridge University Press, Cambridge, United Kingdom.

Brooks R.F. and Shields R. 1985. Cell growth, cell division and cell size homeostasis in Swiss 3T3 cells. *Exp. Cell Res.* **156:** 1–6.

Cheng L., Esch F.S., Marchionni M.A., and Mudge A.W. 1998. Control of Schwann cell survival and proliferation: Autocrine factors and neuregulins. *Mol. Cell. Neurosci.* **12:** 141–156.

Conlon I. and Raff M. 1999. Size control in animal development. *Cell* **96:** 235–244.

———. 2003. Differences in the way a mammalian cell and yeast cells coordinate cell growth and cell-cycle progression. *J. Biol.* **2:** 7.1–7.9.

Conlon I.J., Dunn G.A., Mudge A.W., and Raff M.C. 2001. Extracellular control of cell size. *Nat. Cell Biol.* **3:** 918–921.

Deleu S., Pirson I., Coulonval K., Drouin A., Taton M., Clermont F., Roger P.P., Nakamura T., Dumont J.E., and Maenhaut C. 1999. IGF-1 or insulin, and the TSH cyclic AMP cascade separately control dog and human thyroid cell growth and DNA synthesis, and complement each other in inducing mitogenesis. *Mol. Cell. Endocrinol.* **149:** 41–51.

Fantes P. and Nurse P. 1977. Control of cell size at division in fission yeast by a growth-modulated size control over nuclear division. *Exp. Cell Res.* **107:** 377–386.

Fingar D.C., Salama S., Tsou C., Harlow E., and Blenis J. 2002. Mammalian cell size is controlled by mTOR and its downstream targets S6K1 and 4EBP1/eIF4E. *Genes Dev.* **16:** 1472–1487.

Fox T. and Pardee A. 1969. Animal cells: Noncorrelation of length of G1 phase with size after mitosis. *Science* **167:** 80–82.

Franklin J. and Johnson E. 1998. Control of neuronal size homeostasis by trophic factor-mediated coupling of protein degradation to protein synthesis. *J. Cell Biol.* **142:** 1313–1324.

Hutson N.J. and Mortimore G.E. 1982. Suppression of cytoplasmic protein uptake by lysosomes as the mechanism of protein regain in livers of starved-refed mice. *J. Biol. Chem.* **257:** 9548–9554.

Johnston G.C., Pringle J.R., and Hartwell L.H. 1977. Coordination of growth with cell division in the yeast *Saccharomyces cerevisiae*. *Exp. Cell Res.* **105:** 79–98.

Katso R., Okkenhaug K., Ahmadi K., White S., Timms J., and Waterfield M.D. 2001. Cellular function of phosphoinositide 3-kinases: Implications for development, homeostasis, and cancer. *Annu. Rev. Cell Dev. Biol.* **17:** 615–675.

Killander D. and Zetterberg A. 1965. A quantitative cytochemical investigation of the relationship between cell mass and initiation of DNA synthesis in mouse fibroblasts in vitro. *Exp. Cell Res.* **40:** 12–20.

Nurse P. and Thuriaux P. 1977. Controls over the timing of DNA replication during the cell cycle of fission yeast. *Exp. Cell Res.* **107:** 365–375.

Nurse P., Thuriaux P., and Nasmyth K. 1976. Genetic control of the cell division cycle in the fission yeast *Schizosaccharomyces pombe*. *Mol. Gen. Genet.* **146:** 167–178.

Purves D., Snider W.D., and Voyvodic J.T. 1988. Trophic regulation of nerve cell morphology and innervation in the autonomic nervous system. *Nature* **336:** 123–128.

Shields R., Brooks R.F., Riddle P.N., Capellaro D.F., and Delia D. 1978. Cell size, cell cycle and transition probability in mouse fibroblasts. *Cell* **15:** 469–474.

van Heyningen P., Calver A.R., and Richardson W.D. 2001. Control of progenitor cell number by mitogen supply and demand. *Curr. Biol.* **11:** 232–241.

Wells W.A. 2002. Does size matter? *J. Cell Biol.* **158:** 1156–1159.

Zetterberg A., Engstrom W., and Dafgard E. 1984. The relative effects of different types of growth factors on DNA replication, mitosis, and cellular enlargement. *Cytometry* **5:** 368–375.

4

Coordination of Cell Growth and Cell Division

Emmett Vance Schmidt

Massachusetts General Hospital Cancer Research Center
and the Pediatric Service, MassGeneral Hospital for Children
Boston, Massachusetts 02114

MECHANISMS REGULATING CELLULAR REPRODUCTION are the core problems in the life sciences. Virchow observed, "Omnis cellula a cellula" (Virchow 1858). A single cell is a package of heritable information coupled with the resources needed to duplicate that information. Boveri early observed that the size of the nucleus, which we now know contains the cell's heritable information, does not change after reproduction (Boveri 1905). Julius Sachs further showed that the ratio of cytoplasm to nucleus varied little among cells (Sachs 1898). It is probable that the physics of surface tension sets a limit on the potential sizes of cells within a range of possible cell sizes (Thompson 1961). In contrast, limitations on the content of genetic information are likely based on subtler evolutionary parameters because so much DNA is non-coding. Nevertheless, although total cellular DNA content varies by more than five orders of magnitude among all eukaryotes (Gregory 2001), the ratio of cell size to DNA content can be linear across like species; DNA content in the nucleated erythrocytes of a cohort of closely related mammalian species is an intriguing example of this potential linearity (Gregory 2000). It therefore seems likely that evolutionary selection of optimal cellular DNA content adheres to testable physiologic rules.

Early studies of the question of how cell growth is coordinated with cell division to keep a balance between heritable information and cellular resources were obviously limited by the histologic techniques of the time. The next step in these studies required the identification of DNA as the genetic material of the cell and further identification of the biochemical nature of the various machines needed to reproduce a cell. These biochemical advances, coupled with the identification of isotopes required to measure changes in machinery content over time, have basically governed our definition of growth. In contrast, the measurement of cell division has remained quite simple, since DNA measurement is quite simple.

Measurable cell growth is generally considered to be continuous, whereas DNA synthesis is discontinuous (Fig. 1A). Yet, this point is obviously based heavily on the unit of measurement employed. Cell size con-

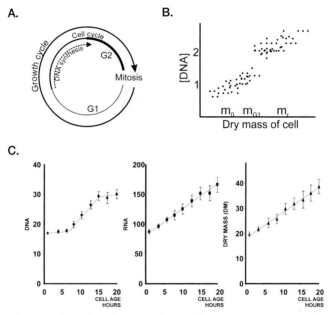

Figure 1. The growth cycle of mammalian cells is continuous (Mitchison 1971; Baserga 1985). (A) Diagram of the cell cycle identifying the discrete interval where DNA synthesis occurs. Cell growth is continuous throughout the cell cycle. (B) Comparison of the dry mass of the cell, measured by interferometry, to DNA content (Killander and Zetterberg 1965). m_0 identifies the mass of cells in G_0 and G_1 when cellular DNA has not replicated. A critical size threshold is reached (m_{G1}) late in G_1 when DNA synthesis starts. Mass continues to accumulate as DNA synthesis continues, including the G_2 phase of the cell cycle (m_r). (C) An increase in cell mass is seen throughout the cell division cycle. DNA content is plotted on the y axis and compared with cell age in hours after initiation of cell division in the first panel. RNA content is similarly plotted on the y axis in the second panel. Dry mass is plotted on the y axis in the third panel. (B, C, Redrawn, with permission, from Killander and Zetterberg 1965 [copyright Elsevier Science].)

tinues to be considered as nearly synonymous with cell growth, probably because it is readily observed and measured. Unfortunately, it is equally obvious that cell size is quite sensitive to the water transport and swelling mechanisms required for the maintenance of an intact lipid bilayer (Morris 2002). Elements that control cell swelling and shrinkage include cell volume regulatory anion channels and organic osmolytes, whose levels can be influenced by mitogenic signals (Lang et al. 2000). Cell volume control by osmotic mechanisms can significantly influence cell division, because osmotic shrinkage can delay S phase (Michea et al. 2000). Nevertheless, most consideration of cell growth looks at changes in a stable osmotic environment over longer time periods than are considered in studies of volume regulation. Therefore, knowing that cell size measurements should be interpreted with the caveat that water measurement could be involved, cell size may be a first approximation of cell growth.

A better approach may be to define growth as an increase in cell mass, which was classically measured by interferometry (Killander and Zetterberg 1965; Zetterberg and Killander 1965a,b; Zetterberg 1966). Intriguingly, dry mass plotted against DNA content suggests that mass doubles with the doubling of DNA and that there is a critical mass at which DNA synthesis is initiated (Fig. 1B). Additionally, mass increases continuously when plotted throughout one cell division cycle in a continuously proliferating population of cells (Fig. 1C) (Killander and Zetterberg 1965). This suggests the possibility that the cell division machinery monitors cell size in order to coordinate continuous growth with discontinuous DNA replication. Mechanisms coordinating growth with cell division must be subject to intense selective pressures, since failure of integration would rapidly produce cells that are either too small or too large to survive. Evidence for such selective pressures is found in the conservation of the size threshold for cell division in prokaryotes (Donachie 1968) and yeast (Johnston et al. 1977).

The concept of a size threshold carries the implication that growth regulates cell division. Before agreeing with that proposition, however, it is worth considering the opposite. The key engine of growth in a cell must be the ribosome, since all enzymatic functions are dependent on its production of new proteins. Moreover, because the ribosome is essentially a ribozyme, with proteins added to preserve its structure, the ribosome might be considered as nearly purely RNA, one synthetic step away from DNA (Puglisi et al. 2000). Importantly, 85% of the RNA in a cell is rRNA, allowing the simple estimation of ribosomes in a cell by measuring its total RNA content. Indeed, the early literature on growth and cell division employed RNA content as a ready measure for growth. One such study demonstrated that RNA content directly tracks the amount of DNA in a cell (Sogin et al. 1974). This result suggests a simple regulatory mechanism for integration of growth and development. It implies that synthesis of rRNA might be constitutive and dependent on rDNA gene dose for its regulation. This simple model would guarantee a balance between growth and division where division regulates growth! Alas, life is not so simple; ribosomal transcription is very sensitive to control by growth stimuli and does not simply respond constitutively to gene dose (Stefanovsky et al. 2001). Considering the two extremes (that growth might drive division or division might drive growth), the solution to how cells coordinate growth with cell division is not unlike the problem of the chicken and the egg. Survival of the cell requires that cell growth and division be inseparable, except in temporary and limited circumstances, and the balance between the two must be continuously monitored and corrected. The monitoring and corrective mechanisms are the subject of this review.

COORDINATING CELL DIVISION BY MONITORING RIBOSOMAL CONTENT AND TRANSLATION INITIATION: CELL GROWTH REGULATING CELL DIVISION

How might cell size generate a signal that would be interpretable by the cell division machinery? In his 1985 review, Renato Baserga noted that ribosomal RNA constituted about 85% of all cellular RNA (Baserga 1985). Its function as a ribozyme suggests that it is the simplest evolutionary unit regulating growth and has likely been subjected to intense selective pressures since the birth of the first unicellular organisms (Szostak et al. 2001). As an alternative key measure of growth that reflects net ribosomal content, cellular RNA is indeed tightly correlated with entry into S (Johnston and Singer 1978; Darzynkiewicz et al. 1979a,b). Nevertheless, although microinjection of antibodies against RNA polymerase I can decrease cellular RNA synthesis by 80%, this does not result in any delay in S-phase entry (Mercer et al. 1984). Thus, the cell is not simply monitoring ribosomal rRNA as the simplest unit of growth. If the next step in macromolecular synthesis after generation of the ribosomal RNA is total protein synthesis, what is regulating protein synthesis?

Protein synthesis can be broken down into three steps—initiation, elongation, and termination (Hershey 1991). The consensus of those studying translation regulation is that the initiation phase is its rate-limiting component (Hershey 1991; Rhoads et al. 1994; Sonenberg and Gingras 1998). It would seem evolutionarily logical to regulate protein synthesis at the initiation step to save any wasted use of the energy needed for subsequent steps in translation. This view is supported by measurement of the relative abundance of translation initiation factors compared to other protein-synthetic functions (Duncan and Hershey 1983; Duncan and Hershey 1985), their responsiveness to growth-signaling pathways (Rhoads 1999), genetic evidence (Hanic-Joyce et al. 1987a; Brenner et al. 1988), and the neoplastic consequence of de-repression of translation initiation (De Benedetti and Rhoads 1990; Lazaris-Karatzas et al. 1990; Koromilas et al. 1992). In particular, all of the signaling pathways known to regulate cell growth in *Drosophila* culminate in effects on the translation initiation complex: (1) Nutrient sensing through the TOR pathway, (2) survival pathways mediated by Akt, and (3) growth factor signaling through the insulin/PI3 kinase pathway (Fig. 2) (Saucedo and Edgar 2002). Although this postulate has been the basis of a number of successful lines of investigation, studies that address the effects of limiting net translation initiation in vivo are comparatively rare (Tsung et al. 1989). Nevertheless, if one assumes that translation initiation is the rate-limiting step in protein synthesis, then how might the rate of translation initiation be coupled to cell division?

A classic genetic demonstration that translation initiation can regulate cell division began with the first identification of a yeast mutant, *prt-1*, that exhibited a rapid cessation of protein synthesis when shifted to its restrictive temperature (Hartwell and McLaughlin 1968, 1969). In keeping with the idea that translation initiation would likely be the rate-limiting step in protein synthesis, this first mutant was indeed shown to be defective in translation initiation. Cloning revealed that *prt-1* encoded a component of the eukaryotic initiation factor 3 complex (Keierleber et al. 1986), and further investigation showed that cell division in a particular *prt-1* mutant, *cdc63-1*, stopped before protein synthesis stopped upon shift to the restrictive temperature (Hanic-Joyce et al. 1987a,b). This mutant arrested at START, the decision point for cell division in yeast, but remained competent to mate. These genetic studies gave strong credence to a view that a key component of growth regulation is translation initiation that may regulate cell division before it affects protein synthesis. Indeed, *cdc63* was an important test case because it continued to grow at the restrictive temperature even in the absence of S-phase entry. That this factor regulated translation initiation suggested that a critical component of the cell division machinery might be translationally regulated. Although translational regulatory mechanisms may be wasteful of cellular resources, since they are active only after the cell has expended energy on mRNA synthesis, such controls may be particularly useful in situations where a rapid response to environmental change is needed (Matthews et

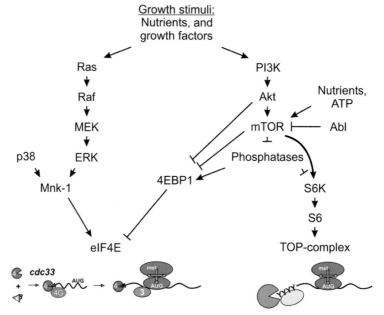

Figure 2. Signaling pathways that regulate protein synthesis converge separately on regulation of 5′-mRNA cap-dependent mRNAs and on 5′-terminal oligopyrimidine (TOP)-dependent mRNAs (Meric and Hunt 2002). Regulation of translation initiation by signal transduction pathways is illustrated. Indicated are the ras-raf pathways (*left* column) and the phosphoinositide 3-kinase (PI3K) pathways (*right* column). The ras-raf pathway controls activation of the 5′ mRNA cap-binding protein (eIF4E) as mediated by the MAP-kinase signal integrating kinase (Mnk) (Scheper and Proud 2002). Activated eIF4E then forms the core of a cap-binding complex that alters the secondary structure of the 5′UTR so that the 40S ribosomal subunit is bound to eIF3, and the ternary complex consisting of eIF2, GTP, and the Met-tRNA is recruited to the mRNA. The ribosome scans the mRNA in a 5′–3′ direction until an AUG start codon is found in the appropriate sequence context. A separable pathway signals through the mammalian target of rapamycin (mTOR). This pathway may regulate cap-dependent translation through its regulation of the eIF4E-binding protein (4EBP1) that inactivates eIF4E. The 4E-BPs are hyperphosphorylated to release eIF4E so that it can interact with the 5′ cap, and assemble the eIF4F initiation complex. mRNAs involved in ribosomal biogenesis contain a terminal oligopyrimidine sequence (TOP) (Meyuhas 2000). The TOR pathway also regulates these mRNAs through its regulation of the ribosomal protein subunit S6 kinase (S6K).

al. 2000). Certainly, the perilous choice to divide represents one such situation.

Identification of cell-size mutants provided a first global approach to identifying situations in which cell growth and division were not coordinated. An early study used an α-factor arrest protocol to identify a small cell-size mutant called *whi-1* (Carter and Sudbery 1980). The gene encoding *whi-1* was then shown to be a cyclin, *CLN3*, that controlled cell size, pheromone arrest, and cell-cycle progression (Cross 1988; Nash et al. 1988). CLN3p is very unstable and its half-life is shorter than the cell cycle (Tyers et al. 1992). It also functions as an upstream regulator of other G_1-specific cyclins (Tyers et al. 1993; Dirick et al. 1995). Cyclins are the master regulators of cell division (Bell and Dutta 2002). They were initially identified as regulators of mitosis with a conserved protein domain, the cyclin box. However, cyclin functions in the transition from G_1 to S phases may be particularly important if there is indeed a critical size threshold for cell division to start (Sherr 1993, 1994). Thus, a linkage between protein synthesis rates and cyclin synthesis might present a logical target for a mechanism that coordinates cell growth and cell division.

Translational regulation occurs when mRNAs are differentially translated (Lodish 1974). The translational machinery in actively growing mammalian cells continuously engages many mRNAs (Endo and Nadal-Ginard 1987). Separating mRNAs engaged by multiple ribosomes from those bound to few or no ribosomes best separates highly from poorly translated species. In practice, such a separation is accomplished by polysomal density gradient separation (Meyuhas et al. 1996). In contrast to the highly translated species that are entirely engaged by multiple ribosomes, as much as 30% of the cell's mRNA is present in free mRNP particles (Geoghegan et al. 1979; Kinniburgh et al. 1979). The different translational efficiency of these two groups of mRNAs is dependent on a variety of *cis*-acting elements, positioned either in the sequences 5′ to the ATG initiation codon or 3′ to the termination codon. Examples of motifs involved in translational regulation include upstream open reading frames (uORFs), 5′-terminal oligopyrimidine tracts (5′TOP), and internal ribosomal entry sequences (IRES). The complexity of the 5′-untranslated sequences (5′UTR) and their GC content also play less specific roles (Lee et al. 1983; Pelletier and Sonenberg 1985; Pyronnet et al. 1996; Rousseau et al. 1996; Lorenzini and Scheffler 1997; Svitkin et al. 2001). As expected, the *cdc63* mutation can exhibit strong translational effects on specific target genes, those including heat shock proteins (Barnes et al. 1993).

How might translation initiation and the control of protein synthesis be coupled to key regulators of the cell cycle like CLN3? The 5′UTR of the critical G_1 cyclin CLN3 contains both an unusually long 5′UTR and an uORF, *cis*-elements common in translational regulation (Polymenis and Schmidt 1997). The effects of these elements can be tracked through a growth cycle by following *CLN3* mRNA's position in polysomal profiles.

CLN3 mRNA was highly translated in actively growing cells but not in starved cells. An uORF identified at −315 of the 364 nucleotide 5′UTR was critical to this regulation, because this cell-cycle-specific translational control was lost when the uORF ATG was mutated to TTG. The A to T mutation caused A-315T/CLN3 to continue to undergo cell division even in starvation conditions, and the resulting cells were smaller in size. As the uORF could be bypassed by leaky scanning to produce CLN3p, the function of the CLN3 uORF appeared to be that of a generalized translational repressor. Moreover, the uORF played a regulatory role in response to the translational inhibitor rapamycin and in response to any effect that changed the net ribosomal content of the cell. Here is a clear mechanism linking cell division to the growth control machinery, as predicted by the historical correlation between ribosomal content and cell division.

A remarkable genetic screen recently examined all of the genes in the budding yeast genome to identify functions that might help coordinate cell division and cell growth (Jorgensen et al. 2002). Cell size distributions were analyzed for 6,000 *Saccharomyces* deletion mutants. All haploid strains deficient in nonessential functions and 1,142 heterozygous diploid strains mutant in essential genes were studied. In this screen, large-cell (*lge*) mutants would be expected to represent cell division regulators, with cell division lagging behind cell growth. START is the moment in the G_1 phase of the yeast cell cycle when the critical size threshold is reached and DNA synthesis is initiated. The primary controllers of the transition from G_1 to S include BCK2 and WHI3→CLN3, which operate in the two parallel pathways (Fig. 3). BCK2 and CLN3 then regulate a series of transcription elements that further regulate CLN1/2 and PRS3 (Fig. 3). As pre-

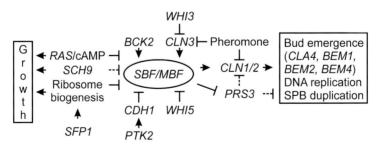

Figure 3. Cell-size control in budding yeast—a genetic overview. Summary of cell-size control genes that act at START. Epistatic interactions were evaluated in Jorgensen et al. (2002). Dashed lines indicate interactions pending further evaluation. (Redrawn, with permission, from Jorgensen et al. 2002 [copyright American Association for the Advancement of Science].)

dicted, several large-cell mutants contained mutations in these START-specific transcriptional controls, and several contained mutations in genes regulating bud emergence.

The small-cell mutants (*whi*) encompassed a diverse set of genes, which mostly did not genetically interact with the START pathway. Thus, they present an unbiased look at the genes that regulate cell growth. The major pathway that emerged in this group regulated ribosomal biogenesis. One gene, *SFP1*, was a new transcriptional controller that regulated transcription of several genes encoding ribosomal proteins. Five known nonessential and ten essential genes in ribosomal biogenesis were found to give small cells. Twenty-one additional genes were known components of the nucleolar protein network involved in ribosome assembly. Thus, in a comprehensive look at the budding yeast genome, the integration of cell division and cell growth is found to be the product of the G_1 cyclin network that regulates initiation of DNA synthesis interacting with genes that regulate ribosomal biogenesis (Jorgensen et al. 2002).

Importantly, tests of this paradigm in more complex organisms reveal conservation of some of these mechanisms, but not all. The mammalian analog of CLN3, cyclin D1, was initially discovered in three separate scenarios: (1) as a translocation breakpoint in cancer; (2) as a human cDNA that complemented a CLN 1,2,3-deficient strain of yeast; and (3) as a cDNA induced when macrophages are stimulated to divide (Matsushime et al. 1991; Motokura et al. 1991; Xiong et al. 1991). As with its yeast counterpart, dominant expression of mammalian cyclin D1 causes a small-cell phenotype (Quelle et al. 1993). Thus, cell division is accelerated by this G_1 cyclin, without an integrated acceleration of cell growth.

The picture becomes substantially more complex, however, when G_1 regulators are considered in other species. The *Drosophila* cell clone assay (see Chapter 2) is an alternative paradigm for identifying genes with features that preferentially alter growth or division. Here, cell division, clonal size, and cell size are readily observed after experimental manipulation of the gene of interest, all within individual mutant clones (Neufeld et al. 1998). In this assay, both known cyclin D1 regulatory targets, dE2F and RBF, function as division regulators, as anticipated from mammalian studies. Surprisingly, overexpression of cyclin D1 and its kinase partner, Cdk4, had no effect on cell size or division rates in proliferating imaginal disc cells (Datar et al. 2000). In contrast, both caused cellular hypertrophy in the postmitotic cells of the fly eye. Finally, in proliferating wing disc cells, cyclin D1 accelerated cell growth, which secondarily increased cell division. It is remarkable to note the tissue specificity of these activities in the fly. Moreover, loss of cyclin D1 also did not alter cell division, but did decrease cell growth rates (Meyer et al. 2000). All of these effects differed profoundly from the effects of RBF, E2F, and fly CycE, which fit the expected paradigms for their proposed functions in mammalian cells. Perhaps the best explanation for these results is that cyclin D regulates targets outside the pRb pathway in flies. This would not be unexpected, given the multiplicity of cyclin D1 functions in human cells (Bernards 1999). Whether these additional functions in flies are similar to those in mammalian cells is unknown. They are, however, a warning that the coordination of cell growth and cell division may not follow the same detailed rules across all species. Intriguingly, budding yeast cell G_1 cyclins better modeled the functions of cyclin D_1 in mammalian cells than did *Drosophila* cyclin D_1!

What about regulation of cyclin D1 in mammalian cells? Is it posttranscriptionally responsive to the regulation of cell growth as in yeast? The signal transduction pathway which regulates the translation initiation pathway that regulates mammalian cell growth has been clearly elucidated (Fig. 2) (Martin and Blenis 2002). Growth signals travel through a rapamycin-inhibitable pathway, which signals through PI3 kinase and Akt. These signals are ultimately interpreted through changes in the eIF4F complex, which binds to the 5′ 7-methyl cap structure required for translation of mammalian mRNAs (Gingras et al. 1999). The critical factors in this complex include the mRNA cap-binding protein eIF4E and its antagonistic binding partner, the 4E-binding protein (4EBP). Phosphorylation of eIF4E stimulates its function; phosphorylation of 4EBP inhibits its function. The stochiometric balance between phosphorylated and unphosphorylated forms of the eIF4E/4EBP pair determines how active the eIF4F translation initiation complex will be. It is particularly noteworthy that *Saccharomyces* mutants in eIF4E also exhibit a G_1 arrest phenotype, identifying eIF4E as the cell-cycle regulator *CDC33* (Brenner et al. 1988).

Several steps in the growth regulatory pathway, which culminates in eIF4E and its antagonist 4EBP, control cyclin D1 through posttranscriptional mechanisms. Rapamycin itself alters both cyclin D1 mRNA and protein stability (Hashemolhosseini et al. 1998). Further downstream, both Akt/PKB and PI3 kinase mediate loss of cyclin D1 translation after tyrosine kinase signaling is inhibited (Muise-Helmericks et al. 1998). However, the most intriguing connection between translation initiation and cell-cycle regulation is seen in mammalian cells overexpressing eIF4E (Lazaris-Karatzas et al. 1990), which become malignantly transformed. This result indicates that not all mRNAs are equally translatable and that those which promote malignancy are more affected by eIF4E levels. Moreover, the ability of translation initiation controls to cause malignancy includes eIF2α, which regulates assembly of the methionyl-tRNA into the initiation complex, and other translation factors (Tatsuka et al. 1992; Meurs et al. 1993; Donze et al. 1995).

Cyclin D1 may be an important mediating target of translational regulation by eIF4E, because cyclin D1 protein levels increase in eIF4E-overexpressing cells without an accompanying increase in cyclin D1 mRNA (Rosenwald et al. 1993a). This regulation is reminiscent of that in yeast. However, the mechanism is quite different. Although cyclin D1 mRNA has a complex 5′UTR, reminiscent of that found in its yeast counterpart, this element is unlikely to account for its translational control, because an expression construct that contains only the coding sequences of cyclin D1 was similarly translationally controlled. Instead, the export of cyclin D1 mRNA appeared to be the target of this control in more comprehensive studies (Rosenwald et al. 1995; Rousseau et al. 1996). A proline-rich homeodomain protein, PRH, was recently shown to bind to nuclear eIF4E and mediate changes in export of the cyclin D1 mRNA (Topisirovic et al. 2003). Moreover, a substantial fraction of eIF4E is located in the nucleus (Lejbkowicz et al. 1992), and the cap structure is required for export of mRNAs from the nucleus. These mRNA export mechanisms might be connected to cell growth if nuclear eIF4E reflects the translational initiation potential in the cytoplasm, but no investigations have shown such potential. In the meantime, although the connection between eIF4E and cyclin D1 in mammalian cells is reminiscent of the situation in yeast, the regulatory mechanisms are different.

Translational control of other cell-cycle proteins is also seen. Translational controls of various cyclins during early frog embryonic development are an intriguing example of this kind of translational control (Groisman et al. 2002). However, these controls do not serve to coordinate cell growth and division but serve more as a timing mechanism for the early intrinsic cell cycle. Interestingly, fission yeast have a second point in the cell cycle where translation of a critical cell-division control determines whether the cells enter mitosis (Daga and Jimenez 1999). This mechanism is extraordinarily similar to the G_1 controls in budding yeast discussed above. Indeed, the cdc25 phosphatase is rate-limiting for entry into mitosis, and its 5′ leader is regulated by levels of the RNA helicase molecule, eIF4A, which is an integral part of the eIF4F initiation complex containing eIF4E. Thus, a critical size threshold for entry into mitosis may be assessed by the eIF4A component of eIF4E, which is rate-limiting for the translation of cdc25.

The G_1 cyclin inhibitor p27 is also controlled at the translational level (Hengst and Reed 1996), but this may play more of a role in counting cell divisions than in integrating growth and cell division. An intrinsic increase in p27 mRNA in the polysomal fraction appears to contribute to cell division arrest in some cells (Millard et al. 1997), and this is regulated by a translational U-rich suppressor element in the 5'UTR of p27 mRNA (Millard et al. 2000). The relevance of this element in cell growth-division integration is untested, but the factors that bind to the U-rich element include several involved in translation initiation control, including HuR, as well as hnRNP C1 and C2. It will be interesting to see whether these components react to cell-size controls in future experiments.

A number of screens to identify the whole spectrum of translationally regulated genes have been attempted. These screens have generally depended on the power of polysomal profiles to distinguish between highly translated mRNAs and less well translated mRNAs. One such screen revealed ribonucleotide reductase, a rate-limiting component of the DNA synthetic machinery, as a target mRNA regulated by altering levels of eIF4E in the cell (Abid et al. 1999). Two additional screens identified translationally regulated genes during T-cell activation and monocyte differentiation, rather larger perturbations in cell physiology than simply changing translation initiation levels (Krichevsky et al. 1999; Mikulits et al. 2000). These screens revealed rather more genes involved in ribosomal metabolism than those involved in cell division control, although both the oncoprotein myb and cyclin G_1 were translational targets in the latter screen. These screens may have looked at global perturbations in cell physiology and not specifically at genes regulated by translation initiation factors. In contrast, a specific look at genes regulated in limiting eIF4F concentrations used poliovirus to reduce cap-binding complexes (Johannes et al. 1999). This study benefited from advances in array technology, and the known translational target *pim-1* was revealed (Hoover et al. 1997; Fernandez et al. 2002). In turn, *pim-1* has known cell-cycle regulatory targets, but its overall significance in cell-cycle regulation has received less attention than many other oncogenes (Mochizuki et al. 1999; Wang et al. 2002). As reagents to perturb translation initiation develop and array technology improves, it will be interesting to see whether these early analyses can be refined to give us a better look at the interface between translational regulation and regulation of cell division.

CROSS-TALK BETWEEN CELL-CYCLE REGULATORS AND RIBOSOMAL BIOGENESIS CONTROLS

The ribosomal RNA genes are the basic element controlling cell growth. Both energy supplies and amino acid availability control rRNA synthesis, connecting growth conditions to the cell growth cycle (Grummt and Grummt 1976; Grummt et al. 1976). rRNA synthesis is therefore interpreting the basic metabolic milieu in which a cell must grow. Ribosomal biogenesis occurs in the nucleolus and is regulated by transcriptional initiation and elongation, in addition to rRNA processing, subunit phosphorylation, acetylation, and trafficking (Leary and Huang 2001). Simplistically, the transcriptional control of rRNA synthesis seems to be a key control element in ribosome biogenesis. The rDNA promoter has been extensively analyzed, and factors regulating rRNA transcription are well-characterized (Grummt 2003). A unique RNA polymerase evolved into the rRNA-specific RNA polymerase I (pol I). RNA pol-I-associated transcription factors include the upstream binding factor (UBF), a promoter selectivity factor (TIF-IB), a regulatory factor associated with initiation-competent pol I (TIF-IA), and a rate-limiting component (TIF-IC) (Fig. 4) (Seither et al. 1998). As noted above, limiting quantities of RNA pol I are not themselves immediately rate-limiting for cell division (Mercer et al. 1984). Furthermore, no mutation in any of the components of the pol I holoenzyme demonstrated cell-size phenotypes when they were systematically surveyed (Jorgensen et al. 2002)! This suggests that transcriptional controls of ribosomal RNA synthesis might be intimately shared with the transcriptional controls of cell-cycle regulators, and would therefore have equivalent effects on both cell growth and cell division. Biochemical approaches have therefore been required to identify key signals coupling regulation of rRNA synthesis to cell division.

In the absence of cell-size effects, it has become necessary to evaluate the known regulators of RNA pol I to see whether they regulate genes involved in cell division and to evaluate genes involved in cell division to determine whether they regulate RNA pol I. No evidence has yet emerged to suggest that yeast pol I–associated transcription factors (the Rrns) act outside the pol I holoenzyme. However, this is a relatively unexplored area of research. The opposite direction has proven more interesting in yeast studies. The nucleolus contains 150 repeats of rDNA and is therefore quite susceptible to damaging recombination events. Perhaps to prevent recombination, RNA pol II is silenced in the nucleolus. A RENT complex that contains Cdc14, Sir2, and Net1 both controls mitotic exit and participates in pol II silencing. The Cdc14 component of this complex is a protein phosphatase that participates in mitotic exit by promoting the degradation of B-type mitotic cyclins. The RENT complex also directly regulates pol I activity, and dominant mutations in Cdc14 can uncouple cell-cycle progression from nucleolar function (Shou et al. 2001). Mechanistically, normal release from mitosis appears to require release of Cdc14 from its

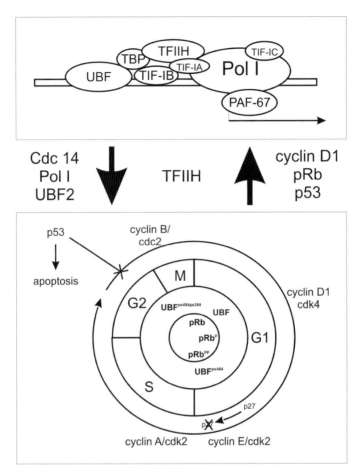

Figure 4. Coordinating mechanisms linking RNA pol I function to cell-cycle controls. Shown are the transcriptional complex that regulates ribosomal RNA synthesis in the upper panel and a diagram of the cell cycle identifying key regulators in the lower. Molecules individually involved in RNA pol I activity that affect cell-cycle regulation are summarized beside the down arrow. Cell-cycle regulators with effects on RNA pol I function are identified beside the up arrow.

tethering function in the nucleolus to function in the mitotic exit network. The physiologic significance of this interaction awaits further work.

The literature becomes richer when this cross-talk is explored in metazoans. The machinery involved in cell growth can directly participate in cell division regulation in at least two cases. First, the machinery of rRNA synthesis has been shown to directly participate in at least one complex synthesizing DNA—the nucleotide excision repair system (Hannan et al. 1999; Grummt 2003). Initial reports described the presence of shared polymerase subunits Ku70/80, DNA topoisomerase I, and PCNA in both complexes (Hannan et al. 1999). More recently, the general transcription factor TFIIH was found to link RNA pol I, RNA pol II, and DNA excision repair (Hoogstraten et al. 2002). Indeed, TFIIH moves rapidly between the nucleolus, where it is required for rRNA transcription, and the additional functional locations, where it participates in excision repair and RNA pol II transcription. To perform these roles, TFIIH acts as a helix opener and promotes transcriptional elongation of rRNA transcripts (Iben et al. 2002). The sharing of TFIIH may play a role in repairing rDNA itself and as a generalized factor useful in all three functions (Conconi et al. 2002). However, the very breadth of the specificity of TFIIH makes it an unlikely coupler of cell division and growth in the same sense as described above. It merely shows how some growth machinery could be shared with DNA synthetic machinery. The second case where the cell growth machinery may regulate cell division is where one of the transcription factors that interacts with pol I has additional transcriptional effects relevant to cell division control. An alternative splicing version of the upstream binding factor (UBF2) associates with the lymphocyte enhancer factor (LEF) that is part of the β-catenin signaling pathway (Shtutman et al. 1999; Tetsu and McCormick 1999; Grueneberg et al. 2003). Although this interaction regulated artificial cyclin D1 promoter-reporters, effects on endogenous genes were not evaluated. Since a number of signaling pathways converge on UBF, this might be a mechanism for growth regulation to affect cyclin D1 levels through its transcriptional regulation.

Cell division controls that can regulate RNA pol I reveal by far the richest connections between rRNA synthesis and cell-cycle control (White 1997). Progression through the cell cycle requires an ordered set of cyclin–cyclin-dependent kinase pairs acting on downstream target molecules (Fig. 4). The retinoblastoma gene product (pRb) is one key target of the cyclins that regulate passage from G_1 into S phase. Since G_1 is essentially defined as the critical growth phase of the cell cycle, one would anticipate a rich interaction between growth and initiation of DNA synthesis. Signaling from cyclin D1 to pol I would be a logical place to look for this connection, because cyclin D1 is a key regulator of this transition. Tryptic peptides of UBF revealed phosphorylation of its Ser-484 during G_1, and exogenously supplied cyclin D1/Cdk 4 increased phosphorylation at this site (Voit et al. 1999). This phosphorylation event and the Ser-484 itself were required for UBF to functionally increase pol I activity. A second phosphorylation at Ser-388, apparently regulated by cyclin E/Cdk2, then modulates the direct interaction between UBF and pol I (Voit and Grummt 2001). Finally, cyclin B/Cdk1 phosphorylation of the rDNA transcription apparatus leads to disassembly of the nucleolus prior to entry into mitosis (Heix et al. 1998; Sirri et al. 2000). Reassembly of the nucleolus then requires inhibition of the cyclin B complex and loss of the UBF phosphorylations (Sirri et al. 2002). It is unclear whether these interactions actually provide feedback between cell growth and cell division, however. Cyclin D1–overexpressing cells are clearly small, suggesting that D1's effects on cell division far outweigh its effects on RNA pol I regulation. Unfortunately, failure to investigate the biological effects of many of these biochemical interactions between rRNA synthesis and cell division control has been a common problem in these studies.

Phosphorylation of pRb by cyclin D1 is the next step in the central regulation of cell division. The interactions of pRb with RNA pol I are well-studied. Nucleolar accumulation of pRb during monocyte differentiation inhibits rRNA synthesis (Cavanaugh et al. 1995). This nucleolar localization also occurs during confluence-induced arrest of fibroblasts (Hannan et al. 2000b). Thus, some element of the decrease in protein synthesis during growth arrest may be regulated by a key interaction between pRb and RNA pol I. One concern about the importance of this interaction has not been addressed, however, which is whether nucleolar localization is key to pRb's tumor suppressor activity. One would need to evaluate a pRb that fails to localize to the nucleolus, and this has not yet been done. The interaction between pRb and RNA pol I is probably biologically significant, however, since Rb$^{-/-}$ cells are resistant to the protein synthesis inhibitor cycloheximide (Herrera et al. 1996).

Subsequent studies have further elucidated mechanisms governing the pRb–UBF interaction. This interaction does not alter binding to DNA but rather results in recruitment of repressors that inactivate UBF by acetylation (Hannan et al. 2000a; Pelletier et al. 2000). Interestingly, both pRb and its "pocket protein" relative p130 can alter RNA pol I activity. This redundancy might also explain some of the lack of pRb's overt phenotypic effects on cell growth (Hannan et al. 2000a; Ciarmatori et al. 2001). Given the importance of pRb in cancer, the significance of these interactions in malignant transformation when pRb is lost is worth further consideration. Unfortunately, the lack of involvement of p130 in human tumors argues that the interaction between pRb and UBF may be interesting but not relevant to tumorigenesis. In addition, it is unclear how the pRb effects on pol I relate to effects on initiation of DNA synthesis, since cell size, protein synthesis, or G$_1$ phenotypes have not been attributed to this biochemical interaction. More extensive characterization is needed of the importance of the pRb-UBF interaction to the cross-talk between cell growth and cell division.

The elusive and fascinatingly pleiotropic oncogene/tumor suppressor p53 also emerges as a coupler between cell cycle regulation and growth control. p53 mediates the response to multiple signals that promote malignancy, including DNA damage, oncogene expression, telomere erosion, hypoxia, ribonucleotide reduction, and decreased cell death. It then passes those signals into response pathways that drive cell cycle arrest, DNA repair, apoptosis, senescence, and differentiation (Vousden and Lu 2002). Although its extensive interactions with RNA pol III are considered below, a brief mention of its interaction with pol I is needed at this point. p53 decreases rRNA synthesis, and this repression could be monitored in reporter gene assays (Budde and Grummt 1999). Neither the mechanism nor the biological significance of this candidate interaction has received much attention since this initial report.

INTERACTIONS BETWEEN CELL-CYCLE REGULATION AND RNA POL III FUNCTION

RNA pol III transcribes a diverse collection of genes that encode structural or catalytic RNAs shorter than 400 bp in length (Schramm and Hernandez 2002). These include the tRNAs that are transcribed from a type 2 RNA pol III promoter (Fig. 5). Genetic evidence shows that synthesis of the tRNAs is linked to cell division, because synthesis of some tRNAs can be rate-limiting for cell division. At least three yeast *cdc* genes are tRNA synthetases, demonstrating the relevance of RNA pol III-driven transcripts to the integration of growth with the cell cycle (Hohmann and Thevelein 1992; Murray et al. 1998; Wrobel et al. 1999). These *cdc* mutants arrest with a different phenotype, however, from the phenotype of yeasts mutant in translation initiation factors. Their arrest occurs before START at a less well-defined point during G_1. Thus, the proteins they encode seem to be rate-limiting for progression through G_1 rather than for entry into the S phase. Remarkably, mutations in the RPC53 subunit of RNA pol III also arrest in early G_1, as large, round, unbudded cells (Mann et al. 1992).

RNA pol III activity varies during the cell cycle. It is shut off during the mitotic shutoff of protein synthesis, as seen for other components of cell division. This shutoff could be reproduced in vitro by *Xenopus* mitotic extracts and could be reproduced by purified maturation-promoting factor (Hartl et al. 1993; Wolf et al. 1994). Furthermore, purified cyclin B/cdc2 directly phosphorylated the TFIIIB component of the RNA pol III

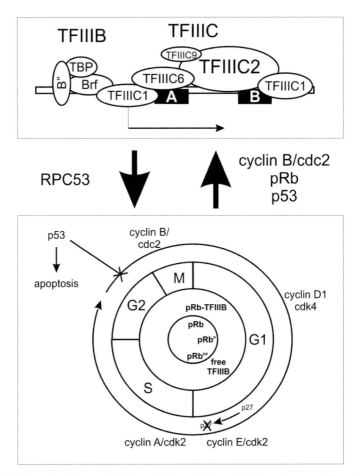

Figure 5. Coordinating mechanisms linking RNA pol III functions to cell-cycle control. Shown are the transcriptional complex that regulates transfer RNA synthesis in the upper panel and a diagram of the cell cycle identifying key regulators in the lower panel. Molecules individually involved in RNA pol III activity that affect cell cycle regulation are summarized beside the down arrow. Cell-cycle regulators with effects on RNA pol III function are identified beside the up arrow.

initiation complex, leading to pol III shutoff (Gottesfeld et al. 1994). These results provided incentive for further studies of cell-cycle regulators that could alter RNA pol III activity, resulting in a very fruitful line of investigation of pRb's effects on pol III.

Levels of tRNA synthesis vary through G_1 in mammalian cells, as predicted by the yeast mutants discussed above (Mauck and Green 1974). Addition of pRb to in vitro extracts of pol III resulted in strong repression of transcription (White et al. 1996). The significance of this result was reinforced by the demonstration that Rb-deficient cells demonstrated elevated pol III activity, and the repression was reversed by mutated pRbs that are associated with cancers. A direct interaction between pRb and both TFIIIB and TFIIIC2 could be shown in GST pull-down experiments, and these components could be coimmunoprecipitated in vivo (Chu et al. 1997; Larminie et al. 1997). These interactions are shared with the other pocket proteins, p130 and p107 (Sutcliffe et al. 1999). pRb directly blocks the interaction between TFIIIB and TFIIIC2 and does not need to recruit any corepressor activity to achieve its inhibition of protein synthesis (Sutcliffe et al. 2000). pRb binding to pol III components increases during both growth arrest and early G_1 (Scott et al. 2001). The pRb–pol III interaction was further regulated by phosphorylation of pRb, especially in response to overexpression of cyclin D1-Cdk4. Thus, a tight connection between pRb and RNA pol III activity exists at all levels of testing, from in vitro extracts to direct evaluation of oncogenic mutants of pRb. Still, a lingering doubt must exist about the priority of these interactions compared to the cell-cycle effects of pRb, again because cyclin D1-overexpressing cells are so clearly smaller than controls, suggesting that the pol III effect lags behind the cell-cycle effects.

Equally robust interactions have been shown for p53 and RNA pol III (White 1997). Loss of p53 results in increased pol III activity, and p53 directly inhibits synthesis of tRNA, as well as other pol III-dependent transcripts (Cairns and White 1998). Oncogenic mutations in p53 directly correlate with derepression of pol III, suggesting that this interaction is biologically significant in tumors with mutated or lost p53 (Stein et al. 2002a,b). This interaction has now been attributed to direct interactions between p53 and the TATA-binding protein (TBP) or blocking a TBP-associated factor, TAF3B2 (Eichhorn and Jackson 2001; Felton-Edkins et al. 2003).

THE *MYC* ONCOGENE AND CELL GROWTH

The c-*myc* proto-oncogene is potentially the most interesting partner in this dance between cell growth and cell division. Some of its interest may be romantic, since it was a gene discovered through an African safari by Dennis Burkitt (Burkitt 1965). The fact that Burkitt's lymphomas caused by *myc* are the fastest-growing tumor studied also must contribute to its interest to studies of the coordination of cell growth and cell division (Iversen et al. 1974). Indeed, Burkitt's lymphomas double in size in 66 hours—a dramatic and awful fact to contemplate during treatment. An encounter with a child suffering from Burkitt's lymphoma is an unforgettable experience in pediatric medicine. Additional interest may come from its status as the first true oncogene linked to a human cancer (Taub et al. 1982). Some of its interest may arise from the bewildering array of functions ascribed to Myc (Oster et al. 2002), and its potential role in a wide array of human cancers, including Burkitt's lymphomas, lung carcinoma, breast carcinoma, and rare cases of colon carcinoma. Elevated expression of the c-*myc* gene occurs in almost one-third of breast and colon carcinomas, and activation of c-*myc* gene expression by too many signal transduction pathways to fully enumerate (Dang 1999). Which of its many functions is essential to *myc*'s role in human cancer remains the critical question. A remarkably diverse set of *myc* target genes has emerged (Eisenman 2001), including those that regulate cell-cycle progression and cell growth (Oster et al. 2002). Importantly, the cell-cycle and growth effects of c-*myc* are genetically separable in p27 null cells (Beier et al. 2000).

Myc is a transcription factor that activates gene expression through its binding to E-box sequences, primarily CACGTG (Eisenman 2001). It also exhibits an inverse capacity to repress transcription, which is coupled to its capacity to activate transcription (Oster et al. 2003). The number of candidate genes regulated by Myc's activation potential has now grown large enough to require documentation on the World Wide Web (http://www.myc-cancer-gene.org/index.asp). Perhaps the greatest dilemma in understanding Myc is that, despite the broad array of target genes, it is a weak activator (Kretzner et al. 1992), and its binding site, CACGTG, could be mathematically present at an average of three copies per gene. New attempts to bring order out of this chaos recently evaluated promoter occupancy by Myc in vivo in chromatin-binding screens (Fernandez et al. 2003; Orian et al. 2003). In *Drosophila*, Myc binds to nearly 15% of available loci, dwarfing even the already large list of current Myc target genes (Orian et al. 2003). The broadest categories of Myc-associated loci were those related to growth and biosynthetic functions of the cell. Chromosome immunoprecipitation studies in human cells evaluated 6,541 promoter regions and found that 1,630 of them contained candidate Myc E boxes (Fernandez et al. 2003). Remarkably, a chromatin immunoprecipitation assay showed that Myc bound 58% of the candidate E boxes, although not all sites showed high-affinity binding. These studies paint a broad picture of Myc as a comparatively weak transcriptional activator that binds a specific DNA sequence present in many genes and weakly activates expression of a large number of genes at any time point. Reconciling these data with Myc's potent oncogenic effects remains an intriguing challenge.

The growth functions of c-Myc were first identified in cells overex-pessing c-*myc*, which were then confirmed in *c-myc* null cells (Dang 1999; Schmidt 1999). Using chimeric proteins that fused c-Myc with domains of the estradiol receptor, Myc function could be temporally regulated (Eilers et al. 1989). Although these cells were initially used to show that Myc induced cell division, subsequent analysis showed that increased protein synthesis clearly precedes S-phase entry (Rosenwald et al. 1993b; Schmidt 1999). This was the first suggestion that c-Myc influenced cell growth, contrasting with a long line of experiments linking Myc to cell division (Pelengaris et al. 2002). Few of the candidate Myc-target genes described at the time were known to play any part in growth regulation. Reasoning that the rate-limiting component of protein synthesis, translation initia-tion factor eIF4E, was likely to play a role in any event involving increased protein synthesis, the transcriptional response of eIF4E to Myc induction was assessed (Rosenwald et al. 1993b). The presence of an essential Myc-binding site in the eIF4E promoter demonstrated a direct role for c-Myc as an activator of eIF4E transcription (Jones et al. 1996). Two additional studies have also identified eIF4E as a Myc-regulated target gene (Schuhmacher et al. 1999; Kim et al. 2000). The most recent chromatin immunoprecipitation experiments confirm that the E boxes in the eIF4E promoter are high-affinity Myc-binding sites in all of the tested cell lines (Fernandez et al. 2003). This line of reasoning suggested that c-Myc worked in part by promoting cell growth. Translation initiation factor eIF4E is a particularly interesting c-Myc target gene because it transforms NIH-3T3 cells, and mutations in eIF4E in budding yeast cause a G_1 arrest phenotype, as discussed above (Brenner et al. 1988; Lazaris-Karatzas et al. 1990).

A broad array of differential expression screens have now identified both additional translation initiation factors and ribosomal proteins as candidate c-Myc target genes in a wide variety of cell lines and conditions (Coller et al. 2000; Greasley et al. 2000; Guo et al. 2000; Pajic et al. 2000; Schuhmacher et al. 2001). Of the 647 analyzed Myc target genes in the Internet database (http://www.myc-cancer-gene.org/index.asp), 81 play a direct role in translation or protein synthesis. These genes include trans-lation elongation factors, translation initiation factors, tRNA synthetases, nucleolar assembly components, and ribosomal proteins. This view of normal Myc function was strongly reinforced when *myc* null rat fibro-blasts were created by somatic recombination techniques (Mateyak et al. 1997). Indeed, the delayed cell division rates demonstrated in these fibroblasts precisely matched their decreased protein synthesis rates, implying that decreased protein synthesis was the primary defect in the *myc* null cells (Mateyak et al. 1997). c-Myc-induced proliferation further correlated with an increase in protein synthesis in the B cells of transgenic mice overexpressing c-*myc* (Iritani and Eisenman 1999). This view is reinforced by evidence for an additional interaction of c-Myc with the TFIIIB component of RNA pol III (Gomez-Roman et al. 2003). This interaction is thought to use Myc as a bridge between the TFIIIB component of RNA pol III and transcriptional coactivators, thus activating RNA pol III. Together, these experiments pushed forward the view that c-Myc regulates the global biosynthetic capacity of the cell.

Myc has far longer been considered a stimulator of cell-cycle progression (Pelengaris et al. 2002), and it is directly involved in regulating several cell-cycle regulators. Myc transcriptionally regulates both cyclin D2 and Cdk4 (Hermeking et al. 2000; Bouchard et al. 2001). This up-regulation may then contribute to enhanced cyclin E/Cdk2 activity by titrating the p27 G_1 cyclin inhibitor (Hermeking et al. 1995; Steiner et al. 1995). In addition, *cdc25* has been proposed as a Myc target gene (Galaktionov et al. 1996). Finally, Myc's repression function has been implicated as a regulator of the cyclin inhibitors p15INK4b and p21 (Staller et al. 2001; Wu et al. 2003). This effect is mediated by Myc's releasing the Miz1 activator from the promoters of *p15* and *p21* genes, thus shutting off their transcription. Given the lines of reasoning linking Myc to both cell growth and cell-cycle progression, which is the critical function? One first step might be to determine whether the two pathways can be independently regulated. By manipulating both serum concentration and Myc levels, Myc can be shown to increase growth without affecting cell division (Schuhmacher et al. 1999). More definitive results should emerge from genetic experiments. For example, a long search for the *Drosophila* homolog of *myc* finally revealed that the *diminutive* mutant of *Drosophila* was mutant in what was then termed *dmyc* (Gallant et al. 1996). Arguing for the priority of cell growth regulation, loss of *dmyc* caused a small-cell phenotype and increased *dmyc* caused a large-cell phenotype (Johnston et al. 1999). This should have closed out the story, but mammalian experiments in c-*myc*-deficient mice provide strong evidence that flies are not merely small flying mice (Trumpp et al. 2001). A variety of *myc* alleles with decreased function were generated in embryonic mouse cells using homologous recombination approaches. As Myc levels decreased, so did the size of the mice harboring the Myc variants. In mice, this was due to a loss of cell numbers, however, not a decrease in cell mass. Importantly, *dmyc* works as a functional *myc* in *myc*-deficient murine cells and vice versa, so the appropriate homologs have been identified. One can conclude that *myc* is the best example of a gene with clear and distinguishable effects on both cell growth and cell division. It therefore now remains for us to determine what their relative contributions are for Myc's role in neoplastic transformation. It would therefore be particularly interesting to see what the effect of an increase in Myc would be when cell growth is inhibited.

GEDANKEN EXPERIMENT

Why might mechanisms integrating cell growth and cell division be of particular interest? Are we merely looking at the chicken and egg problem, where the answer has not led to much progress in our understanding of chicken biology? This summary of signaling pathways that cross between cell-growth regulation and cell-division regulation shows that many mechanisms are shared; rarely have pathways that genuinely coordinate cell growth and division been studied. Cell size remains one of the very few readouts of the results of varying cell division independently of cell growth (or vice versa). Cell size alterations have been genuinely shown for comparatively few alterations in the signaling pathways discussed above. So what are we missing? How might one identify coordinating mechanisms, short of finding cell-size alterations? A gedanken experiment suggests that a perturbation analysis of these two crude aspects of cell proliferation only need investigate eight possible conditions (Fig. 6).

In condition A in Figure 6, growth is stimulated without an accompanying division stimulus. This is often found in nature as a natural response mechanism in terminally arrested cells subjected to stress. Classic examples include cardiac hypertrophy in response to increased peripheral vascular resistance and any exercise-induced hypertrophy of muscles. The event(s) regulating terminal differentiation so effectively interferes with the cell-cycle machinery that potent growth stimuli cannot override the arrest signal (Francis-West et al. 2003). Great excitement has accompanied the elucidation of alterations in the cell-cycle machinery in the last decade and progress in the field led to the idea that a single nonredundant mutation deregulating the cell cycle is a necessary step in the development of all malignancies (Sherr 1996). In contrast, a primary cancer-related growth-stimulating event that is sufficient to cause cancer has not been demonstrated (Ruggero and Pandolfi 2003). The growth regulators described above would be difficult to pin down in such a mechanism, since all those examined to date exhibit dual functions. The translation initiation factor eIF3-p48, which is the integration target of mouse mammary tumor viruses that cause mammary adenocarcinomas, might be one such example (Asano et al. 1997). Unfortunately, it is also clear from the discussions above that translation initiation factors seem predisposed to

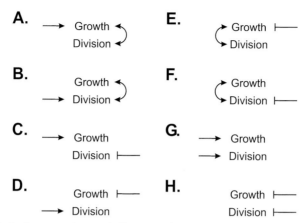

Figure 6. Gedanken experiment. A diagram of potential perturbations of growth and cell-division control and their various combinations.

preferentially regulate cell division regulators, so this example is in thought only, until one looks to see what eIF3-p48's effects are on cell division. The existence of the 5′TOP mechanism hints that one might preferentially drive growth by overexpression of the 5′TOP interacting machinery. Discovery of molecules binding to and regulating 5′TOPs might be of significant interest in this regard (Meyuhas 2000). Finally, perhaps studies of the coordination of cell division and cell growth should be seeking the reactive signals that accompany a growth stimulus (see the double arrow in A). For example, what might happen to the G_1 cell-division controls if one could overexpress RNA pol I or its constituents? The fact that cancer cells exhibit high rates of protein synthesis is not evidence that high rates of protein synthesis are a key to the cancer problem. What we are lacking is evidence that the increase in protein synthesis is an essential and primary problem in cancer.

In condition B in Figure 6, division is stimulated without an accompanying growth stimulus. This is the classic case of cyclin D1 and the other G_1 cyclins that indeed produce small-cell phenotypes (Quelle et al. 1993). However, pure acceleration of the cell cycle is insufficient to cause sustained cell proliferation (Sherr 1994). When G_1 is shortened, a compensatory lengthening of other phases of the cell cycle occurs. Glial growth factor is another rare example of accelerated cell division without growth stimulation (Conlon et al. 2001). Again, we might consider the double arrow in condition B. Do any growth-regulatory signals contribute to the compensatory lengthening of other components of the cell cycle? What signals are sent to the cell growth machinery when cyclin D1 is overexpressed and cell size decreases? Why might there be a lag in mammalian and yeast cells in this signal transfer when *Drosophila* cyclin D preferentially affects growth over division?

Although relatively few studies have identified cross-over signals, the alternative approach of experimentally opposing a growth stimulus against a division inhibitor might yield similar information. This has been experimentally achieved by simultaneously activating a conditional *myc* construct and removing growth factors, resulting in accelerated growth independent of cell division (Schuhmacher et al. 1999). Although Myc clearly regulated growth independently of cell division, the authors did not push the question further. It is interesting to speculate what might limit growth in this scenario. At what point do the cells stop growing and why?

Condition E in Figure 6 is easier to consider experimentally. Inhibition of cell growth stimuli was readily achieved by classic cycloheximide experiments. Like amino acid or growth factor starvation, limitations on translation elongation by cycloheximide led to proliferation arrest in G_1 at a point then termed the restriction point of the cell cycle (Pardee 1974; Zetterberg et al. 1995). Many candidates have been proposed to be the molecule that limits passage through the restriction point, including several key G_1 cyclins (Ekholm et al. 2001; Blagosklonny and Pardee 2002). Certainly, overexpression of cyclin D1 in mammals allows cell proliferation in otherwise limiting amounts of growth factors (Quelle et al. 1993). Similarly, overexpression of Cln3p overcomes the G_1 arrest in *cdc63* yeast that are mutant in translation initiation, and the yeast then arrest in G_2 at their restrictive temperature (M. Polymenis and E.V. Schmidt, unpubl.). These experiments reveal a partial answer to condition D in Figure 6 as well. They represent cases where growth was inhibited simultaneously with acceleration of cell division. It would again be interesting to explore the extreme case of this manipulation. Questions that might be addressed include identification of the second stopping point in G_2 phase of the cell cycle when CLN3p is overexpressed, and clarification of the final limit on cell proliferation in mouse fibroblasts overexpressing cyclin D1.

Overexpression of the binding protein (4EBP, PHAS) that inhibits translation initiation factor eIF4E provides an alternative approach to the same conceptual experiments (Hu et al. 1994; Lin et al. 1994; Pause et al. 1994). These experiments must be interpreted with caution, as eIF4E is already notable for its translational effects on cell division regulators. Nevertheless, pure inhibition of eIF4E by a 4EBP peptide mimic containing an intracellular protein transduction domain causes apoptotic cell death (Herbert et al. 2000). The pro-apoptotic function of 4EBP is also regulated by phosphorylation on multiple sites (Li et al. 2002). It would be interesting to see whether the combination of cell growth inhibition and cell-cycle acceleration could be used to manipulate cancer cells into apoptotic pathways, given the essential nature of cell division mutations in cancer cells (Sherr 1996).

Reagents that purely inhibit cell division (e.g., hydroxyurea) have also been available for many years (condition F, Fig. 6). The p16 inhibitor of Cdk4 might also serve as a biologically relevant block to cell division (Serrano et al. 1993). Both types of division blockade arrest cells quite specifically in the cell cycle. How the growth regulatory pathway then interprets this arrest might provide insight into the double arrow in condition F and help to reveal the mechanisms integrating a division arrest with a growth arrest. Certainly, the idea that pol I and pol III are targets of the Rb pathway suggests that addition of p16 should alter rRNA and tRNA synthesis, and the timing of this change might give a sense of how important pRb is to the actual integration of cell growth and division.

Simultaneous stimulation of both cell growth and cell division is obviously necessary for sustained cell proliferation. The most primitive observations of cell replication described in the opening paragraphs of this chapter revealed this fact. The discovery that many of the genes simultaneously regulating cell growth and cell division are key causes of cancer probably answers the question, "Why is it important to understand the mechanisms that coordinate cell growth and cell division?" Further inquiries might reveal novel pathways and mechanisms that will be helpful as we target therapies more specifically against unique vulnerabilities of the cancer cell. The hope is that some growth or division integrative mechanisms are more specific to cancer cells than to normal cells.

ACKNOWLEDGMENTS

Dr. Schmidt gratefully acknowledges support for his work and for his laboratory from grants RO1 CA-63117 and CA-69069 from the National Cancer Institute, the Harvard Breast Cancer SPORE grant 89393-03, and grant RO1 DK-57857 from the National Institute of Diabetes and Digestive and Kidney Diseases.

REFERENCES

Abid M.R., Li Y., Anthony C., and De Benedetti A. 1999. Translational regulation of ribonucleotide reductase by eukaryotic initiation factor 4E links protein synthesis to the control of DNA replication. *J. Biol. Chem.* **274:** 35991–35998.

Asano K., Merrick W.C., and Hershey J.W. 1997. The translation initiation factor eIF3-p48 subunit is encoded by int-6, a site of frequent integration by the mouse mammary tumor virus genome. *J. Biol. Chem.* **272:** 23477–23480.

Barnes C.A., Singer R.A., and Johnston G.C. 1993. Yeast prt1 mutations alter heat-shock gene expression through transcript fragmentation. *EMBO J.* **12:** 3323–3332.

Baserga R. 1985. *The biology of cell reproduction.* Harvard University Press, Cambridge, Massachusetts.

Beier R., Burgin A., Kiermaier A., Fero M., Karsunky H., Saffrich R., Moroy T., Ansorge W., Roberts J., and Eilers M. 2000. Induction of cyclin E-cdk2 kinase activity, E2F-dependent transcription and cell growth by Myc are genetically separable events. *EMBO J.* **19:** 5813–5823.

Bell S.P. and Dutta A. 2002. DNA replication in eukaryotic cells. *Annu. Rev. Biochem.* **71:** 333–374.

Bernards R. 1999. CDK-independent activities of D type cyclins. *Biochim. Biophys. Acta* **1424:** M17–M22.

Blagosklonny M.V. and Pardee A.B. 2002. The restriction point of the cell cycle. *Cell Cycle* **1:** 103–110.

Bouchard C., Dittrich O., Kiermaier A., Dohmann K., Menkel A., Eilers M., and Luscher B. 2001. Regulation of cyclin D2 gene expression by the Myc/Max/Mad network: Myc-dependent TRRAP recruitment and histone acetylation at the cyclin D2 promoter. *Genes Dev.* **15:** 2042–2047.

Boveri T. 1905. V: Ueber die Abhangigkeit, der Kerngrosse und Zellenzahl von der Chromosomenzal der Ausgangszellen. In *Zellenstudien.* G. Fischer, Jena, Germany.

Brenner C., Nakayama N., Goebl M., Tanaka K., Toh-e A., and Matsumoto K. 1988. CDC33 encodes mRNA cap-binding protein eIF-4E of *Saccharomyces cerevisiae. Mol. Cell. Biol.* **8:** 3556–3559.

Budde A. and Grummt I. 1999. p53 represses ribosomal gene transcription. *Oncogene* **18:** 1119–1124.

Burkitt D. 1965. Malignant lymphomata involving the jaws in Africa. *J. Laryngol. Otol.* **79:** 929–939.

Cairns C.A. and White R.J. 1998. p53 is a general repressor of RNA polymerase III transcription. *EMBO J.* **17:** 3112–3123.

Carter B.L.A. and Sudbery P.E. 1980. Small-sized mutants of *Saccharomyces cerevisiae. Genetics* **96:** 561–566.

Cavanaugh A.H., Hempel W.M., Taylor L.J., Rogalsky V., Todorov G., and Rothblum L.I. 1995. Activity of RNA polymerase I transcription factor UBF blocked by Rb gene product. *Nature* **374:** 177–180.

Chu W.M., Wang Z., Roeder R.G., and Schmid C.W. 1997. RNA polymerase III transcription repressed by Rb through its interactions with TFIIIB and TFIIIC2. *J. Biol. Chem.* **272:** 14755–14761.

Ciarmatori S., Scott P.H., Sutcliffe J.E., McLees A., Alzuherri H.M., Dannenberg J.H., te Riele H., Grummt I., Voit R., and White R.J. 2001. Overlapping functions of the pRb family in the regulation of rRNA synthesis. *Mol. Cell. Biol.* **21:** 5806–5814.

Coller H.A., Grandori C., Tamayo P., Colbert T., Lander E.S., Eisenman R.N., and Golub T.R. 2000. Expression analysis with oligonucleotide microarrays reveals that MYC regulates genes involved in growth, cell cycle, signaling, and adhesion. *Proc. Natl. Acad. Sci.* **97:** 3260–3265.

Conconi A., Bespalov V.A., and Smerdon M.J. 2002. Transcription-coupled repair in RNA polymerase I-transcribed genes of yeast. *Proc. Natl. Acad. Sci.* **99:** 649–654.

Conlon I.J., Dunn G.A., Mudge A.W., and Raff M.C. 2001. Extracellular control of cell size. *Nat. Cell Biol.* **3:** 918–921.

Cross F.R. 1988. *DAF1*, a mutant gene affecting size control, pheromone arrest, and cell cycle kinetics of *Saccharomyces cerevisiae. Mol. Cell. Biol.* **8:** 4675–4684.

Daga R.R. and Jimenez J. 1999. Translational control of the cdc25 cell cycle phosphatase: A molecular mechanism coupling mitosis to cell growth. *J. Cell Sci.* **112:** 3137–3146.

Dang C.V. 1999. c-Myc target genes involved in cell growth, apoptosis, and metabolism. *Mol. Cell. Biol.* **19:** 1–11.

Darzynkiewicz Z., Evenson D., Staiano-Coico L., Sharpless T., and Melamed M.R. 1979a. Relationship between RNA content and progression of lymphocytes through S phase of cell cycle. *Proc. Natl. Acad. Sci.* **76:** 358–362.

———. 1979b. Correlation between cell cycle duration and RNA content. *J. Cell. Physiol.* **100:** 425–438.

Datar S.A., Jacobs H.W., de la Cruz A.F., Lehner C.F., and Edgar B.A. 2000. The *Drosophila* cyclin D-Cdk4 complex promotes cellular growth. *EMBO J.* **19:** 4543–4554.

De Benedetti A. and Rhoads R.E. 1990. Overexpression of eukaryotic protein synthesis initiation factor 4E in HeLa cells results in aberrant growth and morphology. *Proc. Natl. Acad. Sci.* **87:** 8212–8216.

Dirick L., Bohm T., and Nasmyth K. 1995. Roles and regulation of Cln-Cdc28 kinases at the start of the cell cycle of *Saccharomyces cerevisiae. EMBO J.* **14:** 4803–4813.

Donachie W.D. 1968. Relationship between cell size and time of initiation of DNA replication. *Nature* **219:** 1077–1079.

Donze O., Jagus R., Koromilas A.E., Hershey J.W., and Sonenberg N. 1995. Abrogation of translation initiation factor eIF-2 phosphorylation causes malignant transformation of NIH 3T3 cells. *EMBO J.* **14:** 3828–3834.

Duncan R. and Hershey J.W. 1983. Identification and quantitation of levels of protein synthesis initiation factors in crude HeLa cell lysates by two-dimensional polyacrylamide gel electrophoresis. *J. Biol. Chem.* **258:** 7228–7235.

———. 1985. Regulation of initiation factors during translational repression caused by serum depletion. Abundance, synthesis, and turnover rates. *J. Biol. Chem.* **260:** 5486–5492.

Eichhorn K. and Jackson S.P. 2001. A role for TAF3B2 in the repression of human RNA polymerase III transcription in nonproliferating cells. *J. Biol. Chem.* **276:** 21158–21165.

Eilers M., Picard D., Yamamoto K.R., and Bishop J.M. 1989. Chimaeras of myc oncoprotein and steroid receptors cause hormone-dependent transformation of cells. *Nature* **340:** 66–68.

Eisenman R.N. 2001. Deconstructing myc. *Genes Dev.* **15:** 2023–2030.

Ekholm S.V., Zickert P., Reed S.I., and Zetterberg A. 2001. Accumulation of cyclin E is not a prerequisite for passage through the restriction point. *Mol. Cell. Biol.* **21:** 3256–3265.

Endo T. and Nadal-Ginard B. 1987. Three types of muscle-specific gene expression in fusion-blocked rat skeletal muscle cells: Translational control in EGTA-treated cells. *Cell* **49:** 515–526.

Felton-Edkins Z.A., Kenneth N.S., Brown T.R., Daly N.L., Gomez-Roman N., Grandori C., Eisenman R.N., and White R.J. 2003. Direct regulation of RNA polymerase III transcription by RB, p53 and c-Myc. *Cell Cycle* **2:** 181–184.

Fernandez J., Yaman I., Sarnow P., Snider M.D., and Hatzoglou M. 2002. Regulation of internal ribosomal entry site-mediated translation by phosphorylation of the translation initiation factor eIF2alpha. *J. Biol. Chem.* **277:** 19198–19205.

Fernandez P.C., Frank S.R., Wang L., Schroeder M., Liu S., Greene J., Cocito A., and Amati B. 2003. Genomic targets of the human c-Myc protein. *Genes Dev.* **17:** 1115–1129.

Francis-West P.H., Antoni L., and Anakwe K. 2003. Regulation of myogenic differentiation in the developing limb bud. *J. Anat.* **202:** 69–81.

Galaktionov K., Chen X., and Beach D. 1996. Cdc25 cell-cycle phosphatase as a target of c-myc. *Nature* **382:** 511–517.

Gallant P., Shiio Y., Cheng P.F., Parkhurst S.M., and Eisenman R.N. 1996. Myc and Max homologs in *Drosophila*. *Science* **274:** 1523–1527.

Geoghegan T., Cereghini S., and Brawerman G. 1979. Inactive mRNA-protein complexes from mouse sarcoma-180 ascites cells. *Proc. Natl. Acad. Sci.* **76:** 5587–5591.

Gingras A.C., Raught B., and Sonenberg N. 1999. eIF4 initiation factors: Effectors of mRNA recruitment to ribosomes and regulators of translation. *Annu. Rev. Biochem.* **68:** 913–963.

Gomez-Roman N., Grandori C., Eisenman R.N., and White R.J. 2003. Direct activation of RNA polymerase III transcription by c-Myc. *Nature* **421:** 290–294.

Gottesfeld J.M., Wolf V.J., Dang T., Forbes D.J., and Hartl P. 1994. Mitotic repression of RNA polymerase III transcription in vitro mediated by phosphorylation of a TFIIIB component. *Science* **263:** 81–84.

Greasley P.J., Bonnard C., and Amati B. 2000. Myc induces the nucleolin and BN51 genes: Possible implications in ribosome biogenesis. *Nucleic Acids Res.* **28:** 446–453.

Gregory T.R. 2000. Nucleotypic effects without nuclei: Genome size and erythrocyte size in mammals. *Genome* **43:** 895–901.

———. 2001. Coincidence, coevolution, or causation? DNA content, cell size, and the C-value enigma. *Biol. Rev. Camb. Philos. Soc.* **76:** 65–101.

Groisman I., Jung M.Y., Sarkissian M., Cao Q., and Richter J.D. 2002. Translational control of the embryonic cell cycle. *Cell* **109:** 473–483.

Grueneberg D.A., Pablo L., Hu K.Q., August P., Weng Z., and Papkoff J. 2003. A functional screen in human cells identifies UBF2 as an RNA polymerase II transcription factor that enhances the beta-catenin signaling pathway. *Mol. Cell. Biol.* **23:** 3936–3950.

Grummt I. 2003. Life on a planet of its own: Regulation of RNA polymerase I transcription in the nucleolus. *Genes Dev.* **17:** 1691–1702.

Grummt I. and Grummt F. 1976. Control of nucleolar RNA synthesis by the intracellular pool sizes of ATP and GTP. *Cell* **7:** 447–453.

Grummt I., Smith V.A., and Grummt F. 1976. Amino acid starvation affects the initiation frequency of nucleolar RNA polymerase. *Cell* **7:** 439–445.

Guo Q.M., Malek R.L., Kim S., Chiao C., He M., Ruffy M., Sanka K., Lee N.H., Dang C.V., and Liu E.T. 2000. Identification of c-myc responsive genes using rat cDNA microarray. *Cancer Res.* **60:** 5922–5928.

Hanic-Joyce P.J., Johnston G.C., and Singer R.A. 1987a. Regulated arrest of cell proliferation mediated by yeast prt1 mutations. *Exp. Cell Res.* **172:** 134–145.

Hanic-Joyce P.J., Singer R.A., and Johnston G.C. 1987b. Molecular characterization of the yeast *PRT1* gene in which mutations affect translation initiation and regulation of cell proliferation. *J. Biol. Chem.* **262:** 2845–2851.

Hannan K.M., Hannan R.D., Smith S.D., Jefferson L.S., Lun M., and Rothblum L.I. 2000a. Rb and p130 regulate RNA polymerase I transcription: Rb disrupts the interaction between UBF and SL-1. *Oncogene* **19:** 4988–4999.

Hannan K.M., Kennedy B.K., Cavanaugh A.H., Hannan R.D., Hirschler-Laszkiewicz I., Jefferson L.S., and Rothblum L.I. 2000b. RNA polymerase I transcription in confluent cells: Rb downregulates rDNA transcription during confluence-induced cell cycle arrest. *Oncogene* **19:** 3487–3497.

Hannan R.D., Cavanaugh A., Hempel W.M., Moss T., and Rothblum L. 1999. Identification of a mammalian RNA polymerase I holoenzyme containing components of the DNA repair/replication system. *Nucleic Acids Res.* **27:** 3720–3727.

Hartl P., Gottesfeld J., and Forbes D.J. 1993. Mitotic repression of transcription in vitro. *J. Cell Biol.* **120:** 613–624.

Hartwell L.H. and McLaughlin C.S. 1968. Temperature-sensitive mutants of yeast exhibiting a rapid inhibition of protein synthesis. *J. Bacteriol.* **96:** 1664–1671.

———. 1969. A mutant of yeast apparently defective in the initiation of protein synthesis. *Proc. Natl. Acad. Sci.* **62:** 468–474.

Hashemolhosseini S., Nagamine Y., Morley S.J., Desrivieres S., Mercep L., and Ferrari S. 1998. Rapamycin inhibition of the G1 to S transition is mediated by effects on cyclin D1 mRNA and protein stability. *J. Biol. Chem.* **273:** 14424–14429.

Heix J., Vente A., Voit R., Budde A., Michaelidis T.M., and Grummt I. 1998. Mitotic silencing of human rRNA synthesis: Inactivation of the promoter selectivity factor SL1 by cdc2/cyclin B-mediated phosphorylation. *EMBO J.* **17:** 7373–7381.

Hengst L. and Reed S.I. 1996. Translational control of p27Kip1 accumulation during the cell cycle. *Science* **271:** 1861–1864.

Herbert T.P., Fahraeus R., Prescott A., Lane D.P., and Proud C.G. 2000. Rapid induction of apoptosis mediated by peptides that bind initiation factor eIF4E. *Curr. Biol.* **10:** 793–796.

Hermeking H., Funk J.O., Reichert M., Ellwart J.W., and Eick D. 1995. Abrogation of p53-induced cell cycle arrest by c-Myc: Evidence for an inhibitor of p21WAF1/CIP1/SDI1. *Oncogene* **11:** 1409–1415.

Hermeking H., Rago C., Schuhmacher M., Li Q., Barrett J.F., Obaya A.J., O′Connell B.C., Mateyak M.K., Tam W., Kohlhuber F., Dang C.V., Sedivy J.M., Eick D., Vogelstein B., and Kinzler K.W. 2000. Identification of CDK4 as a target of c-MYC. *Proc. Natl. Acad. Sci.* **97:** 2229–2234.

Herrera R.E., Sah V.P., Williams B.O., Makela T.P., Weinberg R.A., and Jacks T. 1996. Altered cell cycle kinetics, gene expression, and G1 restriction point regulation in Rb-deficient fibroblasts. *Mol. Cell. Biol.* **16:** 2402–2407.

Hershey J.W. 1991. Translational control in mammalian cells. *Annu. Rev. Biochem.* **60:** 717–755.

Hohmann S. and Thevelein J.M. 1992. The cell division cycle gene CDC60 encodes cytosolic leucyl-tRNA synthetase in *Saccharomyces cerevisiae*. *Gene* **120:** 43–49.

Hoogstraten D., Nigg A.L., Heath H., Mullenders L.H., van Driel R., Hoeijmakers J.H., Vermeulen W., and Houtsmuller A.B. 2002. Rapid switching of TFIIH between RNA polymerase I and II transcription and DNA repair in vivo. *Mol. Cell* **10:** 1163–1174.

Hoover D.S., Wingett D.G., Zhang J., Reeves R., and Magnuson N.S. 1997. Pim-1 protein expression is regulated by its 5′-untranslated region and translation initiation factor eIF-4E. *Cell Growth Differ.* **8:** 1371–1380.

Hu C., Pang S., Kong X., Velleca M., and Lawrence J.C., Jr. 1994. Molecular cloning and tissue distribution of PHAS-I, an intracellular target for insulin and growth factors. *Proc. Natl. Acad. Sci.* **91:** 3730–3734.

Iben S., Tschochner H., Bier M., Hoogstraten D., Hozak P., Egly J.M., and Grummt I. 2002. TFIIH plays an essential role in RNA polymerase I transcription. *Cell* **109:** 297–306.

Iritani B.M. and Eisenman R.N. 1999. c-Myc enhances protein synthesis and cell size during B lymphocyte development. *Proc. Natl. Acad. Sci.* **96:** 13180–13185.

Iversen O.H., Iversen U., Ziegler J.L., and Bluming A.Z. 1974. Cell kinetics in Burkitt lymphoma. *Eur. J. Cancer* **10:** 155–163.

Johannes G., Carter M.S., Eisen M.B., Brown P.O., and Sarnow P. 1999. Identification of eukaryotic mRNAs that are translated at reduced cap binding complex eIF4F concentrations using a cDNA microarray. *Proc. Natl. Acad. Sci.* **96:** 13118–13123.

Johnston G.C. and Singer R.A. 1978. RNA synthesis and control of cell division in the yeast *S. cerevisiae*. *Cell* **14:** 951–958.

Johnston G.C., Pringle J.R., and Hartwell L.H. 1977. Coordination of growth with cell division in the yeast *Saccharomyces cerevisiae*. *Exp. Cell Res.* **105:** 79–98.

Johnston L.A., Prober D.A., Edgar B.A., Eisenman R.N., and Gallant P. 1999. *Drosophila* myc regulates cellular growth during development. *Cell* **98:** 779–790.

Jones R.M., Branda J., Johnston K.A., Polymenis M., Gadd M., Rustgi A., Callanan L., and Schmidt E.V. 1996. An essential E box in the promoter of the gene encoding the mRNA cap-binding protein (eukaryotic initiation factor 4E) is a target for activation by c-myc. *Mol. Cell. Biol.* **16:** 4754–4764.

Jorgensen P., Nishikawa J.L., Breitkreutz B.J., and Tyers M. 2002. Systematic identification of pathways that couple cell growth and division in yeast. *Science* **297:** 395–400.

Keierleber C., Wittekind M., Qin S.L., and McLaughlin C.S. 1986. Isolation and characterization of PRT1, a gene required for the initiation of protein biosynthesis in *Saccharomyces cerevisiae*. *Mol. Cell. Biol.* **6:** 4419–4424.

Killander D. and Zetterberg A. 1965. A quantitative cytochemical investigation of the relationship between cell mass and initiation of DNA synthesis in mouse fibroblasts in vitro. *Exp. Cell Res.* **40:** 12–20.

Kim S., Li Q., Dang C.V., and Lee L.A. 2000. Induction of ribosomal genes and hepatocyte hypertrophy by adenovirus-mediated expression of c-Myc in vivo. *Proc. Natl. Acad. Sci.* **97:** 11198–11202.

Kinniburgh A.J., McMullen M.D., and Martin T.E. 1979. Distribution of cytoplasmic poly(A+)RNA sequences in free messenger ribonucleoprotein and polysomes of mouse ascites cells. *J. Mol. Biol.* **132:** 695–708.

Koromilas A.E., Roy S., Barber G.N., Katze M.G., and Sonenberg N. 1992. Malignant transformation by a mutant of the IFN-inducible dsRNA-dependent protein kinase. *Science* **257:** 1685–1689.

Kretzner L., Blackwood E.M., and Eisenman R.N. 1992. Myc and Max proteins possess distinct transcriptional activities. *Nature* **359:** 426–429.

Krichevsky A.M., Metzer E., and Rosen H. 1999. Translational control of specific genes during differentiation of HL-60 cells. *J. Biol. Chem.* **274:** 14295–14305.

Lang F., Ritter M., Gamper N., Huber S., Fillon S., Tanneur V., Lepple-Wienhues A., Szabo I., and Gulbins E. 2000. Cell volume in the regulation of cell proliferation and apoptotic cell death. *Cell. Physiol. Biochem.* **10:** 417–428.

Larminie C.G., Cairns C.A., Mital R., Martin K., Kouzarides T., Jackson S.P., and White R.J. 1997. Mechanistic analysis of RNA polymerase III regulation by the retinoblastoma protein. *EMBO J.* **16:** 2061–2071.

Lazaris-Karatzas A., Montine K.S., and Sonenberg N. 1990. Malignant transformation by a eukaryotic initiation factor subunit that binds to mRNA 5′ cap. *Nature* **345:** 544–547.

Leary D.J. and Huang S. 2001. Regulation of ribosome biogenesis within the nucleolus. *FEBS Lett.* **509:** 145–150.

Lee K.A., Guertin D., and Sonenberg N. 1983. mRna secondary structure as a determinant in cap recognition and initiation complex formation. Atp-Mg2+ independent cross-linking of cap binding proteins to m7i-capped inosine-substituted reovirus mRna. *J. Biol. Chem.* **258:** 707–710.

Lejbkowicz F., Goyer C., Darveau A., Neron S., Lemieux R., and Sonenberg N. 1992. A fraction of the mRNA 5′ cap-binding protein, eukaryotic initiation factor 4E, localizes to the nucleus. *Proc. Natl. Acad. Sci.* **89:** 9612–9616.

Li S., Sonenberg N., Gingras A.C., Peterson M., Avdulov S., Polunovsky V.A., and Bitterman P.B. 2002. Translational control of cell fate: Availability of phosphorylation sites on translational repressor 4E-BP1 governs its proapoptotic potency. *Mol. Cell. Biol.* **22:** 2853–2861.

Lin T.A., Kong X., Haystead T.A., Pause A., Belsham G., Sonenberg N., and Lawrence J.C., Jr. 1994. PHAS-I as a link between mitogen-activated protein kinase and translation initiation. *Science* **266:** 653–656.

Lodish H.F. 1974. Model for the regulation of mRNA translation applied to haemoglobin synthesis. *Nature* **251:** 385–388.

Lorenzini E.C. and Scheffler I.E. 1997. Co-operation of the 5′ and 3′ untranslated regions of ornithine decarboxylase mRNA and inhibitory role of its 3′ untranslated region in regulating the translational efficiency of hybrid RNA species via cellular factor. *Biochem. J.* **326:** 361–367.

Mann C., Micouin J.Y., Chiannilkulchai N., Treich I., Buhler J.M., and Sentenac A. 1992. RPC53 encodes a subunit of *Saccharomyces cerevisiae* RNA polymerase C (III) whose inactivation leads to a predominantly G1 arrest. *Mol. Cell. Biol.* **12:** 4314–4326.

Martin K.A. and Blenis J. 2002. Coordinate regulation of translation by the PI 3-kinase and mTOR pathways. *Adv. Cancer Res.* **86:** 1–39.

Mateyak M.K., Obaya A.J., Adachi S., and Sedivy J.M. 1997. Phenotypes of c-Myc-deficient rat fibroblasts isolated by targeted homologous recombination. *Cell Growth Differ.* **8:** 1039–1048.

Matsushime H., Roussel M.F., Ashmun R.A., and Sherr C.J. 1991. Colony-stimulating factor 1 regulates novel cyclins during the G1 phase of the cell cycle. *Cell* **65:** 701–713.

Matthews M.B., Sonenberg N., and Hershey J.W.B. 2000. Origins and principles of translational control. In *Translational control of gene expression* (ed. N. Sonenberg et al.), pp. 1-31. Cold Spring Harbor Laboratory Press, Cold Spring Harbor, New York.

Mauck J.C. and Green H. 1974. Regulation of pre-transfer RNA synthesis during transition from resting to growing state. *Cell* **3:** 171–177.

Mercer W.E., Avignolo C., Galanti N., Rose K.M., Hyland J.K., Jacob S.T., and Baserga R. 1984. Cellular DNA replication is independent of the synthesis or accumulation of ribosomal RNA. *Exp. Cell Res.* **150:** 118–130.

Meric F. and Hunt K.K. 2002. Translation initiation in cancer: A novel target for therapy. *Mol. Cancer Ther.* **1:** 971–979.

Meurs E.F., Galabru J., Barber G.N., Katze M.G., and Hovanessian A.G. 1993. Tumor suppressor function of the interferon-induced double-stranded RNA-activated protein kinase. *Proc. Natl. Acad. Sci.* **90:** 232–236.

Meyer C.A., Jacobs H.W., Datar S.A., Du W., Edgar B.A., and Lehner C.F. 2000. *Drosophila* Cdk4 is required for normal growth and is dispensable for cell cycle progression. *EMBO J.* **19:** 4533–4542.

Meyuhas O. 2000. Synthesis of the translational apparatus is regulated at the translational level. *Eur. J. Biochem.* **267:** 6321–6330.

Meyuhas O., Blierman Y., Pierandrei-Amaldi P., and Amaldi F. 1996. Analysis of polysomal

RNA. In *A laboratory guide to RNA: Isolation, analysis and synthesis* (ed. P. Krieg), pp. 65-81. Wiley Liss, New York.

Michea L., Ferguson D.R., Peters E.M., Andrews P.M., Kirby M.R., and Burg M.B. 2000. Cell cycle delay and apoptosis are induced by high salt and urea in renal medullary cells. *Am. J. Physiol. Renal Physiol.* **278:** F209–F218.

Mikulits W., Pradet-Balade B., Habermann B., Beug H., Garcia-Sanz J.A., and Mullner E.W. 2000. Isolation of translationally controlled mRNAs by differential screening. *FASEB J.* **14:** 1641–1652.

Millard S.S., Vidal A., Markus M., and Koff A. 2000. A U-rich element in the 5′ untranslated region is necessary for the translation of p27 mRNA. *Mol. Cell. Biol.* **20:** 5947–5959.

Millard S.S., Yan J.S., Nguyen H., Pagano M., Kiyokawa H., and Koff A. 1997. Enhanced ribosomal association of p27(Kip1) mRNA is a mechanism contributing to accumulation during growth arrest. *J. Biol. Chem.* **272:** 7093–7098.

Mitchison J.M. 1971. *The biology of the cell cycle.* Cambridge University Press, Cambridge, United Kingdom.

Mochizuki T., Kitanaka C., Noguchi K., Muramatsu T., Asai A., and Kuchino Y. 1999. Physical and functional interactions between Pim-1 kinase and Cdc25A phosphatase. Implications for the Pim-1-mediated activation of the c-Myc signaling pathway. *J. Biol. Chem.* **274:** 18659–18666.

Morris C.E. 2002. How did cells get their size? *Anat. Rec.* **268:** 239–251.

Motokura T., Bloom T., Kim H.G., Juppner H., Ruderman J.V., Kronenberg H.M., and Arnold A. 1991. A novel cyclin encoded by a bcl1-linked candidate oncogene. *Nature* **350:** 512–515.

Muise-Helmericks R.C., Grimes H.L., Bellacosa A., Malstrom S.E., Tsichlis P.N., and Rosen N. 1998. Cyclin D expression is controlled post-transcriptionally via a phosphatidylinositol 3-kinase/Akt-dependent pathway. *J. Biol. Chem.* **273:** 29864–29872.

Murray L.E., Rowley N., Dawes I.W., Johnston G.C., and Singer R.A. 1998. A yeast glutamine tRNA signals nitrogen status for regulation of dimorphic growth and sporulation. *Proc. Natl. Acad. Sci.* **95:** 8619–8624.

Nash R., Tokiwa G., Anand S., Erickson K., and Futcher A.B. 1988. The *WHI1+* gene of *Saccharomyces cerevisiae* tethers cell division to cell size and is a cyclin homolog. *EMBO J.* **7:** 4335–4346.

Neufeld T.P., de la Cruz A.F., Johnston L.A., and Edgar B.A. 1998. Coordination of growth and cell division in the *Drosophila* wing. *Cell* **93:** 1183–1193.

Orian A., van Steensel B., Delrow J., Bussemaker H.J., Li L., Sawado T., Williams E., Loo L.W., Cowley S.M., Yost C., Pierce S., Edgar B.A., Parkhurst S.M., and Eisenman R.N. 2003. Genomic binding by the *Drosophila* Myc, Max, Mad/Mnt transcription factor network. *Genes Dev.* **17:** 1101–1114.

Oster S.K., Ho C.S., Soucie E.L., and Penn L.Z. 2002. The myc oncogene: MarvelouslY Complex. *Adv. Cancer Res.* **84:** 81–154.

Oster S.K., Mao D.Y., Kennedy J., and Penn L.Z. 2003. Functional analysis of the N-terminal domain of the Myc oncoprotein. *Oncogene* **22:** 1998–2010.

Pajic A., Spitkovsky D., Christoph B., Kempkes B., Schuhmacher M., Staege M.S., Brielmeier M., Ellwart J., Kohlhuber F., Bornkamm G.W., Polack A., and Eick D. 2000. Cell cycle activation by c-myc in a Burkitt lymphoma model cell line. *Int. J. Cancer* **87:** 787–793.

Pardee A.B. 1974. A restriction point for control of normal animal cell proliferation. *Proc. Natl. Acad. Sci.* **71:** 1286–1290.

Pause A., Belsham G.J., Gingras A.C., Donze O., Lin T.A., Lawrence J.C., Jr., and Sonenberg N. 1994. Insulin-dependent stimulation of protein synthesis by phosphorylation of a regulator of 5′-cap function. *Nature* **371:** 762–767.

Pelengaris S., Khan M., and Evan G. 2002. c-MYC: More than just a matter of life and death. *Nat. Rev. Cancer* **2:** 764–776.

Pelletier G., Stefanovsky V.Y., Faubladier M., Hirschler-Laszkiewicz I., Savard J., Rothblum L.I., Cote J., and Moss T. 2000. Competitive recruitment of CBP and Rb-HDAC regulates UBF acetylation and ribosomal transcription. *Mol. Cell* **6:** 1059–1066.

Pelletier J. and Sonenberg N. 1985. Insertion mutagenesis to increase secondary structure within the 5′ noncoding region of a eukaryotic mRNA reduces translational efficiency. *Cell* **40:** 515–526.

Polymenis M. and Schmidt E.V. 1997. Coupling of cell division to cell growth by translational control of the G1 cyclin CLN3 in yeast. *Genes Dev.* **11:** 2522–2531.

Puglisi J.D., Blanchard C., and Green R. 2000. Approaching translation at atomic resolution. *Nat. Struct. Biol.* **7:** 855–861.

Pyronnet S., Vagner S., Bouisson M., Prats A.C., Vaysse N., and Pradayrol L. 1996. Relief of ornithine decarboxylase messenger RNA translational repression induced by alternative splicing of its 5′ untranslated region. *Cancer Res.* **56:** 1742–1745.

Quelle D.E., Ashmun R.A., Shurtleff S.A., Kato J.Y., Bar-Sagi D., Roussel M.F., and Sherr C.J. 1993. Overexpression of mouse D-type cyclins accelerates G1 phase in rodent fibroblasts. *Genes Dev.* **7:** 1559–1571.

Rhoads R.E. 1999. Signal transduction pathways that regulate eukaryotic protein synthesis. *J. Biol. Chem.* **274:** 30337–30340.

Rhoads R.E., Joshi B., and Minich W.B. 1994. Participation of initiation factors in the recruitment of mRNA to ribosomes. *Biochimie* **76:** 831–838.

Rosenwald I.B., Lazaris-Karatzas A., Sonenberg N., and Schmidt E.V. 1993a. Elevated levels of cyclin D1 protein in response to increased expression of eukaryotic initiation factor 4E. *Mol. Cell. Biol.* **13:** 7358–7363.

Rosenwald I.B., Rhoads D.B., Callanan L.D., Isselbacher K.J., and Schmidt E.V. 1993b. Increased expression of eukaryotic translation initiation factors eIF-4E and eIF-2 alpha in response to growth induction by c-myc. *Proc. Natl. Acad. Sci.* **90:** 6175–6178.

Rosenwald I.B., Kaspar R., Rousseau D., Gehrke L., Leboulch P., Chen J.J., Schmidt E.V., Sonenberg N., and London I.M. 1995. Eukaryotic translation initiation factor 4E regulates expression of cyclin D1 at transcriptional and post-transcriptional levels. *J. Biol. Chem.* **270:** 21176–21180.

Rousseau D., Kaspar R., Rosenwald I., Gehrke L., and Sonenberg N. 1996. Translation initiation of ornithine decarboxylase and nucleocytoplasmic transport of cyclin D1 mRNA are increased in cells overexpressing eukaryotic initiation factor 4E. *Proc. Natl. Acad. Sci.* **93:** 1065–1070.

Ruggero D. and Pandolfi P.P. 2003. Does the ribosome translate cancer? *Nat. Rev. Cancer* **3:** 179–192.

Sachs J. 1898. *Physiologische Notizen.* N.G. Elwert, Marburg, Germany.

Saucedo L.J. and B.A. Edgar B.A. 2002. Why size matters: Altering cell size. *Curr. Opin. Genet. Dev.* **12:** 565–571.

Scheper G.C. and Proud C.G. 2002. Does phosphorylation of the cap-binding protein eIF4E play a role in translation initiation? *Eur. J. Biochem.* **269:** 5350–5359.

Schmidt E.V. 1999. The role of c-myc in cellular growth control. *Oncogene* **18:** 2988–2996.

Schramm L. and Hernandez N. 2002. Recruitment of RNA polymerase III to its target promoters. *Genes Dev.* **16:** 2593–2620.

Schuhmacher M., Staege M.S., Pajic A., Polack A., Weidle U.H., Bornkamm G.W., Eick D., and Kohlhuber F. 1999. Control of cell growth by c-Myc in the absence of cell division. *Curr. Biol.* **9:** 1255–1258.

Schuhmacher M., Kohlhuber F., Holzel M., Kaiser C., Burtscher H., Jarsch M., Bornkamm G.W., Laux G., Polack A., Weidle U.H., and Eick D. 2001. The transcriptional program of a human B cell line in response to Myc. *Nucleic Acids Res.* **29:** 397–406.

Scott P.H., Cairns C.A., Sutcliffe J.E., Alzuherri H.M., McLees A., Winter A.G., and White R.J. 2001. Regulation of RNA polymerase III transcription during cell cycle entry. *J. Biol. Chem.* **276:** 1005–1014.

Seither P., Iben S., and Grummt I. 1998. Mammalian RNA polymerase I exists as a holoenzyme with associated basal transcription factors. *J. Mol. Biol.* **275:** 43–53.

Serrano M., Hannon G.J., and Beach D. 1993. A new regulatory motif in cell-cycle control causing specific inhibition of cyclin D/CDK4. *Nature* **366:** 704–707.

Sherr C.J. 1993. Mammalian G1 cyclins. *Cell* **73:** 1059–1065.

———. 1994. G1 phase progression: Cycling on cue. *Cell* **79:** 551–555.

———. 1996. Cancer cell cycles. *Science* **274:** 1672–1677.

Shou W., Sakamoto K.M., Keener J., Morimoto K.W., Traverso E.E., Azzam R., Hoppe G.J., Feldman R.M., DeModena J., Moazed D., Charbonneau H., Nomura M., and Deshaies R.J. 2001. Net1 stimulates RNA polymerase I transcription and regulates nucleolar structure independently of controlling mitotic exit. *Mol. Cell* **8:** 45–55.

Shtutman M., Zhurinsky J., Simcha I., Albanese C., D'Amico M., Pestell R., and Ben-Ze'ev A. 1999. The cyclin D1 gene is a target of the beta-catenin/LEF-1 pathway. *Proc. Natl. Acad. Sci.* **96:** 5522–5527.

Sirri V., Hernandez-Verdun D., and Roussel P. 2002. Cyclin-dependent kinases govern formation and maintenance of the nucleolus. *J. Cell Biol.* **156:** 969–981.

Sirri V., Roussel P., and Hernandez-Verdun D. 2000. In vivo release of mitotic silencing of ribosomal gene transcription does not give rise to precursor ribosomal RNA processing. *J. Cell Biol.* **148:** 259–270.

Sogin S.J., Carter B.L., and Halvorson H.O. 1974. Changes in the rate of ribosomal RNA synthesis during the cell cycle in *Saccharomyces cerevisiae. Exp. Cell Res.* **89:** 127–138.

Sonenberg N. and Gingras A.C. 1998. The mRNA 5′ cap-binding protein eIF4E and control of cell growth. *Curr. Opin. Cell Biol.* **10:** 268–275.

Staller P., Peukert K., Kiermaier A., Seoane J., Lukas J., Karsunky H., Moroy T., Bartek J., Massague J., Hanel F., and Eilers M. 2001. Repression of p15INK4b expression by Myc through association with Miz-1. *Nat. Cell Biol.* **3:** 392–399.

Stefanovsky V.Y., Pelletier G., Hannan R., Gagnon-Kugler T., Rothblum L.I., and Moss T. 2001. An immediate response of ribosomal transcription to growth factor stimulation in mammals is mediated by ERK phosphorylation of UBF. *Mol. Cell* **8:** 1063–1073.

Stein T., Crighton D., Boyle J.M., Varley J.M., and White R.J. 2002a. RNA polymerase III transcription can be derepressed by oncogenes or mutations that compromise p53 function in tumours and Li-Fraumeni syndrome. *Oncogene* **21:** 2961–2970.

Stein T., Crighton D., Warnock L.J., Milner J., and White R.J. 2002b. Several regions of p53 are involved in repression of RNA polymerase III transcription. *Oncogene* **21:** 5540–5547.

Steiner P., Philipp A., Lukas J., Godden-Kent D., Pagano M., Mittnacht S., Bartek J., and Eilers M. 1995. Identification of a Myc-dependent step during the formation of active G1 cyclin-cdk complexes. *EMBO J.* **14:** 4814–4826.

Sutcliffe J.E., Brown T.R., Allison S.J., Scott P.H., and White R.J. 2000. Retinoblastoma protein disrupts interactions required for RNA polymerase III transcription. *Mol. Cell. Biol.* **20:** 9192–9202.

Sutcliffe J.E., Cairns C.A., McLees A., Allison S.J., Tosh K., and White R.J. 1999. RNA polymerase III transcription factor IIIB is a target for repression by pocket proteins p107 and p130. *Mol. Cell. Biol.* **19:** 4255–4261.

Svitkin Y.V., Pause A., Haghighat A., Pyronnet S., Witherell G., Belsham G.J., and Sonenberg N. 2001. The requirement for eukaryotic initiation factor 4A (eIF4A) in translation is in direct proportion to the degree of mRNA 5′ secondary structure. *RNA* **7:** 382–394.

Szostak J.W., Bartel D.P., and Luisi P.L. 2001. Synthesizing life. *Nature* **409:** 387–390.

Tatsuka M., Mitsui H., Wada M., Nagata A., Nojima H., and Okayama H. 1992. Elongation factor-1 alpha gene determines susceptibility to transformation. *Nature* **359:** 333–336.

Taub R., Kirsch I., Morton C., Lenoir G., Swan D., Tronick S., Aaronson S., and Leder P. 1982. Translocation of the c-myc gene into the immunoglobulin heavy chain locus in human Burkitt lymphoma and murine plasmacytoma cells. *Proc. Natl. Acad. Sci.* **79:** 7837–7841.

Tetsu O. and McCormick F. 1999. Beta-catenin regulates expression of cyclin D1 in colon carcinoma cells. *Nature* **398:** 422–426.

Thompson D.A.W. 1961. *On growth and form.* Cambridge University Press, Cambridge, United Kingdom.

Topisirovic I., Culjkovic B., Cohen N., Perez J.M., Skrabanek L., and Borden K.L. 2003. The proline-rich homeodomain protein, PRH, is a tissue-specific inhibitor of eIF4E-dependent cyclin D1 mRNA transport and growth. *EMBO J.* **22:** 689–703.

Trumpp A., Refaeli Y., Oskarsson T., Gasser S., Murphy M., Martin G.R., and Bishop J.M. 2001. c-Myc regulates mammalian body size by controlling cell number but not cell size. *Nature* **414:** 768–773.

Tsung K., Inouye S., and Inouye M. 1989. Factors affecting the efficiency of protein synthesis in *Escherichia coli.* Production of a polypeptide of more than 6000 amino acid residues. *J. Biol. Chem.* **264:** 4428–4433.

Tyers M., Tokiwa G., and Futcher B. 1993. Comparison of the *Saccharomyces cerevisiae* G1 cyclins: Cln3 may be an upstream activator of Cln1, Cln2 and other cyclins. *EMBO J.* **12:** 1955–1968.

Tyers M., Tokiwa G., Nash R., and Futcher B. 1992. The Cln3-Cdc28 kinase complex of *S. cerevisiae* is regulated by proteolysis and phosphorylation. *EMBO J.* **11:** 1773–1784.

Virchow R.L.K. 1858. *Die Cellularpathologie in ihrer Begründung auf physiologische und pathologische Gewebelehre.* Verlag von August Hirschwald, Berlin, Germany.

Voit R. and Grummt I. 2001. Phosphorylation of UBF at serine 388 is required for interaction with RNA polymerase I and activation of rDNA transcription. *Proc. Natl. Acad. Sci.* **98:** 13631–13636.

Voit R., Hoffmann M., and Grummt I. 1999. Phosphorylation by G1-specific cdk-cyclin complexes activates the nucleolar transcription factor UBF. *EMBO J.* **18:** 1891–1899.

Vousden K.H. and Lu X. 2002. Live or let die: The cell's response to p53. *Nat. Rev. Cancer* **2:** 594–604.

Wang Z., Bhattacharya N., Mixter P.F., Wei W., Sedivy J., and Magnuson N.S. 2002. Phosphorylation of the cell cycle inhibitor p21Cip1/WAF1 by Pim-1 kinase. *Biochim. Biophys. Acta* **1593:** 45–55.

White R.J. 1997. Regulation of RNA polymerases I and III by the retinoblastoma protein: A mechanism for growth control? *Trends Biochem. Sci.* **22:** 77–80.

White R.J., Trouche D., Martin K., Jackson S.P., and Kouzarides T. 1996. Repression of RNA polymerase III transcription by the retinoblastoma protein. *Nature* **382:** 88–90.

Wolf V.J., Dang T., Hartl P., and Gottesfeld J.M. 1994. Role of maturation-promoting factor (p34cdc2-cyclin B) in differential expression of the *Xenopus* oocyte and somatic-type 5S RNA genes. *Mol. Cell. Biol.* **14:** 4704–4711.

Wrobel C., Schmidt E.V., and Polymenis M. 1999. CDC64 encodes cytoplasmic alanyl-tRNA synthetase, Ala1p, of *Saccharomyces cerevisiae. J. Bacteriol.* **181:** 7618–7620.

Wu S., Cetinkaya C., Munoz-Alonso M.J., von der Lehr N., Bahram F., Beuger V., Eilers M., Leon J., and Larsson L.G. 2003. Myc represses differentiation-induced p21CIP1 expression via Miz-1-dependent interaction with the p21 core promoter. *Oncogene* **22:** 351–360.

Xiong Y., Connolly T., Futcher B., and Beach D. 1991. Human D-type cyclin. *Cell* **65:** 691–699.

Zetterberg A. 1966. Synthesis and accumulation of nuclear and cytoplasmic proteins during interphase in mouse fibroblasts in vitro. *Exp. Cell. Res.* **42:** 500–511.

Zetterberg A. and Killander D. 1965a. Quantitative cytochemical studies on interphase growth. II. Derivation of synthesis curves from the distribution of DNA, RNA and mass values of individual mouse fibroblasts in vitro. *Exp. Cell Res.* **39:** 22–32.

———. 1965b. Quantitative cytophotometric and autoradiographic studies on the rate of protein synthesis during interphase in mouse fibroblasts in vitro. *Exp. Cell Res.* **40:** 1–11.

Zetterberg A., Larsson O., and Wiman K.G. 1995. What is the restriction point? *Curr. Opin. Cell Biol.* **7:** 835–842.

5

TOR Signaling in Yeast: Temporal and Spatial Control of Cell Growth

Robbie Loewith and Michael N. Hall
Division of Biochemistry
Biozentrum, University of Basel
CH4056 Basel, Switzerland

CELL GROWTH IS HIGHLY REGULATED. Cells respond to nutrients or other appropriate growth stimuli by up-regulating macromolecular synthesis, and thereby increasing in size. Conversely, cells respond to nutrient limitation or other types of stress by down-regulating macromolecular synthesis and enhancing turnover of excess mass. Thus, the control of cell growth involves balancing positive regulation of anabolic processes with negative regulation of catabolic processes. Growth is also controlled relative to cell division. In proliferating cells, growth is linked to the cell cycle such that cells generally double their mass before dividing. In other physiological contexts, such as load-induced muscle hypertrophy or growth factor–induced neuronal growth, cell growth can occur postmitotically. Furthermore, in addition to the temporal control of cell growth described above, cell growth can be subject to spatial constraints. For example, budding yeast and neurons grow in a polarized manner as a result of new mass being laid down only at one end of the cell. Finally, in multicellular organisms, growth of individual cells is controlled relative to overall body growth such that the organs constituting the organism are properly proportioned.

What are the mechanisms that mediate and integrate the many parameters of cell growth? In other words, what determines that a cell grows only at the right time and at the right place? Remarkably, the study of these mechanisms has been largely neglected, despite their clinical relevance and despite cell growth being, along with cell division and cell death, one of the most fundamental aspects of cell behavior. Also remarkable is the recent finding that cell growth control, regardless of the eukaryotic organism or the physiological context, often (if not always) involves the same protein—the target of rapamycin TOR protein—and its namesake signaling network (Jacinto and Hall 2003). Here we review the two TORs, TOR1 and TOR2, of *Saccharomyces cerevisiae* and the numerous growth-controlling signaling pathways they mediate. The multiple TOR signaling pathways fall into one of two major signaling branches that integrate the temporal and spatial control of yeast cell growth (Figs. 1 and 2). The branch that determines when a cell grows (temporal control) utilizes TOR1 or TOR2 and is rapamycin-sensitive. The branch that determines where a cell grows (spatial control) contains TOR2, but not TOR1, and is rapamycin-insensitive. Each signaling branch contains a structurally distinct TOR complex.

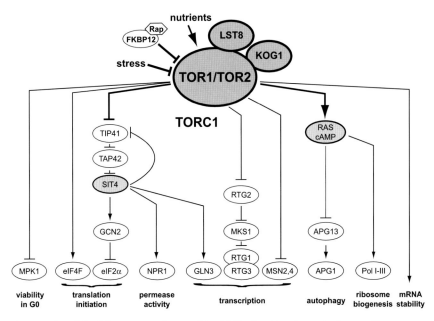

Figure 1. Model of the rapamycin-sensitive TOR signaling branch that mediates temporal control of cell growth. Schematic representation of TOR complex 1 (TORC1), upstream regulators, effector pathways, and rapamycin-sensitive readouts.

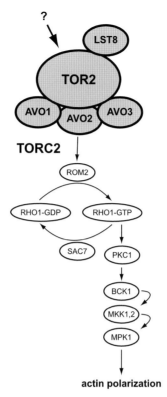

Figure 2. Model of the rapamycin-insensitive TOR2 signaling branch that mediates spatial control of cell growth. Schematic representation of TOR complex 2 (TORC2) and the downstream signaling pathway leading to actin polarization.

GENETIC DISSECTION OF RAPAMYCIN ACTION IN
S. cerevisiae: TWO ESSENTIAL TOR FUNCTIONS

The immunosuppressive and anticancer drug rapamycin binds with high affinity to the highly conserved FKBP12 protein, a peptidylprolyl isomerase encoded by the *FPR1* gene in *S. cerevisiae* (Heitman et al. 1991a; Schreiber 1991). Treatment of yeast cells with rapamycin arrests the cells within one generation and in the G_1 phase of the cell cycle (Heitman et al. 1991a; Kunz et al. 1993; Barbet et al. 1996). However, *fpr1* mutant cells are viable (Heitman et al. 1991b; Wiederrecht et al. 1991) and resistant to rapamycin (Heitman et al. 1991a; Koltin et al. 1991), demonstrating that FPR1 is not the growth-related target of rapamycin but rather a cofactor required for toxicity. The growth-related target of the FPR1–rapamycin complex was discovered with the identification of rapamycin resistance–conferring mutations in the previously unknown genes *TOR1* and *TOR2*. The original rapamycin resistance–conferring *TOR1* and *TOR2* alleles contain the single missense mutations Ser1972Arg and Ser1975Ile, respectively. These alleles confer complete resistance to the antifungal activity of rapamycin by preventing the binding of FPR1–rapamycin to TOR (Heitman et al. 1991a; Cafferkey et al. 1993; Stan et al. 1994; Chen et al. 1995; Lorenz and Heitman 1995; Choi et al. 1996). Intriguingly, the mechanism by which FPR1–rapamycin inhibits TOR is not known (McMahon et al. 2002 and references therein). Cloning of *TOR1* and *TOR2* revealed that the TOR proteins are highly homologous (67%) and related to phosphatidylinositol (PI) kinases, in particular PI 3- and PI 4-kinases (Cafferkey et al. 1993; Kunz et al. 1993; Helliwell et al. 1994). However, the TOR proteins have turned out to be the founding members of a family of phosphatidylinositol kinase–related kinases (PIKKs) which, despite their homology with lipid kinases, are serine/threonine-specific protein kinases. Other members of the PIKK family include ATM, DNA-PK, MEC1, TEL1, and RAD3 (Keith and Schreiber 1995). Integrity of the kinase domains of TOR1 and TOR2 is essential for the function of these proteins (Kunz et al. 1993; Helliwell et al. 1994; Zheng et al. 1995; Schmidt et al. 1996).

In yeast, depletion of TOR1 and TOR2 individually and in combination results in distinct phenotypes, suggesting that the two TORs are functionally related but not identical. Loss of TOR1 results in a slightly reduced growth rate and sensitivity to salts and temperature extremes (Cafferkey et al. 1993; Kunz et al. 1993; Helliwell et al. 1994; Crespo et al. 2001). In contrast, loss of TOR2 is lethal, causing a near-random cell cycle arrest (Kunz et al. 1993; Helliwell et al. 1998a). Simultaneous loss of TOR1 and TOR2 is also lethal, but cells arrest primarily in the G_1 phase of the cell cycle like rapamycin-treated cells (Kunz et al. 1993; Helliwell et al. 1998a). These early observations suggested that the homologous TOR1 and TOR2 proteins perform a redundant, rapamycin-sensitive function required for G_1 progression, and in addition, TOR2 performs an essential function that TOR1 cannot perform. Furthermore, the observation that *TOR1* alleles (e.g., *TOR1-1*) can support growth in the presence of rapamycin suggested that the TOR2-unique function is insensitive to rapamycin (Kunz et al. 1993; Helliwell et al. 1994; Zheng et al. 1995). As discussed below, it is now known that the two so-called functions of TOR correspond to two distinct signaling branches, each mediated by a structurally distinct TOR complex.

Subsequent to the identification of TOR in yeast, TORs have been found in all eukaryotes examined, including plants, worms, flies, birds, and mammals (Brown et al. 1994; Sabatini et al. 1994; Sabers et al. 1995; Crespo and Hall 2002). However, unlike yeasts, which contain two *TOR* genes, higher eukaryotes possess only a single *TOR* gene. As suggested by the clinical efficacy of rapamycin and rapamycin resistance–conferring *mTOR* alleles (Abraham and Wiederrecht 1996; Hosoi et al. 1998; Hidalgo and Rowinsky 2000; Elit 2002; Guba et al. 2002), mammalian TOR (mTOR) signaling is also sensitive to rapamycin (for review, see Schmelzle and Hall 2000; Gingras et al. 2001). Although a single *TOR* gene in higher eukaryotes does not preclude the existence of a second, rapamycin-insensitive mTOR signaling branch, such a branch has yet to be reported in metazoans.

TOR RESPONDS TO NUTRIENTS

Nutrient availability is an important determinant in the control of cell growth. Unicellular organisms in particular must be able to reprogram metabolic pathways to adapt to rapidly changing nutrient conditions. In response to starvation conditions, cells minimize energy expenditure and ultimately arrest growth. Inactivation of TOR, genetically or pharmacologically with rapamycin, mimics a starvation response, suggesting that TOR participates in a signal transduction pathway that links nutrient sensing to cell growth (Barbet et al. 1996; Schmelzle and Hall 2000; Rohde et al. 2001). TOR signaling appears to respond to both the quality and abundance of nitrogen, and possibly carbon (Schmidt et al. 1998; Beck and Hall 1999; Shamji et al. 2000; Kuruvilla et al. 2001; Crespo et al. 2002). Indeed, Schreiber and coworkers have proposed that TOR acts as a multi-channel processor which elicits different responses to distinct nutrient conditions (Shamji et al. 2000; Kuruvilla et al. 2001). In support of this hypothesis, starvation for the preferred nitrogen source glutamine affects only a subset of TOR readouts—those involved in glutamine synthesis (Crespo et al. 2002). Thus, TOR must monitor, directly or indirectly, the quantity and quality of numerous metabolites. However, the identity of these metabolites (other than glutamine) and how they are sensed are presently unknown.

RAPAMYCIN-SENSITIVE TOR SIGNALING BRANCH MEDIATES TEMPORAL CONTROL OF CELL GROWTH

Growth of *S. cerevisiae* cells is temporally controlled. Macromolecular synthesis (and thus cell growth) occurs when nutrients are available. TOR signaling pathways couple nutrient availability to macromolecular synthesis and thus mediate the temporal control of cell growth. In particular, TOR promotes translation initiation, ribosome biogenesis, mRNA stability, and nutrient transport, to maintain a high rate of translation. In contrast, inhibition of TOR function, either through starvation or rapamycin treatment, results in a down-regulation of anabolic processes and an up-regulation of catabolic processes. Specifically, in addition to a general down-regulation of macromolecular synthesis, loss of TOR signaling results in increased transcription of many stress-responsive genes and increased turnover of excess mass through autophagy (Noda and Ohsumi 1998) and other degradative pathways (see below). Thus, a rapamycin-sensitive TOR signaling branch positively controls a growth program and negatively controls a growth-arrest program.

Readouts of the Rapamycin-sensitive TOR Signaling Branch

Readouts of the rapamycin-sensitive TOR signaling branch, which collectively determines the mass of the cell, are diagramed in Figure 1 and discussed below.

TOR Promotes Translation Initiation

Treatment of yeast cells with rapamycin results in a rapid and dramatic inhibition of translation initiation (Barbet et al. 1996). In eukaryotes, initiation of protein synthesis is a highly regulated process (for review, see Hershey and Merrick 2000), and the manner by which TOR signals maintain a robust rate of translation initiation is not clear. Possible primary translation targets of TOR include the translation initiation factors eIF4F and eIF2. Secondary translation targets of TOR include unknown cellular components involved in ribosome biogenesis (see next section) and mRNA stability (see below), and amino acid permeases (see below). Thus, TOR likely controls translation initiation at several levels.

In *S. cerevisiae*, the heterotrimeric initiation factor 4F (eIF4F) is composed of the cap-binding subunit eIF4E (encoded by *CDC33*), the adapter protein eIF4G (encoded by the redundant genes *TIF4631* and *TIF463*), and the DEAD box helicase eIF4A (encoded by the redundant genes *TIF1* and *TIF2*). *cdc33* and *tor* mutants display similar phenotypes, suggesting that TOR may control translation via eIF4E. Supporting this hypothesis is the observation that cap-independent expression of the G_1 cyclin CLN3 suppresses the rapamycin-induced G_1 arrest (Barbet et al. 1996; Danaie et al. 1999). Furthermore, depletion of EAP1, an eIF4E-binding protein and an inhibitor of cap-dependent translation, confers partial resistance to rapamycin (Cosentino et al. 2000). TOR inactivation also results in the degradation of eIF4G (Berset et al. 1998); however, this could be a late and indirect consequence of enhanced proteolysis upon TOR inactivation, and not a primary mechanism by which TOR regulates translation.

TOR has additionally been shown to regulate translation initiation via eIF2. GTP-loaded eIF2 is required to deliver an initiator methionyl-tRNA to the 40S ribosomal subunit and eventually to the translation initiation codon of an mRNA. Phosphorylation of the α-subunit of eIF2 (eIF2α) converts eIF2 from a substrate to an inhibitor of its guanine nucleotide exchange factor eIF2B and consequently reduces general protein synthesis (Hinnebusch 2000). eIF2α is phosphorylated by the protein kinase GCN2. GCN2 is itself activated by the binding of uncharged tRNA to a histidyl-tRNA synthetase-related domain located carboxy-terminal to the kinase domain. GCN2 is also negatively regulated by phosphorylation of Ser-577, and phosphorylation of this residue is controlled by TOR. Rapamycin treatment elicits dephosphorylation of Ser-577, resulting in activation of GCN2 kinase activity, a subsequent increase in eIF2α phosphorylation, and ultimately decreased translation initiation (Cherkasova and Hinnebusch 2003). It will be of interest to determine how much this regulatory mechanism contributes to TOR-mediated control of translation, as all translation is eIF2-dependent, but not all translation is TOR-dependent.

TOR Promotes Ribosome Biogenesis

Ribosome production is energetically a very costly process. In rapidly growing yeast cells, the synthesis of large rRNAs by RNA polymerase (Pol) I accounts for 60% of total nuclear transcription, and ribosomal protein genes account for as many as half of RNA Pol II-dependent initiation events (Warner 1999). Consequently, the production of ribosome components and tRNA is very tightly coupled to growth conditions, to match but not exceed protein synthesis requirements.

TOR signaling regulates ribosome biosynthesis at both transcriptional and translational levels. Inactivation of TOR by rapamycin or nutrient limitation results in a dramatic down-regulation of RNA Pol II-dependent transcription of ribosomal protein genes (Cardenas et al. 1999; Hardwick et al. 1999; Powers and Walter 1999), a down-regulation of RNA Pol I-dependent transcription of rRNA genes, and a down-regulation of RNA Pol III-dependent transcription of tRNA genes (Zaragoza et al. 1998; Powers and Walter 1999). Additionally, inhibition of TOR also reduces the processing of the rRNA 35S precursor (Powers and Walter 1999).

The mechanisms by which TOR signaling regulates and coordinates transcription of ribosomal protein, rRNA, or tRNA genes are not well understood. The RSC chromatin-remodeling complex has been observed to be recruited to the promoters of TOR-regulated ribosomal protein, rRNA, and tRNA genes in a rapamycin-sensitive fashion (Damelin et al. 2002; Ng et al. 2002). However, it remains to be determined whether RSC is a direct target of TOR signaling or a cofactor recruited to TOR-responsive promoters subsequent to the action of a more direct TOR target. Regulation of histone acetylation/deacetylation has also been suggested to play a role in the control of ribosomal protein gene expression by TOR (Rohde and Cardenas 2003).

TOR Promotes mRNA Stability

mRNA turnover is another way by which cells regulate protein synthesis. In yeast, there are three known mRNA decay pathways, the deadenylation-dependent decapping decay pathway, the 3′-to-5′ exonucleolytic decay pathway, and the nonsense-mediated decay pathway (Decker 1998; Czaplinski et al. 1999; Steiger and Parker 2002). In rapidly growing cells, the deadenylation-dependent decapping decay pathway appears to play the most prominent role in degrading mRNAs. Rapamycin treatment accelerates this major mRNA decay pathway, resulting in an enhanced turnover of specific messages (Albig and Decker 2001). However, the control of mRNA turnover by TOR is complex. Some messages are destabilized very rapidly in rapamycin-treated cells, whereas other mRNAs become unstable only after prolonged treatment. The relevant targets of TOR affecting mRNA stability are unknown, and it remains formally possible that rapamycin-induced changes in mRNA turnover are an indirect result of decreased translation (Albig and Decker 2001).

TOR Regulates Amino Acid Permease Activity and Expression

Yeast cells utilize a wide variety of compounds as carbon or nitrogen sources. Amino acids are a particularly important nutrient as they are accumulated by yeast cells for both catabolic and anabolic purposes. Yeast amino acid permeases that transport amino acids from the extracellular medium into the cytoplasm fall into two classes (Sophianopoulou and Diallinas 1995). High-affinity, specific amino acid permeases transport one or a few related L-amino acids. In contrast, the low-affinity general amino acid permeases possess broad specificity and transport most L- and D-amino acids and related compounds. In starved cells, the few general amino acid permeases are up-regulated and the several specific permeases are down-regulated, most likely as a mechanism to avoid maintaining several energetically costly permeases.

Rapamycin-sensitive TOR signaling plays a prominent role in the regulation of both classes of amino acid permeases. Yeast cells growing rapidly in media containing preferred nitrogen sources express specific amino acid permeases and sort these proteins to the plasma membrane where they are believed to transport amino acids for immediate use in protein synthesis (for review, see Magasanik and Kaiser 2002). Inhibition of TOR function by rapamycin or nitrogen starvation induces ubiquitination and subsequent degradation of the tryptophan permease TAT2 and likely the histidine permease HIP1 as well as other specific amino acid permeases (Schmidt et al. 1998; Beck et al. 1999). Thus, TOR function couples amino acid import to protein synthesis.

How does TOR prevent the ubiquitination and degradation of TAT2 and possibly other specific amino acid permeases? TOR may prevent the turnover of TAT2 by inhibiting the serine/threonine kinase NPR1 (Schmidt et al. 1998). TOR maintains NPR1 in a highly phosphorylated state, and overexpression of NPR1 causes a decrease in the amount of TAT2 protein. Thus, it has been proposed that TOR maintains NPR1 in a phosphorylated and thereby inactive state. Upon TOR inactivation, NPR1 is rapidly dephosphorylated and activated. Activated NPR1 in turn leads to the ubiquitination, internalization, and degradation of TAT2. However, the direct substrate of NPR1 kinase activity is not known.

TOR also regulates the general amino acid permease GAP1 and the ammonia permease MEP2. However, unlike the regulation of TAT2 described above, the regulation of GAP1 and MEP2 is negative and via transcription of the *GAP1* and *MEP2* genes (Beck and Hall 1999). The expression of *GAP1* and *MEP2* is derepressed during growth in media containing a poor nitrogen source, suggesting that these permeases are primarily required to import compounds for use as nitrogen sources. Inhibition of TOR also leads to derepression of the *GAP1* and *MEP2* genes, demonstrating that TOR limits expression of nitrogen-scavenging permeases to periods of nitrogen stress. TOR negatively regulates the transcription of the *GAP1* and *MEP2* genes by maintaining the transcription activator GLN3 phosphorylated and thereby sequestered in the cytoplasm (described below).

GAP1 is also controlled posttranscriptionally. The posttranscriptional control of GAP1 involves the kinase NPR1. Consistent with the inverse regulation of TAT2 and GAP1, NPR1 is required for GAP1 transport activity (De Craene et al. 2001). These observations predict that TOR, as a consequence of maintaining NPR1 phosphorylated and inactive, may negatively control GAP1 sorting to the plasma membrane. However, Chen and Kaiser (2002) have presented evidence that TOR does not affect GAP1 posttranscriptionally. Thus, there may be a TOR-insensitive pool of NPR1, or GAP1 is not regulated by NPR1 in the Chen and Kaiser strain background.

Amino acid import in *S. cerevisiae* is also regulated by RHB1, a ras family GTPase (Urano et al. 2000). In mammals and flies, the RHB1-related GTPase Rheb is an upstream activator of TOR (Garami et al. 2003; Inoki et al. 2003; Saucedo et al. 2003; Stocker et al. 2003; Tee et al. 2003; Zhang et al. 2003). It is unlikely that RHB1 functions upstream of TOR in *S. cerevisiae*. Whereas TOR influences the activity of a number of permeases, RHB1 specifically regulates the uptake of arginine and, to a lesser extent, lysine. Furthermore, *rhb1* mutants are viable and not rapamycin hypersensitive, contrary to what would be expected if RHB1 were an upstream activator of TOR (S. Wullschleger and M.N. Hall, unpubl.). Interestingly, *rhb1* mutations in *Schizosaccharomyces pombe* cause a growth arrest similar to that caused by nitrogen starvation and are suppressed by a mammalian *Rheb* gene, suggesting that RHB1 in this yeast might be upstream of TOR (Mach et al. 2000; Yang et al. 2001).

TOR Controls Transcription

The yeast transcriptome is finely adjusted to meet the demands of the environment. In addition to promoting transcription of ribosomal protein, rRNA and tRNA genes under good nutrient conditions (see above), TOR also inhibits the expression of many nutrient-regulated genes (Cardenas et al. 1999; Hardwick et al. 1999; Komeili et al. 2000; Shamji et al. 2000; Rohde et al. 2001; Gasch and Werner-Washburne 2002). Rapamycin enhances expression of several hundred genes. The products of the majority of these genes function in metabolic pathways, nitrogen scavenging, and the tricarboxylic acid cycle. As outlined below, TOR negatively regulates gene expression primarily via cytoplasmic sequestration of several stress-responsive transcription factors (Beck and Hall 1999; Komeili et al. 2000). Thus, TOR prevents transcription of stress combative (but often growth-inhibitory) genes during periods of favorable growth conditions.

TOR inhibits the GATA transcription factors GLN3 and GAT1. Genes whose products are involved in the uptake and assimilation of alternative nitrogen sources are those most strikingly affected by rapamycin treatment.

These genes are regulated in response to the quantity and quality of the nitrogen source via the GATA transcription factors GLN3 and GAT1 (Magasanik and Kaiser 2002). Yeast cells growing in the presence of a high-quality nitrogen source (glutamine, glutamate, or ammonia) maintain GLN3 and GAT1 in the cytoplasm. Shift to a poor nitrogen source (urea or proline), or treatment with rapamycin, results in dephosphorylation of both GLN3 and its cytoplasmic anchor protein URE2. Subsequently, GLN3 dissociates from URE2, accumulates in the nucleus, and induces expression of target genes (Beck and Hall 1999; Cardenas et al. 1999; Hardwick et al. 1999; Magasanik and Kaiser 2002). GAT1 is also negatively regulated by TOR, but the mechanistic details of this regulation are less clear (Beck and Hall 1999). Thus, TOR signaling prevents use of non-favored nitrogen sources by inhibiting expression of the permeases and enzymes required to import and metabolize these compounds.

TOR inhibits the bHLH/Zip transcription factor composed of RTG1 and RTG3. RTG1/3 regulates the expression of specific tricarboxylic acid and glyoxylate cycle genes. The products of these genes promote the synthesis of metabolic intermediates, primarily α-ketoglutarate, that are required for the de novo synthesis of amino acids such as glutamine and glutamate. In the presence of glutamine or glutamate, preferred nitrogen sources, RTG1/3 remains in the cytoplasm. Upon shift to a poor nitrogen source, treatment with rapamycin, or loss of mitochondrial function, RTG1/3 rapidly accumulates in the nucleus and induces expression of target genes (Komeili et al. 2000). Thus, TOR signaling prevents the energetically costly de novo synthesis of nitrogen sources when these compounds can be acquired from the environment.

The regulation of RTG1/3 subcellular localization and activity is not well understood. Nuclear accumulation of RTG1/3 requires RTG2 and is inhibited by MKS1. Epistasis studies suggest that RTG2 and MKS1 function downstream of TOR (Dilova et al. 2002). MKS1 and RTG3 are phosphoproteins, and RTG3 is hyperphosphorylated upon TOR inactivation (Komeili et al. 2000). However, MKS1 phosphorylation is regulated by RTG2 and RTG1/3 as well as by TOR (Dilova et al. 2002; Sekito et al. 2002). Thus, the regulation of the RTGs by TOR appears to be complex.

TOR inhibits the Zn^{++} finger transcription factors MSN2 and MSN4. The partially redundant Zn^{++} finger transcription factors MSN2 and MSN4 are general stress transcription factors, and have been implicated in the up-regulation of many genes in response to nutrient and environmental stresses (Görner et al. 1998; Smith et al. 1998; Gasch and Werner-

Washburne 2002). Under optimal growth conditions, MSN2 and MSN4 are sequestered in the cytoplasm. Starvation for glucose (and possibly nitrogen), or rapamycin treatment, releases the MSNs from the 14-3-3 proteins BMH1 and BMH2 and results in the nuclear accumulation and activation of MSN2 and MSN4 (Beck and Hall 1999). The rapamycin hypersensitivity of *bmh1* and *bmh2* cells further suggests that the BMHs may have a role in maintaining the MSNs in the cytoplasm (Bertram et al. 1998). Seemingly, the MSNs are also controlled in parallel by the RAS/cAMP pathway (Görner et al. 1998). TOR signaling therefore has input into the general stress response in addition to input into the starvation response.

TOR activates transcription of ribosomal protein, rRNA and tRNA genes. In contrast to negative regulation of transcription described above, TOR also regulates transcription positively. Most importantly, TOR activates transcription of genes involved in ribosome and tRNA biogenesis. This includes RNA Pol I-dependent transcription of rRNA genes, RNA Pol II-dependent transcription of ribosomal protein genes, and RNA Pol III-dependent transcription of tRNA genes (see above).

TOR Represses Autophagy

When challenged with a sudden loss of nutrients, yeast cells induce a catabolic membrane-trafficking phenomenon known as autophagy (for review, see Abeliovich and Klionsky 2001; Khalfan and Klionsky 2002; Noda et al. 2002). During autophagy, portions of the cytoplasm are non-selectively engulfed and sequestered in double-membrane structures known as autophagosomes. Autophagosomes subsequently fuse with the vacuolar membrane to release their inner membrane structure into the lumen of the vacuole. These intravacuolar vesicles, called autophagic bodies, are degraded along with their cytoplasmic content. Autophagy serves two important functions. First, it turns over large structures and organelles such as ribosomes and mitochondria. This serves to arrest growth, reduce cellular energy consumption, and protect cells from the toxic effects of mitochondria functioning during suboptimal growth conditions. Second, autophagy recycles unneeded cytoplasmic mass. This provides building blocks and alternative nutrient sources until nutrient conditions improve.

Genetic screens in yeast have identified several *APG* genes encoding the structural elements and signaling components governing autophagy (Tsukada and Ohsumi 1993; Thumm et al. 1994; Harding et al. 1995).

However, the mechanistic details of autophagosome formation and regulation remain sketchy. In nutrient-rich media, TOR signaling limits autophagy. Rapamycin treatment (or starvation) leads to an induction of autophagy (Noda and Ohsumi 1998). The mechanism by which TOR signaling inhibits autophagy appears to be via regulation of the phosphoprotein APG13 and the serine/threonine-specific protein kinase APG1 (Kamada et al. 2000). When TOR is active, APG13 is hyperphosphorylated, APG1 kinase activity is low, and autophagy is inhibited. Inactivation of TOR results in APG13 dephosphorylation. Hypophosphorylated APG13 binds and activates APG1. Active APG1 in turn triggers autophagy (Kamada et al. 2000; Khalfan and Klionsky 2002). The mechanism by which TOR controls APG13 phosphorylation is unknown, but appears to involve signaling through the RAS/cAMP pathway (Noda and Ohsumi 1998; Schmelzle et al. 2004) (see below).

TOR Represses Signaling through the Cell Integrity Pathway

Rapamycin-sensitive TOR signaling negatively regulates the cell integrity pathway (Ai et al. 2002; Krause and Gray 2002; Torres et al. 2002). This pathway consists of the WSC family (including MID2) of cell surface sensors, the RHO1-guanine nucleotide exchange factor (GEF) ROM2, the small GTPase RHO1, the protein kinase C homolog PKC1, and a PKC1-controlled MAP kinase cascade composed of BCK1, MKK1 and 2, and MPK1/SLT2 (Heinisch et al. 1999; Beck et al. 2001). This pathway is important in a number of biological processes, including cell-wall synthesis and actin organization (Heinisch et al. 1999). The level at which rapamycin-sensitive TOR signaling impinges on the cell-integrity pathway is unknown. A current hypothesis proposes that inhibition of TOR by rapamycin leads to plasma membrane stress that is recognized by WSC1 and MID2, which in turn activates the cell-integrity pathway (Torres et al. 2002). However, TOR may repress the cell-integrity pathway more directly, as a PKC1-dependent thickening of the cell wall is a normal physiological response to starvation and is required for viability of cells in G_0. It is also important to note that rapamycin-insensitive TOR2 signaling controls the organization of the actin cytoskeleton via activation of the cell integrity pathway (see below).

EFFECTOR PATHWAYS IN THE RAPAMYCIN-SENSITIVE BRANCH

The rapamycin-sensitive TOR signaling branch, which mediates the temporal control of cell growth, comprises several seemingly separate signaling pathways, as the various readouts that constitute this branch are controlled by different TOR effectors. The best-characterized effector pathway of this signaling branch involves the negative regulation of type 2A phosphatase (PP2A) and the type 2A-related phosphatase SIT4.

Control of Phosphatases by TOR

PP2A, which acts on phosphoserine or phosphothreonine, is a trimer consisting of a catalytic subunit (C) and two regulatory subunits (A and B). The catalytic subunit has multiple targets with substrate specificity and subcellular localization determined by the associated regulatory subunits. In yeast, the catalytic subunit PPH22 or PPH21 associates with the A subunit TPD3 and one of the two B subunits CDC55 and RTS1. The PP2A-related phosphatase, which is also specific for phosphoserine and phosphothreonine, is a dimer consisting of the catalytic subunit SIT4 and one of the four positive regulatory subunits SAP4, SAP155, SAP185, and SAP190.

TOR negatively controls PP2A and SIT4 activity by promoting the binding of the essential, conserved protein TAP42 to PPH21/22 and SIT4, thereby displacing the regulatory subunits of each phosphatase (Di Como and Arndt 1996). TOR may promote the binding of TAP42 to PPH21/22 by phosphorylating TAP42 directly (Jiang and Broach 1999). However, the mechanism by which TOR controls the binding of TAP42 to SIT4 appears to be independent of TAP42 phosphorylation (Jacinto et al. 2001). Rather, TOR controls the binding of TAP42 to SIT4 via the TAP42-binding protein TIP41. TIP41 binds and releases TAP42 from SIT4, allowing SIT4, in turn, to bind its regulatory subunits. TOR controls TIP41 by maintaining it in a phosphorylated state. Phosphorylated TIP41 is unable to bind TAP42. Thus, TOR promotes TAP42 binding to SIT4 by antagonizing the TAP42 inhibitor TIP41. Interestingly, TIP41 is dephosphorylated by SIT4, and is therefore part of a feedback loop. The role of this feedback loop is to rapidly amplify SIT4 activity upon TOR-inactivating conditions. The rapid activation of SIT4 allows cells to adjust quickly to nutrient limitation or other types of TOR-responsive stress.

Which readouts does TOR control via PP2A and SIT4? It remains to be determined which TOR readouts are controlled via PP2A. However, a *sit4* mutation prevents rapamycin-induced dephosphorylation and activation of the transcription factor GLN3, indicating that TOR negatively controls GLN3 by inhibiting SIT4 (Beck and Hall 1999). Similarly, a *sit4* mutation prevents rapamycin-induced dephosphorylation of the kinase

NPR1, suggesting that TOR controls amino acid permease sorting via SIT4 inhibition (Jacinto et al. 2001). A *sit4* mutation also prevents, at least partly, rapamycin-induced dephosphorylation of GCN2 (Cherkasova and Hinnebusch 2003). Finally, a *sit4* mutation also prevents TOR-mediated repression of the cell-integrity pathway (Torres et al. 2002). This last finding is surprising because it suggests that SIT4 is active, rather than inactive, when TOR is active. Although TOR-mediated binding of TAP42 to SIT4 inhibits SIT4 toward GLN3 and NPR1, TAP42 may activate SIT4 toward a so-far-unknown substrate in the cell-integrity pathway. Thus, TAP42 may determine the target specificity of SIT4 rather than act only as a negative regulator of this phosphatase. This model is consistent with the finding that only a small fraction of phosphatase catalytic subunit is bound by TAP42 (Di Como and Arndt 1996).

Several TOR readouts have been shown to be SIT4-independent. A *sit4* mutation does not affect rapamycin-induced activation of the RTG1/3 transcription factor (Liu et al. 2001) or the general stress transcription factors MSN2 and MSN4 (Beck and Hall 1999). Furthermore, a *sit4* mutation does not alter rapamycin-induced down-regulation of genes encoding ribosomal proteins, rRNA (Pol I), or tRNA (Pol III), suggesting that TOR controls ribosome biogenesis independently of SIT4 (Schmelzle et al. 2004). What is the effector pathway of the rapamycin-sensitive TOR signaling branch that controls RTG1/3, MSN2/4, and ribosomal biogenesis? Recent evidence suggests that at least MSN2/4 and ribosome biogenesis are controlled via the RAS/cAMP pathway, which may constitute a new TOR effector pathway in the rapamycin-sensitive branch (Schmelzle et al. 2004). RTG1/3 may be controlled by yet another unknown TOR effector pathway.

TOR and the RAS/cAMP Pathway

The RAS/cAMP pathway, like the TOR pathway, controls cell growth in response to nutrients. The core of this pathway consists of the redundant, small GTPases RAS1 and RAS2, the adenylate cyclase CDC35, the second messenger cAMP, and the cAMP-dependent protein kinase PKA. In yeast, the cAMP-sensitive regulatory subunit of PKA is BCY1. The PKA catalytic subunit is TPK1, TPK2, or TPK3, each encoded by a separate, redundant gene.

TOR may control rapamycin-sensitive, SIT4-independent readouts via the RAS/cAMP pathway (Schmelzle et al. 2004). First, several of the well-characterized readouts of the RAS/cAMP pathway, including ribosome biogenesis (see above) and nuclear localization of the general stress transcription factor MSN2 (see above), are rapamycin-sensitive and SIT4-independent. Conversely, rapamycin-sensitive readouts that are not affected by the RAS/cAMP pathway are SIT4-dependent. Second, constitutive activation of the RAS/cAMP pathway renders ribosome biogenesis and MSN2 localization rapamycin-insensitive. Third, starvation or rapamycin treatment results in nuclear accumulation of the PKA catalytic subunit TPK1 (Griffioen et al. 2000; Schmelzle et al. 2004). Thus, the RAS/cAMP pathway may be a distinct TOR effector pathway. The mechanism by which TOR impinges on the RAS/cAMP pathway is unknown.

It is also worth noting that the RAS/cAMP pathway has been proposed to control the MSNs in parallel to TOR (Görner et al. 1998, 2002). Görner et al. (2002) suggest that the RAS/cAMP pathway inhibits MSN nuclear import, whereas TOR stimulates MSN nuclear export. According to this model, RAS/cAMP and TOR control the MSNs independently. The nature of the relationship between the RAS/cAMP and TOR signaling pathways in the control of the MSNs remains to be determined.

TOR may control autophagy via both SIT4 and the RAS/cAMP pathway. Constitutive activation of the RAS/cAMP pathway or addition of exogenous cAMP renders autophagy largely rapamycin-insensitive (Noda and Ohsumi 1998; Schmelzle et al. 2004). However, although Kamada et al. (2000) report that the induction of autophagy is not controlled by TAP42 or GLN3, a *sit4* mutation partly reduces autophagy (Schmelzle et al. 2004). A joint role for SIT4 and RAS/cAMP in the control of autophagy is reasonable, as autophagy is a complex process that involves many events which ultimately lead to the assembly and trafficking of large membranous structures.

RAPAMYCIN-INSENSITIVE TOR2 SIGNALING BRANCH MEDIATES SPATIAL CONTROL OF CELL GROWTH

Growth of *S. cerevisiae* cells is also spatially controlled. Growth of a daughter cell occurs at a discrete site (the bud site) on the surface of a mother cell. This polarized growth pattern is directed by a polarization of the underlying actin cytoskeleton. Actin cables orient toward, and actin cortical patches concentrate in, the growth site (Chant 1996; Madden and Snyder 1998). This polarization of the actin cytoskeleton orients the secretory pathway and thus targets newly synthesized proteins and other cellular constituents to the bud.

The rapamycin-insensitive TOR2 signaling branch mediates spatial control of cell growth by regulating polarization of the actin cytoskeleton

(Fig. 2). Depletion of TOR2 results in a depolarization of the actin cytoskeleton and loss of viability (Schmidt et al. 1996, 1997; Helliwell et al. 1998a). These phenotypes can be suppressed by overexpression of components of the cell integrity pathway, including ROM2, RHO1, PKC1, BCK1, MKK1/2, and MPK1 (Schmidt et al. 1996, 1997; Bickle et al. 1998; Helliwell et al. 1998a,b). Furthermore, GDP/GTP exchange activity toward RHO1 is reduced in a *tor2* mutant (Schmidt et al. 1997). These findings suggest that TOR2 activates the RHO1 GDP/GTP exchange factor ROM2. Activated ROM2 in turn converts RHO1 to an active, GTP-bound state. GTP-bound RHO1 binds and activates PKC1, which then signals to the actin cytoskeleton via the MAP kinase cascade composed of BCK1, MKK1, and MPK1/SLT2 (Schmidt et al. 1997; Helliwell et al. 1998b). The mechanism by which TOR2 activates ROM2 is unknown.

It is important to note that the rapamycin-sensitive TOR signaling branch represses the cell-integrity pathway, to prevent excessive buildup of the cell wall during vegetative growth (see above). How the rapamycin-sensitive TOR signaling branch represses the cell-integrity pathway while the rapamycin-insensitive TOR2 signaling branch activates the cell-integrity pathway to polarize the actin cytoskeleton is not known. A mechanism for this inverse regulation of the cell-integrity pathway by TOR2 could involve differential compartmentalization of the relevant signaling components.

TORC1 AND TORC2: A DISTINCT TOR COMPLEX FOR EACH SIGNALING BRANCH

The rapamycin-sensitive branch contains both TOR1 and TOR2, whereas the rapamycin-insensitive branch contains only TOR2. What determines the signaling specificity of the TORs? In other words, why is TOR2 more versatile than TOR1 and able to function in both signaling branches? Furthermore, why is TOR2 rapamycin-sensitive only in one of the two branches? The answers to these questions were provided by the identification of two functionally and structurally distinct TOR complexes, TORC1 and TORC2 (Loewith et al. 2002; Wedaman et al. 2003).

Modular Structure of TOR

The first evidence that the TORs assemble into complexes was provided by the sequence of TOR. TOR proteins are large, ~280 kD, and contain a number of known or putative protein–protein interaction domains (Fig. 3). The amino-terminal half of TOR contains ≥ 20 tandemly repeated

HEAT motifs (named after the four proteins in which this motif was originally identified, *H*untingtin, *e*longation factor 3, the *A* subunit of PP2A, and *TOR*) (Hemmings et al. 1990; Andrade and Bork 1995; Perry and Kleckner 2003). HEAT motifs are composed of ~40 amino acids that form antiparallel α helices (Groves and Barford 1999). HEAT repeats mediate protein–protein interactions in multiprotein complexes and are required for the proper subcellular localization of TOR (Kunz et al. 2000). Following the HEAT domain is a so-called FAT domain that has been proposed to serve as a scaffold (Alarcon et al. 1999; Bosotti et al. 2000). FAT domains, found in all PIKKs, are accompanied by a small FATC domain found at the extreme carboxyl terminus of the PIKKs. Following the FAT domain is the FKBP-rapamycin-binding domain (FRB). As mentioned above (see page 3), specific mutations in this region prevent the binding of the FKBP–rapamycin complex and confer complete resistance to rapamycin.

TORC1

TOR complex 1 (TORC1) contains KOG1, LST8, and either TOR1 or TOR2. KOG1, LST8, and, indeed, TORC1 are conserved from yeast to human (Hara et al. 2002; Kim et al. 2002; Loewith et al. 2002; Wedaman et al. 2003). KOG1 (*k*ontroller *o*f *g*rowth) is an essential protein. KOG1 depletion mimics many, if not all, the phenotypes observed in rapamycin-treated cells, including a pronounced drop in translation and derepression of GLN3, RTG1/3, and MSN2/MSN4-regulated genes (Loewith et al. 2002). KOG1 contains a number of protein–protein interaction domains (4 HEAT repeats and 7 WD40 repeats). Characterization of the mammalian ortholog of KOG1 (known as raptor) suggests it performs a scaffolding function coupling TOR to its substrates (Hara et al. 2002). LST8 is

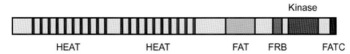

Figure 3. Schematic representation of the domains found in a generic TOR molecule (see text). TOR1 (2,470 amino acids) and TOR2 (2,473 amino acids) are 67% identical. (HEAT) Domain found in *H*untingtin, *e*longation factor 3, the *A* subunit of type 2A protein phosphatase (PP2A), and *TOR*; FAT and FATC are found in all PIKK family members; (FRB) FKBP12-rapamycin-binding domain; (Kinase) protein kinase catalytic domain.

also an essential protein; however, depletion of LST8 recapitulates only a subset of phenotypes observed in rapamycin-treated cells (Roberg et al. 1997; Liu et al. 2001; Loewith et al. 2002). LST8 is a Gβ-like protein consisting essentially entirely of WD40 repeats. LST8 and its mammalian ortholog (mLST8, also known as GβL) appear to have a positive role in TORC1 and mTORC1, respectively (Loewith et al. 2002; Chen and Kaiser 2003; Kim et al. 2003). However, LST8 is more abundant than the TORs or KOG1, and a pool of LST8 biochemically fractionates separately from the TORs (R. Loewith et al., unpubl.), suggesting that LST8 performs additional, TOR-independent functions. The integrity of TORC1 is not affected by rapamycin treatment or nitrogen starvation. TORC1 associates with FPR1 in the presence of rapamycin (Loewith et al. 2002). The findings that TORC1 contains either TOR1 or TOR2 and binds FPR1-rapamycin, and that TORC1 disruption mimics rapamycin treatment, suggest that TORC1 mediates the rapamycin-sensitive TOR signaling branch.

TOR1, TOR2, and LST8, and thus TORC1 (and possibly TORC2), are peripherally associated with the plasma membrane and with an intracellular membranous structure of unknown identity, as determined by subcellular fractionation, immunogold electron microscopy, or indirect immunofluorescence (Kunz et al. 2000; Wedaman et al. 2003). The functional significance of the double localization of TOR is not known. The localization of TOR is not affected by rapamycin or starvation.

TORC2

TOR complex 2 (TORC2) contains TOR2 (but not TOR1), AVO1, AVO2, AVO3, and LST8 (Loewith et al. 2002; Wedaman, et al. 2003). AVO1 (*a*dheres *vo*raciously to TOR2) is an essential protein that contains a putative Ras-binding domain, but whose molecular function is unknown. AVO2 is the only nonessential protein among the subunits of the TOR complexes. AVO2 contains four ankyrin repeats, but no molecular function is known. Indeed, no phenotypes are known for *avo2* cells. AVO3 is an essential protein that contains a RasGEFN domain. RasGEFN domains are found amino-terminal to the catalytic GDP/GTP exchange domain in some Ras-family GEFs. The molecular function of AVO3 is unknown, but the RasGEFN domain predicts that AVO3 may couple TOR2 to downstream effectors such as the GEF ROM2 (see above). Depletion of LST8, AVO1, or AVO3, like depletion of TOR2, results in depolarization of the actin cytoskeleton (Loewith et al. 2002; R. Shioda and M.N. Hall. The integrity of TORC2 is not affected by rapamycin treatment or nitrogen starvation. Furthermore, TORC2 is not bound by FPR1–rapamycin. The findings that TORC2 contains TOR2 (but not TOR1) and is not bound by FPR1-rapamycin, and that TORC2 disruption causes an actin defect, suggest that TORC2 mediates the rapamycin-insensitive TOR2 signaling branch. It is not yet clear whether the AVOs, TORC2, or even a rapamycin-insensitive signaling branch is conserved in mammals (Loewith et al. 2002).

Both AVO3 (also known as TSC11) and TOR2 were identified in a selection for temperature-sensitive suppressors of the calcium sensitivity of *csg2* mutants (*tsc* alleles; Beeler et al. 1998). CSG2 is required for the mannosylation of inositolphosphoceramide, and *csg2* mutants accumulate high levels of sphingolipid intermediates. Most *tsc* mutations are in sphingolipid synthesis genes, suggesting that the accumulation of sphingolipid intermediates renders cells sensitive to Ca^{++}. These findings suggest that TORC2 and sphingolipids may be functionally linked. *avo3* and *tor2* mutants have little to no alteration in sphingolipid levels, suggesting that TORC2 may function downstream or parallel to sphingolipid signaling (R. Shioda et al., unpubl.). The previous findings that overexpression of the PI4P 5-kinase MSS4 or phospholipase C (PLC1) suppresses a *tor2* mutation (Helliwell et al. 1998a), and that TOR itself is related to lipid kinases, also suggest a functional link between TOR2 and a lipid second messenger.

CONCLUSIONS

A decade of genetic studies demonstrated that TOR regulates both temporal and spatial aspects of cell growth via different signaling branches. Biochemical studies that led to the identification of two distinct TOR complexes have now provided a molecular basis for these signaling branches. Future studies will define the mechanisms by which the TORs sense and signal nutrient levels to regulate cell growth. The effector pathways or mechanisms by which the TORs control a number of their readouts, including mRNA stability, ribosome biogenesis, and RTG- and MSN-dependent transcription, also remain to be determined.

Like TOR in yeast, TOR in mammalian cells also controls cell growth in response to nutrients (Jacinto and Hall 2003). Indeed, the TORC1 (and possibly TORC2) signaling branch(es) appears to be part of a primordial signaling network that has been conserved throughout eukaryotic evolution to control the fundamental process of cell growth. Thus, future studies in yeast will continue to contribute to understanding cell growth control in other eukaryotes.

Rapamycin is a drug used in four therapeutic areas: transplantation, cardiovascular disease, rheumatoid arthritis, and cancer (Abraham and Wiederrecht 1996; Marx and Marks 2001; Huang and Houghton 2002). Future studies on yeast TOR and the conserved TOR partner proteins may lead to further understanding of rapamycin action and to the development of new clinically relevant drugs.

ACKNOWLEDGMENTS

We thank Jose-Luis Crespo, Kelly Tatchell, and Tobias Schmelzle for critical reading of the manuscript. R.L. was the recipient of a long-term European Molecular Biology Organization (EMBO) Fellowship. We acknowledge support from the Canton of Basel and the Swiss National Science Foundation (M.N.H.).

REFERENCES

Abeliovich H. and Klionsky D.J. 2001. Autophagy in yeast: Mechanistic insights and physiological function. *Microbiol. Mol. Biol. Rev.* **65:** 463–479.

Abraham R.T. and Wiederrecht G.J. 1996. Immunopharmacology of rapamycin. *Annu. Rev. Immunol.* **14:** 483–510.

Ai W., Bertram P.G., Tsang C.K., Chan T.F., and Zheng X.F. 2002. Regulation of subtelomeric silencing during stress response. *Mol. Cell* **10:** 1295–1305.

Alarcon C.M., Heitman J., and Cardenas M.E. 1999. Protein kinase activity and identification of a toxic effector domain of the target of rapamycin TOR proteins in yeast. *Mol. Biol. Cell* **10:** 2531–2546.

Albig A.R. and Decker C.J. 2001. The target of rapamycin signaling pathway regulates mRNA turnover in the yeast *Saccharomyces cerevisiae*. *Mol. Biol. Cell* **12:** 3428–3438.

Andrade M.A. and Bork P. 1995. HEAT repeats in the Huntington's disease protein. *Nat. Genet.* **11:** 115–116.

Barbet N.C., Schneider U., Helliwell S.B., Stansfield I., Tuite M.F., and Hall M.N. 1996. TOR controls translation initiation and early G1 progression in yeast. *Mol. Biol. Cell* **7:** 25–42.

Beck T. and Hall M.N. 1999. The TOR signalling pathway controls nuclear localization of nutrient-regulated transcription factors. *Nature* **402:** 689–692.

Beck T., Delley P.A., and Hall M.N. 2001. Control of the actin cytoskeleton by extracellular signals. *Results Probl. Cell Differ.* **32:** 231–262.

Beck T., Schmidt A., and Hall M.N. 1999. Starvation induces vacuolar targeting and degradation of the tryptophan permease in yeast. *J. Cell Biol.* **146:** 1227–1238.

Beeler T., Bacikova D., Gable K., Hopkins L., Johnson C., Slife H., and Dunn T. 1998. The *Saccharomyces cerevisiae* TSC10/YBR265w gene encoding 3-ketosphinganine reductase is identified in a screen for temperature-sensitive suppressors of the Ca2+-sensitive csg2Delta mutant. *J. Biol. Chem.* **273:** 30688–30694.

Berset C., Trachsel H., and Altmann M. 1998. The TOR (target of rapamycin) signal transduction pathway regulates the stability of translation initiation factor eIF4G in the yeast *Saccharomyces cerevisiae. Proc. Natl. Acad. Sci.* **95:** 4264–4269.

Bertram P.G., Zeng C., Thorson J., Shaw A.S., and Zheng X.F. 1998. The 14-3-3 proteins positively regulate rapamycin-sensitive signaling. *Curr. Biol.* **8:** 1259–1267.

Bickle M., Delley P.A., Schmidt A., and Hall M.N. 1998. Cell wall integrity modulates RHO1 activity via the exchange factor ROM2. *EMBO J.* **17:** 2235–2245.

Bosotti R., Isacchi A., and Sonnhammer E.L. 2000. FAT: A novel domain in PIK-related kinases. *Trends Biochem. Sci.* **25:** 225–227.

Brown E.J., Albers M.W., Shin T.B., Ichikawa K., Keith C.T., Lane W.S., and Schreiber S.L. 1994. A mammalian protein targeted by G1-arresting rapamycin-receptor complex. *Nature* **369:** 756–758.

Cafferkey R., Young P.R., McLaughlin M.M., Bergsma D.J., Koltin Y., Sathe G.M., Faucette L., Eng W.K., Johnson R.K., and Livi G.P. 1993. Dominant missense mutations in a novel yeast protein related to mammalian phosphatidylinositol 3-kinase and VPS34 abrogate rapamycin cytotoxicity. *Mol. Cell. Biol.* **13:** 6012–6023.

Cardenas M.E., Cutler N.S., Lorenz M.C., Di Como C.J., and Heitman J. 1999. The TOR signaling cascade regulates gene expression in response to nutrients. *Genes Dev.* **13:** 3271–3279.

Chant J. 1996. Generation of cell polarity in yeast. *Curr. Opin. Cell Biol.* **8:** 557–565.

Chen E.J. and Kaiser C.A. 2002. Amino acids regulate the intracellular trafficking of the general amino acid permease of *Saccharomycescerevisiae. Proc. Natl. Acad. Sci.* **99:** 14837–14842.

———. 2003. LST8 negatively regulates amino acid biosynthesis as a component of the TOR pathway. *J. Cell Biol.* **161:** 333–347.

Chen J., Zheng X.F., Brown E.J., and Schreiber S.L. 1995. Identification of an 11-kDa FKBP12-rapamycin-binding domain within the 289-kDa FKBP12-rapamycin-associated protein and characterization of a critical serine residue. *Proc. Natl. Acad. Sci.* **92:** 4947–4951.

Cherkasova V.A. and Hinnebusch A.G. 2003. Translational control by TOR and TAP42 through dephosphorylation of eIF2{alpha} kinase GCN2. *Genes Dev.* **17:** 859–872.

Choi J., Chen J., Schreiber S.L., and Clardy J. 1996. Structure of the FKBP12-rapamycin complex interacting with the binding domain of human FRAP. *Science* **273:** 239–242.

Cosentino G.P., Schmelzle T., Haghighat A., Helliwell S.B., Hall M.N., and Sonenberg N. 2000. Eap1p, a novel eukaryotic translation initiation factor 4E-associated protein in *Saccharomyces cerevisiae. Mol. Cell. Biol.* **20:** 4604–4613.

Crespo J.L. and Hall M.N. 2002. Elucidating TOR signaling and rapamycin action: Lessons from *Saccharomyces cerevisiae. Microbiol. Mol. Biol. Rev.* **66:** 579–591.

Crespo J.L., Daicho K., Ushimaru T., and Hall M.N. 2001. The GATA transcription factors GLN3 and GAT1 link TOR to salt stress in *Saccharomyces cerevisiae. J. Biol. Chem.* **276:** 34441–34444.

Crespo J.L., Powers T., Fowler B., and Hall M.N. 2002. The TOR-controlled transcription activators GLN3, RTG1, and RTG3 are regulated in response to intracellular levels of glutamine. *Proc. Natl. Acad. Sci.* **99:** 6784–6789.

Czaplinski K., Ruiz-Echevarria M.J., Gonzalez C.I., and Peltz S.W. 1999. Should we kill the messenger? The role of the surveillance complex in translation termination and mRNA turnover. *Bioessays* **21:** 685–696.

Damelin M., Simon I., Moy T.I., Wilson B., Komili S., Tempst P., Roth F.P., Young R.A., Cairns B.R., and Silver P.A. 2002. The genome-wide localization of Rsc9, a component of the RSC chromatin-remodeling complex, changes in response to stress. *Mol. Cell* **9:**

563–573.

Danaie P., Altmann M., Hall M.N., Trachsel H., and Helliwell S.B. 1999. CLN3 expression is sufficient to restore G1-to-S-phase progression in *Saccharomyces cerevisiae* mutants defective in translation initiation factor eIF4E. *Biochem J.* **340:** 135–141.

Decker C.J. 1998. The exosome: A versatile RNA processing machine. *Curr. Biol.* **8:** R238–R240.

De Craene J.O., Soetens O., and Andre B. 2001. The Npr1 kinase controls biosynthetic and endocytic sorting of the yeast Gap1 permease. *J. Biol. Chem.* **276:** 43939–43948.

Di Como C.J. and Arndt K.T. 1996. Nutrients, via the Tor proteins, stimulate the association of Tap42 with type 2A phosphatases. *Genes Dev.* **10:** 1904–1916.

Dilova I., Chen C.Y., and Powers T. 2002. Mks1 in concert with TOR signaling negatively regulates RTG target gene expression in *S. cerevisiae. Curr. Biol.* **12:** 389–395.

Elit L. 2002. CCI-779 Wyeth. *Curr. Opin. Investig. Drugs* **3:** 1249–1253.

Garami A., Zwartkruis F.J., Nobukuni T., Joaquin M., Roccio M., Stocker H., Kozma S.C., Hafen E., Bos J.L., and Thomas G. 2003. Insulin activation of Rheb, a mediator of mTOR/S6K/4E-BP signaling, is inhibited by TSC1 and 2. *Mol. Cell* **11:** 1457–1466.

Gasch A.P. and Werner-Washburne M. 2002. The genomics of yeast responses to environmental stress and starvation. *Funct. Integr. Genomics* **2:** 181–192.

Gingras A.C., Raught B., and Sonenberg N. 2001. Regulation of translation initiation by FRAP/mTOR. *Genes Dev.* **15:** 807–826.

Görner W., Durchschlag E., Wolf J., Brown E.L., Ammerer G., Ruis H., and Schuller C. 2002. Acute glucose starvation activates the nuclear localization signal of a stress-specific yeast transcription factor. *EMBO J.* **21:** 135–144.

Görner W., Durchschlag E., Martinez-Pastor M.T., Estruch F., Ammerer G., Hamilton B., Ruis H., and Schuller C. 1998. Nuclear localization of the C2H2 zinc finger protein Msn2p is regulated by stress and protein kinase A activity. *Genes Dev.* **12:** 586–597.

Griffioen G., Anghileri P., Imre E., Baroni M.D., and Ruis H. 2000. Nutritional control of nucleocytoplasmic localization of cAMP-dependent protein kinase catalytic and regulatory subunits in *Saccharomyces cerevisiae. J. Biol. Chem.* **275:** 1449–1456.

Groves M.R. and Barford D. 1999. Topological characteristics of helical repeat proteins. *Curr. Opin. Struct. Biol.* **9:** 383–389.

Guba M., von Breitenbuch P., Steinbauer M., Koehl G., Flegel S., Hornung M., Bruns C.J., Zuelke C., Farkas S., Anthuber M., Jauch K.W., and Geissler E.K. 2002. Rapamycin inhibits primary and metastatic tumor growth by antiangiogenesis: Involvement of vascular endothelial growth factor. *Nat. Med.* **8:** 128–135.

Hara K., Maruki Y., Long X., Yoshino K., Oshiro N., Hidayat S., Tokunaga C., Avruch J., and Yonezawa K. 2002. Raptor, a binding partner of target of rapamycin (TOR), mediates TOR action. *Cell* **110:** 177–189.

Harding T.M., Morano K.A., Scott S.V., and Klionsky D.J. 1995. Isolation and characterization of yeast mutants in the cytoplasm to vacuole protein targeting pathway. *J. Cell Biol.* **131:** 591–602.

Hardwick J.S., Kuruvilla F.G., Tong J.K., Shamji A.F., and Schreiber S.L. 1999. Rapamycin-modulated transcription defines the subset of nutrient-sensitive signaling pathways directly controlled by the Tor proteins. *Proc. Natl. Acad. Sci.* **96:** 14866–14870.

Heinisch J.J., Lorberg A., Schmitz H.P., and Jacoby J.J. 1999. The protein kinase C-mediated MAP kinase pathway involved in the maintenance of cellular integrity in *Saccharomyces cerevisiae. Mol. Microbiol.* **32:** 671–680.

Heitman J., Movva N.R., and Hall M.N. 1991a. Targets for cell cycle arrest by the immunosuppressant rapamycin in yeast. *Science* **253:** 905–909.

Heitman J., Movva N.R., Hiestand P.C., and Hall M.N. 1991b. FK 506-binding protein proline rotamase is a target for the immunosuppressive agent FK 506 in *Saccharomyces cerevisiae. Proc. Natl. Acad. Sci.* **88:** 1948–1952.

Helliwell S.B., Howald I., Barbet N., and Hall M.N. 1998a. TOR2 is part of two related signaling pathways coordinating cell growth in *Saccharomyces cerevisiae. Genetics* **148:** 99–112.

Helliwell S.B., Schmidt A., Ohya Y., and Hall M.N. 1998b. The Rho1 effector Pkc1, but not Bni1, mediates signalling from Tor2 to the actin cytoskeleton. *Curr. Biol.* **8:** 1211–1214.

Helliwell S.B., Wagner P., Kunz J., Deuter-Reinhard M., Henriquez R., and Hall M.N. 1994. TOR1 and TOR2 are structurally and functionally similar but not identical phosphatidylinositol kinase homologues in yeast. *Mol. Biol. Cell* **5:** 105–118.

Hemmings B.A., Adams-Pearson C., Maurer F., Muller P., Goris J., Merlevede W., Hofsteenge J., and Stone S.R. 1990. alpha- and beta-forms of the 65-kDa subunit of protein phosphatase 2A have a similar 39 amino acid repeating structure. *Biochemistry* **29:** 3166–3173.

Hershey J.W.B. and Merrick W.C. 2000. The pathway and mechanisn of initiation of protein synthesis. In *Translational control of gene expression* (ed. N. Sonenberg et al.), pp. 33-88. Cold Spring Harbor Laboratory Press, Cold Spring Harbor, New York.

Hidalgo M. and Rowinsky E.K. 2000. The rapamycin-sensitive signal transduction pathway as a target for cancer therapy. *Oncogene* **19:** 6680–6686.

Hinnebusch A.G. 2000. Mechanism and regulation of initiator methionyl-tRNA binding to ribosomes. In *Translational control of gene expression* (ed. N. Sonenberg et al.), pp. 185–243. Cold Spring Harbor Laboratory Press, Cold Spring Harbor, New York.

Hosoi H., Dilling M.B., Liu L.N., Danks M.K., Shikata T., Sekulic A., Abraham R.T., Lawrence J.C., Jr., and Houghton P.J. 1998. Studies on the mechanism of resistance to rapamycin in human cancer cells. *Mol. Pharmacol.* **54:** 815–824.

Huang S. and Houghton P.J. 2002. Inhibitors of mammalian target of rapamycin as novel antitumor agents: From bench to clinic. *Curr. Opin. Investig. Drugs* **3:** 295–304.

Inoki K., Li Y., Xu T., and Guan K.L. 2003. Rheb GTPase is a direct target of TSC2 GAP activity and regulates mTOR signaling. *Genes Dev.* **17:** 1829–1834.

Jacinto E. and Hall M.N. 2003. Tor signalling in bugs, brain and brawn. *Nat. Rev. Mol. Cell. Biol.* **4:** 117–126.

Jacinto E., Guo B., Arndt K.T., Schmelzle T., and Hall M.N. 2001. TIP41 interacts with TAP42 and negatively regulates the TOR signaling pathway. *Mol. Cell* **8:** 1017–1026.

Jiang Y. and Broach J.R. 1999. Tor proteins and protein phosphatase 2A reciprocally regulate Tap42 in controlling cell growth in yeast. *EMBO J.* **18:** 2782–2792.

Kamada Y., Funakoshi T., Shintani T., Nagano K., Ohsumi M., and Ohsumi Y. 2000. Tor-mediated induction of autophagy via an Apg1 protein kinase complex. *J. Cell Biol.* **150:** 1507–1513.

Keith C.T. and Schreiber S.L. 1995. PIK-related kinases: DNA repair, recombination, and cell cycle checkpoints. *Science* **270:** 50–51.

Khalfan W. and Klionsky D. 2002. Molecular machinery required for autophagy and the cytoplasm to vacuole targeting (Cvt) pathway in *S. cerevisiae. Curr. Opin. Cell Biol.* **14:** 468–475.

Kim D.H., Sarbassov D.D., Ali S.M., King J.E., Latek R.R., Erdjument-Bromage H., Tempst P., and Sabatini D.M. 2002. mTOR interacts with raptor to form a nutrient-sensitive complex that signals to the cell growth machinery. *Cell* **110:** 163–175.

Kim D.H., Sarbassov D.D., Ali S.M., Latek R.R., Guntur K.V.P., Erdjument-Bromage H., Tempst P., and Sabatini D.M. 2003. GβL, a positive regulator of the rapamycin-sensitive pathway required for the nutrient-sensitive interaction between raptor and mTOR. *Mol. Cell* **11:** 895–904.

Koltin Y., Faucette L., Bergsma D.J., Levy M.A., Cafferkey R., Koser P.L., Johnson R.K., and Livi G.P. 1991. Rapamycin sensitivity in *Saccharomyces cerevisiae* is mediated by a peptidyl-prolyl cis-trans isomerase related to human FK506-binding protein. *Mol. Cell. Biol.* **11:** 1718–1723.

Komeili A., Wedaman K.P., O'Shea E.K., and Powers T. 2000. Mechanism of metabolic control. Target of rapamycin signaling links nitrogen quality to the activity of the Rtg1 and Rtg3 transcription factors. *J. Cell Biol.* **151:** 863–878.

Krause S.A. and Gray J.V. 2002. The protein kinase C pathway is required for viability in quiescence in *Saccharomyces cerevisiae. Curr. Biol.* **12:** 588–593.

Kunz J., Schneider U., Howald I., Schmidt A., and Hall M.N. 2000. HEAT repeats mediate plasma membrane localization of Tor2p in yeast. *J. Biol. Chem.* **275:** 37011–37020.

Kunz J., Henriquez R., Schneider U., Deuter-Reinhard M., Movva N.R., and Hall M.N. 1993. Target of rapamycin in yeast, TOR2, is an essential phosphatidylinositol kinase homolog required for G1 progression. *Cell* **73:** 585–596.

Kuruvilla F.G., Shamji A.F., and Schreiber S.L. 2001. Carbon- and nitrogen-quality signaling to translation are mediated by distinct GATA-type transcription factors. *Proc. Natl. Acad. Sci.* **98:** 7283–7288.

Liu Z., Sekito T., Epstein C.B., and Butow R.A. 2001. RTG-dependent mitochondria to nucleus signaling is negatively regulated by the seven WD-repeat protein Lst8p. *EMBO J.* **20:** 7209–7219.

Loewith R., Jacinto E., Wullschleger S., Lorberg A., Crespo J.L., Bonenfant D., Oppliger W., Jenoe P., and Hall M.N. 2002. Two TOR complexes, only one of which is rapamycin sensitive, have distinct roles in cell growth control. *Mol. Cell* **10:** 457–468.

Lorenz M.C. and Heitman J. 1995. TOR mutations confer rapamycin resistance by preventing interaction with FKBP12-rapamycin. *J. Biol. Chem.* **270:** 27531–27537.

Mach K.E., Furge K.A., and Albright C.F. 2000. Loss of Rhb1, a Rheb-related GTPase in fission yeast, causes growth arrest with a terminal phenotype similar to that caused by nitrogen starvation. *Genetics* **155:** 611–622.

Madden K. and Snyder M. 1998. Cell polarity and morphogenesis in budding yeast. *Annu. Rev. Microbiol.* **52:** 687–744.

Magasanik B. and Kaiser C.A. 2002. Nitrogen regulation in *Saccharomyces cerevisiae. Gene* **290:** 1–18.

Marx S.O. and Marks A.R. 2001. Bench to bedside: The development of rapamycin and its application to stent restenosis. *Circulation* **104:** 852–855.

McMahon L.P., Choi K.M., Lin T.A., Abraham R.T., and Lawrence J.C., Jr. 2002. The rapamycin-binding domain governs substrate selectivity by the mammalian target of rapamycin. *Mol. Cell. Biol.* **22:** 7428–7438.

Ng H.H., Robert F., Young R.A., and Struhl K. 2002. Genome-wide location and regulated recruitment of the RSC nucleosome-remodeling complex. *Genes Dev.* **16:** 806–819.

Noda T. and Ohsumi Y. 1998. Tor, a phosphatidylinositol kinase homologue, controls autophagy in yeast. *J. Biol. Chem.* **273:** 3963–3966.

Noda T., Suzuki K., and Ohsumi Y. 2002. Yeast autophagosomes: De novo formation of a membrane structure. *Trends Cell Biol.* **12:** 231–235.

Perry J. and Kleckner N. 2003. The ATRs, ATMs, and TORs are giant HEAT repeat proteins.

Cell **112:** 151–155.

Powers T. and Walter P. 1999. Regulation of ribosome biogenesis by the rapamycin-sensitive TOR-signaling pathway in *Saccharomyces cerevisiae*. *Mol. Biol. Cell* **10:** 987–1000.

Roberg K.J., Bickel S., Rowley N., and Kaiser C.A. 1997. Control of amino acid permease sorting in the late secretory pathway of *Saccharomyces cerevisiae* by SEC13, LST4, LST7 and LST8. *Genetics* **147:** 1569–1584.

Rohde J.R. and Cardenas M.E. 2003. The tor pathway regulates gene expression by linking nutrient sensing to histone acetylation. *Mol. Cell. Biol.* **23:** 629–635.

Rohde J., Heitman J., and Cardenas M.E. 2001. The TOR kinases link nutrient sensing to cell growth. *J. Biol. Chem.* **276:** 9583–9586.

Sabatini D.M., Erdjument-Bromage H., Lui M., Tempst P., and Snyder S.H. 1994. RAFT1: A mammalian protein that binds to FKBP12 in a rapamycin-dependent fashion and is homologous to yeast TORs. *Cell* **78:** 35–43.

Sabers C.J., Martin M.M., Brunn G.J., Williams J.M., Dumont F.J., Wiederrecht G., and Abraham R.T. 1995. Isolation of a protein target of the FKBP12-rapamycin complex in mammalian cells. *J. Biol. Chem.* **270:** 815–822.

Saucedo L.J., Gao X., Chiarelli D.A., Li L., Pan D., and Edgar B.A. 2003. Rheb promotes cell growth as a component of the insulin/TOR signalling network (erratum in *Nat. Cell Biol.* [2003] **5:** 680). *Nat. Cell Biol.* **5:** 566–571.

Schmelzle T. and Hall M.N. 2000. TOR, a central controller of cell growth. *Cell* **103:** 253–262.

Schmelzle T., Beck T., Martin D.E., and Hall M.N. 2004. Activation of the RAS/cAMP pathway suppresses a TOR deficiency in yeast. *Mol. Cell. Biol.* **24:** 338–351.

Schmidt A., Kunz J., and Hall M.N. 1996. TOR2 is required for organization of the actin cytoskeleton in yeast. *Proc. Natl. Acad. Sci.* **93:** 13780–13785.

Schmidt A., Bickle M., Beck T., and Hall M.N. 1997. The yeast phosphatidylinositol kinase homolog TOR2 activates RHO1 and RHO2 via the exchange factor ROM2. *Cell* **88:** 531–542.

Schmidt A., Beck T., Koller A., Kunz J., and Hall M.N. 1998. The TOR nutrient signalling pathway phosphorylates NPR1 and inhibits turnover of the tryptophan permease. *EMBO J.* **17:** 6924–6931.

Schreiber S.L. 1991. Chemistry and biology of the immunophilins and their immunosuppressive ligands. *Science* **251:** 283–287.

Sekito T., Liu Z., Thornton J., and Butow R.A. 2002. RTG-dependent mitochondria-to-nucleus signaling is regulated by Mks1 and is linked to formation of yeast prion [URE3]. *Mol. Biol. Cell* **13:** 795–804.

Shamji A.F., Kuruvilla F.G., and Schreiber S.L. 2000. Partitioning the transcriptional program induced by rapamycin among the effectors of the Tor proteins. *Curr. Biol.* **10:** 1574–1581.

Smith A., Ward M.P., and Garrett S. 1998. Yeast PKA represses Msn2p/Msn4p-dependent gene expression to regulate growth, stress response and glycogen accumulation. *EMBO J.* **17:** 3556–3564.

Sophianopoulou V. and Diallinas G. 1995. Amino acid transporters of lower eukaryotes: Regulation, structure and topogenesis. *FEMS Microbiol. Rev.* **16:** 53–75.

Stan R., McLaughlin M.M., Cafferkey R., Johnson R.K., Rosenberg M., and Livi G.P. 1994. Interaction between FKBP12-rapamycin and TOR involves a conserved serine residue. *J. Biol. Chem.* **269:** 32027–32030.

Steiger M.A. and Parker R. 2002. Analyzing mRNA decay in *Saccharomyces cerevisiae*.

Methods Enzymol. **351:** 648–660.

Stocker H., Radimerski T., Schindelholz B., Wittwer F., Belawat P., Daram P., Breuer S., Thomas G., and Hafen E. 2003. Rheb is an essential regulator of S6K in controlling cell growth in *Drosophila*. *Nat. Cell Biol.* **5:** 559–565.

Tee A.R., Manning B.D., Roux P.P., Cantley L.C., and Blenis J. 2003. Tuberous sclerosis complex gene products, Tuberin and Hamartin, control mTOR signaling by acting as a GTPase-activating protein complex toward Rheb. *Curr. Biol.* **13:** 1259–1268.

Thumm M., Egner R., Koch B., Schlumpberger M., Straub M., Veenhuis M., and Wolf D.H. 1994. Isolation of autophagocytosis mutants of *Saccharomyces cerevisiae*. *FEBS Lett.* **349:** 275–280.

Torres J., Di Como C.J., Herrero E., and De La Torre-Ruiz M.A. 2002. Regulation of the cell integrity pathway by rapamycin-sensitive Tor function in budding yeast. *J. Biol. Chem.* **277:** 43495–43504.

Tsukada M. and Ohsumi Y. 1993. Isolation and characterization of autophagy-defective mutants of *Saccharomyces cerevisiae*. *FEBS Lett.* **333:** 169–174.

Urano J., Tabancay A.P., Yang W., and Tamanoi F. 2000. The *Saccharomyces cerevisiae* Rheb G-protein is involved in regulating canavanine resistance and arginine uptake. *J. Biol. Chem.* **275:** 11198–11206.

Warner J.R. 1999. The economics of ribosome biosynthesis in yeast. *Trends Biochem. Sci.* **24:** 437–440.

Wedaman K.P., Reinke A., Anderson S., Yates J., III, McCaffery J.M., and Powers T. 2003. Tor kinases are in distinct membrane-associated protein complexes in *Saccharomyces cerevisiae*. *Mol. Biol. Cell* **14:** 1204–1220.

Wiederrecht G., Brizuela L., Elliston K., Sigal N.H., and Siekierka J.J. 1991. FKB1 encodes a nonessential FK506-binding protein in *Saccharomyces cerevisiae* and contains regions suggesting homology to the cyclophilins. *Proc. Natl. Acad. Sci.* **88:** 1029–1033.

Yang W., Tabancay A.P., Jr., Urano J., and Tamanoi F. 2001. Failure to farnesylate Rheb protein contributes to the enrichment of G0/G1 phase cells in the *Schizosaccharomyces pombe* farnesyltransferase mutant. *Mol. Microbiol.* **41:** 1339–1347.

Zaragoza D., Ghavidel A., Heitman J., and Schultz M.C. 1998. Rapamycin induces the G0 program of transcriptional repression in yeast by interfering with the TOR signaling pathway. *Mol. Cell. Biol.* **18:** 4463–4470.

Zhang Y., Gao X., Saucedo L.J., Ru B., Edgar B.A., and Pan D. 2003. Rheb is a direct target of the tuberous sclerosis tumour suppressor proteins. *Nat. Cell Biol.* **5:** 578–581.

Zheng X.F., Florentino D., Chen J., Crabtree G.R., and Schreiber S.L. 1995. TOR kinase domains are required for two distinct functions, only one of which is inhibited by rapamycin. *Cell* **82:** 121–130.

6

Growth Regulation by Insulin and TOR Signaling in *Drosophila*

Sally J. Leevers
Growth Regulation Laboratory
Cancer Research UK London Research Institute
London WC2A 3PX, United Kingdom

Ernst Hafen
Zoologisches Institut
Universität Zürich
CH-8057, Zürich, Switzerland

Growth, DEFINED AS INCREASE IN MASS, is a fundamental biological process that influences organism, tissue, and cell size. Thus, growth must be regulated to ensure that animals grow to optimal sizes and that their composite tissues are both appropriately sized and correctly proportioned. For example, if the heart were too small or the limbs too long, the circulatory system would fail. Likewise, although cell size can vary considerably, there are both minimum and maximum sizes at which individual cells will become nonviable or unable to perform their biological function. For example, neurons that are too small may fail to make the required connections, and differentiating cells that are too large may contact too many other cells and therefore receive multiple inductive signals. The observation that individuals of the same species and genotype tend to grow to a highly predictable size suggests that growth is under genetic control. However, growth and final body size are also influenced by external factors, including nutrient levels and temperature. Thus, growth is an essential process that is under genetic control and can be modulated in response to environmental influences.

Tissue growth is accompanied by increases in cell size and/or cell number and/or the accumulation of extracellular matrix (e.g., accretion during bone growth). The term cell growth has been defined as an increase in cell mass and can be used to discuss the growth of unicellular organisms or cells in culture. However, during metazoan tissue growth, individual cells within growing tissues can actually become progressively smaller (if the cells within the tissue are dividing faster than they are growing). Thus, using the term cell growth to describe this process seems inappropriate. To avoid this confusion and take into account the fact that increased tissue mass is not necessarily accompanied by increases in cell size, in this chapter we avoid using the term cell growth and instead use the term "growth" to mean "increase in mass."

The relationship between growth and cell division, and the effect that this relationship has on cell size, is both versatile and complex. Although the last 20 years have witnessed a dramatic increase in our knowledge of the molecular networks that control cell division, far less is known about the control of growth. Furthermore, our understanding of the interplay between growth and cell division at this point is in its infancy and is probably best understood in yeast. Early genetic analyses in yeast have revealed that mutations that delay the cell cycle do not impair growth, clearly demonstrating that growth can proceed in the absence of cell division (for recent reviews, see Millar 2002; Wells 2002; Weitzman 2003; Chapter 4). In contrast, mutations that delay growth also delay cell cycle progression (Nurse 1975; Johnston et al. 1977). These experiments imply that growth can influence the cell cycle and have led to the idea that yeast possesses a cell size/growth checkpoint that has to be satisfied in order for cell division to proceed. In addition, when yeast cultures are switched between media containing different nutrient levels, their growth rate adjusts accordingly and the cells rapidly alter their size (Nurse et al. 1976; Fantes and Nurse 1977). Interestingly, the range of growth rates observed upon altering nutrient levels exceeds the range of cell sizes observed. Together, these observations further support the existence of a yeast cell size/growth checkpoint that somehow couples growth and cell division. Furthermore, they suggest that the threshold of this checkpoint can be modulated by nutrient levels. Subsequent research has identified molecules involved in these checkpoints and revealed potential mechanisms that couple the checkpoints with growth via effects on translation (see Chapters 2, 4, and 10).

Growth in *Drosophila*

Not surprisingly, the relationships between growth and division are less well understood in metazoans. However, the fruit fly, *Drosophila*, has recently emerged as an excellent model system for investigating the control of growth and its potential connections with cell division. In *Drosophila*, growth occurs during the larval stage, when larvae eat more or less continuously and increase their mass 200-fold over 4 days. The bulk of the larva is made up of endoreplicating tissues (ERTs) in which growth is accompanied by DNA replication, but not cell division; thus, ERT cell size and ploidy increase dramatically during the larval period. At the end of larval development, following pupation, the ERTs are destroyed, releasing protein, lipid, and glycogen stores that are utilized during metamorphosis to generate the adult fly. The larva also contains mitotic tissues (MTs) in which growth is accompanied by cell division so the cells remain diploid. These tissues include the imaginal discs, epithelial sacs that grow extensively during the larval period, then are reorganized during metamorphosis to generate the epidermal structures of the adult fly.

Imaginal discs provide an attractive system for the analysis of growth for a number of reasons. Versatile genetic techniques can be employed to manipulate gene activity either throughout the larva, within specific imaginal discs, and within specific regions or randomly generated clones of cells in the discs. These manipulations can be induced and analyzed at specific time points, and their impact on growth assessed using various techniques. Imaginal discs and some disc-derived adult structures are relatively flat, so effects on growth can be assessed by microscopy. Furthermore, clones of cells in which gene activity has been altered can be distinguished using markers such as green fluorescent protein (GFP) in the discs, or wing hair and eye color in the adult fly. Thus, effects on clonal growth can be assessed within intact tissues in a normal biological environment. In addition, cell size and cell cycle distribution can be monitored in developing discs by flow cytometry of dissociated imaginal disc cells.

Importantly, experiments with imaginal discs have clearly shown that, as in yeast, growth can occur even when cell division is delayed (Neufeld et al. 1998). Specifically, when cell division is inhibited in clones of imaginal disc cells, although cell number is reduced, clonal growth can proceed at the normal rate, so cell size increases. Furthermore, if cell division is accelerated, even though the number of cells increases, clonal growth proceeds at the normal rate, so cell size is reduced. Thus, cell division can be modulated in the developing imaginal disc without having any effect on growth.

Intensive research over the last 5 years has led to the identification of a number of molecules that are required within the imaginal discs for their growth (for recent reviews, see Saucedo and Edgar 2002; Oldham and Hafen 2003). Again, there are parallels with yeast. When growth is reduced, by mutating these genes, cell division is also delayed. Additionally, these molecules form signaling pathways that are nutrient-responsive. Moreover, these mutations reduce growth more than cell division and thus also reduce cell size. Many of these growth regulators lie on two well-characterized signaling pathways, the insulin/phosphoinositide 3-kinase (PI3K) pathway and the Target of Rapamycin (TOR) pathway. Below, we describe the insulin/PI3K and TOR signaling pathways in *Drosophila* and potential links between these two pathways. The order of molecules on these pathways has been inferred as a result of biochemical, phenotypic, and genetic interaction analyses in *Drosophila* and by making analogies with the mammalian pathways. Once these pathways have been described, their impact on growth during *Drosophila* development is discussed in more detail, together with the potential relationship between growth and cell division in the developing imaginal disc. Then, the way in which imaginal disc growth and these signaling pathways are subject to extrinsic control by nutrition and hormones is also addressed.

DROSOPHILA INSULIN/PI3K SIGNALING—THE PATHWAY

Insulin Receptor and Insulin Receptor Substrate Proteins

The primary function of insulin signaling in mammals is to maintain blood sugar homeostasis by regulating glucose uptake and metabolism (Saltiel and Kahn 2001). In contrast, the major role performed by mammalian insulin-like growth factors 1 and 2 (IGF-1 and IGF-2) is to promote pre- and postnatal growth (Butler and Le Roith 2001). Insulin and IGFs signal via the insulin receptor and IGF receptor-I, respectively. These receptors share common downstream signaling components, and there is some evidence of cross talk between the insulin and IGF signaling systems. For example, the insulin receptor can mediate IGF-2-induced prenatal growth (Nakae et al. 2001). The mammalian insulin and IGF receptors are tyrosine kinases that are activated by ligand-induced transphosphorylation. Once activated, the receptors transduce signals downstream by phosphorylating various intracellular substrates, including insulin receptor substrate (IRS) proteins. IRS proteins bind via their phosphotyrosine-binding (PTB) domains to autophosphorylated tyrosine residues in the juxtamembrane regions of the activated receptors. Once tyrosine-phosphorylated by the insulin or IGF receptors, the IRS proteins then act as adapters that recruit and activate other signaling molecules containing phosphotyrosine-binding Src homology 2 (SH2) domains. These molecules include the Ras/MAP kinase pathway adapters, Grb2 and Shc, and the class IA PI3Ks (Zick 2001).

Drosophila possess one insulin/IGF receptor homolog, DInr (Fernandez et al. 1995; Chen et al. 1996), and seven insulin-like peptides, DILPs (Brogiolo et al. 2001). Together, these molecules cooperate to regulate both growth and carbohydrate levels in the hemolymph (the insect equivalent of blood; see below and Ikeya et al. 2002; Rulifson et al. 2002). The signaling mechanism of DInr may differ slightly from that of its mammalian homologs, as it possesses a carboxy-terminal extension adjacent to its kinase domain that contains various tyrosine phosphorylation sites. Autophosphorylation of this carboxy-terminal extension is predicted to allow DInr to recruit downstream signaling molecules directly without the need for intermediate adapter proteins. However, *Drosophila* also possess an IRS homolog, Chico, which contains a PTB domain through which Chico binds activated DInr (Bohni et al. 1999; Poltilove et al. 2000). The sequence context of putative tyrosine phosphorylation sites suggests that DInr has the potential to recruit both PI3K and the Ras pathway adapter, Shc, whereas Chico might bind to PI3K and the adapter, Drk (the *Drosophila* homolog of the Ras pathway adapter, Grb2).

Although Shc and Drk have the ability to activate Ras signaling, an increasing amount of evidence suggests that it is the activation of PI3K signaling that provides the critical growth-promoting signal downstream of DInr and Chico. For example, when DInr is activated in mammalian cells, it forms a complex with coexpressed Chico, and that complex also contains mammalian PI3K, but not mammalian Shc or Grb2 (Poltilove et al. 2000). Furthermore, the small fly phenotype of *chico* mutants is completely rescued by a version of *chico* in which the Drk-binding site is mutated (Oldham et al. 2002). In contrast, the PI3K-binding sites on Chico are essential for *chico* function in this assay.

PI3K, Its Targets, and Its Antagonist, PTEN

Mammalian class IA PI3Ks form heterodimers with SH2 domain-containing adapter subunits. *Drosophila* possess one class IA PI3K, Dp110, that is thought to be regulated in an analogous manner to the mammalian class IA PI3Ks, p110α, p110β, and p110δ (Vanhaesebroeck et al. 2001). Thus, Dp110 is found bound to an SH2 domain-containing adapter subunit, p60 (see Fig. 1) (Leevers et al. 1996; Weinkove et al. 1997; Leevers 2001). Both DInr and Chico contain tyrosine residues within the Tyr-Xxx-Xxx-Met consensus motif that, when phosphorylated, is recognized by p60's SH2 domains (Weinkove et al. 1997). Thus, Dp110/p60 heterodimers can be recruited to the plasma membrane as a result of the p60 SH2 domains binding to phosphorylated DInr and Chico. This recruitment gives Dp110 access to its phosphoinositide substrates in the plasma membrane, which it phosphorylates specifically on the 3′ OH of their inositol ring. Elevated levels of 3-phosphoinositides (primarily phosphatidylinositol 3,4,5-triphosphate [PIP3]), transduce signals downstream by binding to molecules with PIP3-binding PH domains, thereby inducing their relocalization to the plasma membrane and/or conformational changes.

In mammals, a number of signaling molecules possess PIP3-binding PH domains, and hence have the potential to be regulated by PI3K. In *Drosophila*, however, two serine/threonine kinases have emerged as key downstream targets: the *Drosophila* homologs of phosphoinositide-dependent kinase 1 (PDK1) and its substrate, Akt (also known as protein kinase B [PKB], see Fig. 1). The PH domain of mammalian PDK1 has a high affinity for 3-phosphoinositides, so PDK1 is thought to be membrane-localized by relatively low levels of PI3K activity. In contrast, the membrane localization of Akt requires higher 3-phosphoinositide levels, such that PDK1 and Akt are thought to colocalize at the membrane only when PI3K activity is elevated. Membrane-localized, PIP3-bound Akt then becomes activated as a result of (1) a conformational change induced by PIP3 binding to its PH domain and (2) dual phosphorylation events performed by PDK1 and a second unidentified kinase termed PDK2

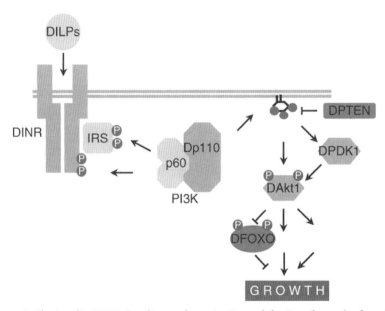

Figure 1. The insulin/PI3K signaling pathway in *Drosophila*. Details can be found in the text.

(Vanhaesebroeck and Alessi 2000). A similar mechanism is thought to operate in *Drosophila*. Consistent with this proposed mechanism, DAkt activity is reduced in larvae lacking Dp110, whereas coexpression of DPDK1 with DAkt in *Drosophila* activates DAkt (Cho et al. 2001; Radimerski et al. 2002b). Furthermore, DPDK1 and DAkt cooperate to induce growth when coexpressed (Rintelen et al. 2001).

Insulin/PI3K signaling in *Drosophila* and mammals is directly antagonized by the tumor suppressor PTEN, a lipid phosphatase that specifically removes the 3′ phosphate from the 3-phosphoinositides generated by PI3K (Goberdhan et al. 1999; Huang et al. 1999; Gao et al. 2000). The conclusion that *Drosophila* PTEN (DPTEN) directly antagonizes PI3K, and the fact that PIP3-induced DAkt activation is the major output of insulin/PI3K signaling, are supported by genetic interactions between *dPTEN* and *dAkt*. Loss of *dPTEN* is lethal and results in larger than normal larvae containing increased levels of PIP3 (Oldham et al. 2002). However, this lethality is suppressed by a mutation in the PIP3-binding PH domain of DAkt that blocks PIP3-mediated recruitment of DAkt to the plasma membrane. These double *dAkt dPTEN* mutants survive despite their elevated PIP3 levels, indicating that DAkt is the primary output of the PI3K pathway (Stocker et al. 2002).

Until recently, the targets that enable DAkt to stimulate growth in *Drosophila* have remained elusive. In mammals, numerous phosphorylation targets of Akt have been identified, many of which might be expected to play a role in promoting growth (Brazil and Hemmings 2001; Scheid and Woodgett 2001). In contrast, in the nematode worm, *Caenorhabditis elegans*, one major target of Akt has emerged, the forkhead domain transcription factor Daf-16. Daf-16 belongs to the FOXO subfamily of forkhead transcription factors (for a recent review, see Tran et al. 2003). FOXO transcription factors contain an amino-terminal forkhead DNA-binding domain and a carboxy-terminal *trans*-activation domain. The FOXO DNA-binding domain is flanked by several conserved Akt phosphorylation sites, and its function is antagonized by Akt phosphorylation (Burgering and Kops 2002). Consistent with this observation, loss of *daf-16* function suppresses all phenotypes associated with loss of insulin signaling in *C. elegans*. Mutation of insulin signaling pathway components can induce dauer larva formation and extend adult life span; both phenotypes are completely reverted by mutations in *daf-16* (Lin et al. 1997; Ogg et al. 1997). These observations suggest that, when insulin signaling is reduced, elevated Daf-16 activity plays a key role in promoting dauer larva formation and life span extension. In mammals, there are four members of the FOXO subfamily, all of which contain several conserved Akt phosphorylation sites. Akt-mediated phosphorylation of FOXO transcription factors inhibits their function by generating 14-3-3 protein-binding sites, thereby leading to the accumulation of 14-3-3-bound FOXO in the cytoplasm (Brunet et al. 1999, 2002; Burgering and Kops 2002).

In *Drosophila*, there is one member of the FOXO transcription factor subfamily, DFOXO. Mutation of the other insulin/PI3K pathway antagonist, *dPTEN*, is lethal, whereas flies homozygous for loss-of-function mutations in *dFOXO* are viable and, although they display an increased sensitivity to oxidative stress, they are normal in size. Despite the fact that mutation of *dFOXO* alone does not promote growth, DFOXO does seem to play a role in inhibiting growth when insulin/PI3K signaling is reduced. The reduced body size of *chico* and other insulin/PI3K pathway mutants is partially suppressed in a *dFOXO* mutant background (Jünger et al. 2003). Interestingly, this effect is almost exclusively due to a suppression of the reduction in cell number rather than the reduction in cell size that is observed in these flies.

The use of microarrays to search for putative target genes of DFOXO has revealed genes involved in the stress response, for example, cytochrome P450 genes, and the gene encoding the translation inhibitor, *Drosophila* 4E-BP (for eukaryotic translation initiator factor 4E (eIF4E) binding protein-1; Jünger et al 2003). Like *dFoxo* mutants, *d4E-BP* mutants do not exhibit a growth phenotype in a wild-type background (Miron et al. 2001). However, the reduced cell number phenotype of a viable *dAkt* mutant combination is partially suppressed by simultaneous mutation of *d4E-BP* (Jünger et al. 2003). Together, these observations indicate that part of the growth inhibitory effect of reduced insulin signaling is mediated by the activation of DFOXO, which accumulates in the nucleus because of reduced phosphorylation by DAkt, and activates *d4E-BP* transcription (Fig. 1). Increased D4E-BP levels then attenuate translation by inhibiting translation initiation (see below).

DROSOPHILA INSULIN/PI3K SIGNALING—ITS EFFECT ON GROWTH

The phenotypes that arise when many of the molecules described above are mutated are similar, consistent with their forming a signaling pathway that regulates growth during *Drosophila* development. For example, clones of cells with loss-of-function mutations in *dInr*, *chico*, *dp110*, *p60*, *dPDK1*, or *dAkt* are smaller than their corresponding twin-spots or sister clones, both in the imaginal discs and in the adult organs that they give rise to (Bohni et al. 1999; Verdu et al. 1999; Weinkove et al. 1999; Brogiolo et al. 2001; Rintelen et al. 2001). This simple observation demonstrates that reducing insulin/PI3K signaling during imaginal disc development reduces growth. These reductions in clonal growth are accompanied by reductions in both cell size and cell number.

When insulin/PI3K signaling is increased, for example, by overexpressing Dp110 or mutating *dPTEN* in clones of cells, clonal growth, cell size, and, in some cases, cell number increase (Leevers et al. 1996; Goberdhan et al. 1999; Huang et al. 1999). Together, these observations of cell clones clearly demonstrate that activation of the insulin/PI3K signaling pathway is required in imaginal disc cells to promote their growth, and that it is sufficient to promote overgrowth. These experiments also provide some insight into the relationship between growth and division in the developing imaginal disc, which is discussed further below.

When insulin/PI3K signaling is modulated throughout an organ or in the entire animal, similar effects on growth are observed. These observations not only reinforce the fact that this pathway promotes growth, but also demonstrate that its modulation can alter final organ and organism size. Null mutations in *dInr*, *dp110*, *p60*, *dPDK1*, or *dAkt* are lethal, whereas hypomorphic mutations in *dInr*, *dp110*, *dPDK1*, or *dAkt*, and null mutations in *chico*, result in small flies that contain small cells and fewer cells (Bohni et al. 1999; Brogiolo et al. 2001; Rintelen et al. 2001). Mutations in each of these genes slow down organismal growth during the larval stage and thus also cause developmental delay. Although development is delayed, the larvae still pupate at a smaller size than wild-type larvae, hence small flies are generated. Since the mitotic imaginal discs only make up a small part of the larvae, these observations of larval growth suggest that insulin/PI3K signaling is also required for growth of the ERT, which make up the bulk of the larva. This conclusion is further supported by the finding that clonal expression of a dominant negative inhibitory version of p60 or of wild-type DPTEN reduces growth and cell size in two ERT, the larval salivary glands and fat bodies (Britton et al. 2002).

Increasing insulin/PI3K signaling is also sufficient to promote growth at the level of the whole organism. For example, certain hypomorphic *dPTEN* mutant combinations increase final larva and fly size, and ubiquitous overexpression of the *Drosophila* insulin-like peptide DILP2 also gives rise to larger than normal flies, which contain larger cells and more cells (Brogiolo et al. 2001; Oldham et al. 2002). Together, these results demonstrate that modulating insulin/PI3K signaling during *Drosophila* development modulates growth, both in clones and in the entire animal, and that these alterations in growth are sufficient to alter final body size.

DROSOPHILA TOR SIGNALING—THE PATHWAY

TOR

TOR is a member of the phosphoinositide kinase-related kinase (PIKK) family of kinases that structurally resemble phosphoinositide kinases but possess serine/threonine rather than lipid kinase activity. TOR was first identified in budding yeast, where gain-of-function mutations in the *TOR1* or *TOR2* genes were shown to confer resistance to rapamycin, a fungal macrolide that inhibits yeast growth. Subsequent experiments in yeast, mammals, and *Drosophila* have established TOR kinases as global regulators of growth that act as nutrient sensors able to couple growth rates to available nutrient (amino acid) levels (see Gingras et al. 2001; Chapters 5 and 7). *Drosophila* possess one TOR homolog, DTOR, which contains a carboxy-terminal kinase domain, an FKBP12-rapamycin-binding domain via which rapamycin is thought to act, and several motifs implicated in protein–protein interactions, including HEAT repeats and a FAT domain (Oldham et al. 2000; Zhang et al. 2000). One way in which TOR signaling has been shown to promote growth is via its ability to stimulate translation (Gingras et al. 2001). Two targets via which DTOR might influence translation have been identified: *Drosophila* S6 kinase (DS6K; Montagne et al. 1999) and the aforementioned D4E-BP (see Fig. 2).

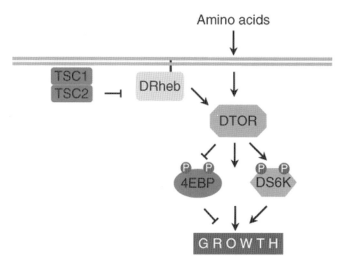

Figure 2. The TOR/S6K signaling pathway in *Drosophila*. Details can be found in the text.

TOR Targets: S6K and 4EBP

Mammalian S6K is a serine/threonine kinase that is activated by a complex series of phosphorylation events, including one controlled by TOR and one performed by PDK1 (Gingras et al. 2001). S6K activation increases the translation of a subclass of mRNAs containing 5′-terminal oligopyrimidine tracts proximal to their 5′ caps (5′TOPs). Many 5′TOP mRNAs encode components of the translational machinery such as ribosomal proteins, elongation factors, and the poly(A)-binding protein. It is not yet clear exactly how S6K activation increases 5′TOP mRNA translation, although its ability to phosphorylate ribosomal protein S6 provides one potential mechanism. Consistent with DS6K being a target of DTOR, the activity and phosphorylation of DS6K is reduced in larvae that have been fed rapamycin, starved of amino acids, or that are mutant for *dTOR* (Oldham et al. 2000; Radimerski et al. 2002b). In addition, the reduced growth of larvae containing *dTOR* hypomorphic mutations can be partially rescued by overexpression of DS6K (Zhang et al. 2000).

A second way in which TOR activation promotes translation is via the phosphorylation and inactivation of 4E-BP1, an inhibitor of eIF4E. eIF4E is a component of eIF4F—the cap-binding complex that directs the translational machinery to the 5′ end of mRNAs so that translation initiation can occur (Gingras et al. 2001). Hypophosphorylated 4E-BP1 binds eIF4E and blocks its interaction with other proteins in the eIF4F complex, thereby preventing translation initiation. Hyperphosphorylation of 4E-BP1 abrogates the 4E-BP1–eIF4E interaction, resulting in the release of eIF4E and subsequent assembly of eIF4F. TOR is one of the kinases responsible for 4E-BP1 hyperphosphorylation. *Drosophila* possess homologs of eIF4F and 4E-BP1 (Lasko 2000), and the up-regulation of translation via TOR-mediated phosphorylation of 4E-BP1 may also be conserved (Fig. 2). In *Drosophila* cells, insulin stimulates D4E-BP phosphorylation and inhibits the association of DeIF4E with D4E-BP, and this effect is inhibited by rapamycin treatment (Miron et al. 2001).

TOR Regulation by TSC and Rheb

Much as insulin/PI3K signaling is kept in check by the tumor suppressor PTEN, TOR signaling is kept in check by the tuberous sclerosis complex (TSC) tumor suppressors, TSC1 and TSC2 (Marygold and Leevers 2002; McManus and Alessi 2002). Mutations in human *TSC1* and *TSC2* are associated with heritable forms of tuberous sclerosis, a disease characterized by benign tumors (hamartomas) in the brain and other tissues

(Montagne et al. 2001; Kandt 2002). One feature of these tumors is that they contain large cells. In *Drosophila*, mutations in *dTSC1* and *dTSC2* have been identified in screens for genes that suppress growth. Loss of *dTSC1* or *dTSC2* accelerates growth and cell division, whereas coexpression of these two genes slows growth and cell cycle progression in a cell-autonomous fashion (Gao and Pan 2001; Potter et al. 2001; Tapon et al. 2001). At least part of the growth inhibitory function of the DTSC complex is dependent on DTOR and DS6K. DS6K activity is dramatically increased in the absence of DTSC function and is suppressed by the presence of excess DTSC1 and DTSC2. In addition, the ability of mutated *dTSC1* or *dTSC2* to increase growth is suppressed by mutations in *dS6K* or *dTOR*. Moreover, the lethality associated with the loss of *dTSC1* is rescued in flies heterozygous for *dTOR* and *dS6K* (Gao et al. 2002; Radimerski et al. 2002a, b). Together, these experiments indicate that one way in which DTSC1 and DTSC2 inhibit growth is by preventing DS6K activation by DTOR.

Recent biochemical and genetic evidence has identified the small GTPase, Rheb, as a TSC target likely to mediate the effect of TSC on S6K activation (Saucedo et al. 2003; Stocker et al. 2003; Zhang et al. 2003). TSC2 contains a carboxy-terminal domain that shares homology with Rap1 GTPase activating protein (GAP) domains. Mammalian TSC2 has been shown to exhibit GAP activity toward the small GTPases, Rap1, and Rab5 in vitro (Wienecke et al. 1997; Xiao et al. 1997). The functional importance of this domain is underscored by the fact that it is often mutated in TSC patients (Maheshwar et al. 1997). These observations suggest that the small GTPase that is the physiological target of TSC2 GAP activity may provide a link between TSC and TOR/S6K activation. Recent searches for loss- and gain-of-function mutations that regulate growth in *Drosophila* identified the small GTPase Rheb as an excellent candidate for this missing link (Saucedo et al. 2003; Stocker et al. 2003). Rheb, which stands for Ras homolog expressed in brain, was first identified in mammalian cells (Yamagata et al. 1994), where its biochemical characterization provided little insight into its function or its role in signaling. However, experiments in *Drosophila* have shown that DRheb might act as a positive component of the TOR signaling pathway.

Mutations in *dRheb* reduce growth and cell size, whereas *dRheb* overexpression increases growth and cell size (Saucedo et al. 2003; Stocker et al. 2003). Genetic epistasis analyses have shown that DRheb function is essential for the overgrowth produced by loss of TSC (Stocker et al. 2003). Furthermore, the fact that DRheb is an essential downstream effector of DTSC is clearly demonstrated by the observation that the lethality of *dTSC* mutants is rescued by a partial reduction in *dRheb* function (Stocker et al. 2003). Other observations place DRheb upstream of DS6K, and possibly DTOR. For example, DS6K activity, but not DAkt activity, is reduced in *dRheb* mutant larvae (Stocker et al. 2003). The hierarchical relationship between DTSC, DRheb, and DS6K is also borne out by gain-of-function experiments performed by several groups. Overexpression of DRheb increases growth and cell size and is in part due to the hyperactivation of DS6K (Patel et al. 2003; Saucedo et al. 2003; Stocker et al. 2003). Finally, work from the Pan and Edgar groups has provided direct biochemical evidence that TSC2 stimulates the GAP activity of Rheb (Zhang et al. 2003). The same relationship has also been shown for human TSC2 and human Rheb (Garami et al. 2003). Thus, the molecular links between DTSC, DRheb, DTOR, DS6K, and D4E-BP provide a second signaling pathway that regulates growth in *Drosophila*.

DROSOPHILA TOR SIGNALING—ITS EFFECT ON GROWTH

Much like many of the genes on the *Drosophila* insulin/PI3K pathway, mutation of *dTOR, dS6K,* or *dRheb* in clones of imaginal disc cells reduces clonal growth and cell size. Likewise *dRheb, dTOR,* and *dS6K* mutant larvae display slowed growth and developmental delay during the larval period (Oldham et al. 2000; Zhang et al. 2000; Saucedo et al. 2003; Stocker et al. 2003). Conversely, mutation of *dTSC1* or *dTSC2* increases growth and cell size, and their co-overexpression inhibits growth and reduces cell size (Gao and Pan 2001; Potter et al. 2001; Tapon et al. 2001). Whereas mutations of *dRheb* or *dTOR* are lethal, *dS6K* mutants are viable and, although their development is delayed, they ultimately develop into flies that are nearly 50% smaller than wild type (Montagne et al. 1999). The size of the cells in these *dS6K* mutant flies is reduced, but the number of cells in these flies is normal, suggesting that loss of DS6K activity reduces growth without affecting the total number of cell divisions that occur during development.

Interestingly, a more detailed characterization of *dTOR* mutant larval phenotypes revealed several features that are also seen in amino acid-starved larvae. For example, cells in the mutant fat body have small nucleoli and aggregated lipid vesicles (Zhang et al. 2000). In addition, ERT growth in *dTOR* and *dRheb* mutants is more severely impaired than imaginal disc growth (Oldham et al. 2000; Stocker et al. 2003). This maintenance of mitotic tissue growth at the expense of ERT growth is reminiscent of the effect of starving larvae of amino acids (Britton and Edgar 1998).

deIF4E is an essential gene, and strong loss-of-function mutations reduce larval growth and result in larval lethality (Lachance et al. 2002). In contrast, flies lacking *d4E-BP1* are fully viable and wild type in size, indicating that the ability of D4E-BP to inhibit DeIF4E is not essential, and that loss of D4E-BP is not sufficient to increase growth (Miron et al. 2001). Consistent with the inability of *d4E-BP* mutation to increase growth, overexpression of DeIF4E is also unable to promote *Drosophila* growth. These observations suggest that other signals, in addition to increased levels of free DeIF4E, must be simultaneously provided to increase growth (Lachance et al. 2002). Indeed, the activity of DeIF4E itself is positively regulated by insulin-induced phosphorylation, and substitution of an acidic residue at the critical site increases fly size (Lachance et al. 2002). It is important to note that, although loss of *d4E-BP* is not sufficient to increase growth in an otherwise wild-type genetic background, *d4E-BP* mutations do suppress the reduced body size of viable *dAkt* mutants (Jünger et al. 2003).

LINKS BETWEEN INSULIN/PI3K AND TOR/S6K SIGNALING

The data described so far have clearly shown that both insulin/PI3K and TOR signaling regulate growth during *Drosophila* development. Several lines of evidence suggest that although these two major growth-regulating pathways act in parallel, there is also considerable cross talk between them at various levels. The ability of these pathways to independently promote growth is supported by the observation that cells in clones that are double-mutant for *dPTEN* and *dTSC1* are twice the size of cells mutant for either *dPTEN* or *dTSC1* (Gao and Pan 2001). However, the observation that the lethality associated with loss of *dInr* function is rescued by a reduction in the gene dosage of either *dTSC1* or *dTSC2* also provides strong evidence for cross talk between these two pathways (Gao and Pan 2001).

To date, genetic and biochemical analyses of the interplay between these two pathways have revealed at least five different links (see Fig. 3).

1. When Akt is overexpressed, it phosphorylates TSC2 and destabilizes the TSC1–TSC2 complex, which should enhance TOR activity (Dennis et al. 2001; Inoki et al. 2002; Jaeschke et al. 2002; Manning et al. 2002; Potter et al. 2002). This observation provides a direct link between the insulin/PI3K pathway and the TOR pathway and is consistent with the fact that insulin-induced DS6K activation can be inhibited by treatment with PI3K inhibitors or double-stranded

RNA-mediated interference of *dAkt* (Lizcano et al. 2003). However, DS6K activity remains unchanged in protein extracts from *dAkt* mutant larvae, suggesting that under physiological conditions the connection between DAkt and DTSC2 may be less important in the majority of larval tissues. The ability of Akt to inhibit TSC function may, however, contribute significantly to the oncogenic potential of Akt in mammals.

2. DAkt activity is increased in *dTor* and *dS6K* mutant larvae and decreased in the absence of a functional TSC complex. These observations suggest that DS6K inhibits DAkt activity via a negative feedback loop, although the molecular mechanism for this negative regulation is unknown (Radimerski et al. 2002a,b; Lizcano et al. 2003).

3. Under physiological conditions, DPDK1 regulates both DAkt and DS6K (Rintelen et al. 2001).

4. Under severe starvation conditions, nuclear DFOXO inhibits translation at least in part by increasing D4E-BP transcription (Jünger et al. 2003). When conditions improve, the insulin/PI3K and TOR pathways can stimulate translation by disrupting the D4E-BP/-DeIF4E complex, via phosphorylation of D4E-BP, and, in parallel, by repressing DFOXO-dependent D4E-BP expression.

5. Under even more severe starvation or stress conditions, full activation of DFOXO up-regulates expression of the insulin receptor itself, thus rendering cells hypersensitive to low insulin levels (Zinke et al. 2002; Puig et al. 2003).

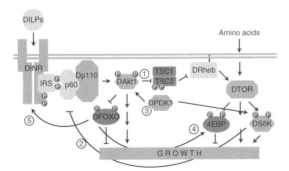

Figure 3. Potential links between insulin/PI3K signaling and TOR signaling in *Drosophila*. A diagram illustrating potential links between insulin/PI3K signaling and TOR signaling in *Drosophila*. Numbering indicates the different mechanisms referred to in the text.

Together, these multiple positive and negative interactions ensure a continuous fine adjustment of the growth rate according to the changing environmental conditions (see below).

RELATIONSHIP BETWEEN GROWTH AND CELL DIVISION IN THE DEVELOPING IMAGINAL DISC

To summarize the phenotypic data discussed above, when insulin/PI3K or TOR signaling is inhibited or activated within developing imaginal discs, growth is reduced or increased, respectively. Although these differences in growth can be accompanied by simultaneous changes in cell number, cell size is also altered, indicating that the magnitude of the effect on growth is greater than the effect on cell division. What do these observations tell us about the possible relationships between growth and cell division in the developing imaginal disc? If growth and division were completely independent processes, the fact that cell size is altered rules out the possibility that these pathways affect growth and division to the same degree, as this would maintain cell size (Fig. 4A). Instead, the altered cell size is consistent with the idea that the effect on growth predominates over the effect on cell division (Fig. 4B). However, as discussed above and elsewhere, there is evidence that growth can directly influence division. If this is the case in the disc, then the ability of insulin and TOR signaling to alter cell size argues that insulin/TOR-mediated reductions in growth do not lead directly to a similarly proportioned reduction in cell division, as this would maintain cell size (Fig. 4C). Instead, the model depicted in Figure 4D seems more likely. In this model, although the reduction in growth caused by reduced insulin/TOR signaling can slow division, the impact on cell division is less than the impact on growth; therefore, cell size is reduced.

At present, it is hard to know whether the models depicted in Figure 4B or 4D represent the situation during imaginal disc growth. Experimental evidence from other systems suggests that both insulin/PI3K signaling and TOR signaling can directly regulate both the growth machinery of the cell and molecules that control cell cycle progression (Fig. 4B) (Schmelzle and Hall 2000; Saltiel and Kahn 2001; Vanhaesebroeck et al. 2001). However, there are also mechanisms through which insulin- or TOR-stimulated growth might influence cell cycle progression (Fig. 4D; see below). If this link were to exist in imaginal discs, it would provide a mechanism to link growth with division, i.e., a growth/cell size checkpoint, and would suggest that altered levels of insulin/PI3K and TOR signaling can modulate that checkpoint.

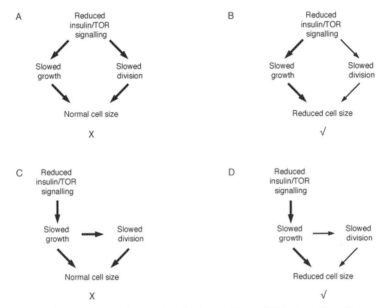

Figure 4. Models depicting the way in which insulin and TOR signaling affect growth and division in the developing imaginal disc. X = results are inconsistent with model; √ = results are consistent with model.

Imaginal disc cell cycles are regulated both at the G_1/S and G_2/M transitions. Progression through G_1/S requires expression of cyclin E, whereas G_2/M is triggered by accumulation of the cdc25 phosphatase homolog, String (Neufeld et al. 1998). What is the effect of modulating insulin/PI3K and TOR/S6k signaling on cell cycle progression? Flow cytometry of dissociated imaginal disc cells has shown that when the activity of certain pathway components is reduced, the proportion of cells in G_1 increases, suggesting that reduced insulin/TOR signaling may delay G_1/S progression (Weinkove et al. 1999). Moreover, when these pathways are activated, the G_1 population is reduced, suggesting that progression through G_1/S, as well as growth, has accelerated (Weinkove et al. 1999; Gao et al. 2000). Together these observations imply that insulin and TOR can positively influence progression through G_1/S.

There is less evidence that these pathways have a direct impact on G_2/M. For example, increasing PI3K activity for 43 hours in clones of cells in the developing wing disc is not sufficient to increase cell number (Weinkove et al. 1999). In contrast, inhibiting insulin/PI3K or TOR signaling can reduce cell number and inhibit cell division, so G_2/M must be delayed. However, this delay in G_2/M may simply reflect the slowed progression through G_1/S rather than the fact that insulin/PI3K and TOR signaling have a direct effect on G_2/M.

How might modulating insulin/PI3K or TOR signaling affect the G_1/S transition at the molecular level? If the effect were independent of growth (Fig. 4B), then one might expect these pathways to control the activity of G_1/S regulators such as cyclin E directly. However, if the effect were an indirect result of the impact of these signaling pathways on growth (Fig. 4D), it is tempting to speculate that translation is involved. There is evidence that both insulin/PI3K and TOR signaling control translation rates, and experiments in other systems have linked rates of translation initiation to progression through G_1/S. For example, in budding yeast, sufficient levels of the G_1 cyclin, Cln3 accumulate only when translation rates are high, because of 5′ upstream ORFs in its transcript. Similarly, insulin/PI3K and TOR signaling might influence progression through G_1/S via translation-mediated effects on cyclin E levels. This idea is supported by the observation that cyclin E levels are strongly reduced in dTOR mutants. Furthermore, quiescent salivary glands lacking dTOR function can be driven into S phase by cyclin E overexpression (Zhang et al. 2000).

It is interesting to note that if a mechanism exists that links growth with cell cycle progression at G_1/S rather than G_2/M, it would provide a means of controlling minimum rather than maximum cell size. Such a mechanism would link growth with cell cycle entry, and hence minimum cell size, rather than with cell division and hence maximum cell size. Consistent with the idea that minimum cell size is more tightly controlled than maximum cell size is the fact that cell size does vary quite considerably during imaginal disc development. In newly hatched *Drosophila* larvae, cell size first increases, then, as cell division kicks in, it remains fairly constant, then, toward the end of the larval period, the disc cells get smaller. These observations suggest that there is some plasticity in the relationship between growth with division during disc development.

It is also important to note that there are other experimental interventions that alter growth *without* affecting cell size (Morata and Ripoll 1975; Neufeld et al. 1998). For example, mutation of *Drosophila* ribosomal protein genes (termed *Minutes*) dominantly slows larval and imaginal disc growth, presumably because ribosome biogenesis is impaired. However, this reduction in growth is often not accompanied by reductions in cell size, implying that reducing global translation rates simultaneously reduces growth and cell division by the same degree. This scenario is more consistent with the model shown in Figure 4C and suggests that there are ways in which growth can be modulated and cell size maintained, perhaps because a growth/cell size checkpoint is left intact. Other experimental interventions can also alter imaginal disc growth without having a significant effect on cell size. For example, during the stimulation of wing imaginal disc growth by an activated version of the Dpp receptor, Thickveins, or by coexpression of cyclin D and CDK4 (Datar et al. 2000; Meyer et al. 2000; Martin-Castellanos and Edgar 2002). In these situations, there is a simultaneous increase in growth and cell number, and cell size remains normal.

Different experimental manipulations that alter growth can thus have very different effects on cell size and, hence, on the relationship between growth and division. Understanding the basis of these differences is likely to require improved knowledge of how different growth regulators promote mass increase. To date, growth has been monitored in the imaginal discs by examining parameters such as clonal area, tissue size, and body mass. Less attention has been paid to effects on the various downstream biosynthetic processes. However, it seems likely that different growth regulators promote growth by targeting different biosynthetic processes, thus, their impact on cell division and cell size may also differ.

NUTRIENT-DEPENDENT REGULATION OF INSULIN/PI3K AND TOR SIGNALING

The two main growth-regulating pathways discussed in this chapter respond to changes in nutrient levels and are likely to be the primary nutrient-responsive growth regulators in multicellular organisms. A brief summary of how nutrients regulate these two pathways is given below.

The phenotypic similarities between insulin/PI3K and TOR pathway mutants and flies that were starved during larval development hints at a connection between nutrient availability and signaling via these pathways. Both interventions reduce fly size, cell number, and cell size (Bohni et al. 1999). A link between nutrient availability and insulin/PI3K signaling has been clearly established by using a PH-domain-GFP reporter protein (tGPH), the membrane localization of which reflects PIP3 levels and thus PI3K activity in the cell (Britton et al. 2002). Overexpression of PI3K increases membrane-bound tGPH, whereas expression of dominant negative forms of the PI3K adapter p60 or starvation decreases levels of membrane-bound tGPH (Britton et al. 2002). Since insulin/PI3K signaling is controlled by the DILP ligands, it seems likely that nutrients somehow affect DILP levels. Indeed, in starved larvae, DILP3 and DILP5 transcription is reduced. *DILP3* and *DILP5* are expressed in the median neurosecretory cells (mNSC) of the larval and adult brain. These bilaterally paired clusters of seven cells project their axon terminals to the corpora cardiaca of the ring gland, the site of juvenile hormone (JH) production, and the aorta from where the DILP peptides can be released into the hemolymph (Ikeya et al. 2002; Rulifson et al. 2002). In addition to the transcriptional regulation of *DILP* expression, nutrition is also likely to modulate DILP levels in the hemolymph via effects on the release of secretory vesicles, as suggested by experiments in other insects (Masumura et al. 1997).

In insects, the growth regulatory function of DILPs is twofold. First, circulating DILPs in the hemolymph activate insulin/PI3K signaling, and hence growth, in the target tissues (Rulifson et al. 2002). This action is complemented by the local production of DILPs in the target tissues in a manner similar to the way in which mammalian IGF-1 is produced by its target tissues (Brogiolo et al. 2001). Second, DILPs exert their effect on growth indirectly. The mNSCs project their axon terminals into the ring gland where ecdysone and JH are synthesized. The stimulation of ecdysone synthesis by insulin-like peptides is well documented in many insects (Hagedorn 1985; Nagasawa 1992; Graf et al. 1997). Furthermore, JH levels are reduced in long-lived insulin receptor mutant flies, suggesting that DILPs also regulate JH synthesis (Tatar et al. 2001). DILPs may regulate growth indirectly by influencing the levels of one or both of these hormones. Under starvation conditions, reduced JH levels may result in the premature increase in ecdysone titer in third-instar larvae, which would lead to earlier initiation of metamorphosis and the production of small flies. Alternatively, starvation may induce an early rise in ecdysone titer by increasing the local concentration of DILPs in the ring gland—starvation has been shown to increase retention of insulin-like peptides in the corpora cardiaca in the silkworm, *Bombyx* (Masumura et al. 1997).

In nematodes, nutrient availability regulates the developmental program and fertility without having a direct effect on cell size or cell number. In the absence of food, the larvae enter the immature, long-lived dauer stage (Riddle 1988). This response is controlled by two signaling pathways; the insulin/PI3K pathway and the daf-4/TGFβ pathway (Finch and Ruvkun 2001). Both these pathways act non-autonomously in the nervous system, and they converge on the nuclear hormone receptor daf-12 (Koga et al. 2001). This implies an intermediate steroid or lipid hormone signal. Indeed, *daf-9*, which acts genetically between *daf-2* and *daf-4*, encodes a cytochrome P450 enzyme involved in steroid and fatty acid metabolism (Gerisch et al. 2001). Therefore, it is likely that the nutrient-dependent growth regulation in nematodes and *Drosophila* is conserved despite the absence of an autonomous requirement of insulin signaling in cell growth in *C. elegans*. Whereas nutrient availability regulates insulin pathway activity indirectly via the control of DILP levels in the hemolymph, nutrients appear to affect TOR pathway activity more directly.

As in yeast, TOR activity in *Drosophila* tissues and in mammalian cells is amino acid-dependent (see Gao et al. 2002; Radimerski et al. 2002a; Chapters 5–7). Although the amino acid-dependence of TOR activity is well-established, the molecular basis for this dependence is currently unclear (Dennis et al. 2001). Interestingly, Rheb, which acts between TSC1/2 and TOR in *Drosophila* and mammalian cells, is also required for nitrogen-regulated growth in *Schizosaccharomyes pombe*. Thus, the amino acid-dependent regulation of the Rheb-TOR function may be a primitive growth regulatory mechanism already present in single-cell organisms such as yeast (Mach et al. 2000).

The coupling between amino acid levels and TOR signaling means that nutrient levels can directly modulate the growth of all larval tissues. In addition, a recent study has shown that TOR signaling in the larval fat body acts as an amino acid sensor that indirectly modulates the growth of other tissues by influencing their insulin/PI3K pathway activity (Bradley and Leevers 2003; Colombani et al. 2003). Lowered amino acid uptake by the fat body reduces TOR signaling in that tissue and modulates the production of a humoral signal that results in lowered insulin/PI3K signaling and growth in other tissues. The *Drosophila* homolog of mammalian acid-labile subunit (DALS) is expressed in the fat body in an amino acid-dependent manner. Mammalian ALS binds to IGFs as part of a ternary complex that extends the ligand's half-life. Thus, DALS may be the TOR-dependent signal that is produced by the fat body to modulate the growth of other tissues.

Surviving periods of low nutrients has been one of the major challenges faced by all organisms during evolution. It is therefore not surprising that mechanisms evolved early in evolution to meet this challenge. The genetic and biochemical studies discussed above highlight the high degree of functional conservation of the TOR pathway from yeast to man and the insulin pathway in multicellular organisms. They support the hypothesis that these pathways represent an essential part of the cellular and organismal mechanisms that allow the attenuation of cellular and organismal growth in response to nutrients.

REFERENCES

Bohni R., Riesgo-Escovar J., Oldham S., Brogiolo W., Stocker H., Andruss B.F., Beckingham K., and Hafen E. 1999. Autonomous control of cell and organ size by CHICO, a *Drosophila* homolog of vertebrate IRS1-4. *Cell* **97:** 865–875.

Bradley G.L. and Leevers S.J. 2003. Amino acids and the humoral regulation of growth: Fat bodies use slimfast. *Cell* **114:** 656–658.

Brazil D.P. and Hemmings B.A. 2001. Ten years of protein kinase B signalling: A hard Akt to follow. *Trends Biochem. Sci.* **26:** 657–664.

Britton J.S. and Edgar B.A. 1998. Environmental control of the cell cycle in *Drosophila:* Nutrition activates mitotic and endoreplicative cells by distinct mechanisms. *Development* **125:** 2149–2158.

Britton J.S., Lockwood W.K., Li L., Cohen S.M., and Edgar B.A. 2002. *Drosophila's* insulin/PI3-kinase pathway coordinates cellular metabolism with nutritional conditions. *Dev. Cell* **2:** 239–249.

Brogiolo W., Stocker H., Ikeya T., Rintelen F., Fernandez R., and Hafen E. 2001. An evolutionarily conserved function of the *Drosophila* insulin receptor and insulin-like peptides in growth control. *Curr. Biol.* **11:** 213–221.

Brunet A., Bonni A., Zigmond M.J., Lin M.Z., Juo P., Hu L.S., Anderson M.J., Arden K.C., Blenis J., and Greenberg M.E. 1999. Akt promotes cell survival by phosphorylating and inhibiting a Forkhead transcription factor. *Cell* **96:** 857–868.

Brunet A., Kanai F., Stehn J., Xu J., Sarbassova D., Frangioni J.V., Dalal S.N., DeCaprio J.A., Greenberg M.E., and Yaffe M.B. 2002. 14-3-3 transits to the nucleus and participates in dynamic nucleocytoplasmic transport [erratum in *J. Cell Biol.* [2002] **157:** 533). *J. Cell Biol.* **156:** 817–828.

Burgering B.M. and Kops G.J. 2002. Cell cycle and death control: Long live Forkheads. *Trends Biochem. Sci.* **27:** 352–360.

Butler A.A. and Le Roith D. 2001. Control of growth by the somatropic axis: Growth hormone and the insulin-like growth factors have related and independent roles. *Annu. Rev. Physiol.* **63:** 141–164.

Chen C., Jack J., and Garofalo R.S. 1996. The *Drosophila* insulin receptor is required for normal growth. *Endocrinology* **137:** 846–856.

Cho K.S., Lee J.H., Kim S., Kim D., Koh H., Lee J., Kim C., Kim J., and Chung J. 2001. *Drosophila* phosphoinositide-dependent kinase-1 regulates apoptosis and growth via the phosphoinositide 3-kinase-dependent signaling pathway. *Proc. Natl. Acad. Sci.* **98:** 6144–6149.

Colombani J., Raisin S., Pantalacci S., Radimerski T., Montagne J., and Leopold P. 2003. A nutrient sensor mechanism controls *Drosophila* growth. *Cell* **114:** 739–749.

Datar S.A., Jacobs H.W., de la Cruz A.F., Lehner C.F., and Edgar B.A. 2000. The *Drosophila* cyclin D-Cdk4 complex promotes cellular growth. *EMBO J.* **19:** 4543–4554.

Dennis P.B., Jaeschke A., Saitoh M., Fowler B., Kozma S.C., and Thomas G. 2001. Mammalian TOR: A homeostatic ATP sensor. *Science* **294:** 1102–1105.

Fantes P. and Nurse P. 1977. Control of cell size at division in fission yeast by a growth-modulated size control over nuclear division. *Exp. Cell Res.* **107:** 377–386.

Fernandez R., Tabarini D., Azpiazu N., Frasch M., and Schlessinger J. 1995. The *Drosophila* insulin receptor homolog: A gene essential for embryonic development encodes two receptor isoforms with different signaling potential. *EMBO J.* **14:** 3373–3384.

Finch C.E. and Ruvkun G. 2001. The genetics of aging. *Annu. Rev. Genomics Hum. Genet.* **2:** 435–462.

Gao X. and Pan D. 2001. TSC1 and TSC2 tumor suppressors antagonize insulin signaling in cell growth. *Genes Dev.* **15:** 1383–1392.

Gao X., Neufeld T.P., and Pan D. 2000. *Drosophila* PTEN regulates cell growth and proliferation through PI3K-dependent and -independent pathways. *Dev. Biol.* **221:** 404–418.

Gao X., Zhang Y., Arrazola P., Hino O., Kobayashi T., Yeung R.S., Ru B., and Pan D. 2002. Tsc tumour suppressor proteins antagonize amino-acid-TOR signalling. *Nat. Cell Biol.* **4:** 699–704.

Garami A., Zwartkruis F.J.T., Joaquin M., Roccia M., Stocker H., Kozma S.C., Hafen E., Bos J.L., and Thomas G. 2003. TSC1/2 tumor suppressor complex functions as a GAP to repress mTOR/S6K/4E-BP signaling through the small GTPase Rheb. *Mol. Cell* **11:** 1457–1466.

Gerisch B., Weitzel C., Kober-Eisermann C., Rootiers V., and Antebi A. 2001. A hormonal signaling pathway influencing *C. elegans* metabolism, reproductive development, and life span. *Dev. Cell* **1:** 841–851.

Gingras A.C., Raught B., and Sonenberg N. 2001. Regulation of translation initiation by FRAP/mTOR. *Genes Dev.* **15:** 807–826.

Goberdhan D.C., Paricio N., Goodman E.C., Mlodzik M., and Wilson C. 1999. *Drosophila* tumor suppressor PTEN controls cell size and number by antagonizing the Chico/PI3-kinase signaling pathway. *Genes Dev.* **13:** 3244–3258.

Graf R., Neuenschwander S., Brown M.R., and Ackermann U. 1997. Insulin-mediated secretion of ecdysteroids from mosquito ovaries and molecular cloning of the insulin receptor homologue from ovaries of bloodfed *Aedes aegypti*. *Insect Mol. Biol.* **6:** 151–163.

Hagedorn H.H. 1985. The role of ecdysteroids in reproduction. In *Comparative insect physiology, biochemistry and pharmacology* (ed. G.A. Kerkut and L.I. Gilbert), vol. 8, pp. 205–262. Pergamon Press, Oxford.

Huang H., Potter C.J., Tao W., Li D.M., Brogiolo W., Hafen E., Sun H., and Xu T. 1999. PTEN affects cell size, cell proliferation and apoptosis during *Drosophila* eye development. *Development* **126:** 5365–5372.

Ikeya T., Galic M., Belawat P., Nairz K., and Hafen E. 2002. Nutrient-dependent expression of insulin-like peptides from neuroendocrine cells in the CNS contributes to growth regulation in *Drosophila*. *Curr. Biol.* **12:** 1293.

Inoki K., Li Y., Zhu T., Wu J., and Guan K.L. 2002. TSC2 is phosphorylated and inhibited by Akt and suppresses mTOR signalling. *Nat. Cell Biol.* **4:** 648–657.

Jaeschke A., Hartkamp J., Saitoh M., Roworth W., Nobukuni T., Hodges A., Sampson J., Thomas G., and Lamb R. 2002. Tuberous sclerosis complex tumor suppressor-mediated S6 kinase inhibition by phosphatidylinositide-3-OH kinase is mTOR independent. *J. Cell Biol.* **159:** 217–224.

Johnston G.C., Pringle J.R., and Hartwell L.H. 1977. Coordination of growth with cell division in the yeast *Saccharomyces cerevisiae*. *Exp. Cell Res.* **105:** 79–98.

Jünger M.A., Rintelen F., Stocker H., Wasserman J.D., Vegh M., Radimerski T., Greenberg M.E., and Hafen E. 2003. The *Drosophila* Forkhead transcription factor FOXO mediates the reduction in cell number associated with reduced insulin signaling. *J. Biol.* **2:** 20.

Kandt R.S. 2002. Tuberous sclerosis complex and neurofibromatosis type 1: The two most common neurocutaneous diseases. *Neurol. Clin.* **20:** 941–964.

Koga M., Ohshima Y., and Antebi A. 2001. daf-12 encodes a nuclear receptor that regulates the dauer diapause and developmental age in *C. elegans*. *Mech. Dev.* **109:** 27–35.

Lachance P.E., Miron M., Raught B., Sonenberg N., and Lasko P. 2002. Phosphorylation of eukaryotic translation initiation factor 4E is critical for growth. *Mol. Cell. Biol.* **22:** 1656–1663.

Lasko P. 2000. The *Drosophila melanogaster* genome: Translation factors and RNA binding proteins. *J. Cell Biol.* **150:** F51–F56.

Leevers S.J. 2001. Growth control: Invertebrate insulin surprises! *Curr. Biol.* **11:** R209–R212.

Leevers S.J., Weinkove D., MacDougall L.K., Hafen E., and Waterfield M.D. 1996. The *Drosophila* phosphoinositide 3-kinase Dp110 promotes cell growth. *EMBO J.* **15:** 6584–6594.

Lin K., Dorman J.B., Rodan A., and Kenyon C. 1997. daf-16: An HNF-3/forkhead family member that can function to double the life-span of *Caenorhabditis elegans* (comment). *Science* **278:** 1319–1322.

Lizcano J.M., Alrubaie S., Kieloch A., Deak M., Leevers S.J., and Alessi D.R. 2003. Insulin-induced *Drosophila* S6 kinase activation requires phosphoinositide 3-kinase and protein kinase B. *Biochem. J.* **374:** 297–306.

Mach K.E., Furge K.A., and Albright C.F. 2000. Loss of Rhb1, a Rheb-related GTPase in fission yeast, causes growth arrest with a terminal phenotype similar to that caused by nitrogen starvation. *Genetics* **155:** 611–622.

Maheshwar M.M., Cheadle J.P., Jones A.C., Myring J., Fryer A.E., Harris P.C., and Sampson J.R. 1997. The GAP-related domain of tuberin, the product of the TSC2 gene, is a target for missense mutations in tuberous sclerosis. *Hum. Mol. Genet.* **6:** 1991–1996.

Manning B.D., Tee A.R., Logsdon M.N., Blenis J., and Cantley L.C. 2002. Identification of the tuberous sclerosis complex-2 tumor suppressor gene product tuberin as a target of the phosphoinositide 3-kinase/akt pathway. *Mol. Cell* **10:** 151–162.

Martin-Castellanos C. and Edgar B.A. 2002. A characterization of the effects of Dpp signaling on cell growth and proliferation in the *Drosophila* wing. *Development* **129:** 1003–1013.

Marygold S.J. and Leevers S.J. 2002. Growth signaling: TSC takes its place. *Curr. Biol.* **12:** R785–R787.

Masumura M., Ishizaki H., Nagata K., Kataoka H., Suzuki A., and Mizoguchi A. 1997. Glucose stimulates the release of bombyxin, an insulin-related peptide of the silkworm *Bombyx mori. Comp. Biochem. Physiol. B Biochem. Mol. Biol.* **118:** 349–357.

McManus E.J. and Alessi D.R. 2002. TSC1-TSC2: A complex tale of PKB-mediated S6K regulation. *Nat. Cell Biol.* **4:** E214–E216.

Meyer C.A., Jacobs H.W., Datar S.A., Du W., Edgar B.A., and Lehner C.F. 2000. *Drosophila* Cdk4 is required for normal growth and is dispensable for cell cycle progression. *EMBO J.* **19:** 4533–4542.

Millar J.B. 2002. A genomic approach to studying cell-size homeostasis in yeast. *Genome Biol.* **3:** REVIEWS1028.

Miron M., Verdu J., Lachance P.E., Birnbaum M.J., Lasko P.F., and Sonenberg N. 2001. The translational inhibitor 4E-BP is an effector of PI(3)K/Akt signalling and cell growth in *Drosophila. Nat. Cell Biol.* **3:** 596–601.

Montagne J., Radimerski T., and Thomas G. 2001. Insulin signaling: Lessons from the *Drosophila* tuberous sclerosis complex, a tumor suppressor. *Sci. STKE* **2001:** PE36.

Montagne J., Stewart M.J., Stocker H., Hafen E., Kozma S.C., and Thomas G. 1999. *Drosophila* S6 kinase: A regulator of cell size. *Science* **285:** 2126–2129.

Morata G. and Ripoll P. 1975. Minutes: Mutants of *Drosophila* autonomously affecting cell division rate. *Dev. Biol.* **42:** 211–221.

Nagasawa H. 1992. Neuropeptides of the silkworm, *Bombyx mori. Experientia* **48:** 425–430.

Nakae J., Kido Y., and Accili D. 2001. Distinct and overlapping functions of insulin and IGF-I receptors. *Endocr. Rev.* **22:** 818–835.

Neufeld T.P., de la Cruz A.F., Johnston L.A., and Edgar B.A. 1998. Coordination of growth and cell division in the *Drosophila* wing. *Cell* **93:** 1183–1193.

Nurse P. 1975. Genetic control of cell size at cell division in yeast. *Nature* **256:** 451–457.

Nurse P., Thuriaux P., and Nasmyth K. 1976. Genetic control of the cell division cycle in the fission yeast *Schizosaccharomyces pombe. Mol. Gen. Genet.* **146:** 167–178.

Ogg S., Paradis S., Gottlieb S., Patterson G.I., Lee L., Tissenbaum H.A., and Ruvkun G. 1997. The Fork head transcription factor DAF-16 transduces insulin-like metabolic and longevity signals in *C. elegans. Nature* **389:** 994–999.

Oldham S. and Hafen E. 2003. Insulin/IGF and target of rapamycin signaling: A TOR de force in growth control. *Trends Cell Biol.* **13:** 79–85.

Oldham S., Montagne J., Radimerski T., Thomas G., and Hafen E. 2000. Genetic and biochemical characterization of dTOR, the *Drosophila* homolog of the target of rapamycin. *Genes Dev.* **14:** 2689–2694.

Oldham S., Stocker H., Laffargue M., Wittwer F., Wymann M., and Hafen E. 2002. The *Drosophila* insulin/IGF receptor controls growth and size by modulating PtdInsP(3) levels. *Development* **129:** 4103–4109.

Patel P.H., Thapar N., Guo L., Martinez M., Maris J., Gau C.-L., Lengyel J.A., and Tamanoi F. 2003. *Drosophila* Rheb GTPase is required for cell cycle progression and cell growth. *J. Cell Sci.* **116:** 3601–3610.

Poltilove R.M., Jacobs A.R., Haft C.R., Xu P., and Taylor S.I. 2000. Characterization of *Drosophila* insulin receptor substrate. *J. Biol. Chem.* **275:** 23346–23354.

Potter C.J., Huang H., and Xu T. 2001. *Drosophila* Tsc1 functions with Tsc2 to antagonize insulin signaling in regulating cell growth, cell proliferation, and organ size. *Cell* **105:** 357–368.

Potter C.J., Pedraza L.G., and Xu T. 2002. Akt regulates growth by directly phosphorylating Tsc2 (comment). *Nat. Cell Biol.* **4:** 658–665.

Puig O., Marr M.T., Ruhf M.L., and Tjian R. 2003. Control of cell number by *Drosophila* FOXO: Downstream and feedback regulation of the insulin receptor pathway. *Genes Dev.* **17:** 2006–2020.

Radimerski T., Montagne J., Hemmings-Mieszczak M., and Thomas G. 2002a. Lethality of *Drosophila* lacking TSC tumor suppressor function rescued by reducing dS6K signaling. *Genes Dev.* **16:** 2627–2632.

Radimerski T., Montagne J., Rintelen F., Stocker H., van der Kaay J., Downes C.P., Hafen E., and Thomas G. 2002b. dS6K-regulated cell growth is dPKB/dPI(3)K-independent, but requires dPDK1. *Nat. Cell Biol.* **4:** 251–255.

Riddle D.L. 1988. The dauer larva. In *The nematode* Caenorhabditis elegans (ed. W.N. Wood), pp. 393-412. Cold Spring Harbor Laboratory Press, Cold Spring Harbor, New York.

Rintelen F., Stocker H., Thomas G., and Hafen E. 2001. PDK1 regulates growth through Akt and S6K in *Drosophila. Proc. Natl. Acad. Sci.* **98:** 15020–15025.

Rulifson E.J., Kim S.K., and Nusse R. 2002. Ablation of insulin-producing neurons in flies: Growth and diabetic phenotypes. *Science* **296:** 1118–1120.

Saltiel A.R. and Kahn C.R. 2001. Insulin signalling and the regulation of glucose and lipid metabolism. *Nature* **414:** 799–806.

Saucedo L.J. and Edgar B.A. 2002. Why size matters: Altering cell size. *Curr. Opin. Genet. Dev.* **12:** 565–571.

Saucedo L.J., Gao X., Chiarelli D.A., Li L., Pan D., and Edgar B.A. 2003. Rheb promotes cell growth as a component of the insulin/TOR signalling network. *Nat. Cell Biol.* **5:** 566–571.

Scheid M.P. and Woodgett J.R. 2001. PKB/AKT: Functional insights from genetic models. *Nat. Rev. Mol. Cell Biol.* **2:** 760–768.

Schmelzle T. and Hall M.N. 2000. TOR, a central controller of cell growth. *Cell* **103:** 253–262.

Stocker H., Andjelkovic M., Oldham S., Laffargue M., Wymann M.P., Hemmings B.A., and Hafen E. 2002. Living with lethal PIP3 levels: Viability of flies lacking PTEN restored by a PH domain mutation in Akt/PKB. *Science* **295:** 2088–2091.

Stocker H., Radimerski T., Schindelholz B., Wittwer F., Belawat P., Daram P., Breuer S., Thomas G., and Hafen E. 2003. Rheb is an essential regulator of sS6K in controlling cell growth in *Drosophila*. *Nat. Cell Biol.* **5:** 559–565.

Tapon N., Ito N., Dickson B.J., Treisman J.E., and Hariharan I.K. 2001. The *Drosophila* tuberous sclerosis complex gene homologs restrict cell growth and cell proliferation. *Cell* **105:** 345–355.

Tatar M., Kopelman A., Epstein D., Tu M.P., Yin C.M., and Garofalo R.S. 2001. A mutant *Drosophila* insulin receptor homolog that extends life-span and impairs neuroendocrine function. *Science* **292:** 107–110.

Tran H., Brunet A., Griffith E.C., and Greenberg M.E. 2003. The many forks in FOXO's road. *Sci. STKE* **2003:** RE5.

Vanhaesebroeck B. and Alessi D.R. 2000. The PI3K-PDK1 connection: More than just a road to PKB. *Biochem. J.* **346:** 561–576.

Vanhaesebroeck B., Leevers S.J., Ahmadi K., Timms J., Katso R., Driscoll P.C., Woscholski R., Parker P.J., and Waterfield M.D. 2001. Synthesis and function of 3-phosphorylated inositol lipids. *Annu. Rev. Biochem.* **70:** 535–602.

Verdu J., Buratovich M.A., Wilder E.L., and Birnbaum M.J. 1999. Cell-autonomous regulation of cell and organ growth in *Drosophila* by Akt/PKB. *Nat. Cell Biol.* **1:** 500–506.

Weinkove D., Leevers S.J., MacDougall L.K., and Waterfield M.D. 1997. p60 is an adaptor for the *Drosophila* phosphoinositide 3-kinase, Dp110. *J. Biol. Chem.* **272:** 14606–14610.

Weinkove D., Neufeld T.P., Twardzik T., Waterfield M.D., and Leevers S.J. 1999. Regulation of imaginal disc cell size, cell number and organ size by *Drosophila* class I(A) phosphoinositide 3-kinase and its adaptor. *Curr. Biol.* **9:** 1019–1029.

Weitzman J.B. 2003. Growing without a size checkpoint. *J. Biol.* **2:** 3.

Wells W.A. 2002. Does size matter? *J. Cell Biol.* **158:** 1156–1159.

Wienecke R., Maize J.C., Jr., Reed J.A., de Gunzburg J., Yeung R.S., and DeClue J.E. 1997. Expression of the TSC2 product tuberin and its target Rap1 in normal human tissues. *Am. J. Pathol.* **150:** 43–50.

Xiao G.H., Shoarinejad F., Jin F., Golemis E.A., and Yeung R.S. 1997. The tuberous sclerosis 2 gene product, tuberin, functions as a Rab5 GTPase activating protein (GAP) in modulating endocytosis. *J. Biol. Chem.* **272:** 6097–6100.

Yamagata K., Sanders L.K., Kaufmann W.E., Yee W., Barnes C.A., Nathans D., and Worley P.F. 1994. rheb, a growth factor- and synaptic activity-regulated gene, encodes a novel Ras-related protein. *J. Biol. Chem.* **269:** 16333–16339.

Zhang H., Stallock J.P., Ng J.C., Reinhard C., and Neufeld T.P. 2000. Regulation of cellular growth by the *Drosophila* target of rapamycin dTOR. *Genes Dev.* **14:** 2712–2724.

Zhang Y., Gao X., Saucedo L.J., Ru B., Edgar B.A., and Pan D. 2003. Rheb is a direct target of the tuberous sclerosis tumour suppressor proteins. *Nat. Cell Biol.* **5:** 578–581.

Zick Y. 2001. Insulin resistance: A phosphorylation-based uncoupling of insulin signaling. *Trends Cell Biol.* **11:** 437–441.

Zinke I., Schutz C.S., Katzenberger J.D., Bauer M., and Pankratz M.J. 2002. Nutrient control of gene expression in *Drosophila*: Microarray analysis of starvation and sugar-dependent response. *EMBO J.* **21:** 6162–6173.

7

Growth Control through the mTOR Network

David A. Guertin, Do-Hyung Kim, and David M. Sabatini
Whitehead Institute for Biomedical Research and
Massachusetts Institute of Technology Department of Biology
Cambridge, Massachusetts 02142

BEFORE TRAVERSING THROUGH THE CELL CYCLE, most dividing cells accumulate mass and grow in size. In unicellular organisms like budding yeast, environmental nutrients are the main regulators of cell growth, whereas animals employ additional levels of control in the form of growth factor-dependent signaling networks. The conserved serine/threonine protein kinase, Target of Rapamycin (TOR; known in mammals as mTOR, FRAP, or RAFT1), has emerged as a critical regulator of cellular growth in eukaryotes. In mammals, mTOR is the core component of a signaling network that integrates inputs derived from nutrients and growth factors to regulate cell growth, an effect mediated in part through the control of the protein synthesis machinery. In this chapter, we provide an overview of mTOR and its associated proteins and discuss in detail how nutrients and growth factors coordinately regulate the mTOR pathway. To conclude, we examine the wider implications of mTOR signaling in development and disease.

mTOR AND ITS EFFECTORS

mTOR was identified in studies aimed at defining the molecular target of the antiproliferative drug rapamycin (Brown et al. 1994; Chiu et al. 1994; Sabatini et al. 1994; Sabers et al. 1995), a macrolide antibiotic produced by the soil bacterium *Streptomyces hygroscopicus* (Sehgal et al. 1975; Vezina et al. 1975). After entering the cell, rapamycin interacts at varying affinities with several members of the FKBP family of prolyl isomerases.

The complex of rapamycin with one member of this family, FKBP12, gains the capacity to bind and perturb the function of mTOR. Although originally valued for its antifungal properties (Sehgal et al. 1975; Baker et al. 1978), rapamycin is now recognized as having multiple potential clinical uses. It is a potent FDA-approved immunosuppressant (marketed under the trade name Rapamune) used in organ transplantation (for review, see Saunders et al. 2001) and, in phase II clinical trials, rapamycin derivatives show promise for the treatment of certain human cancers (for review, see Hidalgo and Rowinsky 2000; Vogt 2001). In vitro the drug is particularly effective at slowing the proliferation of cancer cells deficient for the tumor suppressor gene PTEN, suggesting that in the future, rapamycin therapy may be targeted to patients with specific tumor genotypes (Aoki et al. 2001; Neshat et al. 2001; Podsypanina et al. 2001). Recent clinical trials show that rapamycin prevents the restenosis of blood vessels that frequently occurs after balloon angioplasty and indicate that rapamycin-coated coronary stents may have major uses in cardiology (Morice et al. 2002; Regar et al. 2002; Sharma et al. 2002). Last, as described below, the recent discovery that the mTOR pathway is hyperactive in the genetic disease tuberous sclerosis suggests that rapamycin may have still other unexpected uses.

TOR orthologs exist and have been shown to be essential in *Saccharomyces cerevisiae, Schizosaccharomyces pombe, Caenorhabditis elegans, Drosophila*, and in the plant *Arabidopsis thaliana* (Cafferkey et al. 1993; Kunz et al. 1993; Oldham et al. 2000; Zhang et al. 2000; Weisman and Choder 2001; Long et al. 2002; Menand et al. 2002). Whereas budding and fission yeast contain two partially redundant TOR genes, TOR1 and TOR2, nonfungal organisms have only one. The domain structure of mTOR is diagramed in Figure 1. TOR is the founding member of a family of large proteins (usually greater than 250 kD in molecular weight) containing carboxy-terminal regions with homology to the catalytic domain of the PI-3 lipid kinase (PI3K). This family, collectively named PI3K-like kinases (PIKKs) (Keith and Schreiber 1995), includes several mammalian proteins involved in sensing and responding to DNA damage, such as ATM, ATR, and DNA-PKcs, as well as hSMG1, a critical component of the nonsense-mediated decay (NMD) pathway (Denning et al. 2001; Yamashita et al. 2001). Despite the sequence similarity to the catalytic domain of PI3K, mTOR, the best biochemically characterized TOR, appears to be a protein kinase, phosphorylating itself as well as targets in *trans* (Brown et al. 1995; Brunn et al. 1996, 1997a,b; Burnett et al. 1998). Structural similarity between mTOR and other PIKKs extends beyond the carboxy-terminal kinase domain, as most of the amino-terminal regions

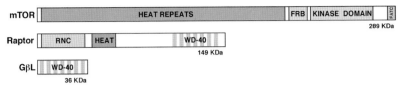

Figure 1. Domain architecture of the mTOR complex components. The core mTOR complex (shown in Fig. 2) is composed of three subunits: mTOR, raptor, and GβL. The 289-kD mTOR kinase is the catalytic subunit of the complex. The kinase domain of mTOR is located in the carboxyl terminus and is structurally related to the PI-3 lipid kinase (PI3K) domain but functions as a Ser/Thr protein kinase. Also in the carboxyl terminus region is the FRB domain, which is the binding site for the rapamycin–FKBP12 complex, and the FATC domain for which a function has not yet been described. Most of the remaining structure of mTOR is composed of HEAT repeats, which are modules that mediate protein–protein interactions. The 149-KD raptor (regulatory associated protein of mTOR) subunit also contains HEAT repeats, in addition to seven WD-40 repeats in the carboxyl terminus, and a highly conserved amino terminal RNC (Raptor N-terminal conserved) domain of unknown function. The 36-kD GβL (G-protein β-subunit like protein) subunit consists almost entirely of seven WD-40 repeats. All three exist in a complex at near-stoichiometric ratios and are required for mTOR signaling. GβL binds the kinase domain near the FRB domain of mTOR, whereas the interaction between Raptor and mTOR is not easily defined and likely encompasses a significant portion of both proteins.

consist of tandemly repeated, loosely conserved HEAT motifs (Andrade and Bork 1995; Perry and Kleckne 2003). These protein modules mediate protein–protein interactions (Chook and Blobel 1999; Cingolani et al. 1999; Groves et al. 1999; Vetter et al. 1999; Marcotrigiano et al. 2001) and were originally discovered in four proteins (Huntingtin, Elongation factor 3, regulatory A subunit of PP2A, and Tor1p) whose names gave rise to the "HEAT" term (Andrade and Bork 1995). Also conserved between PIKK family members is the small FATC domain (Frap, ATM, and Trap C-terminal), an extreme carboxy-terminal region of unknown function (Keith and Schreiber 1995; Bosotti et al. 2000). The FKBP12–rapamycin complex binds directly upstream of the catalytic region of TOR to the so-called FRB (FKBP-rapamycin binding) domain (Chen et al. 1995; Choi et al. 1996). This domain is highly conserved among the TOR proteins but, except for a weakly conserved FRB-like region in the SMG-1 subfamily (Denning et al. 2001), not in other PIKK family members.

As mTOR was discovered by plucking it out of cellular lysates using a rapamycin–FKBP12 affinity reagent, no connections to functionally characterized proteins were known at the time of its identification. Thus, determining its cellular role has been challenging. Until the advent of

RNAi in mammalian cells, rapamycin had been the main tool with which to ask whether mTOR participates in a cellular or molecular process. In early landmark studies, rapamycin was found to affect the phosphorylation state and activity of S6K1 and 4E-BP1 (also known as PHAS-I), implicating mTOR in their control (Chung et al. 1992; Kuo et al. 1992; Price et al. 1992; Lin et al. 1995; Beretta et al. 1996; von Manteuffel et al. 1996). Because S6K1 and 4E-BP1 regulate ribosome biogenesis and cap-dependent mRNA translation, respectively, it became apparent that at least one important function of mTOR is to regulate protein synthesis

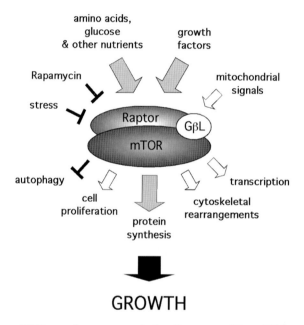

Figure 2. The mTOR complex as a growth signaling center. The mTOR complex integrates diverse input signals into a growth response mediated through regulation of a variety of cellular processes. The core mTOR complex consists of three known subunits: the catalytic subunit, mTOR, and two novel proteins named Raptor and GβL. The best-characterized inputs to the mTOR complex are nutrient and growth-factor-derived signals, but mitochondrial signals and stress may also feed into the complex. The mTOR complex is thought to convert the input signals into a growth response by activating protein synthesis; however, the complex may also regulate transcription, cell proliferation, cytoskeletal rearrangements, and autophagy. The macrolide antibiotic rapamycin perturbs mTOR function, and its ability to prevent growth by inhibiting mTOR signaling has made it a valuable drug with important clinical applications. Rapamycin is an FDA-approved immunosuppressant, and recently it has shown great promise in the treatment of certain cancers and in preventing restenosis after balloon angioplasty.

(Figs. 2 and 3). Activation of S6K1 triggers the subsequent phosphorylation of the 40S ribosomal protein S6 (Jeno et al. 1989) that seems necessary for the translation of mRNAs containing 5′ TOP structures (terminal oligopyrimidine tracts) (Jefferies et al. 1994, 1997). These mRNAs encode the bulk of translational machinery components, including most ribosomal proteins, and constitute about 20% of total cellular mRNAs (Schibler et al. 1977). More recently, however, the importance of S6K1 and S6 in 5′ TOP mRNA translation in some cell types has been challenged (Stolovich et al. 2002). The other well-characterized mTOR effector, 4E-BP1, is an inhibitor of mRNA translation that, in the hypophosphorylated state, interacts with and represses the function of eIF-4E, the translation initiation factor that recognizes the m7GpppN cap structure on the 5′ ends of mRNAs (Koromilas et al. 1992; Pause et al. 1994). Upon mTOR-dependent phosphorylation (Brunn et al. 1997b; Burnett et al. 1998), 4E-BP1 dissociates from eIF-4E, allowing cap-dependent mRNA translation to initiate. It has been suggested that, through 4E-BP1, mTOR regulates the translation of several growth-promoting factors, including c-myc, cyclin D1, and IGF-II (Morice et al. 1993a,b; Nourse et al. 1994; Nielsen et al. 1995; Hashemolhosseini et al. 1998; West et al. 1998; Nelsen et al. 2003).

mTOR appears to control the activation of S6K1 and the inhibition of 4E-BP1 by regulating the phosphorylation of conserved rapamycin-sensitive serine and threonine residues. However, the network that signals to these effectors is not a simple linear pathway, and both proteins also receive upstream inputs independently of mTOR (Fig. 3; see also section on the mTOR complex as a signaling center for growth control) (for review, see Jacinto and Hall 2003; Oldham and Hafen 2003). Although mTOR phosphorylates S6K1 and 4E-BP1 in vitro, it still remains controversial whether this is the case within cells (Brunn et al. 1997b; Burnett et al. 1998; Gingras et al. 1999; Isotani et al. 1999). It is possible that in vivo mTOR activates an unidentified intermediate kinase, and/or inactivates a phosphatase (Ferrari et al. 1993; Peterson et al. 1999). Consistent with a role for a phosphatase in mTOR signaling, treatment of mammalian cells with calyculin A (a preferential inhibitor of PP1, PP2A, PP4, and PP5-like phosphatases) attenuates the dephosphorylation of S6K1 and 4E-BP1 induced by rapamycin or amino acid starvation (Peterson et al. 2000). A weak interaction between PP2A and S6K1 has been detected (Peterson et al. 1999; Westphal et al. 1999), but the connections and regulatory mechanisms between PP2A-like phosphatase activity and mTOR signaling is not known. A balance between mTOR-dependent phosphorylation and mTOR-independent dephosphorylation may be critical for setting the activity of downstream effectors. Alternatively, and perhaps more likely, mTOR may regulate the phosphorylation *and* dephosphorylation of S6K1 and 4E-BP1, by serving as their in vivo kinase and by controlling a still unidentified phosphatase. This notion is consistent with early work in yeast indicating that a critical function of TOR is to directly regulate a phosphatase (DiComo and Arndt 1996; Jiang and Broach 1999).

In addition to S6K1 and 4E-BP1, mTOR activity also regulates the phosphorylation of two additional proteins not apparently involved in protein synthesis. Lipin, the product of the fatty liver dystrophy (*fld*) gene in mice, is required for adipocyte development and, when mutated, results in mice having characteristics of the human disease lipodystrophy (Peterfy et al. 2001). In adipocytes, insulin or amino acids stimulate the hyperphosphorylation of lipin in a rapamycin-sensitive manner (Huffman et al. 2002). Although the effects of phosphorylation on lipin function are not yet known, this work suggests that mTOR may have a role in adipocyte development. Interestingly, rapamycin inhibits the adipogenic differentiation of both primary adipocytes and cell lines in culture (Yeh et al. 1995; Bell et al. 2000; Gagnon et al. 2001). Another target of mTOR signaling is CLIP-170, a microtubule-associated protein that physically interacts with mTOR (Choi et al. 2002). CLIP-170 contains both rapamycin-sensitive and -insensitive phosphorylation sites, and mTOR phosphorylates the rapamycin-sensitive ones in vitro. Moreover, rapamycin treatment inhibits CLIP-170 binding to microtubules, suggesting a role for mTOR in regulating microtubule organization.

Substantial evidence from model organisms and human cell lines implicates TOR signaling in the regulation of a variety of cellular processes besides protein synthesis (Fig. 2). For most of these, the molecular mechanisms involved are still unknown (for review, see Schmelzle and Hall 2000). In yeast, TOR regulates the transcription of genes involved in, for example, the starvation response and metabolism (Barbet et al. 1996; Beck and Hall 1999; Cardenas et al. 1999; Hardwick et al. 1999; Komeili et al. 2000; Crespo et al. 2001). mTOR also regulates the expression of metabolic genes; in human BJAB B-lymphoma cells and mouse CTLL-2 T lymphocytes, rapamycin treatment and amino acid starvation generate similar transcription profiles with marked changes in the levels of mRNAs encoding proteins involved in nutrient and protein metabolism (Peng et al. 2002). Similar to yeast TOR (Noda and Ohsumi 1998), mTOR also appears to be a negative regulator of autophagy in mammals. Autophagy is a nonselective catabolic pathway that degrades proteins and organelles through the lysosomal system, and is suppressed by environmental amino acids (for review, see Ogier-Denis and Codogno 2003). In hepatocytes, nutrient conditions that inhibit autophagy also activate S6, the target of

S6K1, and rapamycin prevents the suppression of autophagy normally elicited by amino acids (Blommaart et al 1995, 1997). As mentioned earlier, yeast have two TOR genes, with TOR2 having a unique rapamycin-resistant role in signaling the reorganization of the actin cytoskeleton during the cell cycle (Schmidt et al. 1996). Through its regulation of CLIP-170 (discussed above), mTOR may also have a role in cytoskeletal organization, but, in contrast to yeast, this mTOR function is clearly rapamycin-sensitive. These observations suggest that both in yeast and mammals protein synthesis regulation is not the only function of the TOR pathway.

mTOR FUNCTIONS IN A COMPLEX

The fact that mTOR is a large protein and contains multiple HEAT repeats suggests that it may function in a complex with other proteins. In support of this notion, mTOR migrates as a 1–2 MD species in size exclusion chromatography (Fang et al. 2001; Kim et al. 2002). Moreover, cellular conditions that block mTOR signaling do not affect the in vitro kinase activity of mTOR, suggesting that within cells, mTOR might interact with unidentified regulatory proteins that are lost during biochemical purification. Nearly a decade after the discovery of mTOR, two groups identified a conserved 149-kD protein named raptor (regulatory associated protein of mTOR) that forms a near stoichiometric complex with mTOR and is required for mTOR signaling (Fig. 1; discussed below) (Hara et al. 2002; Kim et al. 2002). Raptor contains a novel highly conserved amino-terminal region called the RNC (raptor N-terminal conserved) domain, three HEAT repeats, and seven WD-40 repeats (Kim et al. 2002). Raptor interacts with the amino-terminal HEAT repeats of mTOR, but the mTOR binding domain of raptor is not easily defined and probably encompasses a significant fraction of the protein (Kim et al. 2002). In addition to its role in regulating mTOR signaling, raptor may also be important for determining mTOR substrate specificity. Because the sites of phosphorylation of mTOR targets show little homology, it has been puzzling as to how substrate discrimination is achieved within the cell. Raptor can directly interact with 4E-BP1 through a short amino acid sequence called the TOS (mTOR signaling) motif that is found in the carboxyl terminus of 4E-BP1, as well as in the amino termini of all S6 kinases (Schalm and Blenis 2002). The TOS motif is required for efficient mTOR-dependent phosphorylation of both S6K1 and 4E-BP1 within cells, suggesting that raptor may serve as a scaffold for the interaction of the mTOR complex with certain targets (Choi et al. 2003; Nojima et al. 2003; Schalm et al. 2003).

Soon after the discovery of raptor, the mTOR signaling complex was found to contain another protein, GβL (G-protein β-subunit like protein, also known as mLST8) (Loewith et al. 2002; Kim et al. 2003). GβL had been discovered previously as an insulin-inducible protein of unknown function, and, as its name suggests, is similar to the β subunit of heterotrimeric G proteins (Rodgers et al. 2001). The 36-kD protein consists almost entirely of seven WD-40 repeats, binds and activates the kinase domain of mTOR, and, like raptor, is required for mTOR signaling (Fig. 1; discussed below). The GβL binding site on mTOR is adjacent to the FRB domain, suggesting that the FKBP12–rapamycin complex may perturb GβL function. GβL is also important for strengthening the raptor–mTOR interaction, but the mechanism is unclear, as a direct raptor–GβL interaction has not been seen (Kim et al. 2003). Both raptor and GβL are conserved in *S. pombe, S. cerevisiae, Drosophila, C. elegans,* and *A. thaliana* (Kim et al. 2002, 2003; Loewith et al. 2002). In *S. cerevisiae,* homologs of raptor and GβL (Kog1p and Lst8p, respectively), as well as three proteins of unknown function, Avo1p, Avo2p, and Avo3p, exist in TOR complexes (Loewith et al. 2002; Wedaman et al. 2003). Interestingly, *S. cerevisiae* Tor1p and Tor2p interact with different sets of these proteins, defining two TOR complexes, TORC1 and TORC2, of which only TORC1 is sensitive to rapamycin (Loewith et al. 2002). TORC1 contains Kog1p, Lst8p, and either Tor1p or Tor2p, whereas TORC2 has Tor2p exclusively, and also Lst8p, Avo1p, Avo2p, and Avo3p. Whether proteins similar to Avo1p, Avo2p, and Avo3p bind mTOR, or whether multiple mTOR complexes exist in mammals, is not yet known. Ongoing studies of the mTOR complex will undoubtedly shed light on some of these issues in the near future.

Two additional proteins, CLIP-170 and gephyrin, interact with mTOR (Sabatini et al. 1999; Choi et al. 2002) but at lower stoichiometries than raptor or GβL. As discussed above, the interaction with CLIP-170, a microtubule-associated protein, suggests that mTOR may be involved in cytoskeletal dynamics. Gephyrin is a scaffold protein necessary for the clustering of glycine receptors at postsynaptic sites on the dendrites of neurons (Kirsch et al. 1993), and mTOR mutants that cannot associate with gephryin are also defective in signaling to S6K1 and 4E-BP1 (Sabatini et al. 1999). Thus, mTOR could have a role in the control of localized protein synthesis in dendrites, a process emerging as important for the regulation of synaptic plasticity and memory formation (for review, see Tang and Schuman 2002). Interestingly, rapamycin treatment of rat hippocampal neurons, either in dissociated or slice culture, reduces the synaptic potentiation induced by high-frequency stimulation (Tang et al. 2002). Furthermore, in *Aplysia,* serotonin induces synapse-specific facilitation and S6 kinase phosphorylation in a rapamycin-sensitive manner at synaptosomes (Khan et al. 2001). Localization studies in cultured hippocampal neurons indicate that phosphorylated S6K1 is concentrated at the growth cones and filopodia of extending neurites and in the soma (Raymond et al. 2002). Taken together, these observations raise the speculative possibility that mTOR-dependent protein synthesis, perhaps in a cellular subcompartment, may play a role in neurite growth and synaptic plasticity.

THE mTOR COMPLEX AS A SIGNALING CENTER FOR GROWTH CONTROL

mTOR Regulates Growth

Early studies in budding yeast showed that TOR depletion or rapamycin treatment causes an early G_1 arrest and a starvation-like physiological state, suggesting that TOR functions in cell growth control (Barbet et al. 1996). More recently, studies in *Drosophila* have also provided insights on the role of TOR in growth control. Larvae with a loss in dTOR function grow slowly and eventually arrest during larval development, reaching only a fraction of the mass of wild-type animals (Oldham et al. 2000; Zhang et al. 2000). Clonal analysis of homozygous mutant dTOR cells indicates that cells without dTOR function have cell-autonomous defects in size and proliferation. Overexpression of dS6K restores viability to certain hypomorphic dTOR mutants (but not complete loss-of-function animals) (Zhang et al. 2000), consistent with dS6K being one important downstream effector of dTOR. Unlike loss of dTOR, loss of dS6K is only semi-lethal, again suggesting that dS6K does not mediate all of dTOR's effects. Loss of dS6K is semi-lethal because a small percentage of flies lacking dS6K survive. Such flies have a severe reduction in body size caused by smaller rather than fewer cells (Montagne et al. 1999). Similarly, overexpression of a mutant d4E-BP that strongly binds deIF4E also causes a specific cell size defect, but, in addition, affects cell number (Miron et al. 2001).

In mammals and flies, the insulin growth factor signaling pathway regulates organism growth through PI3K-dependent signaling, and insulin regulates the TOR effectors S6K1 and 4E-BP1 in both organisms as insulin stimulation triggers rapamycin-dependent hyperphosphorylation of both S6K1 and 4E-BP1 (Fig. 3) (for review, see Oldham and Hafen 2003). The exact relationship in flies and mammals of TOR to the insulin pathway has been a subject of much interest, but current evidence suggests that a simple upstream/downstream relationship does not exist.

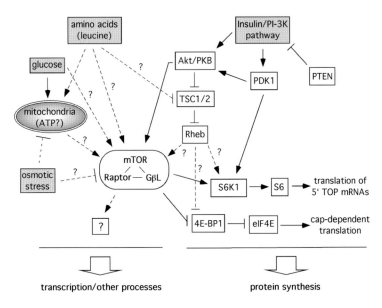

Figure 3. The mTOR signaling network. In this model, the mTOR complex is at the center of a network of diverse input signals that are coordinated into a growth response through regulating cellular processes such as protein synthesis and transcription. The major input signals are represented as shaded figures; solid lines represent known interactions; dashed lines indicate that the important intermediates that regulate these interactions are not known. The best-characterized inputs to the mTOR complex are growth factors that signal through the PI3K branch of the insulin signaling pathway, and nutrients, particularly amino acids such as leucine. Question marks imply that uncertainty still exists as to how these interactions might feed into the mTOR network. For instance, it has not yet been resolved whether TSC1-TSC2/Rheb branch functions through the mTOR complex to regulate S6K1 and 4E-BP1, or whether it functions in a parallel pathway. In the case of the nutrient and metabolic inputs, it is not known whether the described inputs feed into the mTOR complex independently, or whether they converge on a common upstream organizer (such as the mitochondria) that then mediates a nutrient-derived signal to mTOR. Activation of mTOR triggers protein synthesis by phosphorylating S6K1 and 4E-BP1, which triggers subsequent activation and derepression of S6 and eIF-4E, respectively. The PI3K pathway can also regulate the protein synthesis machinery independently of mTOR through PDK1. How mTOR regulates transcription and other cell processes (see Fig. 2) is not known.

Mutations in flies in the genes encoding conserved components of the mammalian insulin pathway, including a receptor with similarity to the insulin receptor, PI3K, Chico (homolog of mammalian IRS1-IRS4), dPTEN, and dAkt/dPKB, cause defects in cell growth, proliferation, and survival, but are not necessarily lethal (Fernandez et al. 1995; Chen et al. 1996; Leevers et al. 1996; Bohni et al. 1999; Goberdhan et al. 1999; Huang et al. 1999; Verdu et al. 1999; Weinkove et al. 1999; Gao et al. 2000). On the other hand, as mentioned above, *Drosophila* TOR is required for viability. dTOR seems to exist in a pathway parallel or downstream from the insulin pathway. For instance, flies with mutations in both dPTEN (a negative regulator of P13K) and dTOR have a dTOR mutant phenotype (Zhang et al. 2000). Likewise, the *Drosophila* homologs of the tumor suppressors TSC1 and TSC2 (downstream effectors of the insulin pathway that negatively regulate growth) have a similar genetic relationship with dTOR (Gao et al. 2002) (discussed below). Complicating the issue further is the fact that in mammals the PI3K pathway clearly has an mTOR-independent signaling input to S6K1 and 4E-BP1. In particular, PDK1 can directly signal to S6K1 (Alessi et al. 1998; Pullen et al. 1998) and, surprisingly, a hypomorphic allele of PDK1, expressing ~10% of normal PDK1 levels, results in viable and fertile mice that are nearly half the size of control mice (Lawlor et al. 2002). How this occurs is unclear, because the ability of insulin to activate Akt/PKB and S6K1 is normal in these mice. Evidence from mammalian tissue culture further supports the notion that S6K1 receives mTOR-dependent and -independent inputs, as expression of an amino- and carboxy-terminal truncated version of S6K1 is insensitive to the effects of rapamycin, but still sensitive to PI3K inhibition with wortmannin (Cheatham et al. 1995; Dennis et al. 1996; Mahalingam and Templeton 1996; Romanelli et al. 2002). Another reason it has been difficult to characterize the growth factor inputs to mTOR is that nutrients, independently of growth factors, can regulate TOR signaling (see below on the role of nutrients in mTOR signaling). In fact, the defects caused by loss of dTOR in *Drosophila* resemble those caused by starvation and are more severe than those found in flies with mutations in insulin pathway components (Britton et al. 2002; Zhang et al. 2000). A similar situation is seen in *C. elegans,* as recent findings show that animals with a decrease in ceTOR function have phenotypes reminiscent of amino-acid-starved animals rather than of mutants in components of the insulin pathway (Long et al. 2002). Thus, the control of growth through TOR signaling molecules, although clearly respondent through some mechanism to growth factors, is also mediated by additional, diverse inputs.

Despite the complicated genetic relationship between insulin and TOR signaling, there have been intense efforts to identify molecular links between components of the insulin pathway and mTOR. Whereas TSC1 and TSC2 function may be emerging as one potential connection between them (see below on mTOR and disease), another mechanism by which the

PI3K pathway might feed into the mTOR network is through direct modification of mTOR by Akt/PKB. Studies using a polyclonal antibody that binds an epitope (amino acids 2433–2450) in the carboxyl terminus of mTOR (Brunn et al. 1997a; Scott et al. 1998) led to the discovery that Akt/PKB directly phosphorylates mTOR. Insulin stimulation or artificial activation of Akt/PKB decreases the binding of this antibody to mTOR, which correlates with an increase in mTOR phosphorylation and in mTOR in vitro activity toward 4E-BP1. The sequence of the epitope recognized by the antibody contains Akt/PKB consensus phosphorylation sites, and, indeed, Akt/PKB phosphorylates mTOR at Ser-2448, a site that is constitutively phosphorylated in some prostate cancer cell lines with deregulated Akt/PKB signaling (Nave et al. 1999; Sekulic et al. 2000). Unexpectedly, Akt/PKB-dependent phosphorylation of Ser-2448 is not essential for S6K1 or 4E-BP1 phosphorylation within cells, because an mTOR variant with a S2448A mutation signals insulin stimulation as effectively as wild-type mTOR (Sekulic et al. 2000). Interestingly, deletion of the domain containing the Akt/PKB phosphorylation site activates mTOR both in vitro and in vivo, suggesting that this region may act as a repressor domain, but the function of the Akt/PKB-mediated phosphorylation remains poorly understood. This repressor domain of mTOR is absent in yeast, raising the possibility that this region of the protein evolved more recently, perhaps to receive inputs from growth factor signaling. Akt/PKB and the mTOR pathway also intersect in a more subtle way. The pro-survival function of Akt/PKB requires its capacity to phosphorylate and inactivate mediators of apoptosis (for review, see Datta et al. 1999) and to stimulate metabolism, which it does in part by maintaining the surface expression of transporters for glucose, amino acids, and iron (Gottlob et al. 2001; Plas et al. 2001; Edinger and Thompson 2002). Importantly, a constitutively active mutant of Akt/PKB promotes survival in the absence of growth factors and induces the cell surface expression of transporters in a rapamycin-sensitive manner, implicating mTOR in this response (Edinger and Thompson 2002).

The insulin growth factor signaling pathway clearly mediates a "growth" signal to mTOR, but what about the role for mTOR itself in mammalian growth control and development? Insight into this question came unexpectedly from a phenotypic screen in mice for mutations that disrupt forebrain development (Hentges et al. 1999). This screen identified a lethal mutation in the *mTOR* gene that leads to a *flat-top* mouse embryo characterized by defective proliferation in the telencephalon and failures of expression of telencephalon marker genes and of the embryos to rotate around the embryonic body axis (Hentges et al. 2001).

Biochemical analyses show that mTOR activity is reduced in *flat-top* embryos, as judged by S6K1 and 4E-BP1 phosphorylation and activity (Hentges et al. 2001). Both S6K1 and 4E-BP1 are normally expressed in the brain regions with decreased proliferation in *flat-top* embryos. Strikingly, rapamycin treatment of developing embryos phenocopies the *flat-top* mutation, strongly suggesting that the effects of rapamycin are due to a defect in mTOR signaling. Preliminary studies did not find cell size defects in a limited sampling of cells derived from the *flat-top* embryos, but, compared to wild types, mutant embryos have one-third fewer cells. Further analysis will be required to conclusively determine whether *flat-top* mutants have cell size defects. Alternatively, perhaps in mice mTOR is critical for proliferation during embryonic development, in which division occurs relatively rapidly and the growth and proliferation machinery are in constant communication. mTOR's role in mass accumulation may not be obvious until later stages of development, in which differentiated cells have exited the cell cycle. This may be the case in nonproliferating muscle cells, because transgenic mice overexpressing Akt1/PKB1 in the heart have a rapamycin-sensitive increase in heart size caused by cardiomyocyte enlargement (Shioi et al. 2002). Disruption of S6K1 in mice, similar to flies, does not affect viability or fertility, but significantly reduces body weight and animal size (Shima et al. 1998). Surprisingly, S6 is still efficiently phosphorylated in these animals, a result that led to the discovery of the partially redundant S6K2 (Shima et al. 1998; Lee-Fruman et al. 1999). Disruption of 4E-BP1 in mice results in viable animals and, in one study, produced male mice weighing roughly 10% less than control mice (Blackshear et al. 1997; Tsukiyama-Kohara et al. 2001). The fact that mice defective for S6K1 or 4E-BP1 have much less severe defects than those without mTOR function suggests that mTOR has essential functions apart from regulating these two translational regulators. To study these roles at the organismal level, mice with conditional loss-of-function alleles of mTOR are needed.

Studies in mammalian tissue culture cells reinforce a role for mTOR in growth control. Treatment of U2OS (human osteosarcoma) cells with rapamycin or LY294002 (an inhibitor of PI3K and mTOR as well as other PIKKs) reduces cell size and number (Fingar et al. 2002; Kim et al. 2002). The LY294002-induced size defect is more pronounced than that caused by rapamycin, consistent with PI3K contributing additional mTOR-independent signals to the growth machinery. Expression of a rapamycin-resistant mTOR mutant rescues the size defects induced by rapamycin, confirming that the drug acts by perturbing mTOR function. Conversely, overexpression of the mTOR effector S6K1, or eIF4E, the target of 4E-BP1 inhibition, synergistically increases cell size (Fingar et al. 2002). Similar to rapamycin treatment, siRNAs directed against the mRNAs encoding mTOR, raptor, or GβL decrease the size of human cells in culture (Kim et al. 2002, 2003). Thus, based on the available evidence, it is likely that mTOR and its associated proteins have a conserved role in growth control.

Role of Nutrients in mTOR Signaling

Genetic studies in model organisms clearly suggest a central role for mTOR in growth regulation, but the identity of the primary input to mTOR is still controversial. As discussed in the previous section, growth factors likely provide an important input signal, but a key aspect of TOR signaling, especially in yeast, is its role in nutrient sensing, and available evidence suggests that mTOR participates in an analogous system in mammals. Like rapamycin treatment, depriving cells for amino acids or other nutrients in culture causes the rapid dephosphorylation of S6K1 and 4E-BP1 (Hara et al. 1998; Lynch et al. 2000; Lynch 2001). Changes in the levels of leucine are sufficient to affect mTOR effectors (Hara et al. 1998; Lynch et al. 2000), and amino acids structurally related to leucine, such as isoleucine and valine, can also stimulate S6K1 and 4E-BP1 phosphorylation, but less potently (Lynch 2001). Initial studies were undertaken in CHO-IR, HEK293T, Swiss 3T3, and adipocyte cells, but similar findings have since been reported in diverse cell types (Patti et al. 1998; Xu et al. 1998; Kimball et al. 1999). It is not yet clear how leucine regulates the mTOR pathway, but it may signal by affecting tRNA aminoacylation, mitochondrial metabolism, and/or through a cell surface receptor (Iiboshi et al. 1999; Shigemitsu et al. 1999; Xu et al. 2001).

In pancreatic β cells, the importance of mitochondrial metabolism in stimulating the mTOR pathway has been well established (Xu et al. 1998, 2001). For exogenous insulin to induce the phosphorylation of 4E-BP1 and S6K1 in β cells in culture, leucine must be present in the media (Xu et al. 1998, 2001). Interestingly, this effect of leucine requires signals derived from the mitochondrial metabolism of the amino acid (Sener and Malaisse 1980; Fahien et al. 1988), and these same signals are also needed to activate the mTOR pathway in β cells (Xu et al. 2001). Rapamycin or several mitochondrial inhibitors block leucine-mediated activation of the pathway without reducing global cellular ATP levels. Thus, it appears that mTOR has an important role in β-cell function, perhaps by coordinating nutrient inputs with the protein synthesis machinery to maintain β-cell mass and proliferative potential. At physiological concentrations, leucine is both a substrate for oxidative decarboxylation and an activator of glutamate dehydrogenase (GDH) (Sener and Malaisse 1980; Fahien et al. 1988), metabolic functions that may provide leads for the eventual identification of the mitochondrial-derived secondary messengers that regulate the mTOR pathway.

How do changes in amino acid levels and mitochondrial signals affect signaling by the mTOR complex? The recent identification of raptor provides some interesting clues (Hara et al. 2002; Kim et al. 2002). A decrease in raptor expression induced by RNA interference (RNAi) attenuates leucine-stimulated phosphorylation of S6K1 and reduces cell size, indicating that raptor has a positive role in mTOR signaling (Kim et al. 2002). Previous studies had failed to identify raptor as an mTOR-associated protein because the association is unstable and broken in commonly used detergents (Hara et al. 2002; Kim et al. 2002). Close examination of the stability of the mTOR–raptor interaction indicates that nutrient deprivation strengthens the association and correlates with a decrease in the in vitro kinase activity of mTOR. Treatment with valinomycin, a mitochondrial uncoupler; antimycin, an electron transport inhibitor; 2-deoxyglucose, a glycolysis inhibitor; or H_2O_2 to induce mitochondrial stress has similar effects. In contrast, a brief stimulation of leucine-deprived cells with leucine reduces the amount of raptor associated with mTOR and correlates with an increase in mTOR in vitro kinase activity (Kim et al. 2002). This same phenomenon occurs with glucose deprivation and re-addition. Thus, mTOR and raptor may exist in a two-state nutrient-sensitive interaction: mTOR and raptor are tightly bound in a stable, inactive complex when nutrient levels are low, and loosely associated but active in an unstable complex when nutrient levels are sufficient for growth. Formation of this two-state complex requires the WD-40-containing protein GβL, because in its absence the interaction between mTOR and raptor is no longer nutrient-sensitive (Kim et al. 2003). Long-term insulin withdrawal or addition does not affect the stability of the mTOR–raptor complex (Kim et al. 2002), suggesting that the function of raptor may be to integrate nutrient-derived signals to mTOR independently of growth factors. In support of such an idea, raptor and GβL are evolutionarily conserved in fission and budding yeast, organisms without growth factor-mediated signaling (Shinozaki-Yabana et al. 2000; Loewith et al. 2002; Wedaman et al. 2003).

What exactly constitutes the nutrient sensor is still a puzzling question. As both mTOR and raptor are relatively large proteins that are part of a big complex, one model is that the complex itself contains the sensor. Such a large complex could accommodate multiple inputs from a variety of sources, such as transiently interacting proteins, or small molecules or metabolites that bind directly to mTOR or its associated proteins. It is also possible that the sensor is upstream of mTOR, receiving nutrient-derived signals and modifying the mTOR complex accordingly, perhaps through posttranslational modifications. One potential candidate is the TSC1–TSC2 complex, which some reports suggest functions upstream of mTOR and whose absence renders S6K1 and 4E-BP1 resistant to amino acid deprivation (Gao et al. 2002; Inoki et al. 2002; Jaeschke et al. 2002; Kenerson et al. 2002; Kwiatkowski et al. 2002; Tee et al. 2002). Alternatively, mTOR may be a direct sensor of intracellular ATP concentrations, which, independently of amino acid levels, reflect the metabolic state of the cell (Dennis et al. 2001). It is tantalizing that inhibition of mitochondrial function has such a pronounced effect on mTOR activity (Dennis et al. 2001; Kim et al. 2002). Because mitochondria are the energy factories of cells, and protein synthesis is a major consumer of cellular energy, it seems appropriate that signaling networks have evolved to relay the energetic state of the cell to the protein synthesis machinery. The challenge now is to determine the nature of the metabolic signals that report the levels of amino acids and other nutrients to the mTOR pathway.

What Is the Logic of mTOR Inputs?

In animals, evolution has apparently added growth factor control to the conserved nutrient-derived signals that regulate cell growth through mTOR. The convergence points and cross-talk mechanisms between nutrient and growth factor signaling on the mTOR complex are still a mystery. Tight control at both levels is surely important to properly regulate cell, organ, and organism size, but many questions remain unanswered. What is the major stimulatory input to mTOR, amino acids or other nutrients, or growth factors? Furthermore, why have a nutrient-sensitive control mechanism in a multicellular organism, in which presumably nutrients are not limiting in the context of individual cellular environments? What are nutrient levels like in a tumorigenic environment? The last question is of particular importance, because losing sensitivity to low nutrient levels may give cancer cells a selective advantage and contribute to the development or progression of neoplasia. These questions may be better understood once the primary signaling mechanisms to mTOR are revealed.

Two models can be proposed to explain the relationship between nutrients and growth factors in regulating mTOR signaling in mammals (Fig. 4). In the first model, nutrient signaling serves as the primary stimulus to mTOR, as is the case in simpler organisms like yeast that respond

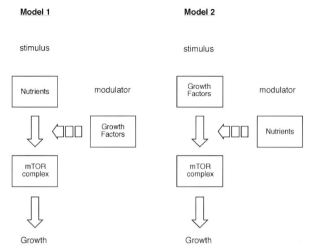

Figure 4. What is the logic of mTOR inputs? It is still a mystery as to what is the primary stimulus to the mTOR complex that is important for regulating growth. Regulation of mTOR by nutrients is conserved from yeast to man, thus Model 1 suggests that nutrients are the primary stimulus and perhaps the pathway has come under control of a growth factor signal in higher eukaryotes to modulate the intensity of the signal in certain tissues or during specific developmental stages. According to Model 2, growth factors have succeeded nutrient signaling as the primary stimulus to the mTOR complex, and the nutrient signals modulate the intensity of signaling. In this scenario, nutrients would serve a checkpoint function only, allowing growth to proceed when sufficient nutrients are available to support both basal cell functions and growth.

only to environmental nutrients. In animals, growth factor control may have been grafted onto the ancient nutrient-signaling pathway to allow a modification of the intensity of signaling in certain tissues or during certain developmental stages. In development, there are instances when cells grow without dividing, and other cases where cells divide without growing, both phenomena perhaps utilizing growth factor modulation of a nutrient-stimulated growth pathway. In an alternative model, growth factors have succeeded nutrient signaling as the primary stimulus to mTOR. In this case, nutrients (or stress) may serve a checkpoint function, only allowing mTOR signaling when nutrient levels are high enough to support both cell growth and basal cellular functions. Much of the work on mTOR signaling has been done in tissue culture cells, and it is not clear what aspects of in vivo metabolism these cells truly represent. It will be important in the future to work out the primary signals to mTOR in a whole-organism model.

Of course, one can imagine that many other conditions besides growth factor and nutrient adequacy must be met for dividing cells to accumulate the mass needed for proliferation. If mTOR is a major integrator of signals that convey the overall status of the cellular environment to the protein synthesis machinery, it seems reasonable to predict that many signals may converge on mTOR. In fact, one report suggests that phosphatidic acid (PA), a lipid second messenger in some mitogenic pathways, positively regulates mTOR (Fang et al. 2001). PA interacts with mTOR in the FRB domain, suggesting that FKBP12–rapamycin may perturb mTOR function by disrupting the PA–mTOR interaction. Another signal, osmotic stress, negatively regulates mTOR signaling (Parrott and Templeton 1999) and, like rapamycin, causes dephosphorylation of S6K1 and 4E-BP1. Consistently, rapamycin-resistant S6K1 mutants are likewise resistant to the dephosphorylation triggered by osmotic stress. This form of stress is not mediated by traditional stress-activated pathways, and may signal by perturbing a mitochondrially derived signal (Desai et al. 2002). These observations are consistent with mTOR serving as an integrator of multiple inputs.

LINK BETWEEN mTOR AND THE CELL CYCLE

In most dividing cells, growth and division are linked processes in which mass accumulation in G_1 is required for progression into S phase. Pioneering studies in yeast indicate that cells must reach a critical size before entering the cell cycle (Fantes and Nurse 1977; Johnston et al. 1977; Barbet et al. 1996; Polymenis and Schmidt 1997; Daga and Jimenez 1999). Division must be coordinated with cell growth to efficiently maintain an appropriate size over generations. Observations that pro-proliferative oncogenes such as Myc, Ras, and cyclin D promote aspects of cell growth suggest that these genes may have roles in linking growth and proliferative control (Johnston et al. 1999; Schuhmacher et al. 1999; Prober and Edgar 2000, 2002). Myc promotes growth at least in part via the transcriptional activation of a number of genes involved in protein synthesis and metabolism (Guo et al. 2000; Schuhmacher et al. 2001). Ras activation can activate the PI3K pathway (Rodriguez-Viciana et al. 1994) as well as cause phosphorylation of eIF-4E, the target of 4E-BP1 inhibition (Pyronnet 2000; Scheper et al. 2001). On the other hand, mTOR may have a role in cell cycle control as rapamycin slows the G_1–S transition in many cell types, and expression profile analysis in mammalian cells shows that inhibition of mTOR by rapamycin reduces the transcription of many genes involved in cell cycle regulation (Peng et al. 2002).

In some cell types, cyclin D1 may be an important mediator of mTOR-dependent growth control. Complexes between cyclin-dependent kinases (cdk) and cyclins control progression through the cell cycle, and the cyclin-D-dependent kinases regulate the G_1/S transition by phosphorylating the retinoblastoma protein (Rb), which leads to its dissociation from the E2F transcription factor (for review, see Sherr and Roberts 1999). Free E2F then induces the transcription of S-phase genes, including G_1/S and S-phase cyclins, which further increase CDK activity toward Rb in a positive feedback loop, and ultimately trigger entry into S phase. Cyclin D1 has a well-established role in cell division as its overexpression in a variety of mammalian tissues (including human mammary, squamous epithelial, and parathyroid endocrine cells) causes hyperplasia and, in some cases, tumor-like masses (Wang et al. 1994; Robles et al. 1996; Imanishi et al. 2001). Treatment of serum-stimulated NIH-3T3 cells with rapamycin inhibits the expression of cyclin D1 (Hashemolhosseini et al. 1998), and similar effects have been observed in other cell types (for review, see Hidalgo and Rowinsky 2000). This suggests that regulation of cyclin D1 expression may be a downstream function of mTOR signaling that links growth regulation to cell proliferation. The situation is not so simple, however, because in *Drosophila*, ovexpression of cyclin D1 has different effects in different tissues (Datar et al. 2000). In dividing wing imaginal cells, increased expression of cyclin D1 triggers increased cell proliferation, whereas in eye and endoreplicating salivary gland cells, increased cyclin D1 expression enhances growth (as well as increasing DNA replication in salivary gland cells). Moreover, mice lacking cyclin D1 are smaller overall in size than control mice and have severe proliferation defects in several tissues, including the retina and mammary epithelium (Fantl et al. 1995; Sicinski et al. 1995). In sum, it appears cyclin D1 may function in both proliferation and growth control, probably in a tissue-specific manner, making it difficult to understand the relationship between mTOR, cyclin D1, growth, and proliferation. It may be that mTOR signaling functions differently in mitotic cells versus nondividing cells, but whether this is a cause or effect of proliferation is not clear, and to fully understand this, mTOR function will need to be studied in a tissue-specific manner.

A particularly clear connection between mTOR and cyclin D1, however, has been discovered in hepatocytes. These cells are highly differentiated but retain the remarkable ability to proliferate in response to a loss in liver mass (Fausto 2000). Transient overexpression of cyclin D1 in hepatocytes generates active cyclin D1–CDK complexes that trigger proliferation and liver growth in mice (Nelsen et al. 2001). Rapamycin, which slows cell cycle progression in a variety of cultured cells, inhibits rat hepatocyte proliferation in response to mitogens like insulin and EGF (Francavilla et al. 1992; Boylan et al. 2001; Nelsen et al. 2003). Interestingly, in cultured rat hepatocytes, expression of either cyclin D1 or E2F, a downstream effector of cyclin D1–CDK, overcomes the rapamycin-induced block in cell cycle progression and triggers protein synthesis. Importantly, rapamycin has similar effects in vivo, as it also blocks liver regeneration and cyclin D1 expression caused by partial hepatectomy in mice (Nelsen et al. 2003). Restoration of cyclin D1 expression in rapamycin-treated mice allows hepatocytes to regenerate liver mass (Nelsen et al. 2003). Thus, at least in hepatocytes, cyclin D1 appears to be a critical downstream target of mTOR signaling for regulating growth.

Rapamycin also inhibits cyclin-A- and cyclin-E-dependent kinase complexes in T cells (Morice et al. 1993a,b). In T cells, rapamycin blocks interleukin-2-dependent down-regulation of the CDK inhibitor p27/Kip1, and up-regulates its expression at both the mRNA and protein levels (Nourse et al. 1994; Kawamata et al. 1998). Moreover, murine fibroblasts and T cells disrupted for the p27/Kip1 gene are less responsive to rapamycin (Luo et al. 1996). Consistent with p27/Kip1 functioning in growth control, mice lacking p27/Kip1 are significantly larger than controls, exhibiting severe hyperplasia in several organs (Kiyokawa et al. 1996; Nakayama et al. 1996). It seems likely that mTOR signaling contributes to cell cycle regulation by a variety of means and that these effects are cell-type-specific. Understanding how growth control and proliferation are linked in the mammalian cell cycle will be an important area of research in the future.

DEREGULATION OF THE mTOR PATHWAY IN HUMAN DISEASE

Tuberous Sclerosis

Until recent work on the disease tuberous sclerosis (TSC), a correlation between deregulated mTOR signaling and human disease had not been found. Several independent reports indicate that the products of the *TSC1* and *TSC2* tumor suppressor genes, which encode hamartin (TSC1) and tuberin (TSC2), respectively, negatively regulate the mTOR pathway (Gao et al. 2002; Inoki et al. 2002; Jaeschke et al. 2002; Kenerson et al. 2002; Kwiatkowski et al. 2002; Tee et al. 2002). Mutations in either *TSC1* or *TSC2* can lead to the onset of tuberous sclerosis, an autosomal dominant disorder affecting about 1 in 6000 newborns in the United States (for review, see Young and Povey 1998). Patients afflicted with the disease develop benign tumors (called harmatomas) and malformations (tubers and hamartias) in several organs, including the brain, kidneys, skin, and heart. Mutations in *TSC2* are also linked to lymphangioleiomyomatosis (LAM), a disease marked by benign lesions of smooth-muscle-like cells in the lung that can lead to pulmonary failure (Carsillo et al. 2000; Franz et al. 2001; Yu et al. 2001). Although tuberous sclerosis rarely results in a malignant cancer, the tubers can lead to a variety of serious neurological complications, including epilepsy, mental retardation, and autism (Gomez et al. 1999). Strikingly, brain lesions from patients with tuberous sclerosis contain abnormally giant cells that are up to 5–10 times larger than a neuron but have a normal DNA content (Mizuguchi and Takashima 2001). The *TSC1* and *TSC2* gene products function together as a stable complex (van Slegtenhorst et al. 1998; Plank et al. 1999), and several disease-associated mutations in *TSC1* and *TSC2* disrupt complex assembly.

Early hints that TSC might have a role in growth control came from studies in *Drosophila* in which the *gigas* mutant containing large cells was found to have a mutation in the fly ortholog of human *TSC2* (Ferrus and Garcia-Bellido 1976; Ito and Rubin 1999). Further analyses in *Drosophila* suggest that dTSC1 and dTSC2 regulate both cell size and proliferation and that they act downstream or in parallel to the dAkt/dPKB and dPTEN components of the insulin signaling pathway (Fig. 3) (Gao and Pan 2001; Potter et al. 2001; Tapon et al. 2001). In addition, in flies, overexpression of dS6K counters the decrease in cell size caused by co-overexpressing dTSC1 and dTSC2, suggesting that the dTSC complex functions either at or upstream of the level of dS6K. Furthermore, the mTOR effectors S6K1 and 4E-BP1 are hyperphosphorylated in mammalian and *Drosophila* tissue culture cells with decreased or absent TSC1 or TSC2 function (Gao et al. 2002; Inoki et al. 2002; Jaeschke et al. 2002; Kenerson et al. 2002; Kwiatkowski et al. 2002; Tee et al. 2002). Conversely, overexpression of TSC1 and TSC2 causes the dephosphorylation of S6K1 and 4E-BP1 and blocks insulin- and amino acid-mediated activation of S6K1 (Goncharova et al. 2002; Inoki et al. 2002; Jaeschke et al. 2002; Tee et al. 2002). Cell size studies show that deletion of *TSC1* results in an increase to nearly two times the size of wild-type cells, whereas in contrast, *dTOR* deletion cells are decreased to roughly one-quarter the size of wild-type cells (Gao et al. 2002). The combination double deletion of *TSC1* and *dTOR* results in cells comparable in size to a *dTOR* deletion alone, and combining a *dS6K* deletion with a *TSC1* null leads to partial suppression of the increased cell size phenotype. Together, these observations implicate TSC1 and TSC2 as upstream negative regulators of S6K1 and positive regulators of 4E-BP1. Interestingly, in *Drosophila*, recombinant TOR interacts with recombinant TSC1 and TSC2 as well as endogenous TSC2 (Gao et al. 2002). However, interactions between endogenous TOR and TSC1 or TSC2 have not been shown in *Drosophila* or in mammals. Moreover, the exact relevance of the detected interaction to the function of TSC1/2 remains to be seen.

As in *Drosophila*, the mammalian TSC complex is also part of the insulin signaling pathway as Akt/PKB negatively regulates it through direct phosphorylation, which leads to ubiquitin-mediated destruction of TSC2 and disruption of the TSC1–TSC2 complex. Most interestingly, loss of TSC function in both *Drosophila* and mammalian cells renders S6K1 and 4E-BP1 phosphorylation resistant to amino acid starvation (Gao et al. 2002). The effect is specific to TSC, as these downstream effectors of mTOR are still sensitive to amino acid withdrawal in cells deficient for PTEN, another upstream negative regulator of the insulin pathway (Gao et al. 2002). Thus, TSC1 and TSC2 appear to have a critical role in regulating the activity of well-characterized effectors of mTOR (Inoki et al. 2002; Manning et al. 2002; Potter et al. 2002).

How does the TSC1–TSC2 complex regulate effectors of mTOR? TSC1 contains an ERM (ezrin-radixin-moesin)-binding domain and may regulate cell adhesion and Rho signaling (Lamb et al. 2000), whereas TSC2 has a GTPase activating protein (GAP) domain (Wienecke et al. 1995; Xiao et al. 1997). Early studies suggested TSC2 may mediate intracellular trafficking (Kleymenova et al. 2001) or cell cycle and transcriptional control (Tsuchiya et al. 1996; Henry et al. 1998; Soucek et al. 2001). However, a recent flurry of papers indicates that a primary function of TSC2 (and thus the TSC1–TSC2 complex) is to regulate the activity of the Ras-related GTPase Rheb (Castro et al. 2003; Garami et al. 2003; Inoki et al. 2003; Saucedo et al. 2003; Stocker et al. 2003; Zhang et al. 2003). Rheb was originally identified as a gene induced by synaptic activity, whose product has an intrinsic GTPase function and is highly expressed in the brain (Yamagata et al. 1994). Using both *Drosophila* and mammalian tissue culture cells, several groups working to identify the GTPase downstream of TSC2 discovered that TSC2 GAP activity specifically targets Rheb (Castro et al. 2003; Garami et al. 2003; Inoki et al. 2003; Zhang et al. 2003). Overexpression of Rheb stimulates the phosphorylation and activation of S6K1 and phosphorylation of 4E-BP1, whereas loss of Rheb function either genetically or by RNAi has the opposite effects (Castro et al. 2003; Garami et al. 2003; Inoki et al. 2003; Saucedo et al. 2003; Stocker et al. 2003; Zhang et al. 2003). Rapamycin blocks the effects of Rheb overexpression on S6K1 (Castro et al. 2003; Garami et al. 2003; Inoki et al. 2003), and taken together, these results suggest a model in which Rheb acts downstream of TSC1–TSC2, and upstream of TOR, but it cannot be ruled out that Rheb functions to inhibit a negative regulator of S6K1 and 4E-BP1, such as a TOR-regulated phosphatase. In *Drosophila*, increasing or decreasing Rheb expression dramatically enlarges or shrinks cell size, respectively, consistent with Rheb being an important regulator of growth downstream of TSC1–TSC2 (Saucedo et al. 2003; Stocker et al. 2003).

As mentioned above, loss of TSC function, which presumably up-regulates Rheb activity, renders S6K1 and 4E-BP1 resistant to amino acid starvation. Likewise, overexpression of Rheb has similar effects, causing S6K1 to become resistant to amino acid starvation and energy depletion, but, interestingly, not to osmotic stress (Garami et al. 2003; Inoki et al. 2003). In fission yeast, deletion of the *Rheb* homolog *rhb1* results in a phenotype similar to nitrogen-starved cells (Mach et al. 2000). These data raise the speculative possibility that nutrient sensing by the TOR network may occur at the level of TSC1–TSC2/Rheb.

The major challenge now is to determine whether the TSC1 and TSC2 tumor suppressors and Rheb function through mTOR or in a parallel pathway, and the relevance of nutrients and growth factors to controlling TSC1–TSC2/Rheb function. A related and equally important issue is to determine the mechanism by which TSC1–TSC2/Rheb affects the phosphorylation state of S6K1 and 4E-BP1. Interestingly, Rheb function, similar to the Ras oncogene, may require farnesylation, a posttranslational lipid modification that allows proteins to attach to membranes (Clark et al. 1997; Yang et al. 2001). Inhibition of farnesylation with farnesyl transferase inhibitors (which were designed to inhibit Ras activity) blocks activation of S6K1 induced by Rheb-overexpression (Cox and Der 2002; Castro et al. 2003); thus, Rheb activity likely requires membrane association and may be a clinically relevant target of such inhibitors. Resolving these issues will not only shed light on the etiology of tuberous sclerosis, but may also uncover important links between nutrient sensitivity, growth signaling, and tumor development.

A Cancer Role for mTOR, PI3K, and PTEN

PTEN is a lipid phosphatase that negatively regulates insulin signaling by dephosphorylating the products of PI3K (phosphatidylinositol-3, 4-biphosphate and phosphatidylinositol-3,4,5-triphosphate; PIP_3) at the D3 position. PTEN is a negative regulator of Akt/PKB, since in the absence of PTEN, cells have elevated levels of PIP_3, which serves as a membrane docking site for a variety of proteins containing PH domains, including Akt/PKB (Haas-Kogan et al. 1998; Stambolic et al. 1998; Whang et al. 1998; Wu et al. 1998). PTEN is a tumor suppressor that is disrupted in many cancers, including glial and prostate tumors, melanomas, endometrial carcinomas, and breast cancer (for review, see Cantley and Neel 1999;

Simpson and Parsons 2001; Vogt 2001). In PTEN-deficient tumors, Akt/PKB activity is significantly elevated, and Akt/PKB itself is overexpressed in several cancers, including gastric, breast, ovarian, pancreatic, and prostate cancer (for review, see Cantley and Neel 1999; Vogt 2001). Thus, aberrant Akt/PKB activation in PTEN-deficient tumors could be the important event with respect to certain cancers. Genetic experiments in *Drosophila* back this hypothesis, because a version of Akt/PKB with an inactivating mutation in the PH domain is sufficient to rescue flies with elevated PIP_3 levels caused by a PTEN deficiency (Stocker et al. 2002). Aberrant Akt/PKB activation could lead to cancer through hyperactivation of the pro-survival response, mTOR signaling, or a combination of these and other effects.

Since substantial evidence suggests that the mTOR and insulin pathways intersect, and since mTOR itself is a direct target of Akt/PKB phosphorylation, mTOR dysfunction may contribute to the development of cancers caused by deregulated PI3K signaling. In support of this possibility, rapamycin blocks the transformation of chicken embryo fibroblasts by P3k (a homolog of the catalytic subunit of class IA PI3K) or activated Akt/PKB (Aoki et al. 2001). The block correlates with a decrease in the constitutive phosphorylation of S6K1 observed in these cells. Moreover, the effects of rapamycin are specific to P3k and Akt/PKB, as the drug has no effect on the transforming ability of several other oncoproteins including Src, Yes, Sea, Abl, Fps, ErbA/ErbB, Crk, Mos, Jun, and Fos (Aoki et al. 2001). Unexpectedly, rapamycin enhances the transforming activity of Myc and Ras, by an unknown mechanism. Studies on the effects of a rapamycin analog, CCI-779, on PTEN-defective tumor cells also suggest an important role for mTOR signaling in transformation mediated by PI3K and Akt/PKB. (Neshat et al. 2001; Podsypanina et al. 2001). Mice heterozygous for *PTEN* frequently develop tumors of the uterus and adrenal medulla (Podsypanina et al. 2001) that are characterized by absent PTEN expression, elevated levels of phosphorylated Akt/PKB, hyperactive S6K1, and increased proliferation. Importantly, inhibition of mTOR signaling with CCI-779 reduces proliferation, tumor size, and S6K1 activity without affecting Akt/PKB phosphorylation. CCI-779 has similar effects in vitro on both mouse and human cells, as *PTEN* null tumor lines have CCI-779-sensitive S6K1 activity (Neshat et al. 2001). Consistently, CCI-779 blocks the proliferation of *PTEN* null cells and the growth of tumors with wild-type PTEN but activated Akt/PKB. Taken together, these findings suggest that a deregulation in mTOR signaling may be crucial to tumor development in cancers with PTEN inactivation/PI3K hyperactivation. In support of this, many multiple myeloma tumor cells also have an activated PI3K–Akt pathway (Hsu et al. 2001), and these cells are also sensitive to CCI-779 (Shi et al. 2002). These observations indicate that mTOR and mTOR effectors are promising targets for drug development for the treatment of cancers due to aberrant PI3K signaling.

How might deregulation of mTOR contribute to cancer in PTEN-deficient tumors? The prominent role of S6K1 in translation and ribosome biogenesis suggests that protein synthesis may be a major factor. Moreover, 4E-BP1 phosphorylation is also elevated in $PTEN^{-/-}$ mouse embryonic fibroblasts compared to $PTEN^{+/+}$, and this hyperphosphorylation, which correlated with decreased binding to eIF4E, was sensitive to CCI-779 as well (Neshat et al. 2001). In addition, eIF4E can transform NIH-3T3 and Rat2 cells (Lazaris-Karatzas et al. 1990). Whereas a potential link between tumorigenesis of PTEN-deficient cells and elevated mTOR-dependent protein synthesis seems plausible, it is equally possible that there could be particular targets of mTOR whose deregulation may contribute to cancer. Because mTOR-dependent control of protein synthesis regulates expression of many proteins, including growth and cell cycle regulators, and mTOR regulates proteins like Lipin and CLIP-170, which are not apparently involved in protein synthesis, it will be important in the future to identify the relevant targets with respect to tumor development.

A Potential Link between mTOR and Diabetes

Circumstantial evidence suggests that the mTOR pathway may play a role in certain types of diabetes. Mice null for the mTOR effector S6K1 develop a form of diabetes remarkably similar to protein-malnutrition diabetes (Pende et al. 2000). This disease occurs in adults who as children were deprived of adequate protein, is common in certain parts of Africa, and is associated with a decrease in the size of insulin-secreting β cells in the islets of Langerhans (for review, see Rao 1984; Swenne et al. 1992).

An interesting role for mTOR in setting the intensity of the insulin signaling pathway suggests that mTOR could also contribute to the development of insulin resistance. Unexpectedly, mTOR appears to regulate insulin signaling through negative feedback inhibition of IRS1 (Insulin Receptor Substrate 1). A variety of insulin signaling models indicate that Ser/Thr phosphorylation of IRS1 has a role in the development of insulin resistance (Takayama et al. 1988; Chin et al. 1993, 1994; Jullien et al. 1993; Hotamisligil et al. 1994, 1996; Tanti et al. 1994; Kanety et al. 1995; De Fea and Roth 1997a,b; Eldar-Finkelman and Krebs 1997; Ricort et al. 1997; Staubs et al. 1998). In adipocytes, prolonged stimulation with insulin

leads to increased phosphorylation of IRS-1, and its subsequent degradation (Li et al. 1999; Haruta et al. 2000). Rapamycin blocks the phosphorylation and degradation of IRS-1, and causes sustained insulin signaling
and an enhancement of glucose uptake. Consistent with this finding,
amino acids, which are natural activators of mTOR signaling, reduce
insulin-stimulated glucose transport in skeletal muscle (Tremblay and
Marette 2001), and this effect is blocked by rapamycin. Similar results are
seen in adipocytes, in which amino acid deprivation down-regulates
insulin-induced phosphorylation and degradation of IRS-1 and increases
2-deoxyglucose uptake, while excess amino acids enhance insulin-induced
phosphorylation and degradation of IRS-1 and decrease 2-deoxyglucose
uptake (Takano et al. 2001). Thus, mTOR appears to be an important
mediator of a negative feedback loop in which amino acid- and growth
factor-derived signals down-regulate insulin signaling. The mechanism by
which mTOR negatively regulates insulin signaling is not yet clear, but one
possibility is that it may control the subcellular localization of IRS-1
(Takano et al. 2001). Insulin stimulates the translocation of IRS-1 from
low-density microsomes (LDMs) to the cytosol (Takano et al. 2001).
Rapamycin inhibits this redistribution in a dose-dependent manner and,
because it also blocks IRS-1 degradation, causes an increase in the levels
of both LDM and cytoplasmic IRS-1. A similar rapamycin-sensitive negative feedback loop also operates for IRS-2 (Simpson et al. 2001). This
function of mTOR may be necessary for maintaining efficient and controlled insulin signaling, and its potential deregulation may be important
in the pathogenesis of diabetes. Although it remains to be seen whether
this is the case in human tissue, a hyperactive mTOR pathway could affect
insulin sensitivity and glucose transport. Interestingly, data from clinical
trials investigating the immunosuppressive effects of rapamycin indicate
that many patients develop a hyperlipidemia syndrome with characteristics reminiscent of the lipid disturbances found in people with type II diabetes (Morrisett et al. 2002).

SUMMARY

The mTOR complex serves as an important organizing center for integrating diverse inputs into a signal that regulates growth. The growth
response is mediated in large part by the protein synthesis machinery, but
likely through other mechanisms as well, such as regulated transcription.
Much of our insight into the role of TOR signaling in animal growth
comes from studies in *Drosophila* and mice, where mutations in components of this signaling network have striking effects of cell, organ, and

organism size. Tight regulation of cell growth is crucial for normal development of organs and organisms, but the identities of the key inputs that regulate mTOR-dependent growth are still not well known. Both nutrients, such as amino acids, and growth factors, such as insulin, are important signals to mTOR, and an important challenge for the future is to determine how these diverse signaling mechanisms coordinately regulate mTOR. Another critical issue to address in future studies is how growth is coordinated with proliferation. Most dividing cells likely coordinate mass accumulation and division, while nondividing differentiated cells may be under growth control alone. mTOR may help coordinate the growth machinery with the cell cycle machinery, but to date, we know relatively little about the contributions mTOR itself makes to cell cycle control. Increasing evidence suggests that defective mTOR signaling may contribute to progression of several diseases, including tuberous sclerosis, cancers with hyperactivated PI3K signaling, and diabetes, and rapamycin is rapidly emerging as an extremely valuable clinical drug with many applications. Understanding the full mechanisms and regulation of the mTOR-signaling network will be a challenge for many years to come and will undoubtedly reveal a fundamental biological strategy that eukaryotes have evolved to regulate growth.

ACKNOWLEDGMENTS

Our work on the mTOR pathway has been supported by institutional funds from the Whitehead Institute and grants from the National Institutes of Health and the G. Harold and Leila Y. Mathers Charitable, Korea Science and Engineering, and the Anna Fuller Foundations, as well as the Human Frontier Science Program.

REFERENCES

Alessi D.R., Kozlowski M.T., Weng Q.P., Morrice N., and Avruch J. 1998. 3-Phosphoinositide-dependent protein kinase 1 (PDK1) phosphorylates and activates the p70 S6 kinase in vivo and in vitro. *Curr. Biol.* **8:** 69–81.

Andrade M.A. and Bork P. 1995. HEAT repeats in the Huntington's disease protein. *Nat. Genet.* **11:** 115–116.

Aoki M., Blazek E., and Vogt P.K. 2001. A role of the kinase mTOR in cellular transformation induced by the oncoproteins P3k and Akt. *Proc. Natl. Acad. Sci.* **98:** 136–141.

Baker H., Sidorowicz A., Sehgal S.N., and Vezina C. 1978. Rapamycin (AY-22, 989), a new antifungal antibiotic. III. In vitro and in vivo evaluation. *J. Antibiot.* **31:** 539–545.

Barbet N.C., Schneider U., Helliwell S.B., Stansfield I., Tuite M.F., and Hall M.N. 1996. TOR controls translation initiation and early G1 progression in yeast. *Mol. Biol. Cell* **7:** 25–42.

Beck T. and Hall M.N. 1999. The TOR signalling pathway controls nuclear localization of nutrient-regulated transcription factors. *Nature* **402**: 689–692.

Bell A., Grunder L., and Sorisky A. 2000. Rapamycin inhibits human adipocyte differentiation in primary culture. *Obes. Res.* **8**: 249–254.

Beretta L., Gingras A.C., Svitkin Y.V., Hall M.N., and Sonenberg N. 1996. Rapamycin blocks the phosphorylation of 4E-BP1 and inhibits cap-dependent initiation of translation. *EMBO J.* **15**: 658–664.

Blackshear P.J., Stumpo D.J., Carballo E., and Lawrence J.C., Jr. 1997. Disruption of the gene encoding the mitogen-regulated translational modulator PHAS-I in mice. *J. Biol. Chem.* **272**: 31510–31514.

Blommaart E.F., Krause U., Schellens J.P., Vreeling-Sindelarova H., and Meijer A.J. 1997. The phosphatidylinositol 3-kinase inhibitors wortmannin and LY294002 inhibit autophagy in isolated rat hepatocytes. *Eur. J. Biochem.* **243**: 240–246.

Blommaart E.F., Luiken J.J., Blommaart P.J., van Woerkom G.M., and Meijer A.J. 1995. Phosphorylation of ribosomal protein S6 is inhibitory for autophagy in isolated rat hepatocytes. *J. Biol. Chem.* **270**: 2320–2326.

Bohni R., Riesgo-Escovar J., Oldham S., Brogiolo W., Stocker H., Andruss B.F., Beckingham K., and Hafen E. 1999. Autonomous control of cell and organ size by CHICO, a *Drosophila* homolog of vertebrate IRS1-4. *Cell* **97**: 865–875.

Bosotti R., Isacchi A., and Sonnhammer E.L. 2000. FAT: A novel domain in PIK-related kinases. *Trends Biochem. Sci.* **25**: 225–227.

Boylan J.M., Anand P., and Gruppuso P.A. 2001. Ribosomal protein S6 phosphorylation and function during late gestation liver development in the rat. *J. Biol. Chem.* **276**: 44457–44463.

Britton J.S., Lockwood W.K., Li L., Cohen S.M., and Edgar B.A. 2002. *Drosophila's* insulin/PI3-kinase pathway coordinates cellular metabolism with nutritional conditions. *Dev. Cell* **2**: 239–249.

Brown E.J., Beal P.A., Keith C.T., Chen J., Shin T.B., and Schreiber S.L. 1995. Control of p70 s6 kinase by kinase activity of FRAP in vivo. *Nature* **377**: 441–446.

Brown E.J., Albers M.W., Shin T.B., Ichikawa K., Keith C.T., Lane W.S., and Schreiber S.L. 1994. A mammalian protein targeted by G1-arresting rapamycin-receptor complex. *Nature* **369**: 756–758.

Brunn G.J., Fadden P., Haystead T.A.J., and Lawrence J.C., Jr. 1997a. The mammalian target of rapamycin phosphorylates sites having a (Ser/Thr)-Pro motif and is activated by antibodies to a region near its COOH terminus. *J. Biol. Chem.* **272**: 32547–32550.

Brunn G.J., Williams J., Sabers C., Wiederrecht G., Lawrence J.C., and Abraham R.T. 1996. Direct inhibition of the signaling functions of the mammalian target of rapamycin by the phosphoinositide 3-kinase inhibitors, wortmannin and Ly294002. *EMBO J.* **15**: 5256–5267.

Brunn G.J., Hudson C.C., Sekulic A., Williams J.M., Hosoi H., Houghton P.J., Lawrence J.C., Jr., and Abraham R.T. 1997b. Phosphorylation of the translational repressor PHAS-I by the mammalian target of rapamycin. *Science* **277**: 99–101.

Burnett P.E., Barrow R.K., Cohen N.A., Snyder S.H., and Sabatini D.M. 1998. RAFT1 phosphorylation of the translational regulators p70 S6 kinase and 4E-BP1. *Proc. Natl. Acad. Sci.* **95**: 1432–1437.

Cafferkey R., Young P.R., McLaughlin M.M., Bergsma D.J., Koltin Y., Sathe G.M., Faucette L., Eng W.K., Johnson R.K., and Livi G.P. 1993. Dominant missense mutations in a novel yeast protein related to mammalian phosphatidylinositol 3-kinase and VPS34 abrogate rapamycin cytotoxicity. *Mol. Cell. Biol.* **13:** 6012–6023.

Cantley L.C. and Neel B.G. 1999. New insights into tumor suppression: PTEN suppresses tumor formation by restraining the phosphoinositide 3-kinase/AKT pathway. *Proc. Natl. Acad. Sci.* **96:** 4240–4245.

Cardenas M.E., Cutler N.S., Lorenz M.C., Di Como C.J., and Heitman J. 1999. The TOR signaling cascade regulates gene expression in response to nutrients. *Genes Dev.* **13:** 3271–3279.

Carsillo T., Astrinidis A., and Henske E.P. 2000. Mutations in the tuberous sclerosis complex gene TSC2 are a cause of sporadic pulmonary lymphangioleiomyomatosis. *Proc. Natl. Acad. Sci.* **97:** 6085–6090.

Castro A.F., Rebhun J.F., Clark G.G., and Quilliam L.A. 2003. Rheb binds TSC2 and promotes S6 kinase activation in a rapamycin- and farnesylation-dependent manner. *J. Biol. Chem.* **3:** 3.

Cheatham L., Monfar M., Chou M.M., and Blenis J. 1995. Structural and functional analysis of pp70S6k. *Proc. Natl. Acad. Sci.* **92:** 11696–11700.

Chen C., Jack J., and Garofalo R.S. 1996. The *Drosophila* insulin receptor is required for normal growth. *Endocrinology* **137:** 846–856.

Chen J., Zheng X.F., Brown E.J., and Schreiber S.L. 1995. Identification of an 11-kDa FKBP12-rapamycin-binding domain within the 289-kDa FKBP12-rapamycin-associated protein and characterization of a critical serine residue. *Proc. Natl. Acad. Sci.* **92:** 4947–4951.

Chin J.E., Liu F., and Roth R.A. 1994. Activation of protein kinase C alpha inhibits insulin-stimulated tyrosine phosphorylation of insulin receptor substrate-1. *Mol. Endocrinol.* **8:** 51–58.

Chin J.E., Dickens M., Tavare J.M., and Roth R.A. 1993. Overexpression of protein kinase C isoenzymes alpha, beta I, gamma, and epsilon in cells overexpressing the insulin receptor. Effects on receptor phosphorylation and signaling. *J. Biol. Chem.* **268:** 6338–6347.

Chiu M.I., Katz H., and Berlin V. 1994. RAPT1, a mammalian homolog of yeast Tor, interacts with the FKBP12/rapamycin complex. *Proc. Natl. Acad. Sci.* **91:** 12574–12578.

Choi J.H., Bertram P.G., Drenan R., Carvalho J., Zhou H.H., and Zheng X.F. 2002. The FKBP12-rapamycin-associated protein (FRAP) is a CLIP-170 kinase. *EMBO Rep.* **3:** 988–994.

Choi J.W., Chen J., Schreiber S.L., and Clardy J. 1996. Structure of the FKBP12-rapamycin complex interacting with the binding domain of human FRAP. *Science* **273:** 239–242.

Choi K.M., McMahon L.P., and Lawrence J.C., Jr. 2003. Two motifs in the translational repressor PHAS-I required for efficient phosphorylation by mTOR and recognition by raptor. *J. Biol. Chem.* **278:**19667–19673.

Chook Y.M. and Blobel G. 1999. Structure of the nuclear transport complex karyopherin-beta2-Ran x GppNHp. *Nature* **399:** 230–237.

Chung J., Kuo C.J., Crabtree G.R., and Blenis J. 1992. Rapamycin-FKBP specifically blocks growth-dependent activation of and signaling by the 70 kD S6 protein kinases. *Cell* **69:** 1227–1236.

Cingolani G., Petosa C., Weis K., and Muller C.W. 1999. Structure of importin-beta bound to the IBB domain of importin-alpha. *Nature* **399:** 221–229.

Clark G.J., Kinch M.S., Rogers-Graham K., Sebti S.M., Hamilton A.D., and Der C.J. 1997. The Ras-related protein Rheb is farnesylated and antagonizes Ras signaling and transformation. *J. Biol. Chem.* **272:** 10608–10615.

Cox A.D. and Der C.J. 2002. Ras family signaling: Therapeutic targeting. *Cancer Biol. Ther.* **1:** 599–606.

Crespo J.L., Daicho K., Ushimaru T., and Hall M.N. 2001. The GATA transcription factors GLN3 and GAT1 link TOR to salt stress in *Saccharomyces cerevisiae. J. Biol. Chem.* **276:** 34441–34444.

Daga R.R. and Jimenez J. 1999. Translational control of the cdc25 cell cycle phosphatase: A molecular mechanism coupling mitosis to cell growth. *J. Cell Sci.* **112:** 3137–3146.

Datar S.A., Jacobs H.W., de la Cruz A.F., Lehner C.F., and Edgar B.A. 2000. The *Drosophila* cyclin D-Cdk4 complex promotes cellular growth. *EMBO J.* **19:** 4543–4554.

Datta S.R., Brunet A., and Greenberg M.E. 1999. Cellular survival: A play in three Akts. *Genes Dev.* **13:** 2905–2927.

De Fea K. and Roth R.A. 1997a. Modulation of insulin receptor substrate-1 tyrosine phosphorylation and function by mitogen-activated protein kinase. *J. Biol. Chem.* **272:** 31400–31406.

———. 1997b. Protein kinase C modulation of insulin receptor substrate-1 tyrosine phosphorylation requires serine 612. *Biochemistry* **36:** 12939–12947.

Denning G., Jamieson L., Maquat L.E., Thompson E.A., and Fields A.P. 2001. Cloning of a novel phosphatidylinositol kinase-related kinase: Characterization of the human SMG-1 RNA surveillance protein. *J. Biol. Chem.* **276:** 22709–22714.

Dennis P.B., Pullen N., Kozma S.C., and Thomas G. 1996. The principal rapamycin-sensitive P70(S6k) phosphorylation sites, T-229 and T-389, are differentially regulated by rapamycin-insensitive kinase kinases. *Mol. Cell. Biol.* **16:** 6242–6251.

Dennis P.B., Jaeschke A., Saitoh M., Fowler B., Kozma S.C., and Thomas G. 2001. Mammalian TOR: A homeostatic ATP sensor. *Science* **294:** 1102–1105.

Desai B.N., Myers B.R., and Schreiber S.L. 2002. FKBP12-rapamycin-associated protein associates with mitochondria and senses osmotic stress via mitochondrial dysfunction. *Proc. Natl. Acad. Sci.* **99:** 4319–4324.

DiComo C.J. and Arndt K.T. 1996. Nutrients, via the Tor proteins, stimulate the association of Tap42 with type 2a phosphatases. *Genes Dev.* **10:** 1904–1916.

Edinger A.L. and Thompson C.B. 2002. Akt maintains cell size and survival by increasing mTOR-dependent nutrient uptake. *Mol. Biol. Cell* **13:** 2276–2288.

Eldar-Finkelman H. and Krebs E.G. 1997. Phosphorylation of insulin receptor substrate 1 by glycogen synthase kinase 3 impairs insulin action. *Proc. Natl. Acad. Sci.* **94:** 9660–9664.

Fahien L.A., MacDonald M.J., Kmiotek E.H., Mertz R.J., and Fahien C.M. 1988. Regulation of insulin release by factors that also modify glutamate dehydrogenase. *J. Biol. Chem.* **263:** 13610–13614.

Fang Y., Vilella-Bach M., Bachmann R., Flanigan A., and Chen J. 2001. Phosphatidic acid-mediated mitogenic activation of mTOR signaling. *Science* **294:** 1942–1945.

Fantes P. and Nurse P. 1977. Control of cell size at division in fission yeast by a growth-modulated size control over nuclear division. *Exp. Cell Res.* **107:** 377–386.

Fantl V., Stamp G., Andrews A., Rosewell I., and Dickson C. 1995. Mice lacking cyclin D1 are small and show defects in eye and mammary gland development. *Genes Dev.* **9:** 2364–2372.

Fausto N. 2000. Liver regeneration. *J. Hepatol.* (suppl. 1) **32:**19–31.

Fernandez R., Tabarini D., Azpiazu N., Frasch M., and Schlessinger J. 1995. The *Drosophila* insulin receptor homolog: A gene essential for embryonic development encodes two receptor isoforms with different signaling potential. *EMBO J.* **14:** 3373–3384.

Ferrari S., Pearson R.B., Siegmann M., Kozma S.C., and Thomas G. 1993. The immuno-suppressant rapamycin induces inactivation of p70s6k through dephosphorylation of a novel set of sites. *J. Biol. Chem.* **268:** 16091–16094.

Ferrus A. and Garcia-Bellido A. 1976. Morphogenetic mutants detected in mitotic recombination clones. *Nature* **260:** 425–426.

Fingar D.C., Salama S., Tsou C., Harlow E., and Blenis J. 2002. Mammalian cell size is controlled by mTOR and its downstream targets S6K1 and 4EBP1/eIF4E. *Genes Dev.* **16:** 1472–1487.

Francavilla A., Carr B.I., Starzl T.E., Azzarone A., Carrieri G., and Zeng Q.H. 1992. Effects of rapamycin on cultured hepatocyte proliferation and gene expression. *Hepatology* **15:** 871–877.

Franz D.N., Brody A., Meyer C., Leonard J., Chuck G., Dabora S., Sethuraman G., Colby T.V., Kwiatkowski D.J., and McCormack F.X. 2001. Mutational and radiographic analysis of pulmonary disease consistent with lymphangioleiomyomatosis and micronodular pneumocyte hyperplasia in women with tuberous sclerosis. *Am. J. Respir. Crit. Care Med.* **164:** 661–668.

Gagnon A., Lau S., and Sorisky A. 2001. Rapamycin-sensitive phase of 3T3-L1 preadipocyte differentiation after clonal expansion. *J. Cell. Physiol.* **189:** 14–22.

Gao X. and Pan D. 2001. TSC1 and TSC2 tumor suppressors antagonize insulin signaling in cell growth. *Genes Dev.* **15:** 1383–1392.

Gao X., Neufeld T.P., and Pan D. 2000. *Drosophila* PTEN regulates cell growth and proliferation through PI3K-dependent and -independent pathways. *Dev. Biol.* **221:** 404–418.

Gao X., Zhang Y., Arrazola P., Hino O., Kobayashi T., Yeung R.S., Ru B., and Pan D. 2002. Tsc tumour suppressor proteins antagonize amino-acid-TOR signalling. *Nat. Cell Biol.* **4:** 699–704.

Garami A., Zwartkruis F.J., Nobukuni T., Joaquin M., Roccio M., Stocker H., Kozma S.C., Hafen E., Bos J.L., and Thomas G. 2003. Insulin activation of Rheb, a mediator of mTOR/S6K/4E-BP signaling, is inhibited by TSC1 and 2. *Mol. Cell.* **11:** 1457–1466.

Gingras A.C., Gygi S.P., Raught B., Polakiewicz R.D., Abraham R.T., Hoekstra M.F., Aebersold R., and Sonenberg N. 1999. Regulation of 4E-BP1 phosphorylation: A novel two-step mechanism. *Genes Dev.* **13:** 1422–1437.

Goberdhan D.C., Paricio N., Goodman E.C., Mlodzik M., and Wilson C. 1999. *Drosophila* tumor suppressor PTEN controls cell size and number by antagonizing the Chico/PI3-kinase signaling pathway. *Genes Dev.* **13:** 3244–3258.

Gomez M.R., Sampson J.R., and Whittemore V.H. 1999. *Tuberous sclerosis complex,* 3rd edition. Oxford University Press, New York.

Goncharova E.A., Goncharov D.A., Eszterhas A., Hunter D.S., Glassberg M.K., Yeung R.S., Walker C.L., Noonan D., Kwiatkowski D.J., Chou M.M., Panettieri R.A., Jr., and Krymskaya V.P. 2002. Tuberin regulates p70 S6 kinase activation and ribosomal protein S6 phosphorylation. A role for the TSC2 tumor suppressor gene in pulmonary lymphangioleiomyomatosis (LAM). *J. Biol. Chem.* **277:** 30958–30967.

Gottlob K., Majewski N., Kennedy S., Kandel E., Robey R.B., and Hay N. 2001. Inhibition of early apoptotic events by Akt/PKB is dependent on the first committed step of glycolysis and mitochondrial hexokinase. *Genes Dev.* **15:** 1406–1418.

Groves M.R., Hanlon N., Turowski P., Hemmings B.A., and Barford D. 1999. The structure of the protein phosphatase 2A PR65/A subunit reveals the conformation of its 15 tandemly repeated HEAT motifs. *Cell* **96:** 99–110.

Guo Q.M., Malek R.L., Kim S., Chiao C., He M., Ruffy M., Sanka K., Lee N.H., Dang C.V., and Liu E.T. 2000. Identification of c-myc responsive genes using rat cDNA microarray. *Cancer Res.* **60:** 5922–5928.

Haas-Kogan D., Shalev N., Wong M., Mills G., Yount G., and Stokoe D. 1998. Protein kinase B (PKB/Akt) activity is elevated in glioblastoma cells due to mutation of the tumor suppressor PTEN/MMAC. *Curr. Biol.* **8:** 1195–1198.

Hara K., Yonezawa K., Weng Q.P., Kozlowski M.T., Belham C., and Avruch J. 1998. Amino acid sufficiency and mTOR regulate p70 S6 kinase and eIF-4E BP1 through a common effector mechanism. *J. Biol. Chem.* **273:** 14484–14494.

Hara K., Maruki Y., Long X., Yoshino K., Oshiro N., Hidayat S., Tokunaga C., Avruch J., and Yonezawa K. 2002. Raptor, a binding partner of target of rapamycin (TOR), mediates TOR action. *Cell* **110:** 177–189.

Hardwick J.S., Kuruvilla F.G., Tong J.K., Shamji A.F., and Schreiber S.L. 1999. Rapamycin-modulated transcription defines the subset of nutrient-sensitive signaling pathways directly controlled by the Tor proteins. *Proc. Natl. Acad. Sci.* **96:** 14866–14870.

Haruta T., Uno T., Kawahara J., Takano A., Egawa K., Sharma P.M., Olefsky J.M., and Kobayashi M. 2000. A rapamycin-sensitive pathway down-regulates insulin signaling via phosphorylation and proteasomal degradation of insulin receptor substrate-1. *Mol. Endocrinol.* **14:** 783–794.

Hashemolhosseini S., Nagamine Y., Morley S.J., Desrivieres S., Mercep L., and Ferrari S. 1998. Rapamycin inhibition of the G1 to S transition is mediated by effects on cyclin D1 mRNA and protein stability. *J. Biol. Chem.* **273:** 14424–14429.

Henry K.W., Yuan X., Koszewski N.J., Onda H., Kwiatkowski D.J., and Noonan D.J. 1998. Tuberous sclerosis gene 2 product modulates transcription mediated by steroid hormone receptor family members. *J. Biol. Chem.* **273:** 20535–20539.

Hentges K., Thompson K., and Peterson A. 1999. The flat-top gene is required for the expansion and regionalization of the telencephalic primordium. *Development* **126:** 1601–1609.

Hentges K.E., Sirry B., Gingeras A.C., Sarbassov D., Sonenberg N., Sabatini D., and Peterson A.S. 2001. FRAP/mTOR is required for proliferation and patterning during embryonic development in the mouse. *Proc. Natl. Acad. Sci.* **98:** 13796–13801.

Hidalgo M. and Rowinsky E.K. 2000. The rapamycin-sensitive signal transduction pathway as a target for cancer therapy. *Oncogene* **19:** 6680–6686.

Hotamisligil G.S., Murray D.L., Choy L.N., and Spiegelman B.M. 1994. Tumor necrosis factor alpha inhibits signaling from the insulin receptor. *Proc. Natl. Acad. Sci.* **91:** 4854–4858.

Hotamisligil G.S., Peraldi P., Budavari A., Ellis R., White M.F., and Spiegelman B.M. 1996. IRS-1-mediated inhibition of insulin receptor tyrosine kinase activity in TNF-alpha- and obesity-induced insulin resistance. *Science* **271:** 665–668.

Hsu J., Shi Y., Krajewski S., Renner S., Fisher M., Reed J.C., Franke T.F., and Lichtenstein A. 2001. The AKT kinase is activated in multiple myeloma tumor cells. *Blood* **98:** 2853–2855.

Huang H., Potter C.J., Tao W., Li D.M., Brogiolo W., Hafen E., Sun H., and Xu T. 1999. PTEN affects cell size, cell proliferation and apoptosis during *Drosophila* eye development. *Development* **126:** 5365–5372.

Huffman T.A., Mothe-Satney I., and Lawrence J.C., Jr. 2002. Insulin-stimulated phosphorylation of lipin mediated by the mammalian target of rapamycin. *Proc. Natl. Acad. Sci.* **99:** 1047–1052.

Iiboshi Y., Papst P.J., Kawasome H., Hosoi H., Abraham R.T., Houghton P.J., and Terada N.

1999. Amino acid-dependent control of p70(s6k). Involvement of tRNA aminoacylation in the regulation. *J. Biol. Chem.* **274:** 1092–1099.

Imanishi Y., Hosokawa Y., Yoshimoto K., Schipani E., Mallya S., Papanikolaou A., Kifor O., Tokura T., Sablosky M., Ledgard F., Gronowicz G., Wang T.C., Schmidt E.V., Hall C., Brown E.M., Bronson R., and Arnold A. 2001. Primary hyperparathyroidism caused by parathyroid-targeted overexpression of cyclin D1 in transgenic mice. *J. Clin. Invest.* **107:** 1093–1102.

Inoki K., Li Y., Xu T., and Guan K.L. 2003. Rheb GTPase is a direct target of TSC2 GAP activity and regulates mTOR signaling. *Genes Dev.* **17:** 17.

Inoki K., Li Y., Zhu T., Wu J., and Guan K.L. 2002. TSC2 is phosphorylated and inhibited by Akt and suppresses mTOR signalling. *Nat. Cell Biol.* **4:** 648–657.

Isotani S., Hara K., Tokunaga C., Inoue H., Avruch J., and Yonezawa K. 1999. Immunopurified mammalian target of rapamycin phosphorylates and activates p70 S6 kinase alpha in vitro. *J. Biol. Chem.* **274:** 34493–34498.

Ito N. and Rubin G.M. 1999. gigas, a *Drosophila* homolog of tuberous sclerosis gene product-2, regulates the cell cycle. *Cell* **96:** 529–539.

Jacinto E. and Hall M.N. 2003. Tor signalling in bugs, brain and brawn. *Nat. Rev. Mol. Cell Biol.* **4:** 117–126.

Jaeschke A., Hartkamp J., Saitoh M., Roworth W., Nobukuni T., Hodges A., Sampson J., Thomas G., and Lamb R. 2002. Tuberous sclerosis complex tumor suppressor-mediated S6 kinase inhibition by phosphatidylinositide-3-OH kinase is mTOR independent. *J. Cell Biol.* **159:** 217–224.

Jefferies H.B., Reinhard C., Kozma S.C., and Thomas G. 1994. Rapamycin selectively represses translation of the "polypyrimidine tract" mRNA family. *Proc. Natl. Acad. Sci.* **91:** 4441–4445.

Jefferies H.B., Fumagalli S., Dennis P.B., Reinhard C., Pearson R.B., and Thomas G. 1997. Rapamycin suppresses 5′TOP mRNA translation through inhibition of p70s6k. *EMBO J.* **16:** 3693–3704.

Jeno P., Jaggi N., Luther H., Siegmann M., and Thomas G. 1989. Purification and characterization of a 40 S ribosomal protein S6 kinase from vanadate-stimulated Swiss 3T3 cells. *J. Biol. Chem.* **264:** 1293–1297.

Jiang Y. and Broach J.R. 1999. Tor proteins and protein phosphatase 2A reciprocally regulate Tap42 in controlling cell growth in yeast. *EMBO J.* **18:** 2782–2792.

Johnston G.C., Pringle J.R., and Hartwell L.H. 1977. Coordination of growth with cell division in the yeast *Saccharomyces cerevisiae*. *Exp. Cell. Res.* **105:** 79–98.

Johnston L.A., Prober D.A., Edgar B.A., Eisenman R.N., and Gallant P. 1999. *Drosophila* myc regulates cellular growth during development. *Cell* **98:** 779–790.

Jullien D., Tanti J.F., Heydrick S.J., Gautier N., Gremeaux T., Van Obberghen E., and Le Marchand-Brustel Y. 1993. Differential effects of okadaic acid on insulin-stimulated glucose and amino acid uptake and phosphatidylinositol 3-kinase activity. *J. Biol. Chem.* **268:** 15246–15251.

Kanety H., Feinstein R., Papa M.Z., Hemi R., and Karasik A. 1995. Tumor necrosis factor alpha-induced phosphorylation of insulin receptor substrate-1 (IRS-1). Possible mechanism for suppression of insulin-stimulated tyrosine phosphorylation of IRS-1. *J. Biol. Chem.* **270:** 23780–23784.

Kawamata S., Sakaida H., Hori T., Maeda M., and Uchiyama T. 1998. The upregulation of p27(Kip1) by rapamycin results in G1 arrest in exponentially growing T-cell lines. *Blood* **91:** 561–569.

Keith C.T. and Schreiber S.L. 1995. PIK-related kinases: DNA repair, recombination, and cell cycle checkpoints. *Science* **270:** 50–51.

Kenerson H.L., Aicher L.D., True L.D., and Yeung R.S. 2002. Activated mammalian target of rapamycin pathway in the pathogenesis of tuberous sclerosis complex renal tumors. *Cancer Res.* **62:** 5645–5650.

Khan A., Pepio A.M., and Sossin W.S. 2001. Serotonin activates S6 kinase in a rapamycin-sensitive manner in Aplysia synaptosomes. *J. Neurosci.* **21:** 382–391.

Kim D.-H., Sarbassov D.D., Ali S.M., King J.E., Latek R.R., Erdjument-Bromage H., Tempst P., and Sabatini D.M. 2002. mTOR interacts with raptor to form a nutrient-sensitive complex that signals to the cell growth machinery. *Cell* **110:** 163–175.

Kim D.-H., Sarbassov D.D., Ali S.M., Latek R.R., Guntur K.V.P., Erdjument-Bromage H., Tempst P., and Sabatini D.M. 2003. GβL: A positive regulator of the rapamycin-sensitive pathway required for a nutrient-sensitive interaction between mTOR and raptor. *Mol. Cell* **11:** 895–904.

Kimball S.R., Shantz L.M., Horetsky R.L., and Jefferson L.S. 1999. Leucine regulates translation of specific mRNAs in L6 myoblasts through mTOR-mediated changes in availability of eIF4E and phosphorylation of ribosomal protein S6. *J. Biol. Chem.* **274:** 11647–11652.

Kirsch J., Wolters I., Triller A., and Betz H. 1993. Gephyrin antisense oligonucleotides prevent glycine receptor clustering in spinal neurons. *Nature* **366:** 745–748.

Kiyokawa H., Kineman R.D., Manova-Todorova K.O., Soares V.C., Hoffman E.S., Ono M., Khanam D., Hayday A.C., Frohman L.A., and Koff A. 1996. Enhanced growth of mice lacking the cyclin-dependent kinase inhibitor function of p27(Kip1). *Cell* **85:** 721–732.

Kleymenova E., Ibragimov-Beskrovnaya O., Kugoh H., Everitt J., Xu H., Kiguchi K., Landes G., Harris P., and Walker C. 2001. Tuberin-dependent membrane localization of polycystin-1: A functional link between polycystic kidney disease and the TSC2 tumor suppressor gene. *Mol. Cell* **7:** 823–832.

Komeili A., Wedaman K.P., O'Shea E.K., and Powers T. 2000. Mechanism of metabolic control. Target of rapamycin signaling links nitrogen quality to the activity of the Rtg1 and Rtg3 transcription factors. *J. Cell Biol.* **151:** 863–878.

Koromilas A.E., Lazaris-Karatzas A., and Sonenberg N. 1992. mRNAs containing extensive secondary structure in their 5′ non-coding region translate efficiently in cells overexpressing eIF-4E. *EMBO J.* **11:** 4153–4158.

Kunz J., Henriquez R., Schneider U., Deuter-Reinhard M., Movva N.R., and Hall M.N. 1993. Target of rapamycin in yeast, TOR2, is an essential phosphatidylinositol kinase homolog required for G1 progression. *Cell* **73:** 585–596.

Kuo C.J., Chung J., Fiorentino D.F., Flanagan W.M., Blenis J., and Crabtree G.R. 1992. Rapamycin selectively inhibits interleukin-2 activation of p70 S6 kinase. *Nature* **358:** 70–73.

Kwiatkowski D.J., Zhang H., Bandura J.L., Heiberger K.M., Glogauer M., el-Hashemite N., and Onda H. 2002. A mouse model of TSC1 reveals sex-dependent lethality from liver hemangiomas, and up-regulation of p70S6 kinase activity in Tsc1 null cells. *Hum. Mol. Genet.* **11:** 525–534.

Lamb R.F., Roy C., Diefenbach T.J., Vinters H.V., Johnson M.W., Jay D.G., and Hall A. 2000. The TSC1 tumour suppressor hamartin regulates cell adhesion through ERM proteins and the GTPase Rho. *Nat. Cell Biol.* **2:** 281–287.

Lawlor M.A., Mora A., Ashby P.R., Williams M.R., Murray-Tait V., Malone L., Prescott A.R., Lucocq J.M., and Alessi D.R. 2002. Essential role of PDK1 in regulating cell size and development in mice. *EMBO J.* **21:** 3728–3738.

Lazaris-Karatzas A., Montine K.S., and Sonenberg N. 1990. Malignant transformation by a eukaryotic initiation factor subunit that binds to mRNA 5′ cap. *Nature* **345:** 544–547.

Lee-Fruman K.K., Kuo C.J., Lippincott J., Terada N., and Blenis J. 1999. Characterization of S6K2, a novel kinase homologous to S6K1. *Oncogene* **18:** 5108–5114.

Leevers S.J., Weinkove D., MacDougall L.K., Hafen E., and Waterfield M.D. 1996. The *Drosophila* phosphoinositide 3-kinase Dp110 promotes cell growth. *EMBO J.* **15:** 6584–6594.

Li J., DeFea K., and Roth R.A. 1999. Modulation of insulin receptor substrate-1 tyrosine phosphorylation by an Akt/phosphatidylinositol 3-kinase pathway. *J. Biol. Chem.* **274:** 9351–9356.

Lin T.A., Kong X., Saltiel A.R., Blackshear P.J., and Lawrence J.C., Jr. 1995. Control of PHAS-I by insulin in 3T3-L1 adipocytes. Synthesis, degradation, and phosphorylation by a rapamycin-sensitive and mitogen-activated protein kinase-independent pathway. *J. Biol. Chem.* **270:** 18531–18538.

Loewith R., Jacinto E., Wullschleger S., Lorberg A., Crespo J.L., Bonenfant D., Oppliger W., Jenoe P., and Hall M.N. 2002. Two TOR complexes, only one of which is rapamycin sensitive, have distinct roles in cell growth control. *Mol. Cell* **10:** 457–468.

Long X., Spycher C., Han Z.S., Rose A.M., Muller F., and Avruch J. 2002. TOR deficiency in *C. elegans* causes developmental arrest and intestinal atrophy by inhibition of mRNA translation. *Curr. Biol.* **12:** 1448–1461.

Luo Y., Marx S.O., Kiyokawa H., Koff A., Massague J., and Marks A.R. 1996. Rapamycin resistance tied to defective regulation of p27Kip1. *Mol. Cell. Biol.* **16:** 6744–6751.

Lynch C.J. 2001. Role of leucine in the regulation of mTOR by amino acids: Revelations from structure-activity studies. *J. Nutr.* **131:** 861S–865S.

Lynch C.J., Fox H.L., Vary T.C., Jefferson L.S., and Kimball S.R. 2000. Regulation of amino acid-sensitive TOR signaling by leucine analogues in adipocytes. *J. Cell. Biochem.* **77:** 234–251.

Mach K.E., Furge K.A., and Albright C.F. 2000. Loss of Rhb1, a Rheb-related GTPase in fission yeast, causes growth arrest with a terminal phenotype similar to that caused by nitrogen starvation. *Genetics* **155:** 611–622.

Mahalingam M. and Templeton D.J. 1996. Constitutive activation of S6 kinase by deletion of amino-terminal autoinhibitory and rapamycin sensitivity domains. *Mol. Cell. Biol.* **16:** 405–413.

Manning B.D., Tee A.R., Logsdon M.N., Blenis J., and Cantley L.C. 2002. Identification of the tuberous sclerosis complex-2 tumor suppressor gene product tuberin as a target of the phosphoinositide 3-kinase/akt pathway. *Mol. Cell* **10:** 151–162.

Marcotrigiano J., Lomakin I.B., Sonenberg N., Pestova T.V., Hellen C.U., and Burley S.K. 2001. A conserved HEAT domain within eIF4G directs assembly of the translation initiation machinery. *Mol. Cell* **7:** 193–203.

Menand B., Desnos T., Nussaume L., Berger F., Bouchez D., Meyer C., and Robaglia C. 2002. Expression and disruption of the *Arabidopsis* TOR (target of rapamycin) gene. *Proc. Natl. Acad. Sci.* **99:** 6422–6427.

Miron M., Verdu J., Lachance P.E., Birnbaum M.J., Lasko P.F., and Sonenberg N. 2001. The translational inhibitor 4E-BP is an effector of PI(3)K/Akt signalling and cell growth in *Drosophila*. *Nat. Cell Biol.* **3:** 596–601.

Mizuguchi M. and Takashima S. 2001. Neuropathology of tuberous sclerosis. *Brain Dev.* **23:** 508–515.

Montagne J., Stewart M.J., Stocker H., Hafen E., Kozma S.C., and Thomas G. 1999. *Drosophila* S6 kinase: A regulator of cell size. *Science* **285:** 2126–2129.

Morice M.C., Serruys P.W., Sousa J.E., Fajadet J., Ban Hayashi E., Perin M., Colombo A., Schuler G., Barragan P., Guagliumi G., Molnar F., and Falotico R. 2002. A randomized comparison of a sirolimus-eluting stent with a standard stent for coronary revascularization. *N. Engl. J. Med.* **346:** 1773–1780.

Morice W.G., Brunn G.J., Wiederrecht G., Siekierka J.J., and Abraham R.T. 1993a. Rapamycin-induced inhibition of p34cdc2 kinase activation is associated with G1/S-phase growth arrest in T lymphocytes. *J. Biol. Chem.* **268:** 3734–3738.

Morice W.G., Wiederrecht G., Brunn G.J., Siekierka J.J., and Abraham R.T. 1993b. Rapamycin inhibition of interleukin-2-dependent p33cdk2 and p34cdc2 kinase activation in T lymphocytes. *J. Biol. Chem.* **268:** 22737–22745.

Morrisett J.D., Abdel-Fattah G., Hoogeveen R., Mitchell E., Ballantyne C.M., Pownall H.J., Opekun A.R., Jaffe J.S., Oppermann S., and Kahan B.D. 2002. Effects of sirolimus on plasma lipids, lipoprotein levels, and fatty acid metabolism in renal transplant patients. *J. Lipid Res.* **43:** 1170–1180.

Nakayama K., Ishida N., Shirane M., Inomata A., Inoue T., Shishido N., Horii I., Loh D.Y., and Nakayama K. 1996. Mice lacking p27(Kip1) display increased body size, multiple organ hyperplasia, retinal dysplasia, and pituitary tumors. *Cell* **85:** 707–720.

Nave B.T., Ouwens M., Withers D.J., Alessi D.R., and Shepherd P.R. 1999. Mammalian target of rapamycin is a direct target for protein kinase B: Identification of a convergence point for opposing effects of insulin and amino-acid deficiency on protein translation. *Biochem J.* **344:** 427–431.

Nelsen C.J., Rickheim D.G., Timchenko N.A., Stanley M.W., and Albrecht J.H. 2001. Transient expression of cyclin D1 is sufficient to promote hepatocyte replication and liver growth in vivo. *Cancer Res.* **61:** 8564–8568.

Nelsen C.J., Rickheim D.G., Tucker M.M., Hansen L.K., and Albrecht J.H. 2003. Evidence that cyclin D1 mediates both growth and proliferation downstream of TOR in hepatocytes. *J. Biol. Chem.* **278:** 3656–3663.

Neshat M.S., Mellinghoff I.K., Tran C., Stiles B., Thomas G., Petersen R., Frost P., Gibbons J.J., Wu H., and Sawyers C.L. 2001. Enhanced sensitivity of PTEN-deficient tumors to inhibition of FRAP/mTOR. *Proc. Natl. Acad. Sci.* **98:** 10314–10319.

Nielsen F.C., Ostergaard L., Nielsen J., and Christiansen J. 1995. Growth-dependent translation of IGF-II mRNA by a rapamycin-sensitive pathway. *Nature* **377:** 358–362.

Noda T. and Ohsumi Y. 1998. Tor, a phosphatidylinositol kinase homologue, controls autophagy in yeast. *J. Biol. Chem.* **273:** 3963–3966.

Nojima H., Tokunaga C., Eguchi S., Oshiro N., Hidayat S., Yoshino K.I., Hara K., Tanaka N., Avruch J., and Yonezawa K. 2003. The mammalian target of rapamycin (mTOR) partner, raptor binds the mTOR substrates p70 S6 kinase and 4E-BP1 through their TOR signaling (TOS) motif. *J. Biol. Chem.* **278:** 15461–15464.

Nourse J., Firpo E., Flanagan W.M., Coats S., Polyak K., Lee M.H., Massague J., Crabtree G.R., and Roberts J.M. 1994. Interleukin-2-mediated elimination of the p27Kip1 cyclin-dependent kinase inhibitor prevented by rapamycin. *Nature* **372:** 570–573.

Ogier-Denis E. and Codogno P. 2003. Autophagy: A barrier or an adaptive response to cancer. *Biochim. Biophys. Acta* **1603:** 113–128.

Oldham S. and Hafen E. 2003. Insulin/IGF and target of rapamycin signaling: A TOR de force in growth control. *Trends Cell Biol.* **13:** 79–85.

Oldham S., Montagne J., Radimerski T., Thomas G., and Hafen E. 2000. Genetic and biochemical characterization of dTOR, the *Drosophila* homolog of the target of rapamycin. *Genes Dev.* **14:** 2689–2694.

Parrott L.A. and Templeton D.J. 1999. Osmotic stress inhibits p70/85 S6 kinase through activation of a protein phosphatase. *J. Biol. Chem.* **274:** 24731–24736.

Patti M.E., Brambilla E., Luzi L., Landaker E.J., and Kahn C.R. 1998. Bidirectional modulation of insulin action by amino acids. *J. Clin. Invest.* **101:** 1519–1529.

Pause A., Belsham G.J., Gingras A.C., Donze O., Lin T.A., Lawrence J.C., and Sonenberg N. 1994. Insulin-dependent stimulation of protein synthesis by phosphorylation of a regulator of 5′-cap function. *Nature* **371:** 762–767.

Pende M., Kozma S.C., Jaquet M., Oorschot V., Burcelin R., Le Marchand-Brustel Y., Klumperman J., Thorens B., and Thomas G. 2000. Hypoinsulinaemia, glucose intolerance and diminished beta-cell size in S6K1-deficient mice. *Nature* **408:** 994–997.

Peng T., Golub T.R., and Sabatini D.M. 2002. The immunosuppressant rapamycin mimics a starvation-like signal distinct from amino acid and glucose deprivation. *Mol. Cell. Biol.* **22:** 5575–5584.

Perry J. and Kleckner N. 2003. The ATRs, ATMs, and TORs are giant HEAT repeat proteins. *Cell* **112:** 151–155.

Peterfy M., Phan J., Xu P., and Reue K. 2001. Lipodystrophy in the fld mouse results from mutation of a new gene encoding a nuclear protein, lipin. *Nat. Genet.* **27:** 121–124.

Peterson R.T., Beal P.A., Comb M.J., and Schreiber S.L. 2000. FKBP12-rapamycin-associated protein (FRAP) autophosphorylates at serine 2481 under translationally repressive conditions. *J. Biol. Chem.* **275:** 7416–7423.

Peterson R.T., Desai B.N., Hardwick J.S., and Schreiber S.L. 1999. Protein phosphatase 2A interacts with the 70-kDa S6 kinase and is activated by inhibition of FKBP12-rapamycin associated protein. *Proc. Natl. Acad. Sci.* **96:** 4438–4442.

Plank T.L., Logginidou H., Klein-Szanto A., and Henske E.P. 1999. The expression of hamartin, the product of the TSC1 gene, in normal human tissues and in TSC1- and TSC2-linked angiomyolipomas. *Mod. Pathol.* **12:** 539–545.

Plas D.R., Talapatra S., Edinger A.L., Rathmell J.C., and Thompson C.B. 2001. Akt and Bcl-xL promote growth factor-independent survival through distinct effects on mitochondrial physiology. *J. Biol. Chem.* **276:** 12041–12048.

Podsypanina K., Lee R.T., Politis C., Hennessy I., Crane A., Puc J., Neshat M., Wang H., Yang L., Gibbons J., Frost P., Dreisbach V., Blenis J., Gaciong Z., Fisher P., Sawyers C., Hedrick-Ellenson L., and Parsons R. 2001. An inhibitor of mTOR reduces neoplasia and normalizes p70/S6 kinase activity in Pten+/− mice. *Proc. Natl. Acad. Sci.* **98:** 10320–10325.

Polymenis M. and Schmidt E.V. 1997. Coupling of cell division to cell growth by translational control of the G1 cyclin CLN3 in yeast. *Genes Dev.* **11:** 2522–2531.

Potter C.J., Huang H., and Xu T. 2001. *Drosophila* Tsc1 functions with Tsc2 to antagonize insulin signaling in regulating cell growth, cell proliferation, and organ size. *Cell* **105:** 357–368.

Potter C.J., Pedraza L.G., and Xu T. 2002. Akt regulates growth by directly phosphorylating Tsc2. *Nat. Cell Biol.* **4:** 658–665.

Price D.J., Grove J.R., Calvo V., Avruch J., and Bierer B.E. 1992. Rapamycin-induced inhibition of the 70-kilodalton S6 protein kinase. *Science* **257:** 973–977.

Prober D.A. and Edgar B.A. 2000. Ras1 promotes cellular growth in the *Drosophila* wing. *Cell* **100:** 435–446.

———. 2002. Interactions between Ras1, dMyc, and dPI3K signaling in the developing

Drosophila wing. *Genes Dev.* **16:** 2286–2299.

Pullen N., Dennis P.B., Andjelkovic M., Dufner A., Kozma S.C., Hemmings B.A., and Thomas G. 1998. Phosphorylation and activation of p70s6k by PDK1. *Science* **279:** 707–710.

Pyronnet S. 2000. Phosphorylation of the cap-binding protein eIF4E by the MAPK-activated protein kinase Mnk1. *Biochem. Pharmacol.* **60:** 1237–1243.

Rao R.H. 1984. The role of undernutrition in the pathogenesis of diabetes mellitus. *Diabetes Care* **7:** 595–601.

Raymond C.R., Redman S.J., and Crouch M.F. 2002. The phosphoinositide 3-kinase and p70 S6 kinase regulate long-term potentiation in hippocampal neurons. *Neuroscience* **109:** 531–536.

Regar E., Serruys P.W., Bode C., Holubarsch C., Guermonprez J.L., Wijns W., Bartorelli A., Constantini C., Degertekin M., Tanabe K., Disco C., Wuelfert E., Morice M.C., and RAVEL Study Group. 2002. Angiographic findings of the multicenter Randomized Study With the Sirolimus-Eluting Bx Velocity Balloon-Expandable Stent (RAVEL): Sirolimus-eluting stents inhibit restenosis irrespective of the vessel size. *Circulation* **106:** 1949–1956.

Ricort J.M., Tanti J.F., Van Obberghen E., and Le Marchand-Brustel Y. 1997. Cross-talk between the platelet-derived growth factor and the insulin signaling pathways in 3T3-L1 adipocytes. *J. Biol. Chem.* **272:** 19814–19818.

Robles A.I., Larcher F., Whalin R.B., Murillas R., Richie E., Gimenez-Conti I.B., Jorcano J.L., and Conti C.J. 1996. Expression of cyclin D1 in epithelial tissues of transgenic mice results in epidermal hyperproliferation and severe thymic hyperplasia. *Proc. Natl. Acad. Sci.* **93:** 7634–7638.

Rodgers B.D., Levine M.A., Bernier M., and Montrose-Rafizadeh C. 2001. Insulin regulation of a novel WD-40 repeat protein in adipocytes. *J. Endocrinol.* **168:** 325–332.

Rodriguez-Viciana P., Warne P.H., Dhand R., Vanhaesebroeck B., Gout I., Fry M.J., Waterfield M.D., and Downward J. 1994. Phosphatidylinositol-3-OH kinase as a direct target of Ras. *Nature* **370:** 527–532.

Romanelli A., Dreisbach V.C., and Blenis J. 2002. Characterization of phosphatidylinositol 3-kinase-dependent phosphorylation of the hydrophobic motif site Thr(389) in p70 S6 kinase 1 (S6K1). *J. Biol. Chem.* **277:** 40281–40289.

Sabatini D.M., Erdjument-Bromage H., Lui M., Tempst P., and Snyder S.H. 1994. RAFT1: A mammalian protein that binds to FKBP12 in a rapamycin-dependent fashion and is homologous to yeast TORs. *Cell* **78:** 35–43.

Sabatini D.M., Barrow R.K., Blackshaw S., Burnett P.E., Lai M.M., Field M.E., Bahr B.A., Kirsch J., Betz H., and Snyder S.H. 1999. Interaction of RAFT1 with gephyrin required for rapamycin-sensitive signaling. *Science* **284:** 1161–1164.

Sabers C.J., Martin M.M., Brunn G.J., Williams J.M., Dumont F.J., Wiederrecht G., and Abraham R.T. 1995. Isolation of a protein target of the FKBP12-rapamycin complex in mammalian cells. *J. Biol. Chem.* **270:** 815–822.

Saucedo L.J., Gao X., Chiarelli D.A., Li L., Pan D., and Edgar B.A. 2003. Rheb promotes cell growth as a component of the insulin/TOR signalling network. *Nat. Cell. Biol.* **5:** 566–571.

Saunders R.N., Metcalfe M.S., and Nicholson M.L. 2001. Rapamycin in transplantation: A review of the evidence. *Kidney Int.* **59:** 3–16.

Schalm S.S. and Blenis J. 2002. Identification of a conserved motif required for mTOR signaling. *Curr. Biol.* **12:** 632–639.

Schalm S.S., Fingar D.C., Sabatini D.M., and Blenis J. 2003. TOS motif-mediated raptor binding regulates 4E-BP1 multisite phosphorylation and function. *Curr. Biol.* **13:** 797–806.

Scheper G.C., Morrice N.A., Kleijn M., and Proud C.G. 2001. The mitogen-activated protein kinase signal-integrating kinase Mnk2 is a eukaryotic initiation factor 4E kinase with high levels of basal activity in mammalian cells. *Mol. Cell. Biol.* **21:** 743–754.

Schibler U., Kelley D.E., and Perry R.P. 1977. Comparison of methylated sequences in messenger RNA and heterogeneous nuclear RNA from mouse L cells. *J. Mol. Biol.* **115:** 695–714.

Schmelzle T. and Hall M.N. 2000. TOR, a central controller of cell growth. *Cell* **103:** 253–262.

Schmidt A., Kunz J., and Hall M.N. 1996. TOR2 is required for organization of the actin cytoskeleton in yeast. *Proc. Natl. Acad. Sci.* **93:** 13780–13785.

Schuhmacher M., Staege M.S., Pajic A., Polack A., Weidle U.H., Bornkamm G.W., Eick D., and Kohlhuber F. 1999. Control of cell growth by c-Myc in the absence of cell division. *Curr. Biol.* **9:** 1255–1258.

Schuhmacher M., Kohlhuber F., Holzel M., Kaiser C., Burtscher H., Jarsch M., Bornkamm G.W., Laux G., Polack A., Weidle U.H., and Eick D. 2001. The transcriptional program of a human B cell line in response to Myc. *Nucleic Acids Res.* **29:** 397–406.

Scott P.H., Brunn G.J., Kohn A.D., Roth R.A., and Lawrence J.C., Jr. 1998. Evidence of insulin-stimulated phosphorylation and activation of the mammalian target of rapamycin mediated by a protein kinase B signaling pathway. *Proc. Natl. Acad. Sci.* **95:** 7772–7777.

Sehgal S.N., Baker H., and Vezina C. 1975. Rapamycin (AY-22, 989), a new antifungal antibiotic. II. Fermentation, isolation and characterization. *J. Antibiot.* **28:** 727–732.

Sekulic A., Hudson C.C., Homme J.L., Yin P., Otterness D.M., Karnitz L.M., and Abraham R.T. 2000. A direct linkage between the phosphoinositide 3-kinase-AKT signaling pathway and the mammalian target of rapamycin in mitogen-stimulated and transformed cells. *Cancer Res.* **60:** 3504–3513.

Sener A. and Malaisse W.J. 1980. L-leucine and a nonmetabolized analogue activate pancreatic islet glutamate dehydrogenase. *Nature* **288:** 187–189.

Sharma S., Bhambi B., and Nyitray W. 2002. Sirolimus-eluting coronary stents (discussion). *N. Engl. J. Med.* **347:** 1285.

Sherr C.J. and Roberts J.M. 1999. CDK inhibitors: Positive and negative regulators of G1-phase progression. *Genes Dev.* **13:** 1501–1512.

Shi Y., Gera J., Hu L., Hsu J.H., Bookstein R., Li W., and Lichtenstein A. 2002. Enhanced sensitivity of multiple myeloma cells containing PTEN mutations to CCI-779. *Cancer Res.* **62:** 5027–5034.

Shigemitsu K., Tsujishita Y., Miyake H., Hidayat S., Tanaka N., Hara K., and Yonezawa K. 1999. Structural requirement of leucine for activation of p70 S6 kinase. *FEBS Lett.* **447:** 303–306.

Shima H., Pende M., Chen Y., Fumagalli S., Thomas G., and Kozma S.C. 1998. Disruption of the p70(s6k)/p85(s6k) gene reveals a small mouse phenotype and a new functional S6 kinase. *EMBO J.* **17:** 6649–6659.

Shinozaki-Yabana S., Watanabe Y., and Yamamoto M. 2000. Novel WD-repeat protein Mip1p facilitates function of the meiotic regulator Mei2p in fission yeast. *Mol. Cell. Biol.* **20:** 1234–1242.

Shioi T., McMullen J.R., Kang P.M., Douglas P.S., Obata T., Franke T.F., Cantley L.C., and

Izumo S. 2002. Akt/protein kinase B promotes organ growth in transgenic mice. *Mol. Cell. Biol.* **22:** 2799–2809.

Sicinski P., Donaher J.L., Parker S.B., Li T., Fazeli A., Gardner H., Haslam S.Z., Bronson R.T., Elledge S.J., and Weinberg R.A. 1995. Cyclin D1 provides a link between development and oncogenesis in the retina and breast. *Cell* **82:** 621–630.

Simpson L. and Parsons R. 2001. PTEN: Life as a tumor suppressor. *Exp. Cell Res.* **264:** 29–41.

Simpson L., Li J., Liaw D., Hennessy I., Oliner J., Christians F., and Parsons R. 2001. PTEN expression causes feedback upregulation of insulin receptor substrate 2. *Mol. Cell. Biol.* **21:** 3947–3958.

Soucek T., Rosner M., Miloloza A., Kubista M., Cheadle J.P., Sampson J.R., and Hengstschlager M. 2001. Tuberous sclerosis causing mutants of the TSC2 gene product affect proliferation and p27 expression. *Oncogene* **20:** 4904–4909.

Stambolic V., Suzuki A., de la Pompa J.L., Brothers G.M., Mirtsos C., Sasaki T., Ruland J., Penninger J.M., Siderovski D.P., and Mak T.W. 1998. Negative regulation of PKB/Akt-dependent cell survival by the tumor suppressor PTEN. *Cell* **95:** 29–39.

Staubs P.A., Nelson J.G., Reichart D.R., and Olefsky J.M. 1998. Platelet-derived growth factor inhibits insulin stimulation of insulin receptor substrate-1-associated phosphatidylinositol 3-kinase in 3T3-L1 adipocytes without affecting glucose transport. *J. Biol. Chem.* **273:** 25139–25147.

Stocker H., Andjelkovic M., Oldham S., Laffargue M., Wymann M.P., Hemmings B.A., and Hafen E. 2002. Living with lethal PIP3 levels: Viability of flies lacking PTEN restored by a PH domain mutation in Akt/PKB. *Science* **295:** 2088–2091.

Stocker H., Radimerski T., Schindelholz B., Wittwer F., Belawat P., Daram P., Breuer S., Thomas G., and Hafen E. 2003. Rheb is an essential regulator of S6K in controlling cell growth in *Drosophila*. *Nat. Cell. Biol.* **5:** 559–565.

Stolovich M., Tang H., Hornstein E., Levy G., Cohen R., Bae S.S., Birnbaum M.J., and Meyuhas O. 2002. Transduction of growth or mitogenic signals into translational activation of TOP mRNAs is fully reliant on the phosphatidylinositol 3-kinase-mediated pathway but requires neither S6K1 nor rpS6 phosphorylation. *Mol. Cell. Biol.* **22:** 8101–8113.

Swenne I., Borg L.A., Crace C.J., and Schnell Landstrom A. 1992. Persistent reduction of pancreatic beta-cell mass after a limited period of protein-energy malnutrition in the young rat. *Diabetologia* **35:** 939–945.

Takano A., Usui I., Haruta T., Kawahara J., Uno T., Iwata M., and Kobayashi M. 2001. Mammalian target of rapamycin pathway regulates insulin signaling via subcellular redistribution of insulin receptor substrate 1 and integrates nutritional signals and metabolic signals of insulin. *Mol. Cell. Biol.* **21:** 5050–5062.

Takayama S., White M.F., and Kahn C.R. 1988. Phorbol ester-induced serine phosphorylation of the insulin receptor decreases its tyrosine kinase activity. *J. Biol. Chem.* **263:** 3440–3447.

Tang S.J. and Schuman E.M. 2002. Protein synthesis in the dendrite. *Philos. Trans. R. Soc. Lond. B Biol. Sci.* **357:** 521–529.

Tang S.J., Reis G., Kang H., Gingras A.C., Sonenberg N., and Schuman E.M. 2002. A rapamycin-sensitive signaling pathway contributes to long-term synaptic plasticity in the hippocampus. *Proc. Natl. Acad. Sci.* **99:** 467–472.

Tanti J.F., Gremeaux T., van Obberghen E., and Le Marchand-Brustel Y. 1994. Serine/threonine phosphorylation of insulin receptor substrate 1 modulates insulin receptor signaling. *J. Biol. Chem.* **269:** 6051–6057.

Tapon N., Ito N., Dickson B.J., Treisman J.E., and Hariharan I.K. 2001. The *Drosophila* tuberous sclerosis complex gene homologs restrict cell growth and cell proliferation. *Cell* **105:** 345–355.

Tee A.R., Fingar D.C., Manning B.D., Kwiatkowski D.J., Cantley L.C., and Blenis J. 2002. Tuberous sclerosis complex-1 and -2 gene products function together to inhibit mammalian target of rapamycin (mTOR)-mediated downstream signaling. *Proc. Natl. Acad. Sci.* **99:** 13571–13576.

Tremblay F. and Marette A. 2001. Amino acid and insulin signaling via the mTOR/p70 S6 kinase pathway. A negative feedback mechanism leading to insulin resistance in skeletal muscle cells. *J. Biol. Chem.* **276:** 38052–38060.

Tsuchiya H., Orimoto K., Kobayashi K., and Hino O. 1996. Presence of potent transcriptional activation domains in the predisposing tuberous sclerosis (Tsc2) gene product of the Eker rat model. *Cancer Res.* **56:** 429–433.

Tsukiyama-Kohara K., Poulin F., Kohara M., DeMaria C.T., Cheng A., Wu Z., Gingras A.C., Katsume A., Elchebly M., Spiegelman B.M., Harper M.E., Tremblay M.L., and Sonenberg N. 2001. Adipose tissue reduction in mice lacking the translational inhibitor 4E-BP1. *Nat. Med.* **7:** 1128–1132.

van Slegtenhorst M., Nellist M., Nagelkerken B., Cheadle J., Snell R., van den Ouweland A., Reuser A., Sampson J., Halley D., and van der Sluijs P. 1998. Interaction between hamartin and tuberin, the TSC1 and TSC2 gene products. *Hum. Mol. Genet.* **7:** 1053–1057.

Verdu J., Buratovich M.A., Wilder E.L., and Birnbaum M.J. 1999. Cell-autonomous regulation of cell and organ growth in *Drosophila* by Akt/PKB. *Nat. Cell Biol.* **1:** 500–506.

Vetter I.R., Arndt A., Kutay U., Gorlich D., and Wittinghofer A. 1999. Structural view of the Ran-Importin beta interaction at 2.3 Å resolution. *Cell* **97:** 635–646.

Vezina C., Kudelski A., and Sehgal S.N. 1975. Rapamycin (AY-22, 989), a new antifungal antibiotic. I. Taxonomy of the producing streptomycete and isolation of the active principle. *J. Antibiot.* **28:** 721–726.

Vogt P.K. 2001. PI 3-kinase, mTOR, protein synthesis and cancer. *Trends Mol. Med.* **7:** 482–484.

von Manteuffel S.R., Gingras A.C., Ming X.F., Sonenberg N., and Thomas G. 1996. 4E-BP1 phosphorylation is mediated by the FRAP-p70s6k pathway and is independent of mitogen-activated protein kinase. *Proc. Natl. Acad. Sci.* **93:** 4076–4080.

Wang T.C., Cardiff R.D., Zukerberg L., Lees E., Arnold A., and Schmidt E.V. 1994. Mammary hyperplasia and carcinoma in MMTV-cyclin D1 transgenic mice. *Nature* **369:** 669–671.

Wedaman K.P., Reinke A., Anderson S., Yates J., III, McCaffery J.M., and Powers T. 2003. Tor kinases are in distinct membrane-associated protein complexes in *Saccharomyces cerevisiae. Mol. Biol. Cell* **14:** 1204–1220.

Weinkove D., Neufeld T.P., Twardzik T., Waterfield M.D., and Leevers S.J. 1999. Regulation of imaginal disc cell size, cell number and organ size by *Drosophila* class I(A) phosphoinositide 3-kinase and its adaptor. *Curr. Biol.* **9:** 1019–1029.

Weisman R. and Choder M. 2001. The fission yeast TOR homolog, tor1+, is required for the response to starvation and other stresses via a conserved serine. *J. Biol. Chem.* **276:** 7027–7032.

West M.J., Stoneley M., and Willis A.E. 1998. Translational induction of the *c-myc* oncogene via activation of the FRAP/TOR signalling pathway. *Oncogene* **17:** 769–780.

Westphal R.S., Coffee R.L., Jr., Marotta A., Pelech S.L., and Wadzinski B.E. 1999. Identification of kinase-phosphatase signaling modules composed of p70 S6 kinase-protein phosphatase 2A (PP2A) and p21-activated kinase-PP2A. *J. Biol. Chem.* **274:** 687–692.

Whang Y.E., Wu X., Suzuki H., Reiter R.E., Tran C., Vessella R.L., Said, J.W., Isaacs W.B., and Sawyers C.L. 1998. Inactivation of the tumor suppressor PTEN/MMAC1 in advanced human prostate cancer through loss of expression. *Proc. Natl. Acad. Sci.* **95:** 5246–5250.

Wienecke R., Konig A., and DeClue J.E. 1995. Identification of tuberin, the tuberous sclerosis-2 product. Tuberin possesses specific Rap1GAP activity. *J. Biol. Chem.* **270:** 16409–16414.

Wu X., Senechal K., Neshat M.S., Whang Y.E., and Sawyers C.L. 1998. The PTEN/MMAC1 tumor suppressor phosphatase functions as a negative regulator of the phosphoinositide 3-kinase/Akt pathway. *Proc. Natl. Acad. Sci.* **95:** 15587–15591.

Xiao G.H., Shoarinejad F., Jin F., Golemis E.A., and Yeung R.S. 1997. The tuberous sclerosis 2 gene product, tuberin, functions as a Rab5 GTPase activating protein (GAP) in modulating endocytosis. *J. Biol. Chem.* **272:** 6097–6100.

Xu G., Kwon G., Cruz W.S., Marshall C.A., and McDaniel M.L. 2001. Metabolic regulation by leucine of translation initiation through the mTOR-signaling pathway by pancreatic beta-cells. *Diabetes* **50:** 353–360.

Xu G., Kwon G., Marshall C.A., Lin T.A., Lawrence J.C., Jr., and McDaniel M.L. 1998. Branched-chain amino acids are essential in the regulation of PHAS-I and p70 S6 kinase by pancreatic beta-cells. A possible role in protein translation and mitogenic signaling. *J. Biol. Chem.* **273:** 28178–28184.

Yamagata K., Sanders L.K., Kaufmann W.E., Yee W., Barnes C.A., Nathans D., and Worley P.F. 1994. rheb, a growth factor- and synaptic activity-regulated gene, encodes a novel Ras-related protein. *J. Biol. Chem.* **269:** 16333–16339.

Yamashita A., Ohnishi T., Kashima I., Taya Y., and Ohno S. 2001. Human SMG-1, a novel phosphatidylinositol 3-kinase-related protein kinase, associates with components of the mRNA surveillance complex and is involved in the regulation of nonsense-mediated mRNA decay. *Genes Dev.* **15:** 2215–2228.

Yang W., Tabancay A.P., Jr., Urano J., and Tamanoi F. 2001. Failure to farnesylate Rheb protein contributes to the enrichment of G0/G1 phase cells in the *Schizosaccharomyces pombe* farnesyltransferase mutant. *Mol. Microbiol.* **41:** 1339–1347.

Yeh W.C., Bierer B.E., and McKnight S.L. 1995. Rapamycin inhibits clonal expansion and adipogenic differentiation of 3T3-L1 cells. *Proc. Natl. Acad. Sci.* **92:** 11086–11090.

Young J. and Povey S. 1998. The genetic basis of tuberous sclerosis. *Mol. Med. Today* **4:** 313–319.

Yu J., Astrinidis A., and Henske E.P. 2001. Chromosome 16 loss of heterozygosity in tuberous sclerosis and sporadic lymphangiomyomatosis. *Am. J. Respir. Crit. Care Med.* **164:** 1537–1540.

Zhang H., Stallock J.P., Ng J.C., Reinhard C., and Neufeld T.P. 2000. Regulation of cellular growth by the *Drosophila* target of rapamycin dTOR. *Genes Dev.* **14:** 2712–2724.

Zhang Y., Gao X., Saucedo L.J., Ru B., Edgar B.A., and Pan D. 2003. Rheb is a direct target of the tuberous sclerosis tumour suppressor proteins. *Nat. Cell. Biol.* **5:** 578–581.

8

IGF-I Receptor Signaling in Cell Growth and Proliferation

Renato Baserga
Kimmel Cancer Institute
Thomas Jefferson University
Philadelphia, Pennsylvania 19107

THIS CHAPTER FOCUSES ON THE ROLE of the type 1 insulin-like growth factor receptor (IGF-IR) in the control of cell growth (cell size) and proliferation. I present a model of cell growth and proliferation built around the IGF-IR and its signaling pathways. This, of course, does not mean that cell growth and proliferation are exclusive prerogatives of the IGF-IR. It means that, using the IGF-IR and its pathways, one can build a model of cell growth and proliferation, whose features can be extended to and combined with other growth factor models. The IGF-IR is not an absolute requirement for growth and proliferation, either in vivo or in vitro, but it plays a significant role, especially in vivo. That cells do not require the IGF-IR for growth in culture is demonstrated by the ability of mouse embryo fibroblasts (MEFs) derived from mice with a targeted disruption of the IGF-IR genes (Liu et al. 1993) to grow and proliferate in serum-supplemented medium (Sell et al. 1993). The same is true in vivo, as discussed below. In this perspective, the IGF-IR is one of several growth factor receptors that play an important role in the physiology of cell growth and proliferation. I begin with a brief background on the IGF-IR itself. For detailed information on the basic aspects of the IGF-IR—its structure, functions, and signaling pathways—the reader is referred to the reviews by Blakesley et al. (1999) and Baserga (2000), and to the book edited by Rosenfeld and Roberts (1999).

TYPE 1 INSULIN-LIKE GROWTH FACTOR RECEPTOR

The full-length human IGF-IR cDNA was cloned by Ullrich et al. in 1986. It is a tyrosine kinase receptor, having 70% homology with the insulin receptor (IR) and, like the IR, it functions as a dimer. It is activated mainly by two ligands, IGF-I and IGF-II, but also by insulin at supraphysiological concentrations. The IGF-IR or its homologs are expressed in essentially all metazoans, from *Caenorhabditis elegans* to humans. It is also expressed in most mammalian cell types, the only known exceptions being hepatocytes and mature B lymphocytes. Because the IGF-IR is found ubiquitously, it participates in the growth regulation of many cell types, even when the primary growth factor is not an IGF. Epithelial cells, for instance, use epidermal growth factor (EGF) as their primary growth factor, whereas hematopoietic and neuronal cells use constellations of specific growth factors. These cells, however, also express the IGF-IR, and IGF-I can modulate their growth and/or survival. The social position of the IGF-IR has increased considerably in the past few years, progressing from a redundant IR, to a receptor with a life of its own, to a receptor important in cell and body growth, cell survival, and malignant transformation. At present, the IGF-IR, activated by its ligands, is known to control the growth and proliferation of cells in a variety of ways: (1) It sends a mitogenic signal; (2) it promotes the growth of cells (a requirement for sustained cell proliferation); (3) it plays a crucial role in the establishment and maintenance of the transformed phenotype; (4) it protects cells from a variety of apoptotic stimuli; (5) it participates in the regulation of cell adhesion and cell motility; and (6) it can induce terminal differentiation. Thus, the IGF-IR, like other receptors, can send contradictory signals (Baserga 2000). The first four functions contribute to an increase in total cell mass, but induction of differentiation does the opposite, as terminally differentiated cells stop dividing and, in many cases, eventually die. We show below that the outcome depends on the availability of appropriate substrates, the so-called "cell context." An intriguing characteristic of the IGF-IR is its role in cell transformation. Cells devoid of IGF-IR are refractory to transformation by cellular and viral oncogenes, and down-regulation of the IGF-IR in cancer cells causes not just reduced growth and proliferation, but also massive apoptosis (Baserga 1999). The IGF axis also plays a prominent role in longevity, from *C. elegans* to man (for review, see Baserga et al. 2002).

The IGF-IR signal transduction pathways have been discussed in many reviews (see above for general references). Briefly, the survival, growth, and proliferation signals originating from the IGF-IR can be summarized as follows. The main pathway goes through the insulin receptor substrates, IRS-1 and the other IRS proteins (White 1998). The IRS proteins, but especially IRS-1, activate a pathway that goes through PI3K (Cantley 2002) and S6 kinase 1, S6K1 (Dufner and Thomas 1999). PI3K activity after IGF-I or insulin stimulation is very low in cells that do not express IRS-1. However, it increases dramatically when IRS-1 is ectopically expressed (Soon et al. 1999; Belletti et al. 2001). There are at least two other intracellular signaling pathways that are operative and become especially important in cells that do not express IRS-1. One is the ERK pathway originating mostly from tyrosine 950 of the IGF-IR (Romano et al. 1999), and the other requires the integrity of serines 1280–1283 of the receptor (Peruzzi et al. 1999). This serine quartet binds 14.3.3 (Craparo et al. 1997), and the interaction induces Raf-1 activation and the mitochondrial translocation of both Raf-1 and Nedd4, a target of caspases (Peruzzi et al. 2001). These three pathways are diagramed in Figure 1. Needless to say, this figure is a drastic simplification of the pathways originating from the IGF-IR, as the receptor interacts directly or indirectly with many other transducing molecules, generating a variety of alternative pathways, to which are entrusted other functions and the fine-tuning of its signaling. These three pathways, however, are the most relevant to our discussion, since they are sufficient for IGF-1-mediated cell growth, proliferation, and survival. Other functions of the IGF-IR require other pathways.

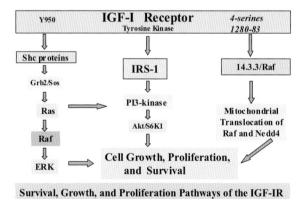

Figure 1. Main pathways of IGF-IR signaling for cell growth, proliferation, and survival. There are three main pathways: one through the insulin receptor substrate-1 (IRS-1), one through the Shc proteins, and one through 14.3.3, which connects with mitochondria. A combination of any two of these three pathways is sufficient for growth, proliferation, and survival. Other functions of the IGF-IR require other pathways and other domains of the receptor.

The Y950 pathway begins with an interaction with the Shc proteins (Craparo et al. 1995; Tartare-Deckert et al. 1996), which connect through Ras and Raf with the ERK proteins (Webster et al. 1994). In turn, the activated ERKs translocate to the nucleus, where they participate in the implementation of the cell-cycle program (Lenormand et al. 1993). The activation of Raf by 14.3.3 seems, at first, a duplication of the Shc pathway, but the evidence has been accumulating that Raf activation by 14.3.3 leads to another pathway, which results in the mitochondrial translocation of Raf (Peruzzi et al. 1999). This mitochondrial translocation has an important role in prevention of apoptosis and probably involves the mitochondrial colocalization of Nedd4. Nedd4, in turn, could be responsible for mitochondrial integrity. A fourth mitogenic pathway may involve c-Src. Wang and coworkers (Zong et al. 1998) have shown that Stat3 plays an important role in IGF-IR-dependent transformation of fibroblasts. However, in other cell lines, Stat3 does not promote proliferation and actually induces differentiation (Prisco et al. 2001).

An interesting aspect of the three main pathways outlined in Figure 1 is their limited redundancy. At least two pathways must be operative for protection from apoptosis, but any two pathways are sufficient (Navarro and Baserga 2001). This explains why the IR cannot protect 32D murine hematopoietic cells from apoptosis caused by IL-3 withdrawal (Peruzzi et al. 1999). A widely used cell line, 32D cells do not express IRS-1 (or IRS-2), and the IR does not have the serine quartet in its carboxyl terminus. There is therefore only one pathway in 32D cells, the equivalent of the Y950 pathway (which, in the IR, is Y960), not enough for a survival signal. Ectopic expression of IRS-1, as expected, restores the ability of the IR to protect 32D cells from apoptosis. In MEFs, which express IRS-1, the difference between IR and IGF-IR in protection from apoptosis (using anoikis to induce apoptosis) is much less dramatic, although a small difference is still detectable (Prisco et al. 1999). For the purpose of this chapter, it is important to remember that the anti-apoptotic pathways of the IGF-IR overlap extensively with its growth and mitogenic pathways, and, therefore, all these pathways contribute to cell mass accumulation.

CELL GROWTH AND RIBOSOMAL RNA

Although this topic is the central theme of the whole book, it is necessary for us to make a digression here, in order to establish the connection between cell size and the IGF-IR. A population of cells in culture, tissues, or animals grows by increasing both cell number and cell size. There is no question that an increase in cell number plays a major role in normal growth (after all, the adult animal originates from a single fertilized egg), but cells also grow in size. An illustration is provided by the postnatal rat liver, where both cell number and cell size keep increasing (Fukuda and Sibatani 1953). Cell size also increases during the cell cycle. Since the early days of the cell cycle, a doubling of cell size from G_1 to G_2 seemed obvious a priori. If a cell did not double in size from G_1 to G_2, its offspring would become smaller and smaller at every division and eventually would vanish. To quote from Fraser and Nurse (1978): "The two daughter cells produced at each division are identical to the parent at the same time in the preceding cycle: this requires that all components are doubled during the course of each cell cycle." The doubling in size from G_1 to G_2 was confirmed experimentally by numerous investigators, including Hartwell (1978) and Skog et al. (1979). This is a very important point. It says that the fundamental growth process is growth in cell size, because cell proliferation depends also on an increase in cell size.

Cell size can be measured directly (FACS analysis), but a good measurement of cell size is given by the amounts of protein and ribosomal RNA (rRNA) per cell. Proteins constitute most of the dry mass of a cell (Gaub et al. 1975), and the amount of protein per cell is influenced by the number of ribosomes; i.e., the amount of rRNA (Cohen and Studzinski 1967; Baxter and Stanners 1978; Darzynkiewicz et al. 1979). The connection between rRNA amounts and cell size has been confirmed in more recent studies. Isolation of size mutants from *Saccharomyces cerevisiae* has shown that many small-size mutants are defective in ribosome biogenesis (Jorgensen et al. 2002). To quote from Sudbery (2002): "...ribosome biogenesis appears to be directly linked to the commitment to cell division," a phenomenon that seems to be evolutionarily conserved, from yeast (which do not have growth factor receptors) to mammalian cells. In turn, ribosome biogenesis is controlled by the rate of rRNA synthesis (Moss and Stefanovsky 1995), which is dependent on the activity of RNA polymerase I (Grummt 1999; Reeder 1999; Comai et al. 2000). Thus, in the last analysis, cell size is regulated by RNA polymerase I and the proteins that modulate its activity. The rRNA synthesizing machinery and the rDNA genes are assembled in the nucleolus. Deletion of the nucleolus (as in the anucleolate mutant of *Xenopus laevis*) or of the RNA polymerase I gene (in yeast) results in nonviable animals or cells. The point that rRNA synthesis is an absolute requirement for cell growth is important. Terminally differentiated cells can survive without a nucleolus, but, as the term implies, they cannot proliferate and have a limited life span. As an interesting aside, successful nuclear transplantation in animal cloning requires the disassembly and subsequent reassembly of nucleoli (Gonda et al. 2003).

ROLE OF THE IGF-IR AND ITS SIGNALING PATHWAYS ON CELL AND BODY SIZE

The evidence for a role of the IGF axis in the regulation of cell size and body size is substantial. Interestingly, all experiments agree that ablation of IGF-IR signaling does not lead to cessation of cell or body growth. Rather, it leads to a decrease in growth, as if the IGF axis were responsible for about 50–60% of normal body growth in vivo, but no more. The conclusion is that there are other pathways for cell and body growth but that the IGF-IR signaling pathways are important, and a fraction of them cannot be replaced by other mechanisms.

IGF Signaling Regulates Cell Size and Body Size in Lower Animals

Drosophila has a receptor that is similar to both the IGF-IR and the IR: Its homology with either one or the other receptor in mammals depends on whether you work with insulin or with IGF-I. From my biased point of view, the *Drosophila* receptor is obviously an IGF-IR. Whether we call it an IGF-IR or an IR, it has signaling pathways that are similar to those in mammalian cells. *Drosophila* has homologs of IRS proteins, Akt, and S6K1. All these homologs have been reported to regulate cell size and body growth in *Drosophila*. This is true of IRS (Bohni et al. 1999), S6K1 (Montagne et al. 1999), and Akt (Verdu et al. 1999). Particularly instructive in the context of this chapter is the *Drosophila* IRS homolog, called *chico*. Deletion of *chico* reduces fly weight by 65% in females and 55% in males. The reduction in body and organ size is due to a reduction in both cell number and cell size. *Chico* is an IRS protein, the only IRS protein of *Drosophila*, and it interacts with the *Drosophila* homologs of the IGF-IR and PI3K (Bohni et al. 1999). S6K1 is a downstream effector of IRS-1 (Myers et al. 1994; and see above), and it is therefore not surprising that it may also regulate body size in *Drosophila*. The position of Akt in the pathway is less clear, as Akt can also be activated in the absence of IRS-1 (Navarro et al. 2001), but a dependence, at least in part, on PI3K is well established (Cantley 2002). In *Drosophila*, therefore, the evidence is convincing that IGF-IR signaling regulates cell size and body growth.

The IGF-1R Knockout Mouse Has a Growth Phenotype

The pioneering work of Estratiadis and collaborators has established the rules on the role of the IGF axis in mouse embryo growth. The main

Table 1. The IGF system in the development of mouse embryos

Genotype			Phenotype
IGF-IR	IGF-2R	IGF-II	(body size)
++	++	++	normal
– –	++	++	46%
++	– –	++	140%
++	++	+ p–	61.5%
– –	– –	++	105%
– –	++	+ p–	34%
++	– –	+ p–	74%
– –	– –	+ p –	34%

Each + or – indicates the presence or absence of the specific alleles; p– means that the paternal allele has been deleted. Mouse embryos produce IGF-II (coded by the paternal allele) but not IGF-I. The percentages indicate the animal weight in percent of normal controls. (Adapted from Ludwig et al. 1996.)

results are summarized in Table 1, which is taken (with modifications) from the paper by Ludwig et al. (1996). To better understand this table, one must keep in mind two additional pieces of information. The first is that mouse embryos produce IGF-II, but not IGF-I. At birth, the genes' transcriptions are reversed, and the adult mouse produces IGF-I but not IGF-II (this is not true of humans). The second thing to remember is that the IGF-II gene is imprinted, and IGF-II is transcribed only from the paternal allele. Thus, deletion of the paternal allele totally suppresses IGF-II production in the mouse embryo.

The data summarized in Table 1 clearly show that the IGF-II/IGF-IR interaction accounts for 66% of mouse embryo growth. When only the IGF-IR is deleted, the mice at birth are 46% of normal in size, suggesting that another receptor may partially substitute for the IGF-IR. Subsequent experiments, in vivo and in vitro, have shown that the residual growth occurring in the absence of the IGF-IR, but in the presence of IGF-II, is largely due to the activation of the IR by IGF-II (Morrione et al. 1997; Louvi et al. 1997), specifically the A isoform of the IR (Sciacca et al. 1999). The IGF-II receptor down-regulates IGF-II, and its deletion actually results in mice that are 150% in size at birth compared to wild-type littermates. A role of IGF-IR signaling in determining cell and body size in mammals is supported by other findings. Mice with a targeted disruption of the IRS-1 (Pete et al. 1999) or S6K1 (Shima et al. 1998) genes are smaller than their wild-type littermates. As mentioned above, IRS-1 is one of the major substrates of the IGF-IR, and S6K1 is a downstream effector of IRS-1 (Myers and White 1996). For other genes regulating mouse embryonic growth, see the review by Efstratiadis (1998), who points out that the IGF-IR is the only growth factor receptor, thus far, whose knockout has a growth phenotype. At least in the mouse embryo, a deficiency in the EGF receptor has no growth phenotype, whereas deletion of the PDGF receptor has a lethal phenotype in early embryonic development (Efstratiadis 1998).

These are the most rigorous experiments defining the role of IGF-IR signaling in *Drosophila* and mouse embryo growth, but there are other data in the literature indirectly supporting a role of the IGF axis in the control of body size in other mammals. There is a good correlation between plasma IGF-I levels and body size in different breeds of dogs (Eigenman et al. 1988) and mice. Plasma IGF-I concentrations vary from 40 ng/ml in the dachshund to 400 ng/ml in the Newfoundland, a giant breed. Similarly, mice with a deficiency in growth-hormone-binding protein, which causes a marked decrease in IGF-I levels, are smaller than their wild-type littermates (Coschigano et al. 2000). Incidentally, higher plasma levels of IGF-I negatively correlate with longevity. As an illustration, Eigenman et al. (1988) have shown that life expectancy has an inverse correlation to body size, with the giant breeds of dogs dying mostly in the first 7 years of life, whereas toy poodles and Chihuahuas often live to be 15 years of age, and sometimes older. The correlation of reduced life expectancy with increasing body size extends to other mammals and even to man (for review, see Baserga et al. 2002).

Transgenic mice confirm the importance of the IGF system in body growth. D'Ercole and coworkers (Mathews et al. 1988) were the first to report that transgenic mice overexpressing IGF-I have an increased body weight. However, the experiments of LeRoith and coworkers with conditional IGF-I mutants (Yakar et al. 1999), in which IGF-I production by the liver was abrogated, clearly show that the plasma levels of IGF-I are less important than the local levels (LeRoith et al. 2001). Despite a large drop in plasma IGF-I levels, these mice are essentially normal in size.

IGF Signaling and Cell Size in Cells in Culture

Since the amount of rRNA per cell must double from G_1 to G_2, inhibition of rRNA synthesis effectively inhibits cell-cycle progression (Lieberman et al. 1963). As mentioned above, increased synthesis of ribosomal RNA leads to a larger number of ribosomes, hence more protein synthesis, and an enlargement of cells (for review, see Baserga 1985). An effect of IGF-IR signaling on rRNA synthesis in cells in culture was suggested by the experiments of Surmacz et al. (1987), who showed that IGF-I could activate the ribosomal DNA promoter, driving a reporter gene. These findings were confirmed in murine hematopoietic 32D cells (Valentinis et al. 2000). 32D cells expressing

IRS-1 are larger than 32D cells not expressing IRS-1, even when both cell lines are proliferating exponentially (although at different rates). Specifically, the size difference between 32D IGF-IR/IRS-1 cells (transformed) and 32D IGF-IR cells (still proliferating but programed for differentiation) was similar to the difference between G_1 and G_2 cells in each cell line. The size difference correlated with the sustained activation of S6K1, in agreement with the findings by Shima et al. (1998) in knockout mice. The size of 32D IGF-IR/IRS-1 cells decreased when the cells were treated with rapamycin. Rapamycin inhibits the S6K1 pathway and induces differentiation of 32D IGF-IR/IRS-1 cells (Valentinis et al. 2000). In turn, differentiation is characterized by inhibition of rRNA transcription (see below).

Significance of the Data on IGF-IR Signaling and Cell or Body Size

In cells in culture, the IGF-IR and IRS-1 are NOT an absolute requirement for cell growth and proliferation. The best demonstration, as mentioned above, is provided by R$^-$ cells (Sell et al. 1993). R$^-$ cells are 3T3-like fibroblasts derived from mouse embryos with a targeted disruption of the IGF-IR genes (Liu et al. 1993). These cells do not proliferate in medium supplemented with the usual growth factors (EGF, PDGF, FGF, and, of course, IGF-I), but they do so in 10% serum, and even in serum-free medium supplemented with a purified growth factor, progranulin (Xu et al. 1998). Progranulin (Bhandari et al. 1993) is not the only purified growth factor that can sustain the proliferation of R$^-$ cells. We have seen that if the IR is overexpressed, IGF-II can sustain the proliferation of R$^-$ cells (Morrione et al. 1997). R$^-$ cells expressing the oncogene v-*src* grow in serum-free medium, without producing any detectable growth factor (Valentinis et al. 1997). As for a requirement for IRS-1, suffice it to say that 32D cells, so often mentioned in this discussion, do not express IRS-1 (or IRS-2) (Wang et al. 1993). These reports make an important point, because they indicate that, in cell cultures, there are alternative pathways that are IGF-independent and fully sufficient for normal cell growth and proliferation in vitro. As we have shown, this is not true of mice and *Drosophila*.

All the data in the previous sections indicate that the IGF axis is very important, but it is not an absolute necessity for cell and body growth. Mice and *Drosophila* devoid of IRS-1, S6K1, or the IGF-IR are smaller, but they do exist. Yet, the IGF axis remains a major component in vivo, so much so as to lead Lupu et al. (2001) to state that " ...the growth control pathway in which the components of the GH/IGF-1 signaling systems participate constitutes the major determinant of body size."

A legitimate question at this point is why I keep referring to the IGF axis, when the downstream signaling pathways of the IGF-IR and the IR are overlapping. Indeed, as the name indicates, IRS-1 was discovered as the major substrate of the IR (Sun et al. 1991) and was only subsequently found to be also a substrate of the IGF-IR. The answer is that mice homozygous for a null allele of the IR gene are of normal size at birth (Accili et al. 1996). True, they die at birth, but due to metabolic abnormalities associated with diabetic ketoacidosis. Thus, it seems that it is IGF-IR and not IR signaling that is most important for mouse embryonic growth. However, one should not forget that the IR can also contribute to mouse embryo growth (see above). The next question is obviously how the IGF-1/insulin system regulates growth. Before attempting to answer this question, we first consider the role of IRS-1 in differentiation.

IRS-1 AND THE INHIBITION OF DIFFERENTIATION

The experiments of Valentinis et al. (1999, 2000) have established that, in 32D murine hematopoietic cells, IRS-1 is a strong inhibitor of the differentiation program. 32D cells expressing a wild-type human IGF-IR (32D IGF-IR cells) proliferate exponentially for about 48 hours after withdrawal of IL-3 and supplementation with IGF-I (Peruzzi et al. 1999). After 48 hours, 32D IGF-IR cells (that do not express IRS-1 or IRS-2) undergo granulocytic differentiation (Valentinis et al. 1999). Ectopic expression of IRS-1 in 32D IGF-IR cells (32D IGF-IR/IRS-1 cells) inhibits granulocytic differentiation; the cells are IL-3-independent and form tumors in mice (Valentinis et al. 2000). The inhibitory effect of IRS-1 on differentiation has been confirmed by examining the expression of Id2 mRNA and protein. The Id proteins, especially Id1 and Id2, are strong inhibitors of differentiation (Benezra et al. 1990; Kreider et al. 1992). The name Id actually derives from "inhibitors of differentiation." In 32D IGF-IR cells, Id2 gene expression is very low, almost undetectable. Ectopic expression of IRS-1 in these cells causes a dramatic increase in Id2 gene expression (Belletti et al. 2001). An inactive mutant of IRS-1, with a deletion of the PTB domain, fails to increase Id2 gene expression (Navarro et al. 2001).

Incidentally, the lack of expression of IRS-1 is not a bizarre characteristic of parental 32D cells. Cell types prone to differentiation often do not express IRS-1, or express very low amounts. This is true not only of 32D cells (Wang et al. 1993) and other hematopoietic cell lines (Sun et al. 1997), but also of neuronal cells (Kim et al. 1998) and myoblasts (Sadowski et al. 2001). In cells with low levels of IRS-1, induction of differentiation causes a further decrease in IRS-1 expression (Sarbassov and

Peterson 1998; Morrione et al. 2001). Reduced levels or absence of IRS-1 in cells prone to differentiate is in itself an indirect confirmation of the inhibitory effect of IRS-1 on differentiation programs.

Differentiation in some cells results in an inhibition of RNA polymerase I (Larson et al. 1993; Comai et al. 2000). When the differentiation program is implemented, cell size decreases (Valentinis et al. 2000), and this decrease is accompanied by a marked reduction in rRNA synthesis and production (Comai et al. 2000; Tu et al. 2002). In such terminally differentiated cells, the nucleolus (the site of rRNA synthesis) disappears (Ringertz and Savage 1976). The integrity of the nucleolus may depend on the activity of the CDK1–cyclin B complex (Sirri et al. 2002), and one could speculate that, in the absence of a mitogenic stimulus, CDK1–cyclin B activity decreases, the nucleolus begins involution, and rRNA synthesis decreases sharply and eventually ceases.

The inhibition of rRNA synthesis in differentiating cells is accompanied by the translocation of the retinoblastoma proteins Rb and p130 to the nucleolus (Hannan et al. 2000a,b). Rb inhibits RNA polymerase I transcription by binding and inactivating the Upstream Binding Factor, UBF1 (Cavanaugh et al. 1995; Voit et al. 1997; Ciarmatori et al. 2001). UBF1 is a key regulator of RNA polymerase I activity and, therefore, of rRNA synthesis (Grummt 1999). Inhibition of differentiation, as shown above, is accompanied by a dramatic increase in Id gene expression. Id proteins, in turn, interact directly with the Rb proteins (Iavarone et al. 1994; Lasorella et al. 2000). A reasonable hypothesis is that IRS-1 does not interact directly with Rb, but simply sets up the Id proteins for inhibition of differentiation, with the Id proteins presumably sequestering the Rb proteins from entering the nucleolus. IRS-1, however, may still be needed for activation of UBF1. This hypothesis is expanded and discussed in a later section.

Interestingly, a similar scheme can apply to the SV40 T antigen, the product of an established oncogene. T antigen binds Rb directly (Fanning 1992), and it also stimulates rRNA synthesis (Soprano et al. 1979, 1983; Zhai et al. 1997). It interacts directly with IRS-1, both in MEFs (Zhou-Li et al. 1995) and in 32D cells expressing ectopic IRS-1 (Zhou-Li et al. 1997). The significance of T-antigen interaction with IRS-1 is not clear. IRS-1 is not required for the nuclear localization of T antigen, which is nuclear even in cells that do not express IRS-1 (Prisco et al. 2002). However, without IRS-1, T antigen cannot protect 32D cells from rapid apoptosis (Zhou-Li et al. 1997). In fact, 32D cells expressing only the T antigen die even faster than parental 32D cells after IL-3 withdrawal (Zhou-Li et al. 1997). A combination of T antigen and IRS-1 protects 32D cells from apoptosis. Therefore, at least in these cells, the association between T antigen and IRS-1 has a functional significance. It should also be mentioned that nucleolar translocation and inhibition of RNA polymerase I have been demonstrated for p53, another protein that, like Rb, binds the SV40 T antigen (Budde and Grummt 1999; Zhai and Comai 2000).

Whatever the mechanism, IRS-1 activated by the IGF-IR is a strong inhibitor of differentiation and a strong activator of rRNA synthesis. To summarize the situation at this point, it is reasonable to accept that cell growth and cell proliferation in general depend on ribosome biogenesis, which means rRNA production, whether it is rRNA synthesis, processing, or stability. I have also made the point that cell growth receives a substantial input by the activated IGF axis, especially through IRS-1. I next discuss the mechanism(s) by which IGF-IR/IRS-1 signaling regulates rRNA synthesis and cell growth.

NUCLEAR TRANSLOCATION OF TRANSDUCING MOLECULES

Signal transduction molecules can translocate to the nucleus. The long list includes ERKs (Lenormand et al. 1993), S6K1, TOR (Reinhard et al. 1994; Kim and Chen 2000), STAT proteins (Bromberg and Darnell 2000; Reddy et al. 2000), PI3K (Neri et al. 2002), Akt (Vandromme et al. 2001; Oh et al. 2002), β-catenin (Morali et al. 2001), the EGF receptor (Lin et al. 2001), certain phosphatases (Bollen and Beullens 2002), the ErbB-3 receptor (Offerdinger et al. 2002), IGFBP-3 (Schedlich et al. 1998), and a cleaved ErbB-4 receptor (Ni et al. 2001). Jans and Hassan (1998) have reviewed the accumulating evidence that growth factors, their receptors, and downstream effectors can translocate to the nucleus. The IRS proteins were not included in that review, but they should be now. The first observation that IRS-1 can translocate to nuclei should be credited to Lassak et al. (2002), who reported a nuclear IRS-1 in medulloblastoma cells expressing the JCV T antigen. At about the same time, Kabuta et al. (2002) claimed nuclear translocation of IRS-3 (but had negative results with IRS-1).

IRS-1 interacts with both the insulin and the IGF-I receptors, and the domains required for their interaction have been identified (Yenush et al. 1998). Because of its interaction with the receptors, its size, and its downstream signaling, it has been generally assumed that IRS-1 is an exclusively cytosolic or plasma membrane protein (Razzini et al. 2000; Jacobs et al. 2001). As already mentioned, IRS-1 is known to interact with the SV40 T antigen (Zhou-Li et al. 1995, 1997) and also with nucleolin (Burks et al.

1998). These two proteins are predominantly nuclear proteins, although minor fractions of either protein can be found in the cytosol (Santos and Butel 1982; Hovanessian et al. 2000). It was therefore tacitly assumed that IRS-1, anchored to the receptors, interacted with the minor cytosolic fractions of either T antigen or nucleolin. There is now convincing evidence that IRS-1 can translocate to the nucleus (Lassak et al. 2002; Prisco et al. 2002; Tu et al. 2002; Sun et al. 2003). The evidence presented in these papers rested on a variety of techniques, including subcellular fractionation, confocal microscopy, immunohistochemistry, and the use of appropriate mutant IRS proteins. Briefly, IRS-1 can be found in the nuclei of cells that express the SV40 T antigen, or the human JCV T antigen, or v-src, of cells in which the IGF-IR is stimulated by IGF-I, and even in adult rat liver (Boylan and Gruppuso 2002).

The finding of IRS-1 in the nuclei of cells expressing the SV40 T antigen makes sense, as it actually explains their known interaction. It is more difficult to explain its nuclear localization in v-src and R+ cells. IRS-1 interacts with both v-Src and c-Src and is constitutively phosphorylated in cells expressing these two proteins (Valentinis et al. 1997). There is no substantial evidence that v-Src localizes to the nucleus (Abram and Courtneidge 2000). A possible explanation is that v-Src may induce translocation of IRS-1, in the same way as it induces translocation of STAT3 (Bromberg and Darnell 2000; Reddy et al. 2000). As for IGF-I-mediated translocation, the obvious candidate for the protein that may chaperon IRS-1 into the nucleus would have been nucleolin, which interacts with IRS-1 (Burks et al. 1998). Unfortunately, there is now evidence against a role for nucleolin in the translocation of IRS-1 to the nucleus (see below). Another possible candidate is c-Myc, which has a nuclear targeting signal (Dang and Lee 1989) and is induced by IGF-I (Sumantran and Feldman 1993; Reiss et al. 1998). Incidentally, nuclear translocation of IRS-1 should not be considered an artifact of tissue culture, as it has been reported in tissue sections of human breast cancer (Schnarr et al. 2000), human medulloblastoma (Lassak et al. 2002), and rat liver (Boylan and Gruppuso 2002).

MECHANISMS OF IRS-1 TRANSLOCATION

We have evidence that IRS-2 can also translocate to the nucleus, although there are some differences between IRS-1 and IRS-2 (Sun et al. 2003). Since IRS-3 has been reported to be capable of nuclear translocation (Kabuta et al. 2002), it seems that at least three of the four IRS proteins can move to the nuclei of mammalian cells. At first sight, it seems that the

Pleckstrin (PH) domain of the IRS proteins should be the determining factor in nuclear translocation. It contains at least two sequences that could be considered as nuclear localization signals (Keller et al. 1993) or even nucleolar localization signals (Yang et al. 2002). To test this possibility, we determined which domains of IRS-1 are required for nuclear translocation. The results (Prisco et al. 2002) can be summarized as follows. A mutant IRS-1 with a deletion of the phosphotyrosine binding domain (δPTB) does not translocate. This mutant has very little activity (Yenush et al. 1996). A mutant IRS-1 with a deletion of the PH domain (δPH) translocates as effectively as wild-type IRS-1. Despite our predictions, this is not a complete surprise, since the δPH IRS-1 is as effective as wild-type IRS-1 in transforming 32D IGF-IR cells to IL-3 independence and in stimulating S6K1 activation (Valentinis and Baserga 2001). In fact, the δPH mutant stabilizes both S6K1 and its nuclear isoform p85 (Navarro et al. 2001), which may account for its growth- and proliferation-promoting abilities. Finally, an IRS-1 comprising only the PH and the PTB domains (PH/PTB) translocates to the nucleus, although not as effectively as the wild-type IRS-1 (Prisco et al. 2002). This last mutant (comprising only the first 300 amino-terminal amino acids of IRS-1) has partial survival-promoting function (Yenush et al. 1998) but cannot stimulate Id2 gene expression and does not inhibit differentiation of 32D IGF-IR cells (Belletti et al. 2001). The finding that the δPH mutant translocates as well as wild-type IRS-1 suggests that nucleolin is not required for translocation, as nucleolin has been reported to bind to the PH domain of IRS-1 (Burks et al. 1998). These results have been obtained in MEFs overexpressing the IGF-IR (R+ cells; Prisco et al. 2002). The results are different with medulloblastoma cells expressing the JCV T antigen (Lassak et al. 2002) and in MEFs expressing the SV40 T antigen (Sun et al. 2003). With T antigens, the PH domain is necessary for nuclear translocation of IRS-1. It seems, therefore, that the PH domain is required for binding of IRS proteins to T antigens but not for IGF-I-mediated translocation. Clearly, other mechanisms must be involved in this process.

BIOLOGICAL SIGNIFICANCE OF IRS-1 TRANSLOCATION

It has been reported that the EGF receptor in the nucleus may act as a transcription factor (Lin et al. 2001). IRS-1 is a potent inducer of Id2 gene expression (Belletti et al. 2001), and Id proteins are, in general, inhibitors of genes required for differentiation (see above). It will be interesting to test whether IRS-1 in the nucleus acts directly or indirectly on Id proteins in modifying the differentiation and proliferation programs. One should also remember the analogies between the IGF-IR and the T antigen, both of which induce nuclear translocation of IRS-1, with which they interact directly. Whereas IRS-1 induces Id2 protein, which interacts with Rb (Lasorella et al. 2000), T antigen interacts directly with Rb (Fanning 1992). The crucial event, however, is the stimulation of rRNA synthesis by either T antigen or IRS-1. T antigen localizes to the nucleus, but not to the nucleoli, whereas IRS-1 localizes to the nucleoli, as well as to the nucleus (Tu et al. 2002; Sun et al. 2003).

The nucleolus, the site of rRNA synthesis, is a very complex structure. Proteomic analysis of the human nucleolus has revealed at least 350 proteins, of which a significant fraction are encoded by novel or uncharacterized genes (Andersen et al. 2002; Scherl et al. 2002). It is the site not only of rRNA synthesis, but also of rRNA processing, through the formation of complexes, one of which has been characterized and designated as a processome (Dragon et al. 2002). The molecular biology of rRNA transcription and processing has been discussed in detail in two recent reviews (Grummt 1999; Reeder 1999). The activity of RNA polymerase I is the crucial event in rRNA synthesis and is regulated by several proteins, among which are UBF1 and nucleolin (Ginisty et al. 1999). UBF1 is the active isoform of UBF, UBF2 being completely inactive in rRNA synthesis (Grummt 1999). It is, therefore, of considerable interest that an antibody to IRS-1 coprecipitates UBF1 from nuclear extracts of cells (Tu et al. 2002; Sun et al. 2003), as illustrated in Figure 2. The same antibody fails to coprecipitate UBF1 from cytoplasmic extracts, as one would expect from the fact that UBF1 is an exclusively nucleolar protein that does not change location regardless of growth conditions (Voit et al. 1995). No UBF is coprecipitated in parental 32D cells that do not express IRS-1. A mutant IRS-1, with a deletion in the PTB domain, also fails to coprecipitate UBF1 (Tu et al. 2002). The significance of this finding is evident from the previous section. We propose that one important function of nuclear IRS-1 is to regulate closely the activity of RNA polymerase I and, therefore, the rate of rRNA synthesis. This can be confirmed by comparing rRNA synthesis of 32D IGF-IR and 32D IGF-IR/IRS-1 cells (Fig. 2). In this experiment, rRNA synthesis was determined in both cell lines while they were both proliferating, albeit at different rates. Expression of IRS-1 in 32D IGF-IR cells results in a threefold increase in rRNA synthesis, in agreement with the findings of Valentinis et al. (2000) that 32D IGF-IR/IRS-1 cells are larger than parental 32D IGF-IR cells. It is obviously not an accident that rRNA synthesis still goes on in proliferating 32D IGF-IR cells. As discussed above, cells can proliferate without IRS-1 (or even without the

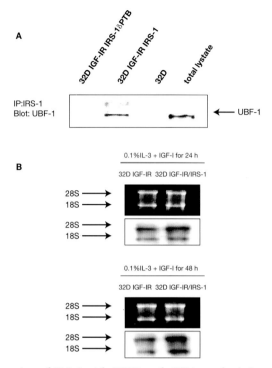

Figure 2. Interaction of IRS-1 with UBF1 and rRNA synthesis in proliferating and nonproliferating 32D cells. (*A*) The upstream binding factor (UBF1) is coprecipitated by an antibody to IRS-1 in 32D IGF-IR/IRS-1 cells, but not in parental 32 cells or in 32D IGF-IR cells expressing an inactive mutant of IRS-1 (deletion of the PTB domain). UBF1 is a key regulator of RNA polymerase I activity. (*B*) rRNA synthesis was determined in 32D IGF-IR and 32D IGF-IR/IRS-1 cells at 24 and 48 hours after withdrawal of IL-3 and supplementation with IGF-I. At this time, both cell lines were proliferating exponentially, but only 32D IGF-IR/IRS-1 cells continue to proliferate. 32D IGF-IR cells eventually differentiate to granulocytes. These experiments directly connect IRS-1 with the rRNA synthesizing machinery. (Modified from Tu et al. 2002.)

IGF-IR). Cells can grow in the absence of IRS-1, but IRS-1 and, more specifically, nuclear IRS-1, increase rRNA synthesis and ribosome biogenesis, and, ultimately, cell proliferation. Nuclear localization, in other words, results in different functions for IRS-1.

The importance of the interaction between IRS-1 and UBF1 should not be underestimated. Given that IGF-IR signaling (especially through IRS-1) is responsible for about 50% of animal growth (from *Drosophila* to mouse), the IRS-1/UBF interaction establishes for the first time a direct molecular link between IRS-1 and the rRNA synthesizing machinery.

A possible model, largely based on the use of 32D murine hematopoietic cells, is outlined in Figure 3. In this model, IRS-1 plays a double role:

inhibition of the differentiation program (by inducing Id gene expression) and stimulation of rRNA synthesis (resulting in increased cell growth). The activation of the IGF-IR/IRS-1 pathway induces Id gene expression, which results in inhibition of differentiation. Id2 proteins may inhibit differentiation of 32D IGF-IR cells by interacting and neutralizing Rb, which cannot move to the nucleolus and inhibit RNA polymerase I by binding to UBF (Cavanaugh et al. 1995; Hannan et al. 2000a). However, inhibition of differentiation is not necessarily a signal for cell proliferation. Overexpression of Id2 proteins inhibits differentiation but does not render 32D cells IL-3-independent (Florio et al. 1998; Prisco et al. 2001). For increased cell growth and cell proliferation, IRS-1 is translocated to the nucleus, where it stimulates rRNA synthesis by its interaction with UBF1 (Tu et al. 2002; Sun et al. 2003). In this model, T antigen inhibits Rb nucleolar translocation by binding directly to Rb. Still, even T antigen would need the cooperation of IRS-1, because, in its absence, T antigen cannot sustain the proliferation of 32D cells (Zhou-Li et al. 1997).

This model could explain the mechanism by which the activated IGF-IR increases cell size. But IGF-I, by itself, can be mitogenic for some cell types and, for proliferation, the cell-cycle program (immediate early genes, cyclins, and cyclin-dependent kinases) must also be activated (see below). At the moment, there is little information on how IGF-I activates the full

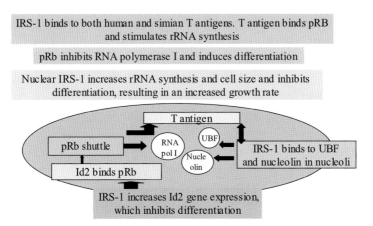

Figure 3. Interactions of IRS-1 with T antigen and nucleolar proteins. In this model, IRS-1 performs two key functions. The first is the induction of Id proteins, which, by binding pRb, inhibit pRb-induced differentiation. The second function is the direct binding to UBF1, a key regulator of rRNA synthesis, which is required for cell growth. In this model, the SV40 T antigen, which binds pRb, will also inhibit differentiation, but it would still require IRS-1 for full-scale cell proliferation.

cell-cycle program. This also depends on IRS-1, but, in addition, it requires signals originating from other domains in the IGF-IR (see Fig. 1). When both the Y950 residue and the 4-serine sequence of IGF-IR are mutated, IRS-1 is incapable of stimulating IGF-I-dependent cell proliferation (Navarro and Baserga 2001). Both of these domains converge on the activation of the Ras/Raf pathway (Peruzzi et al. 2001). A legitimate question at this point is how IL-3 makes parental 32D cells grow in the absence of either IRS-1 or T antigen. The answer again is that the IRS-1 pathway is not an absolute requirement for cell growth. It is a requirement for normal cell growth, but some cell growth and proliferation can be sustained in the absence of the IGF-IR/IRS-1 pathway. IL-3 must maximize alternative pathways in 32D and other IL-3-dependent cell lines. Interestingly, UBF is phosphorylated and activated by ERK in response to growth factors (Stefanovsky et al. 2001). We have mentioned that combined mutations at Y950 and the 4-serine domain of the IGF-IR abrogate the anti-differentiation effect of IRS-1 (Navarro et al. 2001). These mutations affect the subsequent activation of the Ras/Raf/ERK pathways and cause an attenuation of ERK activation, an observation reported several times from our laboratory (Peruzzi et al. 1999; Romano et al. 1999; Navarro and Baserga 2001; Navarro et al. 2001). This explanation is in agreement with our data (Sun et al. 2003) indicating that attenuation of the ERK pathway in this model reduces rRNA synthesis, even when IRS proteins are present. It suggests that full activation of rRNA synthesis requires both a nuclear IRS-1 and a signal from the activated ERK proteins (which also translocate to the nucleus). Since IRS-1 is not a kinase, the activation of rRNA synthesis by IRS-1 must depend on a kinase capable of phosphorylating UBF, phosphorylation of UBF being necessary for its activation (Grummt 1999). We have mentioned above that PI3K can also be found in the nucleus (Neri et al. 2002), and IRS-1 is a potent activator of PI3K (Soon et al. 1999). Although direct evidence is still lacking, it seems reasonable to hypothesize that IRS-1 may activate UBF through the intermediary of PI3K.

COORDINATION BETWEEN CELL-CYCLE PROGRAM AND CELL GROWTH

An increase in cell size is necessary for sustained cell proliferation. The question here is how the two programs, increase in cell size and cell-cycle progression, are coordinated. Clearly, without the implementation of the cell-cycle program, activation of RNA polymerase I could make cells larger but would fail to induce cell division. There must be a coordination between the two programs, so that cell size is usually maintained during cell proliferation. A few recent reports have indicated certain proteins involved in both the cell-cycle program and rRNA synthesis. They include pescadillo (also called PES-1 and Yph1p), involved in DNA replication and ribosome biogenesis (Lerch-Gagg et al. 2002); TAF II/250, which is required for G_1 progression (O'Brien and Tjian 2000) and binds to UBF (Lin et al. 2002); TIID, involved in both RNA polymerase II and RNA polymerase I transcription (Iben et al. 2002); and Cdk4-cyclin D1 and Cdk2-cyclin E, which phosphorylate UBF on Ser-484 and activate it (Voit et al. 1999). Pescadillo, incidentally, is the same protein as Yph1p, isolated from yeast nucleoli by Du and Stillman (2002) and connecting DNA replication to ribosome biogenesis. There are probably other connections among the 350-plus proteins identified in the nucleolus (Andersen et al. 2002; Scherl et al. 2002), but the ones cited above are the first described connections between the two programs. They suggest that the cell that has started the cell-cycle program makes sure, before mitosis, that the size increase is properly coordinated with cell division. A summary of these possibilities is diagramed in Figure 4, which includes other growth factors and other signaling pathways independent of the IGF-IR. These other pathways presumably account for that percentage of cell growth that is completely independent from IGF-IR/IRS-1 signaling. Interestingly, the combined overexpression of Ha-Ras and IRS-1 in 32D cells results in IL-3 independence and a dramatic increase in ERK activity (Cristofanelli et al. 2000).

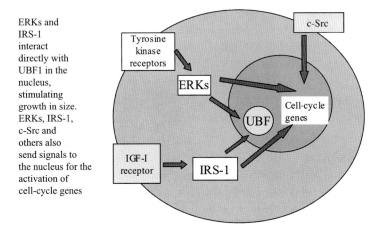

Figure 4. A diagram of interactions for cell growth and proliferation. Cell growth is required but not sufficient for cell proliferation. The cell-cycle program must also be activated. The IGF-I receptor, through IRS-1 and other transducing molecules, can initiate both programs. However, other growth factors and signaling pathways can substitute for the IGF-I receptor, especially in cells in culture.

SUMMARY: A MODEL OF IGF-IR-REGULATED CELL GROWTH

Although cell and body growth is regulated by a constellation of growth factors, certain conclusions can be reached about the part played by the IGF-IR in these processes.

1. The IGF-IR plays a major role in mouse embryonic and postnatal growth (Lupu et al. 2001), with substantial evidence that it does so in lower animals, and in other mammals, including humans. Efstratiadis (1998) has defined the role of the IGF-IR in simple terms: "Thus far, among the classic growth factors, only the IGFs have turned out to be a major growth signaling system for the entire mouse embryo...." Efstratiadis himself, however, sounds a word of caution by reminding us that there are differences in hormonal influences between humans and mice. After making allowances, it still remains that the IGF-IR plays a significant role in determining cell and body size.

2. The essential contribution of the IGF axis to cell growth and proliferation is not as strictly crucial for cells in culture. Cells without the IGF-IR or IRS-1 grow and proliferate in culture when provided with serum and/or the appropriate growth factors. However, absence of the IGF-IR severely limits the ability of cultured cells to proliferate in anchorage-independent conditions (Sell et al. 1993; Baserga 2000). These observations raise some interesting questions. Is the ability of cells in culture to dispense with the IGF axis (at least in monolayer cultures) a reflection of the fact that most cell lines are immortalized? Gene expression changes dramatically when mouse cells in culture become immortalized (Denhardt et al. 1991). One could visualize a situation where the environment of cells in culture may elicit the activation of pathways that are not normally activated in vivo.

3. Several signal transducing pathways originate from the IGF-IR. But the major pathway controlling cell and body size is through IRS-1 and its downstream effectors, such as Akt and S6K1. Again, in metazoans, like *Drosophila*, for instance, IRS-1 is a major contributor, since inhibition of the IRS pathway results in a decreased cell and body size.

4. Cell size is largely dependent on ribosome biogenesis (Jorgensen et al. 2002), which means rRNA synthesis and processing. Present evidence (Tu et al. 2002; Sun et al. 2003) indicates that nuclear IRS-1 interacts directly with components of rRNA synthesis and process-

ing, like UBF1 and nucleolin. UBF1 is the most important link between IRS-1 and the rRNA synthesizing machinery. The inhibition of differentiation and the increase in cell size that are dependent on IRS-1 may be in large part dependent on the nuclear translocation of IRS-1.

5. During the differentiation of hematopoietic cells, rRNA synthesis decreases, the nucleolus tends to involute and disappear, and IRS-1 is again cytoplasmic (Tu et al. 2002).

6. The activated IGF-IR is unequivocally mitogenic in several but not all cell types. The inability of the IGF axis to stimulate cell proliferation in certain types of cells presumably depends on the "cell context." As we have shown, in the absence of IRS-1, the IGF-IR sends a differentiation signal. Interestingly, the cells induced to differentiate survive, indicating that survival signals may occur independently of mitogenic signals.

In conclusion, it seems that the IGF-IR pathways play an important role in cell growth and proliferation. IRS-1 is the main effector of the IGF-IR in the process of cell growth. Ribosomal RNA synthesis and processing are a main target of IRS-1, which explains growth in size of the cells with an active IRS-1 pathway. IGF-I-mediated cell proliferation requires other signals, which activate the cell-cycle program.

ACKNOWLEDGMENTS

This work was supported by grants AG-16291 and CA-089640, from the National Institutes of Health.

REFERENCES

Abram C.L. and Courtneidge S.A. 2000. Src family tyrosine kinases and growth factor signaling. *Exp. Cell Res.* **254:** 1–13.

Accili D., Drago J., Lee E.J., Johnson M.D., Cool M.H., Salvatore P., Asico L.D., José P.A., Taylor S.I., and Westphal H. 1996. Early neonatal death in mice homozygous for a null allele of the insulin receptor gene. *Nat. Genet.* **12:** 106–109.

Andersen J.S., Lyon C.E., Fox A.H., Leung A.K.L., Lam Y.W., Steen H., Mann M., and Lamond A.I. 2002. Directed proteomic analysis of the human nucleolus. *Curr. Biol.* **12:** 1–11.

Baserga R. 1985. *The biology of cell reproduction.* Harvard University Press, Cambridge, Massachusetts.

———. 1999. The IGF-I receptor in cancer research. *Exp. Cell Res.* **253:** 1–6.

———. 2000. The contradictions of the insulin-like growth factor 1 receptor. *Oncogene* **19:** 5574–5581.

Baserga R., Prisco M., and Yuan T. 2002. IGF-I receptor signaling in health and disease. In

Insulin-like growth factors (ed. D. LeRoith et al.). Landes Bioscience, Georgetown, Texas. (EUREKA.com) pp. 104–120.

Baxter G.C. and Stanners C.P. 1978. The effect of protein degradation on cellular growth characteristics. *J. Cell. Physiol.* **96:** 139–146.

Belletti B., Prisco M., Morrione A., Valentinis B., Navarro M., and Baserga R. 2001. Regulation of Id2 gene expression by the IGF-I receptor requires signaling by phosphatidylinositol-3 kinase. *J. Biol. Chem.* **276:** 13867–13874.

Benezra R., Davis R.L., Lockshon D., Turner D.L., and Weintraub H. 1990. The protein Id: A negative regulator of helix-loop-helix DNA binding proteins. *Cell* **61:** 49–59.

Bhandari V., Giaid A., and Bateman A. 1993. The complementary deoxyribonucleic acid sequence, tissue distribution, and cellular localization of the rat granulin precursor. *Endocrinology* **133:** 2682–2689.

Blakesley V.A., Butler A.A., Koval A.P., Okubo Y., and LeRoith D. 1999. IGF-I receptor function: Transducing the IGF-I signal into intracellular events. In *The IGF system* (ed. R.G. Rosenfeld and C.T. Roberts, Jr.), pp. 143–163. Humana Press, Totowa, New Jersey.

Bohni R., Riesco-Escovar J., Oldham S., Brogiolo W., Stocker H., Andruss B.F., Beckingham K., and Hafen E. 1999. Autonomous control of cell and organ size by CHICO, a *Drosophila* homolog of vertebrate IRS-1. *Cell* **97:** 865–875.

Bollen M. and Beullens M. 2002. Signaling by protein phosphatases in the nucleus. *Trends Cell Biol.* **12:** 138–145.

Boylan J.M. and Gruppuso P.A. 2002. Insulin receptor substrate-1 is present in hepatocyte nuclei from intact rats. *Endocrinology* **143:** 4178–4183.

Bromberg J. and Darnell J.E., Jr. 2000. The role of STATs in transcriptional control and their impact on cellular function. *Oncogene* **19:** 2468–2473.

Budde A. and Grummt I. 1999. p53 represses ribosomal gene transcription. *Oncogene* **18:** 1119–1124.

Burks D.J., Wang J., Towery H., Ishibashi O., Lowe D., Riedel H., and White M.F. 1998. IRS pleckstrin homology domains bind to acidic motifs in proteins. *J. Biol. Chem.* **273:** 31061–31067.

Cantley L. 2002. The phosphoinositide 3-kinase pathway. *Science* **206:** 1655–1677.

Cavanaugh A.H., Hempel W.M., Taylor L.J., Rogalsky V., Todorov G., and Rothblum L.I. 1995. Activity of RNA polymerase I transcription factor UBF blocked by Rb gene product. *Nature* **374:** 177–180.

Ciarmatori S., Scott P.H., Sutcliffe J.E., McLees A., Alzuherri H.M., Dannenberg J.H., Riele H., Grummt I., Voit R., and White R.J. 2001. Overlapping function of the pRb family in the regulation of rRNA synthesis. *Mol. Cell. Biol.* **21:** 5806–5814.

Cohen L.S. and Studzinski G.P. 1967. Correlation between cell enlargement and nucleic acid and protein content of HeLa cells in unbalanced growth produced by inhibitors of DNA synthesis. *J. Cell. Physiol.* **69:** 331–340.

Comai L., Song Y., Tan C., and Bui T. 2000. Inhibition of RNA polymerase I transcription in differentiated myeloid leukemia cells by inactivation of selectivity factor-1. *Cell Growth Differ.* **11:** 63–70.

Coschigano K.T., Clemmons D., Bellush L.L., and Kopchick J.J. 2000. Assessment of growth parameters and life span of GHR/BP gene-disrupted mice. *Endocrinology* **141:** 2608–2613.

Craparo A., Freund R., and Gustafson T.A. 1997. 14.3.3 interacts with the insulin-like growth factor I receptor and insulin receptor substrate 1 in a phosphoserine-dependent manner. *J. Biol. Chem.* **272:** 11663–11669.

Craparo A., O'Neill T.J., and Gustafson T.A. 1995. Non-SH2 domains within insulin receptor substrate-1 and SHC mediate their phosphotyrosine-dependent interaction with the NPEY motif of the insulin-like growth factor-I receptor. *J. Biol. Chem.* **270:** 15639–15643.

Cristofanelli B., Valentinis B., Soddu S., Rizzo M.G., Marchetti A., Bossi G., Morena A.R., Dews M., Baserga R., and Sacchi A. 2000. Co-operative transformation of 32D cells by the combined expression of IRS-1 and v-Ha-Ras. *Oncogene* **19:** 3245–3255.

Dang C.V. and Lee W.M.F. 1989. Nuclear and nucleolar targeting sequences of c-erb-A, c-myb, N-myc, p53, HSP70 and HIV tat proteins. *J. Biol. Chem.* **264:** 18019–18023.

Darzynkiewicz D., Evenson D.P., Staiano-Coico L., Sharpless T.K., and Melamed M.R. 1979. Correlation between cell cycle duration and RNA content. *J. Cell. Physiol.* **100:** 425–438.

Denhardt D.T., Edwards D.R., McLeod M., Norton G., Parfett C.L.J., and Zimmer M. 1991. Spontaneous immortalization of mouse embryo cells: Strain differences and changes in gene expression with particular reference to retroviral gag-pol genes. *Exp. Cell Res.* **192:** 128–136.

Dragon F., Gallagher J.E.G., Compagnone-Post P.A., Mitchell B.M., Porwancher K.A., Wehner K.A., Wormsley S., Settlage R.E., Shabanowitz J., Osheim Y., Beyer A.L., Hunt D.F., and Baserga S.J. 2002. A large nucleolar U3 ribonucleoprotein required for 18S ribosomal RNA biogenesis. *Nature* **417:** 967–970.

Du Y.C.N. and Stillman B. 2002. Yph1p, an ORC-interacting protein: Potential link between cell proliferation control, DNA replication and ribosome biogenesis. *Cell* **109:** 835–848.

Dufner A. and Thomas G. 1999. Ribosomal S6 kinase signaling and the control of translation. *Exp. Cell Res.* **253:** 100–109.

Efstratiadis A. 1998. Genetics of mouse growth. *Int. J. Dev. Biol.* **42:** 955–976.

Eigenman J.E., Amador A., and Patterson D.F. 1988. Insulin-like growth factor I levels in proportionate dogs, chondrodystrophic dogs and in giant dogs. *Acta Endocrinol.* **118:** 105–108.

Fanning E. 1992. Simian virus 40 large T antigen: The puzzle, the pieces, and the emerging picture. *J. Virol.* **66:** 1289–1293.

Florio M., Hernandez M.C., Yang H., Shu H.K., Cleveland J.L., and Israel M.A. 1998. Id2 promotes apoptosis by a novel mechanism independent of dimerization to basic helix-loop-helix factors. *Mol. Cell. Biol.* **18:** 5435–5444.

Fraser R.S.S. and Nurse P. 1978. Novel cell cycle control of RNA synthesis in yeasts. *Nature* **27:** 726–730.

Fukuda M. and Sibatani A. 1953. Biochemical studies on the number and the composition of liver cells in the postnatal growth of the rat. *J. Biochem.* **40:** 95–110.

Gaub J., Auer G., and Zetterberg A. 1975. Quantitative cytochemical aspects of a combined Feulgen naphthol yellow S staining procedure for the simultaneous determination of nuclear and cytoplasmic proteins and DNA in mammalian cells. *Exp. Cell Res.* **92:** 323–332.

Ginisty H., Sicard H., Roger B., and Bouvet P. 1999. Structure and function of nucleolin. *J. Cell Sci.* **112:** 761–772.

Gonda K., Fowler J., Katoku-Kikyo N., Haroldson J., Wudel J., and Kikyo N. 2003. Reversible disassembly of somatic nucleoli by the germ cell proteins FRGY2a and FRGY2b. *Nat. Cell Biol.* **5:** 205–210.

Grummt I. 1999. Regulation of mammalian ribosomal gene transcription by RNA polymerase I. *Prog. Nucleic Acid Res. Mol. Biol.* **62**: 109–153.

Hannan K.M., Hannan R.D., Smith S.D., Jefferson L.S., Lun M., and Rothblum L.I. 2000a. Rb and p130 regulate RNA polymerase I transcription: Rb disrupts the interaction between UBF and SL-1. *Oncogene* **19**: 4988–4999.

Hannan K.M., Kennedy B.K., Cavanaugh A.H., Hannan R.D., Hirschler-Laszkiewicz I., Jefferson L.S., and Rothblum L.I. 2000b. RNA polymerase I transcription in confluent cells: Rb downregulates rDNA transcription during confluence-induced cell cycle arrest. *Oncogene* **19**: 3487–3497.

Hartwell L.H. 1978. Cell division from a genetic perspective. *J. Cell Biol.* **77**: 627–637.

Hovanessian A.G., Puvion-Dutilleul F., Nisole S., Svab J., Perret E., Deng J.S., and Krust B. 2000. The cell surface-expressed nucleolin is associated with the actin cytoskeleton. *Exp. Cell Res.* **261**: 312–328.

Iavarone A., Garg P., Lasorella A., Hsu J., and Israel M.A. 1994. The helix-loop-helix protein Id-2 enhances cell proliferation and binds to the retinoblastoma protein. *Genes Dev.* **8**: 1270–1284.

Iben S., Tschochner H., Bier M., Hoogstraten D., Hozak P., Egly J.M., and Grummt I. 2002. TFIIH plays an essential role in RNA polymerase I transcription. *Cell* **109**: 297–306.

Jacobs A.R., LeRoith D., and Taylor S.I. 2001. Insulin receptor substrate-1 pleckstrin homology and phosphotyrosine-binding domains are both involved in plasma membrane targeting. *J. Biol. Chem.* **276**: 40795–40802.

Jans D.A. and Hassan G. 1998. Nuclear targeting by growth factors, cytokines and their receptors: A role in signaling? *BioEssays* **20**: 400–411.

Jorgensen P., Nishikawa J.L., Breitkreutz B.J., and Tyers M. 2002. Systematic identification of pathways that couple cell growth and division in yeast. *Science* **297**: 395–400.

Kabuta T., Hakuno F., Asano T., and Takahashi S. 2002. Insulin receptor substrate 3 functions as transcriptional activator in the nucleus. *J. Biol. Chem.* **277**: 6846–6851.

Keller S.R., Aebersold R., Garner C.W., and Lienhard G.E. 1993. The insulin-elicited 160 kDa phosphotyrosine protein in mouse adipocytes is an insulin receptor substrate-1: Identification by cloning. *Biochim. Biophys. Acta* **1172**: 323–326.

Kim B., Cheng H.-L., Margolis B., and Feldman E.L. 1998. Insulin receptor substrate 2 and Shc play different roles in insulin-like growth factor I signaling. *J. Biol. Chem.* **273**: 34543–34550.

Kim J.E. and Chen J. 2000. Cytoplasmic-nuclear shuttling of FKBP12-rapamycin-associated protein is involved in rapamycin-sensitive signaling and translation initiation. *Proc. Natl. Acad. Sci.* **97**: 14340–14345.

Kreider B.L., Benezra R., Rovera G., and Kadesch T. 1992. Inhibition of myeloid differentiation by the helix-loop-helix protein Id. *Science* **255**: 1700–1702.

Larson D.E., Xie W.Q., Glibetic M., O'Mahony D., Sells B.H., and Rothblum L.I. 1993. Coordinated decrease in rRNA gene transcription factors and rRNA synthesis during muscle differentiation. *Proc. Natl. Acad. Sci.* **90**: 7933–7936.

Lasorella A., Noseda M., Beyna M., and Iavarone A. 2000. Id2 is a retinoblasoma protein target and mediates signaling by Myc oncoproteins. *Nature* **407**: 592–598.

Lassak A., DelValle L., Peruzzi F., Wang J.Y., Enam S., Croul S., Khalili K., and Reiss K. 2002. Insulin receptor substrate-1 translocation to the nucleus by the human JC virus T antigen. *J. Biol. Chem.* **277**: 17231–17238.

Lenormand P., Sardet C., Pages G., L'Allemain G., Brunet A., and Pousseygur J. 1993. Growth factors induce nuclear translocation of MAP kinases (p42[mpk] and p44[mpk]) but not of their activator MAP kinase (p45[mpkk]) in fibroblasts. *J. Cell Biol.* **122**: 1079–1088.

Lerch-Gagg A., Haque J., Li J., Ning G., Traktman P., and Duncan S.A. 2002. Pescadillo is essential for nucleolar assembly, ribosome biogenesis, and mammalian cell proliferation. *J. Biol. Chem.* **277:** 45347–45355

LeRoith D., Bondy C., Yakar S., Liu J.L., and Butler A. 2001. The somatomedin hypothesis 2001. *Endocr. Rev.* **22:** 53–74.

Lieberman I., Abrams R., and Ove P. 1963. Changes in the metabolism of ribonucleic acid preceding the synthesis of deoxyribonucleic acid in mammalian cells cultured from the animal. *J. Biol. Chem.* **238:** 2141–2149.

Lin C.Y., Tuan J., Scalia P., Bui T., and Comai L. 2002. The cell cycle regulatory factor TAF1 stimulates ribosomal DNA transcription by binding to the activator UBF. *Curr. Biol.* **12:** 2142–2146.

Lin S.Y., Makino K., Xia W., Matin A., Wen Y., Kwong K.Y., Bourguignon L., and Hung M.C. 2001. Nuclear localization of EGF receptor and its potential new role as a transcription factor. *Nat. Cell Biol.* **3:** 802–808.

Liu J.-P., Baker J., Perkins A.S., Robertson E.J., and Efstratiadis A. 1993. Mice carrying null mutations of the genes encoding insulin-like growth factor I (igf-1) and type 1 IGF receptor (Igf1r). *Cell* **75:** 59–72.

Louvi A., Accili D., and Efstratiadis A. 1997. Growth-promoting interaction of IGF-II with the insulin receptor during mouse embryonic development. *Dev. Biol.* **189:** 33–48.

Ludwig T., Eggenschwiler J., Fisher P., D'Ercole J.P., Davenport M.L., and Efstratiadis A. 1996. Mouse mutants lacking the type 2 IGF receptor (IGF2R) are rescued from perinatal lethality in Igf2 and Igf1r null backgrounds. *Dev. Biol.* **177:** 517–535.

Lupu F., Terwilliger J.D., Lee K., Segre G.V., and Efstratiadis A. 2001. Roles of growth hormone and insulin-like growth factor 1 in mouse postnatal growth. *Dev. Biol.* **229:** 141–162.

Mathews L.S., Hammer R.E., Behringer R.R., D'Ercole A.J., Bell G.I., Brinster R.L., and Palmiter R.D. 1988. Growth enhancement of transgenic mice expressing human insulin-like growth factor I. *Endocrinology* **123:** 2827–2833.

Montagne J., Stewart M.J., Stocker H., Hafen E., Kozma S.C., and Thomas G. 1999. *Drosophila* 6 kinase: A regulator of cell size. *Science* **285:** 2126–2129.

Morali O.G., Delmas V., Moore R., Jeanney C., Thiery J.P., and Larue L. 2001. IGF-II induces rapid β-catenin relocation to the nucleus during epithelium to mesenchyme transition. *Oncogene* **20:** 4942–4950.

Morrione A., Valentinis B., Xu S.Q., Yumet G., Louvi A., Efstratiadis A., and Baserga R. 1997. Insulin-like growth factor II stimulates cell proliferation through the insulin receptor. *Proc. Natl. Acad. Sci.* **94:** 3777–3782.

Morrione M., Navarro M., Romano G., Dews M., Reiss K., Valentinis B., Belletti B., and Baserga R. 2001. The role of the insulin receptor substrate-1 in the differentiation of rat hippocampal neuronal cells. *Oncogene* **20:** 4842–4852.

Moss T. and Stefanovsky V.Y. 1995. Promotion and regulation of ribosomal transcription in eukaryotes by RNA polymerase I. *Prog. Nucleic Acid Res. Mol. Biol.* **50:** 25–66.

Myers M.G., Jr. and White M.F. 1996. Insulin signal transduction and the IRS proteins. *Annu. Rev. Pharmacol. Toxicol.* **36:** 616–658.

Myers M.G., Jr., Grammer T.C., Wang L.M., Sun X.J., Pierce J.H., Blenis J., and White M.F. 1994. Insulin receptor substrate-1 mediates phosphatidylinositol 3′-kinase and p70S6K signaling during insulin, insulin-like growth factor-I and interleukin-4 stimulation. *J. Biol. Chem.* **269:** 28783–28789.

Navarro M. and Baserga R. 2001 Limited redundancy of survival signals from the type 1 insulin-like growth factor receptor. *Endocrinology* **142:** 1073–1081.

Navarro M., Valentinis B., Belletti B., Romano G., Reiss K., and Baserga R. 2001. Regulation of Id2 gene expression by the type 1 IGF receptor and the insulin receptor substrate-1. *Endocrinology* **142:** 5149–5157.

Neri L.M., Borgatti P., Capitani S., and Martelli A.M. 2002. The nuclear phosphoinositide 3-kinase/Akt pathway: A new second messenger system. *Biochim. Biophys. Acta* **1584:** 73–80.

Ni C.Y., Murphy M.P., Golde T.E., and Carpenter G. 2001. γ-Secretase cleavage and nuclear localization of ErbB-4 receptor tyrosine kinase. *Science* **294:** 2179–2181.

O'Brien T. and Tjian R. 2000. Different functional domains of $TAF_{II}250$ modulate expression of distinct subsets of mammalian genes. *Proc. Natl. Acad. Sci.* **97:** 2456–2461.

Offerdinger M., Schoefer C., Weipoltshammer K., and Grunt T.W. 2002. c-erbB-3: A nuclear protein in mammary epithelial cells. *J. Cell Biol.* **157:** 929–939.

Oh J.S., Kucab J.E., Bushel P.R., Martin K., Bennett L., Collins J., DiAugustine R.P., Barrett J.C., Afshan C.A., and Dunn S.E. 2002. Insulin-like growth factor 1 inscribes a gene expression profile for angiogenic factors and cancer progression in breast epithelial cells. *Neoplasia* **4:** 204–217.

Peruzzi F., Prisco M., Morrione A., Valentinis B., and Baserga R. 2001. Anti-apoptotic signaling of the insulin-like growth factor-I receptor through mitochondrial translocation of c-Raf and Nedd4. *J. Biol. Chem.* **276:** 25990–25996.

Peruzzi F., Prisco M., Dews M., Salomoni P., Grassilli E., Romano G., Calabretta B., and Baserga R. 1999. Multiple signaling pathways of the IGF-I receptor in protection from apoptosis. *Mol. Cell. Biol.* **19:** 7203–7215.

Pete G., Fuller G.R., Oldham J.M., Smith D.R., D'Ercole A.J., Kahn C.R., and Lund P.K. 1999. Postnatal growth responses to insulin-like growth factor 1 in insulin receptor substrate-1 mice. *Endocrinology* **140:** 5478–5487.

Prisco M., Peruzzi F., Belletti B., and Baserga R. 2001. Regulation of Id gene expression by the type 1 insulin-like growth factor: Roles of Stat3 and the tyrosine 950 residue of the receptor. *Mol. Cell. Biol.* **21:** 5447–5458.

Prisco M., Romano G., Peruzzi F., Valentinis B., and Baserga R. 1999. Insulin and IGF-I receptor signaling in protection from apoptosis. *Horm. Metab. Res.* **31:** 81–89.

Prisco M., Santini F., Baffa R., Liu M., Drakas R., Wu A., and Baserga R. 2002. Nuclear translocation of IRS-1 by the SV40 T antigen and the activated IGF-I receptor. *J. Biol. Chem.* **277:** 32078–32085.

Razzini G., Ingrosso A., Brancaccio A., Sciacchitano S., Esposito D.L., and Falasca M. 2000. Different subcellular localization and phosphoinosites binding of insulin receptor substrate protein pleckstrin homology domain. *Mol. Endocrinol.* **14:** 823–836.

Reddy E.P., Korapati A.S., Chaturvedi P., and Rane S. 2000. IL-3 signaling and the role of Src kinases, JAKs and STATs: A covert liaison unveiled. *Oncogene* **19:** 2532–2547.

Reeder R.H. 1999. Regulation of RNA polymerase transcription in yeast and vertebrates. *Prog. Nucleic Acid Res. Mol. Biol.* **62:** 293–327.

Reinhard C., Fernandez A., Lamb N.J.C., and Thomas G. 1994. Nuclear localization of p85S6K: Functional requirement for entry into S phase. *EMBO J.* **13:** 1557–1565.

Reiss K., Valentinis B., Tu X., Xu S.Q., and Baserga R. 1998. Molecular markers of IGF-I-mediated mitogenesis. *Exp. Cell Res.* **242:** 361–372.

Ringertz N.R. and Savage R.E. 1976. *Cell hybrids.* Academic Press, New York.

Romano G., Prisco M., Zanocco-Marani T., Peruzzi F., Valentinis B., and Baserga R. 1999. Dissociation between resistance to apoptosis and the transformed phenotype in IGF-I receptor signaling. *J. Cell. Biochem.* **72:** 294–310.

Rosenfeld R.G. and Roberts C.T., Jr., Eds. 1999. *The IGF system.* Humana Press, Totowa, New Jersey.

Sadowski C.L., Choi T.S., Le M., Wheeler T.T., Wang L.H., and Sadowski H.B. 2001. Insulin induction of SOCS-2 and SOCS-3 mRNA expression in C2C12 skeletal muscle cells is mediated by Stat5. *J. Biol. Chem.* **276:** 20703–20710.

Santos M. and Butel J. 1982. Detection of a complex of SV40 large tumor antigen and 53K cellular protein on the surface of SV40-transformed cells. *J. Cell. Biochem.* **19:** 127–144.

Sarbassov D.D. and Peterson C.A. 1998. Insulin receptor substrate-1 and phosphatidyl-inositol3-kinase regulate extracellular signal-regulated kinase-dependent and -independent signaling pathways during differentiation. *Mol. Endocrinol.* **12:** 1870–1878.

Schedlich I.J., Young T.Y., Firth S.M., and Baxter R.C. 1998. Insulin-like growth factor binding protein (IGFBP)-3 and IGFBP-5 share a common nuclear transport pathway in T47D human breast carcinoma cells. *J. Biol. Chem.* **273:** 18347–18352.

Scherl A., Couté Y., Deon C., Callé A., Kinbeiter K., Sanches J.C., Greco A., Hochstrasser D., and Diaz J.J. 2002. Functional proteomic analysis of human nucleolus. *Mol. Biol. Cell* **13:** 4100–4109.

Schnarr B., Strunz K., Ohsam J., Benner A., Wacker J., and Mayer D. 2000. Down regulation of insulin-like growth factor-1 receptor and insulin receptor substrate-1 expression in advanced human breast cancer. *Int. J. Cancer* **89:** 506–513.

Sciacca L., Costantino A., Pandini G., Mineo R., Frasca F., Scalia P., Sbraccia P., Goldfine I.D., Vigneri R., and Belfiore A. 1999. Insulin receptor activation by IGF-II in breast cancer: Evidence for a new autocrine/paracrine mechanism. *Oncogene* **18:** 2471–2479.

Sell C., Rubini M., Rubin R., Liu J.-P., Efstratiadis A., and Baserga R. 1993. Simian virus 40 large tumor antigen is unable to transform mouse embryonic fibroblasts lacking type-1 IGF receptor. *Proc. Natl. Acad. Sci.* **90:** 11217–11221.

Shima H., Pende M., Chen Y., Fumagalli S., Thomas G., and Kozma S.C. 1998. Disruption of the p70^{S6K}/p85^{S6K} gene reveals a small mouse phenotype and a new functional S6 kinase. *EMBO J.* **17:** 6649–6659.

Sirri V., Hernandez-Verdun D., and Roussel P. 2002. Cyclin-dependent kinases govern formation and maintenance of the nucleolus. *J. Cell Biol.* **156:** 969–981.

Skog S., Eliasson E., and Eliasson E. 1979. Correlation between size and position within the division cycle in suspension cultures of Chang liver cells. *Cell Tissue Kinet.* **12:** 501–511.

Soon L., Flechner L., Gutkind J.S., Wang L.H., Baserga R., Pierce J.H., and Li W. 1999. Insulin-like growth factor 1 synergizes with interleukin 4 for hematopoietic cell proliferation independent of insulin receptor substrate expression. *Mol. Cell. Biol.* **19:** 3816–3828.

Soprano K.J., Dev V.G., Croce C., and Baserga R. 1979. Reactivation of silent rRNA genes by simian virus 40 in human-mouse hybrid cells. *Proc. Natl. Acad. Sci.* **76:** 3885–3889.

Soprano K.J., Galanti N., Jonak G.J., McKercher S., Pipas J.M., Peden K.W.C., and Baserga R. 1983. Mutational analysis of simian virus 40 T antigen: Stimulation of cellular DNA synthesis and activation of rRNA genes by mutants with deletions in the T antigen gene. *Mol. Cell. Biol.* **3:** 214–219.

Stefanovsky V.V., Pelletier G., Hannan R., Gagnon-Kugler T., Rothblum L.I., and Moss T. 2001. An immediate response of ribosomal transcription to growth factor stimulation in mammals is mediated by ERK phosphorylation of UBF. *Mol. Cell* **8:** 1063–1073.

Sudbery P. 2002. When Wee meets Whi. *Science* **297:** 351–352.

Sumantran V.N. and Feldman E.L. 1993. Insulin-like growth factor 1 regulates c-myc and

GAP-43 messenger ribonucleic acid expression in SH-SY5Y human neuroblastoma cells. *Endocrinology* **132:** 2017–2023.

Sun H., Tu X., Prisco M., Wu A., Casiburi I., and Baserga R. 2003. Insulin-like growth factor I receptor signaling and nuclear translocation of insulin receptor substrates 1 and 2. *Mol. Endocrinol.* **17:** 472–486.

Sun X.J., Rothenberg P.L., Kahn C.R., Backer J.M., Araki E., Wilden P.A., Cahill D.A., Goldstein B.J., and White M.F. 1991. The structure of the insulin receptor substrate IRS-1 defines a unique signal transduction protein. *Nature* **352:** 73–77.

Sun X.J., Pons S., Wang L.M., Zhang Y., Yenush L., Burks D., Myers M.G., Jr., Glasheen E., Copeland N.G., Jenkins N.A., and White M.F. 1997. The IRS-2 gene on murine chromosome 8 encodes a unique signaling adapter for insulin and cytokine action. *Mol. Endocrinol.* **11:** 251–262.

Surmacz E., Kaczmarek L., Ronning O., and Baserga R. 1987. Activation of the ribosomal DNA promoter in cells exposed to insulin-like growth factor 1. *Mol. Cell. Biol.* **7:** 657–663.

Tartare-Deckert S., Sawka-Verhelle D., Murdaca J., and van Obberghen E. 1996. Evidence for a differential interaction of SHC and the insulin receptor substrate-1 (IRS-1) with the insulin-like growth factor-I (IGF-I) receptor in the yeast two-hybrid system. *J. Biol. Chem.* **271:** 23456–23460.

Tu X., Batta P., Innocent N., Prisco M., Casaburi I., Belletti B., and Baserga R. 2002. Nuclear translocation of insulin receptor substrate-1 by oncogenes and IGF-I: Effect on ribosomal RNA synthesis. *J. Biol. Chem.* **277:** 44357–44365.

Ullrich A., Gray A., Tam A.W., Yang-Feng T., Tsubokawa M., Collins C., Henzel W., Le Bon T., Kahuria S., Chen E., Jakobs S., Francke U., Ramachandran J., and Fujita-Yamaguchi Y. 1986. Insulin-like growth factor I receptor primary structure: Comparison with insulin receptor suggests structural determinants that define functional specificity. *EMBO J.* **5:** 503–512.

Valentinis B. and Baserga R. 2001. IGF-I receptor signalling in transformation and differentiation. *J. Clin. Pathol. Mol. Pathol.* **54:** 133–137.

Valentinis B., Morrione A., Taylor S.J., and Baserga R. 1997. Insulin-like growth factor 1 receptor signaling in transformation by src oncogenes. *Mol. Cell. Biol.* **17:** 3744–3754.

Valentinis B., Romano G., Peruzzi F., Morrione A., Prisco M., Soddu S., Cristofanelli B., Sacchi A., and Baserga R. 1999. Growth and differentiation signals by the insulin-like growth factor 1 receptor in hemopoietic cells are mediated through different pathways. *J. Biol. Chem.* **274:** 12423–12430.

Valentinis B., Navarro M., Zanocco-Marani T., Edmonds P., McCormick J., Morrione A., Sacchi A., Romano G., Reiss K., and Baserga R. 2000. Insulin receptor substrate-1, p70S6K and cell size in transformation and differentiation of hemopoietic cells. *J. Biol. Chem.* **275:** 25451–25459.

Vandromme M., Rocha A., Meier R., Carnac G., Besser D., Hemmings B.A., Fernandez A., and Lamb N.J.C. 2001. Protein kinase Bβ/Akt2 plays a specific role in muscle differentiation. *J. Biol. Chem.* **276:** 8173–8179.

Verdu J., Buratovich M.A., Wilder E.L., and Birnbaum M.J. 1999. Cell-autonomous regulation of cell and organ growth in *Drosophila* by Akt/PKB. *Nat. Cell Biol.* **1:** 500–506.

Voit R., Hoffmann M., and Grummt I. 1999. Phosphorylation by G1-specific cdk-cyclin complexes activates the nuclear transcription factor UBF. *EMBO J.* **18:** 1891–1899.

Voit R., Schafer K., and Grummt I. 1997. Mechanism of repression of RNA polymerase I transcription by the retinoblastoma protein. *Mol. Cell. Biol.* **17:** 4230–4237.

Voit R., Kuhn A., Sander E.E., and Grummt I. 1995. Activation of mammalian ribosomal gene transcription requires phosphorylation of the nucleolar transcription factor UBF. *Nucleic Acids Res.* **23:** 2593–2599.

Wang L.M., Myers M.G., Jr., Sun X.J., Aaronson S.A., White M., and Pierce J.H. 1993. IRS-1: Essential for insulin- and IL-4-stimulated mitogenesis in hemopoietic cells. *Science* **261:** 1591–1594.

Webster J., Prager D., and Melmed S. 1994. Insulin-like growth factor 1 activation of extracellular signal-related kinase-1 and -2 in growth hormone-secreting cells. *Mol. Endocrinol.* **8:** 539–544.

White M.F. 1998. The IRS-signalling system: A network of docking proteins that mediate insulin action. *Mol. Cell. Biochem.* **182:** 3–11.

Xu S.Q., Tang D., Chamberlain S., Orink G., Masiarz F.R., Kaur S., Prisco M., Zanocco-Marani T., and Baserga R. 1998. The granulin/epithelin precursor abrogates the requirement for the insulin-like growth factor 1 receptor for growth in vitro. *J. Biol. Chem.* **273:** 20078–20083.

Yakar S., Liu J.L., Stannard B., Butler A., Accili D., Sauer B., and LeRoith D. 1999. Normal growth and development in the absence of hepatic insulin-like growth factor 1. *Proc. Natl. Acad. Sci.* **96:** 7324–7329.

Yang Y., Chen Y., Zhang C., Huang H., and Weissman S.M. 2002. Nucleolar localization of hTERT protein is associated with telomerase function. *Exp. Cell Res.* **277:** 201–209.

Yenush L., Zanella C., Uchida T., Bernal D., and White M.F. 1998. The pleckstrin homology and phosphotyrosine binding domains of insulin receptor substrate 1 mediate inhibition of apoptosis by insulin. *Mol. Cell. Biol.* **18:** 6784–6794.

Yenush L., Fernandez R., Myers M.G., Jr., Grammer T.C., Sun X.J., Blenis J., Pierce J.H., Schlessinger J., and White M.F. 1996. The *Drosophila* insulin receptor activates multiple signaling pathways but requires insulin receptor substrate proteins for DNA synthesis. *Mol. Cell. Biol.* **16:** 2509–2517.

Zhai W. and Comai L. 2001. Repression of RNA polymerase I transcription by the tumor suppressor p53. *Mol. Cell. Biol.* **21:** 5930–5938.

Zhai W., Tuan J., and Comai L. 1997. SV40 large T antigen binds to the TBP-TAF complex SL1 and coactivates ribosomal RNA transcription. *Genes Dev.* **11:** 1605–1617.

Zhou-Li F., Xu S.-Q, Dews M., and Baserga R. 1997. Co-operation of simian virus 40 T antigen and insulin receptor substrate-1 in protection from apoptosis induced by interleukin-3 withdrawal. *Oncogene* **15:** 961–970.

Zhou-Li F., D'Ambrosio C., Li S., Surmacz E., and Baserga R. 1995. Association of insulin receptor substrate 1 with simian virus 40 large T antigen. *Mol. Cell. Biol.* **15:** 4232–4239.

Zong C.S., Zong L., Jiang Y., Sadowski H.B., and Wang L.H. 1998. Stat3 plays an important role in oncogenic ros- and insulin-like growth factor 1 receptor-induced anchorage-independent growth. *J. Biol. Chem.* **273:** 28065–28072.

9

S6K Integrates Nutrient and Mitogen Signals to Control Cell Growth

Jacques Montagne and George Thomas
Friedrich Miescher Institute for Biomedical Research
Basel, Switzerland

FOR ALL CELLS IN A SPECIFIC ORGAN TO ACHIEVE their mature size and thereby to control the ultimate size of the organism itself, they must coordinate cell growth (increase in cell mass) with the rate of cell proliferation (increase in cell number) (Conlon and Raff 1999; Edgar 1999; Lehner 1999; Polymenis and Schmidt 1999). However, there are examples in biology where these two phenomena have been separated. For example, without cell division, the growing oocyte of *Xenopus laevis* increases in size 17,000-fold during oogenesis. In contrast, shortly after fertilization, the ensuing embryo passes through 12 synchronous rounds of cell division, dramatically increasing in cell number with no associated growth. Nevertheless, these are exceptions, as growth and proliferation must normally be coordinated within a single cell cycle, such that the new daughter cell will faithfully obtain a full complement of cellular organelles and physiologically important proteins required to carry out its differentiated cellular task (Montagne 2000; Stocker and Hafen 2000; Kozma and Thomas 2002). To achieve these goals, a number of critical growth-promoting anabolic processes must be up-regulated; chief among these is protein synthesis (Thomas 2000).

Rates of protein synthesis must increase not only to generate new cellular organelles and proteins, but also to increase translational capacity (Pardee 1989; Thomas and Hall 1997; Gingras et al. 2001). Indeed, one of the major energy-consuming processes in a growing cell is the generation of nascent ribosomes, with each mature daughter cell containing ~5 million copies of each ribosomal subunit (Schmidt 1999; Warner 1999). This means that ribosomes make up more than 80% of cellular RNA and 5–10% of cellular protein. Earlier studies in higher eukaryotes showed that, although rRNA synthesis is controlled at the transcriptional level, the regulation of ribosomal protein transcripts is largely controlled at the translational level. This translational regulation is through a stretch of polypyrimidines located at the transcriptional start site termed a 5′ terminal oligopyrimidine tract or 5′TOP (Meyuhas et al. 1996; Meyuhas 2000). It has also been known for some time that the translational up-regulation of 5′TOP mRNAs closely follows the multiple phosphorylation of the 40S ribosomal protein S6 (Jefferies and Thomas 1996) and the activation of p70[S6K], later termed S6K1 (see below). Because both events were inhibited by rapamycin, a bacterial macrolide from *Streptomyces hygroscopicus*, activation of p70[S6K] and multiple phosphorylation of S6 were suspected to be the mediators of increased translational capacity. In seeming agreement with this hypothesis, p70[S6K]-deficient mice and *Drosophila* deficient for the fly homolog of p70[S6K], dS6K, have severe defects in growth (Shima et al. 1998; Montagne et al. 1999; Pende et al. 2000). Here we discuss the role of the S6Ks in growth, the signaling pathways involved in controlling S6K activation, and finally the role of S6K in development and human disease.

S6 PHOSPHORYLATION AND THE DISCOVERY OF S6K1 AND S6K2

Identification of S6Kinase

The S6 gene encodes one of the 33 proteins, which in a complex with one molecule of 18S rRNA, comprise the mature 40S ribosomal subunit (Wool et al. 1996). Multiple phosphorylation of S6 was first observed following partial hepatectomy in rats (Gressner and Wool 1974). Since partial hepatectomy induces the activation of protein synthesis and a single round of hepatocyte proliferation to replace the lost liver mass, the above observation prompted a number of subsequent studies which demonstrated that S6 phosphorylation is a ubiquitous response to growth-factor stimulation (Nielsen et al. 1981; Kozma et al. 1989a). Further studies in vivo and in vitro showed that the phosphorylation sites in S6 are clustered at the carboxyl terminus of the protein and that multiple phosphorylation proceeds in an ordered manner: Ser-236 > Ser-235 > Ser-240 > Ser-244 > Ser-247 (Krieg et al. 1988; Bandi et al. 1993). Recently, the conservation of this event with respect to the sites and order of phosphorylation was demonstrated in *Drosophila* Kc 167 tissue culture cells (Radimerski et al.

2000). Consistent with a role in translation, the most highly phosphory-lated derivatives of S6 from growth factor–stimulated mammalian cells were found in heavy polysomes (Duncan and McConkey 1982; Thomas et al. 1982). In addition, chemical cross-linking and protection studies with specific components of the translational apparatus localized S6 to the small head region of the 40S ribosomal subunit at the mRNA–tRNA-binding site (Nygard and Nika 1982; Nygard and Nilsson 1990). This area of the 40S ribosomal subunit is at the interface with the larger 60S ribo-somal subunit, where S6 apparently comes into direct contact with 28S rRNA. The location of S6 in the 40S ribosomal subunit places it in a posi-tion where it could potentially have a functional impact on translation (Nygard and Nilsson 1990).

These initial observations prompted the search for the kinase(s) that links mitogen stimulation to S6 phosphorylation. S6 kinase was first described as an activity from mitogen-stimulated Swiss mouse 3T3 cells responsible for the phosphorylation of S6 in vitro (Novak-Hofer and Thomas 1984). Its eventual purification from this source led to the iden-tification of a kinase with an apparent electrophoretic mobility of 70 kD, termed p70[S6K] (Jenö et al. 1988, 1989; Kozma et al. 1989b). In vitro, p70[S6K] was capable of phosphorylating all the same phosphorylation sites in the equivalent sequential manner as previously observed in vivo (Ferrari et al. 1991). Subsequent sequencing of the enzyme, followed by cloning (Banerjee et al. 1990; Kozma et al. 1990) and expression studies, led to the identification of two isoforms of p70[S6K] (Grove et al. 1991; Reinhard et al. 1992) that are produced from the same transcript by alternative transla-tional start sites. The larger isoform of p70[S6K], termed p85[S6K], is distinct from the shorter form in that it contains an amino-terminal 23–amino acid extension (Grove et al. 1991; Reinhard et al. 1992), generated from an alternative, weak translational initiation site, whereas the shorter form has a strong Kozak initiation site (Kozak 1991). The additional 23 amino acids in the p85[S6K] isoform contain a nuclear localization signal (NLS), which localizes p85[S6K] to the nucleus, whereas the shorter p70[S6K] appears to be largely localized to the cytoplasm (Reinhard et al. 1994).

Unexpectedly, it was recently found that deletion of the p70[S6K]/p85[S6K] gene in mice did not impair either S6 phosphorylation or 5′TOP mRNA translation, leading to the identification of a second S6 kinase gene (Shima et al. 1998), which is highly homologous to the first gene (Fig. 1). For sim-plicity, this new gene has been referred to as either S6K2 or S6Kβ, and the p70[S6K]/p85[S6K] gene is now collectively referred to as S6K1 or S6Kα (Gout et al. 1998; Saitoh et al. 1998; Shima et al. 1998). Although it has largely been argued from studies with rapamycin in mammalian cells that S6K1 and

S6K2 are responsible for the multiple phosphorylation of S6 in vivo, recent studies in *X. laevis* oocytes revealed that phosphorylation of S6 may be carried out by p90[rsk] during meiotic maturation (Schwab et al. 1999).

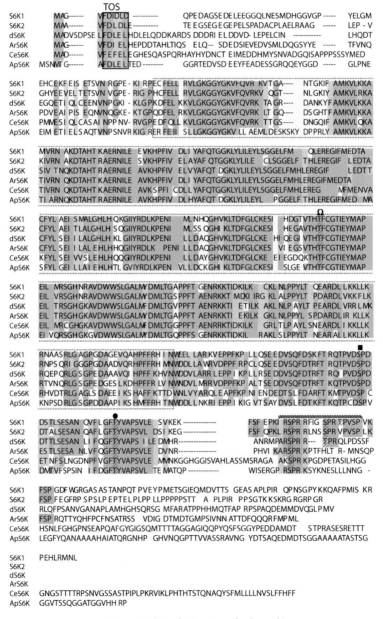

Figure 1. (*See facing page for legend.*)

S6K1 and S6K2 belong to the A, G, and C family of serine/threonine protein kinases, the largest family of serine/threonine protein kinases (Hanks and Hunter 1995). A comparison of the S6K1 sequence with that of S6K2 reveals that they are highly homologous, sharing 82% identity in the catalytic domain and almost the same degree of homology in the linker and autoinhibitory domains (Gout et al. 1998; Saitoh et al. 1998; Shima et al. 1998). In addition, the acidic amino-terminal domain, as well as critical phosphorylation sites, including those equivalent to T229, S371, T389, and the S/TP sites in the autoinhibitory domain of S6K1 (see below), are all conserved (Fig. 1). Although a longer form of S6K2 has been reported, we have not detected this form with specific antibodies, nor would the genome structure of *S6K2* appear to support the existence of the longer form (M. Pende et al., unpubl.). However, a potential NLS resides at the carboxyl terminus of S6K2. S6K1 is identical in amino acid sequence in all mammalian species examined to date, and S6K2 also displays a high degree of identity. This homology extends to invertebrates where single-copy orthologs for the *S6K* gene have been found in *Drosophila melanogaster* (Stewart et al. 1996; Watson et al. 1996), *Caenorhabditis elegans* (Long et al. 2002), *Aplysia californica* (Khan et al. 2001), and *Artemia franciscana* (Fig. 1) (Santiago and Sturgill 2001). In addition, two S6K homologs have also been found in *Arabidopsis thaliana* (Turck et al. 1998), and a novel S6K was recently described in *Schizosaccharomyces pombe* (Matsuo et al. 2003), indicating that the S6Ks are evolutionarily conserved in eukaryotes.

S6 Phosphorylation and 5´TOP Messengers

As mentioned above, it is known that translation of 5´TOP mRNAs is selectively up-regulated following mitogenic stimulation. The 5´TOP

Figure 1. Sequence comparison of S6 kinases from human (S6K1 and S6K2), *Drosophila melanogaster* (dS6K), *Artemia franciscana* (ArS6K), *Caenorhabditis elegans* (CeS6K), and *Aplysia californica* (ApS6K). The highly conserved residues are marked by a deep-blue background, whereas the partially conserved residues between species are marked by a light-blue background. The amino-terminal TOS domain is boxed. The catalytic domain from residue 66 to 333 of S6K1 (*pink dotted boxed*) contains the conserved T-loop kinase (Ω). The linker domain between the catalytic domain and the autoinhibitory domain (*bold red line*) contains the entirely conserved S371 (■) and T389 (●) with respect to S6K1 primary sequence. In contrast, the regulatory S/TP sites within the autoinhibitory domain are partially conserved among species (*hatched pink background*).

transcripts can represent up to 20% of the total mRNA in the cell and encode ribosomal proteins, translational elongation factors, and a number of initiation factors (Meyuhas et al. 1996). The 5′TOP motif invariably begins with a cytidine followed by a stretch of 5–15 pyrimidines. However, for efficient translation of these transcripts, the integrity of the region immediately downstream of the 5′TOP is also required. The 5′TOP itself appears to act as a repressor of translation (Mariottini and Amaldi 1990; Hammond et al. 1991; Levy et al. 1991), such that mutation or removal of the 5′TOP leads to the immediate recruitment of these mRNAs into polysomes (Avni et al. 1994). Recent studies show that these effects may be mediated by three proteins; the La antigen, which binds to the 5′TOP, the cellular nucleic binding protein (CNBP), which binds to the sequence immediately downstream of the polypyrimidine tract, and the Ro60 autoantigen, which mediates the binding of La antigen and CNBP to their respective sequences (Pellizzoni et al. 1998). It appears that CNBP acts as a suppressor, whereas La can promote the translation of 5′TOP mRNAs, with the binding of either protein to the 5′TOP mRNA excluding the binding of the other (Crosio et al. 2000).

The notion that S6K1 and S6 phosphorylation may be involved in mediating the up-regulation of 5′TOP mRNAs was originally derived from the observation that rapamycin selectively represses recruitment of 5′TOP mRNAs into large polysomes without affecting global translation (Jefferies et al. 1994; Terada et al. 1994). In support of this model, it was further shown that (1) an intact 5′TOP tract is required for rapamycin to elicit an inhibitory effect on the translation of these mRNAs, (2) a dominant interfering allele of S6K1 represses the mitogen-induced translational up-regulation of 5′TOP mRNAs to the same extent as rapamycin, and (3) expression of a rapamycin-resistant S6K1 mutant largely negates the inhibitory effects of rapamycin on 5′TOP mRNA translation (Jefferies et al. 1994, 1997; Schwab et al. 1999). Consistent with this, it was reported that in S6K1-deficient embryonic stem (ES) cells, in contrast to wild-type ES cells, S6 phosphorylation is abolished and rapamycin no longer affects the translation of 5′TOP mRNAs (Kawasome et al. 1998). However, it has been difficult to interpret this latter finding, as S6 phosphorylation is largely intact in all tissues analyzed from mice deficient for S6K1, apparently due to the presence of S6K2, which is up-regulated in all these same tissues (Shima et al. 1998). In addition, it has been later reported that S6 becomes phosphorylated in S6K1-deficient ES cells in response to mitogen stimulation and that S6K2 is present and, in parallel, activated by mitogen stimulation (Lee-Fruman et al. 1999).

Recently, there has been a report arguing that S6K1, S6 phosphorylation, and mTOR are not required for translation of 5′TOP mRNAs in ligand-induced ES cells or in NGF- and serum-stimulated PC12 cells (Stolovich et al. 2002). However, others have reported that up-regulation of 5′TOP mRNAs in NGF and serum-stimulated PC12 cells is totally repressed by rapamycin (Petroulakis and Wang 2002). In part, these discrepancies may be explained by the different way in which these experiments were performed. In the first case, the authors compared total polysomes to the sub-polysome fractions, whereas the second group monitored the movement of the 5′TOP mRNA along the polysome profile, from small to large polysomes. It may also be that differences in specific cell types reflect the potential of different pathways to converge on a common mitogenic response. Indeed, the phosphorylation of GSK3β and BAD are regulated not only by PKB, but also by S6K1 (Peyrollier et al. 2000; Harada et al. 2001). Further investigations will be necessary to understand the role of the S6Ks and S6 phosphorylation in the regulation of 5′TOP mRNA. Finally, it should also be noted that S6K1 has other biological targets which can affect cell growth, including protein synthesis elongation factor 2 kinase (eEF2K) (Wang et al. 2001) and initiation factor 4B (eIF4B) (Raught et al. 2001). Most likely, other targets exist, and their identification will be of great help in understanding the mechanisms by which S6K1 and S6K2 mediate cell growth.

S6K ACTIVATION

Multiple Residues Are Phosphorylated in S6K1

Our understanding of the mechanism of kinase activation was preceded by the identification of the signaling components that mediate S6K1 activation, which in turn have been instructive in elucidating the relative importance of S6Ks in cell growth. Initial studies showed that recovery of S6K1 activity from cells required the presence of phosphatase inhibitors in the extraction buffer, giving credence to the concept of mitogen-induced kinase cascades (Novak-Hofer and Thomas 1984, 1985). These studies subsequently led to the realization that S6K1 activation involves the sequential interplay of five distinct domains and the hierarchical phosphorylation of at least seven specific sites in S6K1 (Fig. 1) (Pullen and Thomas 1997). The first domain is represented by a short segment at the amino terminus, which is quite acidic in nature and confers rapamycin sensitivity to S6K1 (Cheatham et al. 1995; Weng et al. 1995b; Dennis et al. 1996). Adjacent to this domain is the highly conserved catalytic domain, which contains a key site of phosphorylation in the activation loop, T229, which lies in a hydrophobic motif (Pearson et al. 1995; Weng et al. 1995b). The catalytic domain is coupled to the carboxy-terminal autoinhibitory domain through a linker domain. The linker domain contains two essential phosphorylation sites, S371, an S/TP site, and T389, which lies in a hydrophobic motif (Pearson et al. 1995; Dennis et al. 1996; Moser et al. 1997; Saitoh et al. 2002). At the end of the linker domain there is an additional rapamycin-sensitive phosphorylation site, at position 404, although mutation of this site to either an acidic or an alanine residue does not affect basal or serum-stimulated S6K1 activity (Pearson et al. 1995). In contrast to most members of the AGC family, which end with the linker domain, S6K1 and S6K2 have two additional carboxy-terminal domains (Fig. 1). The first domain, which contains a sequence that is homologous to the amino acid sequence surrounding phosphorylation sites in S6, serves as an autoinhibitory domain (Cheatham et al. 1995; Weng et al. 1995b; Dennis et al. 1998). This domain contains four S/T-P phosphorylation sites: S411, S418, T421, and S424, whose phosphorylation facilitates the phosphorylation of T389 (Dennis et al. 1998), which is critical for T229 phosphorylation in the activation loop (see below). Finally, the extreme carboxyl terminus of S6K1 contains a PDZ-binding domain, whereas S6K2 contains a number of potential SH3-binding motifs. These two domains may permit differential interactions with potential upstream activators or downstream targets. Based on these findings, the current model of S6K1 activation is as follows. The first step is the phosphorylation of the S/T-P sites in the autoinhibitory domain. The phosphorylation of the S/T-P sites then functions in conjunction with the amino terminus to allow phosphorylation of T389 (Fig. 2). These two phosphorylation events disrupt a presumed interaction between the carboxyl and amino termini, allowing phosphorylation of T229 and resulting in S6K1 activation.

In parallel with establishing the mechanism of S6K1 activation, two key observations were made with regard to upstream signaling components leading to S6K1 activation. First, it was initially presumed that activated growth factor receptors mediated S6K1 activation through p21ras, as oncogenic alleles of p21ras lead to constitutive S6 phosphorylation (Blenis and Erikson 1984). However, by using a combination of receptor docking site mutants, dominant interfering alleles of specific signaling components, and pharmacological agents, which selectively inhibit distinct signaling pathways, it became evident that S6K1 resides on a pathway distinct from that of p21ras (Ballou et al. 1991; Ming et al. 1994). The pathway leading to S6K1 activation appeared in many cases to be triggered by activation of phosphatidylinositide-3OH kinase (PI3K) (Cheatham et al. 1994; Chung et al. 1994; Weng et al. 1995b). Second, it was discovered that rapamycin selectively blocks S6K1 activation (Calvo et al. 1992;

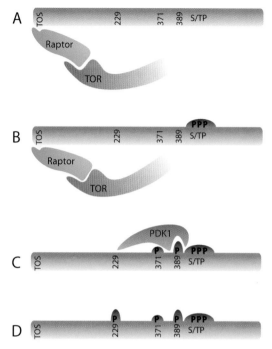

Figure 2. Molecular mechanism of S6K1 sequential activation. (*A*) Raptor is able to make a bridge between TOR and S6K1. The S/TP sites within the autoinhibitory domain have first to be phosphorylated to facilitate TOR-mediated phosphorylation of T389 (*B*). Phosphorylated T389 constitutes a docking site for PDK1 (*C*), which will ultimately phosphorylate the T-loop, T229, within the catalytic domain. (*D*) S6K1 activity always paralleled 371 phosphorylation.

Chung et al. 1992; Kuo et al. 1992; Price et al. 1992), eventually leading to the identification of the mammalian target of rapamycin (TOR) as a key regulator of S6K1 activation (Brown et al. 1995). The theoretical division of S6K1 into functional domains (Figs. 1 and 2) and the identification of a signaling pathway distinct from that of p21ras (Fig. 3), constituted the basis for the further investigations reported below.

PDK1 is the Activation-loop Kinase

An important step toward the understanding of the mechanism of S6K regulation was the discovery that an S6K1 variant, having acidic amino acid substitutions in the autoinhibitory domain phosphorylation sites and at T389, has elevated T229 phosphorylation and high basal kinase activity (Dennis et al. 1996). In addition, this mutant was largely

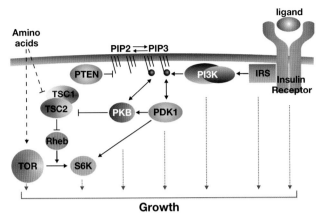

Figure 3. Nutrients and insulin receptor signaling to S6 kinase activation. Positive regulators of growth are indicated by green/blue colors, whereas negative regulators are indicated by red/orange colors. The specific effects for each intermediate in regulating S6K1 activity are described in the text.

rapamycin- and wortmannin-resistant. These observations, in combination with mutagenesis studies, implied that phosphorylation of T229 was the final step in kinase activation and also suggested that the kinase responsible for regulating phosphorylation at this site was independent of mitogen stimulation (Dennis et al. 1996). In parallel studies, it was reported that the newly discovered 3-phosphoinositide-dependent protein kinase 1 (PDK1) was responsible for phosphorylation of the PKB activation-loop site 308 (Alessi et al. 1997a,b; Stokoe et al. 1998), which was highly homologous to that of T229 in S6K1. Moreover, PDK1 appeared to be constitutively active (Alessi et al. 1997a), consistent with the characteristics associated with the T229 kinase (Dennis et al. 1996). Subsequent in vitro and in vivo studies confirmed that PDK1 is the kinase responsible for T229 phosphorylation, and that its activity toward T229 is resistant to wortmannin and rapamycin when employing the S6K1 acidic mutant as a substrate (Alessi et al. 1998; Pullen et al. 1998). This observation is not readily noted, since PDK1 is thought to be activated by the increased production of phosphatidylinositide-3, 4,5-P3 (PIP3) by PI3K (Fig. 3) (Alessi et al. 1997a,b; Stokoe et al. 1998). However, rather than PIP3 inducing PDK1 activation, it instead appears to play two distinct functions with regard to PKB T308 phosphorylation. First it binds to the pleckstrin-homology (PH) domain of PKB, presumably disrupting its interaction with T308 and allowing PDK1 to assess this phosphorylation site (Biondi et al. 2001). Second, through the PH domain, PIP3 recruits PKB to the membrane where it can be activated by PDK1. On the basis of these observations, it has been proposed that PDK1 is a constitutively active AGC kinase that selectively activates downstream target kinases whose activation is dependent on a prior step (Fig. 2), which allows PDK1 to access the activation-loop site of the target kinase (Pullen et al. 1998; Biondi et al. 2001; Frodin et al. 2002). Consistent with PDK1 being the T229 kinase, it was subsequently demonstrated that S6K1 activity is absent in ES cells derived from mice deficient for PDK1 (Williams et al. 2000) and that the activity of its *Drosophila* ortholog, dS6K, is absent in larvae which carry an inactive dPDK1 (Radimerski et al. 2002b).

In mouse ES cells deficient for PDK1, not only T229 phosphorylation is impaired, but also phosphorylation of T389 is absent (Williams et al. 2000). The simplest explanation for this latter observation is that the T389 kinase is also dependent on PDK1 for activity (Balendran et al. 1999). However, another possibility has been suggested from recent studies showing that, once phosphorylated, T389 serves as a docking site for PDK1 (Fig. 2), allowing the phosphorylation of T229 (Biondi et al. 2001; Frodin et al. 2002). In brief, a carboxy-terminal truncated version of S6K1 serves as an excellent substrate for PDK1, whereas full-length S6K1 is a poor substrate. Absence of the carboxyl terminus, or conversion of the four phosphorylation sites within the autoinhibitory domain to acidic residues, raises basal levels of T389 phosphorylation such that these S6K1 variants bind PDK1. In contrast, the wild-type S6K1 has low T389 phosphorylation and fails to interact with PDK1 (Biondi et al. 2001). Consistent with these findings, interaction and activation of the carboxy-terminal truncated S6K1 variant was greatly facilitated by conversion of T389 to an acidic residue (Dennis et al. 1998). Interestingly, dephosphorylation at T389 occurs rapidly in vivo, but not in vitro when the critical lysine at position 100 in the ATP-binding site is mutated to an arginine to render S6K1 inactive (Cheatham et al. 1995). This lysine would be necessary to create a hydrophobic pocket that can interact with and protect the T389 hydrophobic motif. Thus, an alternative explanation for lack of T389 phosphorylation in PDK1$^{-/-}$ cells may be the failure of PDK1 to bind to the phosphorylated hydrophobic motif, which would protect it from dephosphorylation by a phosphatase.

mTOR, the T389 Kinase, and More

As mentioned above, rapamycin is a potent inhibitor of both basal and mitogen-stimulated S6K1 activity (Calvo et al. 1992; Chung et al. 1992; Kuo et al. 1992; Price et al. 1992). Rapamycin inhibits growth and proliferation by forming a gain-of-function inhibitory complex with the cellular protein FKBP12, which binds to and blocks the activity of mTOR (Chen et al. 1995). In yeast, two *TOR* genes were originally identified as mutations that confer resistance to rapamycin (Heitman et al. 1991; Kunz et al. 1993; Helliwell et al. 1994) and later shown to be involved in nutrient responsive cell growth (Barbet et al. 1996). In mammalian cells, there is only one *mTOR* gene. mTOR was initially found to be an upstream regulator of S6K1 with the demonstration that an mTOR mutant unable to bind FKBP12–rapamycin protects S6K1 activity from inhibition by rapamycin (Brown et al. 1995). Subsequent studies showed that mTOR also regulates the activity of the initiation factor 4E binding protein (4E-BP1) (Beretta et al. 1996; von Manteuffel et al. 1996, 1997; Brunn et al. 1997), a repressor of initiation factor 4E. Because it has been difficult to detect changes in mTOR activity, S6K and 4EBP1 activities and phosphorylation have been used as reporters for mTOR activity. Consistent with studies in yeast, S6K1 activity and 4EBP1 phosphorylation are modulated by nutrient availability through mTOR (Hara et al. 1998; Xu et al. 1998; Iiboshi et al. 1999; Shigemitsu et al. 1999). In addition, it has been shown recently that homeostatic levels of ATP (Dennis et al. 2001) as well as phosphatidic acid (Fang et al. 2001) directly control mTOR activity. Thus, mTOR is emerging as a permissive factor, charged with integrating nutrient and energy homeostasis with a mitogenic input (Chapter 7).

mTOR contains a PI3K-like catalytic domain that is closely related to the catalytic domains of the DNA damage-repair protein kinases ATM, ATR, and DNA-dependent PK (Dennis et al. 1999). The mTOR catalytic domain possesses protein kinase activity that mediates the phosphorylation of T389 as well as that of S411, T421, S424 (Burnett et al. 1998; Isotani et al. 1999; Saitoh et al. 2002), and, as more recently shown, S371 (Saitoh et al. 2002). Neither replacement of the S/TP sites by acidic residues nor deletion of the carboxy-terminal autoregulatory domain leads to significant rapamycin resistance (Cheatham et al. 1995; Pearson et al. 1995; Dennis et al. 1996). Although deletion of the amino-terminal domain or fusion to glutathione transferase produces an inactive S6K1, both variants are activated by mitogens when the carboxyl terminus is also deleted, in a wortmannin-sensitive and rapamycin-resistant manner (Cheatham et al. 1995; Weng et al. 1995a; Dennis et al. 1996, 2001). These findings suggested that the amino terminus of S6K constitutes a target for mTOR-mediated regulation.

That S6K1 interacts physically with mTOR was initially suggested by the fact that overexpression of wild-type, kinase-dead, or activated S6K1 is able to inhibit mTOR signaling toward 4EBP1 (von Manteuffel et al. 1997). Recent studies suggest that mTOR binds S6K1 and 4E-BP via the mTOR partner protein raptor (Hara et al. 2002; Kim et al. 2002; Nojima et al. 2003). Within the highly divergent amino termini of S6K1, S6K2, and S6 kinases of other species resides a consensus FDIDL amino acid sequence (Fig. 1), which has recently been shown to be critical for S6K1 activation (Schalm and Blenis 2002). This motif is also found in 4EBP1 and has been termed the TOR signaling (TOS) motif (Fig. 2). Mutation of the S6K1 TOS motif in combination with deletion of the carboxy-terminal domain partially restores kinase activity in a rapamycin-resistant manner. Interestingly, mutations in the TOS domain inhibit phosphorylation of T389 and T229 but not of S371 or the other S/TP phosphorylation sites (Schalm and Blenis 2002). This suggests that the only critical phosphorylation site for mTOR signaling through the TOS domain is T389, with PDK1 phosphorylation of T229 dependent on mTOR phosphorylation of T389. Thus, it is intriguing that the amino- and carboxy-terminally truncated S6K1 is resistant to rapamycin. The simplest explanation is that S6K1 in the absence of the TOS motif may be phosphorylated by an alternative wortmannin-sensitive kinase. Another possibility is that in the absence of the TOS domain, T389 phosphorylation is still regulated by mTOR, but in a rapamycin-independent manner. Consistent with such a model, recent studies with S6K1 and 4EBP1 (McMahon et al. 2002) suggest that the rapamycin–FKBP12 complex alters substrate specificity rather than blocking kinase activity. Thus, T389, like a number of the SP sites in 4EBP1, may be regulated by mTOR in a rapamycin-independent manner. If so, it would be interesting to know whether a dominant interfering allele of mTOR blocks activation of the double truncated S6K1, and if so, whether this is paralleled by loss of wild-type mTOR activity in vitro.

Given that the double truncated S6K1 is wortmannin-sensitive, the possibility was also raised that mTOR instead represses a phosphatase (Peterson et al. 1999) and that a novel mitogen-activated kinase, which is wortmannin-sensitive, may be responsible for mediating this response. Seemingly consistent with this model, the NIMA-related kinases NEK6 and NEK7 were recently reported to be mitogen-induced, wortmannin-sensitive T389 kinases (Belham et al. 2001). They were purified from rat liver, employing S6K1 as a substrate (Belham et al. 2001). However, a subsequent report aimed at characterizing the substrate specificity of NEK6, as well as its activation by insulin and its sensitivity to wortmannin, has led others to conclude that NEK6 is not an in vivo T389 kinase (Lizcano et al. 2002). Since atypical protein kinase Cζ (PKCζ) has been identified as a downstream target of PI3K and as a potential activator of S6K1, PKCζ has been proposed to be a PI3K mediator for S6K1 activation (Nakanishi et al. 1993; Herrera-Velit et al. 1997). More recently, it has been shown that PDK1 and PKCζ cooperate to induce T389 phosphorylation of either a wild-type or an amino- carboxy-terminally truncated S6K1 (Romanelli et al. 1999, 2002). In contrast, PDK1-induced phosphorylation of T389 is weak and not synergistic with PKCζ if the assay is conducted with an inactive S6K1. Since the inactive S6K1 is also rapidly dephosphorylated on T389, the authors suggested that PDK1 and PKCζ cooperate to induce the autophosphorylation of S6K1 (Romanelli et al. 1999, 2002). However, the mutation employed has been reported to modify the structure of the kinase as well as T389 dephosphorylation. Hence the apparent requirement of a functional S6K1 for a cooperative effect between PDK1 and PKCζ might rather reflect some conformational change in S6K1 structure. Nonetheless, PKCζ is probably not the wortmannin-sensitive kinase that mediates S/TP phosphorylation downstream of PI3K, and although a few other kinases have been proposed to carry out this reaction, the identity of this kinase remains elusive.

PI3K Signaling in Fly Versus Mouse

Genetic studies in *Drosophila* have underscored the importance of dS6K in the insulin-signaling pathway in controlling cell, organ, and animal size (Montagne et al. 1999). Because of its relatively small genome size, conserved signaling pathways, and ease of genetic manipulation, *Drosophila* has also proven powerful in establishing epistatic relationships and, more recently, in providing direct biochemical readouts. Components of the insulin-signaling pathway include Chico (the insulin receptor substrate homolog), dPI3K, dPKB, dPTEN, and dS6K (Bohni et al. 1999; Goberdhan et al. 1999; Huang et al. 1999; Montagne et al. 1999; Verdu et al. 1999; Weinkove et al. 1999). Mutations in any of these components have a striking effect on cell size and number, with the exception of *dS6K* (Montagne et al. 1999). Mutations in *dS6K* affect cell size but not cell number, seemingly consistent with arguments that dS6K is a distal effector in the signaling pathway, directly controlled by dTOR (Oldham et al. 2000; Zhang et al. 2000), a downstream effector of dPI3K and dPKB. However, recent studies showed that dS6K activity is unimpaired in *chico*-deficient larvae (Oldham et al. 2000), suggesting that dS6K activation may be mediated through dPI3K docking directly to the *Drosophila* insulin receptor. Biochemical, genetic, and pharmacological studies have argued that dS6K resides on an insulin-signaling pathway distinct from that of dPKB (Radimerski et al. 2002b). However, more striking, dS6K activation was not dependent on dPI3K. Indeed, dS6K immunoprecipitated from either mutant Dp110 or dPKB larvae was as active as dS6K from wild-type larvae (Radimerski et al. 2002b). Epistasis experiments also revealed that dP110 is able to trigger growth in the absence of dS6K, whereas this growth was totally reliant on dPKB. Despite being independent of dPKB and dPI3K, dS6K activity is dependent on the *Drosophila* homolog of PDK1, dPDK1 (Cho et al. 2001; Rintelen et al. 2001), as well as dTOR, demonstrating that dPDK1-mediated dS6K activation is phosphatidyl-inositide-3, 4,5-P_3 (PIP$_3$) independent (Radimerski et al. 2002b).

Seemingly consistent with PKB independence, neither Ca^{++}, amino acids, nor TPA treatment of mammalian cells affects PKB activation despite leading to acute stimulation of S6K1 in a rapamycin-dependent manner (Conus et al. 1998). Depending on the cell type, the ability of some of these agents to activate S6K1 is also wortmannin-resistant, suggesting that under certain conditions S6K1 activation is PI3K-independent (Iiboshi et al. 1999). Such observations suggest that with the additional doubling of the mammalian genome, mammalian receptors may have evolved more complex downstream signaling networks than those found in invertebrates, with several branch points leading to S6K activation, some of which bifurcate upstream and downstream of both PI3K and PKB. However, in *Drosophila*, the signaling pathways and the number of genes involved are less complex, such that the dTOR/dS6K and dPI3K/dPKB signaling pathways may be largely parallel and independent of one another (Radimerski et al. 2002b). However, despite these observations, recent studies demonstrate that dS6K can act negatively on the dPI3K/dPKB pathway (see below). In addition, what remains unresolved is the mechanism by which insulin induces dS6K activation in *Drosophila*, if not through dPI3K (Radimerski et al. 2002b). One possibility is that insulin stimulation of amino acid uptake drives the dTOR/dS6K pathway, consistent with the exquisite sensitivity of dS6K to essential amino acids, especially branch chain amino acids (Xu et al. 1998). Knowledge derived from *Drosophila* will be important in establishing the role of S6Ks in controlling the growth and development of higher eukaryotes.

TSC, the S6K1 Upstream Tumor Suppressor

Recent studies underscored the importance of S6K signaling in tuberous sclerosis complex (TSC) syndrome (Gao and Pan 2001; Potter et al. 2001;

Tapon et al. 2001). In humans, TSC has been characterized as an autosomal dominant inherited disease, whose clinical manifestations include cutaneous benign tumors, neurological dysfunction, and bilateral renal carcinomas (Young and Povey 1998). TSC is caused by mutations in either one of the two tumor suppressor genes *TSC1* and *TSC2*, which encode hamartin (TSC1) and tuberin (TSC2), with 1 in 6000 born with a mutation in one of the two genes (Sparagana and Roach 2000; Montagne et al. 2001). Hamartomatous lesions also have been observed in patients with one of three related autosomal dominant disorders associated with germline *PTEN* mutations: Cowden disease (CD), Lhermitte-Duclos disease (LDD), and Bannayan-Zonana syndrome (BZS) (Liaw et al. 1997). These syndromes share specific developmental defects, hamartomatous lesions in the colonic mucosa, and increased susceptibility to cancer (for review, see Cantley and Neel 1999; Di Cristofano and Pandolfi 2000).

The role of TSC1 and TSC2 in S6K signaling first emerged in studies in *Drosophila* (Gao and Pan 2001; Potter et al. 2001; Tapon et al. 2001). Initial observations indicated that mitotic clones lacking either one of the fly TSC homologs (dTsc1 or dTsc2) were enlarged. Conversely, overexpression of dTsc1 and dTsc2 in the *Drosophila* eye, but neither alone, led to a reduction in the size of the eye. Strikingly, the growth-suppressive effect could be rescued upon coexpression of dS6K, and the overgrowth phenotype, caused by loss of dTsc1 or dTsc2 function, was reversed in a dS6K-deficient background (Potter et al. 2001; Tapon et al. 2001). That dS6K was epistatic to dTsc1/2 suggested that dS6K acts downstream of dTsc1/2. Indeed, subsequent studies have recently shown that overexpression of dTsc1/2 in larvae represses dS6K activity, whereas the kinase activity was increased in dTsc1 mutant larvae or in *Drosophila Kc167* cultured cells treated with siRNAs to either dTsc1 or dTsc2 (Gao et al. 2002; Inoki et al. 2002; Radimerski et al. 2002a). Similarly, S6K1 and 4EBP1 phosphorylation was found elevated in mouse embryonic fibroblasts derived from TSC2-deficient mice or in cell lines derived from humans having lesions in either TSC1 or TSC2 (Gao et al. 2002; Goncharova et al. 2002; Inoki et al. 2002; Jaeschke et al. 2002; Manning et al. 2002; Tee et al. 2002). Conversely, overexpression of TSC1 and TSC2 suppresses insulin-induced S6K1 phosphorylation and activation, whereas mutants of TSC2 fail to elicit this response. Although it is clear that TSC1 and TSC2 suppress S6K1 signaling, the molecular mechanism through which this repression occurs is yet to be established (Fig. 3).

Based on the fact that dTsc2 and dTOR co-immunoprecipitate (Gao and Pan 2001), and that in cells depleted of either dTsc1 or dTsc2, dS6K activity is resistant to amino acid depletion (Gao et al. 2002), it has been proposed that dTsc1/2 exerts its tumor suppressor effects through dTOR. To address the issue in mammalian cells, advantage was taken of the two rapamycin-resistant alleles of *S6K1* described above (see p. 276), whose activities are still mediated by mitogens. The prediction was that if the effects of TSC1/2 are through mTOR, then rapamycin-insensitive S6Ks would be resistant to inhibition by TSC1/2, whereas if the effects were elicited downstream of mTOR, they would be inhibited. Surprisingly, opposite results were obtained with these two constructs, largely discrediting their use for such studies at this point (Inoki et al. 2002; Jaeschke et al. 2002; Tee et al. 2002). The differential findings also suggest that distinct molecular mechanisms can render S6K1 rapamycin-resistant, suggesting unique routes of S6K1 activation.

A parallel aim of the above studies was to identify the mechanisms by which TSC1/2 are regulated. Collectively, it appears that in mammalian cells TSC1/2 constitute a missing link between PI3K signaling and S6K1 activation (Inoki et al. 2002; Jaeschke et al. 2002; Manning et al. 2002; Potter et al. 2002). Again the details are not clear, but it appears that activation of PI3K induces either destabilization or degradation of the TSC1/2 complex. In both cases, this appears to be mediated by the phosphorylation of TSC2, which either disrupts the interaction of TSC2 with TSC1 (Dan et al. 2002; Potter et al. 2002) or creates a docking site for 14-3-3-protein-mediated transport to the proteasome (Li et al. 2002; Liu et al. 2002; Nellist et al. 2002). Three studies have concluded that TSC2 phosphorylation is mediated by PKB (Dan et al. 2002; Inoki et al. 2002; Manning et al. 2002). This finding would seem in contradiction to earlier results that a dominant interfering allele of PKB blocks wild-type reporter PKB activation, but has no effect on S6K1 reporter activation (Dufner et al. 1999). One possibility that has been suggested to explain this difference is that more than one kinase is involved in regulating mitogen-induced TSC2 phosphorylation (Manning et al. 2002). Evidence that an additional mechanism or kinase is involved is that dS6K activation in *Drosophila* is dependent on inhibition of dTsc1/2 function and independent of dPKB signaling (Radimerski et al. 2002a,b). Despite these observations, it has been reported in S2 *Drosophila* cells that dTcs2 phosphorylation by ectopic overexpression of dPKB mediates the dissociation of the dTsc complex (Potter et al. 2002). That dPKB resides upstream of dTsc1/2 is further supported by the fact that the enhanced growth in the *Drosophila* eye caused by either PKB overexpression or loss of dTsc1/2 function is not additive (Potter et al. 2002). However, the first experiment is dependent on ectopic overexpression of dPKB, and the second shows that loss of dTsc1/2 function leads to dPKB inhibition, an effect mediated by dS6K

(Radimerski et al. 2002a). Thus, in the latter case, the fact that dTsc1/2 appears epistatic to dPKB could be explained by the repression of dPKB by hyperactivated dS6K. *Drosophila* should prove useful in identifying the PKB-independent mechanism of S6K activation. In addition, it will be critical to determine whether TSC2 degradation occurs only in experimental conditions, or whether such an antagonism to the tumor suppressor TSC1/2 also occurs in a pathological situation, such as in PTEN-transformed cells.

How then does TSC1/2 regulate S6K1 signaling? Recent studies in both *Drosophila* and mammals have shown that the target of TSC1/2 is the small GTPase Rheb (<u>R</u>as <u>h</u>omolog <u>e</u>nriched in <u>b</u>rain). In overexpression-based genetic screens, the Rheb *Drosophila* homolog, *dRheb*, has been identified as a promoter of growth, which is epistatic to *dTsc1/2* loss-of-function mutants (Saucedo et al. 2003; Stocker et al. 2003). Consistent with this finding, dS6K activity is suppressed and dPKB activity is elevated in *dRheb* mutants (Saucedo et al. 2003; Stocker et al. 2003). Likewise in mammalian cells, overexpression of Rheb drives S6K1 activation and suppresses PKB activation (Garami et al. 2003). This activation can be counteracted by coexpression of TSC1/2 and is dependent in both mammalian and *Drosophila* cells on the GAP domain of TSC2 (Garami et al. 2003; Zhang et al. 2003). More importantly, insulin stimulation of quiescent cells in culture increases the amount of Rheb in the GTP-bound state, and Rheb siRNA blocks insulin-induced activation of S6K1 (Garami et al. 2003), arguing that insulin mediates S6K1 activation via Rheb. Further investigations designed to elucidate the mechanisms through which TSC/Rheb controls mTOR/S6K1 signaling will also help to establish the molecular mechanisms leading to S6K1 activation.

S6 KINASES: BIOLOGICAL FUNCTIONS AND PATHOLOGICAL ASPECTS

S6K in the Control of Growth and Cell Size

Gene disruption is a very informative approach to study a biological function, and reduction of S6K expression has been achieved in *C. elegans* by double-strand RNA interference (RNAi), and in *D. melanogaster* and mouse by gene deletion. The phenotype observed in each case was distinct, as was the effect of each mutation on growth and development. In the case of *C. elegans*, RNAi-mediated depletion of S6K (Cep70) led to a very subtle phenotype with minor growth defects (Long et al. 2002). The more severe phenotype was the one reported in *Drosophila* (Fig. 4), where

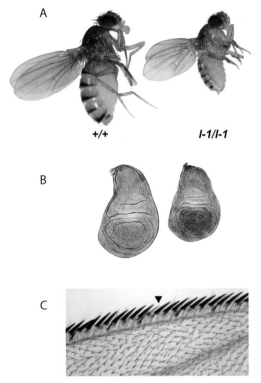

Figure 4. Phenotypic features of *D. melanogaster* S6K (*dS6K*) mutants. (*A*) Null mutant flies for *dS6K* (*l-1/l-1*) are semi-lethal, but the few escapers that emerge have a strong development delay and are smaller in size than wild-type flies (*+/+*). (*B*) Size reduction is observed during larval life as depicted for imaginal discs dissected just prior to metamorphosis onset, from either wild-type (*left*) or *dS6k* mutant larvae (*right*). (*C*) The cell growth and cell size effect is cell-autonomous, since a *dS6K*-lacking bristle (*arrowhead*) induced by mitotic recombination is extremely reduced in size as compared to the surrounding bristles that express dS6K. The yellow marker that is used to label the somatic recombinant *dS6K* mutant cells is neutral with respect to growth.

most of the *dS6K* null mutants die during late larval or early pupal stages, and only a few escapers emerge as adults with a severe developmental delay and an acute reduction in body size (Montagne et al. 1999). The few flies that escape live for only 2 weeks, are lethargic, and do not fly. Unexpectedly, the reduction in body size is due to a decrease in cell size rather than in cell number. In contrast, S6K1-deficient mice are viable (Fig. 5A,B), although they are 15–30% smaller than wild-type mice (Shima et al. 1998). However, such mice still have a copy of S6K2, which

A

B

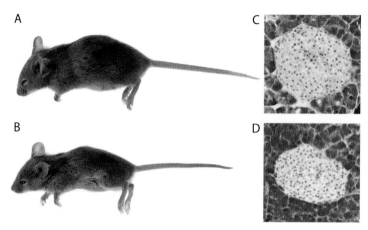

C

D

Figure 5. Disruption of S6K1 in mouse leads to animal growth defect (*B*) and reduction of β-cell size of Langerhans islets (*D*) as compared to wild type (*A* and *C*).

is up-regulated in all tissues examined, suggesting a compensatory response to the loss of S6K1. Consistent with this, most mice deficient for both *S6K1* and *S6K2* die at birth (Pende et al. 2004). However, despite the presence of S6K2, S6K1-deficient mice are hypoinsulinemic and mildly glucose intolerant (Pende et al. 2000). Both phenotypes arise from reduced insulin pancreatic content and a severe impairment in insulin secretion, due to a selective decrease in size of β-cells of the Langerhans islets (Fig. 5C,D).

The phenotype observed for *dS6K* mutant flies suggests that the kinase plays an important role in regulating cell size (Montagne et al. 1999). Maintaining cell size is critical for the physiological integrity of differentiated cells, as cytoplasmic volume has profound effects on intracellular transport and on the cell surface to volume ratio, a key parameter in the transfer of molecules across the plasma membrane (Szarski 1983; Agutter and Wheatley 2000). However, the few *dS6K* escapers are also severely delayed at adult eclosion, with cell cycle times in mutant larvae twice as long as those in wild-type larvae (Montagne et al. 1999). Since ribosome biogenesis is a rate-limiting factor for cell growth, the delay in development observed in *Drosophila dS6K* mutants is consistent with dS6K affecting translational capacity. However, a defect in ribosome biogenesis is not sufficient to explain the reduction in cell size, as *Drosophila* harboring hypomorphic mutations in ribosomal proteins, termed *Minutes*, are delayed in development but have normal cell size (Morata and Ripoll 1975; Neufeld et al. 1998; Montagne et al. 1999). One model that has been proposed to explain the difference in cell size between *dS6K* mutants and *Minutes* is that *dS6K* mutants may affect the translation of all 5′TOP mRNAs, whereas lowering the expression of one ribosomal protein gene does not affect the translation of the other ribosomal protein genes (Thomas 2000). As 5′TOP mRNAs represent up to 20% of cell mRNA and in most cell types mRNA is in excess of ribosomes, a decrease in the amount of these transcripts recruited to polysomes would not only inhibit the production of nascent ribosomes, but would also lead to a proportional increase in the translation of other transcripts. If these transcripts encoded cell cycle regulators, such as E2F (Neufeld et al. 1998), they could drive cells through the cell cycle at a smaller cell size. Indeed, this model would largely satisfy the hypothesis, derived from studies on proliferative cells, that final cell size essentially results from the integration of growth and proliferation rates (Stocker and Hafen 2000; Thomas 2000). However, that differentiated cells have distinct sizes suggests the existence of cell size control mechanisms, which would act on the cell growth and proliferation machineries (Schmidt and Schibler 1995; Montagne 2000). Since ribosomes represent a significant portion of the cytoplasmic volume and are required to generate organelles (Warner 1999), cell size is linked to translational capacity. Thus, in part, cell size control depends on the adjustment of the ribosome content to genetic information specific for each cell type. dS6K may participate as described above in the control of cell size, or it may play a more profound role, as discussed below.

Although a number of *Drosophila* mutations have been shown to affect both cell size and cell number (Coelho and Leevers 2000; Montagne 2000; Stocker and Hafen 2000), *dS6K* mutants are the only ones whose body size reduction is exclusively due to a decrease in cell size. The normal number of cells observed in dS6K mutants suggests the potential existence of a mechanism for counting cell number; this is in contradiction to the classical hypothesis, which proposes that the control of body size solely relies on cell mass rather than cell number (Conlon and Raff 1999). The normal cell number in *dS6K* mutants may be a consequence of a mechanism that would delay the developmental process, allowing the normal number of cell divisions to proceed. This mechanism would not appear to be dependent on a critical cell mass threshold, since some *Drosophila* mutants, such as *dPKB* hypomorphs or *chico*, have fewer cells, are smaller at adult eclosion, and exhibit a shorter developmental delay (Bohni et al. 1999; Stocker et al. 2002). Hence, if a mechanism for counting cell number exists, it would be perturbed in *dPKB* mutants but not in *dS6K* mutants. Indeed, the model that a developmental counter measures only cell mass and not cell number is based largely on experiments that modulate cell cycle regulators, and a mechanism for counting cell number would be proposed to be perturbed in this case. Interestingly, a counter for cell number has been described in *Dictyostelium* (Brock and Gomer 1999). Thus, understanding the reason *dS6K* mutants have a normal number of cells may lead to the discovery of a new limiting event for the developmental program, or potentially reveal whether a mechanism for counting cell number acts in parallel to a cell mass counter.

Integrating Nutrient Availability with Growth Factor Signaling

As pointed out above, S6K1-deficient mice are smaller in size, hypoinsulinemic, and mildly glucose intolerant (Shima et al. 1998; Pende et al. 2000). The latter effects appear to be a consequence of a reduction in β-cell size (Fig. 5C,D), as β-cell size has been shown to have a profound effect on the ability of these cells to secrete insulin (Swenne et al. 1988; Giordano et al. 1993). Because insulin is argued to play an important role in the late development of the embryo, the growth defect observed in S6K1-deficient mice may also be a consequence of a defect in insulin production during development. However, preliminary data suggest this is not the case, because the effects on growth retardation appear to emerge much earlier in embryogenesis than can be attributed to a defect in insulin production (S.H. Um et al., unpubl.). In addition, replacement of the *S6K1* gene in β-cells deficient for the kinase appears to rescue the defect in cell size (Y.G. Gangloff et al., unpubl.). Consistent with this, analysis of somatic mutant clones for dS6K in *Drosophila* indicates that the growth and size defect are also cell autonomous (Fig. 4C) (Montagne et al. 1999). Unexpectedly, the effect on cell size appears to be largely specific for β-cells, despite the ubiquitous expression of S6K1 and the presence of S6K2 in β-cells. It is known that β-cell growth, like S6K1 activity, is extremely sensitive to mitogens (Hugl et al. 1998) such as insulin and insulin-like growth factors, as well as to nutrients, including glucose and amino acids, and energy. Thus, β-cells lacking S6K1 may be reduced in size because they fail to efficiently integrate the parallel growth factor and nutrient stimulatory inputs required for β-cell development. This can have profound pathological consequences, as decreased insulin secretion and reduced β-cell mass are usually associated with later stages of Type 2 diabetes. However, they are thought to be the primary causal factors for the disease when induced by protein malnutrition during gestation or childhood (Phillips 1996). Corroborative in vivo data for this hypothesis have come from studies in rats, where a limited period of protein malnutrition leads to glucose intolerance arising from a persistent decrease in β-cell size and insulin secretion (Swenne et al. 1987, 1992). That impaired

S6K1 function in a cell-autonomous manner can lead to the development in a mouse model of a specific form of Type 2 diabetes underscores the importance of the kinase as a nutrient effector.

Despite the fact that S6K1-deficient mice are hypoinsulinemic and mildly glucose intolerant, they exhibit normal fasting glucose levels (Pende et al. 2000). Interestingly, in the studies described above where rats were placed on a limited protein diet, the effects of hypoinsulinemia were partially attenuated by mild insulin hypersensitivity in peripheral tissues (Swenne et al. 1992). Such an explanation may also account for the normal fasting glucose levels in S6K1-deficient mice. In fact, in recent studies it has been shown that rapamycin treatment of 3T3 pseudo-adipocytes leads to potentiation of insulin signaling and increased glucose uptake (Berg et al. 2002). It appears that mTOR acts in a negative feedback loop to dampen insulin signaling. Consistent with these findings, in *Drosophila* larvae deficient for dTOR, dPKB activity levels are elevated (Radimerski et al. 2002b). Similar results are obtained with dS6K-deficient larvae, suggesting that the effects of the loss-of-function *dTOR* alleles on dPKB activation are through inhibition of dS6K signaling (Radimerski et al. 2002a). Furthermore, mutations in either *dTsc1* or *dTsc2* lead to elevated dS6K activity and suppression of dPKB activity. The effect on dPKB is rescued in a double *dTsc1 dS6K* mutant (Radimerski et al. 2002a). Thus, elevation of dS6K activity can inhibit dPKB signaling. Similarly in mice, it may be that the loss of S6K1 potentiates signaling in the insulin-signaling pathway.

Binding of insulin to the insulin receptor triggers the activation of two major signaling cascades: the MAPK and the PI3K pathways (Skolnik et al. 1993). Recruitment of the proteins Grb-2 and Sos to Shc activates the MAPK cascade, whereas association of PI3K with the IRS proteins results in production of PIP3, which serves as a second messenger in the activation of PKB, S6K, and atypical isoforms of PKC (Kotani et al. 1994). Together, these kinase cascades mediate the metabolic and growth-promoting functions of insulin, such as the translocation of GLUT4 glucose transporter containing vesicles from intracellular pools to the plasma membrane (Foran et al. 1999), stimulation of glycogen and protein synthesis (Cross et al. 1995), uptake of amino acids, and initiation of specific gene transcription programs (Hajduch et al. 1998). The coordinate response triggered by insulin suggests that nutritional and mitogenic pathways have been integrated during evolution. The negative feedback described above leads to down-regulation of glucose uptake and other metabolic actions of insulin. It may be that as amino acids and energy levels decrease, this will eventually lead to an inhibition of mTOR signaling and inactivation of the S6Ks. Thus, with time, the impact of the negative feedback loop on PI3K will diminish, leading again to an increase in glucose and amino acid uptake. In this way, it is hypothesized that S6Ks could contribute in maintaining tissue homeostasis as a function of insulin signaling and nutrient levels.

REFERENCES

Agutter P.S. and Wheatley D.N. 2000. Random walks and cell size. *Bioessays* **22**: 1018–1023.

Alessi D.R., Kozlowski M.T., Weng Q.P., Morrice N., and Avruch J. 1998. 3-Phosphoinositide-dependent protein kinase 1 (PDK1) phosphorylates and activates the p70 S6 kinase in vivo and in vitro. *Curr. Biol.* **8**: 69–81.

Alessi D.R., James S.R., Downes C.P., Holmes A.B., Gaffney P.R., Reese C.B., and Cohen P. 1997a. Characterization of a 3-phosphoinositide-dependent protein kinase which phosphorylates and activates protein kinase Bα. *Curr. Biol.* **7**: 261–269.

Alessi D.R., Deak M., Casamayor A., Caudwell F.B., Morrice N., Norman D.G., Gaffney P., Reese C.B., MacDougall C.N., Harbison D., Ashworth A., and Bownes M. 1997b. 3-Phosphoinositide-dependent protein kinase-1 (PDK1): Structural and functional homology with the *Drosophila* DSTPK61 kinase. *Curr. Biol.* **7**: 776–789.

Avni D., Shama S., Loreni F., and Meyuhas O. 1994. Vertebrate mRNAs with a 5′-terminal pyrimidine tract are candidates for translation repression in quiescent cells. Characterization of the translational *cis*-regulatory elememt (TLRE). *Mol. Cell. Biol.* **14**: 3822–3833.

Balendran A., Currie R., Armstrong C.G., Avruch J., and Alessi D.R. 1999. Evidence that 3-phosphoinositide-dependent protein kinase-1 mediates phosphorylation of p70 S6 kinase in vivo at Thr-412 as well as Thr-252. *J. Biol. Chem.* **274**: 37400–37406.

Ballou L.M., Luther H., and Thomas G. 1991. MAP2 kinase and 70k S6 kinase lie on distinct signalling pathways. *Nature* **349**: 348–350.

Bandi H.R., Ferrari S., Krieg J., Meyer H.E., and Thomas G. 1993. Identification of 40 S ribosomal protein S6 phosphorylation sites in Swiss mouse 3T3 fibroblasts stimulated with serum. *J. Biol. Chem.* **268**: 4530–4533.

Banerjee P., Ahamad M.F., Grove J.R., Kozlosky C., Price D.J., and Avruch J. 1990. Molecular structure of a major insulin/mitogen-activated 70kDa S6 protein kinase. *Proc. Natl. Acad. Sci.* **87**: 8550–8554.

Barbet N.C., Schneider U., Helliwell S.B., Stansfield I., Tuite M.F., and Hall M.N. 1996. TOR controls translation initiation and early G1 progression in yeast. *Mol. Biol. Cell* **7**: 25–42.

Belham C., Comb M.J., and Avruch J. 2001. Identification of the NIMA family kinases NEK6/7 as regulators of the p70 ribosomal S6 kinase. *Curr. Biol.* **11**: 1155–1167.

Beretta L., Gingras A.C., Svitkin Y.V., Hall M.N., and Sonenberg N. 1996. Rapamycin blocks the phosphorylation of 4E-BP1 and inhibits cap-dependent initiation of translation. *EMBO J.* **15**: 658–664.

Berg C.E., Lavan B.E., and Rondinone C.M. 2002. Rapamycin partially prevents insulin resistance induced by chronic insulin treatment. *Biochem. Biophys. Res. Commun.* **293**: 1021–1027.

Biondi R.M., Kieloch A., Currie R.A., Deak M., and Alessi D.R. 2001. The PIF-binding pocket in PDK1 is essential for activation of S6K and SGK, but not PKB. *EMBO J.* **20**: 4380–4390.

Blenis J. and Erikson R.L. 1984. Phosphorylation of the ribosomal protein S6 is elevated in cells transformed by a variety of tumor viruses. *J. Virol.* **50:** 966–969.

Bohni R., Riesgo-Escovar J., Oldham S., Brogiolo W., Stocker H., Andruss B.F., Beckingham K., and Hafen E. 1999. Autonomous control of cell and organ size by CHICO, a *Drosophila* homolog of vertebrate IRS1-4. *Cell* **97:** 865–875.

Brock D.A. and Gomer R.H. 1999. A cell-counting factor regulating structure size in *Dictyostelium*. *Genes Dev.* **13:** 1960–1969.

Brown E.J., Beal P.A., Keith C.T., Chen J., Shin T.B., and Schreiber S.L. 1995. Control of p70 S6 kinase by kinase activity of FRAP in vivo. *Nature* **377:** 441–446.

Brunn G.J., Hudson C.C., Sekulic A., Williams J.M., Hosoi H., Houghton P.J., Lawrence J.C., Jr., and Abraham R.T. 1997. Phosphorylation of the translational repressor PHAS-I by the mammalain target of rapamycin. *Science* **277:** 99–101.

Burnett P.E., Barrow R.K., Cohen N.A., Snyder S.H., and Sabatini D.M. 1998. RAFT1 phosphorylation of the translational regulators p70 S6 kinase and 4E-BP1. *Proc. Natl. Acad. Sci.* **95:** 1432–1437.

Calvo V., Crews C.M., Vik T.A., and Bierer B.E. 1992. Interleukin 2 stimulation of p70 S6 kinase activity is inhibited by the immunosuppressant rapamycin. *Proc. Natl. Acad. Sci.* **89:** 7571–7575.

Cantley L.C. and Neel B.G. 1999. New insights into tumor suppression: PTEN suppresses tumor formation by restraining the phosphoinositide 3-kinase/AKT pathway. *Proc. Natl. Acad. Sci.* **96:** 4240–4245.

Cheatham B., Vlahos C.J., Cheatham L., Wang L., Blenis J., and Kahn C.R. 1994. Phosphatidylinositol 3-kinase activation is required for insulin stimulation of pp70 S6 kinase, DNA synthesis, and glucose transporter translocation. *Mol. Cell. Biol.* **14:** 4902–4911.

Cheatham L., Monfar M., Chou M.M., and Blenis J. 1995. Structural and functional analysis of pp70[s6k]. *Proc. Natl. Acad. Sci.* **92:** 11696–11700.

Chen J., Zheng X.-F., Brown E.J., and Schreiber S.L. 1995. Identification of an 11-kDa FKBP12-rapamycin-binding domain within the 289-kDa FKBP12-rapamycin-associated protein and characterization of a critical serine residue. *Proc. Natl. Acad. Sci.* **92:** 4947–4951.

Cho K.S., Lee J.H., Kim S., Kim D., Koh H., Lee J., Kim C., Kim J., and Chung J. 2001. *Drosophila* phosphoinositide-dependent kinase-1 regulates apoptosis and growth via the phosphoinositide 3-kinase-dependent signaling pathway. *Proc. Natl. Acad. Sci.* **98:** 6144–6149.

Chung J., Kuo C.J., Crabtree G.R., and Blenis J. 1992. Rapamycin-FKBP specifically blocks growth-dependent activation of and signaling by the 70 kd S6 protein kinases. *Cell* **69:** 1227–1236.

Chung J., Grammer T.C., Lemon K.P., Kazlauskas A., and Blenis J. 1994. PDGF- and insulin-dependent pp70[s6k] activation mediated by phosphatidylinositol-3-OH kinase. *Nature* **370:** 71–75.

Coelho C.M. and Leevers S.J. 2000. Do growth and cell division rates determine cell size in multicellular organisms? *J. Cell Sci.* **113:** 2927–2934.

Conlon I. and Raff M. 1999. Size control in animal development. *Cell* **96:** 235–244.

Conus N.M., Hemmings B.A., and Pearson R.B. 1998. Differential regulation by calcium reveals distinct signaling requirements for the activation of Akt and p70[s6k]. *J. Biol. Chem.* **273:** 4776–4782.

Crosio C., Boyl P.P., Loreni F., Pierandrei-Amaldi P., and Amaldi F. 2000. La protein has a positive effect on the translation of TOP mRNAs in vivo. *Nucleic Acids Res.* **28:** 2927–2934.

Cross D.A.E., Alessi D.R., Cohen P., Andjelkovic M., and Hemmings B.A. 1995. Inhibition of glycogen synthase kinase-3 by insulin mediated by protein kinase B. *Nature* **378:** 785–789.

Dan H.C., Sun M., Yang L., Feldman R.I., Sui X.M., Ou C.C., Nellist M., Yeung R.S., Halley D.J., Nicosia S.V., Pledger W.J., and Cheng J.Q. 2002. Phosphatidylinositol 3-kinase/Akt pathway regulates tuberous sclerosis tumor suppressor complex by phosphorylation of tuberin. *J. Biol. Chem.* **277:** 35364–35370.

Dennis P.B., Fumagalli S., and Thomas G. 1999. Target of rapamycin (TOR): Balancing the opposing forces of protein synthesis and degradation. *Curr. Opin. Genet. Dev.* **9:** 49–54.

Dennis P.B., Pullen N., Kozma S.C., and Thomas G. 1996. The principal rapamycin-sensitive p70^{s6k} phosphorylation sites T_{229} and T_{389} are differentially regulated by rapamycin-insensitive kinase-kinases. *Mol. Cell. Biol.* **16:** 6242–6251.

Dennis P.B., Pullen N., Pearson R.B., Kozma S.C., and Thomas G. 1998. Phosphorylation sites in the autoinhibitory domain participate in p70^{s6k} activation loop phosphorylation. *J. Biol. Chem.* **273:** 14845–14852.

Dennis P.B., Jaeschke A., Saitoh M., Fowler B., Kozma S.C., and Thomas G. 2001. Mammalian TOR: A homeostatic ATP sensor. *Science* **294:** 1102–1105.

Di Cristofano A. and Pandolfi P.P. 2000. The multiple roles of PTEN in tumor suppression. *Cell* **100:** 387–390.

Dufner A., Andjelkovic M., Burgering B.M.T., Hemmings B.A., and Thomas G. 1999. Protein kinase B localization and activation differentially affect S6 kinase 1 activity and eukaryotic translation initiation factor 4E-binding protein1 phosphorylation. *Mol. Cell. Biol.* **19:** 4525–4534.

Duncan R. and McConkey E.H. 1982. Preferential utilization of phosphorylated 40-S ribosomal subunits during initiation complex formation. *Eur. J. Biochem.* **123:** 535–538.

Edgar B.A. 1999. From small flies come big discoveries about size control. *Nat. Cell Biol.* **1:** E191–E193.

Fang Y., Vilella-Bach M., Bachmann R., Flanigan A., and Chen J. 2001. Phosphatidic acid-mediated mitogenic activation of mTOR signaling. *Science* **294:** 1942–1945.

Ferrari S., Bandi H.R., Bussian B.M., and Thomas G. 1991. Mitogen-activated 70K S6 kinase. *J. Biol. Chem.* **266:** 22770–22775.

Foran P.G., Fletcher L.M., Oatey P.B., Mohammed N., Dolly J.O., and Tavare J.M. 1999. Protein kinase B stimulates the translocation of GLUT4 but not GLUT1 or transferrin receptors in 3T3-L1 adipocytes by a pathway involving SNAP-23, synaptobrevin-2, and/or cellubrevin. *J. Biol. Chem.* **274:** 28087–28095.

Frodin M., Antal T.L., Dummler B.A., Jensen C.J., Deak M., Gammeltoft S., and Biondi R.M. 2002. A phosphoserine/threonine-binding pocket in AGC kinases and PDK1 mediates activation by hydrophobic motif phosphorylation. *EMBO J.* **21:** 5396–5407.

Gao X. and Pan D. 2001. TSC1 and TSC2 tumor suppressors antagonize insulin signaling in cell growth. *Genes Dev.* **15:** 1383–1392.

Gao X., Zhang Y., Arrazola P., Hino O., Kobayashi T., Yeung R.S., Ru B., and Pan D. 2002. Tsc tumour suppressor proteins antagonize amino-acid-TOR signalling. *Nat. Cell Biol.* **4:** 699–704.

Garami A., Zwartkruis F.J.T., Nobukuni T., Joaquin M., Roccio M., Stocker H., Kozma S.C., Hafen E., Bos J.L., and Thomas G. 2003. Insulin activation of Rheb, a mediator of mTOR/S6K/4E-BP signaling, is inhibited by TSC1 and 2. *Mol. Cell* **11:** 1457–1466.

Gingras A. C., Raught B., and Sonenberg N. 2001. Regulation of translation initiation by FRAP/mTOR. *Genes Dev.* **15:** 807–826.

Giordano E., Cirulli V., Bosco D., Rouiller D., Halban P., and Meda P. 1993. B-cell size influences glucose-stimulated insulin secretion. *Am. J. Physiol.* **265:** C358–C364.

Goberdhan D.C., Paricio N., Goodman E.C., Mlodzik M., and Wilson C. 1999. *Drosophila* tumor suppressor PTEN controls cell size and number by antagonizing the Chico/PI3-kinase signaling pathway. *Genes Dev.* **13:** 3244–3258.

Goncharova E.A., Goncharov D.A., Eszterhas A., Hunter D.S., Glassberg M.K., Yeung R.S., Walker C.L., Noonan D., Kwiatkowski D.J., Chou M.M., Panettieri R.A., Jr., and Krymskaya V.P. 2002. Tuberin regulates p70 S6 kinase activation and ribosomal protein S6 phosphorylation. A role for the TSC2 tumor suppressor gene in pulmonary lymphangioleiomyomatosis (LAM). *J. Biol. Chem.* **277:** 30958–30967.

Gout I., Minami T., Hara K., Tsujishita Y., Filonenko V., Waterfield M.D., and Yonezawa K. 1998. Molecular cloning and characterization of a novel p70 S6 kinase, p70 S6 kinase β containing a proline-rich region. *J. Biol. Chem.* **273:** 30061–30064.

Gressner A.M. and Wool I.G. 1974. The phosphorylation of liver ribosomal proteins in vivo. Evidence that only a single small subunit (S6) is phosphorylated. *J. Biol. Chem.* **249:** 6917–6925.

Grove J.R., Banerjee P., Balasubramanyam A., Coffer P.J., Price D.J., Avruch J., and Woodgett J.R. 1991. Cloning and expression of two human p70 S6 kinase polypeptides differing only at their amino termini. *Mol. Cell. Biol.* **11:** 5541–5550.

Hajduch E., Alessi D.R., Hemmings B.A., and Hundal H.S. 1998. Constitutive activation of protein kinase B alpha by membrane targeting promotes glucose and system A amino acid transport, protein synthesis, and inactivation of glycogen synthase kinase 3 in L6 muscle cells. *Diabetes* **47:** 1006–1013.

Hammond M.L., Merrick W., and Bowman L.H. 1991. Sequences mediating the translation of mouse S16 ribosomal protein mRNA during myoblast differentiation and in vitro and possible control points for the in vitro translation. *Genes Dev.* **5:** 1723–1736.

Hanks S.K. and Hunter T. 1995. The eukaryotic protein kinase superfamily: Kinase (catalytic) domain structure and classification. *FASEB J.* **9:** 576–596.

Hara K., Yonezawa K., Weng Q.-P., Kozlowski M.T., Belham C., and Avruch J. 1998. Amino acid sufficiency and mTOR regulate p70 S6 kinase and eIF-4E BP1 through a common effector mechanism. *J. Biol. Chem.* **273:** 14484–14494.

Hara K., Maruki Y., Long X., Yoshino K., Oshiro N., Hidayat S., Tokunaga C., Avruch J., and Yonezawa K. 2002. Raptor, a binding partner of target of rapamycin (TOR), mediates TOR action. *Cell* **110:** 177–189.

Harada H., Andersen J.S., Mann M., Terada N., and Korsmeyer S.J. 2001. p70S6 kinase signals cell survival as well as growth, inactivating the pro-apoptotic molecule BAD. *Proc. Natl. Acad. Sci.* **98:** 9666–9670.

Heitman J., Movva N.R., and Hall M.N. 1991. Targets for cell cycle arrest by the immunosupressant rapamycin in yeast. *Science* **253:** 905–909.

Helliwell S.B., Wagner P., Kunz J., Deuter-Reinhars M., Henriquez R., and Hall M.N. 1994. TOR1 and TOR2 are structurally and functionally similar but not identical phosphotidylinositol kinase homologues in yeast. *Mol. Biol. Cell* **5:** 105–118.

Herrera-Velit P., Knutson K.L., and Reiner N.E. 1997. Phosphatidylinositol 3-kinase-dependent activation of protein kinase C-zeta in bacterial lipopolysaccharide-treated human monocytes. *J. Biol. Chem.* **272:** 16445–16452.

Huang H., Potter C.J., Tao W., Li D.M., Brogiolo W., Hafen E., Sun H., and Xu T. 1999.

PTEN affects cell size, cell proliferation and apoptosis during *Drosophila* eye development. *Development* **126:** 5365–5372.

Hugl S.R., White M.F., and Rhodes C.J. 1998. Insulin-like growth factor I (IGF-I)-stimulated pancreatic beta-cell growth is glucose-dependent. Synergistic activation of insulin receptor substrate-mediated signal transduction pathways by glucose and IGF-I in INS-1 cells. *J. Biol. Chem.* **273:** 17771–17779.

Iiboshi Y., Papst P.J., Kawasome H., Hosoi H., Abraham R.T., Houghton P.J., and Terada N. 1999. Amino acid-dependent control of p70(s6k). Involvement of tRNA aminoacylation in the regulation. *J. Biol. Chem.* **274:** 1092–1099.

Inoki K., Li Y., Zhu T., Wu J., and Guan K.L. 2002. TSC2 is phosphorylated and inhibited by Akt and suppresses mTOR signalling. *Nat. Cell Biol.* **4:** 648–657.

Isotani S., Hara K., Tokunaga C., Inoue H., Avruch J., and Yonezawa K. 1999. Immunopurified mammalian target of rapamycin phosphorylates and activates p70 S6 kinase alpha in vitro. *J. Biol. Chem.* **274:** 34493–34498.

Jaeschke A., Hartkamp J., Saitoh M., Roworth W., Nobukuni T., Hodges A., Sampson J., Thomas G., and Lamb R. 2002. Tuberous sclerosis complex tumor suppressor-mediated S6 kinase inhibition by phosphatidylinositide-3-OH kinase is mTOR independent. *J. Cell Biol.* **159:** 217–224.

Jefferies H.B.J. and Thomas G. 1996. Ribosomal protein S6 phosphorylation and signal transduction. In *Translational control* (ed. J.W.B. Hershey et al.), pp. 389–409. Cold Spring Harbor Laboratory Press, Cold Spring Harbor, New York.

Jefferies H.B.J., Reinhard C., Kozma S.C., and Thomas G. 1994. Rapamycin selectively represses translation of the "polypyrimidine tract" mRNA family. *Proc. Natl. Acad. Sci.* **91:** 4441–4445.

Jefferies H.B.J., Fumagalli S., Dennis P.B., Reinhard C., Pearson R.B., and Thomas G. 1997. Rapamycin suppresses 5′TOP mRNA translation through inhibition of p70[s6k]. *EMBO J.* **12:** 3693–3704.

Jenö P., Ballou L.M., Novak-Hofer I., and Thomas G. 1988. Identification and characterization of a mitogenic-activated S6 kinase. *Proc. Natl. Acad. Sci.* **85:** 406–410.

Jenö P., Jäggi N., Luther H., Siegmann M., and Thomas G. 1989. Purification and characterization of a 40 S ribosomal protein S6 kinase from vanadate-stimulated Swiss 3T3 cells. *J. Biol. Chem.* **264:** 1293–1297.

Kawasome K., Papst P., Webb S., Keller G.M., Johnson G.L., Gelfand E.W., and Terada N. 1998. Targeted disruption of p70[s6k] defines its role in protein synthesis and rapamycin sensitivity. *Proc. Natl. Acad. Sci.* **95:** 5033–5038.

Khan A., Pepio A.M., and Sossin W.S. 2001. Serotonin activates S6 kinase in a rapamycin-sensitive manner in *Aplysia* synaptosomes. *J. Neurosci.* **21:** 382–391.

Kim D.H., Sarbassov D.D., Ali S.M., King J.E., Latek R.R., Erdjument-Bromage H., Tempst P., and Sabatini D.M. 2002. mTOR interacts with raptor to form a nutrient-sensitive complex that signals to the cell growth machinery. *Cell* **110:** 163–175.

Kotani K., Yonezawa K., Hara K., Ueda H., Kitamura Y., Sakaue H., Ando A., Chavanieu A., Calas B., and Grigorescu F., et al. 1994. Involvement of phosphoinositide 3-kinase in insulin- or IGF-1-induced membrane ruffling. *EMBO J.* **13:** 2313–2321.

Kozak M. 1991. An analysis of vertebrate mRNA sequences: Intimations of translational control. *J. Cell Biol.* **115:** 887–903.

Kozma S.C. and Thomas G. 2002. Regulation of cell size in growth, development and human disease: PI3K, PKB and S6K. *Bioessays* **24:** 65–71.

Kozma S.C., Ferrari S., and Thomas G. 1989a. Unmasking a growth factor/oncogene-activated S6 phosphorylation cascade. *Cell. Signal.* **1:** 219–225.

Kozma S.C., Ferrari S., Bassand P., Siegmann M., Totty N., and Thomas G. 1990. Cloning of the mitogen-activated S6 kinase from rat liver reveals an enzyme of the second messenger subfamily. *Proc. Natl. Acad. Sci.* **87:** 7365–7369.

Kozma S.C., Lane H.A., Ferrari S., Luther H., Siegmann M., and Thomas G. 1989b. A stimulated S6 kinase from rat liver: Identity with the mitogen-activated S6 kinase from 3T3 cells. *EMBO J.* **8:** 4125–4132.

Krieg J., Hofsteenge J., and Thomas G. 1988. Identification of the 40 S ribosomal protein S6 phosphorylation sites induced by cycloheximide. *J. Biol. Chem.* **263:** 11473–11477.

Kunz J., Henriquez R., Scheider U., Deuter-Reinhard M., Movva N.R., and Hall M.N. 1993. Target of rapamycin in yeast, TOR2 is an essential phosphatidylinosito kinase homolog required for G_1 progression. *Cell* **73:** 585–596.

Kuo C.J., Chung J., Fiorentino D.F., Flanagan W.M., Blenis J., and Crabtree G.R. 1992. Rapamycin selectively inhibits interleukin-2 activation of p70 S6 kinase. *Nature* **358:** 70–73.

Lee-Fruman K.K., Kuo C.J., Lippincott J., Terada N., and Blenis J. 1999. Characterization of S6K2, a novel kinase homologous to S6K1. *Oncogene* **18:** 5108–5114.

Lehner C.F. 1999. The beauty of small flies. *Nat. Cell Biol.* **1:** E129–E130.

Levy S., Avni D., Hariharan N., Perry R.P., and Meyuhas O. 1991. Oligopyrimidine tract at the 5′ end of mammalian ribosomal protein mRNAs is required for their translational control. *Proc. Natl. Acad. Sci.* **88:** 3319–3323.

Li Y., Inoki K., Yeung R., and Guan K.L. 2002. Regulation of TSC2 by 14-3-3 binding. *J. Biol. Chem.* **277:** 44593–44596.

Liaw D., Marsh D.J., Li J., Dahia P.L., Wang S.I., Zheng Z., Bose S., Call K.M., Tsou H.C., Peacocke M., Eng C., and Parsons R. 1997. Germline mutations of the PTEN gene in Cowden disease, an inherited breast and thyroid cancer syndrome. *Nat. Genet.* **16:** 64–67.

Liu M.Y., Cai S., Espejo A., Bedford M.T., and Walker C.L. 2002. 14-3-3 interacts with the tumor suppressor tuberin at Akt phosphorylation site(s). *Cancer Res.* **62:** 6475–6480.

Lizcano J.M., Deak M., Morrice N., Kieloch A., Hastie C.J., Dong L., Schutkowski M., Reimer U., and Alessi D.R. 2002. Molecular basis for the substrate specificity of NIMA-related kinase-6 (NEK6). Evidence that NEK6 does not phosphorylate the hydrophobic motif of ribosomal S6 protein kinase and serum- and glucocorticoid-induced protein kinase in vivo. *J. Biol. Chem.* **277:** 27839–27849.

Long X., Spycher C., Han Z.S., Rose A.M., Muller F., and Avruch J. 2002. TOR deficiency in *C. elegans* causes developmental arrest and intestinal atrophy by inhibition of mRNA translation. *Curr. Biol.* **12:** 1448–1461.

Manning B.D., Tee A.R., Logsdon M.N., Blenis J., and Cantley L.C. 2002. Identification of the tuberous sclerosis complex-2 tumor suppressor gene product tuberin as a target of the phosphoinositide 3-kinase/akt pathway. *Mol. Cell* **10:** 151–162.

Matsuo T., Kubo Y., Watanabe Y., and Yamamoto M. 2003. *Schizosaccharomyces pombe* AGC family kinase Gad8p forms a conserved signaling module with TOR and PDK1-like kinases. *EMBO J.* **22:** 3073–3083.

Mariottini P. and Amaldi F. 1990. The 5′ untranslated region of mRNA for ribosomal protein S19 is involved in its translational regulation during *Xenopus* development. *Mol. Cell. Biol.* **10:** 816–822.

McMahon L.P., Choi K.M., Lin T.A., Abraham R.T., and Lawrence J.C., Jr. 2002. The rapamycin-binding domain governs substrate selectivity by the mammalian target of rapamycin. *Mol. Cell. Biol.* **22:** 7428–7438.

Meyuhas O. 2000. Synthesis of the translational apparatus is regulated at the translational level. *Eur. J. Biochem.* **267:** 6321–6330.

Meyuhas O., Avni D., and Shama S. 1996. Translational control of ribosomal protein mRNAs in eukaryotes. In *Translational control* (ed. J.W.B. Hershey et al.), pp. 363–388. Cold Spring Harbor Laboratory Press, Cold Spring Harbor, New York.

Ming X.F., Burgering B.M.T., Wennström S., Claesson-Welsh L., Heldin C.H., Bos J.L., Kozma S.C., and Thomas G. 1994. Activation of p70/p85 S6 kinase by a pathway independent of p21ras. *Nature* **371:** 426–429.

Montagne J. 2000. Genetic and molecular mechanisms of cell size control. *Mol. Cell. Biol. Res. Commun.* **4:** 195–202.

Montagne J., Radimerski T., and Thomas G. 2001. Insulin signaling: Lessons from the *Drosophila* tuberous sclerosis complex, a tumor suppressor. Sci. STKE **2001:** PE36.

Montagne J., Stewart M.J., Stocker H., Hafen E., Kozma S.C., and Thomas G. 1999. *Drosophila* S6 kinase: A regulator of cell size. *Science* **285:** 2126–2129.

Morata G. and Ripoll P. 1975. Minutes: Mutants of *Drosophila* autonomously affecting cell division rate. *Dev. Biol.* **42:** 211–221.

Moser B.A., Dennis P.B., Pullen N., Pearson R.B., Williamson N.A., Wettenhall E.H., Kozma S.C., and Thomas G. 1997. Dual requirement for a newly identified phosphorylation site in p70s6k. *Mol. Cell. Biol.* **17:** 5648–5655.

Nakanishi H., Brewer K.A., and Exton J.H. 1993. Activation of the zeta isozyme of protein kinase C by phosphatidylinositol 3,4,5-trisphosphate. *J. Biol. Chem.* **268:** 13–16.

Nellist M., Goedbloed M.A., de Winter C., Verhaaf B., Jankie A., Reuser A.J., van den Ouweland A.M., van der Sluijs P., and Halley D.J. 2002. Identification and characterization of the interaction between tuberin and 14-3-3zeta. *J. Biol. Chem.* **277:** 39417–39424.

Neufeld T.P., de la Cruz A.F., Johnston L.A., and Edgar B.A. 1998. Coordination of growth and cell division in the *Drosophila* wing. *Cell* **93:** 1183–1193.

Nielsen P.J., Duncan R., and McConkey E.H. 1981. Phosphorylation of ribosomal protein S6: Relationship to protein synthesis in HeLa cells. *Eur. J. Biochem.* **120:** 523–527.

Nojima H., Tokunaga C., Eguchi S., Oshiro N., Hidayat S., Yoshino K., Hara K., Tanaka N., Avruch J., and Yonezawa K. 2003. The mammalian target of rapamycin (mTOR) partner, raptor, binds the mTOR substrates, p70 S6 kinase and 4E-BP1 through their TOR signaling TOS motif. *J. Biol. Chem.* **278:** 15461–15464.

Novak-Hofer I. and Thomas G. 1984. An activated S6 kinase in extracts from serum- and epidermal growth factor-stimulated Swiss 3T3 cells. *J. Biol. Chem.* **259:** 5995–6000.

———. 1985. Epidermal growth factor-mediated activation of an S6 kinase in Swiss mouse 3T3 cells. *J. Biol. Chem.* **260:** 10314–10319.

Nygard O. and Nika H. 1982. Identification by RNA-protein cross-linking of ribosomal proteins located at the interface between the small and the large subunits of mammalian ribosomes. *EMBO J.* **1:** 357–362.

Nygard O. and Nilsson L. 1990. Translational dynamics. Interactions between the translational factors, tRNA and ribosomes during eukaryotic protein synthesis. *Eur. J. Biochem.* **191:** 1–17.

Oldham S., Montagne J., Radimerski T., Thomas G., and Hafen E. 2000. Genetic and biochemical characterization of dTOR, the *Drosophila* homolog of the target of rapamycin. *Genes Dev.* **14:** 2689–2694.

Pardee A.B. 1989. G_1 events and regulation of cell proliferation. *Science* **246:** 603–608.

Pearson R.B., Dennis P.B., Han J.W., Williamson N.A., Kozma S.C., Wettenhall R.E.H., and Thomas G. 1995. The principal target of rapamycin-induced p70^{s6k} inactivation is a novel phosphorylation site within a conserved hydrophobic domain. *EMBO J.* **21:** 5279–5287.

Pellizzoni L., Lotti F., Rutjes S.A., and Pierandrei-Amaldi P. 1998. Involvement of the *Xenopus laevis* Ro60 autoantigen in the alternative interaction of La and CNBP proteins with the 5′UTR of L4 ribosomal protein mRNA. *J. Mol. Biol.* **281:** 593–608.

Pende M., Um S.H., Mieulet V., Sticker M., Gross V.L., Mestan J., Mueller M., Fumagalli S., Kozma S.C., and Thomas G. 2004. S6Kl$^{-/-}$/S6K2$^{-/-}$ Mice exhibit perinatal lethality, rapamycin sensitive 5′TOP mRNA translation, and uncover A MAP kinase-dependent S6 kinase pathway. *Mol. Cell. Biol.*

Pende M., Kozma S.C., Jaquet M., Oorschot V., Burcelin R., Le Marchand-Brustel Y., Klumperman J., Thorens B., and Thomas G. 2000. Hypoinsulinaemia, glucose intolerance and diminished beta-cell size in S6K1-deficient mice. *Nature* **408:** 994–997.

Peterson R.T., Desai B.N., Hardwick J.S., and Schreiber S.L. 1999. Protein phosphatase 2A interacts with the 70-kDa S6 kinase and is activated by inhibition of FKBP12-rapamycin-associated protein. *Proc. Natl. Acad. Sci.* **96:** 4438–4442.

Petroulakis E. and Wang E. 2002. Nerve growth factor specifically stimulates translation of eukaryotic elongation factor 1A-1 (eEF1A-1) mRNA by recruitment to polyribosomes in PC12 cells. *J. Biol. Chem.* **277:** 18718–18727.

Peyrollier K., Hajduch E., Blair A. S., Hyde R., and Hundal H.S. 2000. L-leucine availability regulates phosphatidylinositol 3-kinase, p70 S6 kinase and glycogen synthase kinase-3 activity in L6 muscle cells: Evidence for the involvement of the mammalian target of rapamycin (mTOR) pathway in the L-leucine-induced up-regulation of system A amino acid transport. *Biochem J.* **350:** 361–368.

Phillips D.I. 1996. Insulin resistance as a programmed response to fetal undernutrition (comments). *Diabetologia* **39:** 1119–1122.

Polymenis M. and Schmidt E.V. 1999. Coordination of cell growth with cell division. *Curr. Opin. Genet. Dev.* **9:** 76–80.

Potter C.J., Huang H., and Xu T. 2001. *Drosophila* tsc1 functions with tsc2 to antagonize insulin signaling in regulating cell growth, cell proliferation, and organ size. *Cell* **105:** 357–368.

Potter C.J., Pedraza L.G., and Xu T. 2002. Akt regulates growth by directly phosphorylating Tsc2. *Nat. Cell Biol.* **4:** 658–665.

Price D.J., Grove J.R., Calvo V., Avruch J., and Bierer B.E. 1992. Rapamycin-induced inhibition of the 70-kilodalton S6 protein kinase. *Science* **257:** 973–977.

Pullen N. and Thomas G. 1997. The modular phosphorylation and activation of p70^{s6k}. *FEBS Lett.* **410:** 78–82.

Pullen N., Dennis P.B., Andjelkovic M., Dufner A., Kozma S., Hemmings B.A., and Thomas G. 1998. Phosphorylation and activation of p70^{s6k} by PDK1. *Science* **279:** 707–710.

Radimerski T., Montagne J., Hemmings-Mieszczak M., and Thomas G. 2002a. Lethality of *Drosophila* lacking TSC tumor suppressor function rescued by reducing dS6K signaling. *Genes Dev.* **16:** 2627–2632.

Radimerski T., Mini T., Schneider U., Wettenhall R.E., Thomas G., and Jeno P. 2000. Identification of insulin-induced sites of ribosomal protein S6 phosphorylation in *Drosophila melanogaster. Biochemistry* **39:** 5766–5774.

Radimerski T., Montagne J., Rintelen F., Stocker H., van Der Kaay J., Downes C.P., Hafen E., and Thomas G. 2002b. dS6K-regulated cell growth is dPKB/dPI(3)K-independent, but requires dPDK1. *Nat. Cell Biol.* **4:** 251–255.

Raught B., Gingras A.C., and Sonenberg N. 2001. The target of rapamycin (TOR) proteins. *Proc. Natl. Acad. Sci.* **98:** 7037–7044.

Reinhard C., Thomas G., and Kozma S.C. 1992. A single gene encodes two isoforms of the p70 S6 kinase: Activation upon mitogenic stimulation. *Proc. Natl. Acad. Sci.* **89:** 4052–4056.

Reinhard C., Fernandez A., Lamb N.J.C., and Thomas G. 1994. Nuclear localization of p85[s6k]: Functional requirement for entry into S phase. *EMBO J.* **13:** 1557–1565.

Rintelen F., Stocker H., Thomas G., and Hafen E. 2001. PDK1 regulates growth through Akt and S6K in *Drosophila. Proc. Natl. Acad. Sci.* **98:** 15020–15025.

Romanelli A., Dreisbach V.C., and Blenis J. 2002. Characterization of phosphatidylinositol 3-kinase-dependent phosphorylation of the hydrophobic motif site Thr(389) in p70 S6 kinase 1. *J. Biol. Chem.* **277:** 40281–40289.

Romanelli A., Martin K.A., Toker A., and Blenis J. 1999. p70 S6 kinase is regulated by protein kinase C zeta and participates in a phosphoinositide 3-kinase-regulated signalling complex. *Mol. Cell. Biol.* **19:** 2921–2928.

Saitoh M., ten Dijke P., Miyazono K., and Ichijo H. 1998. Cloning and characterization of p70[s6k]β defines a novel family of p70 S6 kinases. *Biochem. Biophys. Res. Commun.* **253:** 470–476.

Saitoh M., Pullen N., Brennan P., Cantrell D., Dennis P.B., and Thomas G. 2002. Regulation of an activated S6K1 variant reveals a novel mammalian target of rapamycin phosphorylation site. *J. Biol. Chem.* **277:** 20104-20112.

Santiago J. and Sturgill T.W. 2001. Identification of the S6 kinase activity stimulated in quiescent brine shrimp embryos upon entry to preemergence development as p70 ribosomal protein S6 kinase: Isolation of *Artemia franciscana* p70S6k cDNA. *Biochem. Cell Biol.* **79:** 141–152.

Saucedo L.J., Gao X., Chiarelli D.A., Li L., Pan D., and Edgar B.A. 2003. Rheb promotes cell growth as a component of the insulin/TOR signalling network. *Nat. Cell Biol.* **5:** 566–571.

Schalm S.S. and Blenis J. 2002. Identification of a conserved motif required for mTOR signaling. *Curr. Biol.* **12:** 632–639.

Schmidt E.E. and Schibler U. 1995. Cell size regulation, a mechanism that controls cellular RNA accumulation: Consequences on regulation of the ubiquitous transcription factors Oct1 and NF-Y and the liver-enriched transcription factor DBP. *J. Cell Biol.* **128:** 467–483.

Schmidt E.V. 1999. The role of c-myc in cellular growth control. *Oncogene* **18:** 2988–2996.

Schwab M.S., Kim S.H., Terada N., Edfjall C., Kozma S.C., Thomas G., and Maller J.L. 1999. p70(S6K) controls selective mRNA translation during oocyte maturation and early embryogenesis in *Xenopus laevis. Mol. Cell. Biol.* **19:** 2485–2494.

Shigemitsu K., Tsujishita Y., Hara K., Nanahoshi M., Avruch J., and Yonezawa K. 1999. Regulation of translational effectors by amino acid and mammalian target of rapamycin signaling pathways. Possible involvement of autophagy in cultured hepatoma cells. *J. Biol. Chem.* **274:** 1058–1065.

Shima H., Pende M., Chen Y., Fumagalli S., Thomas G., and Kozma S.C. 1998. Disruption of the p70[s6k]/p85[s6k] gene reveals a small mouse phenotype and a new functional S6 kinase. *EMBO J.* **17:** 6649–6659.

Skolnik E.Y., Batzer A., Li N., Lee C.H., Lowenstein E., Mohammadi M., Margolis B., and Schlessinger J. 1993. The function of GRB2 in linking the insulin receptor to Ras signaling pathways. *Science* **260:** 1953–1955.

Sparagana S.P. and Roach E.S. 2000. Tuberous sclerosis complex. *Curr. Opin. Neurol.* **13:** 115–119.

Stewart M.J., Berry C.O.A., Zilberman F., Thomas G., and Kozma S.C. 1996. The *Drosophila* p70[s6k] homolog exhibits conserved regulatory elements and rapamycin sensitivity. *Proc. Natl. Acad. Sci.* **93:** 10791–10796.

Stocker H. and Hafen E. 2000. Genetic control of cell size. *Curr. Opin. Genet. Dev.* **10:** 529–535.

Stocker H., Andjelkovic M., Oldham S., Laffargue M., Wymann M.P., Hemmings B.A., and Hafen E. 2002. Living with lethal PIP3 levels: Viability of flies lacking PTEN restored by a PH domain mutation in Akt/PKB. *Science* **295:** 2088–2091.

Stocker H., Radimerski T., Schindelholz B., Wittwer F., Belawat P., Daram P., Breuer S., Thomas G., and Hafen E. 2003. Rheb is an essential regulator of S6K in controlling cell growth in *Drosophila. Nat. Cell Biol.* **5:** 559–566.

Stokoe D., Stephens L.R., Copeland T., Gaffney P.R.J., Reese C.B., Painter G.F., Holmes A.B., McCormick F., and Hawkins P.T. 1998. Dual role of phosphatidylinositol-3,4,5-trisphosphate in the activation of protein kinase B. *Science* **277:** 567–570.

Stolovich M., Tang H., Hornstein E., Levy G., Cohen R., Bae S.S., Birnbaum M.J., and Meyuhas O. 2002. Transduction of growth or mitogenic signals into translational activation of TOP mRNAs is fully reliant on the phosphatidylinositol 3-kinase-mediated pathway but requires neither S6K1 nor rpS6 phosphorylation. *Mol. Cell. Biol.* **22:** 8101–8113.

Swenne I., Crace C.J., and Jansson L. 1988. Intermittent protein-calorie malnutrition in the young rat causes long-term impairment of the insulin secretory response to glucose in vitro. *J. Endocrinol.* **118:** 295–302.

Swenne I., Crace C.J., and Milner R.D. 1987. Persistent impairment of insulin secretory response to glucose in adult rats after limited period of protein-calorie malnutrition early in life. *Diabetes* **36:** 454–458.

Swenne I., Borg L.A., Crace C.J., and Schnell Landstrom A. 1992. Persistent reduction of pancreatic beta-cell mass after a limited period of protein-energy malnutrition in the young rat. *Diabetologia* **35:** 939–945.

Szarski H. 1983. Cell size and the concept of wasteful and frugal evolutionary strategies. *J. Theor. Biol.* **105:** 201–209.

Tapon N., Ito N., Dickson B.J., Treisman J.E., and Hariharan I.K. 2001. The *Drosophila* tuberous sclerosis complex gene homologs restrict cell growth and cell proliferation. *Cell* **105:** 345–355.

Tee A.R., Fingar D.C., Manning B.D., Kwiatkowski D.J., Cantley L.C., and Blenis J. 2002. Tuberous sclerosis complex-1 and -2 gene products function together to inhibit mammalian target of rapamycin (mTOR)-mediated downstream signaling. *Proc. Natl. Acad. Sci.* **99:** 13571–13576.

Terada N., Patel H.R., Takase K., Kohno K., Narin A.C., and Gelfand E.W. 1994. Rapamycin selectively inhibits translation of mRNAs encoding elongation factors and ribosomal proteins. *Proc. Natl. Acad. Sci.* **91:** 11477–11481.

Thomas G. 2000. An "encore" for ribosome biogenesis in cell proliferation. *Nat. Cell Biol.* **2:** E71–E72.

Thomas G. and Hall M.N. 1997. TOR signalling and control of cell growth. *Curr. Opin. Cell Biol.* **9:** 782–787.

Thomas G., Martin-Pèrez J., Siegmann M., and Otto A.M. 1982. The effect of serum, EGF,

PGF$_{2a}$ and insulin on S6 phosphylation and the initiation of protein and DNA synthesis. *Cell* **30:** 235–242.

Turck F., Kozma S.C., Thomas G., and Nagy F. 1998. A heat-sensitive *Arabidopsis thaliana* kinase substitutes for human p70s6k function in vivo. *Mol. Cell. Biol.* **18:** 2038–2044.

Verdu J., Buratovich M.A., Wilder E.L., and Birnbaum M.J. 1999. Cell-autonomous regulation of cell and organ growth in *Drosophila* by Akt/PKB. *Nat. Cell Biol.* **1:** 500–506.

von Manteuffel S.R., Gingras A.-C., Ming X.-F., Sonenberg N., and Thomas G. 1996. 4E-BP1 phosphorylation is mediated by the FRAP-p70[s6k] pathway and is independent of mitogen-activated protein kinase. *Proc. Natl. Acad. Sci.* **93:** 4076–4080.

von Manteuffel S.R., Dennis P.B., Pullen N., Gingras A.C., Sonenberg N., and Thomas G. 1997. The insulin-induced signalling pathway leading to S6 and initiation factor 4E binding protein 1 phosphorylation bifurcates at a rapamycin-sensitive point immediately upstream of p70s6k. *Mol. Cell. Biol.* **17:** 5426–5436.

Wang X., Li W., Williams M., Terada N., Alessi D.R., and Proud C.G. 2001. Regulation of elongation factor 2 kinase by p90(RSK1) and p70 S6 kinase. *EMBO J.* **20:** 4370–4379.

Warner J.R. 1999. The economics of ribosome biosynthesis in yeast. *Trends Biochem. Sci.* **24:** 437–440.

Watson K.L., Chou M.M., Blenis J., Gelbart W.M., and Erickson R.L. 1996. A *Drosophila* gene structurally and functionally homologous to the mammalian 70-kDa S6 kinase gene. *Proc. Natl. Acad. Sci.* **93:** 13694–13698.

Weinkove D., Neufeld T.P., Twardzik T., Waterfield M.D., and Leevers S.J. 1999. Regulation of imaginal disc cell size, cell number and organ size by *Drosophila* class I(A) phosphoinositide 3-kinase and its adaptor. *Curr. Biol.* **9:** 1019–1029.

Weng Q.-P., Andrabi K., Kozlowski M.T., Grove J.R., and Avruch J. 1995a. Multiple independent inputs are required for activation of the p70 S6 kinase. *Mol. Cell. Biol.* **15:** 2333–2340.

Weng Q.-P., Andrabi K., Klippel A., Kozlowski M.T., Williams L.T., and Avruch J. 1995b. Phosphatidylinositol 3-kinase signals activation of p70 S6 kinase in situ through site specific p70 phosphorylation. *Proc. Natl. Acad. Sci.* **92:** 5744–5748.

Williams M.R., Arthur J.S., Balendran A., van der Kaay J., Poli V., Cohen P., and Alessi D.R. 2000. The role of 3-phosphoinositide-dependent protein kinase 1 in activating AGC kinases defined in embryonic stem cells. *Curr. Biol.* **10:** 439–448.

Wool I.G., Chan Y.-L., and Glück A. 1996. Mammalian ribosomes: The structure and the evolution of the proteins. In *Translational control* (ed. J.W.B. Hershey et al.), pp. 685–732. Cold Spring Harbor Laboratory Press, Cold Spring Harbor, New York.

Xu G., Kwon G., Marshall C.A., Lin T.A., Lawrence J.C., Jr., and McDaniel M.L. 1998. Branched-chain amino acids are essential in the regulation of PHAS-I and p70 S6 kinase by pancreatic beta-cells. A possible role in protein translation and mitogenic signaling. *J. Biol. Chem.* **273:** 28178–28184.

Young J. and Povey S. 1998. The genetic basis of tuberous sclerosis. *Mol. Med. Today* **4:** 313–319.

Zhang H., Stallock J.P., Ng J.C., Reinhard C., and Neufeld T.P. 2000. Regulation of cellular growth by the *Drosophila* target of rapamycin dTOR. *Genes Dev.* **14:** 2712–2724.

Zhang Y., Gao X., Saucedo L.J., Ru B., Edgar B.A., and Pan D. 2003. Rheb is a direct target of the tuberous sclerosis tumour suppressor proteins. *Nat. Cell Biol.* **5:** 578–581.

10

Translation Initiation and Cell Growth Control

Emmanuel Petroulakis and Nahum Sonenberg
Department of Biochemistry and
McGill Cancer Research Centre
McGill University
Montreal, Quebec H3G 1Y6 Canada

A GREAT DEAL OF RESEARCH EFFORT has been directed toward understanding the principles of cell growth. Growth and proliferation are tightly linked under most circumstances. Growth is defined as an increase in cell mass, and proliferation is defined as an increase in cell number. Cell growth consumes a large amount of energy due to the need for protein synthesis, which is the most energy-consuming process in the cell. Early work in yeast as well as more recent studies in mammals and *Drosophila* have established a critical link between cell growth and translational initiation. Furthermore, biological processes that require growth, such as cell division, differentiation, and development, are regulated at the translational level. Translation initiation is the primary target for translational control, and therefore it is implicated directly in the control of cell growth. A plethora of extracellular cues trigger the phosphorylation of several initiation factors in order to recruit ribosomes to the mRNA (Gingras et al. 1999a; Hershey and Merrick 2000). Changes in expression of the translational apparatus and dysregulation of its phosphorylation state cause dramatic changes in cell growth and often lead to diseases such as cancer (Hershey and Miyamoto 2000). This chapter describes how phosphorylation of translation initiation factors governs cell growth.

THE TRANSLATION INITIATION PATHWAY

Translation initiation in eukaryotes requires a large number of translation factors (~12) that promote the recruitment of ribosomes to the mRNA to form a translation initiation complex (for reviews, see Pain 1996; Hershey and Merrick 2000; Raught et al. 2000a). The initiation pathway can be divided into distinct stages (depicted in Fig. 1), the first of which requires the dissociation of the 40S and 60S ribosomal subunits, which is facilitated by eukaryotic initiation factors (eIF) 1A and eIF3. The multimeric complex eIF3 also stabilizes the 43S preinitiation complex (PIC), which is formed through an interaction between the eIF3-bound 40S small ribosomal subunit and the Met•tRNA$_i$•eIF2•GTP ternary complex. The 48S PIC is formed in the third stage via the recruitment of the 43S PIC to the mRNA, which is mediated by the eIF4 family of initiation factors (see below). The 5′ end of nuclear transcribed mRNAs contains the structure m^{7}GpppN (where N is any nucleotide) termed the cap. The majority of eukaryotic mRNAs are translated in a cap-dependent manner (Gingras et al. 1999a). The 5′ cap is bound directly by the eIF4F complex that is responsible for unwinding the secondary structure of the mRNA (see below). Once the 43S PIC is bound to the mRNA, it is thought to move in a 5′ to 3′ direction until an initiation codon is encountered. Generally, the first AUG (or a cognate triplet) that is encountered is utilized as the translation start site (Hershey and Merrick 2000). In the final stage of initiation, eIF5 stimulates GTP hydrolysis by eIF2 in the ternary complex, leading to the release of initiation factors and permitting eIF5B-promoted association of the 60S ribosomal subunit. Subsequent rounds of initiation occur once the GDP in the eIF2•GDP binary complex is exchanged for GTP. This exchange is effected by eIF2B, a guanine nucleotide exchange factor (Hinnebusch 2000).

ROLE OF eIF2α IN GROWTH

Global translation rates are reduced upon phosphorylation of eIF2α resulting in reduced levels of the ternary complex Met•tRNAi•eIF2•GTP. eIF2α is phosphorylated at Ser-51 in response to many environmental factors including growth factors and nutrient deprivation (e.g., glucose, amino acid), heat shock, exposure to heavy metals, and viral infection (Chen 2000; Kaufman 2000; Ron and Harding 2000). Several eIF2α kinases that respond to specific stimuli have been studied in great detail (Chen 2000; Kaufman 2000; Ron and Harding 2000). These include double-stranded RNA (dsRNA)-dependent protein kinase (PKR), which is acti-

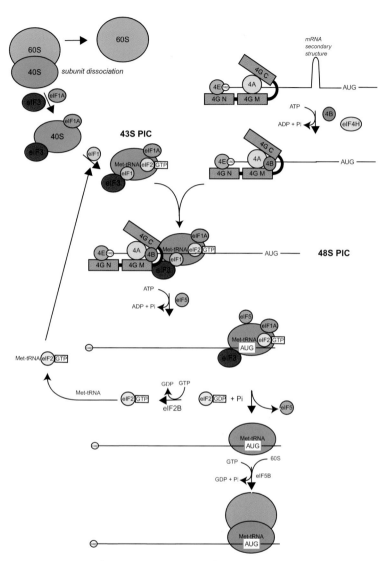

Figure 1. Steps in translation initiation. Translation initiation requires the dissociation of the 40S and 60S ribosomal subunits. Formation of the 43S PIC, which involves the recruitment of the Met•tRNA$_f$•eIF2•GTP complex to the 40S subunit, is facilitated by eIF1, eIF1A, eIF3. Following the binding of the eIF4F complex (eIF4E, eIF4A, and eIF4G) to the 5′ cap of the mRNA, the RNA secondary structure is unwound. The mRNA is then recruited by the 43S PIC to yield the 48S PIC. eIF5 mediates the release of initiation factors. The 60S ribosome is then recruited by eIF4B to form the 80S initiation complex. The three domains of eIF4G are indicated as 4G N (amino-terminal), 4G M (middle), and 4G C (carboxy-terminal). See the text for details. (Adapted, with permission, from Hershey and Merrick 2000.)

vated in response to dsRNA, polyanionic compounds and various cellular and viral proteins (Kaufman 2000), heme-regulated inhibitor kinase (HRI), which is activated in mature red blood cells in response to heme deprivation (Chen 2000); PERK (PKR-like ER-resident kinase) in the lumen of the endoplasmic reticulum, which is activated in response to protein misfolding (Ron and Harding 2000); and GCN2, which is activated in response to amino acid deficiency and UV irradiation (Hinnebusch 2000; Deng et al. 2002; Zhang et al. 2002).

The phosphorylation of eIF2α effects growth control in mammals. Overexpression of a dominant-negative PKR mutant in NIH-3T3 cells causes malignant transformation, implicating PKR in tumor suppression (Koromilas et al. 1992b). Consistent with the importance of eIF2α phosphorylation in growth, overexpression of a Ser51Ala eIF2α mutant causes malignant transformation of NIH-3T3 cells (Donze et al. 1995). PKR also transduces signals to transcription factors NF-κB (Kumar et al. 1994; Williams 2001), IFN-regulatory factor-1 (Kirchhoff et al. 1995), and activating transcription factor-2 (Bandyopadhyay et al. 2000). Thus, given the variety of PKR substrates, it is reasonable to expect that, during malignancy, PKR regulates pathways other than those that affect translation. Phosphorylation of eIF2α is also implicated in the survival of pancreatic cells through the involvement of PERK (Harding et al. 2001; Scheuner et al. 2001). PERK-deficient mice or those harboring a Ser51Ala mutant allele display normal pancreatic development, but severe postnatal pancreatic dysfunction (Harding et al. 2001; Scheuner et al. 2001). PERK-knockout mice develop hyperglycemia by one month of age (Harding et al. 2001), and Ser51Ala mutant animals die of hypoglycemia within the first day after birth (Scheuner et al. 2001). Both animal models display a selective decrease in β-pancreatic cells, a phenotype which resembles Wolcott-Rallison syndrome (WRS) in humans that is caused by mutations in the PERK gene (Delepine et al. 2000). These findings illustrate the importance of translational regulation in cell growth, proliferation, and human disease.

ROLE OF THE CAP IN EUKARYOTIC mRNAs

Translation initiation rates of different mRNAs may vary dramatically. These differences can be explained in many instances by differential requirement for eIF4 initiation factors (see below) in recruiting ribosomes to form the translation initiation complex. The cap is also important in other aspects of mRNA metabolism, including the stability of mRNAs and the efficient splicing of mRNA precursors via the activity of the *n*uclear *c*ap *b*inding *c*omplex (nCBC) (Izaurralde et al. 1995).

THE eIF4F COMPLEX

The cap is bound in the cytoplasm by the eIF4F complex, which recruits the 40S ribosomal subunit to the 5′ end of the mRNA (Fig. 2) (Gingras et al. 1999a). Although the cap facilitates the translation of mRNAs, residual translation may still occur in its absence (Palmer et al. 1993; Gunnery and Mathews 1995). In the most general mechanism of translation initiation, the assembly of the eIF4F initiation complex on the 5′ cap is required for the recruitment of the 40S ribosomal subunit (Gingras et al. 1999a). The eIF4F complex contains eIF4E, eIF4A, and eIF4G. eIF4E, which binds the cap directly, was initially purified by affinity chromatography using immobilized m⁷GDP (Sonenberg et al. 1978, 1979) and stimulates the translation of capped mRNAs in vitro (Sonenberg et al. 1980). eIF4E binds to eIF4G, a large modular scaffolding protein that also contains binding sites for eIF4A (a member of the DEAD box family of proteins) (Rogers et al. 2002), which is thought to act as an ATP-dependent RNA helicase to unwind the mRNA secondary structure (Rozen et al. 1990; Raught et al. 2000a). eIF4A activity is enhanced dramatically by the two

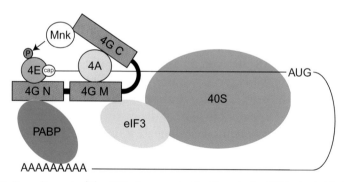

Figure 2. Recruitment of the 40S ribosome to the 5′ end of mRNAs by initiation factors. The eIF4F complex coordinates the recruitment of the 40S ribosomal subunit to the 5′end of eukaryotic mRNAs. eIF4E (4E) binds the 5′ m⁷G cap structure (cap) of the mRNA. The three modular domains of eIF4G (amino-terminal, 4G N; middle, 4G M; carboxy-terminal, 4G C) serve as docking sites for PABP, eIF4E, eIF4A, eIF3, and Mnk1. eIF3 functions as a bridge between eIF4G and the 40S subunit. PABP interacts with the amino-terminal domain of eIF4G (Tarun and Sachs 1996; Imataka et al. 1998) to mediate circularization of the mRNA (Wells et al. 1998).

eIF4G isoforms, eIF4GI and eIF4GII (Rozen et al. 1990; Gradi et al. 1998), and two RNA-binding initiation factors, eIF4B and eIF4H (Rozen et al. 1990; Rogers et al. 2001). eIF4G contains additional binding sites for eIF3 (Lamphear et al. 1995; Imataka and Sonenberg 1997) and the eIF4E kinase, Mnk1 (see Fig. 1) (Pyronnet et al. 1999). The poly(A)-binding protein (PABP) also binds eIF4G to mediate mRNA circularization by bringing the 5′ and 3′ ends of the mRNA in close proximity (Tarun and Sachs 1996; Imataka et al. 1998; Wells et al. 1998). The circularization of the mRNA is thought to act synergistically to stimulate translation initiation (Sachs 2000). The importance of the PABP-eIF4G interaction is underscored by the observation that expression in *Xenopus* oocytes of a eIF4G variant, which is deficient in PABP binding, inhibits progesterone-induced oocyte maturation and the translation of polyadenylated mRNAs (Wakiyama et al. 2000).

THE eIF4E-BINDING PROTEINS

eIF4F complex formation is negatively controlled by eIF4E-binding proteins (4E-BPs), originally identified as a family of three translational repressors in mammals (4E-BP1, 4E-BP2, and 4E-BP3) that bind eIF4E and prevent formation of the eIF4F initiation complex (for review, see Gingras et al. 1999a; Raught et al. 2000a). Consequently, cap-dependent but not cap-independent translation is inhibited (Pause et al. 1994; Poulin et al. 1998). The 4E-BPs share a high degree of structural similarity in that 4E-BP1 and 4E-BP2 are 56% identical, whereas 4E-BP3 shares 57% and 59% identity with 4E-BP1 and 4E-BP2, respectively (Pause et al. 1994; Poulin et al. 1998). 4E-BPs contain an eIF4E-binding site that consists of the motif, YXXXXLΦ (where X is any amino acid and Φ is a hydrophobic amino acid: L, M, or F). The importance of this motif in the control of translation became apparent in light of the discovery that it is also present in all eIF4E-binding proteins, including eIF4GI and eIF4GII (Mader et al. 1995) as well as the yeast homologs of eIF4G (Tif4631 and Tif4632) (Raught et al. 2000a). Thus, the motif is phylogenetically highly conserved. Mutations in this motif in 4E-BPs or eIF4G strongly diminish the interactions of these proteins with eIF4E. Synthetic peptides containing the motif also inhibit cap-dependent translation in vitro (Fletcher et al. 1998; Marcotrigiano et al. 1999) and, interestingly, cause cell death of nontransformed fibroblasts (Herbert et al. 2000). The 4E-BPs are unstructured molecules in solution, and 4E-BP1- or eIF4GII-derived peptides undergo a "disorder-to-order" transition upon binding to the convex dorsal surface of eIF4E (Marcotrigiano et al. 1999). The intermolecular contacts of eIF4G and 4E-BP1 with eIF4E are strikingly similar (Marcotrigiano et al. 1999). Thus, 4E-BP1 impedes eIF4F formation by mimicking the interaction between eIF4E and eIF4G (Marcotrigiano et al. 1999).

The phosphorylation status of the 4E-BPs on several serine and threonine residues is a critical factor in regulating their interaction with eIF4E. In the hypophosphorylated state the 4E-BPs bind tightly to eIF4E, whereas hyperphosphorylation of the 4E-BPs by numerous stimuli (described below) weakens this interaction, resulting in the release of 4E-BP from eIF4E and enhanced cap-dependent translation (Fig. 3).

The phosphorylation of 4E-BP1 is the best characterized of the three 4E-BPs (Gingras et al. 1998, 1999a,b, 2001; Wang et al. 2003). 4E-BP1 phosphorylation is regulated by numerous stimuli, including hormones (such as insulin), growth factors (EGF, NGF, FGF), cytokines (IL-3, TGFβ), mitogens (TPA), G-protein-coupled receptor ligands (DAMGO), and virus infection (adenovirus and picornavirus). 4E-BP1 is phosphorylated on seven sites (numbering according to human 4E-BP1): Thr-37, Thr-46, Ser-65, Thr-70, Ser-83, Ser-101, and Ser-112 (Raught et al. 2000a; Yang and Kastan 2000; Gingras et al. 2001b; Wang et al. 2003). The eIF4E-binding motif spans residues 54 to 60 and resides between two critical sets of phosphorylation sites (Thr-37/Thr-46 and Ser-65/Thr-70). Using phosphopeptide mapping combined with phospho-specific antibodies, 4E-BP1 phosphorylation in serum-stimulated human embryonic kidney 293 cells was shown to occur in an ordered, hierarchical manner (Gingras et al. 1999b, 2001a,b; McMahon et al. 2002). Following serum stimulation, the phosphorylation of Thr-37 and Thr-46 is mildly enhanced, whereas that of Ser-65 and Thr-70 is dramatically increased (Gingras et al. 1999b). Strikingly, phosphorylation at Ser-65 and Thr-70 requires the prior phosphorylation at Thr-37 and Thr-46 (Gingras et al. 2001b). Substitution of Thr-37 and/or Thr-46 with an alanine prevents phosphorylation of Ser-65 and Thr-70, whereas substitution for glutamic acid (a phospho-mimetic side chain) permits increased phosphorylation at these same sites (Gingras et al. 2001b). In turn, phosphorylation at Thr-70 precedes phosphorylation at Ser-65, since the latter is phosphorylated only after the other three sites become phosphorylated (Gingras et al. 1999b). Interestingly, the phosphorylation of 4E-BP1 depends on an amino-terminal RAIP sequence that is cleaved off during apoptosis (Tee and Proud 2002). Cleavage of 4E-BP1 during caspase-dependent cell death yields a truncated 4E-BP1 that cannot be phosphorylated and therefore remains tightly bound to eIF4E, suggesting a mechanism by which cap-dependent translation is repressed during apoptosis (Tee and Proud 2002).

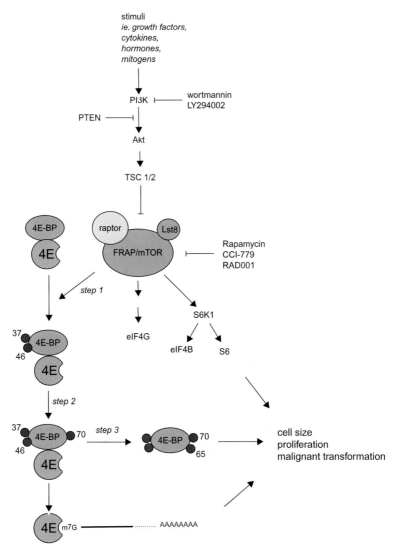

Figure 3. Signaling pathways leading to cap binding, 4E-BP phosphorylation, and cell growth. Stimuli (i.e., growth factors) activate the phosphatidylinositol-3 kinase (PI3K)-Akt-FRAP/mTOR signaling pathway. 4E-BP1 is phosphorylated through a multistep mechanism: FRAP/mTOR signaling mediates phosphorylation at Thr-37 and Thr-46 prior to phosphorylation of Ser-65 and Thr-70. Hyperphosphorylation of 4E-BP1 causes its release from eIF4E, allowing eIF4E to bind to the 5′ cap of mRNA. Mutations in PTEN and TSC 1/2 (see Chapter 7), elevate FRAP/mTOR activity, increase levels of eIF4F, and enhance translation of a subset of mRNAs that exhibit extensive secondary structure. FRAP/mTOR activity is inhibited by rapamycin and its derivatives (CCI-779, RAD-001). Enhanced mRNA translation correlates with increased cell size, growth, and malignant transformation.

The phosphorylation status of 4E-BP1 also changes during the cell cycle. 4E-BP1 is hyperphosphorylated during G_1 and is dephosphorylated during G_2/M (Pyronnet et al. 2001). The reduction in 4E-BP1 phosphorylation during mitosis coincides with a decrease in cap-dependent translation, dephosphorylation of eIF4E, a reduction in eIF4F complex formation, and enhanced binding of 4E-BP1 to eIF4E, resulting in the release of eIF4E from the eIF4F complex (Pyronnet et al. 2001). However, this finding is at variance with another study showing that 4E-BP1 becomes hyperphosphorylated during mitosis (Heesom et al. 2001). Denton's group used phosphopeptide-specific antibodies to demonstrate that phosphorylation at Ser-65 and Thr-70 increases in mitotic cells and that Thr-70 is more efficiently phosphorylated when 4E-BP1 is complexed with eIF4E, in comparison to unbound 4E-BP1 which is poorly phosphorylated (Heesom et al. 2001). They further show that phosphorylation of Thr-70 is mediated by the cdc2–cyclin B complex, the level of which increases at the onset of mitosis (Heesom et al. 2001). Whereas Pyronnet et al. (2001) used nocodazole treatment to enrich for mitotic cells, Heesom et al. (2001) isolated mitotic cells following release from an aphidicolin block (Heesom et al. 2001). A possible reason for the discrepancy between these two reports is that Heesom et al. (2001) examined cells at a late stage of telophase or at anaphase, at which time translation might be enhanced such that cells may enter G_1 (Heesom et al. 2001). Thus, the stage at which cap-dependent translation is inactivated during mitosis needs to be defined.

eIF4GI AND eIF4GII

eIF4GI is a large protein that functions as a scaffold to recruit translation initiation factors to assemble the ribosome complex at the 5′ end of the mRNA (Gingras et al. 1999a). eIF4GI is a modular protein that can be divided into three functional and structural domains, which were initially defined through cleavage by viral proteases (Lamphear et al. 1995). The amino-terminal region provides binding sites for eIF4E (Lamphear et al. 1995; Mader et al. 1995) and PABP (Imataka et al. 1998). The middle domain binds eIF3, eIF4A, and mRNA (Lamphear et al. 1995; Imataka and Sonenberg 1997). The carboxy-terminal domain contains a second binding site for eIF4A (Lamphear et al. 1995; Imataka and Sonenberg 1997) and a site for Mnk1 and Mnk2 (MAP kinase signal integrated kinase 1 and 2), which phosphorylate eIF4E (Waskiewicz et al. 1997, 1999; Pyronnet et al. 1999; Scheper et al. 2001). An eIF4GI-related gene product, termed eIF4GII, is 46% identical to eIF4GI and shares the same modular features (Gradi et al. 1998). It is not immediately clear why there are two mammalian eIF4G isoforms, raising the possibility that they are involved in the translation of different subsets of mRNAs (Gradi et al. 1998). eIF4G is implicated in cell growth as its overexpression induces malignant transformation of NIH-3T3 cells (Fukuchi-Shimogori et al. 1997). The malignant phenotype induced by eIF4G is similar to that caused by eIF4E overexpression (Fukuchi-Shimogori et al. 1997). Interestingly, eIF4G is also overexpressed in squamous lung cell carcinoma (Bauer et al. 2002).

DAP-5 (*d*eath *a*ssociated *p*rotein-5)/p97/NAT1 is an eIF4G-related protein. It is 63% similar and 39% identical to the middle portion of eIF4G (Imataka et al. 1997; Levy-Strumpf et al. 1997; Yamanaka et al. 1997). DAP-5/p97 lacks the amino-terminal domain of eIF4G and therefore does not contain an eIF4E-binding site. Low-level ectopic expression of DAP-5/p97 elicits resistance to cell death, whereas higher expression of DAP-5 is toxic and inhibits cell growth (Levy-Strumpf et al. 1997). DAP-5/p97 has been implicated in apoptosis through cleavage of its carboxyl terminus. The apoptosis-induced cleavage product supports translation via the DAP-5/p97 internal ribosome entry site (IRES) (Henis-Korenblit et al. 2002). In contrast, cleaved eIF4G does not support IRES translation in vivo, suggesting that DAP5/p97 is an important mediator of IRES-dependent translation during apoptosis (Henis-Korenblit et al. 2002). Thus, DAP-5/p97 may be important for efficient translation of mRNAs that are activated during apoptosis (Henis-Korenblit et al. 2002).

The eIF4Gs are phosphoproteins. eIF4GI contains two sets of phosphorylation sites. One set resides in the "hinge" region between the middle and the carboxy-terminal domains, and the second set is found within the amino-terminal domain. Examination of eIF4G using phosphopeptide mapping and tandem mass spectrometry revealed that serines 1148, 1188, and 1232 (numbering is according to the extended eIF4GI isoform; Bradley et al. 2002; Zamora et al. 2002), in the hinge region, become phosphorylated in 293 cells in response to serum stimulation (Raught et al. 2000b). The PI3K-FRAP/mTOR signaling pathway (see below) is implicated in eIF4GI phosphorylation, since its phosphorylation is inhibited by PI3K inhibitors (wortmannin and LY294002) and the inhibitor of FRAP/mTOR (rapamycin) (Raught et al. 2000b). The signaling pathway that leads to amino-terminal phosphorylation has not been characterized. Because amino-terminal deletion mutants of eIF4G are constitutively phosphorylated within the hinge region, it was proposed that the amino-terminal phosphorylation serves to prime the phosphorylation at the hinge region. Phosphorylation of the amino-terminal site is suggested to relieve an intramolecular interaction that exposes the hinge region to yet unidentified kinases (Raught et al. 2000b). One possible kinase involved is the Ca^{++}/calmodulin-dependent protein kinase I, which phosphorylates eIF4GII at Ser-1156 (Qin et al. 2003). Thus, an ordered eIF4G phosphorylation pathway may work in concert with that of 4E-BP1 to modulate efficient cap-dependent translation (Gingras et al. 1999a).

eIF4B

eIF4B stimulates the ATP-dependent helicase activity of eIF4A by increasing both the affinity of eIF4A for mRNA (Abramson et al. 1987, 1988) and the utilization of ATP by eIF4A (Rogers et al. 1999; Bi et al. 2000). Phosphorylation of eIF4B may play a role in regulating its function, as it is enhanced in response to insulin in mouse fibroblasts overexpressing insulin receptors (Manzella et al. 1991), in 3T3-L1 cells in response to several stimuli (serum, insulin, phorbol esters) (Morley and Traugh 1993), and by several kinases in reticulocyte lysate (Morley and Traugh 1989; 1990; Ochs et al. 2002). Phosphorylation of eIF4B by serum stimulation is also inhibited by rapamycin, indicating that FRAP/mTOR is implicated in this pathway (F. Peiretti et al., in prep.). Furthermore, S6K1 phosphorylates eIF4B directly in vivo, since expression of a rapamycin-resistant S6K1 mutant results in eIF4B phosphorylation that is refractory to rapamycin treatment. The S6K1 phosphorylation site on eIF4B is Ser-422. However, the biological significance of S6K1 in eIF4B phosphorylation is not clear (F. Peiretti et al., in prep.). The translation of mRNAs bearing a 5′TOP (terminal oligopyrimidine) element is enhanced, coinciding with S6K1 activation. Since efficient 5′TOP-containing mRNA translation requires PI3K signaling, but is not dependent on rpS6 phosphorylation in response to amino acids (Tang et al. 2001) or growth factors (Stolovich et al. 2002), the phosphorylation of eIF4B may prove to be an important target in S6K-mediated control of cell growth.

eIF4E

In metazoans, eIF4E is phosphorylated on a single serine residue, Ser-209 in mammals and Ser-251 in *Drosophila*. Phosphorylation of eIF4E on Ser-209 is tightly regulated, and its importance in translation and growth is bolstered by the finding that *Drosophila* harboring a Ser251Ala mutation exhibit decreased body size and slow development (Lachance et al. 2002). Phosphorylation of eIF4E correlates with growth status and overall mRNA translation rates in mammalian cells. During mitosis, eIF4E becomes dephosphorylated concomitant with the decrease in cap-dependent translation (Bonneau and Sonenberg 1987). The MAP kinase signaling pathway is implicated in the phosphorylation of eIF4E in response to many treatments, including serum, growth factors, and phorbol esters (Raught and Gingras 1999; Scheper and Proud 2002). The p38 MAP kinase signaling pathway is also implicated in eIF4E phosphorylation, albeit in response to cytokine treatment and several environmental stresses such as exposure to arsenite or anisomycin (Wang et al. 1998). However, other stresses such as heat shock or viral infection lead to the disassembly of the eIF4F complex, recruitment of eIF4E by 4E-BPs, a decrease in mRNA translation, and dephosphorylation of eIF4E (for review, see Scheper and Proud 2002). It appears that eIF4E phosphorylation may be important for only a subset of mRNAs since it does appear to be required for general translation in vitro (Svitkin et al. 1996; Morley and Naegele 2002; Scheper and Proud 2002) and is not required for its binding to the cap or cap analogs (Edery et al. 1988; Carberry et al. 1989).

The MAP kinase interacting kinases, Mnk1 and Mnk2, phosphorylate eIF4E at Ser-209 (Waskiewicz et al. 1997; Pyronnet et al. 1999). Mnk1 was identified as an Erk2-interacting protein in a yeast two-hybrid screen (Waskiewicz et al. 1997). Mnk1 and Mnk2 bind to Erk1, Erk2, and p38 MAPK (only the latter two phosphorylate Mnk1 and Mnk2). The interaction between Erk2 and Mnk1 is also weakened by phosphorylation on Erk2 (Waskiewicz et al. 1997). The basal activity of Mnk2 is much higher than that of Mnk1, and therefore the effect of extracellular signals on eIF4E phosphorylation may depend on the ratio of Mnk1 to Mnk2 expression levels (Scheper and Proud 2002).

SIGNALING PATHWAYS

Extracellular signals activate several signaling pathways that affect mRNA translation rates. 4E-BP1 is a well-characterized phosphorylation target of the phosphatidylinositol-3 kinase (PI3K) signaling pathway, through the serine/threonine kinase Akt/protein kinase B (Akt/PKB) and the FKBP12-rapamycin-associated protein/mammalian target of rapamycin (FRAP/mTOR). PI3K, Akt/PKB, and FRAP/mTOR and their downstream targets (including 4E-BP1, S6K1, S6K2, eIF4B, and eIF4GI) are implicated as critical regulators of cell growth (Gingras et al. 2001a). The following is a short summary of the components of the PI3K–Akt/PKB–FRAP/mTOR signaling pathway.

PI3K

The PI3K family of lipid-phosphorylating enzymes controls the metabolism of 3-phosphoinositides that play critical roles in many biological processes, including cell proliferation, growth, apoptosis, and protein synthesis (Fruman et al. 1998; Katso et al. 2001). PI3K signaling is also implicated in human diseases, including cancers and inherited disorders such as Cowden disease, Lhermite-Duclos disease (LDD), and Bannayan-Zonana syndrome (Cantley and Neel 1999; Backman et al. 2001; Maehama et al. 2001). Extracellular signals activate the regulatory subunit of PI3K, which then recruits its catalytic subunit to the membrane. At the membrane, activated PI3K phosphorylates the hydroxyl group at the D3 position in the inositol ring of phosphatidylinositol (PtdIns)-3,4-biphosphate to generate PtdIns-3,4,5-triphosphate, a prerequisite for the activation of PKB/Akt (Fruman et al. 1998). Inhibition of PI3K by pharmacological means leads to a reduction in 4E-BP1 phosphorylation (Gingras et al. 1999a). The activity of the catalytic subunit of PI3K is inhibited irreversibly by low concentrations of wortmannin (Ui et al. 1995), and reversibly by LY294002 (Vlahos et al. 1994). Consistent with the finding that PI3K is an upstream activator of 4E-BP1 phosphorylation, overexpression of a membrane-targeted PI3K catalytic subunit gives rise to a constitutively phosphorylated 4E-BP1 (Gingras et al. 1998).

PTEN

PTEN (phosphatase and tensin homolog) deleted on chromosome 10, also referred to as MMAC/TEP1 (mutated in multiple advanced cancers/transforming growth factor β enhanced protein 1), is a tumor suppressor gene whose protein product negatively regulates the activation of PKB/Akt by dephosphorylating PI3K-generated PtdIns-3,4,5-triphosphate (Maehama and Dixon 1999; Di Cristofano and Pandolfi 2000). PTEN is mutated in many cancers (only p53 is more frequently mutated), including glioblastomas, breast and prostate cancers, and melanomas (Li et al. 1997; Steck et al. 1997). PTEN inactivation is commonly found in advanced metastatic cancers having poor prognosis (Li et al. 1997; Steck et al. 1997).

PTEN and the downstream effectors of PI3K signaling regulate cell size in flies and mammals (Backman et al. 2002). Loss-of-function mutations in *Drosophila* PTEN (dPTEN) cause a significant increase in cell size and organs, whereas dPTEN overexpression antagonizes these phenotypes

(Gao et al. 2000). Inactivation of mouse PTEN in the cerebellum and den-date gyrus results in brain enlargement that coincides with a twofold increase in neuronal cell size and increased phosphorylation of Akt/PKB (Backman et al. 2001; Kwon et al. 2001).

PTEN conceivably controls growth by regulating protein translation through Akt-mediated activation of FRAP/mTOR (see below), a key reg-ulator of cap-dependent translation (Gingras et al. 2001a). FRAP/mTOR-mediated phosphorylation activates S6K1 and inactivates 4E-BP1 (result-ing in the release of 4E-BP1 from eIF4E) (Gingras et al. 2001a). S6K1 is also phosphorylated directly by PDK1 (*phosphoinositide-dependent kinase* 1) (Pullen et al. 1998; Dufner and Thomas 1999). Activated S6K1 phosphorylates the 40S ribosomal protein, S6. Activation of S6K1 was suggested to play a predominant role in regulating cell size through trans-lational control of TOP-bearing mRNAs, including ribosomal proteins and elongation factors (Dufner and Thomas 1999; Gingras et al. 2001a), although recent data question the role of both S6K1 and mTOR in this process (Tang et al. 2001; Stolovich et al. 2002). Deletion of S6K in *Drosophila* or mice results in decreased body and organ size as a direct result of reduced cell size (Shima et al. 1998; Montagne et al. 1999; Pende et al. 2000). Deletion of S6K1 in mice causes reduced pancreatic cell mass that results from a decrease in β-cell size, leading to hypoinsulinemia and glucose intolerance (Pende et al. 2000). Genetic experiments showing the involvement of S6K1 in the regulation of cell size strongly implicate S6K1 in cell growth (Pende et al. 2000).

Akt/PKB

Akt/PKB was first described as a viral oncogene, v-akt (Staal et al. 1977; Staal and Hartley 1988). The cellular homolog of v-akt is implicated in many human neoplasias (Cheng et al. 1992; Bellacosa et al. 1995). Akt/PKB consists of a family of three Ser/Thr protein kinases (Akt1/PKBα, Akt2/PKBβ, Akt3/PKBγ) that differ in their tissue expres-sion (Chan et al. 1999; Brazil et al. 2002). PI3K-generated D3-phosphory-lated lipid products bind to the pleckstrin homology (PH) domain of Akt/PKB and mediate its phosphorylation by PDK1. Akt/PKB promotes cell survival and cell growth (Datta et al. 1999). Its expression is elevated in many tumor types, and it is thought to promote an aggressive malig-nant phenotype by inhibiting the apoptotic death of malignant cells. The role of Akt/PKB in cell signaling was first described by its activation in growth-factor-treated cells (Burgering and Coffer 1995; Franke et al.

1995). Activation of Akt/PKB is inhibited by wortmannin, and its activity is correlated with the activation of many growth-related downstream targets. Interestingly, dAkt may regulate d4E-BP through regulation of the FOXO transcription factor (Puig et al. 2003). In addition to regulation of d4E-BP phosphorylation by dAkt signaling, dAkt can also sequester dFOXO and prevent d4E-BP transcription, thus providing alternative signaling strategies to regulate cell growth (Puig et al. 2003). Such studies have sparked much interest in understanding the role of Akt/PKB and downstream translational targets in cell growth as described in Chapter 7.

MAMMALIAN TARGET OF RAPAMYCIN

TOR (target of rapamycin), also referred to as FRAP (FKBP12 and rapamycin-associated protein) or RAFT1 (rapamycin and FKBP12 target), is an important downstream effector molecule in PI3K signaling. Two TOR proteins (TOR1 and TOR2) were first discovered in budding yeast during a screen for mutant strains that are resistant to rapamycin, a microbial macrolide (Vezina et al. 1975). The sequence of mammalian TOR (mTOR) is very similar (~45% identical) to the yeast homolog, suggesting that their functions and signaling pathways are conserved. Rapamycin is also a potent immunosuppressant that causes G_1 arrest in T cells. It acts by forming a "gain of function" complex with the 12-kD immunophilin FKBP12 (FK506-binding protein) to inhibit TOR signaling (Abraham and Wiederrecht 1996; Hall 1996). The growth-inhibitory effect of rapamycin prompted the development of related analogs such as CCI-779 (Wyeth-Ayerst) and RAD001 (Novartis) that show promise in the treatment of various types of cancers (Huang and Houghton 2002; see below). TOR is a serine/threonine protein kinase that belongs to the PIKK (phosphoinositide 3-kinase related kinase) family of checkpoint kinases, which include ATM (ataxia telangiectasia mutated protein), ATR/FRP (ATM and Rad3-related protein/FRAP-related protein), and DNA-PK (DNA-dependent protein kinase). The consensus phosphorylation site sequence for PIKKs is serine or threonine followed by glutamine at the +1 position. The consensus sequence for the 4E-BPs or S6K TOR targets is an exception to this rule, since the +1 position usually contains proline or a hydrophobic residue. This is explained by the finding that FRAP/mTOR substrates are presented through an adapter protein (raptor; see below).

   Two groups independently identified the mammalian raptor protein (*regulatory associated protein of mTOR*), a 150-kD FRAP/mTOR-interacting protein that acts as a nutrient sensor to modulate the phosphorylation of 4E-BP1 and S6K1, and cell size (Hara et al. 2002; Kim et al. 2002). Raptor is the yeast homolog of Kog1 (*kontroller* of *growth* 1), and along with Lst-8, which is required for trafficking of amino acid permeases to the plasma membrane in yeast (Roberg et al. 1997), is a component of TOR complex 1 (TORC1) (Loewith et al. 2002). A second TOR complex, TORC2, also associates with Lst8, but not with raptor (Loewith et al. 2002). TORC1 mediates rapamycin-sensitive signaling whereas TORC2 mediates rapamycin-insensitive signaling to the cytoskeleton (Loewith et al. 2002).

Raptor interacts with both 4E-BP1 and S6K1 (Hara et al. 2002). The stimulatory effect of raptor on cell size in HEK293T cells parallels that of FRAP/mTOR (Kim et al. 2002). A reduction in the expression of either raptor (Ce-Raptor) or Ce-TOR via RNA interference (RNAi) in *Caenorhabditis elegans* results in similar phenotypes, characterized by delayed gonadal development, increased gonadal degeneration, and an enlarged intestinal lumen that is associated with compromised nutrient absorption (Hara et al. 2002; Long et al. 2002). These RNAi-induced phenotypes are similar to those caused by RNAi against initiation factors Ce-eIF4G, Ce-eIF2α, and Ce-eIF2β subunits, and are therefore thought to be caused by perturbations in overall mRNA translation in *C. elegans* (Long et al. 2002).

The phosphorylation and regulation of S6K1 and 4E-BP1 by FRAP/mTOR is dependent on a TOR signaling (TOS) motif (Schalm and Blenis 2002). The TOS motif is a five-residue sequence present at the amino terminus of the two S6 kinases (FDIDL) and at the carboxyl terminus of the 4E-BPs (FEMDI) (Schalm and Blenis 2002). The TOS motif found in 4E-BP1 is conserved in all three human 4E-BPs and in *Drosophila* 4E-BP (Schalm and Blenis 2002). Phosphorylation of S6K1 at Thr-389 by amino acid stimulation is inhibited by rapamycin, and a point mutation (F5A) in the TOS motif of S6K prevents its phosphorylation (Schalm and Blenis 2002). Mutation of the 4E-BP1 TOS motif (F114A) inhibits insulin-stimulated phosphorylation of 4E-BP1 (Schalm and Blenis 2002). The TOS motif of 4E-BP1 and S6K1 is also required to bind raptor, and point mutations in the TOS motif prevent their interaction with raptor (Nojima et al. 2003).

REGULATION OF CELL SIZE BY eIF4E AND 4E-BPs

eIF4E and 4E-BPs regulate cell size in mammalian cells and in *Drosophila*. Targeted expression of d4E-BP to the *Drosophila* wing-imaginal disc causes a significant reduction in wing size (Miron et al. 2001). Cell numbers are reduced only upon overexpression of a d4E-BP mutant that binds very

tightly to eIF4E, suggesting that the reduced proliferation is a consequence rather than a direct cause of reduced cell size (Miron et al. 2001).
d4E-BP also antagonizes the increase in cell size which is induced by dAkt
or the dPI3K subunit, Dp110 (Verdu et al. 1999; Miron et al. 2001).
Furthermore, when d4E-BP is co-expressed with a Dp110 mutant or in
flies that are deficient for the p60 PI3K subunit, there is a further reduction in cell size (Miron et al. 2001). Therefore, d4E-BP is a regulator of cell
growth, which is mediated by the dPI3K signaling pathway. The effects of
d4E-BP are also independent of dS6K, indicating that signaling downstream of dPI3K bifurcates at some point to differentially regulate d4E-BP
and dS6K1 activity (Miron et al. 2001; Radimerski et al. 2002a,b).

4E-BP1 and S6K1 cooperate via FRAP/mTOR signaling to regulate
cell size (Fingar et al. 2002). Overexpression of eIF4E or S6K1 causes an
increase in cell size, and their co-expression has an additive effect. The
increase is dependent on PI3K-mTOR signaling through 4E-BP1 and
S6K1 and is also modulated by eIF4E. Overexpression of eIF4E counteracts the decrease in cell size caused by rapamycin. Most likely, the effect on
cell size is manifested through regulation of cap-dependent translation,
since a dominant mutant of 4E-BP1 (T37A/T46A double mutant) prevents an increase in cell size when cells are grown in the presence of serum
(Fingar et al. 2002).

TRANSLATION INITIATION AND CANCER

The importance of translational control in regulating cell growth and proliferation is underscored by the finding that many of the initiation factors
are overexpressed in transformed cells and tumors (Hershey and Miyamoto
2000). Moreover, several translation factors exhibit oncogenic properties, as
they promote transformation in rodent cells. The oncogenic activity of
some initiation factors is also consistent with their function as downstream
targets of signaling pathways that are intimately connected to cancer.

eIF4E is perhaps the best characterized of all the initiation factors
with respect to cellular transformation (Raught and Gingras 1999; Raught
et al. 2000a). eIF4E was first ascribed oncogenic properties based on the
finding that overexpression of murine eIF4E in NIH-3T3 cells causes
malignant transformation (Lazaris-Karatzas et al. 1990). eIF4E also transforms mouse embryo fibroblasts (MEFs) in concert with an immortaliz-
ing gene such as *E1A* or *myc* (Lazaris-Karatzas and Sonenberg 1992).
Overexpression of eIF4E results in anchorage-independent cell growth
and the formation of foci in soft agar, and also promotes tumor formation
in nude mice (Lazaris-Karatzas et al. 1990). Neutralizing antibodies to Ras

as well as a dominant negative Ras mutant inhibit the oncogenicity of eIF4E, indicating that the mitogenic and oncogenic effects of eIF4E are mediated at least partly through Ras (Lazaris-Karatzas et al. 1992). These findings prompted studies to determine the expression status of eIF4E in tumors. Indeed, it was found that eIF4E expression is elevated in many types of cancers, including breast (Kerekatte et al. 1995; Anthony et al. 1996), head and neck squamous carcinomas (Haydon et al. 2000), and non-Hodgkin's lymphomas (Wang et al. 1999). The amount of eIF4E is limiting in cells, and therefore, its overexpression is thought to enhance translation via an increase in eIF4F complex formation (Koromilas et al. 1992a). Consistent with this hypothesis, mRNAs with highly structured 5′UTRs are more efficiently translated in eIF4E-overexpressing cells (Koromilas et al. 1992a). Complex 5′UTRs are present in mRNAs whose translation is tightly regulated, such as ornithine decarboxylase (ODC) (Grens and Scheffler 1990; Manzella and Blackshear 1990; Manzella et al. 1991), fibroblast growth factor 2 (FGF2) (Kevil et al. 1995), and vascular endothelial growth factor (VEGF) (Cohen et al. 1996; Kevil et al. 1996), all of which enhance growth, proliferation, and angiogenesis (Kozak 1987; van der Velden and Thomas 1999).

Overexpression of eIF4E in NIH-3T3 cells promotes cell survival under serum-free conditions (Polunovsky et al. 1996). eIF4E also prevents cytostatic drug-induced apoptosis in c-myc-expressing rat embryo fibroblasts (REFs) through a mechanism that requires cyclin D1 (Tan et al. 2000). Treatment of cells with antisense cyclin D1 oligomers or transfection with dominant negative cyclin D1 counteracts the anti-apoptotic effect of eIF4E in myc-transfected REFs (Tan et al. 2000). In oncogenic Ras-bearing fibroblasts, the apoptotic effect of cytostatic drugs is enhanced by exposure to rapamycin (Polunovsky et al. 2000). Consistent with this result, overexpression of 4E-BP1 enhances the sensitivity of ras-transfected fibroblasts to cytostatic drug-induced apoptosis (Polunovsky et al. 2000). These results highlight the importance of cap-dependent translation in cell survival.

Inhibition of eIF4E would be expected to negate its transformation activity and perhaps transformation by other oncogenes. The transformation of NIH-3T3 cells by eIF4E, Ha-Ras, or v-Src can be partially reversed through overexpression of 4E-BP1 or 4E-BP2 (Rousseau et al. 1996). In this sense, the 4E-BPs function as tumor suppressors, although mutations in 4E-BP have not been found in any type of cancer. Overexpression of 4E-BP1 also inhibits the proliferation of MCF-7 human breast cancer cells (Jiang et al. 2003). Constitutive expression of 4E-BP1 in MCF-7 cells results in decreased cyclin D1 expression (Jiang et al. 2003). Cyclin D1 is implicated as a key regulator of cell proliferation in eIF4E-overexpressing cells (Long et al. 2001). When a non-phosphorylatable mutant of 4E-BP1 (five phosphorylation sites were mutated to alanine) was expressed, p27kip mRNA translation increased via a cap-independent mechanism (Jiang et al. 2003).

TRANSLATION INITIATION AND GROWTH IN YEAST

The budding yeast *Saccharomyces cerevisiae* reaches a critical size before cell division (Rupes 2002). Cell division proceeds once a restriction point is reached in the G_1 phase of the cell cycle, referred to as "Start," which is defined as the point where protein synthesis rates are sufficient to proceed to DNA replication (S-phase) (see, e.g., Popolo et al. 1982). The mechanism of Start is linked to Sfp1, a transcription factor that regulates the expression of at least 60 genes implicated in ribosome assembly (Jorgensen et al. 2002). *Sfp1Δ* mutants exhibit decreased cell size, and restoration of *Sfp1* expression correlates with enhanced expression of ribosomal proteins and translation factors (Jorgensen et al. 2002). However, the small size of *Sfp1Δ* mutant cells does not accompany alterations in cell division (Jorgensen et al. 2002). These results indicate that ribosome biogenesis and the translation machinery in budding yeast are implicated more specifically in regulating cell size rather than the cell cycle (Jorgensen et al. 2002).

Translation initiation does play a role in regulating the cell cycle in budding yeast. Cdc33 (the yeast homolog of eIF4E) is required for viability (Brenner et al. 1988) and regulates the cell cycle (Anthony et al. 2001) by playing a critical role in Start to regulate the translation of CLN3, a G_1 cyclin (Brenner et al. 1988). Overexpression of Cdc33 stimulates the recruitment of CLN3 mRNA to polysomes and is thought to be a direct cause of the increase in the amount of Cln1 and Cln2 that form complexes with Cdc28 to regulate budding, duplication of the spindle pole, and DNA replication.

There are no structural homologs of the 4E-BPs in yeast. However, cap-dependent translation in *S. cerevisiae* is inhibited by p20 (also referred to as Caf20p), a small protein that contains the canonical eIF4E-binding motif, YXXXXLΦ (Altmann et al. 1989, 1997; Lanker et al. 1992). Unlike mammalian and *Drosophila* cells in which the 4E-BP1 interaction with eIF4E is regulated by 4E-BP phosphorylation, the yeast p20–eIF4E interaction is not regulated by phosphorylation, although p20 is a phosphoprotein (Altmann et al. 1997). Deletion of p20 reduces the generation time of exponentially growing cells, and p20-transformed yeast strains have a reduced growth rate, implicating p20 in yeast growth (Altmann et al. 1997). The PAS kinase enzymes, PSK1 and PSK2, are serine/threonine kinases that utilize their PAS domains as metabolic sensors (Taylor and Zhulin 1999; Rutter et al. 2002). PSK2 directly phosphorylates p20 in its carboxy-terminal region at serines 58 and 59 (Rutter et al. 2002). Thus, PAS kinases are potential regulators of translational initiation in yeast that link nutrient availability to translation rates. The mammalian ortholog, PASKIN, is expressed highly in testes and is thought to play a role in spermatogenesis (Hofer et al. 2001). Unlike PSK1 and PSK2 in yeast, a role for PASKIN in metabolic sensing has not been demonstrated.

CONCLUSIONS

It is abundantly clear that the regulation of translation initiation is intimately linked to cell growth. The FRAP/mTOR signaling pathway mediates the phosphorylation of many translation targets such as 4E-BP1, S6K, eIFGI, and eIF4B. Some of these targets play critical roles in growth processes by controlling the assembly of the eIF4F initiation complex. A critical outcome is the enhanced translation of certain mRNAs that affect growth and proliferation. A pertinent question to consider in understanding this process is how FRAP/mTOR discriminates between its various substrates. The identification of the FRAP/mTOR-interacting proteins raptor and Lst8 is likely to be key to delineating how FRAP/mTOR functions as a nutrient sensor. Another important finding is that FRAP/mTOR activity is specifically inhibited by rapamycin. Since it is thought that FRAP/mTOR inhibitors act primarily by inhibiting the translational apparatus, the initiation pathway is an attractive target for the development of novel inhibitors to inhibit cancer growth. Many cancers are caused by mutations in genes that encode tumor suppressors such as PTEN. In such cases, the net increase in the output of PI3K signaling has a general effect on the phosphorylation of FRAP/mTOR targets. The importance of the control of translation initiation in cell growth and disease will become more compelling with further understanding of the mechanisms of initiation factor phosphorylation and their effect on mRNA translation.

ACKNOWLEDGMENTS

We thank Mathieu Miron and Andrea Brueschke for suggestions and for proofreading the manuscript. This work was supported by grants from the National Cancer Institute and Canadian Institutes of Health Research of Canada. N.S. is a James McGill professor and Howard Hughes International Scholar.

REFERENCES

Abraham R.T. and Wiederrecht G.J. 1996. Immunopharmacology of rapamycin. *Annu. Rev. Immunol.* **14:** 483–510.

Abramson R.D., Dever T.E., and Merrick W.C. 1988. Biochemical evidence supporting a mechanism for cap-independent and internal initiation of eukaryotic mRNA. *J. Biol. Chem.* **263:** 6016–6019.

Abramson R.D., Dever T.E., Lawson T.G., Ray B.K., Thach R.E., and Merrick W.C. 1987. The ATP-dependent interaction of eukaryotic initiation factors with mRNA. *J. Biol. Chem.* **262:** 3826–3832.

Altmann M., Krieger M., and Trachsel H. 1989. Nucleotide sequence of the gene encoding a 20 kDa protein associated with the cap binding protein eIF-4E from *Saccharomyces cerevisiae*. *Nucleic Acids Res.* **17:** 7520.

Altmann M., Schmitz N., Berset C., and Trachsel H. 1997. A novel inhibitor of cap-dependent translation initiation in yeast: p20 competes with eIF4G for binding to eIF4E. *EMBO J.* **16:** 1114–1121.

Anthony B., Carter P., and De Benedetti A. 1996. Overexpression of the proto-oncogene/translation factor 4E in breast-carcinoma cell lines. *Int. J. Cancer* **65:** 858–863.

Anthony C., Zong Q., and De Benedetti A. 2001. Overexpression of eIF4E in *Saccharomyces cerevisiae* causes slow growth and decreased alpha-factor response through alterations in CLN3 expression. *J. Biol. Chem.* **276:** 39645–39652.

Backman S., Stambolic V., and Mak T. 2002. PTEN function in mammalian cell size regulation. *Curr. Opin. Neurobiol.* **12:** 516–522.

Backman S.A., Stambolic V., Suzuki A., Haight J., Elia A., Pretorius J., Tsao M.S., Shannon P., Bolon B., Ivy G.O., and Mak T.W. 2001. Deletion of Pten in mouse brain causes seizures, ataxia and defects in soma size resembling Lhermitte-Duclos disease. *Nat. Genet.* **29:** 396–403.

Bandyopadhyay S.K., de La Motte C.A., and Williams B.R. 2000. Induction of E-selectin expression by double-stranded RNA and TNF-alpha is attenuated in murine aortic endothelial cells derived from double-stranded RNA-activated kinase (PKR)-null mice. *J. Immunol.* **164:** 2077–2083.

Bauer C., Brass N., Diesinger I., Kayser K., Grasser F.A., and Meese E. 2002. Overexpression of the eukaryotic translation initiation factor 4G (eIF4G-1) in squamous cell lung carcinoma. *Int. J. Cancer* **98:** 181–185.

Bellacosa A., de Feo D., Godwin A.K., Bell D.W., Cheng J.Q., Altomare D.A., Wan M., Dubeau L., Scambia G., and Masciullo V., et al. 1995. Molecular alterations of the AKT2 oncogene in ovarian and breast carcinomas. *Int. J. Cancer* **64:** 280–285.

Bi X., Ren J., and Goss D.J. 2000. Wheat germ translation initiation factor eIF4B affects eIF4A and eIFiso4F helicase activity by increasing the ATP binding affinity of eIF4A. *Biochemistry* **39:** 5758–5765.

Bonneau A.M. and Sonenberg N. 1987. Involvement of the 24-kDa cap-binding protein in regulation of protein synthesis in mitosis. *J. Biol. Chem.* **262:** 11134–11139.

Bradley C.A., Padovan J.C., Thompson T.L., Benoit C.A., Chait B.T., and Rhoads R.E. 2002. Mass spectrometric analysis of the N terminus of translational initiation factor eIF4G-1 reveals novel isoforms. *J. Biol. Chem.* **277:** 12559–12571.

Brazil D.P., Park J., and Hemmings B.A. 2002. PKB binding proteins. Getting in on the Akt. *Cell* **111:** 293–303.

Brenner C., Nakayama N., Goebl M., Tanaka K., Toh-e A., and Matsumoto K. 1988. CDC33 encodes mRNA cap-binding protein eIF-4E of *Saccharomyces cerevisiae*. *Mol. Cell. Biol.* **8:** 3556–3559.

Burgering B.M. and Coffer P.J. 1995. Protein kinase B (c-Akt) in phosphatidylinositol-3-OH kinase signal transduction. *Nature* **376:** 599–602.

Cantley L.C. and Neel B.G. 1999. New insights into tumor suppression: PTEN suppresses tumor formation by restraining the phosphoinositide 3-kinase/AKT pathway. *Proc. Natl. Acad. Sci.* **96:** 4240–4245.

Carberry S.E., Rhoads R.E., and Goss D.J. 1989. A spectroscopic study of the binding of m7GTP and m7GpppG to human protein synthesis initiation factor 4E. *Biochemistry* **28:** 8078–8083.

Chan T.O., Rittenhouse S.E., and Tsichlis P.N. 1999. AKT/PKB and other D3 phosphoinositide-regulated kinases: Kinase activation by phosphoinositide-dependent phosphorylation. *Annu. Rev. Biochem.* **68:** 965–1014.

Chen J.-J. 2000. Heme-regulated eIF2a kinase. In *Translational control of gene expression* (ed. N. Sonenberg et al.), pp. 529-546. Cold Spring Harbor Laboratory Press, Cold Spring Harbor, New York.

Cheng J.Q., Godwin A.K., Bellacosa A., Taguchi T., Franke T.F., Hamilton T.C., Tsichlis P.N., and Testa J.R. 1992. AKT2, a putative oncogene encoding a member of a subfamily of protein-serine/threonine kinases, is amplified in human ovarian carcinomas. *Proc. Natl. Acad. Sci.* **89:** 9267–9271.

Cohen T., Nahari D., Cerem L.W., Neufeld G., and Levi B.Z. 1996. Interleukin 6 induces the expression of vascular endothelial growth factor. *J. Biol. Chem.* **271:** 736–741.

Datta S.R., Brunet A., and Greenberg M.E. 1999. Cellular survival: A play in three Akts. *Genes Dev.* **13:** 2905–2927.

Delepine M., Nicolino M., Barrett T., Golamaully M., Lathrop G.M., and Julier C. 2000. EIF2AK3, encoding translation initiation factor 2-alpha kinase 3, is mutated in patients with Wolcott-Rallison syndrome. *Nat. Genet.* **25:** 406–409.

Deng J., Harding H.P., Raught B., Gingras A.C., Berlanga J.J., Scheuner D., Kaufman R.J., Ron D., and Sonenberg N. 2002. Activation of GCN2 in UV-irradiated cells inhibits translation. *Curr. Biol.* **12:** 1279–1286.

Di Cristofano A. and Pandolfi P.P. 2000. The multiple roles of PTEN in tumor suppression. *Cell* **100:** 387–390.

Donze O., Jagus R., Koromilas A.E., Hershey J.W., and Sonenberg N. 1995. Abrogation of translation initiation factor eIF-2 phosphorylation causes malignant transformation of NIH 3T3 cells. *EMBO J.* **14:** 3828–3834.

Dufner A. and Thomas G. 1999. Ribosomal S6 kinase signaling and the control of translation. *Exp. Cell Res.* **253:** 100–109.

Edery I., Altmann M., and Sonenberg N. 1988. High-level synthesis in *Escherichia coli* of functional cap-binding eukaryotic initiation factor eIF-4E and affinity purification using a simplified cap-analog resin. *Gene* **74:** 517–525.

Fingar D.C., Salama S., Tsou C., Harlow E., and Blenis J. 2002. Mammalian cell size is controlled by mTOR and its downstream targets S6K1 and 4EBP1/eIF4E. *Genes Dev.* **16:** 1472–1487.

Fletcher C.M., McGuire A.M., Gingras A.C., Li H., Matsuo H., Sonenberg N., and Wagner G. 1998. 4E binding proteins inhibit the translation factor eIF4E without folded structure. *Biochemistry* **37:** 9–15.

Franke T.F., Yang S.I., Chan T.O., Datta K., Kazlauskas A., Morrison D.K., Kaplan D.R., and Tsichlis P.N. 1995. The protein kinase encoded by the Akt proto-oncogene is a target of the PDGF-activated phosphatidylinositol 3-kinase. *Cell* **81:** 727–736.

Fruman D.A., Meyers R.E., and Cantley L.C. 1998. Phosphoinositide kinases. *Annu. Rev. Biochem.* **67:** 481–507.

Fukuchi-Shimogori T., Ishii I., Kashiwagi K., Mashiba H., Ekimoto H., and Igarashi K. 1997. Malignant transformation by overproduction of translation initiation factor eIF4G. *Cancer Res.* **57:** 5041–5044.

Gao X., Neufeld T.P., and Pan D. 2000. *Drosophila* PTEN regulates cell growth and proliferation through PI3K-dependent and -independent pathways. *Dev. Biol.* **221:** 404–418.

Gingras A.C., Raught B., and Sonenberg N. 1999a. eIF4 initiation factors: Effectors of mRNA recruitment to ribosomes and regulators of translation. *Annu. Rev. Biochem.* **68:** 913–963.

———. 2001a. Regulation of translation initiation by FRAP/mTOR. *Genes Dev.* **15:** 807–826.

Gingras A.C., Kennedy S.G., O'Leary M.A., Sonenberg N., and Hay N. 1998. 4E-BP1, a repressor of mRNA translation, is phosphorylated and inactivated by the Akt(PKB) signaling pathway. *Genes Dev.* **12:** 502–513.

Gingras A.C., Gygi S.P., Raught B., Polakiewicz R.D., Abraham R.T., Hoekstra M.F., Aebersold R., and Sonenberg N. 1999b. Regulation of 4E-BP1 phosphorylation: A novel two-step mechanism. *Genes Dev.* **13:** 1422–1437.

Gingras A.C., Raught B., Gygi S.P., Niedzwiecka A., Miron M., Burley S.K., Polakiewicz R.D., Wyslouch-Cieszynska A., Aebersold R., and Sonenberg N. 2001b. Hierarchical phosphorylation of the translation inhibitor 4E-BP1. *Genes Dev.* **15:** 2852–2864.

Gradi A., Imataka H., Svitkin Y.V., Rom E., Raught B., Morino S., and Sonenberg N. 1998. A novel functional human eukaryotic translation initiation factor 4G. *Mol. Cell. Biol.* **18:** 334–342.

Grens A. and Scheffler I.E. 1990. The 5′- and 3′-untranslated regions of ornithine decarboxylase mRNA affect the translational efficiency. *J. Biol. Chem.* **265:** 11810–11816.

Gunnery S. and Mathews M.B. 1995. Functional mRNA can be generated by RNA polymerase III. *Mol. Cell. Biol.* **15:** 3597–3607.

Hall M.N. 1996. The TOR signalling pathway and growth control in yeast. *Biochem. Soc. Trans.* **24:** 234–239.

Hara K., Maruki Y., Long X., Yoshino K., Oshiro N., Hidayat S., Tokunaga C., Avruch J., and Yonezawa K. 2002. Raptor, a binding partner of target of rapamycin (TOR), mediates TOR action. *Cell* **110:** 177–189.

Harding H.P., Zeng H., Zhang Y., Jungries R., Chung P., Plesken H., Sabatini D.D., and Ron D. 2001. Diabetes mellitus and exocrine pancreatic dysfunction in perk-/- mice reveals a role for translational control in secretory cell survival. *Mol. Cell* **7:** 1153–1163.

Haydon M.S., Googe J.D., Sorrells D.S., Ghali G.E., and Li B.D. 2000. Progression of eIF4e gene amplification and overexpression in benign and malignant tumors of the head and neck. *Cancer* **88:** 2803–2810.

Heesom K.J., Gampel A., Mellor H., and Denton R.M. 2001. Cell cycle-dependent phosphorylation of the translational repressor eIF- 4E binding protein-1 (4E-BP1). *Curr. Biol.* **11:** 1374–1379.

Henis-Korenblit S., Shani G., Sines T., Marash L., Shohat G., and Kimchi A. 2002. The caspase-cleaved DAP5 protein supports internal ribosome entry site-mediated translation of death proteins. *Proc. Natl. Acad. Sci.* **99:** 5400–5405.

Herbert T.P., Fahraeus R., Prescott A., Lane D.P., and Proud C.G. 2000. Rapid induction of apoptosis mediated by peptides that bind initiation factor eIF4E. *Curr. Biol.* **10:** 793–796.

Hershey J.W.B. and Merrick W.C. 2000. Pathway and mechanism of initiation of protein synthesis. In *Translational control of gene expression* (ed. N. Sonenberg et al.), pp. 33-88. Cold Spring Harbor Laboratory Press, Cold Spring Harbor, New York.

Hershey J.W.B. and Miyamoto S. 2000. Translational control and cancer. In *Translational control of gene expression* (ed. N. Sonenberg et al.), pp. 637-654. Cold Spring Harbor Laboratory Press, Cold Spring Harbor, New York.

Hinnebusch A.G. 2000. Mechanism and regulation of initiator methionyl-tRNA binding to ribosomes. In *Translational control of gene expression* (ed. N. Sonenberg et al.), pp. 185-243. Cold Spring Harbor Laboratory Press, Cold Spring Harbor, New York.

Hofer T., Spielmann P., Stengel P., Stier B., Katschinski D.M., Desbbaillets I., Gassmann M., and Wenger R.H. 2001. Mammalian PASKIN, a PAS-serine/threonine kinase related to bacterial oxygen sensors. *Biochem. Biophys. Res. Commun.* **288:** 757–764.

Huang S. and Houghton P.J. 2002. Inhibitors of mammalian target of rapamycin as novel antitumor agents: From bench to clinic. *Curr. Opin. Investig. Drugs* **3:** 295–304.

Imataka H. and Sonenberg N. 1997. Human eukaryotic translation initiation factor 4G (eIF4G) possesses two separate and independent binding sites for eIF4A. *Mol. Cell. Biol.* **17:** 6940–6947.

Imataka H., Gradi A., and Sonenberg N. 1998. A newly identified N-terminal amino acid sequence of human eIF4G binds poly(A)-binding protein and functions in poly(A)-dependent translation. *EMBO J.* **17:** 7480–7489.

Imataka H., Olsen H.S., and Sonenberg N. 1997. A new translational regulator with homology to eukaryotic translation initiation factor 4G. *EMBO J.* **16:** 817–825.

Izaurralde E., Lewis J., Gamberi C., Jarmolowski A., McGuigan C., and Mattaj I.W. 1995. A cap-binding protein complex mediating U snRNA export. *Nature* **376:** 709–712.

Jiang H., Coleman J., Miskimins R., and Miskimins W.K. 2003. Expression of constitutively active 4EBP-1 enhances p27Kip1 expression and inhibits proliferation of MCF7 breast cancer cells. *Cancer Cell Int.* **3:** 2.

Jorgensen P., Nishikawa J.L., Breitkreutz B.J., and Tyers M. 2002. Systematic identification of pathways that couple cell growth and division in yeast. *Science* **297:** 395–400.

Katso R., Okkenhaug K., Ahmadi K., White S., Timms J., and Waterfield M.D. 2001. Cellular function of phosphoinositide 3-kinases: Implications for development, homeostasis, and cancer. *Annu. Rev. Cell Dev. Biol.* **17:** 615–675.

Kaufman R.J. 2000. Double stranded RNA-activated protein kinase PKR. In *Translational control of gene expression* (ed. N. Sonenberg et al.), pp. 503–528. Cold Spring Harbor Laboratory Press, Cold Spring Harbor, New York.

Kerekatte V., Smiley K., Hu B., Smith A., Gelder F., and De Benedetti A. 1995. The proto-oncogene/translation factor eIF4E: A survey of its expression in breast carcinomas. *Int. J. Cancer* **64:** 27–31.

Kevil C., Carter P., Hu B., and DeBenedetti A. 1995. Translational enhancement of FGF-2 by eIF-4 factors, and alternate utilization of CUG and AUG codons for translation initiation. *Oncogene* **11:** 2339–2348.

Kevil C.G., De Benedetti A., Payne D.K., Coe L.L., Laroux F.S., and Alexander J.S. 1996. Translational regulation of vascular permeability factor by eukaryotic initiation factor 4E: Implications for tumor angiogenesis. *Int. J. Cancer* **65:** 785–790.

Kim D.H., Sarbassov D.D., Ali S.M., King J.E., Latek R.R., Erdjument-Bromage H., Tempst

P., and Sabatini D.M. 2002. mTOR interacts with raptor to form a nutrient-sensitive complex that signals to the cell growth machinery. *Cell* **110:** 163–175.

Kirchhoff S., Koromilas A.E., Schaper F., Grashoff M., Sonenberg N., and Hauser H. 1995. IRF-1 induced cell growth inhibition and interferon induction requires the activity of the protein kinase PKR. *Oncogene* **11:** 439–445.

Koromilas A.E., Lazaris-Karatzas A., and Sonenberg N. 1992b. mRNAs containing extensive secondary structure in their 5′ non-coding region translate efficiently in cells overexpressing initiation factor eIF-4E. *EMBO J.* **11:** 4153–4158.

Koromilas A.E., Roy S., Barber G.N., Katze M.G., and Sonenberg N. 1992b. Malignant transformation by a mutant of the IFN-inducible dsRNA-dependent protein kinase. *Science* **257:** 1685–1689.

Kozak M. 1987. An analysis of 5′-noncoding sequences from 699 vertebrate messenger RNAs. *Nucleic Acids Res.* **15:** 8125–8148.

Kumar A., Haque J., Lacoste J., Hiscott J., and Williams B.R. 1994. Double-stranded RNA-dependent protein kinase activates transcription factor NF-kappa B by phosphorylating I kappa B. *Proc. Natl. Acad. Sci.* **91:** 6288–6292.

Kwon C.H., Zhu X., Zhang J., Knoop L.L., Tharp R., Smeyne R.J., Eberhart C.G., Burger P.C., and Baker S.J. 2001. Pten regulates neuronal soma size: A mouse model of Lhermitte-Duclos disease. *Nat. Genet.* **29:** 404–411.

Lachance P.E., Miron M., Raught B., Sonenberg N., and Lasko P. 2002. Phosphorylation of eukaryotic translation initiation factor 4E is critical for growth. *Mol. Cell. Biol.* **22:** 1656–1663.

Lamphear B.J., Kirchweger R., Skern T., and Rhoads R.E. 1995. Mapping of functional domains in eukaryotic protein synthesis initiation factor 4G (eIF4G) with picornaviral proteases. Implications for cap-dependent and cap-independent translational initiation. *J. Biol. Chem.* **270:** 21975–21983.

Lanker S., Muller P.P., Altmann M., Goyer C., Sonenberg N., and Trachsel H. 1992. Interactions of the eIF-4F subunits in the yeast *Saccharomyces cerevisiae*. *J. Biol. Chem.* **267:** 21167–21171.

Lazaris-Karatzas A. and Sonenberg N. 1992. The mRNA 5′ cap-binding protein, eIF-4E, cooperates with v-myc or E1A in the transformation of primary rodent fibroblasts. *Mol. Cell. Biol.* **12:** 1234–1238.

Lazaris-Karatzas A., Montine K.S., and Sonenberg N. 1990. Malignant transformation by a eukaryotic initiation factor subunit that binds to mRNA 5′ cap. *Nature* **345:** 544–547.

Lazaris-Karatzas A., Smith M.R., Frederickson R.M., Jaramillo M.L., Liu Y.L., Kung H.F., and Sonenberg N. 1992. Ras mediates translation initiation factor 4E-induced malignant transformation. *Genes Dev.* **6:** 1631–1642.

Levy-Strumpf N., Deiss L.P., Berissi H., and Kimchi A. 1997. DAP-5, a novel homolog of eukaryotic translation initiation factor 4G isolated as a putative modulator of gamma interferon-induced programmed cell death. *Mol. Cell. Biol.* **17:** 1615–1625.

Li J., Yen C., Liaw D., Podsypanina K., Bose S., Wang S.I., Puc J., Miliaresis C., Rodgers L., McCombie R., Bigner S.H., Giovanella B.C., Ittmann M., Tycko B., Hibshoosh H., Wigler M.H., and Parsons R. 1997. PTEN, a putative protein tyrosine phosphatase gene mutated in human brain, breast, and prostate cancer. *Science* **275:** 1943–1947.

Loewith R., Jacinto E., Wullschleger S., Lorberg A., Crespo J.L., Bonenfant D., Oppliger W., Jenoe P., and Hall M.N. 2002. Two TOR complexes, only one of which is rapamycin sensitive, have distinct roles in cell growth control. *Mol. Cell* **10:** 457–468.

Long E., Lazaris-Karatzas A., Karatzas C., and Zhao X. 2001. Overexpressing eukaryotic translation initiation factor 4E stimulates bovine mammary epithelial cell proliferation. *Int. J. Biochem. Cell Biol.* **33:** 133–141.

Long X., Spycher C., Han Z.S., Rose A.M., Muller F., and Avruch J. 2002. TOR deficiency in *C. elegans* causes developmental arrest and intestinal atrophy by inhibition of mRNA translation. *Curr. Biol.* **12:** 1448–1461.

Mader S., Lee H., Pause A., and Sonenberg N. 1995. The translation initiation factor eIF-4E binds to a common motif shared by the translation factor eIF-4 gamma and the translational repressors 4E-binding proteins. *Mol. Cell. Biol.* **15:** 4990–4997.

Maehama T. and Dixon J.E. 1999. PTEN: A tumour suppressor that functions as a phospholipid phosphatase. *Trends Cell Biol.* **9:** 125–128.

Maehama T., Taylor G.S., and Dixon J.E. 2001. PTEN and myotubularin: Novel phosphoinositide phosphatases. *Annu. Rev. Biochem.* **70:** 247–279.

Manzella J.M. and Blackshear P.J. 1990. Regulation of rat ornithine decarboxylase mRNA translation by its 5′-untranslated region. *J. Biol. Chem.* **265:** 11817–11822.

Manzella J.M., Rychlik W., Rhoads R.E., Hershey J.W., and Blackshear P.J. 1991. Insulin induction of ornithine decarboxylase. Importance of mRNA secondary structure and phosphorylation of eucaryotic initiation factors eIF-4B and eIF-4E. *J. Biol. Chem.* **266:** 2383–2389.

Marcotrigiano J., Gingras A.C., Sonenberg N., and Burley S.K. 1999. Cap-dependent translation initiation in eukaryotes is regulated by a molecular mimic of eIF4G. *Mol. Cell* **3:** 707–716.

McMahon L.P., Choi K.M., Lin T.A., Abraham R.T., and Lawrence J.C., Jr. 2002. The rapamycin-binding domain governs substrate selectivity by the mammalian target of rapamycin. *Mol. Cell. Biol.* **22:** 7428–7438.

Miron M., Verdu J., Lachance P.E., Birnbaum M.J., Lasko P.F., and Sonenberg N. 2001. The translational inhibitor 4E-BP is an effector of PI(3)K/Akt signaling and cell growth in *Drosophila. Nat. Cell Biol.* **3:** 596–601.

Montagne J., Stewart M.J., Stocker H., Hafen E., Kozma S.C., and Thomas G. 1999. *Drosophila* S6 kinase: A regulator of cell size. *Science* **285:** 2126–2129.

Morley S.J. and Naegele S. 2002. Phosphorylation of eukaryotic initiation factor (eIF) 4E is not required for de novo protein synthesis following recovery from hypertonic stress in human kidney cells. *J. Biol. Chem.* **277:** 32855–32859.

Morley S.J. and Traugh J.A. 1989. Phorbol esters stimulate phosphorylation of eukaryotic initiation factors 3, 4B, and 4F. *J. Biol. Chem.* **264:** 2401–2404.

———. 1990. Differential stimulation of phosphorylation of initiation factors eIF-4F, eIF-4B, eIF-3, and ribosomal protein S6 by insulin and phorbol esters. *J. Biol. Chem.* **265:** 10611–10616.

———. 1993. Stimulation of translation in 3T3-L1 cells in response to insulin and phorbol ester is directly correlated with increased phosphate labeling of initiation factor (eIF-) 4F and ribosomal protein S6. *Biochimie* **75:** 985–989.

Nojima H., Tokunaga C., Eguchi S., Oshiro N., Hidayat S., Yoshino K., Hara K., Tanaka N., Avruch J., and Yonezawa K. 2003. The mammalian target of rapamycin (mTOR) partner, raptor, binds the mTOR substrates, p70 S6 kinase and 4E-BP1 through their TOR signaling (TOS) motif. *J. Biol. Chem.* **278:** 15461–15464.

Ochs K., Saleh L., Bassili G., Sonntag V.H., Zeller A., and Niepmann M. 2002. Interaction of translation initiation factor eIF4B with the poliovirus internal ribosome entry site. *J. Virol.* **76:** 2113–2122.

Pain V.M. 1996. Initiation of protein synthesis in eukaryotic cells. *Eur. J. Biochem.* **236:** 747–771.

Palmer T.D., Miller A.D., Reeder R.H., and McStay B. 1993. Efficient expression of a protein coding gene under the control of an RNA polymerase I promoter. *Nucleic Acids Res.* **21:** 3451–3457.

Pause A., Belsham G.J., Gingras A.C., Donze O., Lin T.A., Lawrence J.C., Jr., and Sonenberg N. 1994. Insulin-dependent stimulation of protein synthesis by phosphorylation of a regulator of 5′-cap function (see comments). *Nature* **371:** 762–767.

Pende M., Kozma S.C., Jaquet M., Oorschot V., Burcelin R., Le Marchand-Brustel Y., Klumperman J., Thorens B., and Thomas G. 2000. Hypoinsulinaemia, glucose intolerance and diminished beta-cell size in S6K1-deficient mice. *Nature* **408:** 994–997.

Polunovsky V.A., Rosenwald I.B., Tan A.T., White J., Chiang L., Sonenberg N., and Bitterman P.B. 1996. Translational control of programmed cell death: Eukaryotic translation initiation factor 4E blocks apoptosis in growth-factor-restricted fibroblasts with physiologically expressed or deregulated Myc. *Mol. Cell. Biol.* **16:** 6573–6581.

Polunovsky V.A., Gingras A.C., Sonenberg N., Peterson M., Tan A., Rubins J.B., Manivel J.C., and Bitterman P.B. 2000. Translational control of the antiapoptotic function of Ras. *J. Biol. Chem.* **275:** 24776–24780.

Popolo L., Vanoni M., and Alberghina L. 1982. Control of the yeast cell cycle by protein synthesis. *Exp. Cell Res.* **142:** 69–78.

Poulin F., Gingras A.C., Olsen H., Chevalier S., and Sonenberg N. 1998. 4E-BP3, a new member of the eukaryotic initiation factor 4E-binding protein family. *J. Biol. Chem.* **273:** 14002–14007.

Puig O., Marr M.T., Ruhf M.L., and Tjian R. 2003. Control of cell number by *Drosophila* FOXO: Downstream and feedback regulation of the insulin receptor pathway. *Genes Dev.* **17:** 2006–2020.

Pullen N., Dennis P.B., Andjelkovic M., Dufner A., Kozma S.C., Hemmings B.A., and Thomas G. 1998. Phosphorylation and activation of p70s6k by PDK1. *Science* **279:** 707–710.

Pyronnet S., Dostie J., and Sonenberg N. 2001. Suppression of cap-dependent translation in mitosis. *Genes Dev.* **15:** 2083–2093.

Pyronnet S., Imataka H., Gingras A.C., Fukunaga R., Hunter T., and Sonenberg N. 1999. Human eukaryotic translation initiation factor 4G (eIF4G) recruits mnk1 to phosphorylate eIF4E. *EMBO J.* **18:** 270–279.

Qin H., Raught B., Sonenberg N., Goldstein E.G., and Edelman A.M. 2003. Phosphorylation screening identifies translational initiation factor 4GII as an intracellular target of Ca2+/calmodulin-dependent protein kinase I. *J. Biol. Chem.* **278:** 48570–48579.

Radimerski T., Montagne J., Hemmings-Mieszczak M., and Thomas G. 2002a. Lethality of *Drosophila* lacking TSC tumor suppressor function rescued by reducing dS6K signaling. *Genes Dev.* **16:** 2627–2632.

Radimerski T., Montagne J., Rintelen F., Stocker H., van der Kaay J., Downes C.P., Hafen E., and Thomas G. 2002b. dS6K-regulated cell growth is dPKB/dPI(3)K-independent, but requires dPDK1. *Nat. Cell Biol.* **4:** 251–255.

Raught B. and Gingras A.C. 1999. eIF4E activity is regulated at multiple levels. *Int. J. Biochem. Cell Biol.* **31:** 43–57.

Raught B., Gingras A.C., and Sonenberg N. 2000a. Regulation of ribosomal recruitment in eukaryotes. In *Translational control of gene expression* (ed. N. Sonenberg et al.), pp. 245-293. Cold Spring Harbor Laboratory Press, Cold Spring Harbor, New York.

Raught B., Gingras A.C., Gygi S.P., Imataka H., Morino S., Gradi A., Aebersold R., and Sonenberg N. 2000b. Serum-stimulated, rapamycin-sensitive phosphorylation sites in the eukaryotic translation initiation factor 4GI. *EMBO J.* **19:** 434–444.

Roberg K.J., Bickel S., Rowley N., and Kaiser C.A. 1997. Control of amino acid permease sorting in the late secretory pathway of *Saccharomyces cerevisiae* by SEC13, LST4, LST7 and LST8. *Genetics* **147:** 1569–1584.

Rogers G.W., Jr., Komar A.A., and Merrick W.C. 2002. eIF4A: The godfather of the DEAD box helicases. *Prog. Nucleic Acid Res. Mol. Biol.* **72:** 307–331.

Rogers G.W., Jr., Richter N.J., and Merrick W.C. 1999. Biochemical and kinetic characterization of the RNA helicase activity of eukaryotic initiation factor 4A. *J. Biol. Chem.* **274:** 12236–12244.

Rogers G.W., Jr., Richter N.J., Lima W.F., and Merrick W.C. 2001. Modulation of the helicase activity of eIF4A by eIF4B, eIF4H, and eIF4F. *J. Biol. Chem.* **276:** 30914–30922.

Ron D. and Harding H.P. 2000. PERK and translational control by stress in the endoplasmic reticulum. In *Translational control of gene expression* (ed. N. Sonenberg et al.), pp. 547–560. Cold Spring Harbor Laboratory Press, Cold Spring Harbor, New York.

Rousseau D., Gingras A.C., Pause A., and Sonenberg N. 1996. The eIF4E-binding proteins 1 and 2 are negative regulators of cell growth. *Oncogene* **13:** 2415–2420.

Rozen F., Edery I., Meerovitch K., Dever T.E., Merrick W.C., and Sonenberg N. 1990. Bidirectional RNA helicase activity of eucaryotic translation initiation factors 4A and 4F. *Mol. Cell. Biol.* **10:** 1134–1144.

Rupes I. 2002. Checking cell size in yeast. *Trends Genet.* **18:** 479–485.

Rutter J., Probst B.L., and McKnight S.L. 2002. Coordinate regulation of sugar flux and translation by PAS kinase. *Cell* **111:** 17–28.

Sachs A.B. 2000. Physical and functional interactions between the mRNA cap structure and the poly(A) tail. In *Translational control of gene expression* (ed. N. Sonenberg et al.), pp. 447-466. Cold Spring Harbor Laboratory Press, Cold Spring Harbor, New York.

Schalm S.S. and Blenis J. 2002. Identification of a conserved motif required for mTOR signaling. *Curr. Biol.* **12:** 632–639.

Scheper G.C. and Proud C.G. 2002. Does phosphorylation of the cap-binding protein eIF4E play a role in translation initiation? *Eur. J. Biochem.* **269:** 5350–5359.

Scheper G.C., Morrice N.A., Kleijn M., and Proud C.G. 2001. The mitogen-activated protein kinase signal-integrating kinase Mnk2 is a eukaryotic initiation factor 4E kinase with high levels of basal activity in mammalian cells. *Mol. Cell. Biol.* **21:** 743–754.

Scheuner D., Song B., McEwen E., Liu C., Laybutt R., Gillespie P., Saunders T., Bonner-Weir S., and Kaufman R.J. 2001. Translational control is required for the unfolded protein response and in vivo glucose homeostasis. *Mol. Cell* **7:** 1165–1176.

Shima H., Pende M., Chen Y., Fumagalli S., Thomas G., and Kozma S.C. 1998. Disruption of the p70(s6k)/p85(s6k) gene reveals a small mouse phenotype and a new functional S6 kinase. *EMBO J.* **17:** 6649–6659.

Sonenberg N., Morgan M.A., Merrick W.C., and Shatkin A.J. 1978. A polypeptide in eukaryotic initiation factors that crosslinks specifically to the 5′-terminal cap in mRNA. *Proc. Natl. Acad. Sci.* **75:** 4843–4847.

Sonenberg N., Rupprecht K.M., Hecht S.M., and Shatkin A.J. 1979. Eukaryotic mRNA cap binding protein: Purification by affinity chromatography on sepharose-coupled m7GDP. *Proc. Natl. Acad. Sci.* **76:** 4345–4349.

Sonenberg N., Trachsel H., Hecht S., and Shatkin A.J. 1980. Differential stimulation of capped mRNA translation in vitro by cap binding protein. *Nature* **285:** 331–333.

Staal S.P. and Hartley J.W. 1988. Thymic lymphoma induction by the AKT8 murine retrovirus. *J. Exp. Med.* **167:** 1259–1264.

Staal S.P., Hartley J.W., and Rowe W.P. 1977. Isolation of transforming murine leukemia viruses from mice with a high incidence of spontaneous lymphoma. *Proc. Natl. Acad. Sci.* **74:** 3065–3067.

Steck P.A., Pershouse M.A., Jasser S.A., Yung W.K., Lin H., Ligon A.H., Langford L.A., Baumgard M.L., Hattier T., Davis T., Frye C., Hu R., Swedlund B., Teng D.H., and Tavtigian S.V. 1997. Identification of a candidate tumour suppressor gene, MMAC1, at chromosome 10q23.3 that is mutated in multiple advanced cancers. *Nat. Genet.* **15:** 356–362.

Stolovich M., Tang H., Hornstein E., Levy G., Cohen R., Bae S.S., Birnbaum M.J., and Meyuhas O. 2002. Transduction of growth or mitogenic signals into translational activation of TOP mRNAs is fully reliant on the phosphatidylinositol 3-kinase-mediated pathway but requires neither S6K1 nor rpS6 phosphorylation. *Mol. Cell. Biol.* **22:** 8101–8113.

Svitkin Y.V., Ovchinnikov L.P., Dreyfuss G., and Sonenberg N. 1996. General RNA binding proteins render translation cap dependent. *EMBO J.* **15:** 7147–7155.

Tan A., Bitterman P., Sonenberg N., Peterson M., and Polunovsky V. 2000. Inhibition of Myc-dependent apoptosis by eukaryotic translation initiation factor 4E requires cyclin D1. *Oncogene* **19:** 1437–1447.

Tang H., Hornstein E., Stolovich M., Levy G., Livingstone M., Templeton D., Avruch J., and Meyuhas O. 2001. Amino acid-induced translation of TOP mRNAs is fully dependent on phosphatidylinositol 3-kinase-mediated signaling, is partially inhibited by rapamycin, and is independent of S6K1 and rpS6 phosphorylation. *Mol. Cell. Biol.* **21:** 8671–8683.

Tarun S.Z., Jr. and Sachs A.B. 1996. Association of the yeast poly(A) tail binding protein with translation initiation factor eIF-4G. *EMBO J.* **15:** 7168–7177.

Taylor B.L. and Zhulin I.B. 1999. PAS domains: Internal sensors of oxygen, redox potential, and light. *Microbiol. Mol. Biol. Rev.* **63:** 479–506.

Tee A.R. and Proud C.G. 2002. Caspase cleavage of initiation factor 4E-binding protein 1 yields a dominant inhibitor of cap-dependent translation and reveals a novel regulatory motif. *Mol. Cell. Biol.* **22:** 1674–1683.

Ui M., Okada T., Hazeki K., and Hazeki O. 1995. Wortmannin as a unique probe for an intracellular signalling protein, phosphoinositide 3-kinase. *Trends Biochem. Sci.* **20:** 303–307.

van der Velden A.W. and Thomas A.A. 1999. The role of the 5′ untranslated region of an mRNA in translation regulation during development. *Int. J. Biochem. Cell Biol.* **31:** 87–106.

Verdu J., Buratovich M.A., Wilder E.L., and Birnbaum M.J. 1999. Cell-autonomous regulation of cell and organ growth in *Drosophila* by Akt/PKB. *Nat. Cell Biol.* **1:** 500–506.

Vezina C., Kudelski A., and Sehgal S.N. 1975. Rapamycin (AY-22,989), a new antifungal antibiotic. I. Taxonomy of the producing streptomycete and isolation of the active principle. *J. Antibiot.* **28:** 721–726.

Vlahos C.J., Matter W.F., Hui K.Y., and Brown R.F. 1994. A specific inhibitor of phosphatidylinositol 3-kinase, 2-(4-morpholinyl)-8-phenyl-4H-1-benzopyran-4-one (LY294002). *J. Biol. Chem.* **269:** 5241–5248

Wakiyama M., Imataka H., and Sonenberg N. 2000. Interaction of eIF4G with poly(A)-binding protein stimulates translation and is critical for *Xenopus* oocyte maturation. *Curr. Biol.* **10:** 1147–1150.

Wang S., Rosenwald I.B., Hutzler M.J., Pihan G.A., Savas L., Chen J.J., and Woda B.A. 1999.

Expression of the eukaryotic translation initiation factors 4E and 2alpha in non-Hodgkin's lymphomas. *Am. J. Pathol.* **155:** 247–255.

Wang X., Li W., Parra J.L., Beugnet A., and Proud C.G. 2003. The C terminus of initiation factor 4E-binding protein 1 contains multiple regulatory features that influence its function and phosphorylation. *Mol. Cell. Biol.* **23:** 1546–1557.

Wang X., Flynn A., Waskiewicz A.J., Webb B.L., Vries R.G., Baines I.A., Cooper J.A., and Proud C.G. 1998. The phosphorylation of eukaryotic initiation factor eIF4E in response to phorbol esters, cell stresses, and cytokines is mediated by distinct MAP kinase pathways. *J. Biol. Chem.* **273:** 9373–9377.

Waskiewicz A.J., Flynn A., Proud C.G., and Cooper J.A. 1997. Mitogen-activated protein kinases activate the serine/threonine kinases Mnk1 and Mnk2. *EMBO J.* **16:** 1909–1920.

Waskiewicz A.J., Johnson J.C., Penn B., Mahalingam M., Kimball S.R., and Cooper J.A. 1999. Phosphorylation of the cap-binding protein eukaryotic translation initiation factor 4E by protein kinase Mnk1 in vivo. *Mol. Cell. Biol.* **19:** 1871–1880.

Wells S.E., Hillner P.E., Vale R.D., and Sachs A.B. 1998. Circularization of mRNA by eukaryotic translation initiation factors. *Mol. Cell* **2:** 135–140.

Williams B.R. 2001. Signal integration via PKR. *Sci. STKE* **2001:** RE2.

Yamanaka S., Poksay K.S., Arnold K.S., and Innerarity T.L. 1997. A novel translational repressor mRNA is edited extensively in livers containing tumors caused by the transgene expression of the apoB mRNA-editing enzyme. *Genes Dev.* **11:** 321–333.

Yang D.Q. and Kastan M.B. 2000. Participation of ATM in insulin signaling through phosphorylation of eIF-4E-binding protein 1. *Nat. Cell Biol.* **2:** 893–898.

Zamora M., Marissen W.E., and Lloyd R.E. 2002. Multiple eIF4GI-specific protease activities present in uninfected and poliovirus-infected cells. *J. Virol.* **76:** 165–177.

Zhang P., McGrath B.C., Reinert J., Olsen D.S., Lei L., Gill S., Wek S.A., Vattem K.M., Wek R.C., Kimball S.R., Jefferson L.S., and Cavener D.R. 2002. The GCN2 eIF2alpha kinase is required for adaptation to amino acid deprivation in mice. *Mol. Cell. Biol.* **22:** 6681–6688.

11

Forging the Factory: Ribosome Synthesis and Growth Control in Budding Yeast

Paul Jorgensen and Mike Tyers
Department of Medical Genetics and Microbiology, University of Toronto
Toronto, Ontario, Canada M5S 1A8; Samuel Lunenfeld Research Institute
Mount Sinai Hospital, Toronto, Ontario, Canada M5G 1X5

Jonathan R. Warner
Department of Cell Biology, Albert Einstein College of Medicine
Bronx, New York 10461

THE OBSERVATION THAT THE LEVEL OF RNA in a culture of *Escherichia coli* is related to the growth rate originated the study of macromolecular regulation (for review, see Maaloe and Kjeldgaard 1966). As the role of the ribosome in translation became evident, the issue could be clearly framed: *Since the rate of growth depends on the synthesis of proteins, and the synthesis of proteins depends on ribosomes, the regulation of growth must ultimately depend on the regulation of synthesis of new ribosomes, which consumes a large fraction of the cell's resources.* This is undoubtedly as true of eukaryotic cells as of prokaryotic cells.

This chapter deals largely with the regulation of ribosome synthesis in budding yeast. Aspects of the regulation of ribosomal RNA and of ribosomal protein (RP) synthesis in metazoan cells are covered in Chapters 12 and 9, respectively.

At the outset, a short detour into the regulation of ribosome synthesis in *E. coli* can serve as a useful introduction to the problem. More than 20 years ago, it became clear from the work of Nomura, Lindahl, and others that the synthesis of RPs in *E. coli* is regulated by an extraordinary example of tight inventory control. Although the transcription of mRNA encoding RPs is more or less constitutive, the translation of RP messenger RNAs (mRNAs) depends on the continued formation of new rRNA with which the new RPs can associate. If there is insufficient transcription of new rRNA, the free RPs bind specifically to their (usually polycistronic) mRNAs to prevent further translation (for review, see Nomura et al. 1984; Zengel and Lindahl 1994).

The control of ribosome synthesis in *E. coli* thus depends on the control of rRNA transcription. A series of papers over many years, first from the Nomura lab and more recently from the Gourse lab (for review, see Gourse et al. 1996), has culminated with the recent demonstration that transcription at the critical rRNA promoters requires an unusually high concentration of the initiating NTP (iNTP). Should a reduction in iNTP concentration occur due to hyperactive protein synthesis or to declining energy availability, rRNA transcription decreases and ribosome synthesis is reduced; the limited energy is then available for other purposes (Schneider et al. 2002). This simple but elegant system directly couples ribosome synthesis, and thus cell growth, to the activity of the translation apparatus, and to the energy resources available to the cell. Superimposed on this system are other mechanisms; e.g., inhibition of rRNA promoters during the "stringent response" by ppGpp, a secondary messenger produced upon amino acid starvation (Gourse et al. 1996).

One additional point requires emphasis. In contrast to many other regulated systems, the stoichiometry of ribosome synthesis matters. Each ribosome contains a single molecule of each of the RNA and protein components, with a couple of minor exceptions. Thus, the underexpression of a single RP will lead to a corresponding undersynthesis of ribosomes, as well as to the potential accumulation of many defective ribosomal particles. Conversely, the overexpression of a single RP will not only be of no value to the cell, but could easily be deleterious, since it would generate a "rogue" RNA-binding protein that could interfere with normal RNA–protein interactions of the cell. Thus, a consideration of the regulation of ribosome synthesis must include not only the response to the environment, but also the maintenance of an equimolar supply of ribosomal components.

BACKGROUND ON YEAST

We know far less about the mechanisms regulating ribosome synthesis in *Saccharomyces cerevisiae* (Nomura 1999). Yet there can be no doubt that such regulation is a critical aspect of cell growth control, as a consideration of the role of ribosome synthesis in the economy of the cell makes evident (Warner 1999):

- A rapidly growing yeast cell has about 200,000 ribosomes.

- Therefore, with a doubling time of ~100 minutes, the cell must synthesize 2,000 ribosomes/minute in its nucleolus, and export them to the cytoplasm.

- The 100–200 rRNA genes are responsible for about 60% of total transcription.

- The 137 genes encoding 78 RPs are responsible for about 30% of all mRNAs (Holstege et al. 1998), and because these transcripts generally have short half-lives, RP genes account for ~50% of initiation events by RNA polymerase II.

- RP transcripts account for 90% of all mRNA splicing events (Ares et al. 1999).

- The assembly of ribosomes in the nucleus requires the import of nearly 1,000 RPs/minute/nuclear pore.

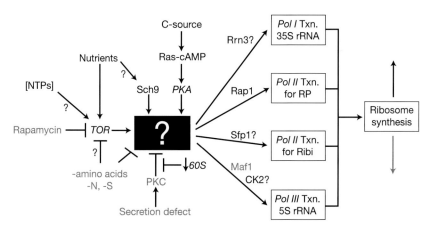

Figure 1. Regulation of ribosome synthesis in *S. cerevisiae*. Although many of the signaling pathways that communicate nutrient status and many of the effector proteins that directly induce transcription of the various ribosomal components have been identified, the connections between these upstream and downstream factors remain mysterious (*black box*). Of particular importance is whether positive (*black*) and negative (*gray*) signals converge at a common point upstream of the four types of transcription (*boxes*) or whether each set of promoters is regulated by multiple signals. See text for details.

Ribosome synthesis requires the coordinated activity of all three RNA polymerases (Fig. 1). Producing the raw materials for ribosome construction involves RNA Pol I transcription of the 35S rRNA genes, RNA Pol II transcription of the RP genes, and RNA Pol III transcription of the 5S rRNA genes. The products of the 35S and 5S rRNA genes should be equimolar, but the products of the RP genes need be equimolar only after the translation step, providing additional points of potential regulation. In addition to these raw materials, more than a hundred other proteins have been implicated in rRNA processing and ribosome assembly (for review, see Fatica and Tollervey 2002). Recent evidence suggests that transcription of these genes is coordinately regulated, being expressed in a manner that is similar to, but distinct from, RP genes (Hughes et al. 2000; Wade et al. 2001; Jorgensen et al. 2002). As in *E. coli*, the energetic costs of ribosome synthesis have selected for a coordinated and tightly harnessed control system.

REGULATION OF RIBOSOMAL RNA TRANSCRIPTION

Befitting the massive rate of ribosome synthesis in budding yeast, the nucleolus dominates this organism's nucleus, occupying a third of the nuclear volume (Wente et al. 1997). The nucleolus is organized around the single ribosomal DNA (rDNA) cluster, found on chromosome XII at the *RDN1* locus, that consists of a tandem array of 100–200 repeats of a standard 9.1-kb unit (Fig. 2). In budding yeast, the 9.1-kb unit encodes both the 35S rRNA, the precursor to 5.8S, 18S, and 25S rRNAs, and the 5S rRNA. Each repeat also contains important sequence elements including those that regulate 35S and 5S rRNA transcription (see below), an origin of replication (ARS element), and a replication fork barrier (Fig. 2). The latter prevents replication forks from colliding with transcribing RNA polymerase I during S phase. Stalled replication forks within the rDNA locus engender high rates of recombination between rDNA repeats, with frequent expansion or contraction of the number of repeats, leading to some variability in repeat number. In addition, autonomously replicating circles of one or more repeats are generated, a process thought to underlie senescence in yeast (for review, see Ivessa and Zakian 2002).

In *S. cerevisiae*, RNA polymerase I has 14 subunits, 5 of which are shared with Pol II and III, and 2 of which are shared only with Pol III (for review, see Reeder 1999). The transcription of 35S rRNA depends on two upstream sequences and the two multisubunit factors that bind to them (Fig. 3A). Core factor (CF) is composed of three proteins (Rrn6, Rrn7, and Rrn11) and binds to the core promoter element (CPE) about 30 bp

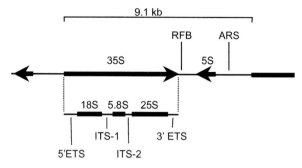

Figure 2. Structure of the *S. cerevisiae* rDNA repeat. The rDNA locus is composed of 100–200 head-to-tail repeats of a 9.1-kb unit. Each repeat contains a gene encoding the 35S rRNA precursor (transcribed by RNA Pol I) and a gene encoding the 5S rRNA (transcribed by RNA Pol III). The 35S rRNA is extensively cleaved and processed, resulting in 18S, 5.8S, and 25S rRNAs that are pseudouridylated and methylated at specific nucleotides. Cleavage and trimming events remove two external transcribed spacers (5′ETS, 3′ETS) and two internal transcribed spacers (ITS-1, ITS-2) from the rRNA. Each rDNA repeat also possesses an origin of replication (ARS) and a replication fork barrier (RFB), which prevents replication forks originating at the ARS from entering the 35S gene and colliding with RNA Pol I. The RFB also influences the expansion and contraction of the number of rDNA repeats that occurs during strain propagation.

upstream of the initiation site, where it is required for basal rates of initiation. Upstream activation factor (UAF) is composed of six proteins (Rrn5, Rrn9, Rrn10, Uaf30, and histones H3 and H4) and binds to the upstream control element (UCE) located 90–100 bp upstream of the initiation site, where it activates high-level transcription. Additionally, transcription requires TBP, and Rrn3, which binds to the polymerase (Siddiqi et al. 2001). These complexes are thought to be sequentially recruited to promoters such that promoter-bound UAF recruits TBP, which in turn recruits CF and a Rrn3/RNA Pol I complex (Steffan et al. 1996; Bordi et al. 2001). Recently, the RNA Pol II general transcription factor, TFIIH, has been shown to bind to RNA Pol I and to play an essential role in RNA Pol I initiation and elongation in the yeast and mouse nucleolus (Iben et al. 2002).

The rate-limiting and growth-regulated step in initiation by RNA Pol I appears to be the formation of an active Rrn3/RNA Pol I complex. Many years ago, it was shown that Pol I from vertebrates was purified in two forms, only one of which was active (Tower and Sollner-Webb 1987) . It has recently become clear that the active Pol I is associated with TIF-IA, whose homolog in *S. cerevisiae* is Rrn3 (Bodem et al. 2000). Similarly, in

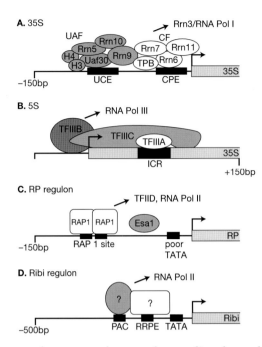

Figure 3. Four sets of promoters that must be coordinately regulated for ribosome synthesis in budding yeast. At some RP gene promoters Abf1 acts in place of Rap1. See text and references for details. (*A*, Adapted from Moss and Stefanovsky 2002; *B*, adapted from Schramm and Hernandez 2002.)

yeast extracts only the small fraction of Pol I that is associated with Rrn3 is active in vitro (Milkereit and Tschochner 1998). Furthermore, only Pol I that is phosphorylated, probably on subunits A43 and A190, can bind Rrn3. Rrn3, in turn, dissociates shortly after initiation of transcription (Milkereit and Tschochner 1998; Fath et al. 2001). Thus, one view (undoubtedly oversimplified) is that the rate of rRNA transcription is controlled simply by the phosphorylation of Pol I, allowing it to bind to the essential transcription factor Rrn3! We need only find the right kinase(s), or phosphatase(s) and the case is solved. Perhaps!

Unlike the situation for most genes, transcription of rRNA can be regulated at two levels, by the opening or closing of individual rRNA genes within the tandem repeat, and by the activation or inactivation of Pol I. By use of psoralen cross-linking, Sogo's group demonstrated that only about 50% of the rRNA genes are open and active, even in the most rapidly growing cells (Dammann et al. 1993). When cells reach stationary phase, the number of open and active genes is reduced by half (Dammann et al. 1993), although the rate of rRNA transcription has been reduced by >90% (Ju and Warner 1994). In mammalian cells there is now strong evidence that rRNA genes are silenced by methylation of specific upstream nucleotides (Santoro et al. 2002). In contrast, in *S. cerevisiae* the histone deacetylase Rpd3 was recently shown to be required to close rRNA genes, even though the histone acetylation of the rDNA chromatin was found to be independent of growth phase (Sandmeier et al. 2002). Interestingly, Sir2, a histone deacetylase that is required for the repression of RNA Pol II transcription within the *RDN1* locus, does not regulate rRNA repeat opening and closing, as measured by psoralen cross-linking (Sandmeier et al. 2002). The authors pose the intriguing suggestion that it is the acetylation of the H3 and/or H4 components of UAF that controls the availability of the rRNA gene (Sandmeier et al. 2002). Elegant analysis of the rDNA genes with electron microscopy revealed that in *rpd3Δ* strains the number of active rRNA genes is not decreased in stationary phase, but the loading of polymerases is greatly reduced; thus *rpd3Δ* strains still effectively repress rRNA transcription in stationary phase (Sandmeier et al. 2002). A recent genome-wide survey of histone acetylation raises the exciting possibility that the Hos1 and Hos3 histone deacetylases may function primarily within the rDNA locus (Robyr et al. 2002).

Manipulation of the number of rRNA genes in *S. cerevisiae* has shown that cells with as few as 40 repeats can grow as rapidly as wild-type cells (Wai et al. 2000). In such cells, all 40 rRNA genes appear to be active and to have twice the polymerase density of active genes in wild-type cells (French et al. 2003). The rDNA locus itself is effectively dispensable for wild-type growth and can be replaced with a plasmid-borne rDNA repeat, suggesting that the tandem nature of the rDNA array and the nucleolus it organizes is not actually essential for ribosome production (Wai et al. 2000). Taken together, these results suggest that although the availability of the rRNA genes may play some role in the regulation of rRNA transcription, it is not the primary one. Rather, the rate of RNA Pol I initiation appears to be the most important variable.

Several other factors have less-defined roles in RNA Pol I transcription. In addition to the UCE and CPE elements, a putative Pol I enhancer element was identified ~100 bp downstream from the preceding 35S rDNA sequence (see Fig. 2) (Elion and Warner 1984). The enhancer element appeared to activate transcription of 35S rDNA located both upstream and downstream from it and to overlap with the region of rRNA termination (Lang et al. 1994). More recent analyses, however, suggest that the enhancer actually functions by binding proteins that partially localize the reporter gene to the nucleolus, where RNA polymerase I is concentrated. Interestingly, the transcription factor and chromatin organizing protein Reb1, the yeast equivalent of the mammalian TTF1, binds both within the enhancer and also just upstream of the promoter, leading to suggestions that it plays a structural role in organizing the rDNA repeats, similar to TTF1 (Johnson and Warner 1989; Planta 1997). Another factor with a role in organizing the nucleolus is Cfi1/Net1, the scaffold for the RENT complex that also contains the silencing protein Sir2 and Cdc14 (see below for the connection of Net1 with the cell cycle) (Straight et al. 1999). Net1 binds to RNA Pol I and stimulates RNA Pol I transcription in vitro (Shou et al. 2001). Where Net1 binds in the rDNA repeats and how it interacts with the UAF and CF complexes remains to be determined. Finally, the HMG-box protein Hmo1 exhibits genetic interactions with deletions of nonessential RNA Pol I subunits and localizes to the nucleolus (Gadal et al. 2002). The HMG-box protein UBF plays a critical role in the activation and growth control of RNA Pol I transcription in vertebrates (Moss and Stefanovsky 2002), and it is tempting to speculate that Hmo1 is functionally related to UBF.

Less is known about the control of 5S rRNA transcription. RNA Pol III is recruited to 5S rRNA promoters by the TBP-containing TFIIIB complex, which is in turn recruited by a TFIIIC/TFIIIA complex that specifically interacts with the internal control region (ICR) located downstream from the transcription start site (see Fig. 3B) (Schramm and Hernandez 2002). 5S rRNA is present at roughly equivalent amounts as the other rRNAs (Neigeborn and Warner 1990), but how this balance is achieved is unknown. If 35S transcription is artificially reduced, 5S transcription continues for many hours, suggesting no immediate links between RNA Pol I and III (Wittekind et al. 1990).

Recent work suggests that Maf1 is a repressor of all RNA Pol III transcription in response to a range of stressful stimuli, including secretory pathway defects, nutrient limitation, and DNA damage (Upadhya et al. 2002). Although this study mostly examined the effects of Maf1 on tRNA transcription by RNA Pol III, deletion of *MAF1* does prevent the repression of 5S transcription in response to a defect in the secretory pathway and presumably other stresses as well (Upadhya et al. 2002). Maf1 appears to act by blocking TFIIIB binding to DNA and has no effect on the transcription of RP genes or the 35S rRNA (Upadhya et al. 2002). Importantly, Maf1 functions downstream of at least three distinct signaling pathways, suggesting that it is a common mediator of RNA Pol III repression.

Casein kinase II (CK2) is an important positive regulator of initiation by RNA Pol III (Hockman and Schultz 1996). CK2 stably associates with the TFIIIB complex, and phosphorylation of TFIIIB subunits by CK2 appears to be required for efficient recruitment of TFIIIB to promoters (Ghavidel and Schultz 2001). Intriguingly, CK2 is associated with RNA Pol I holoenzyme in vertebrates and is found bound to a number of proteins implicated in ribosome biogenesis in yeast, suggesting that CK2 could have additional roles in controlling ribosome production (Hannan et al. 1998; Albert et al. 1999; Leary and Huang 2001; Gavin et al. 2002; Ho et al. 2002). Indeed, crippled alleles of CK2 subunits exhibit decreased Pol I transcription rates in yeast, although CK2 had no effect on Pol I transcription in yeast extracts (Hockman and Schultz 1996). There is some evidence that the repression of RNA Pol I and III transcription caused by DNA damage is relayed by diminished CK2 activity (Ghavidel and Schultz 2001).

Finally, in mammalian cells, RNA Pol I and III activity is repressed during mitosis, probably through the action of the mitotic cyclin-dependent kinase Cdc2 (Gottesfeld et al. 1994; Kuhn et al. 1998). RNA Pol I and III activity is not repressed during mitosis in *S. cerevisiae*, perhaps because of the relatively modest chromosome condensation observed in this organism (Wente et al. 1997).

REGULATION OF RIBOSOMAL PROTEIN SYNTHESIS

At the outset, a word of warning about nomenclature. For many years there were several versions of the names for yeast RPs. We now use the official codification (Mager et al. 1997), but many older references have used a different nomenclature. A translation is available at MIPS (http://www. mips.biochem.mpg.de/proj/yeast/reviews/rib_nomencl.html).

TRANSCRIPTION OF RP GENES

One of the most striking features of the molecular physiology of *S. cerevisiae* is that not only are the RP genes transcribed to very high levels, they are also transcribed in concert. This has been known for many years from the analyses of individual RP genes (for review, see Woolford and Warner 1991), but has become particularly striking with the recent analyses of the yeast transcriptome under many conditions. The RP genes are typically one of the strongest cohorts seen when genome-wide expression data are clustered. Throughout this review, we refer to genes that are transcriptionally co-regulated and appear to be controlled by the same *cis-* and *trans-*acting factors as a regulon (e.g., the RP regulon).

What, then, is the basis of this coordinate regulation? The RP genes bear certain striking similarities. The promoters of most of the RP genes carry two adjacent binding sites for the transcription factor Rap1, in positions –250 to –500 bp upstream of the start codon (Fig. 3C). These sites are essential for high-level transcription. In a minority of genes, Abf1- or Reb1-binding sites substitute for the Rap1 sites (Lascaris et al. 1999). Downstream from these elements are one or two T-rich stretches that appear to be a common structural element of many yeast promoters (Iyer and Struhl 1995). It has recently been suggested that RP genes are characterized by rather unfavorable TATA boxes (Mencia et al. 2002). A new protein, the forkhead transcription factor Fhl1, has recently been shown to bind almost exclusively to RP gene promoters (Lee et al. 2002). A role for Fhl1 in ribosome production was anticipated by the gross defects in rRNA processing observed in an *fhl1* deletion strain. Another factor, *IFH1*, exhibits strong genetic interactions with *FHL1* and so may also regulate RP gene transcription (Hermann-Le Denmat et al. 1994; Cherel and Thuriaux 1995).

Rap1, Abf1, and Reb1 are all multifunctional proteins. Rap1 in particular has numerous important roles in the cell. On the one hand, it is responsible for the very active transcription not only of the RP genes, but also of other translation factors and of most of the abundant glycolytic enzymes. On the other hand, it is also an important transcriptional repressor, binding and initiating silencing both at telomeres and at the silent *MAT* loci (for review, see Morse 2000). Similarly, Abf1 has additional roles at the sites of replication initiation, whereas Reb1 binds to elements in the rDNA repeats and appears to have structural roles within the rDNA locus (see above).

Chromatin immunoprecipitation (ChIP) assays indicate that the Rap1- and Abf1-binding sites are constitutively occupied, whether or not high-level transcription is taking place (Reid et al. 2000). (A word of caution: The RP genes are among the most vigorously transcribed genes in the genome. Even when repressed, they are transcribed at a rate that is higher than that of most of the genes in the cell.) The lone essential histone acetylase of *S. cerevisiae*, Esa1, has been found at RP genes (Reid et al. 2000), but only those that contain Rap1- or Abf1-binding sites. Depletion of Esa1 leads to a substantial reduction of bulk histone acetylation and to a selective reduction in the transcription of RPs (Reid et al. 2000). Thus, it has been suggested that Rap1 recruits Esa1, which acetylates the histones (or something else) around the RP gene promoter, leading to active transcription. There is a certain circularity in this argument, however, because it is not known whether Esa1 activates transcription, or whether it is merely recruited to transcriptionally active regions. As noted above, the Hos family of histone acetylases appears to modify the chromatin of rRNA and of a few RP genes (Robyr et al. 2002), although there is no direct evidence that these activities regulate transcription.

A recent report suggests that the role of Rap1 is to recruit TAFs, which in turn recruit TBP to the suboptimal TATA box, thereby promoting transcription (Mencia et al. 2002). This is an intriguing suggestion, but it does not explain the extraordinarily high rate of transcription at RP genes, nor the means by which transcription can be so readily regulated over an order of magnitude or more. One striking feature of RP transcription is the myriad signaling pathways that affect its regulation (Fig. 1, and see below). Rapamycin, heat shock, a failure of the *sec* pathway, and reduced Ras-cAMP-PKA pathway activity, to name a few, cause a rapid decline in RP gene transcription, even in the presence of excess nutrients. It is not known whether all of these repressive stimuli signal to a common transcription factor, like Rap1, or whether they act via unique repressors that bind to all of the RP gene promoters.

In summary, although we have some idea of the structure of RP genes, and of the proteins that bind to them, we remain woefully ignorant about the way that transcription is activated to such a high level, how transcription is modulated to provide almost the same number of mRNAs for every RP (Holstege et al. 1998), and how each of the 137 genes is turned on, or off, with such intensity.

SPLICING OF RP mRNAs

An intriguing feature of the *S. cerevisiae* genome is that it contains fewer than 250 introns (Spingola et al. 1999). Of these, 95 are in the 137 RP genes (Ares et al. 1999). Indeed, this led to some confusion because a number of the original temperature-sensitive (*ts*) mutants of Hartwell that were classified as defective in ribosome synthesis (Hartwell et al. 1970) were actually mutants of the splicing apparatus, identified as such only a decade later when introns had been discovered (Rosbash et al. 1981).

The strong enrichment for introns in transcripts encoding RPs suggests that regulated splicing could control individual or coordinate expression of these proteins. In two cases, for the RPs L30 and S14, the splicing of the transcripts is subject to negative feedback control by their products. In the case of L30, overproduction of the L30 protein leads to the accumulation of unspliced *RPL30* transcript (Dabeva et al. 1986). This is due to the specific interaction of L30 with a bulged stem-loop structure within the 5′ leader and the first few nucleotides of the intron (Eng and Warner 1991; Vilardell and Warner 1994). The atomic resolution solution of this structure provides perhaps the most detailed view available of a regulated splice event (Mao et al. 1999). The binding surface of L30 has been conserved over two billion years, as L30 from the archaeon, *Sulfolobus acidocaldarius,* can prevent the growth of *S. cerevisiae* cells by inhibiting the productive splicing of the *S. cerevisiae RPL30* transcript (Vilardell et al. 2000a). The regulation of splicing of the *RPL30* transcript contributes substantially to the fitness of the cell; cells unable to regulate *RPL30* splicing are rapidly outgrown by wild-type cells (Li et al. 1996). Excess L30 may also interfere with the normal function of other proteins that have, during evolution, "borrowed" the RNA-binding domain of L30, such as Nhp2, an essential component of the H/ACA snoRNPs (Henras et al. 2001).

The case of S14 is somewhat different. S14 is encoded by two genes, *CRY1* and *CRY2* (for cryptopleurine resistant—this is the same RP that when mutated causes emetine resistance in mammalian cells) (Paulovich et al. 1993). Under normal conditions, 90% of the mRNA encoding S14 is derived from *CRY1* (Li et al. 1995). If *CRY1* is deleted, however, the amount of *CRY2* mRNA increases to provide almost a full complement (Li et al. 1995). In fact, *CRY2* is transcribed at about the same level as *CRY1,* but any S14 that is not incorporated into a ribosome can bind to a complex structure in the *CRY2* transcript, preventing splicing (Fewell and Woolford 1999). Interference with the splicing of the *CRY2* transcript leads to its immediate destruction; no unspliced precursor accumulates, unlike the situation with L30.

A number of attempts, mostly unpublished, have searched for other examples of the feedback control of splicing among the RP transcripts. Thus far, they have been unsuccessful, perhaps because of the rapid turnover of unspliced transcripts. One is left to wonder why the RP genes have selectively maintained their introns. It has recently been demonstrated that introns in at least two yeast genes activate transcription by a factor of two (Furger et al. 2002), an effect that may contribute to the very high rates of transcription exhibited by RP genes. Furthermore, proteins involved in nuclear export are physically coupled to the splicing machinery in budding yeast, suggesting that introns could provide the short-lived RP transcripts a quick exit from the nucleus (Reed and Hurt 2002).

TRANSLATION OF RP mRNAS

An intriguing biological fact is that not only is the synthesis of bacterial RPs regulated at the translational level, as described above, but the synthesis of vertebrate RPs is as well, albeit by an entirely different mechanism (for review, see Meyuhas and Hornstein 2000; see also Chapter 9). The transcripts of mammalian RP genes initiate with several pyrimidine residues, termed a "terminal oligopyrimidine tract," or TOP sequence. In quiescent cells, a large proportion of the RP mRNAs are maintained in cytoplasmic RNPs and are not associated with ribosomes. Upon stimulation with a growth factor, these mRNAs are actively translated in polysomes. Neither the mechanism of inhibition nor the mechanism of activation of the translation of these mRNAs has been fully explained. Although some evidence points toward the phosphorylation of ribosomal protein S6 as being the gatekeeper (Volarevic and Thomas 2001), other experiments provide a contrary view (Meyuhas and Hornstein 2000; Stolovich et al. 2002).

With RP synthesis in both bacterial and mammalian cells being regulated at the level of translation, it is surprising that there is little evidence for this practice in *S. cerevisiae*. Scattered reports offer glimpses of potential regulation of translation of RP mRNAs (Warner et al. 1985; Kuhn et al. 2001), but none has provided any insight into mechanisms. In only one case, that of L30, has translational control been shown directly, but this control seems likely to be an inadvertent by-product of the regulated splicing of this mRNA (Dabeva and Warner 1993; Vilardell et al. 2000b). Nevertheless, recent analysis of the genomes of several species closely related to *S. cerevisiae* shows an unexpectedly high degree of conservation in the 5′ leader sequences of RP transcripts over many tens of millions of years, suggesting that there may be regulatory effects on translation, subtle perhaps, but strong enough to have persisted over evolutionary time (Cliften et al. 2003).

RP AND rRNA STOICHIOMETRY

Various mechanisms exist to maintain a roughly equal concentration of rRNAs and each of the RPs in the cell. As discussed above, the levels of some RPs are kept in check by negative feedback control of their own splicing. The fine-tuning, however, is largely based on the rapid turnover of unused RPs. This has been most dramatically demonstrated using overexpression of individual proteins. In several reports, the half-life of excess RPs has been estimated to be 0.5 to 3 minutes, whereas the half-life of proteins incorporated into the ribosomes of log-phase cells is at least hours (Warner et al. 1985; elBaradi et al. 1986; Maicas et al. 1988). Interestingly, three RP proteins (S31, L40A, and L40B) are encoded in the genome as ubiquitin fusions from which the ubiquitin moiety is rapidly cleaved by ubiquitin hydrolase activity. Removal of the ubiquitin-coding part of the gene results in a dramatic defect in rRNA processing, suggesting a potential chaperone function for ubiquitin in ribosome assembly (Finley et al. 1989). As befits the high level of RP protein expression, the three RP–ubiquitin fusions supply all of the ubiquitin needed for intracellular proteolysis under non-stress conditions (Hershko and Ciechanover 1998).

There seems to be little direct crosstalk between the transcription of rRNA and the synthesis of RPs. When rRNA transcription was steadily decreased by depriving cells of a subunit of RNA Pol I, RPs were synthesized at the same rate but steadily became more unstable to match the declining levels of rRNA (Wittekind et al. 1990), as had been found for mammalian cells some time before (Warner 1977). In the converse situation, the absence of RPs has little effect on rRNA synthesis, at least in the short term, but does lead to impaired rRNA processing and increased rRNA degradation (Warner and Udem 1972). An important overall corollary to these observations is that individually increasing the synthesis of either rRNA or the RP regulon will not lead to increased rates of ribosome synthesis. Rather, the gene sets must be coordinately induced.

RIBOSOME ASSEMBLY

It has recently become apparent that the processing of the rRNA precursor and the assembly of the RNA and protein products into ribosomal subunits is far more complex than hitherto suspected (Fatica and Tollervey 2002). Ribosome assembly is directly coupled with transcription in the nucleolus, as demonstrated by the association of huge particles with the nascent 5′ end of the 35S pre-rRNA in chromatin spreads of transcribing rDNA (Miller and Beatty, 1969; Mougey et al. 1993). Following transcription, the 35S rRNA is covalently modified in numerous positions by addition of methyl groups to 2′ hydroxyls and by conversion of uridines to pseudouridines. These events are directed by several dozen sequence-specific snoRNPs (Fatica and Tollervey 2002). The modified pre-rRNA is also subjected to a partially ordered series of endonuclease cleavages and exonuclease trimmings, leading to the mature 18S rRNA within the 40S subunit and the 25S:5.8S rRNA complex within the 60S subunit (Fatica and Tollervey 2002). Through a process of undoubtedly labyrinthine complexity that is just beginning to be understood, the 32 RPs of the 40S and the 46 RPs of the 60S subunit assemble with the rRNAs.

Recent proteomic analysis has uncovered a plethora of novel proteins in the nucleolus, many of which can be linked directly or indirectly to ribosome assembly. In total, some 350 proteins have been identified by purifying human nucleoli (Andersen et al. 2002; Scherl et al. 2002). For budding yeast, the affinity purification of nascent ribosomes using tagged nucleolar components has revealed hundreds of interacting partners (Dragon et al. 2002; Grandi et al. 2002). The complexity and incomplete understanding of the ribosome assembly process precludes a comprehensive discussion—instead we refer the reader to an excellent recent review on the topic (Fatica and Tollervey 2002). In the first characterized step, ribosome biogenesis factors mostly specific to 40S maturation, including a large U3 snoRNP, assemble with predominantly 40S ribosomal proteins on the 5′ portion of the 35S rRNA, thereby forming a 90S particle (Fig. 4) (Dragon et al. 2002; Grandi et al. 2002). The U3 snoRNP and associated proteins almost certainly correspond to the "terminal balls" at the end of the "Christmas tree" structure formed at transcriptionally active rDNA loci, as visualized by electron microscopy (Miller and Beatty 1969; Mougey et al. 1993; Dragon et al. 2002). After initial cleavage events in the 5′ETS and ITS-1, pre-40S and pre-60S particles appear to progress along distinct routes in a series of dynamic protein complexes, which progressively re-localize from the nucleolus, to the nucleoplasm, and finally into the cytoplasm (Fig. 4) (Nissan et al. 2002). Although the early separation of 40S and 60S assembly was unexpected, it may suggest a coupling of the assembly process with transcription (Fatica and Tollervey 2002). As transcribing the 7-kb 35S pre-rRNA takes ~5 minutes, the assembly of factors and ribosomal proteins presumably begins on the proximal 18S rRNA before RNA Pol I has transcribed the more distal 5.8S and 25S rRNA sequences (Warner 2001). Regardless, the biochemical processing of the rRNA transcript, as assayed by ribose-methylation, does not occur until the transcript is complete (Udem and Warner 1972). It is quite remarkable to consider that the dozens of snoRNAs that specify the sites of methylation and pseudouridylation may line up along the transcript as it is being synthesized, awaiting a signal indicating the successful completion of transcription. The final step in the production of 60S subunits is delivery to the cytoplasm. Several nucleo-cytoplasmic shuttling proteins, such as Tif6 and Nmd3, have been implicated in the export of 60S subunits (Ho and Johnson 1999; Basu et al. 2001; Senger et al. 2001).

Discerning the roles and the dynamics of the many novel proteins in the various particles associated with nascent ribosomes represents an immediate and exciting challenge (Warner 2001; Fatica and Tollervey 2002; Milkereit et al. 2003).

One crucial area of investigation concerns the regulation of the assembly process and whether completion of specific steps is monitored

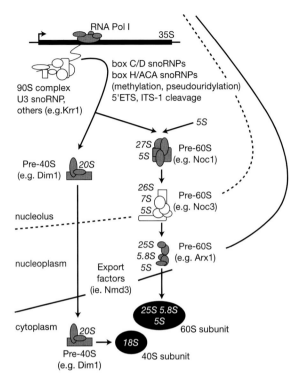

Figure 4. A simple depiction of ribosome biogenesis. Transcription of the nascent 5′ end of the 35S pre-RNA is followed by binding of numerous proteins, mostly specific to 40S subunit assembly and including the U3 snoRNP. During the early stages of assembly, box C/D and box H/ACA snoRNP complexes covalently modify the rRNA at specific nucleotides. Cleavage events in the 5′ ETS and ITS-1 regions separate the pre-40S and pre-60S particles. Pre-40S particles are exported to the cytoplasm where processing of the 20S rRNA to the mature 18S rRNA is completed. A series of pre-60S particles specific to the nucleolus, nucleoplasm, and cytoplasm have been characterized and are greatly simplified here (Nissan et al. 2002). Maturation of the pre-60S particle includes endonuclease cleavages and exonuclease trimmings of the 27S pre-rRNA leading to the mature 25S and 5.8S rRNAs, via 26S and 7S intermediates. Ribosome biogenesis factors that characterize each of the pre-ribosomal particles are noted in brackets. RP binding to the various rRNA intermediates is not depicted but generally occurs very early, within the nucleolus. rRNAs are represented in italics. (Adapted from Fatica and Tollervey 2002.)

by one or more checkpoints. Several kinases (e.g., CK2, Cdc2), GTPases (e.g., Nog1, Nug1), and a very large AAA ATPase family member (Rea1) have been shown to interact with preribosomal particles and are thus likely regulatory factors (Fatica and Tollervey 2002). CK2 and Cdc2 have both been shown to phosphorylate nucleolar factors (Leary and Huang 2001). It will be of particular importance to determine which, if any, steps in ribosome assembly are sensitive to nutrient regulation, and what mechanisms monitor each pre-ribosome for accurate assembly and/or function. For example, recent experiments suggest that failure of ribosome assembly due to the deficiency of a single 60S subunit can not only lead to breakdown of the entire particle, but can also disrupt regulatory signals that feed back on ribosome biogenesis from the secretory machinery (Zhao et al. 2003).

TRANSCRIPTION OF THE RIBI REGULON

The abundance of the machinery that assembles ribosomes is also regulated coordinately at the transcriptional level (Hughes et al. 2000; Wade et al. 2001). More than 140 proteins are currently implicated in rRNA processing and ribosome assembly in yeast (Fatica and Tollervey 2002), and the genes for the great majority of these are part of a large suite of transcriptionally co-regulated genes termed the Ribi regulon (for ribosome biogenesis; also known as the RRB regulon, for ribosome and rRNA biosynthesis) (Hughes et al. 2000; Wade et al. 2001; Jorgensen et al. 2002; Wu et al. 2002). The genes in the Ribi and RP regulons generally show similar reactions to stimuli, but in some instances are differentially expressed (Fig. 5) (Gasch et al. 2000; Hughes et al. 2000; Causton et al. 2001; Wade et al. 2001; Jorgensen et al. 2002; Wu et al. 2002). Fittingly, when the cell decreases synthesis of the raw materials of ribosome synthesis (rRNAs, RPs), the apparatus that assembles the ribosomes also appears to be down-sized.

The Ribi regulon currently encompasses ~100 genes, but it may contain up to 400 genes (Wu et al. 2002; P. Jorgensen and M. Tyers, unpubl.), making it one of the largest co-regulated gene sets in yeast. Indeed, the Ribi regulon appears to be a general regulator of the translation apparatus, because, in addition to containing the genes for ribosome assembly factors, the Ribi regulon includes genes whose products act both downstream and upstream of ribosome assembly (Gasch et al. 2000; Hughes et al. 2000; Wade et al. 2001; Jorgensen et al. 2002; Wu et al. 2002). Genes upstream of ribosome assembly include those that encode the unique and shared components of RNA Pol I and III, but apparently none of the unique subunits of RNA Pol II (Jorgensen et al. 2002). It is unknown whether transcription of the Pol I and III subunits normally limits the rate of rDNA transcription or whether such regulation simply anticipates pos-

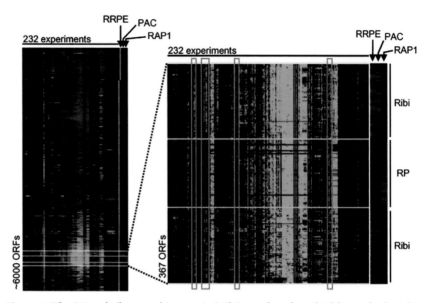

Figure 5. The RP and ribosome biogenesis (Ribi) regulons have highly similar but distinct patterns of expression. 232 microarray experiments measuring the abundance of ~6000 transcripts were subjected to two-dimensional, hierarchical clustering to group experiments and genes with similar expression patterns. On the resulting dendogram, >1.5-fold inductions are colored red and >1.5-fold repressions are colored green. Gene clusters corresponding to the RP (*blue lines*) and Ribi (*yellow lines*) regulons were identified by their strong enrichment for RAP1 and RRPE/PAC promoter elements in upstream regions (*red stripes*), respectively; the RP and Ribi clusters were highly enriched for RP genes and genes implicated in ribosome assembly, respectively. The Ribi regulon is divided into two large clusters, separated by an intervening cluster representing the RP regulon. The separation of the two Ribi clusters is an artifact of the clustering procedure, but does reveal the strong similarity between RP and Ribi expression. The RP and Ribi regulons are distinctly expressed, however, as revealed by nine highlighted experiments (*magenta rectangles*). The data for the 232 experiments are published and represent transcriptional regulation during the cell cycle, sporulation, stress responses, and other conditions.

sible future demand. The regulon also contains numerous genes involved in nucleotide metabolism, which may replenish the pool of ribonucleotides depleted by rapid rRNA transcription. Downstream from ribosome assembly, genes encoding the translation initiation and elongation complexes also form part of the Ribi regulon. In addition, many of the genes encoding tRNA synthetases appear to be co-regulated with the Ribi regulon (Jorgensen et al. 2002).

The Ribi regulon is likely to be regulated by a common set of yet-to-be-identified transcription factors. The majority of genes in the Ribi regulon possess two common sequence motifs in their upstream promoter regions, usually found within 200 bp of the translational start codon: the PAC element and the RRPE element (Fig. 3D)(Gasch et al. 2000; Hughes et al. 2000; Wade et al. 2001; Jorgensen et al. 2002). Eliminating either element appears to diminish the induction of at least one Ribi regulon gene (*EBP2*), suggesting that these sequence motifs are indeed functional (Wade et al. 2001). In addition, the promoters of Ribi regulon genes appear to be enriched for poly-T stretches, a feature of many yeast promoters including those of the RP genes (Iyer and Struhl 1995; see above, P. Jorgensen and M. Tyers, unpubl.). Although the cognate *trans*-acting factors for PAC/RRPE elements remain obscure, one strong candidate is a zinc-finger transcription factor called Sfp1 (Blumberg and Silver 1991; Xu and Norris 1998). Strains that lack *SFP1* are viable but exhibit decreased expression of the Ribi and RP regulons, slow growth, and extremely small cell size (see below) (Jorgensen et al. 2002). Restoration of *SFP1* expression to an *sfp1Δ* strain cell results in a strong, rapid induction of the Ribi regulon (Jorgensen et al. 2002). Intriguingly, the rapid induction of the Ribi regulon is followed by induction of the RP regulon, suggesting direct coordination of the two gene sets. The identification of all components of the transcription factor complexes that control the Ribi regulon will be critical to allow dissection of the upstream regulatory pathways.

Transcriptional control of ribosome biogenesis components also occurs in mammalian cells. Genes encoding several such factors, P120/hNOP2 (Fonagy et al. 1995), SAN5/hNOP5 (Nelson et al. 2000), BN51/hRPC53 (Ittmann 1994), and nucleolin (Herblot et al. 1999), are induced by growth factors. In addition, numerous groups have recently demonstrated that transcription of genes encoding ribosome biogenesis factors and RPs is elevated by the c-Myc protein (Coller et al. 2000; Greasley et al. 2000; Boon et al. 2001; Neiman et al. 2001; Iritani et al. 2002); such genes are often transcriptionally induced in malignant cells (Kinoshita et al. 2001; Nakamoto et al. 2001). At least some mammalian ribosome biogenesis factor mRNAs (e.g., fibrillarin) are also regulated translationally by the presence of the 5′TOP sequence implicated in the translational regulation of ribosomal proteins (Meyuhas and Hornstein 2000; see Chapter 9). Taken together with the observed regulation of RNA Pol I and RNA Pol III activity by c-Myc and the retinoblastoma (Rb) gene product (see Chapter 12) it is evident that the initiation of cell proliferation entails highly coordinated induction of ribosome biogenesis. Whether separate Ribi and RP transcriptional regulons exist in mammalian cells, and which signaling pathways function upstream of such hypothetical regulons, are obviously important questions.

REGULATORY CONDITIONS

As pointed out in Figure 1, there are numerous conditions under which the synthesis of ribosomes is regulated, in either a positive or negative direction, as summarized below.

Environmental Stresses

The earliest example of the coordinate regulation of ribosome synthesis in budding yeast came from the wild-type control in heat shock experiments with *ts* mutants (Warner and Udem 1972). When a wild-type culture is shifted from 23°C to 37°C, there is an almost immediate halt to the transcription of RP genes. The transcript abundance of RP genes rapidly declines as the mRNAs decay with a half-life of 5–10 minutes, and 15 minutes after heat shock, the synthesis of RPs is reduced to 10–20% of the normal rate. Transcription quickly recovers, however, and ~60 minutes after the shock, the level of RP mRNA has returned to the original level. The transcription of rRNA is also reduced after the heat shock, but to a substantially lesser extent, while the Ribi regulon is transiently repressed with kinetics which parallels that of the RP genes (Gasch et al. 2000; Causton et al. 2001; Wade et al. 2001). Surprisingly, the heat shock effect is seen even when cells are switched from 23°C to 30°C, the optimal temperature for growth, although the extent of the effect seems to depend on the magnitude of the temperature shift. The mechanism of this effect is unknown, but recent results suggest that it is triggered by the presence of misfolded proteins (Trotter et al. 2002).

DNA microarray analyses indicate that in addition to heat shock, numerous other extracellular and intracellular stresses appear to coordinately down-regulate both the RP and Ribi regulons. Indeed, the RP and Ribi regulons comprise a large proportion of the so-called environmental stress response (ESR), a group of ~900 genes whose transcripts are regulated, often transiently, by nearly any severe stress, including heat shock, hydrogen peroxide, osmotic shock, and DNA damage (Gasch et al. 2000, 2001; Causton et al. 2001). In most cases, the consequences of these stresses on RNA Pol I and RNA Pol III transcriptional machinery are not known, although DNA damage clearly decreases transcription initiation by both polymerases (Ghavidel and Schultz 2001). In the case of DNA damage, the signal to repress RP and Ribi transcription depends on the canonical DNA damage signaling pathway, as this repression is not observed in the absence of either the Mec1 or Dun1 kinases (Gasch et al. 2001). Diminished CK2 activity in response to DNA damage may in part explain the reduction in RNA Pol I and Pol III activity (Ghavidel and Schultz 2001).The pathways that signal other stresses to the RP and Ribi regulons are not known.

Growth Cycle

Microorganisms must constantly adapt to ever-changing nutrient conditions. (for review, see Werner-Washburne et al. 1996; Herman 2002). For example, *S. cerevisiae* cultures exhibit log-phase growth in rich glucose medium, yet at a density of about 3×10^7 to 4×10^7 cells/ml, when at least one-quarter of the original glucose is still present, the synthesis of ribosomes declines rapidly, as reflected by the reduced transcription of rRNA, RP genes, and the Ribi regulon (Ju and Warner 1994; DeRisi et al. 1997). Therefore, the yeast cell appears to anticipate its growth potential and prepares early for future shortages. As pointed out at the beginning of this chapter, the synthesis of ribosomes consumes a major portion of the cell's resources. Indeed, the synthesis of a ribosome represents a "capital investment" that the yeast makes in future growth. A slight, temporary rebound in RP transcripts occurs after diauxie, with the induction of oxidative metabolism and catabolism of stored carbohydrates (Ju and Warner 1994; Sillje et al. 1997). As cells approach the final density of $\sim 10^8$ cells/ml, there is a substantial degradation of ribosomes, so that in stationary phase the RNA/DNA level of the cells is only about 25% that of a log-phase culture. Nevertheless, within 15 minutes of dilution of a reasonably fresh stationary culture into fresh medium, RP mRNA recovers to log-phase levels, even in the absence of protein synthesis (J.R.Warner, unpubl.).

Effect of Nutrients, the Ras-cAMP-PKA, Sch9, and TOR Pathways

As illustrated during the diauxic shift, RP, Ribi, and rRNA transcription is exquisitely regulated by nutrient availability and the perceived growth potential of the cell. As in bacteria, amino acid deprivation results in a "stringent response" during which RP transcription, Ribi transcription, and rRNA transcription quickly decrease (Warner and Gorenstein 1978; Gasch et al. 2000). How the various nutrients are sensed and signaled to control RP, Ribi, and rRNA transcription remains mysterious, despite much effort.

The resumption of ribosome synthesis on dilution of a stationary-phase culture into fresh glucose-containing medium is likely due to an increase in the intracellular levels of cAMP in response to glucose (Nikawa et al. 1987; Francois et al. 1988). Several studies have shown that increased levels of cAMP preferentially stimulate the transcription of RP genes, an effect that appears to operate through the Rap1-binding promoter elements (Klein and Struhl 1994; Neuman-Silberberg et al. 1995). Thus, a major determinant of ribosome synthesis is activity of the Ras-cAMP-PKA pathway, although no substrate for PKA that interfaces with the transcription of RP genes has as yet been identified. Indeed, induction of RP transcription by the restoration of Ras-cAMP-PKA signaling can be blocked by inhibiting translation with cycloheximide, suggesting that PKA may not act directly on the RP gene promoters (Neuman-Silberberg et al. 1995). It should also be noted that nitrogen regulation of ribosomal protein synthesis (as opposed to carbon regulation) appears to not involve cAMP, nor the Rap1 promoter elements (Neuman-Silberberg et al. 1995), but likely acts through the TOR pathway (see below).

The Ribi regulon is strongly induced upon overexpression of the Sch9 kinase (P. Jorgensen and M. Tyers, unpubl.). Ras-cAMP-PKA and Sch9 appear to stimulate growth via parallel pathways, as hyperactivation of either pathway can suppress the loss of the other (Toda et al. 1988). The mechanisms whereby these two pathways control RP, Ribi, and rRNA transcription remain to be determined. Within the yeast AGC kinase family, Sch9 has the strongest sequence similarity with the oncogene/survival factor PKB/Akt and has similar effects on stress resistance, longevity, and cell size in budding yeast as PKB/Akt has on these processes in higher organisms (Fabrizio et al. 2001; Jorgensen et al. 2002; see Chapters 2 and 8). Because it is a central conduit that is downstream of many growth factor receptors, the mammalian PI3K/Akt pathway may also control the transcription of genes encoding mammalian ribosome biogenesis factors (see above).

The TOR (target of rapamycin) kinase pathway has been studied extensively in yeast and metazoan model systems and is discussed thoroughly in Chapters 5, 6, and 7. That the TOR pathway activates ribosome synthesis in yeast is evident from the observation that addition of rapamycin, a potent inhibitor of TOR, leads to rapid extinction of the transcription of rRNA by Pol I, of RP genes by Pol II (Powers and Walter 1999), of 5S rRNA by Pol III (Zaragoza et al. 1998), and of the Ribi regulon (Shamji et al. 2000). Rapamycin treatment also inhibits translation, but somewhat more slowly and to a lesser degree than it inhibits the transcription of ribosomal components (Cardenas et al. 1999; Hardwick et al. 1999; Powers and Walter 1999). The TOR pathway has been implicated in the response to nitrogen availability, to which it couples a number of readouts (for review, see Schmelzle and Hall 2000; Rohde et al. 2001). One of the critical substrates of the TOR kinase is Tap42, the phosphorylated form of which interacts with and inhibits PP2A catalytic subunits, including Pph21, Pph22, and Sit4 (Di Como and Arndt 1996; Jiang and Broach 1999). Although the transcriptional repression of both the Ribi and RP regulons was initially thought to be mediated by Tap42, presumably via release of the PP2A phosphatases (Shamji et al. 2000), recent studies suggest that this repression is actually Tap42-independent (T. Schmelzle et al., pers. comm.). Do the TOR and Ras-cAMP-PKA pathways converge at a single point above the RP and Ribi regulons or do they regulate *trans*-acting factors at both sets of promoters (Figs. 1, 3)? Recent work suggests that the TOR pathway signals to the RP regulon (as well as to RNA Pol I and III) through the Ras-cAMP-PKA pathway; in particular, the localization and presumably the activation of Tpk1, one of three PKA homologs in yeast, can be manipulated by rapamycin (T. Schmelzle et al., pers. comm.). In any case, TOR signaling impinges on the chromatin status of the RP gene promoters. Rapamycin causes the release of Esa1 from RP genes, leading to histone H4 deacetylation, probably by Rpd3, and concomitant suppression of RP gene transcription without loss of Rap1 and Abf1 binding (Rohde and Cardenas 2003).

Meiosis

As might be expected, since the signal to initiate meiosis is nitrogen starvation, cells placed in sporulation medium promptly cease ribosome synthesis and begin to degrade their ribosomes. Interestingly, however, late in meiosis the prespores resume ribosome synthesis (Emanuel and Magee 1981; Chu et al. 1998), and the bulk of the ribosomes in a mature spore reflect the progeny genotype. Thus, a diploid heterozygous for resistance to cycloheximide will not grow on moderate concentrations of cycloheximide, whereas two of the four spore clones do grow (Emanuel and Magee 1981).

Secretory Pathway

A search for conditional mutants defective in ribosome synthesis surprisingly turned up several genes involved in the secretory pathway (Mizuta and Warner 1994; Li and Warner 1996). It is now evident that any defect in the secretory pathway, from the initial ribosome/ER interaction to the final fusion of vesicles with the plasma membrane, leads to a rapid repression of ribosome synthesis, including transcription of both the rRNA genes and the RP regulon. The repression of ribosome synthesis appears to be a stress response to the stretching of the plasma membrane that occurs when protein synthesis in the cell continues while the plasma membrane (and cell wall) cannot grow because of a defective secretory pathway. The putative stretch sensor, Wsc1, signals to Pkc1, which responds by stimulating the transcription of cell wall proteins through a MAP kinase pathway, and by repressing the synthesis of ribosomes through another as-yet-unidentified pathway (Verna et al. 1997; Li et al. 1999; Li et al. 2000).

Rap1 and RP Gene Expression

Since Rap1 is not only the activator of most RP genes, but also a repressor of many genes, it is natural to ask whether repression of RP genes occurs through Rap1. Indeed, the allele *rap1-17*, a truncation mutant that is unable to silence either telomere-proximal genes or silent *MAT* loci, will not support the repression of RP genes in response to a defect in the secretory pathway (Mizuta et al. 1998). Surprisingly, the lack of repression conferred by *rap1-17* extends not only to genes that lack a Rap1 site, such as *RPL3*, whose transcription is driven by Abf1 (Hamil et al. 1988), but also to the rRNA transcription unit (Miyoshi et al. 2001) and to the Ribi regulon as well (Miyoshi et al. 2003). The repressive effect of Rap1 on the RP regulon differs from that at the *MAT* and telomere loci, however, since it does not require the *SIR* genes (Nierras and Warner 1999).

PROTEIN SYNTHESIS AND START PROGRESSION

To maintain a uniform size, cells must double in mass with each division. (Conlon and Raff 1999). As the predominant determinant of cellular growth, ribosome synthesis may play a key role in this coordination. In budding yeast, growth and division are coupled at Start, the point in late G_1 at which the cell commits to the next round of cell division (Hartwell et al. 1974; for a recent review, see Rupes 2002). To pass Start, yeast must reach a critical cell size. This requirement ensures that sufficient growth has occurred to traverse the cell cycle in the absence of additional nutrients and enforces a growth requirement on daughter cells, which are typically much smaller than mother cells (Johnston et al. 1977). Even yeast that are larger than the critical cell size must also have sufficient nutrients and a threshold level of protein synthesis in order to pass Start (Johnston et al. 1977; Moore 1988). Correspondingly, conditional alleles of translation components (Hanic-Joyce et al. 1987; Brenner et al 1988) and some ribosome biogenesis components cause cell cycle arrest prior to Start (Zimmerman and Kellogg 2001), although others do not (Kinoshita et al. 2001; Du and Stillman 2002).

At a minimum, entry into the cell cycle depends on sufficient ribosome synthesis. But exactly what is the molecular or cellular element that serves as a proxy for cell size? In perhaps the simplest scenario, the cell size threshold may be an indirect readout of overall translation rate, which impacts unstable cell cycle regulators, notably the G_1 cyclins that activate the cyclin-dependent kinase Cdc28 at Start (Fig. 6A) (see Chapter 4). *CLN1, CLN2,* and *CLN3* have partially overlapping activities; in particular, *CLN3* controls the expression of *CLN1* and *CLN2* as part of a suite of some 200 genes that is induced in late G_1 phase (Tyers et al. 1993; Futcher 2002). It has been proposed that *CLN3*, the highly unstable, rate-limiting activator of the Cdc28 kinase at Start, is a sensor of translation rate, which in turn reflects ribosome content and cell size (Futcher 1996; Polymenis and Schmidt 1997; Danaie et al. 1999). Notably, the *CLN3* 5′UTR contains a small upstream open reading frame (uORF) that represses translation of the bona fide *CLN3* reading frame, such that production of Cln3 protein is highly sensitive to translation rate (Polymenis and Schmidt 1997; Danaie et al. 1999). Cln3-Cdc28 concentration in the nucleus would thus be predicted to increase with cytoplasmic ribosome content, until at a critical concentration it activates Start (Futcher 1996; Nash et al. 2001). Indeed, mutation of the *CLN3* uORF reduces critical cell size (Polymenis and Schmidt 1997), whereas replacement of the uORF with the 5′ region of the *UBI4* gene, which is expressed in stationary phase, causes cells to become resistant to the G_1 arrest caused by rapamycin and sensitive to stationary phase conditions (Barbet et al. 1996; see Chapter 5). Intriguingly, the localization of the *CLN3* mRNA is also regulated, via an interaction with the RNA-binding protein Whi3 (Gari et al. 2001). Whi3 colocalizes with the *CLN3* mRNA in cytoplasmic foci, an effect that appears to prevent Cln3 accumulation in the nucleus without altering its overall abundance or translation rate (Gari et al. 2001). Thus, Cln3 abundance is precisely controlled through multiple mechanisms.

Although manipulation of *CLN3* can obviously alter Start progression, whether or not this is the primary physiological mechanism that dictates the critical cell size threshold is not easily determined. In particular, the very low abundance of Cln3 has so far precluded definitive tests of the above model. Where such measurements have been attempted, *CLN3* mRNA and protein both appear to peak at the end of mitosis rather than in late G_1 phase (Tyers et al. 1993; McInerny et al. 1997). Because cells that lack *CLN3* altogether still alter critical cell size in response to physiological signals (Nash et al. 1988), *CLN3* regulation cannot be the entire story. The *CLN3* translational control model also does not explain the long-standing observation that the critical cell size threshold is inversely related to nutrient status, being lowest in cells growing in poor nutrient conditions with slow translation rates and low ribosome content (Johnston et al. 1979; Tyson et al. 1979). Despite the fact that *CLN3* dosage is limiting for Start, the expression and activity of *CLN3* are markedly reduced under poor nutrient conditions, even though these cells pass Start at a small cell size (Tokiwa 1995; Hall et al. 1998). This inverse correlation suggests that some compensating mechanism allows cells growing in poor media to pass Start with much lower levels of Cln3 (Fig. 6B). One possibility is that activity of a Cln3-Cdc28 antagonist is diminished under poor nutrients. Alternatively, it has been suggested that drastically elevated Cln3-Cdc28 activity is needed in rich nutrients in order to drive rapid G_1 progression when cells are growing quickly (Hall et al. 1998). That is, in poor nutrients, the cell cycle machinery and translation may be more or less matched, in which case the cell cycle must necessarily be accelerated in glucose medium. The fact that the *CLN3* promoter contains a glucose-inducible element supports, but does not prove, this idea (Newcomb et al. 2002).

It is important to bear in mind that glucose massively stimulates the rate of protein synthesis but has different effects on Start progression, depending on context. Upon return of stationary-phase cells to proliferative growth, the Ras-cAMP-PKA pathway activates expression of *CLN1/2* (Hubler et al. 1993), whereas in logarithmic growth the Ras-cAMP-PKA pathway mediates glucose repression of *CLN1/2* transcription (Baroni et al. 1994; Tokiwa et al. 1994).

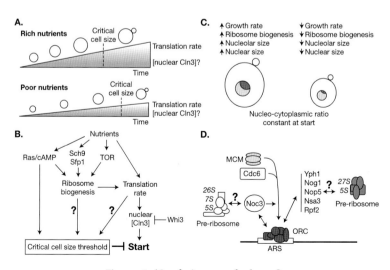

Figure 6. (*See facing page for legend*).

LINKS BETWEEN RIBOSOME BIOGENESIS AND CELL CYCLE CONTROL

The long-established correlation between growth rate and critical cell size has led to the suggestion that translation rates have a repressive influence at Start, manifested as an increased critical cell size threshold. Such effects would partially counter the positive effect of elevated Cln3 synthesis. It

---

Figure 6. Connections between ribosome biogenesis, the nucleolus, and the cell cycle machinery. (A) Cells may gauge their cell size by measuring translation rate. As daughter cells grow in size during G_1 phase of the cell cycle, their translational capacity (*gray slope*) increases. The critical cell size at which Start occurs may therefore reflect a critical translation rate. The concentration of Cln3 in the nucleus may measure translation rate. When cells are grown in rich nutrients, the critical cell size/translation threshold is set much higher than in poor nutrients. Increases in translation rates are shown, for simplicity, as being linear, but both cell size and translation rates increase exponentially in yeast (Elliott and McLaughlin 1978; Nurse and Fantes 1981; Woldringh et al. 1993). (B) The effect of nutrient availability on the critical cell size threshold and on Cln3 concentration are in opposition. A fundamental discrepancy in the understanding of Start is the observation that although cells growing in rich nutrients have much higher concentrations of the rate-limiting Start activator Cln3, they pass Start at a much larger critical cell size. Nutrient-responsive mechanisms appear to compensate for low Cln3 levels in cells grown in poor nutrients. The Ras-cAMP-PKA pathway increases the critical cell size threshold by repressing the *CLN1* promoter (Baroni et al. 1994; Tokiwa et al. 1994). The Ras-cAMP-PKA pathway, Sfp1, Sch9, and TOR may also raise the critical cell size threshold by increasing the rate of ribosome biogenesis, which appears to have a repressive effect on Start. (C) A possible mechanism to explain links between ribosome biogenesis and Start. Critical cell size at Start may be dictated by a nucleo-cytoplasmic volume ratio, potentially mediated by the nuclear accumulation of Cln3 translated in the cytoplasm. As the nucleolus (*stripes*) occupies roughly one-third of the yeast nucleus (*gray*) in cells grown in rich media, changing nucleolar size could affect nuclear volume and thereby the nucleo-cytoplasmic volume ratio. In cells grown in poor media or in cells with defects in ribosome biogenesis, the observed decreases in cell size could be due to hypothetical decreases in nucleolar and nuclear volume. (D) Several proteins implicated in ribosome biogenesis interact with pre-replication complexes. Noc3 is part of a pre-60S particle but is also required to recruit Cdc6 and the MCM complex to origins of replication (ARS element) (Fatica and Tollervey 2002; Zhang et al. 2002). It is not known whether Noc3 performs this replication function in the context of a pre-ribosomal particle. The ORC complex, which binds directly to the ARS element, was found to interact with five additional proteins implicated in ribosome biogenesis, including Yph1 (Du and Stillman 2002). The role of Yph1 at the replication origin is not well characterized, and it is not known whether ORC interacts with pre-ribosomal particles. Note that each of the 100–200 rRNA repeats contains an ARS element (Fig. 2).

should be noted, however, that this correlation was established by altering growth rates by varying nutrient conditions (Johnston et al. 1979; Tyson et al. 1979), which have wide-ranging effects on the cell, aside from increasing overall translation rates. If protein synthetic rate directly determines critical cell size, then the correlation should also exist when growth rate is altered by a genetic lesion in the protein synthesis apparatus. However, recent analysis of a number of mutant strains with defects in ribosome biogenesis contradicts this simple view. A systematic screen for mutants that exhibit partial uncoupling of growth and division revealed that deletion of any of five nonessential genes implicated in ribosome biogenesis causes a small cell size (Jorgensen et al. 2002). In budding yeast, small cell size is referred to as a Whiskey or Whi phenotype for an infamous bet over a bottle of whiskey that such mutants could indeed be isolated (Sudbery et al. 1980). Furthermore, ten essential genes, all involved in ribosome biogenesis, exhibit a haploinsufficient Whi phenotype (Jorgensen et al. 2002). These reductions in cell size are not simply a result of slow growth, because mutations in ribosomal structural genes that similarly retard growth rate have less effect on cell size. Thus, a number of ribosome biogenesis factors appear to couple growth and division. Moreover, two of the strongest *whi* mutants isolated, *sfp1Δ* and *sch9Δ*, correspond to factors that activate transcription of the Ribi regulon and perhaps the RP regulon (Jorgensen et al. 2002; P. Jorgensen and M. Tyers, in prep.). These findings suggest a model in which the process of ribosome biogenesis itself sends an inhibitory signal to the Start machinery (Fig. 6B).

The effects of ribosome biogenesis on Start are consistent with the observation that the critical cell size threshold increases with growth rate, as does the rate of ribosome biogenesis (Johnston et al. 1979). As noted above, as growth slows in late log phase, ribosome synthesis ceases, while cell division continues for one or two additional cycles (Ju and Warner 1994). The critical issue of whether the reduction in ribosome biogenesis is accompanied by a decrease in critical cell size in subsequent divisions has not been examined carefully. In the converse situation, in which cells are shifted from a poor carbon source such as ethanol to a rich carbon source such as glucose, rRNA transcription and RP expression increase almost immediately, whereas translation rate per se does not increase for nearly an hour (Kief and Warner 1981). As critical cell size is reset immediately upon such shifts (Johnston et al. 1979; Lorincz and Carter 1979), the process of ribosome biogenesis may in reality be more closely linked with critical cell size than translation rates.

The nature of the inhibitory signal from ribosome biogenesis remains to be determined, but at least two possible scenarios can be imagined. In the first, certain ribosome biogenesis mutations might compact the nucleolus, and thereby reduce nuclear volume (Fig. 6C). For unknown reasons, the nuclear to cytoplasmic volume ratio often correlates with cell size in many species (Wilson 1925; Conlon and Raff 1999); as with most eukaryotic cells, yeast cell size increases with increasing ploidy (Mortimer 1958). It is difficult to alter nuclear DNA content, and hence nuclear volume, without complications such as altered gene dosage and checkpoint-dependent cell cycle delays. However, when DNA content is increased by 10% by introducing an abundant but otherwise benign 2μ plasmid into yeast, critical cell size is increased by a roughly proportional amount (B. Futcher, pers. comm.). Thus, a possible consequence of decreased nucleolar volume in ribosome biogenesis mutants would be to reduce the critical cell size threshold. In the second scenario, the ribosome biogenesis machinery might be directly linked to Start through one or more rate-limiting factors. The nature of this putative signal is unknown, but interestingly, a dominant negative version of the ribosome biogenesis factor Bop1 in mouse cells (Erb1 in yeast) causes a p53-dependent G_1 arrest without altering protein synthetic rate (Pestov et al. 2001). In this regard, it is not insignificant that p53 and its associated regulatory factors Mdm2 and Arf are controlled in part through localization to the nucleolar compartment. Indeed, recent evidence suggests that Arf may directly influence rates of ribosome biogenesis (Sugimoto et al. 2003). The conservation of the cell cycle and ribosome biogenesis machineries, and the obvious need to couple division with growth, argue that deep links between these two processes may have been forged early in evolution.

A second and quite unanticipated connection between ribosome biogenesis and the cell cycle has recently come to light. That is, proteins involved in ribosome biogenesis appear to physically interact with replication origins (Fig. 6D). In one case, a protein implicated in transport of 60S pre-ribosomal particles out of the nucleus, Noc3 (Milkereit et al. 2001), localizes to replication origins and interacts with both the ORC and MCM complexes (Zhang et al. 2002). Noc3 is required to load Cdc6 and the MCM complex onto chromatin, an essential step in forming pre-replicative origins, such that conditional alleles of *NOC3* arrest with unfired replication origins despite having undergone bud emergence (Zhang et al. 2002). In a parallel study, the ORC complex was found to interact with a number of ribosome biogenesis factors, including Yph1(Nop7), Nog1, Nop5, Nsa3(Cic1), and Rpf2 (Du and Stillman 2002). Yph1 also appears to be necessary for efficient entry into S phase and, by virtue of its depletion as cells enter stationary phase, this requirement may help ensure that replication is not initiated under adverse conditions (Du and Stillman 2002). Whether these findings are a consequence of specific interactions between the many replication origins in the rDNA locus (Fig.

2) and the ribosome biogenesis machinery, or whether they reflect a more general role for ribosome biogenesis factors in DNA replication, remains to be determined. The apparent positive role for ribosome biogenesis factors in S phase contrasts with their repressive function at Start, and underscores the potential complexity of linkages between the ribosome biogenesis and cell cycle machineries.

In a third cell cycle connection, the nucleolus has been shown to play an important role in the M-to-G_1 phase transition, also known as mitotic exit or telophase. For most of the cell cycle, the Cdc14 phosphatase is sequestered in an inactive state in the nucleolus via an interaction with the nucleolar protein Net1 (Shou et al. 1999; Visintin et al. 1999). Once cells have successfully segregated the daughter nucleus into the bud at the end of anaphase, a complex signaling pathway, called the *mitotic exit network* (MEN), triggers release of Cdc14 into the nucleus and cytoplasm, where it reverses numerous Clb-Cdc28 phosphorylation events and thereby switches the cell into the low-CDK state of G_1 phase (Visintin et al. 1998). Net1 plays a ubiquitous role in nucleolar function, as it is required for proper nucleolar structure, binds and cooperates with Sir2 to mediate rDNA silencing, and helps activate RNA Pol I-dependent transcription (Straight et al. 1999; Shou et al. 2001). Further work is required to solidify and further explore the connections between ribosome synthesis and the cell cycle, but it is likely that many of these regulatory connections will be conserved from yeast to humans. The elaboration of these preliminary connections may herald major advances in understanding how cell growth and division are coupled.

CONCLUDING REMARKS

Although steady progress has been made in characterizing the RP and rRNA promoters and the events upstream of transcriptional induction, the physical connections between upstream signals and downstream effectors remain mysterious. Of particular import is whether the numerous stimuli that control the four transcriptional programs important for ribosome synthesis control a common point upstream of all four programs or whether these stimuli impinge separately on the four sets of promoters (Figs. 1 and 3). A flood of genomic and proteomic information has already stimulated numerous advances, such as the identification of new transcriptional regulators (e.g., Fhl1), new co-regulated gene sets (e.g., the Ribi regulon), and massive particles that process rRNA and assemble ribosomes. Because the synthesis of new ribosomes dominates the yeast cell economy and employs hundreds of gene products, the true extent of this process and the pathways that regulate it can only be appreciated in a genomic context, an outlook that is now becoming entrenched in the field (Wade et al. 2001; Dragon et al. 2002; Grandi et al. 2002; Nissan et al. 2002; Milkereit et al. 2003; Peng et al. 2003).

Compared to cell proliferation, the study of eukaryotic cell growth and ribosome synthesis has been rather badly neglected. But the fact that cell growth is typically limiting for cell cycle progression in most organisms portends an inevitable renaissance. In particular, emergent connections between growth control and many oncogenes and tumor suppressors are certain to heighten general interest in ribosome biosynthesis and its attendant regulatory pathways. Application of the unparalleled genomic technologies and molecular genetics of budding yeast will forge the way in illuminating the tightly harnessed and coordinated nature of ribosome synthesis.

ACKNOWLEDGMENTS

Work from the authors' laboratories was supported in part by grants from the National Institutes of Health (GM-25532 to J.R.W. and CAI-3330 to the Albert Einstein Cancer Center) and grants from the Canadian Institutes of Health Research (CIHR) to M.T. P.J. is supported by a CIHR Studentship and M.T. holds a Canada Research Chair in Biochemistry.

REFERENCES

Albert A.C., Denton M., Kermekchiev M., and Pikaard C.S. 1999. Histone acetyltransferase and protein kinase activities copurify with a putative *Xenopus* RNA polymerase I holoenzyme self-sufficient for promoter-dependent transcription. *Mol. Cell. Biol.* **19:** 796–806.

Andersen J.S., Lyon C.E., Fox A.H., Leung A.K., Lam Y.W., Steen H., Mann M., and Lamond A.I. 2002. Directed proteomic analysis of the human nucleolus. *Curr. Biol.* **12:** 1–11.

Ares M., Jr., Grate L., and Pauling M.H. 1999. A handful of intron-containing genes produces the lion's share of yeast mRNA. *RNA* **5:** 1138–1139.

Barbet N.C., Schneider U., Helliwell S.B., Stansfield I., Tuite M.F., and Hall M.N. 1996. TOR controls translation initiation and early G1 progression in yeast. *Mol. Biol. Cell* **7:** 25–42.

Baroni M.D., Monti P., and Alberghina L. 1994. Repression of growth-regulated G1 cyclin expression by cyclic AMP in budding yeast. *Nature* **371:** 339–342.

Basu U., Si K., Warner J.R., and Maitra U. 2001. The *Saccharomyces cerevisiae TIF6* gene encoding translation initiation factor 6 is required for 60S ribosomal subunit biogenesis. *Mol. Cell. Biol.* **21:** 1453–1462.

Blumberg H. and Silver P. 1991. A split zinc-finger protein is required for normal yeast growth. *Gene* **107:** 101–110.

Bodem J., Dobreva G., Hoffmann-Rohrer U., Iben S., Zentgraf H., Delius H., Vingron M., and Grummt I. 2000. TIF-IA, the factor mediating growth-dependent control of ribosomal RNA synthesis, is the mammalian homolog of yeast Rrn3p. *EMBO Rep.* **1:** 171–175.

Boon K., Caron H.N., van Asperen R., Valentijn L., Hermus M.C., van Sluis P., Roobeek I., Weis I., Voute P.A., Schwab M., and Versteeg R. 2001. N-myc enhances the expression of a large set of genes functioning in ribosome biogenesis and protein synthesis. *EMBO J.* **20:** 1383–1393.

Bordi L., Cioci F., and Camilloni G. 2001. In vivo binding and hierarchy of assembly of the yeast RNA polymerase I transcription factors. *Mol. Biol. Cell* **12:** 753–760.

Brenner C., Nakayama N., Goebl M., Tanaka K., Toh-e A., and Matsumoto K. 1988. *CDC33* encodes mRNA cap-binding protein eIF-4E of *Saccharomyces cerevisiae*. *Mol. Cell. Biol.* **8:** 3556–3559.

Cardenas M.E., Cutler N.S., Lorenz M.C., Di Como C.J., and Heitman J. 1999. The TOR signaling cascade regulates gene expression in response to nutrients. *Genes Dev.* **13:** 3271–3279.

Causton H.C., Ren B., Koh S.S., Harbison C.T., Kanin E., Jennings E.G., Lee T.I., True H.L., Lander E.S., and Young R.A. 2001. Remodeling of yeast genome expression in response to environmental changes. *Mol. Biol. Cell* **12:** 323–337.

Cherel I. and Thuriaux P. 1995. The *IFH1* gene product interacts with a fork head protein in *Saccharomyces cerevisiae*. *Yeast* **11:** 261–270.

Chu S., DeRisi J., Eisen M., Mulholland J., Botstein D., Brown P.O., and Herskowitz I. 1998. The transcriptional program of sporulation in budding yeast. *Science* **282:** 699–705.

Cliften P., Sudarsanam P., Desikan A., Fulton L., Fulton B., Majors J., Waterson R., Cohen B.A., and Johnston M. 2003. Finding functional features in *Saccharomyces* genomes by phylogenetic footprinting. *Science* **301:** 71–76.

Coller H.A., Grandori C., Tamayo P., Colbert T., Lander E.S., Eisenman R.N., and Golub T.R. 2000. Expression analysis with oligonucleotide microarrays reveals that MYC regulates genes involved in growth, cell cycle, signaling, and adhesion. *Proc. Natl. Acad. Sci.* **97:** 3260–3265.

Conlon I. and Raff M. 1999. Size control in animal development. *Cell* **96:** 235–244.

Dabeva M.D. and Warner J.R. 1993. Ribosomal protein L32 of *Saccharomyces cerevisiae* regulates both splicing and translation of its own transcript. *J. Biol. Chem.* **268:** 19669–19674.

Dabeva M.D., Post-Beittenmiller M.A., and Warner J.R. 1986. Autogenous regulation of splicing of the transcript of a yeast ribosomal protein gene. *Proc. Natl. Acad. Sci.* **83:** 5854–5857.

Dammann R., Lucchini R., Koller T., and Sogo J.M. 1993. Chromatin structures and transcription of rDNA in yeast *Saccharomyces cerevisiae*. *Nucleic Acids Res.* **21:** 2331–2338.

Danaie P., Altmann M., Hall M.N., Trachsel H., and Helliwell S.B. 1999. *CLN3* expression is sufficient to restore G1-to-S-phase progression in *Saccharomyces cerevisiae* mutants defective in translation initiation factor eIF4E. *Biochem. J.* **340:** 135–141.

DeRisi J.L., Iyer V.R., and Brown P.O. 1997. Exploring the metabolic and genetic control of gene expression on a genomic scale. *Science* **278:** 680–686.

Di Como C.J. and Arndt K.T. 1996. Nutrients, via the Tor proteins, stimulate the association of Tap42 with type 2A phosphatases. *Genes Dev.* **10:** 1904–1916.

Dragon F., Gallagher J.E., Compagnone-Post P.A., Mitchell B.M., Porwancher K.A., Wehner K.A., Wormsley S., Settlage R.E., Shabanowitz J., Osheim Y., Beyer A.L., Hunt D.F., and Baserga S.J. 2002. A large nucleolar U3 ribonucleoprotein required for 18S ribosomal RNA biogenesis. *Nature* **417:** 967–970.

Du Y.C. and Stillman B. 2002. Yph1p, an ORC-interacting protein: Potential links between cell proliferation control, DNA replication, and ribosome biogenesis. *Cell* **109:** 835–848.

elBaradi T.T., van der Sande C.A., Mager W.H., Raue H.A., and Planta R.J. 1986. The cellular level of yeast ribosomal protein L25 is controlled principally by rapid degradation of excess protein. *Curr. Genet.* **10:** 733–739.

Elion E.A. and Warner J.R. 1984. The major promoter element of rRNA transcription in yeast lies 2 kb upstream. *Cell* **39:** 663–673.

Elliott S.G. and McLaughlin C.S. 1978. Rate of macromolecular synthesis through the cell cycle of the yeast *Saccharomyces cerevisiae*. *Proc. Natl. Acad. Sci.* **75:** 4384–4388.

Emanuel J.R. and Magee P.T. 1981. Timing of ribosome synthesis during ascosporogenesis of yeast cells: Evidence for early function of haploid daughter genomes. *J. Bacteriol.* **145:** 1342–1350.

Eng F.J. and Warner J.R. 1991. Structural basis for the regulation of splicing of a yeast messenger RNA. *Cell* **65:** 797–804.

Fabrizio P., Pozza F., Pletcher S.D., Gendron C.M., and Longo V.D. 2001. Regulation of longevity and stress resistance by Sch9 in yeast. *Science* **292:** 288–290.

Fath S., Milkereit P., Peyroche G., Riva M., Carles C., and Tschochner H. 2001. Differential roles of phosphorylation in the formation of transcriptional active RNA polymerase I. *Proc. Natl. Acad. Sci.* **98:** 14334–14339.

Fatica A. and Tollervey D. 2002. Making ribosomes. *Curr. Opin. Cell Biol.* **14:** 313–318.

Fewell S.W. and Woolford J.L., Jr. 1999. Ribosomal protein S14 of *Saccharomyces cerevisiae* regulates its expression by binding to RPS14B pre-mRNA and to 18S rRNA. *Mol. Cell. Biol.* **19:** 826–834.

Finley D., Bartel B., and Varshavsky A. 1989. The tails of ubiquitin precursors are ribosomal proteins whose fusion to ubiquitin facilitates ribosome biogenesis. *Nature* **338:** 394–401.

Fonagy A., Swiderski C., and Freeman J.W. 1995. Nucleolar p120 is expressed as a delayed early response gene and is inducible by DNA-damaging agents. *J. Cell. Physiol.* **164:** 634–643.

Francois J., Villanueva M.E., and Hers H.G. 1988. The control of glycogen metabolism in yeast. 1. Interconversion in vivo of glycogen synthase and glycogen phosphorylase induced by glucose, a nitrogen source or uncouplers. *Eur. J. Biochem.* **174:** 551–559.

French S.L., Osheim Y., Cioci F., Nomura M., and Beyer A. 2003. In exponentially growing *Saccharomyces cerevisiae* cells, ribosomal RNA synthesis is determined by the summed RNA polymerase I loading rate rather than by the number of active genes. *Mol. Cell. Biol.* **23:** 1558–1568.

Furger A., O'Sullivan J.M., Binnie A., Lee B.A., and Proudfoot N.J. 2002. Promoter proximal splice sites enhance transcription. *Genes Dev.* **16:** 2792–2799.

Futcher B. 1996. Cyclins and the wiring of the yeast cell cycle. *Yeast* **12:** 1635–1646.

—-. 2002. Transcriptional regulatory networks and the yeast cell cycle. *Curr. Opin. Cell Biol.* **14:** 676–683.

Gadal O., Labarre S., Boschiero C., and Thuriaux P. 2002. Hmo1, an HMG-box protein, belongs to the yeast ribosomal DNA transcription system. *EMBO J.* **21:** 5498–5507.

Gari E., Volpe T., Wang H., Gallego C., Futcher B., and Aldea M. 2001. Whi3 binds the mRNA of the G1 cyclin *CLN3* to modulate cell fate in budding yeast. *Genes Dev.* **15:** 2803–2808.

Gasch A.P., Huang M., Metzner S., Botstein D., Elledge S.J., and Brown P.O. 2001. Genomic expression responses to DNA-damaging agents and the regulatory role of the yeast ATR homolog Mec1p. *Mol. Biol. Cell* **12:** 2987–3003.

Gasch A.P., Spellman P.T., Kao C.M., Carmel-Harel O., Eisen M.B., Storz G., Botstein D., and Brown P.O. 2000. Genomic expression programs in the response of yeast cells to environmental changes. *Mol. Biol. Cell* **11:** 4241–4257.

Gavin A.C., Bosche M., Krause R., Grandi P., Marzioch M., Bauer A., Schultz J., Rick J.M., Michon A.M., and Cruciat C.M., et al. 2002. Functional organization of the yeast proteome by systematic analysis of protein complexes. *Nature* **415:** 141–147.

Ghavidel A. and Schultz M.C. 2001. TATA binding protein-associated CK2 transduces DNA damage signals to the RNA polymerase III transcriptional machinery. *Cell* **106:** 575–584.

Gottesfeld J.M., Wolf V.J., Dang T., Forbes D.J., and Hartl P. 1994. Mitotic repression of RNA polymerase III transcription in vitro mediated by phosphorylation of a TFIIIB component. *Science* **263:** 81–84.

Gourse R.L., Gaal T., Bartlett M.S., Appleman J.A., and Ross W. 1996. rRNA transcription and growth rate-dependent regulation of ribosome synthesis in *Escherichia coli*. *Annu. Rev. Microbiol.* **50:** 645–677.

Grandi P., Rybin V., Bassler J., Petfalski E., Strauss D., Marzioch M., Schafer T., Kuster B., Tschochner H., Tollervey D., Gavin A.C., and Hurt E. 2002. 90S pre-ribosomes include the 35S pre-rRNA, the U3 snoRNP, and 40S subunit processing factors but predominantly lack 60S synthesis factors. *Mol. Cell* **10:** 105–115.

Greasley P.J., Bonnard C., and Amati B. 2000. Myc induces the nucleolin and BN51 genes: Possible implications in ribosome biogenesis. *Nucleic Acids Res.* **28:** 446–453.

Hall D.D., Markwardt D.D., Parviz F., and Heideman W. 1998. Regulation of the Cln3-Cdc28 kinase by cAMP in *Saccharomyces cerevisiae*. *EMBO J.* **17:** 4370–4378.

Hamil K.G., Nam H.G., and Fried H.M. 1988. Constitutive transcription of yeast ribosomal protein gene *TCM1* is promoted by uncommon *cis*- and *trans*-acting elements. *Mol. Cell. Biol.* **8:** 4328–4341.

Hanic-Joyce P.J., Johnston G.C., and Singer R.A. 1987. Regulated arrest of cell proliferation mediated by yeast *prt1* mutations. *Exp. Cell Res.* **172:** 134–145.

Hannan R.D., Hempel W.M., Cavanaugh A., Arino T., Dimitrov S.I., Moss T., and Rothblum L. 1998. Affinity purification of mammalian RNA polymerase I. Identification of an associated kinase. *J. Biol. Chem.* **273:** 1257–1267.

Hardwick J.S., Kuruvilla F.G., Tong J.K., Shamji A.F., and Schreiber S.L. 1999. Rapamycin-modulated transcription defines the subset of nutrient-sensitive signaling pathways directly controlled by the Tor proteins. *Proc. Natl. Acad. Sci.* **96:** 14866–14870.

Hartwell L.H., McLaughlin C.S., and Warner J.R. 1970. Identification of ten genes that control ribosome formation in yeast. *Mol. Gen. Genet.* **109:** 42–56.

Hartwell L.H., Culotti J., Pringle J.R., and Reid B.J. 1974. Genetic control of the cell division cycle in yeast. *Science* **183:** 46–51.

Henras A., Dez C., Noaillac-Depeyre J., Henry Y., and Caizergues-Ferrer M. 2001. Accumulation of H/ACA snoRNPs depends on the integrity of the conserved central domain of the RNA-binding protein Nhp2p. *Nucleic Acids Res.* **29:** 2733–2746.

Herblot S., Chastagner P., Samady L., Moreau J.L., Demaison C., Froussard P., Liu X., Bonnet J., and Theze J. 1999. IL-2-dependent expression of genes involved in cytoskeleton organization, oncogene regulation, and transcriptional control. *J. Immunol.* **162:** 3280–3288.

Herman P.K. 2002. Stationary phase in yeast. *Curr. Opin. Microbiol.* **5:** 602–607.

Hermann-Le Denmat S., Werner M., Sentenac A., and Thuriaux P. 1994. Suppression of yeast RNA polymerase III mutations by *FHL1*, a gene coding for a fork head protein involved in rRNA processing. *Mol. Cell. Biol.* **14:** 2905–2913.

Hershko A. and Ciechanover A. 1998. The ubiquitin system. *Annu. Rev. Biochem.* **67:** 425–479.

Ho J.H. and Johnson A.W. 1999. *NMD3* encodes an essential cytoplasmic protein required for stable 60S ribosomal subunits in *Saccharomyces cerevisiae. Mol. Cell. Biol.* **19:** 2389–2399.

Ho Y., Gruhler A., Heilbut A., Bader G.D., Moore L., Adams S.L., Millar A., Taylor P., Bennett K., and Boutilier K., et al. 2002. Systematic identification of protein complexes in *Saccharomyces cerevisiae* by mass spectrometry. *Nature* **415:** 180–183.

Hockman D.J. and Schultz M.C. 1996. Casein kinase II is required for efficient transcription by RNA polymerase III. *Mol. Cell. Biol.* **16:** 892–898.

Holstege F.C., Jennings E.G., Wyrick J.J., Lee T.I., Hengartner C.J., Green M.R., Golub T.R., Lander E.S., and Young R.A. 1998. Dissecting the regulatory circuitry of a eukaryotic genome. *Cell* **95:** 717–728.

Hubler L., Bradshaw-Rouse J., and Heideman W. 1993. Connections between the Ras-cyclic AMP pathway and G1 cyclin expression in the budding yeast *Saccharomyces cerevisiae. Mol. Cell. Biol.* **13:** 6274–6282.

Hughes J.D., Estep P.W., Tavazoie S., and Church G.M. 2000. Computational identification of *cis*-regulatory elements associated with groups of functionally related genes in *Saccharomyces cerevisiae. J. Mol. Biol.* **296:** 1205–1214.

Iben S., Tschochner H., Bier M., Hoogstraten D., Hozak P., Egly J.M., and Grummt I. 2002. TFIIH plays an essential role in RNA polymerase I transcription. *Cell* **109:** 297–306.

Iritani B.M., Delrow J., Grandori C., Gomez I., Klacking M., Carlos L.S., and Eisenman R.N. 2002. Modulation of T-lymphocyte development, growth and cell size by the Myc antagonist and transcriptional repressor Mad1. *EMBO J.* **21:** 4820–4830.

Ittmann M.M. 1994. Cell cycle control of the BN51 cell cycle gene which encodes a subunit of RNA polymerase III. *Cell Growth Differ.* **5:** 783–788.

Ivessa A.S. and Zakian V.A. 2002. To fire or not to fire: Origin activation in *Saccharomyces cerevisiae* ribosomal DNA. *Genes Dev.* **16:** 2459–2464.

Iyer V. and Struhl K. 1995. Poly(dA:dT), a ubiquitous promoter element that stimulates transcription via its intrinsic DNA structure. *EMBO J.* **14:** 2570–2579.

Jiang Y. and Broach J.R. 1999. Tor proteins and protein phosphatase 2A reciprocally regulate Tap42 in controlling cell growth in yeast. *EMBO J.* **18:** 2782–2792.

Johnson S.P. and Warner J.R. 1989. Unusual enhancer function in yeast rRNA transcription. *Mol. Cell. Biol.* **9:** 4986–4993.

Johnston G.C., Pringle J.R., and Hartwell L.H. 1977. Coordination of growth with cell division in the yeast *Saccharomyces cerevisiae. Exp. Cell Res.* **105:** 79–98.

Johnston G.C., Ehrhardt C.W., Lorincz A., and Carter B.L. 1979. Regulation of cell size in the yeast *Saccharomyces cerevisiae. J. Bacteriol.* **137:** 1–5.

Jorgensen P., Nishikawa J.L., Breitkreutz B.J., and Tyers M. 2002. Systematic identification of pathways that couple cell growth and division in yeast. *Science* **297:** 395–400.

Ju Q. and Warner J.R. 1994. Ribosome synthesis during the growth cycle of *Saccharomyces cerevisiae. Yeast* **10:** 151–157.

Kief D.R. and Warner J.R. 1981. Coordinate control of syntheses of ribosomal ribonucleic acid and ribosomal proteins during nutritional shift-up in *Saccharomyces cerevisiae. Mol. Cell. Biol.* **1:** 1007–1015.

Kinoshita Y., Jarell A.D., Flaman J.M., Foltz G., Schuster J., Sopher B.L., Irvin D.K., Kanning K., Kornblum H.I., Nelson P. S., Hieter P., and Morrison R.S. 2001. Pescadillo, a novel cell cycle regulatory protein abnormally expressed in malignant cells. *J. Biol. Chem.* **276:** 6656–6665.

Klein C. and Struhl K. 1994. Protein kinase A mediates growth-regulated expression of yeast ribosomal protein genes by modulating RAP1 transcriptional activity. *Mol. Cell. Biol.* **14:** 1920–1928.

Kuhn A., Vente A., Doree M., and Grummt I. 1998. Mitotic phosphorylation of the TBP-containing factor SL1 represses ribosomal gene transcription. *J. Mol. Biol.* **284:** 1–5.

Kuhn K.M., DeRisi J.L., Brown P.O., and Sarnow P. 2001. Global and specific translational regulation in the genomic response of *Saccharomyces cerevisiae* to a rapid transfer from a fermentable to a nonfermentable carbon source. *Mol. Cell. Biol.* **21:** 916–927.

Lang W.H., Morrow B.E., Ju Q., Warner J.R., and Reeder R.H. 1994. A model for transcription termination by RNA polymerase I. *Cell* **79:** 527–534.

Lascaris R.F., Mager W.H., and Planta R.J. 1999. DNA-binding requirements of the yeast protein Rap1p as selected in silico from ribosomal protein gene promoter sequences. *Bioinformatics* **15:** 267–277.

Leary D.J. and Huang S. 2001. Regulation of ribosome biogenesis within the nucleolus. *FEBS Lett.* **509:** 145–150.

Lee T.I., Rinaldi N.J., Robert F., Odom D.T., Bar-Joseph Z., Gerber G.K., Hannett N.M., Harbison C.T., Thompson C.M., and Simon I., et al. 2002. Transcriptional regulatory networks in *Saccharomyces cerevisiae*. *Science* **298:** 799–804.

Li B. and Warner J.R. 1996. Mutation of the Rab6 homologue of *Saccharomyces cerevisiae*, YPT6, inhibits both early Golgi function and ribosome biosynthesis. *J. Biol. Chem.* **271:** 16813–16819.

Li B., Nierras C.R., and Warner J.R. 1999. Transcriptional elements involved in the repression of ribosomal protein synthesis. *Mol. Cell. Biol.* **19:** 5393–5404.

Li B., Vilardell J., and Warner J.R. 1996. An RNA structure involved in feedback regulation of splicing and of translation is critical for biological fitness. *Proc. Natl. Acad. Sci.* **93:** 1596–1600.

Li Y., Moir R.D., Sethy-Coraci I.K., Warner J.R., and Willis I. M. 2000. Repression of ribosome and tRNA synthesis in secretion-defective cells is signaled by a novel branch of the cell integrity pathway. *Mol. Cell. Biol.* **20:** 3843–3851.

Li Z., Paulovich A.G., and Woolford J.L., Jr. 1995. Feedback inhibition of the yeast ribosomal protein gene *CRY2* is mediated by the nucleotide sequence and secondary structure of *CRY2* pre-mRNA. *Mol. Cell. Biol.* **15:** 6454–6464.

Lorincz A. and Carter B.L.A. 1979. Control of cell size at bud initiation in *Saccharomyces cerevisiae*. *J. Gen. Microbiol.* **113:** 287–295.

Maaloe O. and Kjeldgaard N.O. 1966. *Control of macromolecular synthesis.* W.A. Benjamin, New York.

Mager W.H., Planta R.J., Ballesta J.G., Lee J.C., Mizuta K., Suzuki K., Warner J.R., and Woolford J. 1997. A new nomenclature for the cytoplasmic ribosomal proteins of *Saccharomyces cerevisiae*. *Nucleic Acids Res.* **25:** 4872–4875.

Maicas E., Pluthero F.G., and Friesen J.D. 1988. The accumulation of three yeast ribosomal proteins under conditions of excess mRNA is determined primarily by fast protein decay. *Mol. Cell. Biol.* **8:** 169–175.

Mao H., White S.A., and Williamson J.R. 1999. A novel loop-loop recognition motif in the yeast ribosomal protein L30 autoregulatory RNA complex. *Nat. Struct. Biol.* **6:** 1139–1147.

McInerny C.J., Partridge J.F., Mikesell G.E., Creemer D.P., and Breeden L.L. 1997. A novel Mcm1-dependent element in the *SWI4, CLN3, CDC6,* and *CDC47* promoters activates M/G1-specific transcription. *Genes Dev.* **11:** 1277–1288.

Mencia M., Moqtaderi Z., Geisberg J.V., Kuras L., and Struhl K. 2002. Activator-specific recruitment of TFIID and regulation of ribosomal protein genes in yeast. *Mol. Cell* **9:** 823–833.

Meyuhas O. and Hornstein E. 2000. Translational control of TOP mRNAs. In *Translational control of gene expression* (ed. N. Sonenberg et al.), pp. 671–693, Cold Spring Harbor Laboratory Press, Cold Spring Harbor, New York.

Milkereit P. and Tschochner H. 1998. A specialized form of RNA polymerase I, essential for initiation and growth-dependent regulation of rRNA synthesis, is disrupted during transcription. *EMBO J.* **17:** 3692–3703.

Milkereit P., Kuhn H., Gas N., and Tschochner H. 2003. The pre-ribosomal network. *Nucleic Acids Res.* **31:** 799–804.

Milkereit P., Gadal O., Podtelejnikov A., Trumtel S., Gas N., Petfalski E., Tollervey D., Mann M., Hurt E., and Tschochner H. 2001. Maturation and intranuclear transport of pre-ribosomes requires Noc proteins. *Cell* **105:** 499–509.

Miller O.L., Jr. and Beatty B.R. 1969. Visualization of nucleolar genes. *Science* **164:** 955–957.

Miyoshi K., Miyakawa T., and Mizuta K. 2001. Repression of rRNA synthesis due to a secretory defect requires the C-terminal silencing domain of Rap1p in *Saccharomyces cerevisiae. Nucleic Acids Res.* **29:** 3297–3303.

Miyoshi K., Shirai C., and Mizuta K. 2003. Transcription of genes encoding trans-acting factors required for rRNA maturation/ribosomal subunit assembly is coordinately regulated with ribosomal protein genes and involves Rap1 in *Saccharomyces cerevisiae. Nucleic Acids Res.* **31:** 1969–1973.

Mizuta K. and Warner J.R. 1994. Continued functioning of the secretory pathway is essential for ribosome synthesis. *Mol. Cell. Biol.* **14:** 2493–2502.

Mizuta K., Tsujii R., Warner J.R., and Nishiyama M. 1998. The C-terminal silencing domain of Rap1p is essential for the repression of ribosomal protein genes in response to a defect in the secretory pathway. *Nucleic Acids Res.* **26:** 1063–1069.

Moore S.A. 1988. Kinetic evidence for a critical rate of protein synthesis in the *Saccharomyces cerevisiae* yeast cell cycle. *J. Biol. Chem.* **263:** 9674–9681.

Morse R.H. 2000. RAP, RAP, open up! New wrinkles for RAP1 in yeast. *Trends Genet.* **16:** 51–53.

Mortimer R.K. 1958. Radiobiological and genetic studies on a polyploid series (haploid to hexaploid) of *Saccharomyces cerevisiae. Radiat. Res.* **9:** 312–326.

Moss T. and Stefanovsky V.Y. 2002. At the center of eukaryotic life. *Cell* **109:** 545–548.

Mougey E.B., O'Reilly M., Osheim Y., Miller O.L., Jr., Beyer A., and Sollner-Webb B. 1993. The terminal balls characteristic of eukaryotic rRNA transcription units in chromatin spreads are rRNA processing complexes. *Genes Dev.* **7:** 1609–1619.

Nakamoto K., Ito A., Watabe K., Koma Y., Asada H., Yoshikawa K., Shinomura Y., Matsuzawa Y., Nojima H., and Kitamura Y. 2001. Increased expression of a nucleolar Nop5/Sik family member in metastatic melanoma cells: Evidence for its role in nucleolar sizing and function. *Am. J. Pathol.* **159:** 1363–1374.

Nash P., Tang X., Orlicky S., Chen Q., Gertler F.B., Mendenhall M.D., Sicheri F., Pawson T., and Tyers M. 2001. Multisite phosphorylation of a CDK inhibitor sets a threshold for the onset of DNA replication. *Nature* **414:** 514–521.

Nash R., Tokiwa G., Anand S., Erickson K., and Futcher A.B. 1988. The *WHI1+* gene of

Saccharomyces cerevisiae tethers cell division to cell size and is a cyclin homolog. *EMBO J.* **7:** 4335–4346.

Neigeborn L. and Warner J.R. 1990. Expression of yeast 5S RNA is independent of the rDNA enhancer region. *Nucleic Acids Res.* **18:** 4179–4184.

Neiman P.E., Ruddell A., Jasoni C., Loring G., Thomas S.J., Brandvold K.A., Lee R., Burnside J., and Delrow J. 2001. Analysis of gene expression during myc oncogene-induced lymphomagenesis in the bursa of Fabricius. *Proc. Natl. Acad. Sci.* **98:** 6378–6383.

Nelson S.A., Aris J.P., Patel B.K., and LaRochelle W.J. 2000. Multiple growth factor induction of a murine early response gene that complements a lethal defect in yeast ribosome biogenesis. *J. Biol. Chem.* **275:** 13835–13841.

Neuman-Silberberg F.S., Bhattacharya S., and Broach J.R. 1995. Nutrient availability and the RAS/cyclic AMP pathway both induce expression of ribosomal protein genes in *Saccharomyces cerevisiae* but by different mechanisms. *Mol. Cell. Biol.* **15:** 3187–3196.

Newcomb L.L., Hall D.D., and Heideman W. 2002. *AZF1* is a glucose-dependent positive regulator of *CLN3* transcription in *Saccharomyces cerevisiae*. *Mol. Cell. Biol.* **22:** 1607–1614.

Nierras C.R. and Warner J.R. 1999. Protein kinase C enables the regulatory circuit that connects membrane synthesis to ribosome synthesis in *Saccharomyces cerevisiae*. *J. Biol. Chem.* **274:** 13235–13241.

Nikawa J., Cameron S., Toda T., Ferguson K.M., and Wigler M. 1987. Rigorous feedback control of cAMP levels in *Saccharomyces cerevisiae*. *Genes Dev.* **1:** 931–937.

Nissan T.A., Bassler J., Petfalski E., Tollervey D., and Hurt E. 2002. 60S pre-ribosome formation viewed from assembly in the nucleolus until export to the cytoplasm. *EMBO J.* **21:** 5539–5547.

Nomura M. 1999. Regulation of ribosome biosynthesis in *Escherichia coli* and *Saccharomyces cerevisiae:* Diversity and common principles. *J. Bacteriol.* **181:** 6857–6864.

Nomura M., Gourse R., and Baughman G. 1984. Regulation of the synthesis of ribosomes and ribosomal components. *Annu. Rev. Biochem.* **53:** 75–117.

Nurse P. and Fantes P.A. 1981. Cell cycle controls in fission yeast: A genetic analysis. In *The cell cycle* (ed. P.C.L. John), pp. 85-98. Cambridge University Press, London, United Kingdom.

Paulovich A.G., Thompson J.R., Larkin J.C., Li Z., and Woolford J.L., Jr. 1993. Molecular genetics of cryptopleurine resistance in *Saccharomyces cerevisiae:* Expression of a ribosomal protein gene family. *Genetics* **135:** 719–730.

Peng W.T., Robinson M.D., Mnaimneh S., Krogan N.J., Cagney G., Morris Q., Davierwala A.P., Grigull J., Yang X., Zhang W., Mitsakakis N., Ryan O.W., Datta N., Jojic V., Pal C., Canadien V., Richards D., Beattie B., Wu L.F., Altschuler S.J., Roweis S., Frey B.J., Emili A., Greenblatt J.F., and Hughes T.R. 2003. A panoramic view of yeast noncoding RNA processing. *Cell* **113:** 919–933.

Pestov D.G., Strezoska Z., and Lau L.F. 2001. Evidence of p53-dependent cross-talk between ribosome biogenesis and the cell cycle: Effects of nucleolar protein Bop1 on G(1)/S transition. *Mol. Cell. Biol.* **21:** 4246–4255.

Planta R.J. 1997. Regulation of ribosome synthesis in yeast. *Yeast* **13:** 1505–1518.

Polymenis M. and Schmidt E.V. 1997. Coupling of cell division to cell growth by translational control of the G1 cyclin CLN3 in yeast. *Genes Dev.* **11:** 2522–2531.

Powers T. and Walter P. 1999. Regulation of ribosome biogenesis by the rapamycin-sensitive TOR- signaling pathway in *Saccharomyces cerevisiae. Mol. Biol. Cell* **10:** 987–1000.

Reed R. and Hurt E. 2002. A conserved mRNA export machinery coupled to pre-mRNA splicing. *Cell* **108:** 523–531.

Reeder R.H. 1999. Regulation of RNA polymerase I transcription in yeast and vertebrates. *Prog. Nucleic Acid Res. Mol. Biol.* **62:** 293–327.

Reid J.L., Iyer V.R., Brown P.O., and Struhl K. 2000. Coordinate regulation of yeast ribosomal protein genes is associated with targeted recruitment of Esa1 histone acetylase. *Mol. Cell* **6:** 1297–1307.

Robyr D., Suka Y., Xenarios I., Kurdistani S.K., Wang A., Suka N., and Grunstein M. 2002. Microarray deacetylation maps determine genome-wide functions for yeast histone deacetylases. *Cell* **109:** 437–446.

Rohde J.R. and Cardenas M.E. 2003. The TOR pathway regulates gene expression by linking nutrient sensing to histone acetylation. *Mol. Cell. Biol.* **23:** 629–635.

Rohde J., Heitman J., and Cardenas M.E. 2001. The TOR kinases link nutrient sensing to cell growth. *J. Biol. Chem.* **276:** 9583–9586.

Rosbash M., Harris P.K., Woolford J.L., Jr., and Teem J.L. 1981. The effect of temperature-sensitive RNA mutants on the transcription products from cloned ribosomal protein genes of yeast. *Cell* **24:** 679–686.

Rupes I. 2002. Checking cell size in yeast. *Trends Genet.* **18:** 479–485.

Sandmeier J.J., French S., Osheim Y., Cheung W.L., Gallo C.M., Beyer A.L., and Smith J.S. 2002. *RPD3* is required for the inactivation of yeast ribosomal DNA genes in stationary phase. *EMBO J.* **21:** 4959–4968.

Santoro R., Li J., and Grummt I. 2002. The nucleolar remodeling complex NoRC mediates heterochromatin formation and silencing of ribosomal gene transcription. *Nat. Genet.* **32:** 393–396.

Scherl A., Coute Y., Deon C., Calle A., Kindbeiter K., Sanchez J.C., Greco A., Hochstrasser D., and Diaz J.J. 2002. Functional proteomic analysis of human nucleolus. *Mol. Biol. Cell* **13:** 4100–4109.

Schmelzle T. and Hall M.N. 2000. TOR, a central controller of cell growth. *Cell* **103:** 253–262.

Schneider D.A., Gaal T., and Gourse R.L. 2002. NTP-sensing by rRNA promoters in *Escherichia coli* is direct. *Proc. Natl. Acad. Sci.* **99:** 8602–8607.

Schramm L. and Hernandez N. 2002. Recruitment of RNA polymerase III to its target promoters. *Genes Dev.* **16:** 2593–2620.

Senger B., Lafontaine D.L., Graindorge J.S., Gadal O., Camasses A., Sanni A., Garnier J.M., Breitenbach M., Hurt E., and Fasiolo F. 2001. The nucle(ol)ar Tif6p and Efl1p are required for a late cytoplasmic step of ribosome synthesis. *Mol. Cell* **8:** 1363–1373.

Shamji A.F., Kuruvilla F.G., and Schreiber S.L. 2000. Partitioning the transcriptional program induced by rapamycin among the effectors of the Tor proteins. *Curr. Biol.* **10:** 1574–1581.

Shou W., Seol J.H., Shevchenko A., Baskerville C., Moazed D., Chen Z.W., Jang J., Charbonneau H., and Deshaies R.J. 1999. Exit from mitosis is triggered by Tem1-dependent release of the protein phosphatase Cdc14 from nucleolar RENT complex. *Cell* **97:** 233–244.

Shou W., Sakamoto K.M., Keener J., Morimoto K.W., Traverso E.E., Azzam R., Hoppe G.J., Feldman R.M., DeModena J., Moazed D., Charbonneau H., Nomura M., and Deshaies R.J. 2001. Net1 stimulates RNA polymerase I transcription and regulates nucleolar structure independently of controlling mitotic exit. *Mol. Cell* **8:** 45–55.

Siddiqi I., Keener J., Vu L., and Nomura M. 2001. Role of TATA binding protein (TBP) in yeast ribosomal DNA transcription by RNA polymerase I: Defects in the dual functions of transcription factor UAF cannot be suppressed by TBP. *Mol. Cell. Biol.* **21:** 2292–2297.

Sillje H.H., ter Schure E.G., Rommens A.J., Huls P.G., Woldringh C.L., Verkleij A.J., Boonstra J., and Verrips C.T. 1997. Effects of different carbon fluxes on G1 phase duration, cyclin expression, and reserve carbohydrate metabolism in *Saccharomyces cerevisiae*. *J. Bacteriol.* **179:** 6560–6565.

Spingola M., Grate L., Haussler D., and Ares M., Jr. 1999. Genome-wide bioinformatic and molecular analysis of introns in *Saccharomyces cerevisiae*. *RNA* **5:** 221–234.

Steffan J.S., Keys D.A., Dodd J.A., and Nomura M. 1996. The role of TBP in rDNA transcription by RNA polymerase I in *Saccharomyces cerevisiae*: TBP is required for upstream activation factor-dependent recruitment of core factor. *Genes Dev.* **10:** 2551–2563.

Stolovich M., Tang H., Hornstein E., Levy G., Cohen R., Bae S.S., Birnbaum M.J., and Meyuhas O. 2002. Transduction of growth or mitogenic signals into translational activation of TOP mRNAs is fully reliant on the phosphatidylinositol 3′-kinase-mediated pathway but requires neither S6K1 nor rpS6 phosphorylation. *Mol. Cell. Biol.* **22:** 8101–8113.

Straight A.F., Shou W., Dowd G.J., Turck C.W., Deshaies R.J., Johnson A.D., and Moazed D. 1999. Net1, a Sir2-associated nucleolar protein required for rDNA silencing and nucleolar integrity. *Cell* **97:** 245–256.

Sudbery P.E., Goodey A.R., and Carter B.L. 1980. Genes which control cell proliferation in the yeast *Saccharomyces cerevisiae*. *Nature* **288:** 401–404.

Sugimoto M., Kuo M.-L., Roussel M., and Sherr C.J. 2003. Nucleolar Arf tumour suppressor inhibits ribosomal RNA processing. *Mol. Cell* **11:** 415–424.

Toda T., Cameron S., Sass P., and Wigler M. 1988. *SCH9*, a gene of *Saccharomyces cerevisiae* that encodes a protein distinct from, but functionally and structurally related to, cAMP-dependent protein kinase catalytic subunits. *Genes Dev.* **2:** 517–527.

Tokiwa G. 1995. "Cell cycle control in *Saccharomyces cerevisiae*: G1 cyclin regulation and a connection with the Ras/cAMP signaling pathways." Ph.D. thesis, State University of New York, Stony Brook.

Tokiwa G., Tyers M., Volpe T., and Futcher B. 1994. Inhibition of G1 cyclin activity by the Ras/cAMP pathway in yeast. *Nature* **371:** 342–345.

Tower J. and Sollner-Webb B. 1987. Transcription of mouse rDNA is regulated by an activated subform of RNA polymerase I. *Cell* **50:** 873–883.

Trotter E.W., Kao C.M., Berenfeld L., Botstein D., Petsko G.A., and Gray J.V. 2002. Misfolded proteins are competent to mediate a subset of the responses to heat shock in *Saccharomyces cerevisiae*. *J. Biol. Chem.* **277:** 44817–44825.

Tyers M., Tokiwa G., and Futcher B. 1993. Comparison of the *Saccharomyces cerevisiae* G1 cyclins: Cln3 may be an upstream activator of Cln1, Cln2 and other cyclins. *EMBO J.* **12:** 1955–1968.

Tyson C.B., Lord P.G., and Wheals A.E. 1979. Dependency of size of *Saccharomyces cerevisiae* cells on growth rate. *J. Bacteriol.* **138:** 92–98.

Udem S.A. and Warner J.R. 1972. Ribosomal RNA synthesis in *Saccharomyces cerevisiae*. *J. Mol. Biol.* **65:** 227–242.

Upadhya R., Lee J., and Willis I.M. 2002. Maf1 is an essential mediator of diverse signals that repress RNA polymerase III transcription. *Mol. Cell* **10:** 1489–1494.

Verna J., Lodder A., Lee K., Vagts A., and Ballester R. 1997. A family of genes required for maintenance of cell wall integrity and for the stress response in *Saccharomyces cerevisiae*. *Proc. Natl. Acad. Sci.* **94:** 13804–13809.

Vilardell J. and Warner J.R. 1994. Regulation of splicing at an intermediate step in the formation of the spliceosome. *Genes Dev.* **8:** 211–220.

Vilardell J., Yu S.J., and Warner J.R. 2000a. Multiple functions of an evolutionarily conserved RNA binding domain. *Mol. Cell* **5:** 761–766.

Vilardell J., Chartrand P., Singer R.H., and Warner J.R. 2000b. The odyssey of a regulated transcript. *RNA* **6:** 1773–1780.

Visintin R., Hwang E.S., and Amon A. 1999. Cfi1 prevents premature exit from mitosis by anchoring Cdc14 phosphatase in the nucleolus. *Nature* **398:** 818–823.

Visintin R., Craig K., Hwang E.S., Prinz S., Tyers M., and Amon A. 1998. The phosphatase Cdc14 triggers mitotic exit by reversal of Cdk-dependent phosphorylation. *Mol. Cell* **2:** 709–718.

Volarevic S. and Thomas G. 2001. Role of S6 phosphorylation and S6 kinase in cell growth. *Prog. Nucleic Acid Res. Mol. Biol.* **65:** 101–127.

Wade C., Shea K.A., Jensen R.V., and McAlear M.A. 2001. *EBP2* is a member of the yeast RRB regulon, a transcriptionally coregulated set of genes that are required for ribosome and rRNA biosynthesis. *Mol. Cell. Biol.* **21:** 8638–8650.

Wai H.H., Vu L., Oakes M., and Nomura M. 2000. Complete deletion of yeast chromosomal rDNA repeats and integration of a new rDNA repeat: Use of rDNA deletion strains for functional analysis of rDNA promoter elements in vivo. *Nucleic Acids Res.* **28:** 3524–3534.

Warner J.R. 1977. In the absence of ribosomal RNA synthesis, the ribosomal proteins of HeLa cells are synthesized normally and degraded rapidly. *J. Mol. Biol.* **115:** 315–333.

———. 1999. The economics of ribosome biosynthesis in yeast. *Trends Biochem. Sci.* **24:** 437–440.

———. 2001. Nascent ribosomes. *Cell* **107:** 133–136.

Warner J.R. and Gorenstein C. 1978. Yeast has a true stringent response. *Nature* **275:** 338–339.

Warner J.R. and Udem S.A. 1972. Temperature sensitive mutations affecting ribosome synthesis in *Saccharomyces cerevisiae*. *J. Mol. Biol.* **65:** 243–257.

Warner J.R., Mitra G., Schwindinger W.F., Studeny M., and Fried H.M. 1985. *Saccharomyces cerevisiae* coordinates accumulation of yeast ribosomal proteins by modulating mRNA splicing, translational initiation, and protein turnover. *Mol. Cell. Biol.* **5:** 1512–1521.

Wente S.R., Gasser S.M., and Caplan A.J. 1997. The nucleus and nucleocytoplasmic transport in *Saccharomyces cerevisiae*. In *The molecular and cellular biology of the yeast Saccharomyces: Cell cycle and cell biology* (ed. J.R. Pringle et al.), vol. 3, pp. 471–546. Cold Spring Harbor Laboratory Press, Cold Spring Harbor, New York.

Werner-Washburne M., Braun E.L., Crawford M.E., and Peck V.M. 1996. Stationary phase in *Saccharomyces cerevisiae*. *Mol. Microbiol.* **19:** 1159–1166.

Wilson E.B. 1925. *The cell in development and heredity*. Macmillan, New York.

Wittekind M., Kolb J.M., Dodd J., Yamagishi M., Memet S., Buhler J.M., and Nomura M. 1990. Conditional expression of *RPA190*, the gene encoding the largest subunit of yeast RNA polymerase I: Effects of decreased rRNA synthesis on ribosomal protein synthesis. *Mol. Cell. Biol.* **10:** 2049–2059.

Woldringh C.L., Huls P.G., and Vischer N.O. 1993. Volume growth of daughter and parent cells during the cell cycle of *Saccharomyces cerevisiae* a/alpha as determined by image cytometry. *J. Bacteriol.* **175:** 3174–3181.

Woolford J.L. and Warner J.R. 1991. The ribosome and its synthesis. In *The molecular and cellular biology of the yeast* Saccharomyces: *Genome dynamics, protein synthesis, and energetics* (ed. J.R. Broach et al.), vol. 1, pp. 587–626, Cold Spring Harbor Laboratory Press, Cold Spring Harbor, New York.

Wu L.F., Hughes T.R., Davierwala A.P., Robinson M.D., Stoughton R., and Altschuler S.J. 2002. Large-scale prediction of *Saccharomyces cerevisiae* gene function using overlapping transcriptional clusters. *Nat. Genet.* **31:** 255–265.

Xu Z. and Norris D. 1998. The *SFP1* gene product of *Saccharomyces cerevisiae* regulates G2/M transitions during the mitotic cell cycle and DNA-damage response. *Genetics* **150:** 1419–1428.

Zaragoza D., Ghavidel A., Heitman J., and Schultz M.C. 1998. Rapamycin induces the G0 program of transcriptional repression in yeast by interfering with the TOR signaling pathway. *Mol. Cell. Biol.* **18:** 4463–4470.

Zengel J.M. and Lindahl L. 1994. Diverse mechanisms for regulating ribosomal protein synthesis in *Escherichia coli*. *Prog. Nucleic Acid Res. Mol. Biol.* **47:** 331–370.

Zhang Y., Yu Z., Fu X., and Liang C. 2002. Noc3p, a bHLH protein, plays an integral role in the initiation of DNA replication in budding yeast. *Cell* **109:** 849–860.

Zhao Y., Sohn J.H., and Warner J.R. 2003. Autoregulation in the biosynthesis of ribosomes. *Mol. Cell. Biol.* **23:** 699–707.

Zimmerman Z.A. and Kellogg D.R. 2001. The Sda1 protein is required for passage through start. *Mol. Biol. Cell* **12:** 201–219.

12

Control of rRNA and tRNA Production Is Closely Tied to Cell Growth

Robert J. White
Institute of Biomedical and Life Sciences
Division of Biochemistry and Molecular Biology
University of Glasgow
Glasgow, G12 8QQ, United Kingdom

IN ACTIVELY GROWING CELLS, up to 80% of nuclear transcription can be taken up by the synthesis of rRNA and tRNA (Moss and Stefanovsky 2002). Indeed, these stable transcripts constitute ~95% of a cell's RNA content (Warner 1999). The Herculean task of supplying large rRNA is performed in eukaryotes by RNA polymerase (Pol) I, which is dedicated solely to manufacturing a single transcript that is then processed into the 18S, 5.8S, and 28S rRNA products. Pol III is highly specialized for the synthesis of short untranslated transcripts, including tRNA, 5S rRNA, and 7SL RNA (for review, see White 2002). Transcription of the protein-encoding genes is carried out by Pol II, which is smaller than Pols I and III and has fewer subunits (Sentenac 1985).

Whereas the genetic templates for Pols II and III are dispersed throughout the genome, those of Pol I are clustered in highly reiterated tandem arrays (for review, see Paule 1998). These rRNA genes are so intensely transcribed that the resulting dense concentration of macromolecules can be seen under a light microscope as nucleoli, the most prominent visible features of a nucleus. Nucleoli can be regarded as dynamic ribosome factories, in which rRNA is synthesized in the fibrillar centers and then assembled into ribosomes in the surrounding granular regions (for review, see Carmo-Fonseca et al. 2000). Although Pol III transcription occurs in the nucleoplasm, many of its products, including tRNA, move to nucleoli for processing and maturation (Jacobson et al.

1997; Bertrand et al. 1998; Carmo-Fonseca et al. 2000). The reason for this colocalization is unclear, but it may help coordinate production.

Since the output of Pol I is concerned exclusively with protein synthesis, and the most abundant Pol III products (tRNA and 5S rRNA) are dedicated to the same end, one could predict that these two systems should be coregulated; this seems especially important for the 5S rRNA, which is required in equimolar quantities with the other rRNA molecules. In general, coordinate regulation is indeed the norm. For example, the outputs of mammalian Pols I and III increase in parallel following serum stimulation and fluctuate together during passage through the cell cycle (Johnson et al. 1974; Mauck and Green 1974; White et al. 1995a; Klein and Grummt 1999). Such coordination may be of considerable importance to the balance of cellular metabolism, given the amount of energy expended in generating rRNA and tRNA. Indeed, complex mechanisms exist to ensure that the output of Pols I and III is matched to a cell's biosynthetic requirements. Although these transcription systems are often regarded as constitutive housekeeping functions, they are in fact very tightly regulated. However, transformed cells subvert these controls and can synthesize rRNA and tRNA at highly elevated rates. This chapter describes how transcription by Pols I and III can influence cell growth, what is known about how these systems are controlled, and how this regulation goes awry in cancers.

CELL GROWTH AND Pols I AND III

Because of its primordial role in determining ribosome production, rRNA gene transcription can provide a key control point for regulating growth and hence proliferation (Nomura 1999; Warner 1999). The growth rate of cells and the production of rRNA and tRNA are tightly coupled (for review, see Hannan and Rothblum 1995; Grummt 1999; Reeder 1999; Brown et al. 2000). Under starvation conditions, growth rate and the synthesis of rRNA and tRNA decrease, whereas the overall production of mRNA is unchanged (Johnson et al. 1974; Mauck and Green 1974; Grummt et al. 1976). There is a strict inverse relationship between Pol I transcription levels and cell doubling time (Derenzini et al. 1998). Indeed, nucleolar size provides a cytohistological parameter of growth rate (Trere et al. 1989; Derenzini et al. 1998). This is readily explained by the fact that growth, defined as an increase in mass, is directly proportional to the rate of protein accumulation (Baxter and Stanners 1978). This is inevitable, since 80–90% of a cell's dry mass is protein (Zetterberg and Killander

1965). It is well documented that biosynthesis and the attainment of adequate mass are essential prerequisites for cell cycle progression (Killander and Zetterberg 1965; Johnston et al. 1977; for review, see Nasmyth 1996; Neufeld and Edgar 1998; Montagne 2000; Stocker and Hafen 2000; Fingar et al. 2002; Sudbery 2002). Indeed, the cell cycle machinery is subordinate to growth: slowing biosynthesis can cause cell cycle arrest, but blocking cell cycle progression does not prevent growth (Hartwell 1971; Nurse 1975; Johnston 1977). For example, a 50% reduction in translation causes fibroblasts to exit the cell cycle and quiesce, whereas activation of translation can increase cell size (Brooks 1977; Ronning et al. 1981; Kung et al. 1993; Fingar et al. 2002). However, manipulations that accelerate the cell cycle need not promote growth and so can result in smaller cells (Hartwell 1971; Nurse 1975; Johnston 1977; Neufeld et al. 1998). The overall rate of protein synthesis is limited by the total number of ribosomes (Zetterberg and Killander 1965), reflecting the fact that in most cell types, mRNA concentration exceeds that of ribosomes (Kief and Warner 1981). Protein accumulation therefore depends critically on ribosome content (Thomas 2000). Synthesis of rRNA is a limiting step in ribosome production, as there is little or no wastage of rRNA—it all gets incorporated into ribosomes (Liebhaber et al. 1978). Indeed, the rate of protein synthesis is strongly correlated with a cell's RNA content, which is 95% rRNA and tRNA (Zetterberg and Killander 1965). Therefore, the output of Pols I and III must be maintained at a high level for cells to sustain the synthesis of ribosomes and protein that underlie active growth.

A striking demonstration that this is the case came from a recent study of cardiac hypertrophy. Hannan and colleagues examined neonatal cardiomyocytes that had been stimulated to grow by contraction or the α_1-adrenergic agonist phenylephrine (Brandenburger et al. 2001). Such treatment raises the level of the Pol I transcription factor UBF and activates rRNA synthesis, but this could be prevented specifically using an adenoviral antisense vector directed against UBF mRNA (Brandenburger et al. 2001). Not only did this restrict the level of rRNA to that of unstimulated cells, but it also blocked the increases in protein content and cell size that characterize a hypertrophic response (Brandenburger et al. 2001). This provides convincing evidence that an activation of rRNA synthesis is required for cardiomyocytes to grow. Similarly, a dominant negative mutant of the Pol I-specific factor TIF-IA can suppress cell cycle progression in proliferating HEK293T cells (Zhao et al. 2003), presumably by reducing ribosome production and hence growth. Conversely, an activated version of the same factor stimulated proliferation by 43% (Zhao et al. 2003). This striking result argues that Pol I transcription remains limiting for growth and proliferation even in these highly transformed human cells.

Studies in worms and fruit flies also support the contention that growth can be controlled by regulating rRNA synthesis (Frank and Roth 1998; Frank et al. 2002). *Drosophila* brat (brain tumor) and its *Caenorhabditis elegans* homolog ncl-1 specifically inhibit transcription of rRNA genes (large and 5S) by Pols I and III and suppress cell growth (Frank and Roth 1998; Frank et al. 2002). Inactivation of brat or ncl-1 results in enlarged cells with inflated nucleoli and an elevated content of rRNA and ribosomes (Frank and Roth 1998; Frank et al. 2002). In the case of brat, this results in brain tumors, and so brat is classified as a tumor suppressor (Frank et al. 2002). Since brat can rescue the large nucleolus and slow-growth phenotype of ncl-1 mutant worms (Frank et al. 2002), this regulatory mechanism is clearly conserved across a considerable expanse of metazoan evolution. It is unlikely to have disappeared in higher organisms.

Levels of tRNA can also have a major impact on the rates of protein synthesis, growth, and proliferation. For example, a twofold reduction in the level of initiator tRNA can cause a threefold increase in yeast doubling time; this was shown in *Saccharomyces cerevisiae* in which copies of the tRNA gene had been deleted (Francis and Rajbhandary 1990). Similarly, protein synthesis soon reaches a limit in fibroblasts experiencing a tRNA deficit under unbalanced conditions where rRNA and mRNA production continue to rise (Mauck and Green 1974). Such data suggest that tRNA levels are not in vast excess, but may be close to limiting for translation and protein synthesis. This is consistent with the rapid induction of tRNA gene transcription that forms an integral component of the growth response in all organisms examined.

There is strong evidence to link protein synthetic capacity with size homeostasis, the ill-defined mechanism that couples growth to cell division (Polymenis and Schmidt 1997, 1999). Furthermore, Pols I and III have been linked to this process by a systematic screen of the entire *S. cerevisiae* genome carried out to identify mutants with altered cell size (Jorgensen et al. 2002). Although a range of genes were implicated, mutations in ribosomal components were most frequently responsible for reduced cell size (Jorgensen et al. 2002). Ribosome biogenesis was identified as a major pathway for controlling cell size, and mutations in Pol I subunits were found to cause a substantial decrease in size that exceeded the slowing of growth (Jorgensen et al. 2002). Moreover, the most potent effect on size was ascribed to Sfp1, a transcription factor that directly activates the genes encoding Pols I and III, along with translation factors and various nucleolar components. These observations support the idea that the output of Pols I and III influences cell size, presumably by determining protein synthetic capacity.

TRANSCRIPTION MACHINERY OF Pols I AND III

Despite their complexity, eukaryotic RNA polymerases have little specificity for particular DNA sequences. In the case of mammalian Pol I, productive recruitment to the rRNA promoters requires the factor SL1, which binds just upstream of the transcription Start site (Table 1) (for review, see Paule and White 2000; Moss and Stefanovsky 2002; Grummt 2003). SL1 contains the TATA-binding protein (TBP) and three tightly bound TBP-associated factor (TAF) subunits (Comai et al. 1992). The rRNA promoter is also bound by UBF, an architectural protein containing multiple HMG domains that can wrap ~140 bp of DNA around itself in a single turn of ~360° to provide a three-dimensional structure that is recognized by SL1 (Bazett-Jones et al. 1994). Two TAF subunits of SL1 carry binding sites for TIF-IA, a polypeptide that interacts with Pol I and recruits it into the initiation complex (Miller et al. 2001). Additional polymerase-associated factors are also recruited along with Pol I (Paule and White 2000; Moss and Stefanovsky 2002; Grummt 2003). One of these is TFIIH, which until recently was believed to be Pol II specific, but is now known to be required for a postinitiation step in Pol I transcription (Iben et al. 2002).

Pol III recruitment to specific genes requires TFIIIB, which binds just upstream of the transcription start site (Kassavetis et al. 1990). TFIIIB is a complex of three subunits, TBP, Bdp1 (until recently known as B''; Willis 2002), and Brf1 (for review, see Paule and White 2000; Geiduschek and Kassavetis 2001; Schramm and Hernandez 2002). Since most Pol III-transcribed genes have no TATA box, TBP is recruited to the promoter by pro-

---

Table 1. Key components of the mammalian Pol I and III initiation machinery

Pol I	RNA polymerase with 14 subunits
SL1	Composed of TBP, TAF_I110, TAF_I63, and TAF_I48
UBF	Homodimeric DNA-binding factor
TIF-IA	Pol I-associated polypeptide; homolog of yeast Rrn3
TFIIH	Complex of 9 subunits shared with Pol II
Pol III	RNA polymerase with 17 subunits
TFIIIB	Composed of TBP, Brf1, and Bdp1
TFIIIC	DNA-binding factor with 6 subunits

tein–protein interactions between TFIIIB and the DNA-binding factor TFIIIC (White and Jackson 1992a). TFIIIC binds directly to promoter sequences that are present within the coding regions of most Pol III templates, including tRNA genes (Lassar et al. 1983). An exception is provided by the 5S rRNA genes, where the small polypeptide TFIIIA provides a platform for TFIIIC (Lassar et al. 1983). Once assembled, TFIIIB binds the Pol III enzyme and places it over the initiation site so that transcription can commence (Kassavetis et al. 1990). Indeed, the assembly factors TFIIIA and TFIIIC are no longer required after they have brought TFIIIB to a promoter, at least in vitro (Kassavetis et al. 1990). TFIIIB can therefore be regarded as the pivotal initiation factor in the Pol III system. Figure 1 provides a simplified illustration of how the Pol I and III transcription complexes are envisaged. Much greater detail can be obtained from recent reviews (Paule and White 2000; Geiduschek and Kassavetis 2001; Moss and Stefanovsky 2002; Schramm and Hernandez 2002; White 2002; Grummt 2003).

COORDINATION OF TRANSCRIPTION BY Pols I AND III

Actively proliferating eukaryotic cells use ~80% of their energy in producing ribosomes and other components of the protein synthetic apparatus (Volarevic et al. 2000). It therefore seems inevitable that mechanisms

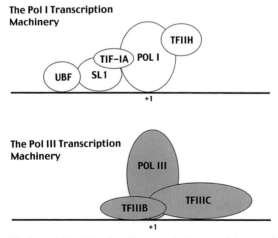

Figure 1. Simplified models of the basal transcription machinery of Pols I and III. Only the factors referred to in this chapter are shown. Note: The yeast Pol I machinery is very different from the mammalian system shown here. The transcription Start site is signified by +1.

have evolved to couple production of the biosynthetic machinery with the needs of the cell and the availability of nutrients and/or growth factors. The TOR signaling pathway is central to this process in diverse organisms (Schmelzle and Hall 2000). In *S. cerevisiae*, blocking this pathway has been shown to mediate a coordinated decrease in both translation and the transcription of genes encoding rRNA, tRNA, and ribosomal proteins (Zaragoza et al. 1998; Cardenas et al. 1999; Powers and Walter 1999; Schmelzle and Hall 2000). Thus, 1 hour after TOR was inactivated by treatment with rapamycin, transcription by Pols I and III had fallen two- to fourfold, while translation had dropped by approximately fivefold (Zaragoza et al. 1998). The catalytic activity of Pol III enzyme was halved in extracts from cells treated for 1 hour with rapamycin (Zaragoza et al. 1998). Pol III transcription was also inactivated when active extracts prepared from a conditional *tor* mutant strain grown at the permissive temperature were then shifted to the nonpermissive temperature in vitro (Zaragoza et al. 1998). This suggests that the effect of TOR on the Pol III system is probably not an indirect response to blocking protein synthesis, since transcription in vitro does not require ongoing translation. Little else is currently known about the mechanistic basis of this Pol III control, or how TOR regulates the Pol I system. However, it seems to be a control point that is conserved through evolution, since rapamycin also inhibits transcription by Pols I and III in mammals (E. Graham and P. Scott, pers. comm.). Indeed, Grummt (2003) has reported unpublished evidence that the mTOR pathway regulates rRNA synthesis by modulating the activity of TIF-IA.

The proto-oncogene product c-Myc has a potent capacity to stimulate cell growth (Dang 1999; Schmidt 1999; Grandori et al. 2000). Under some conditions, c-Myc can drive protein synthesis and growth in the absence of cell cycle progression (Iritani and Eisenman 1999; Schuhmacher et al. 1999; Beier et al. 2000; Kim, et al. 2000). Many putative c-Myc target genes, transcribed by Pol II, encode anabolic enzymes, translation factors, and ribosomal proteins (Coller et al. 2000; Boon et al. 2001; Schuhmacher et al. 2001; Shiio et al. 2002). Furthermore, Pol III transcription of tRNA and 5S rRNA genes is also induced strongly by c-Myc (Gomez-Roman et al. 2003). Indeed, the magnitude of induction exceeds what is characteristically observed for Pol II-transcribed c-Myc targets. That endogenous c-Myc is a natural regulator of Pol III transcription at physiological concentrations was demonstrated by using knockout fibroblasts and by RNA interference (Felton-Edkins et al. 2003b; Gomez-Roman et al. 2003). Furthermore, chromatin immunoprecipitation revealed the presence of endogenous c-Myc at chromosomal tRNA and 5S rRNA genes, both in fibroblasts and in epithelial cells (Felton-Edkins et al. 2003b; Gomez-Roman et al. 2003). Recruitment to these sites appears to involve protein–protein interactions with TFIIIB, rather than direct DNA recognition by c-Myc (Gomez-Roman et al. 2003). In contrast to these Pol III templates, the large rRNA genes contain matches to the E-box DNA sequence that is bound by c-Myc. Indeed, the presence of c-Myc at these loci can be detected by chromatin immunoprecipitation (C. Grandori and R. Eisenman, pers. comm.). Preliminary data suggest that c-Myc interacts with SL1 and activates Pol I transcription directly (C. Grandori et al., unpubl.). This is consistent with the decrease in rRNA observed in *c-myc$^{-/-}$* knockout fibroblasts (Mateyak et al. 1997). The potent effects of c-Myc on cell growth may therefore involve coordinately inducing synthesis of rRNA, tRNA, and mRNAs encoding ribosomal proteins, through parallel effects on the three transcription systems.

Another excellent example of how biosynthetic processes can be balanced is provided in *S. cerevisiae* by the secretory pathway, which is largely devoted to the production of membranes and secretion of protein components of the cell wall. Attempts to grow when the plasma membrane is unable to expand, due to an insufficient supply of lipids and proteins, raises the intracellular turgor pressure and triggers the PKC-dependent cell integrity pathway. This results in a rapid and coordinated repression of rRNA synthesis by Pol I, ribosomal protein mRNA synthesis by Pol II, and tRNA and 5S rRNA production by Pol III (Mizuta and Warner 1994; Nierras and Warner 1999; Li et al. 2000). In all, >85% of nuclear transcription responds to this pathway in *S. cerevisiae* (Li et al. 2000). Inevitably, growth will be quickly curtailed under these conditions. PKC has also been found to influence Pol III transcription in mammals (James and Carter 1992; Wang et al. 1995), although it has yet to be determined whether this forms part of a response to altered turgor.

CK2 is another protein kinase that regulates the output of Pols I and III in both yeast and man. This ubiquitous and highly conserved enzyme forms part of the Wnt signaling pathway in flies and mammals (Willert et al. 1997; Song et al. 2000). Increases in the level, nuclear localization, and/or activity of CK2 are associated with cell growth and proliferation (Klarlund and Czech 1988; Belenguer et al. 1989; Carroll and Marshak 1989; Munstermann et al. 1990; Orlandini et al. 1998; Bosc et al. 1999). Thus, mitogens can raise CK2 expression (Carroll and Marshak 1989; Orlandini et al. 1998), and CK2 is most abundant in cells with high mitotic activity, such as transformed cells and normal colorectal mucosa (Munstermann et al. 1990). Indeed, microinjection of CK2 in the absence of growth factors can induce expression of genes that normally respond rapidly to mitogens (Gauthier-Rouviere et al. 1991). Conversely, inactivation of CK2 using specific antibodies or antisense oligonucleotides can arrest proliferation of primary human fibroblasts (Pepperkok et al. 1991, 1994). Similarly, cell cycle progression is blocked in *S. cerevisiae* when temperature-sensitive CK2 mutants are cultured at the nonpermissive temperature (Hanna et al. 1995). In both yeast and mammals, CK2 induces the synthesis of rRNA and tRNA, a function that may be important for its growth-promoting capacity (Belenguer et al. 1989; Voit et al. 1992; Hockman and Schultz 1996; Ghavidel and Schultz 1997, 2001; Johnston et al. 2002). Indeed, CK2 associates stably with TFIIIB, which it phosphorylates and activates in humans, mice, and yeast (Ghavidel and Schultz 1997, 2001; Johnston et al. 2002). Phosphorylation by CK2 appears to be necessary for TFIIIB to interact efficiently with TFIIIC in vitro and in fibroblasts (Johnston et al. 2002). It therefore regulates assembly of the Pol III preinitiation complex. A Pol-I-specific target for CK2 has yet to be established in yeast, but in rats it interacts stably with Pol I itself and phosphorylates its largest subunit (Hannan et al. 1998). In addition, the carboxy-terminal tail of UBF is heavily phosphorylated by CK2 in mammals (Voit et al. 1992, 1995; Voit and Grummt 2001). Removal of this region or phosphatase treatment of intact UBF blocks transcriptional activation (O' Mahony et al. 1992; Voit et al. 1992; Tuan et al. 1999). Indeed, UBF phosphorylation regulates the ability of its carboxy-terminal tail to bind and recruit SL1 (Kihm et al. 1998; Tuan et al. 1999). Thus, phosphorylation controls promoter recruitment of the TBP-containing initiation factor (SL1 or TFIIIB) in both the Pol I and Pol III systems.

Genotoxic agents such as UV or methane methylsulfonate (MMS) can trigger a rapid and coordinate suppression of rRNA and tRNA synthesis in *S. cerevisiae* that is mediated, at least in part, by CK2 (Ghavidel and Schultz 2001). Thus, the DNA-damage-induced reduction in transcription by Pols I and III is severely attenuated in CK2 mutant strains (Ghavidel and Schultz 2001). The response is also attenuated by alanine substitution of a CK2 phosphoacceptor site in TBP (Ghavidel and Schultz 2001). The catalytic subunit of CK2 dissociates from TBP following UV or MMS treatment, presumably resulting in dephosphorylation and loss of activity (Ghavidel and Schultz 2001). This provides an interesting contrast to most signaling pathways, in which a stimulus provokes the association of a kinase with its substrate. Here, CK2 binds to TFIIIB under benign conditions, ensuring active transcription, and then dissociates in response to genotoxic stress. Reducing the output of Pols I and III may serve to restrict cell growth under conditions in which chromosomal replication is prohibited.

In addition to CK2, other kinases are also involved in regulating assembly of the Pol I and III preinitiation complexes. UBF can be phosphorylated at Ser-484 by cdk4/cyclin D1 and cdk2/cyclin E, both in vitro and in NIH-3T3 cells (Voit et al. 1999). Furthermore, substitution of Ser-484 to alanine weakens the ability of UBF to stimulate Pol I transcription in a reconstituted system or in transfected cells (Voit et al. 1999). As well as a ~twofold reduction in rRNA synthesis, a fibroblast line that stably expressed this Ser484Ala mutant was reported to "grow" ~25% more slowly than a line expressing wild-type UBF (Voit et al. 1999). At the end of G_1 phase, UBF also gets phosphorylated at Scr-388 by cdk2/cyclin E and cdk2/cyclin A (Voit and Grummt 2001). Conversion of this site to glycine inactivates UBF, whereas substitution to aspartate enhances its activity (Voit and Grummt 2001). The status of this residue does not affect SL1 binding, but instead regulates the ability of UBF to interact with Pol I (Voit and Grummt 2001).

UBF is also phosphorylated in response to growth factors by the MAP kinases Erk 1 and 2 (Stefanovsky et al. 2001). This can result in a rapid induction of rRNA synthesis within 10 minutes of serum addition (Stefanovsky et al. 2001). Thr-117 and Thr-201 are the phosphoacceptor sites for Erk, and substitution of these sites to alanine compromises the ability of UBF to activate Pol I transcription (Stefanovsky et al. 2001). These residues lie within the HMG domains 1 and 2 that mediate DNA binding. Erk phosphorylation was shown to reduce interactions with an rRNA gene fragment in vitro; this was postulated to facilitate promoter clearance by allowing passage of the polymerase (Stefanovsky et al. 2001). Thus, UBF modification seems to provide an important control point for regulating the assembly and function of a Pol I transcription complex; its binding to DNA responds to the Erks (Stefanovsky et al. 2001), its tail needs to be phosphorylated for SL1 recruitment (Kihm et al. 1998; Tuan et al. 1999), and it needs to be phosphorylated on Ser-388 by cdk2 in order to bind to Pol I (Voit and Grummt 2001).

TIF-IA is also bound, phosphorylated, and activated by Erk2, as well as the 90-kD ribosomal S6 kinases (Rsk) that are controlled by Erk (Zhao et al. 2003). Indeed, it is TIF-IA, rather than UBF, that mediates the Pol I response to the Erk cascade in FM3A cells (Zhao et al. 2003). Mitogen-stimulated phosphorylation of TIF-IA by Rsk may enhance its interaction with Erk2. Furthermore, substitution to alanine of the Ser-649 phospho-acceptor site creates a mutant TIF-IA that suppresses rRNA synthesis and retards proliferation (Zhao et al. 2003). Thus, TIF-IA provides a target through which Erk2 and Rsk can transduce growth stimuli to the Pol I machinery. This is consistent with the central role played by TIF-IA in controlling rRNA synthesis in response to mitogens and nutrients (Buttgereit et al. 1985; Yuan et al. 2002; Grummt 2003).

Pol III transcription also responds to the Erk signaling pathway. Thus, Erk activates TFIIIB by binding and phosphorylating its Brf1 subunit (Felton-Edkins et al. 2003a). Serum induction of Pol III transcription can be severely compromised by substitutions in Brf1 that disrupt the Erk docking or phosphoacceptor sites. The interaction of TFIIIB with both TFIIIC and Pol III, as determined by coimmunoprecipitation, is sensitive to Erk activity. Consequently, ChIP assays show that promoter occupancy by Brf1 and Pol III at endogenous tRNA and 5S rRNA genes is significantly diminished by treating cells with a specific inhibitor of the Erk-activating kinase MEK. This MAP kinase pathway therefore ensures that initiation complex assembly on Pol III templates responds to mitogenic stimuli. These observations can explain the rapid and coordinated induction of rRNA and tRNA synthesis that follows growth factor addition (Johnson et al. 1974). The Erk pathway has been shown to act on a range of targets in order to promote balanced growth and proliferation (Whitmarsh and Davis 2000). Its effect on Pols I and III is likely to make a substantial contribution to the cellular response to mitogens.

In addition to the immediate-early response that appears to be mediated by Erks, Pol III transcription increases further when fibroblasts progress from G_1 into S phase (Mauck and Green 1974; White et al. 1995a; Scott et al. 2001). This reflects a derepression of TFIIIB, much of which is inactivated during G_0 and early G_1 phases through interaction with the retinoblastoma protein RB (Scott et al. 2001). When bound by RB, TFIIIB is unable to interact with either Pol III or TFIIIC and is therefore sequestered into an inactive complex (Sutcliffe et al. 2000). At the G_1/S transition, the complex dissociates following phosphorylation of RB by cyclin D- and E-dependent kinases; only hypophosphorylated RB can bind and neutralize TFIIIB (Scott et al. 2001). Since the activity of these cdks is mitogen-dependent (Mittnacht 1998), this system plays a major role in ensuring that Pol III output responds to growth factor availability. Consequently, cells derived from RB-knockout mice are substantially compromised in their ability to suppress Pol III transcription following serum withdrawal (Scott et al. 2001). As well as RB, the related pocket proteins p107 and p130 bind and repress TFIIIB (Sutcliffe et al. 1999), and these interactions are also lost at the G_1/S transition under the influence of cdks (Scott et al. 2001). Thus, the cyclin D- and E-dependent kinases can activate transcription by both Pol I and Pol III in mammalian cells (Voit et al. 1999; Scott et al. 2001; Voit and Grummt 2001). However, unlike the direct effects of cdk2 and cdk4 on UBF (Voit et al. 1999; Voit and Grummt 2001), there is currently no evidence that these kinases act directly on the Pol III machinery, their known effects instead being mediated via RB and its relatives.

RB and p130 are also involved in repressing Pol I transcription (Cavanaugh et al. 1995; Voit et al. 1997; Hannan et al. 2000a,b; Pelletier et al. 2000; Ciarmatori et al. 2001). Thus, $Rb^{-/-}$ $p130^{-/-}$ double-knockout fibroblasts express ~50% more 18S and 28S rRNA than matched wild-type cells (Ciarmatori et al. 2001). Despite a high degree of homology and functional overlap with RB and p130 (Mulligan and Jacks 1998), the third pocket protein p107 seems unable to regulate Pol I transcription (Voit et al. 1997; Hannan et al. 2000a; Ciarmatori et al. 2001). UBF is bound by RB and p130, but not by p107 (Hannan et al. 2000a; Ciarmatori et al. 2001). Several mechanisms have been suggested to explain how this results in transcriptional repression. RB can disrupt the DNA-binding activity of UBF in vitro (Voit et al. 1997; Hannan et al. 2000a). It can also interfere with the interaction between UBF and SL1 (Hannan et al. 2000a). In addition, RB competes with the acetyltransferase CBP for UBF binding and thereby suppresses the acetylation of UBF that activates Pol I transcription (Pelletier et al. 2000). These effects are illustrated in Figure 2. The relative importance of the three mechanisms has yet to be determined under physiological circumstances.

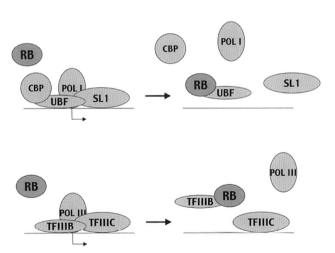

Figure 2. Schematic representations of the effects that RB can have on the Pol I and III initiation complexes. Binding of RB to UBF has been found in vitro to disrupt its interactions with DNA, SL1, and CBP. RB binding to TFIIIB has been shown to block its interactions with TFIIIC and Pol III both in vitro and in vivo.

Synthesis of rRNA decreases about fivefold when 3T6 fibroblasts are grown to confluence (Hannan et al. 2000b). This correlates with an accumulation of hypophosphorylated RB (Hannan et al. 2000b). UBF was found to coimmunoprecipitate exclusively with the hypophosphorylated form of RB (Hannan et al. 2000b), implying that the interaction is regulated by cdk activity. Furthermore, immunofluorescence revealed that the nucleolar content of RB is markedly elevated in density-arrested cells (Hannan et al. 2000b). RB may therefore play a significant role in down-regulating Pol I output when fibroblasts reach confluence.

RB also accumulates in the nucleolus during differentiation along the monocytic lineage of human U937 promyelocytic leukemia cells (Cavanaugh et al. 1995). This coincides with a substantial increase in its binding to UBF and a decrease in rRNA synthesis (Cavanaugh et al. 1995). However, down-regulation of Pol I transcription in this situation is mediated principally by inactivation of SL1, the limiting factor in these cells (Comai et al. 2000). This does not reflect any change in the abundance or association of the SL1 subunits (Comai et al. 2000). The mechanistic basis of the transcriptional decrease remains unclear, although TBP phosphorylation was observed to accompany U937 cell differentiation (Comai et al. 2000).

A loss of active SL1 is also responsible for down-regulating Pol I transcription when F9 embryonal carcinoma cells differentiate into parietal endoderm (Alzuherri and White 1999). This, however, reflects a marked decrease in the levels of its TAF_I95 and TAF_I48 subunits (Alzuherri and White 1999). In contrast, the abundance of the remaining SL1-specific subunit, TAF_I68, shows little or no decrease following differentiation (Alzuherri and White 1999). Since the TBP content of SL1 fractions also changes little (Alzuherri and White 1998), it may be that differentiated F9 cells contain significant amounts of a TBP/TAF_I68 subcomplex. A close parallel is seen for Pol III transcription, which also decreases dramatically during parietal endoderm formation, both in culture and in the developing mouse (Vasseur et al. 1985; White et al. 1989). Biochemical experiments showed that a dramatic loss of TFIIIB activity is responsible for this change (White et al. 1989; Alzuherri and White 1998). This was confirmed by western blotting, which demonstrated a significant decrease in level of the TFIIIB-specific TAF subunit Brf1 (Alzuherri and White 1998). We have since found that levels of the other TFIIIB TAF Bdp1 also drop substantially when F9 cells differentiate (D. Athineos and R.J. White, unpubl.). A decrease in the abundance of specific TAFs therefore serves to reduce the output of Pols I and III in differentiated F9 cells, which have a reduced requirement for ribosome synthesis due to a cessation of growth.

The level of TBP also decreases under these circumstances (Alzuherri and White 1998, 1999; Perletti et al. 2001). Nevertheless, there is no general change in Pol II transcriptional activity, as evidenced by the more or less constant transcription of HPRT and α-tubulin genes (White et al. 1989). Most of the Pol II TAFs are unaffected by differentiation, although TAF$_{II}$135 selectively undergoes proteasome-mediated degradation (Perletti et al. 2001). It will be interesting to determine whether the Pol I and Pol III TAFs that are down-regulated during F9 cell differentiation are also victims of the proteasome.

The TAF components of SL1 and TFIIIB are also targeted when cycling cells pass through mitosis. This phase of the cell cycle is characterized by a generalized repression of nuclear transcription in vertebrates (Taylor 1960; Prescott and Bender 1962; Gottesfeld and Forbes 1997). Both SL1 and TFIIIB undergo specific inactivation through phosphorylation, which can be mediated by the master mitotic kinase cdc2/cyclin B (Gottesfeld et al. 1994; Heix et al. 1998; Kuhn et al. 1998). The amino-terminal domain of TBP is hyperphosphorylated during M phase in both humans and *Xenopus* (White et al. 1995b; Leresche et al. 1996; Segil et al. 1996; Heix et al. 1998), but the functional significance of this remains unclear. Inactivation of SL1 reflects the phosphorylation of TAF$_I$110, which compromises binding to UBF (Heix et al. 1998). Similarly, TFIIIB repression at mitosis is due to a specific loss of TAF activity (White et al. 1995b; Leresche et al. 1996), which correlates with the hyperphosphorylation of Brf1 (J. Fairley and R.J. White, unpubl.). Since TFIID has also been shown to be phosphorylated and inhibited at mitosis (Segil et al. 1996), it seems that the inactivation of TBP-containing complexes is a general mechanism for transcriptional control during M phase.

As TBP is utilized by Pols I, II, and III (White and Jackson 1992b), it would appear to provide an ideal target for coordinating transcriptional output, especially considering its ability to bind a wide variety of regulatory factors. However, there are relatively few examples of its being used in this way. One is provided by the hepatitis B virus (HBV) X oncoprotein, which raises the level of TBP and thereby stimulates transcription by Pols I and III (Wang et al. 1995, 1997, 1998). The large-T antigen of SV40 has been shown to bind directly to both TBP and TAFs in the SL1, TFIID, and TFIIIB complexes (Damania and Alwine 1996; Zhai et al. 1997; Damania et al. 1998). Although T antigen's interaction with SL1 is important for Pol I activation (Zhai et al. 1997), the functional significance of its association with TFIIIB has not been established. Instead, its effect on Pol III transcription seems to be mediated primarily through neutralizing RB and thereby releasing TFIIIB from repression (Larminie et al. 1999).

A TBP-binding protein that can regulate Pols II and III is Dr1 (also called NC2), which was first isolated from HeLa cells because of its ability to repress the adenovirus major late promoter (Inostroza et al. 1992). The importance of Dr1 is indicated by its high degree of conservation through evolution. Thus, *S. cerevisiae* has a homolog of Dr1 that is 37% identical to the human protein and is essential for viability (Kim et al. 1997). Furthermore, human Dr1 is 44% identical to an *Arabidopsis* protein (Kuromori and Yamamoto 1994). Dr1 was found to inactivate TFIIIB by binding to TBP and disrupting its interaction with Brf1 (White et al. 1994). This can repress Pol III transcription both in vitro and in live yeast (White et al. 1994; Kim et al. 1997). In contrast, Pol I transcription is immune to the effects of Dr1 (White et al. 1994; Kim et al. 1997). This reflects an inability to interact with SL1 (R.J. White, unpubl.). Crystallographic analysis has shown that Dr1 binds to the same site as TFIIB on the surface of TBP (Kamada et al. 2001). Since TAF_I48 has been shown to block access of TFIIB to TBP (Beckmann et al. 1995), it is likely that it also prevents Dr1 from reaching TBP in the SL1 complex (Fig. 3). Furthermore, immunofluorescence suggests that Dr1 is absent from nucleoli (A. Mills and R.J. White, unpubl.). Despite its lack of effect on Pol I transcription, Dr1 can inhibit cell growth when it is overexpressed in yeast (Kim et al. 1997). However, little is currently known about the physiological role of Dr1. Although it is expressed constitutively, its cofactor

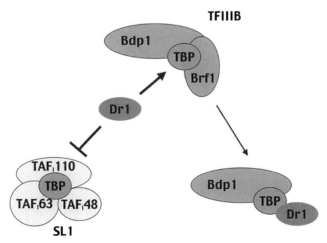

Figure 3. Model showing the proposed effects of Dr1 on transcription by Pols I and III. Binding of Dr1 to TBP can displace Brf1 and thereby disrupt TFIIIB, at least in vitro. Although Bdp1 remains bound in the model, this has never been tested. Dr1 is unable to contact SL1, probably because its access to TBP is blocked by TAF_Is.

DRAP1 is most abundant in nondividing cells (Mermelstein et al. 1996). It may therefore contribute to the reduction of Pol III transcription that characterizes the resting state, since DRAP1 facilitates TFIIIB repression by Dr1 in vitro (R.J. White, unpubl.). No studies have yet been reported that examine the extent to which Pol III is influenced by endogenous Dr1/DRAP1 under physiological conditions.

TBP is also bound directly by p53, one of the very few factors that, like RB and p130, has been shown to regulate transcription by all three nuclear RNA polymerases. The relevance of the TBP interaction for Pol II regulation has been questioned, because a TBP mutant that cannot bind p53 is nevertheless capable of supporting Pol II activation by p53 in vivo (Tansey and Herr 1995). However, interaction with TBP may be of functional significance for the control of Pols I and III. Thus, unlike Dr1, p53 is able to bind to SL1, probably via TBP (Zhai and Comai 2000). This prevents SL1 from binding to UBF and consequently represses rRNA synthesis (Zhai and Comai 2000). TFIIIB is the target for p53 in the Pol III system (Chesnokov et al. 1996; Cairns and White 1998), and again the interaction seems to be mediated through TBP (Crighton et al. 2003). ChIP assays show that TFIIIB is excluded from promoters when bound by p53, apparently because its ability to interact with TFIIIC is compromised (Crighton et al. 2003). As a consequence, tRNA and 5S rRNA synthesis is markedly elevated in fibroblasts from p53-knockout mice (Cairns and White 1998).

In addition to TBP, the three transcription systems also share several subunits of the RNA polymerases themselves. Thus, there are five essential subunits common to Pols I, II, and III, as well as two more that are shared between Pols I and III (for review, see Sentenac 1985; White 2002). Although these would appear to provide further opportunity for coordinate regulation, there is no evidence that this potential is ever exploited. The free-living amoeba *Acanthamoeba castellanii* provides a striking example of how a shared polymerase subunit might have been exploited, but is not (for review, see Paule 1998). When starved of essential nutrients, *Acanthamoeba* differentiates into a dormant cyst, and this is accompanied by a rapid decrease in rRNA synthesis by Pols I and III (Matthews et al. 1995). The down-regulation of large rRNA synthesis is due to specific inactivation of the Pol I enzyme, which correlates closely with a highly specific increase in the electrophoretic mobility of its 39-kD subunit (Bateman and Paule 1986). This change is presumed to reflect posttranslational modification, although its chemical nature remains to be established (Paule 1998). Within 20 minutes of transfer to rich growth medium, the amoebae resume rRNA synthesis in parallel with a decrease in 39-kD subunit mobility (Paule 1998). This polymerase subunit is shared between Pols I and III and so would appear to provide an ideal target for coordinating transcription during *Acanthamoeba* encystment. However, the down-regulation of rRNA synthesis precedes any change in Pol III activity, and 5S rRNA gene regulation is instead achieved through the specific inactivation of TFIIIA (Matthews et al. 1995). Thus, coordinate regulation appears to be achieved through apparently unrelated mechanisms.

A combination of shared and distinct strategies may be involved in controlling Pols I and III during the switch between resting and growing states. As described above, both systems are repressed by RB and p130 in G_0 phase fibroblasts and are activated in response to mitogens through the effects of cdks, Erks, and perhaps CK2. However, in cardiac myocytes, induction of rRNA synthesis by hypertrophic stimuli involves a rise in the level of UBF mRNA and protein (Hannan et al. 1995, 1996; Hannan and Rothblum 1995; Brandenburger et al. 2001). Indeed, if this increase is prevented using an antisense construct directed against the UBF mRNA, activation of Pol I transcription is also blocked (Brandenburger et al. 2001). Most striking of all is the finding that hypertrophic growth is itself prevented by this treatment (Brandenburger et al. 2001). This important result indicates that cardiac myocytes cannot grow under these conditions unless they raise their rate of rRNA production. Amino acid starvation or cycloheximide treatment influences the output of mammalian Pol I through the polymerase-associated factor TIF-IA (Buttgereit et al. 1985; Bodem et al. 2000; Yuan 2002; Grummt 2003). This was originally identified as an activity that restores transcription when added to extracts of stationary-phase mouse cells (Buttgereit et al. 1985). It was subsequently found to correspond to the *Saccharomyces* factor Rrn3, for which it can substitute in vivo (Bodem et al. 2000; Moorefield et al. 2000). Only when it is bound by Rrn3/TIF-IA can Pol I be recruited to an rRNA gene promoter (Milkereit and Tschochner 1998; Miller et al. 2001). The interaction between Pol I and Rrn3/TIF-IA is controlled by phosphorylation in both mammals and yeast (Fath et al. 2001; Cavanaugh et al. 2002). Apart from TBP, Rrn3/TIF-IA is the only known Pol I-specific transcription factor that is conserved between yeast and man, so it is likely providing a very important control point. Yet no comparable factor or mechanism has been identified for Pol III. Nutrient deprivation in *Saccharomyces* results in reduced TFIIIB levels and activity (Sethy et al. 1995; Clarke et al. 1996). Thus, Brf1 is threefold less abundant in stationary-phase yeast when compared to those that are actively cycling (Sethy et al. 1995). Brf1 availability is limiting for Pol III transcription even in logarithmically growing yeast, and the deficit increases in stationary phase (Lopez-de-Leon et al. 1992; Sethy et al. 1995).

Entirely distinct strategies are also used by *Xenopus* oocytes to generate the enormous stockpiles of ribosomal components that sustain them through the initial stages of embryogenesis when biosynthesis is curtailed. In addition to the 400 "somatic" 5S genes that are transcribed in all cell types, the haploid *Xenopus* genome contains 20,000 copies of an auxiliary "major oocyte" 5S rRNA gene family, which is highly expressed prior to fertilization and then silenced permanently as development proceeds (for review, see Wolffe and Brown 1988). Transcription of this enormous number of templates is possible in oocytes because each contains ~1.5 billion TFIIIA molecules, providing ~70,000 per 5S gene (Shastry et al. 1984). During early oogenesis, 5S rRNA is synthesized in a 15- to 20-fold excess over 18S and 28S rRNA (Ford 1971). However, this imbalance is subsequently compensated through the specific amplification of Pol I templates to give 500,000 copies per oocyte, while 5S rRNA gene numbers remain constant (Brown and Dawid 1968).

Infection of HeLa cells with poliovirus results in the rapid inhibition of host-cell RNA synthesis by all three classes of nuclear RNA polymerase (Fradkin et al. 1987; Kliewer and Dasgupta 1988; Rubinstein and Dasgupta 1989). However, a distinct mechanism is used in each case and the shutoff is not coordinated. Pol I transcription is down-regulated most rapidly, within 2–3 hours of infection, followed by Pol II after 3–4 hours and Pol III at 4–5 hours postinfection (Fradkin et al. 1987; Kliewer and Dasgupta 1988; Rubinstein and Dasgupta 1989). Cessation of rRNA synthesis appears to reflect inactivation of Pol I and/or its associated factors (Rubinstein and Dasgupta 1989). The shutoff of Pol II transcription correlates with a specific decrease in the activity of fractions containing TFIID (Kliewer and Dasgupta 1988; Clark et al. 1993). Finally, silencing of Pol III activity is primarily due to the partial degradation of TFIIIC by a viral protease, generating a factor that is transcriptionally inactive, although still able to bind DNA (Shen et al. 1996).

Terminal differentiation of rat L6 myoblasts is accompanied by a decrease in the rate of Pol I transcription (Larson et al. 1993). This occurs in parallel with a fall in the abundance of UBF protein and is preceded by a decrease in the level of UBF mRNA (Larson et al. 1993). In contrast, the transcription rates of genes encoding 5S rRNA and ribosomal protein L32 remain unchanged (Zahradka et al. 1991). The surplus of these molecules is then degraded, thereby preventing the accumulation of a static pool of ribosomal components (Zahradka et al. 1991). The half-life of 5S rRNA falls from 6.2 hours in undifferentiated myoblasts to 1.1 hours in myotubes, whereas the half-lives of ribosomal proteins can drop to 30 minutes from >25 hours (Zahradka et al. 1991). This excessive synthesis and degradation of far more products than are required provides an extraordinary example of cellular profligacy.

DEREGULATION OF Pols I AND III IN TRANSFORMED CELLS

Abnormally high rates of transcription by Pols I and III are a general feature of transformed and tumor cells. For example, an early report of this concerned mice with myelomas, where Pols I and III are both hyperactive while overall Pol II activity remains normal (Schwartz et al. 1974). It has been known for over a century that nucleolar hypertrophy is one of the most consistent cytological features of cancer cells (Pianese 1896). Indeed, enlarged nucleoli are used by pathologists as a strong diagnostic indicator of tumor formation (Busch and Smetana 1970; Derenzini and Ploton 1994; King 1996). This shows that transformation in situ is tightly linked to the deregulation of Pol I, because the size of the nucleolus reflects the level of rRNA synthesis (Kurata et al. 1978; Altmann and Leblond 1982; Moss and Stefanovsky 1995). Many types of tumor cell lines have been found to overexpress the products of Pols I and III, including cells transformed by DNA tumor viruses (e.g., SV40, HBV), RNA tumor viruses (e.g., human T-cell leukemia virus 1), and chemical carcinogens (e.g., methylcholanthrene) (Scott et al. 1983; White et al. 1990; Wang et al. 1995, 1997; Gottesfeld et al. 1996; Ying et al. 1996; Larminie et al. 1999; Zhai and Comai 1999). When different SV40-transformed clones are compared, the highest abundance of Pol III transcripts is found in those which most efficiently induce tumors, whereas lower levels are detected in the less tumorigenic lines (Scott et al. 1983; White et al. 1990). Furthermore, lines transformed using temperature-sensitive mutants of SV40 large-T antigen rapidly down-regulate Pol III products at the nonpermissive temperature while reverting to normal morphology and phenotype (Scott et al. 1983; Luo et al. 1997). Although most of these studies have employed cell lines, their relevance has been validated for tumors in situ (Chen et al. 1997a,b; Winter et al. 2000). For example, RT-PCR showed that rRNA and tRNA are overproduced consistently in human ovarian cancers (Winter et al. 2000). An extensive northern analysis of 80 tumor specimens representing 19 types of cancer revealed that 7SL RNA is abnormally abundant in every tumor analyzed, relative to healthy tissue from the same patient (Chen et al. 1997a). Furthermore, in situ hybridization of breast, lung, and tongue carcinomas revealed increased levels of Pol III transcripts in neoplastic cells relative to the surrounding healthy tissue (Chen et al. 1997a,b).

An obvious explanation for the consistent deregulation of Pols I and III in transformed cells was suggested by the discovery that these systems are subject to repression in healthy cells by both RB and p53 (Cavanaugh et al. 1995; White et al. 1996; Cairns and White 1998; Zhai and Comai 2000). The fact that they are targeted by two major and unrelated tumor suppressors provides a clear indication of the importance of controlling their output. Since the function of p53 and/or RB is compromised in most human cancers, it seems likely that Pols I and III will be released from repression in the majority of tumors. Thus, about half of all sporadic human tumors carry a mutation in one p53 gene combined with a deletion of the second p53 gene (Hollstein et al. 1991,1994). In most cases, the mutation lies in the central core domain of p53, a region which it needs to regulate Pols I and III (Zhai and Comai 2000; Stein et al. 2002a,b). The most common substitution detected in human cancers is the mutation to histidine of p53 residue 175 within the core domain, a change that is associated with high tumorigenicity and an extremely poor prognosis in patients (Goh et al. 1995). This substitution abolishes the ability of p53 to repress Pols I and III, both in vitro and in vivo (Zhai and Comai 2000; Stein et al. 2002a). Two out of three other tumor-derived p53 substitutions tested were also found to severely compromise the capacity to regulate Pol III (Stein et al. 2002a). The exception, mutation Arg181Leu, is an oddity, since it also remains able to activate Pol II reporters (Crook et al. 1994). Inherited mutations in p53 can cause Li-Fraumeni syndrome, a familial cancer predisposition (for review, see Varley et al. 1997). Pol III activity was found to be abnormally elevated in primary cells from eight out of ten tested Li-Fraumeni syndrome patients (Stein et al. 2002a). Many tumors neutralize wild-type p53 by overexpressing the oncoprotein Mdm2 (for review, see Momand et al. 1998). As might be expected, Mdm2 can release Pol III in vivo from the repressive effects of p53 (Stein et al. 2002a). The same is true of E6, an oncoprotein produced by human papillomavirus in most cervical carcinomas (Stein et al. 2002a). Overall, it is estimated that p53 function is compromised in over 80% of human tumors (Lozano and Elledge 2000). Derepression of Pols I and III can be expected in a high proportion of these cases.

It has been suggested that the pathway involving RB may be disrupted in all human malignancies (Weinberg 1995). Many human cancers carry mutations in RB, including retinoblastomas where the gene was first identified (Friend et al. 1986; DiCiommo et al. 2000). In many such cases, RB expression is ablated completely; $Rb^{-/-}$ mice provide a model for this situation and have allowed confirmation that tRNA and 5S rRNA synthesis is elevated in vivo when RB is missing (White et al. 1996; Scott et al.

2001). Other tumors express mutant forms of RB, and in these cases, the mutation generally affects the pocket domain (Hu et al. 1990). Deletion and substitution analyses have shown that the pocket is essential for RB to regulate Pol III activity (White et al. 1996; Chu et al. 1997). Furthermore, TFIIIB is unable to bind a mutant form of RB that is found in the osteosarcoma cell line SAOS2 (Larminie et al. 1999), where a carboxy-terminal truncation has removed part of the pocket (Shew et al. 1990). Several examples have been described of highly localized mutations that inactivate the pocket. For example, in one small-cell lung carcinoma, a point mutation created a stop codon and a novel splice donor site within exon 22, thereby eliminating 38 residues from the pocket domain of the product (Horowitz et al. 1990). In another small-cell lung carcinoma, a single base change in a splice acceptor site gave rise to an RB polypeptide that lacked the 35 amino acids encoded by exon 21 (Horowitz et al. 1990). A third inactivating mutation from the same cancer type resulted in a single residue substitution at codon 706 (Kaye et al. 1990). Each of these subtle naturally occurring mutations abolishes the ability to regulate Pol III transcription (White et al. 1996). The same is true of a substitution at residue 567 of RB, which was identified in the germ line of a child with bilateral retinoblastoma (Brown et al. 2000). This is clearly a limited survey, but it nevertheless demonstrates that mutations which arise in RB in tumors can compromise its ability to regulate Pol III transcription. This is also the case for Pol I, which is not repressed by RB carrying pocket domain substitutions at residues 567 or 706 (Cavanaugh et al. 1995; Hannan et al. 2000a; T.R. Brown and R.J. White, unpubl.). However, it is not clear that this is sufficient to allow increased rRNA synthesis in vivo, due to redundancy of function with p130 (Hannan et al. 2000a; Ciarmatori et al. 2001). Pol I activity appears to be normal in RB-knockout mouse embryonic fibroblasts and only becomes deregulated if p130 is lost as well (Ciarmatori et al. 2001). It will be important to determine whether this also applies to other cell types, especially those of epithelial origin, from which most tumors arise.

Although mutations in p130 are extremely rare in transformed cells, its function is often lost, along with that of RB and p107, due to changes affecting a function upstream of the pocket protein family (for review, see Grana et al. 1998; Mulligan and Jacks 1998). The most common mechanism by which this occurs is through constitutive phosphorylation by cdks, which are hyperactive in many cancers (for review, see Hunter and Pines 1994; Bates and Peters 1995). For example, cyclin D1 is overexpressed in 30–40% of primary breast tumors. In addition to situations in which cyclins are affected directly, many other cancers have lost the function of p16, a specific repressor of the cyclin-D-dependent kinases (Hirama and Koeffler 1995; Rocco and Sidransky 2001). For example, the gene for p16 is deleted in many pancreatic, esophageal, and lung carcinomas. Thus, cyclin-D-dependent kinase activity is abnormally elevated in a broad spectrum of tumors, resulting in hyperphosphorylation of the pocket proteins. Neither UBF nor TFIIIB will bind to RB once it is hyperphosphorylated (Hannan et al. 2000b; Scott et al. 2001). It has yet to be formally confirmed that this is also the case for p130, although TFIIIB was shown to dissociate from p130 as the latter undergoes phosphorylation at the G_1/S transition (Scott et al. 2001). Pol III transcription can be activated in vivo by cotransfection of G_1 cyclin/cdk pairs or by using a ribozyme to deplete endogenous p16 (Scott et al. 2001). In addition to regulating the pocket proteins, cdk2 and cdk4 can also directly phosphorylate and activate UBF (Voit et al. 1999; Voit and Grummt 2001).

Several viruses encode oncoproteins that bind and neutralize the pocket proteins (for review, see Vousden 1995). The most clinically important example of this is the E7 product of malignant HPV strains, which binds and inactivates RB and its relatives (Dyson et al. 1989; Munger et al. 1989). An E7 peptide can dissociate RB from UBF in vitro (Cavanaugh et al. 1995). Furthermore, Pol III transcription can be strongly stimulated in vivo by E7 of the highly malignant strain HPV-16 (Larminie et al. 1999; Sutcliffe et al. 1999). This is not an indirect response to cell transformation, since Pol III is also activated by a nontransforming mutant version of E7 that has retained its ability to neutralize RB (Larminie et al. 1999). However, deletions or substitutions in the pocket-binding domain of E7 abolish its capacity to stimulate a Pol III reporter (Larminie et al. 1999; Sutcliffe et al. 1999). It is therefore highly likely that E7 deregulates Pol III transcription by releasing TFIIIB from repression by RB and its relatives p107 and p130.

The oncoproteins of several other DNA tumor viruses can also bind RB and neutralize its function, including adenoviral E1A (Whyte et al. 1988, 1989) and the large-T antigens of SV40 and polyomavirus (DeCaprio et al. 1988; Moran 1988; Dyson et al. 1990). There is currently little or no evidence linking this to a deregulation of Pol I. However, RB-mediated repression of Pol III transcription has been shown to be reversed by E1A and the large-T antigens of SV40 or polyomavirus, both in vitro and in vivo (White et al. 1996; Larminie et al. 1999; Felton-Edkins and White 2002). Furthermore, the interaction between RB and TFIIIB is diminished substantially in fibroblasts transformed by SV40 or polyomavirus (Larminie et al. 1999; Felton-Edkins and White 2002). Release of TFIIIB from the inhibitory effects of the pocket proteins is therefore a feature of cells transformed by several DNA tumor viruses. Thus, aberrant Pol III activity can result when RB function is compromised through genetic mutation, hyperphosphorylation, or binding of viral oncoproteins (Fig. 4). Since one or other of these mechanisms is thought to apply in most if not all human tumors (Weinberg 1995), these observations seem to go a long way toward explaining the very high incidence of Pol III deregulation in cancers. There is currently less reason to believe that defective RB is a principal contributor to Pol I activation.

Deregulation of c-Myc is found in a range of human neoplasms, such as Burkitt's lymphomas and breast and colon carcinomas, and may contribute to one-seventh of all U.S. cancer deaths (Dang 1999; Nesbit et al. 1999). Overexpression of c-Myc can strongly stimulate transcription by Pols I and III in fibroblasts or B cells (Gomez-Roman et al. 2003; C. Grandori et al., unpubl.). It is likely that this will also be true of other cell types and that abnormal c-Myc activity will often help drive rRNA and tRNA synthesis during malignant transformation. Indeed, elevated c-Myc levels have been found to increase Pol III transcription in an ovarian carcinoma line (N. Gomez-Roman and R.J. White, unpubl.). Furthermore, elevated N-myc levels were shown to stimulate rRNA production in neuroblastoma cells (Boon et al. 2001).

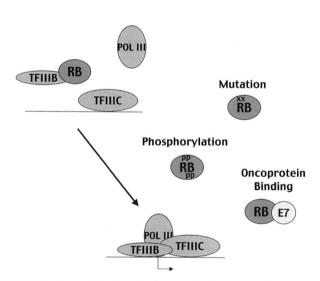

Figure 4. Mutation, hyperphosphorylation, or binding to viral oncoproteins, such as HPV E7, can prevent RB from repressing Pol III transcription both in vitro and in vivo.

As c-Myc, RB, and p53 have multiple targets and pleiotropic effects, it could be argued that elevated Pol I or Pol III activity is an unimportant side effect of their deregulation, with little impact on the transformation process. However, such an argument could not explain an unrelated mechanism that also deregulates Pol III in some situations. TFIIIC, a factor essential for tRNA and 5S rRNA synthesis (Paule and White 2000), is overexpressed specifically at both the mRNA and protein levels in each of four lines tested of fibroblasts transformed by SV40 or polyomavirus (White et al. 1990; Larminie et al. 1999; Felton-Edkins and White 2002). This effect is independent of large-T antigen (Felton-Edkins and White 2002). Furthermore, it is not a secondary response to accelerated proliferation, because TFIIIC levels are unaffected by growth factor availability or cell cycle arrest (Winter et al. 2000; Scott et al. 2001). These viruses therefore achieve very high levels of Pol III transcription by deregulating both TFIIIB and TFIIIC (Fig. 5). Similar deregulation is observed in biopsy samples from cancer patients (Winter et al. 2000). The DNA-binding activity of TFIIIC was measured in extracts of tissue from nine individuals with grade 2 or grade 3 ovarian epithelial carcinomas; in each case, the tumor sample had higher TFIIIC activity than the corresponding healthy ovarian tissue from the same patient (Winter et al. 2000). RT-PCR revealed that the tumors overexpress mRNAs encoding all five subunits of TFIIIC, while control mRNAs remained at normal levels (Winter et al. 2000). Although the number of cases examined so far is small, the results suggest that deregulation of TFIIIC may be a frequent feature of ovarian carcinomas. These observations provided the first evidence that a Pol III-specific factor is overproduced in human tumors. It is extremely unlikely to be a random effect, since as few as ~1.5% of transcripts are significantly deregulated in some tumors (Zhang et al. 1997). These observations suggest strongly that there is selective pressure to up-regulate TFIIIC in ovarian cancers. As TFIIIC is dedicated exclusively to Pol III transcription, this implies a specific drive to increase Pol III output as the tumors develop.

As well as TFIIIC, the Bdp1 subunit of TFIIIB is also overexpressed at the mRNA and protein levels in cells transformed by polyomavirus or SV40 (Felton-Edkins and White 2002). Again, this is a large-T-independent effect (Felton-Edkins and White 2002). Its significance is unclear, since the same lines produce normal levels of TBP and Brf1, the other subunits of TFIIIB (Felton-Edkins and White 2002). However, massive overexpression of Bdp1 mRNA has been found in one cervical carcinoma, out of five that were screened (N. Daly and R.J. White, unpubl.).

Much less is known about how transformation affects the Pol I machinery, especially in tumors in situ. The Pol I response to SV40 is one

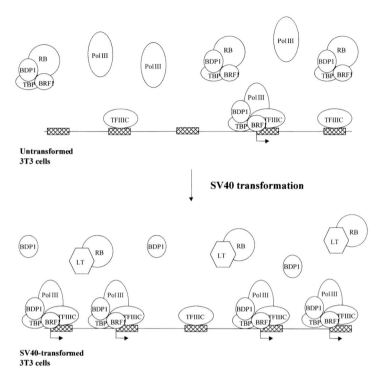

Figure 5. Changes to the Pol III machinery that accompany SV40 transformation of 3T3 fibroblasts. In untransformed 3T3 cells, TFIIIB activity is limiting because much of it is bound and repressed by RB and its relatives during G_0 and early G_1 phases. SV40 transformation triggers a marked increase in the levels of TFIIIC and Bdp1, whereas large-T antigen (LT) binds and neutralizes RB, thereby derepressing TFIIIB. As a consequence of these changes, SV40-transformed cells display very high levels of Pol III transcriptional activity.

of the few situations that has been studied in detail. The most striking effect is on rRNA half-life, which increases from 72 hours in primary human WI38 fibroblasts to 700 hours in SV40-transformed derivatives (Liebhaber et al. 1978). This is not a response to changes in growth rate, since rRNA stability is the same in serum-starved or rapidly growing WI38 cells (Liebhaber et al. 1978). The rate of Pol I transcription also increases in response to SV40 large-T antigen (Liebhaber et al. 1978; Learned et al. 1983; Zhai et al. 1997). This involves binding of large T to SL1 and the phosphorylation of UBF by an unidentified T-associated kinase (Zhai et al. 1997; Zhai and Comai 1999). Very different mechanisms therefore appear to be responsible for the activation of Pols I and III by SV40.

When 16 cases of hepatocellular carcinoma were examined, 11 showed overexpression of UBF mRNA relative to healthy liver from the same patients (Huang et al. 2002). Hepatitis B virus (HBV) is strongly linked to the development of hepatocellular carcinoma, and its X onco-gene can induce liver cancer in transgenic mice (Kim et al. 1991). X has been shown to stimulate transcription by both Pol I and Pol III (Aufiero and Schneider 1990; Wang et al. 1995, 1997, 1998). An increase in the level of TBP can explain this coordinate activation (Wang et al. 1995, 1997, 1998). The effect is likely to be direct, since it can be obtained with recombinant TBP in vitro and also by transfecting cells with a TBP mutant that is specifically defective in supporting Pol II transcription (Wang et al. 1997, 1998). The promoter of the TBP gene is activated by X in a Ras-dependent manner that requires signaling through the Raf/MEK/Erk cascade (Johnson et al. 2000). Thus, in addition to the rapid activation of Pols I and III produced by Erk phosphorylation of UBF and Brf1, this pathway can also elicit a slower response through raising the level of TBP. About 30% of human tumors carry activating mutations in Ras (for review, see Bos 1989). Such mutations can be expected to impact on the Pol I and III systems in several ways, including direct effects of Erk, induction of TBP and cyclin D1, and hyperphosphorylation of the pocket proteins (Downward 1997; Marshall 1999; Johnson et al. 2000; Stefanovsky et al. 2001; Felton-Edkins et al. 2003a; Zhao et al. 2003).

GROWTH SUPPRESSION BY TUMOR SUPPRESSORS MAY BE MEDIATED BY Pols I AND III

Transcription by Pols I and III is inhibited by three unrelated tumor suppressors, RB, p53, and brat. Furthermore, the tumor suppressor Arf can block rRNA production by inhibiting posttranscriptional processing (Sugimoto et al. 2003). These observations serve to underline the importance of controlling the output of these systems. There are reasons to consider that restraining rRNA and tRNA synthesis may be an important component of the growth control function of these tumor suppressors (Nasmyth 1996; White 1997; Neufeld and Edgar 1998). A prevalent view has been that the activity of RB can be explained by its ability to repress E2F. However, as pointed out by Nasmyth (1996), this does not fully make sense. E2F is a potent inducer of cell cycle progression, but it does not promote cell growth (Neufeld et al. 1998; Beier et al. 2000). Alternative interactions must therefore explain the ability of RB to restrain growth. RB is two orders of magnitude more abundant than E2F and has been shown to bind and regulate a variety of targets (for review, see Weinberg 1995;

Grana et al. 1998; Mulligan and Jacks 1998). The relative contributions of its many targets have yet to be clearly elucidated. However, its capacity to repress Pols I and III provides a clear link to growth control (Fig. 6). A similar argument can be made for p53, which inhibits growth under a range of stress conditions, including hypoxia, ribonucleotide depletion, and exposure to genotoxins (for review, see Oren 1999; Vousden and Lu 2002). In such circumstances, curtailing biosynthesis may be important for maintaining cell viability. Reducing the production of rRNA and tRNA may be especially advantageous when ribonucleotide pools are depleted, and indeed, we have found that such conditions do trigger a p53-dependent repression of Pol III transcription (J. Morton and R.J. White, unpubl.). In this regard, it is highly noteworthy that a specific block in rRNA processing and hence ribosome production has been shown to trigger a p53-mediated reversible arrest (Pestov et al. 2001). Indeed, the cell cycle arrest produced in this manner is even more potent than that produced by the paradigm cdk inhibitor p21. This has led to the intriguing hypothesis that p53 senses "nucleolar stress" and thereby helps to ensure that cell cycle progression is coupled to ribosome biogenesis (Pestov et al. 2001). Furthermore, Sherr has argued that the primordial role of Arf was to slow ribosome biogenesis in response to certain stress signals; its subsequent linkage to p53 may have evolved as a more efficient

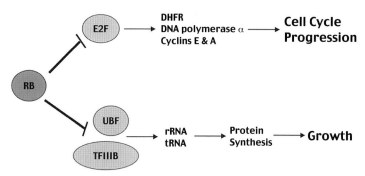

Figure 6. Model proposing how RB might achieve a balanced suppression of cell growth and proliferation. E2F can promote cell cycle progression through its ability to induce genes encoding proteins that drive the cell cycle, such as cyclin E, and proteins involved in DNA synthesis, such as dihydrofolate reductase (DHFR). Its ability to inhibit E2F provides RB with a potent mechanism for arresting the cell cycle. However, E2F does not induce cell growth (mass increase), which results from protein synthesis. The ability of RB to restrain rRNA and tRNA production through its effects on UBF and TFIIIB may provide a brake on the biosynthetic apparatus that restricts cell growth.

checkpoint for coupling ribosome production with p53-dependent inhibitors of cell cycle progression (Sugimoto et al. 2003).

At present there is no direct evidence that tumor suppressors do in fact achieve growth restraint by regulating Pol I and/or Pol III. However, it is a reasonable contention, given the clear data showing that these systems are targeted directly by RB and p53. It is unquestionable that growth can be prevented when the synthesis of rRNA or tRNA becomes limiting. The important issue is to what extent this ever happens under physiological circumstances. Genetic analysis indicates that RB suppresses tRNA levels by 2.4-fold in serum-starved mouse embryonic fibroblasts (Scott et al. 2001). It is not known whether this is sufficient to limit mammalian cell growth. However, such a reduction in yeast is clearly enough to slow growth substantially (Francis and Rajbhandary 1990). As already described, the hypertrophic growth of rat cardiomyocytes can be blocked by specifically preventing the 37% increase in rRNA synthesis that follows treatment with agonists (Brandenburger et al. 2001). The use of knockout mice has shown that RB and p130 together suppress rRNA production by ~40% (Ciarmatori et al. 2001). Such quantitative comparisons make it seem likely that growth rates may indeed be limited through the action of pocket proteins on UBF and TFIIIB. This might only be the case under particular conditions, whereas other pathways could mediate growth restraint in different cell types or circumstances. Nevertheless, the control points provided by Pols I and III may serve crucial regulatory functions in some situations. Loss of such controls during tumor development may constitute an important step toward neoplastic growth.

SUMMARY

A substantial part of the cell's biosynthetic capacity is engaged in producing rRNA and tRNA. Furthermore, the availability of these products is likely to be a principal determinant of a cell's potential to accumulate protein and thereby grow. It is therefore inevitable that the Pol I and Pol III transcription machines are tightly controlled by a complex interplay of regulatory networks which serve to ensure that their output is appropriate for the prevailing environmental conditions. In many situations, the activities of Pols I and III are coordinated by shared mechanisms working in parallel. However, some surprising instances are also known in which approximate coregulation is achieved through apparently unrelated mechanisms. Given the fundamental importance of rRNA and tRNA synthesis in determining potential for growth and hence proliferation, it is not surprising that Pols I and III are both targeted directly by RB and p53 in higher organisms. Indeed, the capacity of these key tumor suppressors to restrain rRNA and tRNA production may be an important component of their growth-restraining functions. When one or both become incapacitated during tumorigenesis, the transcriptional activity of Pols I and III is likely to rise; this is especially clear in the case of Pol III, which has been shown to respond in vivo when RB becomes inactivated through mutation, phosphorylation, or the binding of viral oncoproteins. Apart from the loss of tumor suppressors, a variety of other mechanisms can activate transcription by Pols I and III in transformed cells, such as derangement of the Ras/Erk pathway or overexpression of TFIIIC2. There seems to be considerable pressure to activate these systems in cancers, and this end can be achieved through a diversity of means. The consequences of such deregulation are only beginning to be explored, but they may very well be profound.

REFERENCES

Altmann G.G. and Leblond C.P. 1982. Changes in the size and structure of the nucleolus of columnar cells during their migration from crypt base to villus top in rat jejunum. *J. Cell Sci.* **56:** 83–99.

Alzuherri H.M. and White R.J. 1998. Regulation of a TATA-binding protein-associated factor during cellular differentiation. *J. Biol. Chem.* **273:** 17166–17171.

———. 1999. Regulation of RNA polymerase I transcription in response to F9 embryonal carcinoma stem cell differentiation. *J. Biol. Chem.* **274:** 4328–4334.

Aufiero B. and Schneider R.J. 1990. The hepatitis B virus X-gene product trans-activates both RNA polymerase II and III promoters. *EMBO J.* **9:** 497–504.

Bateman E. and Paule M.R. 1986. Regulation of eukaryotic ribosomal RNA synthesis by RNA polymerase modification. *Cell* **47:** 445–450.

Bates S. and Peters G. 1995. Cyclin D1 as a cellular proto-oncogene. *Semin. Cancer Biol.* **6:** 73–82.

Baxter G.C. and Stanners C.P. 1978. The effect of protein degradation on cellular growth characteristics. *J. Cell. Physiol.* **96:** 139–146.

Bazett-Jones D.P., Leblanc B., Herfort M., and Moss T. 1994. Short-range DNA looping by the *Xenopus* HMG-box transcription factor, xUBF. *Science* **264:** 1134–1137.

Beckmann H., Chen J.-L., O'Brien T., and Tjian R. 1995. Coactivator and promoter-selective properties of RNA polymerase I TAFs. *Science* **270:** 1506–1509.

Beier R., Burgin A., Kiermaier A., Fero M., Karsunky H., Saffrich R., Moroy T., Ansorge W., Roberts J., and Eilers M. 2000. Induction of cyclin E-cdk2 kinase activity, E2F-dependent transcription and cell growth by Myc are genetically separable events. *EMBO J.* **19:** 5813–5823.

Belenguer P., Baldin V., Mathieu C., Prats H., Bensaid M., Bouche G., and Amalric F. 1989. Protein kinase NII and the regulation of rDNA transcription in mammalian cells. *Nucleic Acids Res.* **17:** 6625–6636.

Bertrand E., Houser-Scott F., Kendall A., Singer R.H., and Engelke D.R. 1998. Nucleolar localization of early tRNA processing. *Genes Dev.* **12:** 2463–2468.

Bodem J., Dobreva G., Hoffmann-Rohrer U., Iben S., Zentgraf H., Delius H., Vingron M., and Grummt I. 2000. TIF-IA, the factor mediating growth-dependent control of ribosomal RNA synthesis, is the mammalian homolog of yeast Rrn3p. *EMBO Rep.* **1:** 171–175.

Boon K., Caron H.N., van Asperen R., Valentijn L., Hermus M.-C., van Sluis P., Roobeek I., Weis I., Voute P.A., Schwab M., and Versteeg R. 2001. N-myc enhances the expression of a large set of genes functioning in ribosome biogenesis and protein synthesis. *EMBO J.* **20:** 1383–1393.

Bos J.L. 1989. Ras oncogenes in human cancer: A review. *Cancer Res.* **49:** 4682–4689.

Bosc D.G., Luscher B., and Litchfield D.W. 1999. Expression and regulation of protein kinase CK2 during the cell cycle. *Mol. Cell. Biochem.* **191:** 213–222.

Brandenburger Y., Jenkins A., Autelitano D.J., and Hannan R.D. 2001. Increased expression of UBF is a critical determinant for rRNA synthesis and hypertrophic growth of cardiac myocytes. *FASEB J.* **15:** 2051–2053.

Brooks R.F. 1977. Continuous protein synthesis is required to maintain the probability of entry into S phase. *Cell* **12:** 311–317.

Brown D.D. and Dawid I.B. 1968. Specific gene amplification in oocytes. *Science* **160:** 272–280.

Brown T.R.P., Scott P.H., Stein T., Winter A.G., and White R.J. 2000. RNA polymerase III transcription: Its control by tumor suppressors and its deregulation by transforming agents. *Gene Expr.* **9:** 15–28.

Busch H. and Smetana K. 1970. Nucleoli of tumor cells. In *The nucleolus*, pp. 448–471. Academic Press, New York.

Buttgereit D., Pflugfelder G., and Grummt I. 1985. Growth-dependent regulation of rRNA synthesis is mediated by a transcription initiation factor (TIF-IA). *Nucleic Acids Res.* **13:** 8165–8180.

Cairns C.A. and White R.J. 1998. p53 is a general repressor of RNA polymerase III transcription. *EMBO J.* **17:** 3112–3123.

Cardenas M.E., Cutler N.S., Lorenz M.C., Di Como C.J., and Heitman J. 1999. The TOR signaling cascade regulates gene expression in response to nutrients. *Genes Dev.* **13:** 3271–3279.

Carmo-Fonseca M., Mendes-Soares L., and Campos I. 2000. To be or not to be in the nucleolus. *Nat. Cell Biol.* **2:** E107–E112.

Carroll D. and Marshak D.R. 1989. Serum-stimulated cell growth causes oscillations in casein kinase II activity. *J. Biol. Chem.* **264:** 7345–7348.

Cavanaugh A.H., Hempel W.M., Taylor L.J., Rogalsky V., Todorov G., and Rothblum L.I. 1995. Activity of RNA polymerase I transcription factor UBF blocked by *Rb* gene product. *Nature* **374:** 177–180.

Cavanaugh A.H., Hirschler-Laszkiewicz I., Hu Q., Dundr M., Smink T., Misteli T., and Rothblum L.I. 2002. Rrn3 phosphorylation is a regulatory checkpoint for ribosome biogenesis. *J. Biol. Chem.* **277:** 27423–27432.

Chen W., Bocker W., Brosius J., and Tiedge H. 1997a. Expression of neural BC200 RNA in human tumors. *J. Pathol.* **183:** 345–351.

Chen W., Heierhorst J., Brosius J., and Tiedge H. 1997b. Expression of neural *BC1* RNA: Induction in murine tumors. *Eur. J. Cancer* **33:** 288–292.

Chesnokov I., Chu W.-M., Botchan M.R., and Schmid C.W. 1996. p53 inhibits RNA polymerase III-directed transcription in a promoter-dependent manner. *Mol. Cell. Biol.* **16:** 7084–7088.

Chu W.-M., Wang Z., Roeder R.G., and Schmid C.W. 1997. RNA polymerase III transcription repressed by Rb through its interactions with TFIIIB and TFIIIC2. *J. Biol. Chem.* **272:** 14755–14761.

Ciarmatori S., Scott P.H., Sutcliffe J.E., McLees A., Alzuherri H.M., Dannenberg J.-H., Te Riele H., Grummt I., Voit R., and White R.J. 2001. Overlapping functions of the pRb family in the regulation of rRNA synthesis. *Mol. Cell. Biol.* **21:** 5806–5814.

Clark M.E., Lieberman P.M., Berk A.J., and Dasgupta A. 1993. Direct cleavage of human TATA-binding protein by poliovirus protease 3C in vivo and in vitro. *Mol. Cell. Biol.* **13:** 1232–1237.

Clarke E.M., Peterson C.L., Brainard A.V., and Riggs D.L. 1996. Regulation of the RNA polymerase I and III transcription systems in response to growth conditions. *J. Biol. Chem.* **271:** 22189–22195.

Coller H.A., Grandori C., Tamayo P., Colbert T., Lander E.S., Eisenman R.N., and Golub T.R. 2000. Expression analysis with oligonucleotide microarrays reveals that MYC regulates genes involved in growth, cell cycle, signaling, and adhesion. *Proc. Natl. Acad. Sci.* **97:** 3260–3265.

Comai L., Tanese N., and Tjian R. 1992. The TATA-binding protein and associated factors are integral components of the RNA polymerase I transcription factor, SL1. *Cell* **68:** 965–976.

Comai L., Song Y., Tan C., and Bui T. 2000. Inhibition of RNA polymerase I transcription in differentiated myeloid leukaemia cells by inactivation of selectivity factor 1. *Cell Growth Differ.* **11:** 63–70.

Crighton D., Woiwode A., Zhang C., Mandavia N., Morton J.P., Warnock L.J., Milner J., White R.J., and Johnson D.L. 2003. p53 represses RNA polymerase III transcription by targeting TBP and inhibiting promoter occupancy by TFIIIB. *EMBO J.* **22:** 2810–2820.

Crook T., Marston N.J., Sara E.A., and Vousden K.H. 1994. Transcriptional activation by p53 correlates with suppression of growth but not transformation. *Cell* **79:** 817–827.

Damania B. and Alwine J.C. 1996. TAF-like function of SV40 large T antigen. *Genes Dev.* **10:** 1369–1381.

Damania B., Mital R., and Alwine J.C. 1998. Simian virus 40 large T antigen interacts with human TFIIB-related factor and small nuclear RNA-activating protein complex for transcriptional activation of TATA-containing polymerase III promoters. *Mol. Cell. Biol.* **18:** 1331–1338.

Dang C.V. 1999. c-Myc target genes involved in cell growth, apoptosis, and metabolism. *Mol. Cell. Biol.* **19:** 1–11.

DeCaprio J.A., Ludlow J.W., Figge J., Shew J.-Y., Huang C.-M., Lee W.-H., Marsilio E., Paucha E., and Livingston D.M. 1988. SV40 large tumor antigen forms a specific complex with the product of the retinoblastoma susceptibility gene. *Cell* **54:** 275–283.

Derenzini M. and Ploton D. 1994. Interphase nucleolar organizer regions. In *Molecular biology in histopathology* (ed. J. Crocker), pp. 231–249. Wiley, New York.

Derenzini M., Trere D., Pession A., Montanaro L., Sirri V., and Ochs R.L. 1998. Nucleolar function and size in cancer cells. *Am. J. Pathol.* **152:** 1291–1297.

DiCiommo D., Gallie B.L., and Bremner R. 2000. Retinoblastoma: The disease, gene and protein provide critical leads to understand cancer. *Semin. Cancer Biol.* **10:** 255–269.

Downward J. 1997. Routine role for Ras. *Curr. Biol.* **7:** R258–R260.

Dyson N., Howley P.M., Munger K., and Harlow E. 1989. The human papillomavirus-16

E7 oncoprotein is able to bind to the retinoblastoma gene product. *Science* **243:** 934–937.

Dyson N., Bernards R., Friend S.H., Gooding L.R., Hassel J.A., Major E.O., Pipas J.M., Vandyke T., and Harlow E. 1990. Large T antigens of many polyomaviruses are able to form complexes with the retinoblastoma protein. *J. Virol.* **64:** 1353–1356.

Fath S., Milkereit P., Peyroche G., Riva M., Carles C., and Tschochner H. 2001. Differential roles of phosphorylation in the formation of transcriptional active RNA polymerase I. *Proc. Natl. Acad. Sci.* **98:** 14334–14339.

Felton-Edkins Z.A. and White R.J. 2002. Multiple mechanisms contribute to the activation of RNA polymerase III transcription in cells transformed by papovaviruses. *J. Biol. Chem.* **277:** 48182–48191.

Felton-Edkins Z.A., Fairley J.A., Graham E.L., Johnston I.M., White R.J., and Scott P.H. 2003a. The mitogen-activated protein (MAP) kinase ERK induces tRNA synthesis by phosphorylating TFIIIB. *EMBO J.* **22:** 2422–2432.

Felton-Edkins Z.A., Kenneth N.S., Brown T.R.P., Daly N.L., Gomez-Roman N., Grandori C., Eisenman R.N., and White R.J. 2003b. Direct regulation of RNA polymerase III transcription by RB, p53 and c-Myc. *Cell Cycle* **2:** 181–184.

Fingar D.C., Salama S., Tsou C., Harlow E., and Blenis J. 2002. Mammalian cell size is controlled by mTOR and its downstream targets S6K1 and 4EBP1/eIF4E. *Genes Dev.* **16:** 1472–1487.

Ford P. 1971. Non-coordinated accumulation and synthesis of 5S ribonucleic acid by ovaries of *Xenopus laevis*. *Nature* **233:** 561–564.

Fradkin L.G., Yoshinaga S.K., Berk A.J., and Dasgupta A. 1987. Inhibition of host cell RNA polymerase III-mediated transcription by poliovirus: Inactivation of specific transcription factors. *Mol. Cell. Biol.* **7:** 3880–3887.

Francis M.A. and Rajbhandary U.L. 1990. Expression and function of a human initiator tRNA gene in the yeast *Saccharomyces cerevisiae*. *Mol. Cell. Biol.* **10:** 4486–4494.

Frank D.J. and Roth M.B. 1998. ncl-1 is required for the regulation of cell size and ribosomal RNA synthesis in *Caenorhabditis elegans*. *J. Cell Biol.* **140:** 1321–1329.

Frank D.J., Edgar B.A., and Roth M.B. 2002. The *Drosophila melanogaster* gene *brain tumor* negatively regulates cell growth and ribosomal RNA synthesis. *Development* **129:** 399–407.

Friend S.H., Bernards R., Rogelj S., Weinberg R.A., Rapaport J.M., Alberts D.M., and Dryja T.P. 1986. A human DNA segment with properties of the gene that predisposes to retinoblastoma and osteosarcoma. *Nature* **323:** 643–646.

Gauthier-Rouviere C., Basset M., Blanchard J.-M., Cavadore J.-C., Fernandez A., and Lamb N.J.C. 1991. Casein kinase II induces *c-fos* expression via the serum response element pathway and p67[SRF] phosphorylation in living fibroblasts. *EMBO J.* **10:** 2921–2930.

Geiduschek E.P. and Kassavetis G.A. 2001. The RNA polymerase III transcription apparatus. *J. Mol. Biol.* **310:** 1–26.

Ghavidel A. and Schultz M.C. 1997. Casein kinase II regulation of yeast TFIIIB is mediated by the TATA-binding protein. *Genes Dev.* **11:** 2780–2789.

———. 2001. TATA binding protein-associated CK2 transduces DNA damage signals to the RNA polymerase III transcriptional machinery. *Cell* **106:** 575–584.

Goh H.S., Yao J., and Smith D.R. 1995. p53 point mutation and survival in colorectal cancer patients. *Cancer Res.* **55:** 5217–5221.

Gomez-Roman N., Grandori C., Eisenman R.N., and White R.J. 2003. Direct activation of RNA polymerase III transcription by c-Myc. *Nature* **421:** 290–294.

Gottesfeld J.M. and Forbes D.J. 1997. Mitotic repression of the transcriptional machinery. *Trends Biochem. Sci.* **22:** 197–202.

Gottesfeld J.M., Johnson D.L., and Nyborg J.K. 1996. Transcriptional activation of RNA polymerase III-dependent genes by the human T-cell leukaemia virus type 1 Tax protein. *Mol. Cell. Biol.* **16:** 1777–1785.

Gottesfeld J.M., Wolf V.J., Dang T., Forbes D.J., and Hartl P. 1994. Mitotic repression of RNA polymerase III transcription in vitro mediated by phosphorylation of a TFIIIB component. *Science* **263:** 81–84.

Grana X., Garriga J., and Mayol X. 1998. Role of the retinoblastoma protein family, pRB, p107 and p130 in the negative control of cell growth. *Oncogene* **17:** 3365–3383.

Grandori C., Cowley S.M., James L.P., and Eisenman R.N. 2000. The Myc/Max/Mad network and the transcriptional control of cell behavior. *Annu. Rev. Cell Dev. Biol.* **16:** 653–699.

Grummt I. 1999. Regulation of mammalian ribosomal gene transcription by RNA polymerase I. *Prog. Nucleic Acid Res. Mol. Biol.* **62:** 109–154.

———. 2003. Life on a planet of its own: Regulation of RNA polymerase I transcription in the nucleolus. *Genes Dev.* **17:** 1691–1702.

Grummt I., Smith V.A., and Grummt F. 1976. Amino acid starvation affects the initiation frequency of nucleolar RNA polymerase. *Cell* **7:** 439–445.

Hanna D.E., Rethinaswamy A., and Glover C.V.C. 1995. Casein kinase II is required for cell cycle progression during G_1 and G_2/M in *Saccharomyces cerevisiae*. *J. Biol. Chem.* **270:** 25905–25914.

Hannan R.D. and Rothblum L.I. 1995. Regulation of ribosomal DNA transcription during neonatal cardiomyocyte hypertrophy. *Cardiovasc. Res.* **30:** 501–510.

Hannan R.D., Luyken J., and Rothblum L.I. 1995. Regulation of rDNA transcription factors during cardiomyocyte hypertrophy induced by adrenergic agents. *J. Biol. Chem.* **270:** 8290–8297.

———. 1996. Regulation of ribosomal DNA transcription during contraction-induced hypertrophy of neonatal cardiomyocytes. *J. Biol. Chem.* **271:** 3213–3220.

Hannan R.D., Hempel W.M., Cavanaugh A., Arino T., Dimitrov S.I., Moss T., and Rothblum L. 1998. Affinity purification of mammalian RNA polymerase I. *J. Biol. Chem.* **273:** 1257–1267.

Hannan K.M., Hannan R.D., Smith S.D., Jefferson L.S., Lun M., and Rothblum L.I. 2000a. Rb and p130 regulate RNA polymerase I transcription: Rb disrupts the interaction between UBF and SL-1. *Oncogene* **19:** 4988–4999.

Hannan K.M., Kennedy B.K., Cavanaugh A.H., Hannan R.D., Hirschler-Laszkiewicz I., Jefferson L.S., and Rothblum L.I. 2000b. RNA polymerase I transcription in confluent cells: Rb downregulates rDNA transcription during confluence-induced cell cycle arrest. *Oncogene* **19:** 3487–3497.

Hartwell L.H. 1971. Genetic control of the cell division cycle in yeast. II. Genes controlling DNA replication and its initiation. *J. Mol. Biol.* **59:** 183–194.

Heix J., Vente A., Voit R., Budde A., Michaelidis T.M., and Grummt I. 1998. Mitotic silencing of human rRNA synthesis: Inactivation of the promoter selectivity factor SL1 by cdc2/cyclin B-mediated phosphorylation. *EMBO J.* **17:** 7373–7381.

Hirama T. and Koeffler H.P. 1995. Role of the cyclin-dependent kinase inhibitors in the development of cancer. *Blood* **86:** 841–854.

Hockman D.J. and Schultz M.C. 1996. Casein kinase II is required for efficient transcription by RNA polymerase III. *Mol. Cell. Biol.* **16:** 892–898.

Hollstein M., Sidransky D., Vogelstein B., and Harris C.C. 1991. p53 mutations in human cancers. *Science* **253**: 49–53.

Hollstein M., Rice K., Greenblatt M.S., Soussi T., Fuchs R., Sorlie T., Hovig E., Smith-Sorensen B., Montesano R., and Harris C.C. 1994. Database of p53 gene somatic mutations in human tumors and cell lines. *Nucleic Acids Res.* **22**: 3551–3555.

Horowitz J.M., Park S.-H., Bogenmann E., Cheng J.-C., Yandell D.W., Kaye F.J., Minna J.D., Dryja T.P., and Weinberg R.A. 1990. Frequent inactivation of the retinoblastoma antioncogene is restricted to a subset of human tumor cells. *Proc. Natl. Acad. Sci.* **87**: 2775–2779.

Hu Q., Dyson N., and Harlow E. 1990. The regions of the retinoblastoma protein needed for binding to adenovirus E1A or SV40 large T antigen are common sites for mutations. *EMBO J.* **9**: 1147–1155.

Huang R., Wu T., Xu L., Liu A., Ji Y., and Hu G. 2002. Upstream binding factor up-regulated in hepatocellular carcinoma is related to the survival and cisplatin-sensitivity of cancer cells. *FASEB J.* **16**: 293–301.

Hunter T. and Pines J. 1994. Cyclins and cancer II: Cyclin D and CDK inhibitors come of age. *Cell* **79**: 573–582.

Iben S., Tschochner H., Bier M., Hoogstraten D., Hozak P., Egly J.-M., and Grummt I. 2002. TFIIH plays an essential role in RNA polymerase I transcription. *Cell* **109**: 297–306.

Inostroza J.A., Mermelstein F.H., Ha I., Lane W.S., and Reinberg D. 1992. Dr1, a TATA-binding protein-associated phosphoprotein and inhibitor of class II gene transcription. *Cell* **70**: 477–489.

Iritani B.M. and Eisenman R.N. 1999. c-Myc enhances protein synthesis and cell size during B lymphocyte development. *Proc. Natl. Acad. Sci.* **96**: 13180–13185.

Jacobson M.R., Cao L.-G., Taneja K., Singer R.H., Wang Y.-I., and Pederson T. 1997. Nuclear domains of the RNA subunit of RNase P. *J. Cell Sci.* **110**: 829–837.

James C.B.L. and Carter T.H. 1992. Activation of protein kinase C inhibits adenovirus VA gene transcription *in vitro*. *J. Gen. Virol.* **73**: 3133–3139.

Johnson L.F., Abelson H.T., Green H., and Penman S. 1974. Changes in RNA in relation to growth of the fibroblast. I. Amounts of mRNA, rRNA, and tRNA in resting and growing cells. *Cell* **1**: 95–100.

Johnson S.A.S., Mandavia N., Wang H.-D., and Johnson D.L. 2000. Transcriptional regulation of the TATA-binding protein by Ras cellular signaling. *Mol. Cell. Biol.* **20**: 5000–5009.

Johnston G.C. 1977. Cell size and budding during starvation of the yeast *Saccharomyces cerevisiae*. *J. Bacteriol.* **132**: 738–739.

Johnston G.C., Pringle J.R., and Hartwell L.H. 1977. Coordination of growth with cell division in the yeast *Saccharomyces cerevisiae*. *Exp. Cell Res.* **105**: 79–98.

Johnston I.M., Allison S.J., Morton J.P., Schramm L., Scott P.H., and White R.J. 2002. CK2 forms a stable complex with TFIIIB and activates RNA polymerase III transcription in human cells. *Mol. Cell. Biol.* **22**: 3757–3768.

Jorgensen P., Nishikawa J.L., Breitkreutz B.-J., and Tyers M. 2002. Systematic identification of pathways that couple cell growth and division in yeast. *Science* **297**: 395–400.

Kamada K., Shu F., Chen H., Malik S., Stelzer G., Roeder R.G., Meisterernst M., and Burley S.K. 2001. Crystal structure of negative cofactor 2 recognizing the TBP-DNA transcription complex. *Cell* **106**: 71–81.

Kassavetis G.A., Braun B.R., Nguyen L.H., and Geiduschek E.P. 1990. *S. cerevisiae* TFIIIB is the transcription initiation factor proper of RNA polymerase III, while TFIIIA and TFIIIC are assembly factors. *Cell* **60:** 235–245.

Kaye F.J., Kratzke R.A., Gerster J.L., and Horowitz J.M. 1990. A single amino acid substitution results in a retinoblastoma protein defective in phosphorylation and oncoprotein binding. *Proc. Natl. Acad. Sci.* **87:** 6922–6926.

Kief D.R. and Warner J.R. 1981. Coordinate control of syntheses of ribosomal ribonucleic acid and ribosomal proteins during nutritional shift-up in *Saccharomyces cerevisiae*. *Mol. Cell. Biol.* **1:** 1007–1015.

Kihm A.J., Hershey J.C., Haystead T.A., Madsen C.S., and Owens G.K. 1998. Phosphorylation of the rRNA transcription factor upstream binding factor promotes its association with TATA binding protein. *Proc. Natl. Acad. Sci.* **95:** 14816–14820.

Killander D. and Zetterberg A. 1965. A quantitative cytochemical investigation of the relationship between cell mass and initiation of DNA synthesis in mouse fibroblasts *in vitro*. *Exp. Cell Res.* **40:** 12–20.

Kim C.-M., Koike K., Saito I., Miyamura T., and Jay G. 1991. *HBx* gene of hepatitis B virus induces liver cancer in transgenic mice. *Nature* **351:** 317–320.

Kim S., Li Q., Dang C.V., and Lee L.A. 2000. Induction of ribosomal genes and hepatocyte hypertrophy by adenovirus-mediated expression of c-Myc *in vivo*. *Proc. Natl. Acad. Sci.* **97:** 11198–11202.

Kim S., Na J.G., Hampsey M., and Reinberg D. 1997. The Dr1/DRAP1 heterodimer is a global repressor of transcription *in vivo*. *Proc. Natl. Acad. Sci.* **94:** 820–825.

King R.J.B. 1996. *Cancer biology*. Longman, Harlow, United Kingdom.

Klarlund J.K. and Czech M.P. 1988. Insulin-like growth factor I and insulin rapidly increase casein kinase II activity in BALB/c 3T3 fibroblasts. *J. Biol. Chem.* **263:** 15872–15875.

Klein J. and Grummt I. 1999. Cell cycle-dependent regulation of RNA polymerase I transcription: The nucleolar transcription factor UBF is inactive in mitosis and early G_1. *Proc. Natl. Acad. Sci.* **96:** 6096–6101.

Kliewer S. and Dasgupta A. 1988. An RNA polymerase II transcription factor inactivated in poliovirus-infected cells copurifies with transcription factor TFIID. *Mol. Cell. Biol.* **8:** 3175–3182.

Kuhn A., Vente A., Doree M., and Grummt I. 1998. Mitotic phosphorylation of the TBP-containing factor SL1 represses ribosomal gene transcription. *J. Mol. Biol.* **284:** 1–5.

Kung A.L., Sherwood S.W., and Schimke R.T. 1993. Differences in the regulation of protein synthesis, cyclin B accumulation, and cellular growth in response to the inhibition of DNA synthesis in Chinese hamster ovary and HeLa S3 cells. *J. Biol. Chem.* **268:** 23072–23080.

Kurata S., Kog K., and Sakaguchi B. 1978. Nucleolar size in parallel with ribosomal RNA synthesis at diapause termination in the eggs of *Bombyx mori*. *Chromosoma* **68:** 313–317.

Kuromori T. and Yamamoto M. 1994. Cloning of cDNAs from *Arabidopsis thaliana* that encode putative protein phosphatase 2C and a human Dr1-like protein by transformation of a fission yeast mutant. *Nucleic Acids Res.* **22:** 5296–5301.

Larminie C.G.C., Sutcliffe J.E., Tosh K., Winter A.G., Felton-Edkins Z.A., and White R.J. 1999. Activation of RNA polymerase III transcription in cells transformed by simian virus 40. *Mol. Cell. Biol.* **19:** 4927–4934.

Larson D.E., Xie W., Glibetic M., O'Mahony D., Sells B.H., and Rothblum L.I. 1993. Coordinated decreases in rRNA gene transcription factors and rRNA synthesis during muscle cell differentiation. *Proc. Natl. Acad. Sci.* **90:** 7933–7936.

Lassar A.B., Martin P.L., and Roeder R.G. 1983. Transcription of class III genes: Formation of preinitiation complexes. *Science* **222:** 740–748.

Learned R.M., Smale S., Haltiner M., and Tjian R. 1983. Regulation of human ribosomal RNA transcription. *Proc. Natl. Acad. Sci.* **80:** 3558–3562.

Leresche A., Wolf V.J., and Gottesfeld J.M. 1996. Repression of RNA polymerase II and III transcription during M phase of the cell cycle. *Exp. Cell Res.* **229:** 282–288.

Li Y., Moir R.D., Sethy-Coraci I.K., Warner J.R., and Willis I.M. 2000. Repression of ribosome and tRNA synthesis in secretion-defective cells is signaled by a novel branch of the cell integrity pathway. *Mol. Cell. Biol.* **20:** 3843–3851.

Liebhaber S.A., Wolf S., and Schlessinger D. 1978. Differences in rRNA metabolism of primary and SV40-transformed human fibroblasts. *Cell* **13:** 121–127.

Lopez-de-Leon A., Librizzi M., Tuglia K., and Willis I. 1992. PCF4 encodes an RNA polymerase III transcription factor with homology to TFIIB. *Cell* **71:** 211–220.

Lozano G. and Elledge S. 2000. p53 sends nucleotides to repair DNA. *Nature* **404:** 24–25.

Luo Y., Kurz J., MacAfee N., and Krause M.O. 1997. C-myc deregulation during transformation induction: Involvement of 7S K RNA. *J. Cell. Biochem.* **64:** 313–327.

Marshall C. 1999. How do small GTPase signal transduction pathways regulate cell cycle entry? *Curr. Opin. Cell Biol.* **11:** 732–736.

Mateyak M.K., Obaya A.J., Adachi S., and Sedivy J.M. 1997. Phenotypes of c-Myc-deficient rat fibroblasts isolated by targeted homologous recombination. *Cell Growth Differ.* **8:** 1039–1048.

Matthews J.L., Zwick M.G., and Paule M.R. 1995. Coordinate regulation of ribosomal component synthesis in *Acanthamoeba castellanii:* 5S RNA transcription is down regulated during encystment by alteration of TFIIIA activity. *Mol. Cell. Biol.* **15:** 3327–3335.

Mauck J.C. and Green H. 1974. Regulation of pre-transfer RNA synthesis during transition from resting to growing state. *Cell* **3:** 171–177.

Mermelstein F., Yeung K., Cao J., Inostroza J.A., Erdjument-Bromage H., Eagelson K., Landsman D., Levitt P., Tempst P., and Reinberg D. 1996. Requirement of a corepressor for Dr1-mediated repression of transcription. *Genes Dev.* **10:** 1033–1048.

Milkereit P. and Tschochner H. 1998. A specialized form of RNA polymerase I, essential for initiation and growth-dependent regulation of rRNA synthesis, is disrupted during transcription. *EMBO J.* **17:** 3692–3703.

Miller G., Panov K.I., Friedrich J.K., Trinkle-Mulcahy L., Lamond A.I., and Zomerdijk J.C.B.M. 2001. hRRN3 is essential in the SL1-mediated recruitment of RNA polymerase I to rRNA gene promoters. *EMBO J.* **20:** 1373–1382.

Mittnacht S. 1998. Control of pRB phosphorylation. *Curr. Opin. Genet. Dev.* **8:** 21–27.

Mizuta K. and Warner J.R. 1994. Continued functioning of the secretory pathway is essential for ribosome synthesis. *Mol. Cell. Biol.* **14:** 2493–2502.

Momand J., Jung D., Wilczynski S., and Niland J. 1998. The *MDM2* gene amplification database. *Nucleic Acids Res.* **26:** 3453–3459.

Montagne J. 2000. Genetic and molecular mechanisms of cell size control. *Mol. Cell Biol. Res. Commun.* **4:** 195–202.

Moorefield B., Greene E.A., and Reeder R.H. 2000. RNA polymerase I transcription factor Rrn3 is functionally conserved between yeast and human. *Proc. Natl. Acad. Sci.* **97:** 4724–4729.

Moran E. 1988. A region of SV40 large T antigen can substitute for a transforming domain of the adenovirus E1A products. *Nature* **334:** 168–170.

Moss T. and Stefanovsky V.Y. 1995. Promotion and regulation of ribosomal transcription in eukaryotes by RNA polymerase I. *Prog. Nucleic Acid Res. Mol. Biol.* **50:** 25–66.

———. 2002. At the center of eukaryotic life. *Cell* **109:** 545–548.

Mulligan G. and Jacks T. 1998. The retinoblastoma gene family: Cousins with overlapping interests. *Trends Genet.* **14:** 223–229.

Munger K., Werness B.A., Dyson N., Phelps W.C., Harlow E., and Howley P.M. 1989. Complex formation of human papillomavirus E7 proteins with the retinoblastoma tumor suppressor gene product. *EMBO J.* **8:** 4099–4105.

Munstermann U., Fritz G., Seitz G., Lu Y.P., Schneider H.R., and Issinger O.G. 1990. Casein kinase II is elevated in solid human tumors and rapidly proliferating nonneoplastic tissue. *Eur. J. Biochem.* **189:** 251–257.

Nasmyth K. 1996. Another role rolls in. *Nature* **382:** 28–29.

Nesbit C.E., Tersak J.M., and Prochownik E.V. 1999. MYC oncogenes and human neoplastic disease. *Oncogene* **18:** 3004–3016.

Neufeld T.P. and Edgar B.A. 1998. Connections between growth and the cell cycle. *Curr. Opin. Cell Biol.* **10:** 784–790.

Neufeld T.P., de la Cruz A.F., Johnston L.A., and Edgar B.A. 1998. Coordination of growth and cell division in the *Drosophila* wing. *Cell* **93:** 1183–1193.

Nierras C.R. and Warner J.R. 1999. Protein kinase C enables the regulatory circuit that connects membrane synthesis to ribosome synthesis in *Saccharomyces cerevisiae*. *J. Biol. Chem.* **274:** 13235–13241.

Nomura M. 1999. Regulation of ribosome biosynthesis in *Escherichia coli* and *Saccharomyces cerevisiae*: Diversity and common principles. *J. Bacteriol.* **181:** 6857–6864.

Nurse P. 1975. Genetic control of cell size at cell division in yeast. *Nature* **256:** 547–551.

O' Mahony D.J., Xie W., Smith S.D., Singer H.A., and Rothblum L.I. 1992. Differential phosphorylation and localization of the transcription factor UBF *in vivo* in response to serum deprivation. *J. Biol. Chem.* **267:** 35–38.

Oren M. 1999. Regulation of the p53 tumor suppressor protein. *J. Biol. Chem.* **274:** 36031–36034.

Orlandini M., Semplici F., Ferruzzi R., Meggio F., Pinna L.A., and Oliviero S. 1998. Protein kinase *CK2α′* is induced by serum as a delayed early gene and cooperates with Ha-*ras* in fibroblast transformation. *J. Biol. Chem.* **273:** 21291–21297.

Paule M.R., Ed. 1998. *Transcription of ribosomal RNA genes by eukaryotic RNA polymerase I*. Springer-Verlag, Berlin.

Paule M.R. and White R.J. 2000. Transcription by RNA polymerases I and III. *Nucleic Acids Res.* **28:** 1283–1298.

Pelletier G., Stefanovsky V.Y., Faubladier M., Hirschler-Laszkiewicz I., Savard J., Rothblum L.I., Cote J., and Moss T. 2000. Competitive recruitment of CBP and Rb-HDAC regulates UBF acetylation and ribosomal transcription. *Mol. Cell* **6:** 1059–1066.

Pepperkok R., Lorenz P., Ansorge W., and Pyerin W. 1994. Casein kinase II is required for transition of G_0/G_1, eraly G_1, and G_1/S phases of the cell cycle. *J. Biol. Chem.* **269:** 6986–6991.

Pepperkok R., Lorenz P., Jakobi R., Ansorge W., and Pyerin W. 1991. Cell growth stimulation by EGF: Inhibition through antisense-oligodeoxynucleotides demonstrates important role of casein kinase II. *Exp. Cell Res.* **197:** 245–253.

Perletti L., Kopf E., Carre L., and Davidson I. 2001. Coordinate regulation of RARgamma2, TBP, and TAF$_{II}$135 by targeted proteolysis during retinoic acid-induced differentiation of F9 embryonal carcinoma cells. *BMC Mol. Biol.* **2:** 4 (http://www.biomedcentral.com/1471-2199/2/4).

Pestov D.G., Strezoska Z., and Lau L.F. 2001. Evidence of p53-dependent cross-talk between ribosome biogenesis and the cell cycle: Effects of nucleolar protein Bop1 on G1/S transition. *Mol. Cell. Biol.* **21:** 4246–4255.

Pianese G. 1896. Beitrag zur Histologie und Aetiologie der Carcinoma Histologische und experimentelle Untersuchungen. *Beitr. Pathol. Anat. Allg. Pathol.* **142:** 1–193.

Polymenis M. and Schmidt E.V. 1997. Coupling of cell division to cell growth by translational control of the G_1 cyclin CLN3 in yeast. *Genes Dev.* **11:** 2522–2531.

———. 1999. Coordination of cell growth with cell division. *Curr. Opin. Genet. Dev.* **9:** 76–80.

Powers T. and Walter P. 1999. Regulation of ribosome biogenesis by the rapamycin-sensitive TOR-signaling pathway in *Saccharomyces cerevisiae. Mol. Biol. Cell* **10:** 987–1000.

Prescott D.M. and Bender M.A. 1962. Synthesis of RNA and protein during mitosis in mammalian tissue culture cells. *Exp. Cell Res.* **26:** 260–268.

Reeder R.H. 1999. Regulation of RNA polymerase I transcription in yeast and vertebrates. *Prog. Nucleic Acid Res. Mol. Biol.* **62:** 293–327.

Rocco J.W. and Sidransky D. 2001. p16(MTS-1/CDKN2/INK4a) in cancer progression. *Exp. Cell Res.* **264:** 42–55.

Ronning O.W., Lindmo T., Pettersen E.O., and Seglen P.O. 1981. The role of protein accumulation in the cell cycle control of human NHIK 3035 cells. *J. Cell. Physiol.* **109:** 411–418.

Rubinstein S.J. and Dasgupta A. 1989. Inhibition of rRNA synthesis by poliovirus: Specific inactivation of transcription factors. *J. Virol.* **63:** 4689–4696.

Schmelzle T. and Hall M.N. 2000. TOR, a central controller of cell growth. *Cell* **103:** 253–262.

Schmidt E.V. 1999. The role of c-myc in cellular growth control. *Oncogene* **18:** 2988–2996.

Schramm L. and Hernandez N. 2002. Recruitment of RNA polymerase III to its target promoters. *Genes Dev.* **16:** 2593–2620.

Schuhmacher M., Staege M.S., Pajic A., Polack A., Weidle U.H., Bornkamm G.W., Eick D., and Kohlhuber F. 1999. Control of cell growth by c-Myc in the absence of cell division. *Curr. Biol.* **9:** 1255–1258.

Schuhmacher M., Kohlhuber F., Holzel M., Kaiser C., Burtscher H., Jarsch M., Bornkamm G.W., Laux G., Polack A., Weidle U.H., and Eick D. 2001. The transcriptional program of a human B cell line in response to Myc. *Nucleic Acids Res.* **29:** 397–406.

Schwartz L.B., Sklar V.E.F., Jaehning J.A., Weinmann R., and Roeder R.G. 1974. Isolation and partial characterization of the multiple forms of deoxyribonucleic acid-dependent ribonucleic acid polymerase in mouse myeloma MOPC 315. *J. Biol. Chem.* **249:** 5889–5897.

Scott M.R.D., Westphal K.-H., and Rigby P.W. 1983. Activation of mouse genes in transformed cells. *Cell* **34:** 557–567.

Scott P.H., Cairns C.A., Sutcliffe J.E., Alzuherri H.M., McLees A., Winter A.G., and White R.J. 2001. Regulation of RNA polymerase III transcription during cell cycle entry. *J. Biol. Chem.* **276:** 1005–1014.

Segil N., Guermah M., Hoffmann A., Roeder R.G., and Heintz N. 1996. Mitotic regulation of TFIID: Inhibition of activator-dependent transcription and changes in subcellular localization. *Genes Dev.* **10:** 2389–2400.

Sentenac A. 1985. Eukaryotic RNA polymerases. *CRC Crit. Rev. Biochem.* **18:** 31–90.

Sethy I., Moir R.D., Librizzi M., and Willis I.M. 1995. In vitro evidence for growth regulation of tRNA gene transcription in yeast. *J. Biol. Chem.* **270:** 28463–28470.

Shastry B.S., Honda B.M., and Roeder R.G. 1984. Altered levels of a 5S gene-specific transcription factor (TFIIIA) during oogenesis and embryonic development of *Xenopus laevis. J. Biol. Chem.* **259:** 11373–11382.

Shen Y., Igo M., Yalamanchili P., Berk A.J., and Dasgupta A. 1996. DNA binding domain and subunit interactions of transcription factor IIIC revealed by dissection with poliovirus 3C protease. *Mol. Cell. Biol.* **16:** 4163–4171.

Shew J.-Y., Lin B.T.-Y., Chen P.-L., Tseng B.Y., Yang-Feng T.L., and Lee W.-H. 1990. C-terminal truncation of the retinoblastoma gene product leads to functional inactivation. *Proc. Natl. Acad. Sci.* **87:** 6–10.

Shiio Y., Donohoe S., Yi E.C., Goodlett D.R., Aebersold R., and Eisenman R.N. 2002. Quantitative proteomic analysis of Myc oncoprotein function. *EMBO J.* **21:** 5088–5096.

Song D.H., Sussman D.J., and Seldin D.C. 2000. Endogenous protein kinase CK2 participates in Wnt signaling in mammary epithelial cells. *J. Biol. Chem.* **275:** 23790–23797.

Stefanovsky V.Y., Pelletier G., Hannan R., Gagnon-Kugler T., Rothblum L.I., and Moss T. 2001. An immediate response of ribosomal transcription to growth factor stimulation in mammals is mediated by ERK phosphorylation of UBF. *Mol. Cell* **8:** 1063–1073.

Stein T., Crighton D., Boyle J.M., Varley J.M., and White R.J. 2002a. RNA polymerase III transcription can be derepressed by oncogenes or mutations that compromise p53 function in tumors and Li-Fraumeni syndrome. *Oncogene* **21:** 2961–2970.

Stein T., Crighton D., Warnock L.J., Milner J., and White R.J. 2002b. Several regions of p53 are involved in repression of RNA polymerase III transcription. *Oncogene* **21:** 5540–5547.

Stocker H. and Hafen E. 2000. Genetic control of cell size. *Curr. Opin. Genet. Dev.* **10:** 529–535.

Sudbery P. 2002. When wee meets whi. *Science* **297:** 351–352.

Sugimoto M., Kuo M.-L., Roussel M.F., and Sherr C.J. 2003. Nucleolar Arf tumor suppressor inhibits ribosomal RNA processing. *Mol. Cell* **11:** 415–424.

Sutcliffe J.E., Brown T.R.P., Allison S.J., Scott P.H., and White R.J. 2000. Retinoblastoma protein disrupts interactions required for RNA polymerase III transcription. *Mol. Cell. Biol.* **20:** 9192–9202.

Sutcliffe J.E., Cairns C.A., McLees A., Allison S.J., Tosh K., and White R.J. 1999. RNA polymerase III transcription factor IIIB is a target for repression by pocket proteins p107 and p130. *Mol. Cell. Biol.* **19:** 4255–4261.

Tansey W.P. and Herr W. 1995. The ability to associate with activation domains *in vitro* is not required for the TATA box-binding protein to support activated transcription *in vivo. Proc. Natl. Acad. Sci.* **92:** 10550–10554.

Taylor J. 1960. Nucleic acid synthesis in relation to the cell division cycle. *Ann. N.Y. Acad. Sci.* **90:** 409–421.

Thomas G. 2000. An encore for ribosome biogenesis in the control of cell proliferation. *Nat. Cell Biol.* **2:** E71–E72.

Trere D., Pession A., and Derenzini M. 1989. The silver-stained proteins of interphasic nucleolar organizer regions as a parameter of cell duplication rate. *Exp. Cell Res.* **184:** 131–137.

Tuan J.C., Zhai W., and Comai L. 1999. Recruitment of TATA-binding protein-TAF$_I$ complex SL1 to the human ribosomal DNA promoter is mediated by the carboxy-terminal activation domain of upstream binding factor (UBF) and is regulated by UBF phosphorylation. *Mol. Cell. Biol.* **19:** 2872–2879.

Varley J.M., Evans D.G.R., and Birch J.M. 1997. Li-Fraumeni syndrome—A molecular and clinical review. *Br. J. Cancer* **76:** 1–14.

Vasseur M., Condamine H., and Duprey P. 1985. RNAs containing B2 repeated sequences are transcribed in the early stages of mouse embryogenesis. *EMBO J.* **4:** 1749–1753.

Voit R. and Grummt I. 2001. Phosphorylation of UBF at serine 388 is required for interaction with RNA polymerase I and activation of rDNA transcription. Proc. Natl. *Acad. Sci.* **98:** 13631–13636.

Voit R., Hoffmann M., and Grummt I. 1999. Phosphorylation by G_1-specific cdk-cyclin complexes activates the nucleolar transcription factor UBF. *EMBO J.* **18:** 1891–1899.

Voit R., Schafer K., and Grummt I. 1997. Mechanism of repression of RNA polymerase I transcription by the retinoblastoma protein. *Mol. Cell. Biol.* **17:** 4230–4237.

Voit R., Kuhn A., Sander E.E., and Grummt I. 1995. Activation of mammalian ribosomal gene transcription requires phosphorylation of the nucleolar transcription factor UBF. *Nucleic Acids Res.* **23:** 2593–2599.

Voit R., Schnapp A., Kuhn A., Rosenbauer H., Hirschmann P., Stunnenberg H.G., and Grummt I. 1992. The nucleolar transcription factor mUBF is phosphorylated by casein kinase II in the C-terminal hyperacidic tail which is essential for transactivation. *EMBO J.* **11:** 2211–2218.

Volarevic S., Stewart M.J., Ledermann B., Zilberman F., Terracciano L., Montini E., Grompe M., Kozma S.C., and Thomas G. 2000. Proliferation, but not growth, blocked by conditional deletion of 40S ribosomal protein S6. *Science* **288:** 2045–2047.

Vousden K.H. 1995. Regulation of the cell cycle by viral oncoproteins. *Semin. Cancer Biol.* **6:** 109–116.

Vousden K.H. and Lu X. 2002. Live or let die: The cell's response to p53. *Nat. Rev. Cancer* **2:** 594–604.

Wang H.-D., Trivedi A., and Johnson D.L. 1997. Hepatitis B virus X protein induces RNA polymerase III-dependent gene transcription and increases cellular TATA-binding protein by activating the Ras signalling pathway. *Mol. Cell. Biol.* **17:** 6838–6846.

———. 1998. Regulation of RNA polymerase I-dependent promoters by the hepatitis B virus X protein via activated Ras and TATA-binding protein. *Mol. Cell. Biol.* **18:** 7086–7094.

Wang H.-D., Yuh C.-H., Dang C.V., and Johnson D.L. 1995. The hepatitis B virus X protein increases the cellular level of TATA-binding protein, which mediates transactivation of RNA polymerase III genes. *Mol. Cell. Biol.* **15:** 6720–6728.

Warner J.R. 1999. The economics of ribosome biosynthesis in yeast. *Trends Biochem. Sci.* **24:** 437–440.

Weinberg R.A. 1995. The retinoblastoma protein and cell cycle control. *Cell* **81:** 323–330.

White R.J. 1997. Regulation of RNA polymerases I and III by the retinoblastoma protein: A mechanism for growth control? *Trends Biochem. Sci.* **22:** 77–80.

———. 2002. *RNA polymerase III transcription*, 3rd edition. Landes Bioscience, Austin, Texas (*http://www.eurekah.com*).

White R.J. and Jackson S.P. 1992a. Mechanism of TATA-binding protein recruitment to a TATA-less class III promoter. *Cell* **71:** 1041–1053.

———. 1992b. The TATA-binding protein: A central role in transcription by RNA polymerases I, II and III. *Trends Genet.* **8:** 284–288.

White R.J., Stott D., and Rigby P.W.J. 1989. Regulation of RNA polymerase III transcription in response to F9 embryonal carcinoma stem cell differentiation. *Cell* **59:**

1081–1092.

————. 1990. Regulation of RNA polymerase III transcription in response to simian virus 40 transformation. *EMBO J.* **9:** 3713–3721.

White R.J., Gottlieb T.M., Downes C.S., and Jackson S.P. 1995a. Cell cycle regulation of RNA polymerase III transcription. *Mol. Cell. Biol.* **15:** 6653–6662.

————. 1995b. Mitotic regulation of a TATA-binding-protein-containing complex. *Mol. Cell. Biol.* **15:** 1983–1992.

White R.J., Khoo B.C.-E., Inostroza J.A., Reinberg D., and Jackson S.P. 1994. The TBP-binding repressor Dr1 differentially regulates RNA polymerases I, II and III. *Science* **266:** 448–450.

White R.J., Trouche D., Martin K., Jackson S.P., and Kouzarides T. 1996. Repression of RNA polymerase III transcription by the retinoblastoma protein. *Nature* **382:** 88–90.

Whitmarsh A.J. and Davis R.J. 2000. A central control for cell growth. *Nature* **403:** 255–256.

Willis I.M. 2002. A universal nomenclature for subunits of the RNA polymerase III transcription initiation factor TFIIIB. *Genes Dev.* **16:** 1337–1338.

Whyte P., Williamson N.M., and Harlow E. 1989. Cellular targets for transformation by the adenovirus E1A proteins. *Cell* **56:** 67–75.

Whyte P., Buchkovich K.J., Horowitz J.M., Friend S.H., Raybuck M., Weinberg R.A., and Harlow E. 1988. Association between an oncogene and an anti-oncogene: The adenovirus E1A proteins bind to the retinoblastoma gene product. *Nature* **334:** 124–129.

Willert K., Brink M., Wodarz A., Varmus H., and Nusse R. 1997. Casein kinase 2 associates with and phosphorylates Dishevelled. *EMBO J.* **16:** 3089–3096.

Winter A.G., Sourvinos G., Allison S.J., Tosh K., Scott P.H., Spandidos D.A., and White R.J. 2000. RNA polymerase III transcription factor TFIIIC2 is overexpressed in ovarian tumors. *Proc. Natl. Acad. Sci.* **97:** 12619–12624.

Wolffe A.P. and Brown D.D. 1988. Developmental regulation of two 5S ribosomal RNA genes. *Science* **241:** 1626–1632.

Ying C., Gregg D.W., and Gorski J. 1996. Estrogen-induced changes in rRNA accumulation and RNA polymerase I activity in the rat pituitary: Correlation with pituitary tumor susceptibilty. *Mol. Cell. Endocrinol.* **118:** 207–213.

Yuan X., Zhao J., Zentgraf H., Hoffmann-Rohrer U., and Grummt I. 2002. Multiple interactions between RNA polymerase I, TIF-IA and TAF$_I$ subunits regulate preinitiation complex assembly at the ribosomal gene promoter. *EMBO Rep.* **3:** 1082–1087.

Zahradka P., Larson D.E., and Sells B.H. 1991. Regulation of ribosome biogenesis in differentiated rat myotubes. *Mol. Cell. Biochem.* **104:** 189–194.

Zaragoza D., Ghavidel A., Heitman J., and Schultz M.C. 1998. Rapamycin induces the G$_0$ program of transcriptional repression in yeast by interfering with the TOR signaling pathway. *Mol. Cell. Biol.* **18:** 4463–4470.

Zetterberg A. and Killander D. 1965. Quantitative cytophotometric and autoradiographic studies on the rate of protein synthesis during interphase in mouse fibroblasts in vitro. *Exp. Cell Res.* **40:** 1–11.

Zhai W. and Comai L. 1999. A kinase activity associated with simian virus 40 large T antigen phosphorylates upstream binding factor (UBF) and promotes formation of a stable initiation complex between UBF and SL1. *Mol. Cell. Biol.* **19:** 2791–2802.

————. 2000. Repression of RNA polymerase I transcription by the tumor suppressor p53. *Mol. Cell. Biol.* **20:** 5930–5938.

Zhai W., Tuan J., and Comai L. 1997. SV40 large T antigen binds to the TBP-TAF$_I$ complex

SL1 and coactivates ribosomal RNA transcription. *Genes Dev.* **11:** 1605–1617.

Zhang L., Zhou W., Velculescu V.E., Kern S.E., Hruban R.H., Hamilton S.R., Vogelstein B., and Kinzler K.W. 1997. Gene expression profiles in normal and cancer cells. *Science* **276:** 1268–1272.

Zhao J., Yuan X., Frodin M., and Grummt I. 2003. ERK-dependent phosphorylation of the transcription initiation factor TIF-IA is required for RNA polymerase I transcription and cell growth. *Mol. Cell* **11:** 405–413.

13

Autophagy: Reversing Cell Growth

Yoshinori Ohsumi

Department of Cell Biology
National Institute for Basic Biology
Myodaijicho, Okazaki, 444-8585, Japan

IT IS NOW WELL KNOWN THAT EVERY CELLULAR event requires a balance of synthesis and breakdown of many proteins. Every protein has its own lifetime within a wide range, from a few minutes to more than ten days. We do not yet know the determinants of the lifetime of each protein or the exact purpose of protein turnover, but a dynamic state of equilibrium must be crucial for the control of cell growth.

Proteins are not degraded spontaneously, but rather are degraded by active processes. There are two major pathways of intracellular protein degradation. First, the ubiquitin–proteasome system in the cytoplasm mediates the degradation of short-lived, damaged, or misfolded proteins (Hochstrasser 1996; Hershko and Ciechanover 1998). Proteins targeted for degradation are first tagged with a small protein, ubiquitin, and then digested by a huge proteinase complex, the proteasome. Both the ubiquitination and cleavage processes require ATP hydrolysis and are strictly dependent on recognition of target proteins by the sophisticated ubiquitin ligase system and the proteasome. Short-lived proteins play crucial roles in important cellular events such as transcriptional regulation and cell cycle control.

The second major pathway of intracellular protein degradation involves the degradation of long-lived proteins within a specific compartment, the lysosome/vacuole. So far, several delivery routes to this lytic compartment are proposed. The self-degradation of intracellular components in lysosomes is generally called autophagy, in contrast to heterophagy of extracellular materials (Mortimore and Poso 1987). Macroautophagy, the major autophagic pathway, begins with a so-called isolation membrane engulfing a portion of the cytoplasm to form an autophagosome, a double-membrane sac containing the engulfed cytoplasm (Fig. 1) (Seglen and Bohley 1992). The outer membrane of an autophagosome subsequently fuses with the lysosome, thereby releasing the inner membrane and cytoplasmic contents into the lysosome, where these are digested for reuse. Microautophagy is a process in which the lysosome or vacuole directly engulfs a portion of the cytoplasm. Finally, chaperone-mediated autophagy is the direct transport of proteins across the lysosomal membrane, with the help of the chaperones Hsc73 and

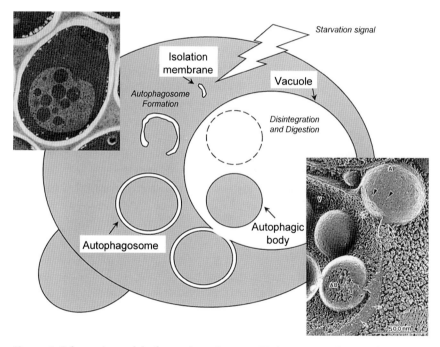

Figure 1. Schematic model of autophagy in yeast. Various types of starvation trigger signal transduction to induce autophagy. A small membrane sac, termed the isolation membrane, appears in the cytoplasm and elongates to enwrap a portion of the cytoplasm. Once a double-membrane structure, the autophagosome (AP), is formed, it immediately fuses with the vacuole. Upon fusion of the outer membrane of the autophagosome with the vacuole (V), the inner membrane vesicle of the autophagosome is released into the vacuolar lumen, giving rise to the autophagic body (AB). In wild-type cells, the autophagic body rapidly disintegrates, and its contents are digested for reuse.

Lgp96 (Terlecky et al. 1992; Cuervo and Dice 1996). The focus of this chapter is on macroautophagy, which is referred to here simply as autophagy.

Autophagy is characterized as the nonselective and bulk degradation of cytoplasmic components, including proteins, large complexes such as ribosomes, and organelles. More than 99% of cellular proteins are long-lived; thus, turnover of long-lived proteins via autophagy is important in the control of cell growth. Bulk protein degradation also plays a role in the cellular starvation response, cell differentiation and remodeling, and some aspects of organelle homeostasis. Of particular importance to cell growth is the role of autophagy in the starvation response. Autophagy mediates the shrinkage of the ribosome pool and thus the slowing or reversal of cell growth when nutrients are limiting (Fig. 2). The ribosome pool is a limiting factor in cell growth.

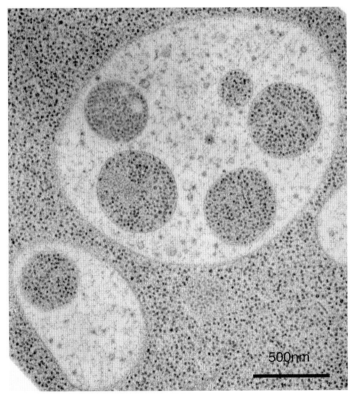

500nm

Figure 2. Electron micrograph of autophagic yeast cells. Note autophagic bodies filled with ribosomes.

Almost half a century has passed since the lytic organelle, the lysosome, was discovered by Christian de Duve using cell fractionation procedures (de Duve 1959). Since then, many electron micrographs have revealed autophagy occurring in a variety of cells from different organs and in cultured cells, and it is now generally accepted that autophagy is a ubiquitous activity of eukaryotic cells. However, elucidating a general picture of the autophagic process was hindered by the dynamic nature of mammalian lysosomes. Moreover, a specific marker protein and a quantitative assay for monitoring autophagy in mammalian cells were lacking. Genes specifically involved in autophagy were discovered only recently, and the molecular basis of autophagy is just now being uncovered. This chapter focuses on recent progress in the molecular dissection of autophagy using yeast as a model system. Molecular biological approaches using the yeast *Saccharomyces cerevisiae* have begun to unravel the mystery of autophagy.

VACUOLAR PROTEIN DEGRADATION IN YEAST

It has been postulated that the yeast vacuole is a lytic organelle since, like mammalian lysosomes, it is an acidic compartment and contains various hydrolytic enzymes (Klionsky et al. 1990). Genetic studies on yeast proteinases were undertaken mainly in the laboratories of Elizabeth Jones and Dieter Wolf (Achstetter and Wolf 1985; Jones et al. 1997). Enzymology and the biosynthetic pathway of vacuolar enzymes have also been studied quite extensively (Raymond et al. 1992; Stack et al. 1995). It was shown that bulk protein turnover is induced upon nitrogen starvation and is dependent on vacuolar enzyme activities (Zubenko and Jones 1981; Egner et al. 1993), suggesting that the vacuole is the site of bulk degradation. However, these findings led to a number of obvious questions. What kinds of substrates are degraded in the vacuole? What is the mechanism of sequestering those substrates into the vacuole? How is the process regulated in response to nutrients?

DISCOVERY OF AUTOPHAGY IN YEAST

In the early 1990s, we found by direct microscopic inspection of yeast cells that various different nutrient starvation conditions induce autophagy. When vacuolar proteinase-deficient mutants grown in a rich medium were transferred to nitrogen-depleted medium, spherical structures appeared in the vacuole after 30–40 minutes, accumulated, and almost filled the vacuole for up to 10 hours (Takeshige et al. 1992). These structures, named autophagic bodies, were mostly single-membrane-bound structures containing a portion of the cytoplasm (Figs. 1 and 2) (Takeshige et al. 1992). Autophagosomes, double-membrane structures, were found in the cytoplasm of the starved cells. Autophagosomes in yeast are about 300–900 nm in diameter and contain cytoplasmic enzymes, ribosomes, and occasionally other cellular structures, including mitochondria and rough endoplasmic reticulum (rER) (Takeshige et al. 1992). Biochemical and immunoelectron microscopic analyses clearly indicated that autophagosomes enclose cytoplasmic components nonselectively. Fusion of the outer membrane of autophagosomes with the vacuolar membrane was shown by freeze-fracture electron microscopy (Baba et al. 1994, 1995).

Quite similar membrane phenomena were induced under not only nitrogen- but also carbon-, sulfate-, phosphate-, and auxotrophic single amino acid-starvation. This strongly indicated that yeast cells, in response to adverse nutrient conditions, transfer a portion of cytoplasm to the lytic compartment via autophagosomes. The membrane dynamics of yeast autophagy are topologically the same as macroautophagy in mammals, although the yeast vacuole is much larger than a mammalian lysosome. A schematic representaiton of the entire autophagic process in yeast is shown in Figure 1.

GENETIC APPROACHES TO YEAST AUTOPHAGY

Genetic approaches were taken to address the molecular mechanism of autophagy. As the first approach, the morphological change of the vacuole under starvation conditions was used to screen for autophagy-defective mutants. Cells were screened for a defect in the vacuolar accumulation of autophagic bodies, and one mutant, *apg1*, was obtained (Tsukada and Ohsumi 1993). This mutant failed to induce protein degradation in response to starvation and could not maintain viability upon prolonged nitrogen starvation. This phenotype, assumed to be due to a defect in autophagy, was then used to screen for more mutants. Fourteen more *apg* mutants were obtained. Another approach, taken by Thumm and colleagues, involved immunoscreening for cells that retained a cytoplasmic enzyme, fatty acid synthase, after starvation (Thumm et al. 1994). By this method, originally six *aut* mutants were obtained. Later, two hybrid screens using Apg proteins as bait identified two more *APG* genes (Mizushima et al. 1999; Kamada et al. 2000). Klionsky's group, isolating mutants defective in maturation of the vacuolar enzyme aminopeptidase I (API), identified the Cvt (cytoplasm to vacuole targeting) pathway (Harding et al. 1995). The *CVT* genes significantly overlapped with the autophagy-related *APG* and *AUT* genes (Harding et al. 1996; Scott et al. 1996). The Cvt and autophagy pathways use quite similar machinery for membrane dynamics (Scott et al. 1997), although the two pathways are quite different. The autophagy pathway is degradative, whereas the Cvt pathway, which is active in non-starved cells, is biosynthetic.

PHENOTYPES OF AUTOPHAGY-DEFECTIVE MUTANTS

The above mutants were selected as nonconditional mutants, assuming that autophagy is not essential for vegetative cell growth. In fact, all *apg* and *aut* mutants display similar growth phenotypes. They grow like wild-type cells despite failing to induce bulk protein degradation under various nutrient-depletion conditions. However, homozygous diploids with any *apg* or *aut* mutation fail to sporulate (Tsukada and Ohsumi 1993), suggesting that this intracellular remodeling process triggered by nitrogen depletion requires bulk protein degradation via autophagy. Another characteristic feature of autophagy-defective mutants is loss of viability during nitrogen starvation. These mutants start to die after two days of starvation, and almost all cells lose viability within one week (Tsukada and Ohsumi 1993). Under carbon starvation, they maintain viability even after prolonged starvation. The *pep4* mutant, which has pleiotropic defects in vacuolar hydrolases, shows loss-of-viability phenotype like *apg* mutants, although an accumulation of autophagosomes is clearly visible. This indicates that loss of viability occurs only when autophagic degradation is completely blocked. Interestingly, all the isolated *apg* mutants behave like *apg* null mutants. However, they do not display any significant differences from wild-type cells in their response to various types of stress such as heat, osmotic, and salt stress. Furthermore, vacuolar functions, secretion, and endocytosis in these mutants are almost completely normal. As a consequence of the mutant isolation strategies, mutants with abnormal vacuole morphology, partially defective mutants, and mutants of genes overlapping with other essential functions were eliminated. These genetic analyses revealed that at least 16 genes (Table 1) are required specifically for the autophagosome formation (APG+AUT10) in yeast. Sixteen is close to the final number of autophagy-specific genes, as the mutant screens were nearly saturated, although it is becoming clear that more than 16 genes are required for normal levels of autophagy.

Table 1. Functions of the Apg, Aut, Cvt, and Gsa proteins and mammalian homologs

Apg	Aut	Cvt	Gsa	Mammalian	Function/localization
Apg1	Aut3	Cvt10	Gsa10	ULK1	protein kinase, localizes to the PAS
Apg2	Aut8		Gsa11	Apg2	localizes to the PAS
Apg3	Aut1		Gsa20	Apg3	Apg8 conjugating enzyme (E2)
Apg4	Aut2			Apg4A, Apg4B	cysteine protease for processing the carboxyl terminus of Apg8
Apg5				Apg5	substrate of Apg12 conjugating reaction, localizes to the PAS
Apg6				Beclin-1	subunit of PI3K complex, involved in protein sorting to the vacuole as Vps30
Apg7		Cvt2	Gsa7	Apg7	activating enzyme (E1) of Apg8 and Apg12
Apg8	Aut7	Cvt5		LC3, GATE16, GABARAP	ubiquitin-like protein, conjugates with PE, localizes to the PAS and autophagosomes
Apg9	Aut9	Cvt7	Gsa14	Apg9?	transmembrane protein, required for PAS formation
Apg10				Apg10	Apg12 conjugating enzyme (E2)
Apg12				Apg12	ubiquitin-like protein, conjugates with Apg5
Apg13				?	subunit of Apg1 kinase, phosphorylated under growing conditions
Apg14		Cvt12		?	subunit of autophagy-specific PI3K complex
Apg16		Cvt11		Apg16L	binds with Apg12–Apg5 and forms tetramer, required for Apg12–Apg5 recruitment to the PAS
Apg17				?	member of Apg1 complex, not required for the Cvt pathway
	Aut10	Cvt18	Gsa12		WD repeat protein
	Aut4				disintegration of autophagic bodies in the vacuole
	Aut5	Cvt17			disintegration of autophagic bodies in the vacuole, putative lipase
		Cvt9	Gsa9		required only for the Cvt pathway, localizes to the PAS
		Cvt19			receptor of API for the Cvt vesicle, localizes to the PAS
		Mai1			required for the Cvt pathway, not for macroautophagy

PROTEINS INVOLVED IN AUTOPHAGY

Cloning and identification of the autophagy genes revealed that almost all were previously uncharacterized genes. The sole exception was *APG6*, which turned out to be allelic to *VPS30*, a gene required for vacuolar protein sorting (Kametaka et al. 1998). The genes related to autophagy are shown in Table 1. In yeast, autophagy is almost completely shut off under nutrient-rich, growth-promoting conditions. Autophagy is strictly induced upon nutrient starvation, but every *APG* gene is expressed in growing conditions. This constitutive expression of the *APG* genes may be in anticipation of emergencies or may be partly due to the Cvt pathway being active in growing cells. Transcription of several *APG* genes, such as *APG8* and *APG14*, is further up-regulated by starvation and is under negative regulation by TOR, a phosphoinositide kinase-related protein kinase (Noda and Ohsumi 1998; Kirisako et al. 1999; Chan et al. 2001). Molecular biological and biochemical analyses of the *APG* genes and their products uncovered genetic and physical interactions among Apg proteins. We now know that the Apg proteins comprise four functional subgroups: the Apg12 conjugation system, the Apg8 lipidation system, the Apg1 protein kinase complex, and a PI 3-kinase complex.

THE Apg12 CONJUGATION SYSTEM

Among the more remarkable discoveries to come out of the analysis of the Apg proteins are two ubiquitin-like conjugation systems (Ohsumi 2001). Actually, more than half of the *APG* genes are involved in these novel conjugation systems. First, Apg12, a small hydrophilic protein of 186 amino acids having no apparent homology with ubiquitin, forms a covalently linked conjugate with Apg5 (Mizushima et al. 1998b). This conjugate formation is essential for autophagy and, like ubiquitination, is mediated by consecutive reactions. The carboxy-terminal Gly residue of Apg12 is activated by an activating enzyme, Apg7 (E1), and then attached via a thioester linkage to a conjugating enzyme, Apg10 (E2) (Shintani et al. 1999; Tanida et al. 1999). Finally, Apg12 forms a conjugate with Apg5 via an isopeptide bond between the carboxy-terminal Gly of Apg12 and Lys-149 of Apg5. Apg12 and Apg5 form a conjugate immediately after synthesis; free forms of Apg12 and Apg5 are hardly detectable. Apg5 seems to be the only target of the Apg12 modification. So far, no protease activity to deconjugate Apg12-Apg5 has been found, suggesting that the conjugation reaction is irreversible. The conjugate further forms a complex with Apg16. Apg16 was originally isolated in a two-hybrid screen using Apg12 as bait, although Apg16 binds Apg5 directly and Apg12 only indirectly (Mizushima et al. 1999). Apg16 has a coiled-coil region in its carboxy-terminal half and forms an oligomer through this region. Four copies of the Apg12–Apg5–Apg16 trimeric complex likely form a supercomplex that is essential for autophagy (Kuma et al. 2002).

THE Apg8 LIPIDATION SYSTEM

The second ubiquitin-like protein essential for autophagy is Aut7/Apg8, a 117-amino-acid protein. Cell fractionation studies indicated that about half the Apg8 pool is peripherally bound to membrane whereas the other half behaves like an integral membrane protein (Kirisako et al. 2000). Furthermore, Apg8 is a good marker for membrane dynamics during autophagy because it resides on the membrane of the isolated membrane, the autophagosome, and the autophagic body (Kirisako et al. 1999).

As revealed by epitope tagging, nascent Apg8 is processed at its carboxyl terminus (Kirisako et al. 2000). Apg4, a novel cysteine protease, removes a terminal Arg, thus exposing a Gly at the carboxyl terminus of Apg8. The processed form of Apg8 is in turn activated by Apg7 and then transferred to a conjugating enzyme, Apg3 (Ichimura et al. 2000). Apg7 is a unique enzyme that activates two different ubiquitin-like proteins, Apg12 and Apg8, and assigns them to proper E2 enzymes, Apg10 and Apg3, respectively. Apg3 is somewhat homologous to Apg10, but has no overall significant homology with E2 enzymes of the ubiquitin system. What is the target of Apg8? Interestingly, Apg8 forms a conjugate not with a protein but with phosphatidylethanolamine (PE), an abundant membrane phospholipid (Ichimura et al. 2000). This lipidation is necessary for the membrane dynamics of autophagy. For normal autophagosome formation, Apg8-PE needs to be deconjugated by the deconjugating enzyme Apg4, suggesting that Apg8 is recycled between a lipidated and a nonlipidated form (Ichimura et al. 2000). The two Apg8-PE and Apg12-Apg5 conjugation reactions are somehow related, as the level of Apg8-PE is significantly reduced in cells lacking the Apg12–Apg5 conjugate.

THE Apg1 PROTEIN KINASE COMPLEX

The third protein complex required for autophagy is the Apg1 protein kinase complex. Apg1 is a serine/threonine protein kinase (Matsuura et al. 1997; Straub et al. 1997). The protein kinase domain of Apg1, whose kinase activity can be detected in vitro, is amino-terminal. Apg1 kinase activity increases during the induction of autophagy, and kinase-negative *apg1* mutations block the induction of autophagy, suggesting that Apg1 kinase activity is essential for the regulation of autophagosome formation (Matsuura et al. 1997; Kamada et al. 2000). Apg1 is physically associated with Apg13, Apg17, and Cvt9.

Apg13 is highly phosphorylated under nutrient-rich conditions. Upon starvation or addition of the TOR inhibitor rapamycin, Apg13 is dephosphorylated by an as-yet-unknown phosphatase (Kamada et al. 2000). Conversely, upon addition of nutrients to starved cells, Apg13 is rapidly hyperphosphorylated. Overproduction of Apg1 partially suppresses the autophagy defect of an *apg13* null mutation (Funakoshi et al. 1997). Consistent with this genetic interaction, Apg13, via a central region, physically associates with Apg1 (Kamada et al. 2000). Under starvation conditions, Apg13 is tightly associated with Apg1, whereas under nutrient-rich conditions, the affinity is lower (Kamada et al. 2000). In addition, in an *apg13Δ* mutant, the kinase activity of Apg1 is low. These results suggest that Apg13 is a positive regulator for the Apg1 protein kinase. The above observations also suggest that Apg13 and ultimately Apg1 kinase activity are controlled by the TOR signaling pathway in response to nutrient conditions.

The transport of API to the vacuole is completely blocked when the *apg13* null mutant is grown in a nutrient-rich medium, but the block can be partially overcome by incubation in starvation conditions (Abeliovich et al. 2000). Furthermore, in an *apg13* mutant in which Apg13 lacks most of the Apg1-binding region, transport of API is normal, but autophagy is completely absent (Kamada et al. 2000). Thus, Apg13 may regulate both autophagy and the Cvt pathway through the Apg1 protein kinase. It is known that the carboxy-terminal region of Apg13 also interacts with Vac8 (Scott et al. 2000).

THE PI 3-KINASE COMPLEX

The fourth complex is an autophagy-specific phosphatidylinositol 3-kinase (PI3K) complex. Cloning and characterization of *APG6* revealed that it is allelic to *VPS30*. Vps30/Apg6 is a dual-function protein involved in vacuolar protein sorting and autophagy. Apg14 is a possibly coiled-coil protein associated with Vps30. Overexpression of Apg14 partially suppresses the autophagic defect of a mutant expressing a truncated form of Vps30 but does not suppress a complete deletion allele of *VPS30*. This suggests that Apg14 binds Vps30 to exert its autophagic function. In contrast to a *vps30* mutant, an *apg14* mutant is not defective in vacuolar protein sorting.

Later, it was found that Vps30 forms two distinct protein complexes (Kihara et al. 2001). One consists of Vps30, Apg14, Vps34, and Vps15, and the other of Vps30, Vps38, Vps34, and Vps15. These two complexes differ based on whether they contain Apg14 or Vps38. Vps34 is the sole PI3K in yeast, and Vps15 is a regulatory protein kinase of Vps34 (Budovskaya et al. 2002). The former PI3K complex is responsible for autophagy, whereas the latter mediates vacuolar protein sorting. Vps30 is a possibly coiled-coil protein and is peripherally membrane associated. Lack of Vps34 or Vps15 results in solubilization of Vps30. Apg14 is unique to the autophagy-specific PI3K complex, and may therefore play an important role in determining the specificity of the PI3K complex (Kihara et al. 2001). Apg14 is peripherally associated with an as-yet-unknown membrane (Kametaka et al. 1998).

FUNCTION OF THE Apg PROTEINS

All *apg* mutants fail to accumulate autophagosomes during starvation, indicating that all the *APG* gene products function at or before the step of autophagosome formation. So far, studies on the Apg proteins have reached the conclusion that all the Apg proteins function at the autophagosome formation step. There are many fundamental problems to be solved. What is the origin of the autophagosomal membranes? How does the autophagosome assemble to form a spherical structure? What is the machinery mediating closure of the autophagosome (intra-autophagosomal membrane fusion) and fusion of the autophagosomal outer membrane with the vacuolar membrane?

The membrane dynamics of autophagy are quite different from classical vesicular membrane trafficking. For a long time, the origin of autophagosomal membrane was thought to be the endoplasmic reticulum (ER). The autophagosomal membranes of yeast seem thinner than any other cellular membranes, and the outer and inner membranes stick together, leaving almost no lumenal space (Takeshige et al. 1992; Baba et al. 1994). In freeze-fracture images, the inner membrane completely lacks intramembrane particles, and the outer membrane contains sparse but significant particles (Baba et al. 1995). This indicates that the two membranes are somehow differentiated and specialized, presumably for delivery of a portion of the cytoplasm to the lytic compartment. So far, nobody has shown that membrane vesicles are involved in the elongation of the isolation membrane which leads to autophagosome formation. We proposed that autophagosome formation is not simply the enwrapping of cytoplasm by preexisting large membrane structures such as the ER, but rather assembly of new membrane from its constituents.

As mentioned above, all Apg proteins function at or just before the autophagosome formation step. Recently, we have shown that many Apg proteins are localized in a small structure, named the preautophagosome structure (PAS), close to the vacuole (Suzuki et al. 2001). PAS, detected by GFP-Aut7, colocalizes with (Apg12-)Apg5, the Apg1 kinase complex, Apg2, and, presumably, Apg14. Thus, PAS seems to be an organizing center of the autophagosome. The lipidation of Apg8 is a prerequisite for the recruitment of Apg8 to PAS. Furthermore, in *apg14* or *apg6* mutants, Apg8 and Apg5 do not form a dot structure in the cytoplasm, suggesting that the autophagy-specific PI3K complex plays an important role in the organization of PAS (Suzuki et al. 2001). In contrast, defects in the Apg1 kinase complex have little effect on PAS structure.

REGULATORS OF AUTOPHAGY

Autophagy is induced not only by nitrogen starvation, but also by starvation for other nutrients, which means that autophagy is a general physiological response to nutrient limitation. In yeast cells growing in nutrient-rich conditions, the autophagic level is negligible. Conditions that promote cell growth, such as high cAMP levels or mutations that activate protein kinase A (PKA), block the induction of autophagy (Noda and Ohsumi 1998), indicating that cell growth and the induction of autophagy are oppositely regulated. Furthermore, when rapamycin is added to cells growing in a nutrient-rich medium, the cells behave as if in starvation medium, and autophagy is induced (Noda and Ohsumi 1998). Thus, TOR kinase negatively regulates autophagy as a master regulator. At present, the upstream regulators and downstream effectors of TOR controlling autophagy are not well understood, but may involve an interaction between the TOR pathway and PKA.

Several enzymes involved in carbon metabolism, such as ADH, G6PDH, and PGK, are nonselectively sequestered in the vacuole via an autophagosome (Takeshige et al. 1992; Baba et al. 1994). Mitochondria and rER are also engulfed by autophagosomes. These findings indicate that autophagy is a nonselective degradation process. However, one cannot exclude the possibility that certain molecules are selectively sequestered by autophagosomes. In fact, API and α-mannosidase are selectively enclosed in autophagosomes under starvation conditions, but as part of a biosynthetic pathway rather than a degradation process. Whether autophagy targets certain molecules for selective degradation is an open question.

Several mutants accumulate autophagosomes in the cytoplasm under starvation conditions. These mutants display aberrant vacuolar morphology, suggesting that the processes of autophagosome–vacuole fusion and vacuolar biogenesis share common components, such as SNARE molecules. In wild-type cells, autophagic bodies effectively disappear within a minute. Aut5/Cvt17 and Aut4 are involved in this process (Suriapranata et al. 2000; Epple et al. 2001). Aut5/Cvt17 contains a putative lipase domain, but it is still unknown why autophagosomal membranes disintegrate so quickly.

PHYSIOLOGICAL ROLES OF AUTOPHAGY IN YEAST

In the yeast *Saccharomyces cerevisiae,* autophagy is not essential for growth. However, autophagy-defective mutants cannot form spores upon nitrogen starvation, indicating that this cell differentiation process requires bulk protein turnover. Another common feature of autophagy-defective mutants is a loss of viability upon prolonged starvation. Mutants start to die after two days of starvation, and most cells lose viability within one week. It is still unclear why autophagy is essential for the maintenance of cell viability upon starvation. A simple explanation is that the minimal nutrient supply provided by autophagy is essential for survival. Another possibility is that a reduction in cellular activity, such as a down-regulation of ribosomal activity and cell growth, is a prerequisite for survival during adverse conditions. The most frequent stress in nature, especially for microorganisms, is likely nutrient limitation. Thus, there must be strong selective pressure to acquire systems, such as the autophagic system, that allow survival under severe starvation conditions. The ubiquitin–proteasome system alone would not be sufficient to satisfy this selective pressure, as it does not mediate bulk protein degradation. Autophagy is a bulk degradation process during which significant amounts of cytoplasm and large structures are degraded en masse, and is therefore energetically less costly than the ubiquitin–proteasome system.

AUTOPHAGY IN HIGHER EUKARYOTES

The identification of autophagy genes in yeast has facilitated the study of autophagy in higher eukaryotic cells. Many Apg orthologs have been found in many eukaryotes. The Apg12 conjugation system is highly conserved from yeast to mammals and plants (Mizushima et al. 1998a, 2002, 2003; Tanida et al. 2001; Doelling et al. 2002). Mouse Apg12-Apg5 is necessary for autophagosome formation, in particular for elongation of the isolation membrane (Mizushima et al. 2001). LC3, an Aut7/Apg8 ortholog, is a good marker for monitoring autophagic membrane dynamics in mammals (Kabeya et al. 2000). The finding that the basic mechanism of autophagy is well conserved from yeast to human suggests that autophagy is a fundamental cellular activity in eukaryotic organisms. However, the physiological role of autophagy must be broader in higher eukaryotes, where autophagy is involved in development, differentiation, and cellular homeostasis. In higher eukaryotes, autophagy may be induced not only upon starvation, but also in response to various other physiological demands. The molecular dissection of autophagy in higher eukaryotes is currently an exciting area of biology.

SUMMARY

Protein turnover is a crucial aspect of the control of cell growth. Autophagy is a major system of bulk degradation of cellular components, especially long-lived proteins and organelles, in the lysosome or vacuole. Autophagy requires a dynamic rearrangement of membranes to form a specific organelle, the autophagosome. The molecular mechanism of autophagy has long remained mysterious. However, using the yeast *S. cerevisiae*, a molecular dissection of the autophagic process has begun, and it is now known that at least 16 genes are required specifically for autophagy. Characterization of these genes has revealed two novel ubiquitin-like conjugation systems and two kinase complexes for protein and lipid. These systems are well conserved from yeast to mammals, suggesting that autophagy is a fundamental activity of eukaryotic cells.

ACKNOWLEDGMENTS

I thank my colleagues who contributed to some of the work discussed in this chapter, particularly Drs. T. Noda and N. Mizushima. Our work was supported by several grants for Scientific Research from Monbukagaku-sho, Japan.

REFERENCES

Abeliovich H., Dunn W.A., Jr., Kim J., and Klionsky D.J. 2000. Dissection of autophagosome biogenesis into distinct nucleation and expansion steps. *J. Cell Biol.* **151:** 1025–1034.
Achstetter T. and Wolf D.H. 1985. Proteinases, proteolysis and biological control in the yeast *Saccharomyces cerevisiae*. *Yeast* **1:** 139–157.

Baba M., Oshumi M., and Ohsumi Y. 1995. Analysis of the membrane structures involved in autophagy in yeast by freeze-replica method. *Cell Struct. Funct.* **20:** 465–471.

Baba M., Takeshige K., Baba N., and Ohsumi Y. 1994. Ultrastructural analysis of the autophagic process in yeast: Detection of autophagosomes and their characterization. *J. Cell Biol.* **124:** 903–913.

Budovskaya Y.V., Hama H., DeWald D.B., and Herman P.K. 2002. The C terminus of the Vps34p phosphoinositide 3-kinase is necessary and sufficient for the interaction with the Vps15p protein kinase. *J. Biol. Chem.* **277:** 287–294.

Chan T.F., Bertram P.G., Ai W., and Zheng X.F. 2001. Regulation of APG14 expression by the GATA-type transcription factor Gln3p. *J. Biol. Chem.* **276:** 6463–6467.

Cuervo A.M. and Dice J.F. 1996. A receptor for the selective uptake and degradation of proteins by lysosomes. *Science* **273:** 501–503.

de Duve C. 1959. Lysomes, a new group of cytoplasmic particles. In *Subcellular particles* (ed. T. Hayashi), pp 128-159. Ronald Press, New York.

Doelling J.H., Walker J.M., Friedman E.M., Thompson A.R., and Vierstra R.D. 2002. The APG8/12-activating enzyme APG7 is required for proper nutrient recycling and senescence in *Arabidopsis thaliana*. *J. Biol. Chem.* **277:** 33105–33114.

Egner R., Thumm M., Straub M., Simeon A., Schuller H.J., and Wolf D.H. 1993. Tracing intracellular proteolytic pathways. Proteolysis of fatty acid synthase and other cytoplasmic proteins in the yeast *Saccharomyces cerevisiae*. *J. Biol. Chem.* **268:** 27269–27276.

Epple U.D., Suriapranata I., Eskelinen E.L., and Thumm M. 2001. Aut5/Cvt17p, a putative lipase essential for disintegration of autophagic bodies inside the vacuole. *J. Bacteriol.* **183:** 5942–5955.

Funakoshi T., Matsuura A., Noda T., and Ohsumi Y. 1997. Analyses of *APG13* gene involved in autophagy in yeast, *Saccharomyces cerevisiae*. *Gene* **192:** 207–213.

Harding T.M., Hefner-Gravink A., Thumm M., and Klionsky D.J. 1996. Genetic and phenotypic overlap between autophagy and the cytoplasm to vacuole protein targeting pathway. *J. Biol. Chem.* **271:** 17621–17624.

Harding T.M., Morano K.A., Scott S.V., and Klionsky D.J. 1995. Isolation and characterization of yeast mutants in the cytoplasm to vacuole protein targeting pathway. *J. Cell Biol.* **131:** 591–602.

Hershko A. and Ciechanover A. 1998. The ubiquitin system. *Annu. Rev. Biochem.* **67:** 425–479.

Hochstrasser M. 1996. Ubiquitin-dependent protein degradation. *Annu. Rev. Genet.* **30:** 405–439.

Ichimura Y., Kirisako T., Takao T., Satomi Y., Shimonishi Y., Ishihara N., Mizushima N., Tanida I., Kominami E., Ohsumi M., Noda T., and Ohsumi Y. 2000. A ubiquitin-like system mediates protein lipidation. *Nature* **408:** 488–492.

Jones E.W., Webb G.C., and Hiller M.A. 1997. Biogenesis and function of the yeast vacuole. In *The molecular and cellular biology of the yeast* Saccharomyces: *Cell cycle and cell biology* (ed. J.R. Pringle et al.), vol. 3, pp. 363-470. Cold Spring Harbor Laboratory Press, Cold Spring Harbor, New York.

Kabeya Y., Mizushima N., Ueno T., Yamamoto A., Kirisako T., Noda T., Kominami E., Ohsumi Y., and Yoshimori T. 2000. LC3, a mammalian homologue of yeast Apg8p, is localized in autophagosome membranes after processing. *EMBO J.* **19:** 5720–5728.

Kamada Y., Funakoshi T., Shintani T., Nagano K., Ohsumi M., and Ohsumi Y. 2000. Tor-mediated induction of autophagy via an Apg1 protein kinase complex. *J. Cell Biol.* **150:** 1507–1513.

Kametaka S., Okano T., Ohsumi M., and Ohsumi Y. 1998. Apg14p and Apg6/Vps30p form a protein complex essential for autophagy in the yeast, *Saccharomyces cerevisiae*. *J. Biol. Chem.* **273:** 22284–22291.

Kihara A., Noda T., Ishihara N., and Ohsumi Y. 2001. Two distinct Vps34 phosphatidyl-inositol 3-kinase complexes function in autophagy and carboxypeptidase Y sorting in *Saccharomyces cerevisiae*. *J. Cell Biol.* **152:** 519–530.

Kirisako T., Baba M., Ishihara N., Miyazawa K., Ohsumi M., Yoshimori T., Noda T., and Ohsumi Y. 1999. Formation process of autophagosome is traced with Apg8/Aut7p in yeast. *J. Cell Biol.* **147:** 435–446.

Kirisako T., Ichimura Y., Okada H., Kabeya Y., Mizushima N., Yoshimori T., Ohsumi M., Takao T., Noda T., and Ohsumi Y. 2000. The reversible modification regulates the mem-brane-binding state of Apg8/Aut7 essential for autophagy and the cytoplasm to vacuole targeting pathway. *J. Cell Biol.* **151:** 263–276.

Klionsky D.J., Herman P.K., and Emr S.D. 1990. The fungal vacuole: Composition, func-tion, and biogenesis. *Microbiol. Rev.* **54:** 266–292.

Kuma A., Mizushima N., Ishihara N., and Ohsumi Y. 2002. Formation of the approxi-mately 350-kDa Apg12-Apg5·Apg16 multimeric complex, mediated by Apg16 oligomerization, is essential for autophagy in yeast. *J. Biol. Chem.* **277:** 18619–18625.

Matsuura A., Tsukada M., Wada Y., and Ohsumi Y. 1997. Apg1p, a novel protein kinase required for the autophagic process in *Saccharomyces cerevisiae*. *Gene* **192:** 245–250.

Mizushima N., Noda T., and Ohsumi Y. 1999. Apg16p is required for the function of the Apg12p-Apg5p conjugate in the yeast autophagy pathway. *EMBO J.* **18:** 3888–3896.

Mizushima N., Yoshimori T., and Ohsumi Y. 2002. Mouse Apg10 as an Apg12 conjugating enzyme: Analysis by the conjugation-mediated yeast two-hybrid method. *FEBS Lett.* **532:** 450–454.

———. 2003. Role of the Apg12 conjugation system in mammalian autophagy. *Int. J. Biochem. Cell Biol.* **35:** 553–561.

Mizushima N., Sugita H., Yoshimori T., and Ohsumi Y. 1998a. A new protein conjugation system in human. The counterpart of the yeast Apg12p conjugation system essential for autophagy. *J. Biol. Chem.* **273:** 33889–33892.

Mizushima N., Noda T., Yoshimori T., Tanaka Y., Ishii T., George M.D., Klionsky D.J., Ohsumi M., and Ohsumi Y. 1998b. A protein conjugation system essential for autophagy. *Nature* **395:** 395–398.

Mizushima N., Yamamoto A., Hatano M., Kobayashi Y., Kabeya Y., Suzuki K., Tokuhisa T., Ohsumi Y., and Yoshimori T. 2001. Dissection of autophagosome formation using Apg5-deficient mouse embryonic stem cells. *J. Cell Biol.* **152:** 657–668.

Mortimore G.E. and Poso A.R. 1987. Intracellular protein catabolism and its control dur-ing nutrient deprivation and supply. *Annu. Rev. Nutr.* **7:** 539–564.

Noda T. and Ohsumi Y. 1998. Tor, a phosphatidylinositol kinase homologue, controls autophagy in yeast. *J. Biol. Chem.* **273:** 3963–3966.

Ohsumi Y. 2001. Molecular dissection of autophagy: Two ubiquitin-like systems. *Nat. Rev. Mol. Cell. Biol.* **2:** 211–216.

Raymond C.K., Roberts C.J., Moore K.E., Howald I., and Stevens T.H. 1992. Biogenesis of the vacuole in *Saccharomyces cerevisiae*. *Int. Rev. Cytol.* **139:** 59–120.

Scott S.V., Baba M., Ohsumi Y., and Klionsky D.J. 1997. Aminopeptidase I is targeted to the vacuole by a nonclassical vesicular mechanism. *J. Cell Biol.* **138:** 37–44.

Scott S.V., Hefner-Gravink A., Morano K.A., Noda T., Ohsumi Y., and Klionsky D.J. 1996. Cytoplasm-to-vacuole targeting and autophagy employ the same machinery to deliver proteins to the yeast vacuole. *Proc. Natl. Acad. Sci.* **93**: 12304–12308.

Scott S.V., Nice D.C., III, Nau J.J., Weisman L.S., Kamada Y., Keizer-Gunnink I., Funakoshi T., Veenhuis M., Ohsumi Y., and Klionsky D.J. 2000. Apg13p and Vac8p are part of a complex of phosphoproteins that are required for cytoplasm to vacuole targeting. *J. Biol. Chem.* **275**: 25840–25849.

Seglen P.O. and Bohley P. 1992. Autophagy and other vacuolar protein degradation mechanisms. *Experientia* **48**: 158–172.

Shintani T., Mizushima N., Ogawa Y., Matsuura A., Noda T., and Ohsumi Y. 1999. Apg10p, a novel protein-conjugating enzyme essential for autophagy in yeast. *EMBO J.* **18**: 5234–5241.

Stack J.H., Horazdovsky B., and Emr S.D. 1995. Receptor-mediated protein sorting to the vacuole in yeast: Roles for a protein kinase, a lipid kinase and GTP-binding proteins. *Annu. Rev. Cell Dev. Biol.* **11**: 1–33.

Straub M., Bredschneider M., and Thumm M. 1997. AUT3, a serine/threonine kinase gene, is essential for autophagocytosis in *Saccharomyces cerevisiae*. *J. Bacteriol.* **179**: 3875–3883.

Suriapranata I., Epple U.D., Bernreuther D., Bredschneider M., Sovarasteanu K., and Thumm M. 2000. The breakdown of autophagic vesicles inside the vacuole depends on Aut4p. *J. Cell. Sci.* **113**: 4025–4033.

Suzuki K., Kirisako T., Kamada Y., Mizushima N., Noda T., and Ohsumi Y. 2001. The preautophagosomal structure organized by concerted functions of *APG* genes is essential for autophagosome formation. *EMBO J.* **20**: 5971–5981.

Takeshige K., Baba M., Tsuboi S., Noda T., and Ohsumi Y. 1992. Autophagy in yeast demonstrated with proteinase-deficient mutants and conditions for its induction. *J. Cell Biol.* **119**: 301–311.

Tanida I., Tanida-Miyake E., Ueno T., and Kominami E. 2001. The human homolog of *Saccharomyces cerevisiae* Apg7p is a protein-activating enzyme for multiple substrates including human Apg12p, GATE-16, GABARAP, and MAP-LC3. *J. Biol. Chem.* **276**: 1701–1706.

Tanida I., Mizushima N., Kiyooka M., Ohsumi M., Ueno T., Ohsumi Y., and Kominami E. 1999. Apg7p/Cvt2p: A novel protein-activating enzyme essential for autophagy. *Mol. Biol. Cell* **10**: 1367–1379.

Terlecky S.R., Chiang H.L., Olson T.S., and Dice J.F. 1992. Protein and peptide binding and stimulation of in vitro lysosomal proteolysis by the 73-kDa heat shock cognate protein. *J. Biol. Chem.* **267**: 9202–9209.

Thumm M., Egner R., Koch B., Schlumpberger M., Straub M., Veenhuis M., and Wolf D.H. 1994. Isolation of autophagocytosis mutants of *Saccharomyces cerevisiae*. *FEBS Lett.* **349**: 275–280.

Tsukada M. and Ohsumi Y. 1993. Isolation and characterization of autophagy-defective mutants of *Saccharomyces cerevisiae*. *FEBS Lett.* **333**: 169–174.

Zubenko G.S. and Jones E.W. 1981. Protein degradation, meiosis and sporulation in proteinase-deficient mutants of *Saccharomyces cerevisiae*. *Genetics* **97**: 45–64.

14

Synaptic Growth, Synaptic Maintenance, and the Persistence of Long-term Memory Storage

Craig H. Bailey[2], Robert D. Hawkins[2], and Eric R. Kandel[1,2]
[1]Howard Hughes Medical Institute and
[2]Center for Neurobiology and Behavior
College of Physicians & Surgeons of
Columbia University
New York State Psychiatric Institute
New York, New York 10032

T HE ELEMENTARY EVENTS THAT UNDERLIE synaptic plasticity, the ability of neurons to modulate the strength of their synapses in response to extra- or intracellular cues, are thought to be fundamental both for the fine-tuning of synaptic connections during development, and for behavioral learning and memory storage in the adult organism. Indeed, activity-dependent modulation of synaptic strength and structure is emerging as one of the key mechanisms by which information is processed and stored within the brain (Kandel 2001).

Earlier behavioral studies in vertebrates and invertebrates have shown that the formation of both explicit (declarative) and implicit (non-declarative) forms of memory consist of two temporally distinct stages: short-term memory lasting minutes to hours and long-term memory lasting days, weeks, or longer. This temporal distinction in behavior is reflected in specific forms of synaptic plasticity that underlie each form of behavioral memory, as well as specific molecular requirements for each of these two forms of synaptic plasticity. In each case, the short-term form involves the covalent modifications of preexisting proteins mediated in part by cAMP and cAMP-dependent protein kinase (PKA) and is expressed as an alteration in the effectiveness of preexisting connections.

In contrast, the long-term forms require PKA, MAPK, and CREB-mediated gene expression, new mRNA and protein synthesis, and are often associated with the growth of new synaptic connections (Bailey et al. 1996). For both implicit and explicit memory storage, the synaptic growth is thought to represent the final and self-sustaining change that stabilizes the long-term process. Initial insights into the molecular mechanisms that underlie this learning-related synaptic remodeling suggest a role for both CREB-activated gene expression and the modulation of NCAM-related cell adhesion molecules (Bailey and Kandel 1993).

The finding that long-term memory is associated with synaptic growth and requires new macromolecular synthesis raises three interesting conceptual issues. First, it raises the question of how the covalent modification of preexisting proteins and the alteration of preexisting connections during short-term memory become transformed into the formation of new synaptic connections. Are the structural changes dependent on, and perhaps induced by, the changes in gene and protein expression in the neurons involved? If so, what are the signaling pathways that trigger gene induction, and what are the molecules and mechanisms that initiate synaptic growth? Second, the finding of synaptic growth provides one way of thinking about the stability of long-term memory storage. Is the stability of long-term memory, which would seem to require some mechanism that can survive molecular turnover, achieved because of the relative stability of synaptic structure? If so, what are the cellular and molecular processes that serve to stabilize synaptic structure, and how do these differ from those that initiate the structural change? Finally, the finding that synaptic growth and the formation of new synaptic connections are the most reliable anatomical markers for the long-term process raises the question of how closely these mechanisms for learning-related structural plasticity in the brain resemble the mechanisms and molecules that govern the universal process of cell growth and cytoplasmic expansion in other systems.

In this review, we consider the degree to which the genes and proteins for synaptic growth and synaptic maintenance are conserved in the two major forms of memory storage by focusing on these issues. We begin by examining the growth of sensory neuron synapses that accompanies an elementary form of implicit memory—long-term sensitization in *Aplysia*. We then consider recent evidence for structural remodeling and synaptic growth associated with long-term potentiation (LTP) in the hippocampus, a type of enduring synaptic plasticity thought to be involved in the storage of long-term memory for explicit forms of learning in the mammalian brain. Finally, we outline some of the molecular insights that have been provided by these neurobiological studies into the mechanisms responsible for both the initiation and stabilization of the synaptic growth associated with memory storage and consider how these may relate to the processes that regulate cellular growth and alterations in cytoplasmic mass in other systems.

GROWTH OF NEW SYNAPSES BY THE SENSORY NEURONS ACCOMPANIES LONG-TERM SENSITIZATION—AN IMPLICIT FORM OF MEMORY IN *APLYSIA*

The marine mollusk *Aplysia californica* has proven advantageous for cellular studies of memory because it has a tractable nervous system, consisting of approximately 20,000 neurons, many of which are large, invariant, and readily identifiable. As a result, the neural circuitry underlying many of its behaviors has been at least partially characterized. One such behavior is the gill-withdrawal reflex, which can undergo several types of behavioral modification, including both nonassociative and associative learning. One elementary form of nonassociative learning exhibited by this reflex is sensitization, a form of learned fear, which can be induced by a strong stimulus to another site such as the neck or tail. As is the case for other defensive reflexes, the behavioral memory for sensitization of the gill-withdrawal reflex is graded, and retention is proportional to the number of training trials. A single stimulus to the tail gives rise to short-term sensitization lasting minutes to hours. Repetition of the stimulus produces long-term sensitization that can last days to weeks (Fig. 1) (Frost et al. 1985).

The memory for both the short- and long-term forms of sensitization is represented on an elementary level by the monosynaptic connections between identified mechanoreceptor sensory neurons and their follower cells. Although this component accounts for only a part of the behavioral modification measured in the intact animal, its simplicity has allowed the reduction of the analysis of the short- and long-term memory of sensitization to the cellular and molecular level. For example, this monosynaptic pathway can be reconstituted in dissociated cell culture (Montarolo et al. 1986), where serotonin (5-HT), a modulatory neurotransmitter normally released by sensitizing stimuli, can substitute for the shock to the neck or tail used during behavioral training in the intact animal (Glanzman et al. 1989). A single application of 5-HT produces short-term changes in synaptic effectiveness, whereas five spaced applications given over a period of 1.5 hours produces long-term changes lasting one or more days.

Biophysical studies of this monosynaptic connection suggest that both the similarities and the differences in memory reflect, at least in part, intrinsic cellular mechanisms of the nerve cells participating in memory

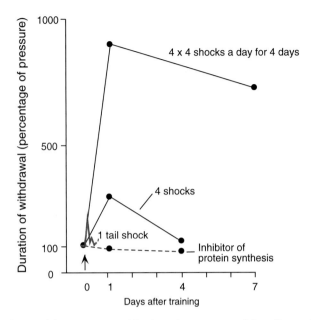

Figure 1. Behavioral long-term sensitization. A summary of the effects of long-term sensitization training on the duration of siphon withdrawal in *Aplysia californica*. The retention of the memory for sensitization is a graded function proportional to the number of training trials. Before sensitization, a weak touch to the siphon causes only a brief siphon- and gill-withdrawal reflex. Following a single noxious, sensitizing shock to the tail, that same weak touch elicits a much larger response that lasts about 1 hour. More tail shocks increase the size and duration of the response. Application of protein synthesis inhibitors blocks the long-term but not the short-term memory for sensitization. (Modified from Frost et al. 1985.)

storage. Thus, studies of the connections between sensory and motor neurons both in the intact animal and in cells in culture indicate that phenotypically the long-term changes are surprisingly similar to the short-term changes. A component of the increase in synaptic strength observed during both the short- and long-term changes is due, in each case, to enhanced release of transmitter by the sensory neuron, accompanied by an increase in the excitability of the sensory neuron, attributable to the depression of a specific potassium channel (Klein and Kandel 1980; Frost et al. 1985; Hochner et al. 1986; Montarolo et al. 1986; Dale et al. 1987; Scholz and Byrne 1987).

   Despite these several similarities, the short-term cellular changes differ from the long-term process in two important ways. First, the short-term change involves only covalent modification of preexisting proteins and an alteration of preexisting connections. Both short-term behavioral sensitization in the animal and short-term facilitation in dissociated cell culture do not require ongoing macromolecular synthesis: The short-term change is not blocked by inhibitors of transcription or translation (Schwartz et al. 1971; Montarolo et al. 1986). In contrast, these inhibitors selectively block the induction of the long-term changes both in the semi-intact animal (Castelluci et al. 1989) and in primary cell culture (Montarolo et al. 1986). Most striking is the finding that the induction of long-term facilitation at this single synapse in *Aplysia* exhibits a critical time window in its requirement for protein and RNA synthesis character-istic of that necessary for other forms of learning in both vertebrates and invertebrates. From a molecular perspective, these studies indicate that both the long-term behavioral and the long-term cellular changes require the expression of genes and proteins not required for the short term.

Second, as described below, the long-term but not the short-term process involves a structural change. Bailey and Chen (1983, 1988a,b, 1989) first demonstrated that long-term sensitization training is associated with the growth of new synaptic connections by the sensory neurons onto their follower cells. This synaptic growth can be induced in the intact ganglion by the intracellular injection of cAMP, a second messenger activated by 5-HT (Nazif et al. 1991), and can be reconstituted in sensory-motor cocul-tures by repeated presentations of 5-HT (Glanzman et al. 1990).

To study possible structural transformations that may underlie the transition from short-term to long-term memory, Bailey and Chen com-bined selective intracellular labeling techniques with the analysis of serial thin sections to study complete reconstructions of identified sensory neu-ron synapses from both control and behaviorally modified animals (for review, see Bailey and Kandel 1993). Their results indicate that long-term memory (lasting several weeks) is accompanied by a family of morpho-logical alterations at identified sensory neuron synapses. These changes reflect structural plasticity at two different levels of synaptic organization: (1) alterations in focal regions of membrane specialization of the synapse that mediate transmitter release—the number, size, and vesicle comple-ment of sensory neuron active zones are larger in sensitized animals than controls and smaller in habituated animals (Bailey and Chen 1983, 1988b)—and (2) a parallel but more pronounced and widespread effect involving modulation of the total number of presynaptic varicosities per sensory neuron (Bailey and Chen 1988a). There is a positive correlation between these changes in synapse number and each behavioral modifica-tion. This effect is bidirectional—sensitization increases behavioral per-formance as well as the total number of sensory neuron varicosities, whereas habituation has an opposing effect, decreasing both behavioral effectiveness and varicosity number. Thus, sensory neurons from long-term habituated animals had on average 35% fewer varicosities (compared to their controls). In contrast, the morphological changes that accompany long-term sensitization involve a doubling in the total number of synaptic varicosities. In addition to an increase in varicosity number, sensory neurons from long-term sensitized animals exhibit further evidence of long-lasting growth; i.e., they display an increase in the total linear extent of their axonal arbor within the ganglion (Fig. 2).

To determine which class of structural change at sensory neuron synapses might be necessary for the retention of long-term sensitization, Bailey and Chen (1989) compared the time course for each morphological change with the behavioral duration of the memory. They found that not all structural changes persist as long as the memory. The increase in the size and vesicle complement of sensory neuron active zones present 24 hours following the completion of behavioral training is back to control levels when tested one week later. These data indicate that, insofar as modulation of active-zone size and associated vesicles is one of the structural mechanisms underlying long-term sensitization, it is associated with the initial expression and not with the persistence of the long-term process. In contrast, the duration of changes in varicosity and active-zone number, which persist unchanged for at least one week and are only partially reversed at the end of three weeks, parallels the behavioral time course of memory, indicating that only the changes in the number of sensory neuron synapses contribute to the maintenance of long-term sensitization. These findings suggest that the growth of new sensory neuron synapses may represent the final and perhaps most stable phase of long-term memory storage and provide evidence for an intriguing notion—that the stability of the long-term process may be achieved, in part, because of the relative stability of synaptic structure.

This long-lasting growth of new synaptic connections between sensory neurons and their follower cells (both interneurons and motor neurons) during long-term sensitization can be reconstituted in dissociated sensory-motor neuron cocultures by repeated applications of 5-HT (Glanzman et al. 1990). In culture, the structural change can be correlated with the long-term (24 hours) enhancement in synaptic effectiveness and depends on the presence of an appropriate target cell similar to the synapse formation that occurs during development. The nature of this interaction between the presynaptic cell and its target is not known. The signal from the postsynaptic neuron may be cell-associated, such as a constituent of the motor cell's membrane, or diffusible, perhaps being released locally from the motor cell's processes. Both in the ganglion and in culture, the long-term increase in the number of sensory neuron varicosities is dependent on new macromolecular synthesis (Bailey et al. 1992b).

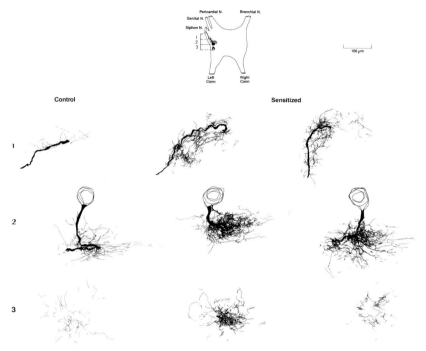

Figure 2. Serial reconstruction of sensory neurons from long-term sensitized and control animals. Total extent of the neuropil arbors of sensory neurons from one control and two long-term-sensitized animals are shown. In each case, the rostral (row *3*) to caudal (row *1*) extent of the arbor is divided roughly into thirds. Each panel was produced by the superimposition of camera lucida tracings of all horseradish peroxidase-labeled processes present in 17 consecutive slab-thick sections and represents a linear segment through the ganglion of roughly 340 μm. For each composite, ventral is up, dorsal is down, lateral is to the left, and medial is to the right. By examining images across each row (rows *1*, *2*, and *3*), the viewer is comparing similar regions of each sensory neuron. In all cases, the arbor of long-term-sensitized cells is markedly increased compared to control. (Reprinted, with permission, from Bailey and Chen 1988a.)

STRUCTURAL REMODELING AND SYNAPSE FORMATION ASSOCIATED WITH LONG-TERM POTENTIATION: INSIGHTS INTO EXPLICIT MEMORY STORAGE IN THE MAMMALIAN BRAIN

Long-term potentiation (LTP) in the hippocampus is a form of synaptic plasticity that is thought to be involved in the storage of long-term memory for explicit forms of learning, forms of learning concerned with acquiring information of places, objects, and other living beings (Baudry and Davis 1991; Malenka and Nicoll 1999). LTP can be induced by the application of a brief, high-frequency electrical tetanus to a hippocampal pathway, and this increase in synaptic strength can last for hours in an anesthetized animal and can, if repeated, last for days and even weeks in an alert, freely moving animal. Evidence for changes in synapse number with LTP in the CA1 region first came from studies by Lynch and his coworkers (Lee et al. 1980), who reported that electrical stimulation that produced LTP either in vivo or in vitro also led to a rapid increase in the number of synapses onto dendritic shafts.

This observation was confirmed and extended in the slice by Chang and Greenough (1984), who also found a rapid (within 10–15 minutes after stimulation) increase in the number of shaft synapses per unit area and, in addition, an increase in the number of sessile spine synapses (presumed to be immature or transitional synaptic contacts) in slices exposed to LTP compared to equivalent low-frequency stimulated control groups. Using serial sections and improved, unbiased methods for synapse quantification, Geinisman et al. (1991) also demonstrated changes in synapse number during the induction of LTP in the dendate gyrus. These studies found that LTP was accompanied by a highly selective increase in a single morphological subtype of synaptic contact-perforated axospinous synapses with a segmented postsynaptic density—which could represent the selective formation of a particular class of synapse or a dynamic transformation of one synaptic form into another. This and other studies (for review, see Geinisman 2000) suggested that potentiation might trigger a physical splitting of the presynaptic active zone and postsynaptic density such that more neurotransmitter could be released following the arrival of an action potential in the terminal, thus providing a larger depolarization of the postsynaptic cell.

Subsequent studies in the mammalian central nervous system have provided additional evidence for changes either in the number of functional synapses (Liao et al. 1995; Durand et al. 1996; Wu et al. 1996; Bolshakov et al. 1997; Isaac et al. 1997; Chavis et al. 1998) or in the structural organization of synapses (Chang and Greenough 1984; Chang et al. 1991; Geinisman et al. 1991, 1993, 1996; Buchs and Muller 1996; Trommald et al. 1996; Rusakov et al. 1997) following various forms of synaptic plasticity including LTP (for review, see Yuste and Bonhoeffer 2001). However, in the mammalian brain, interpretation of changes in the number of functional synapses (Kullman and Siegelbaum 1995; Malenka and Nicoll 1997) and in the structure of synapses remains controversial (Sorra and Harris 1998).

In particular, studies examining the relationship between changes in synapse number or structure and LTP have been hampered by four major problems. First, there is considerable heterogeneity and concomitant variability among different experimental preparations, which hinders comparisons of any changes in pre- or postsynaptic structure between populations of naive and stimulated neurons. Second, until recently, most studies have examined only the early, transient phase of LTP which, because it does not depend on new protein synthesis and does not persist for more than 1 or 2 hours, might not be expected to be accompanied by significant structural plasticity. Third, these studies inevitably look at electrically induced LTP, where the electrical stimulation is likely to induce changes at only a small fraction of the total population of synapses, leaving a high background of unaltered synaptic connections. Finally, excitatory synapses in the mammalian brain can display a variety of morphologies, making it very difficult to analyze selectively the structural remodeling at only potentiated synaptic contacts.

Recent technical advances have begun to address these problems and have taken some important steps in correlating changes in synaptic strength with structural remodeling at the synapse during LTP. Maletic-Savatic et al. (1999) first showed that brief periods of NMDA-receptor activation, which is essential for induction of LTP, can also rapidly induce changes in synaptic structure. Using green fluorescent protein (GFP) and two-photon laser-scanning microscopy to visualize dendritic spines of CA1 pyramidal neurons in organotypic hippocampal slice cultures, they delivered tetanic stimulation, which induced LTP in close proximity to the labeled dendrites. Within 5 minutes of the electrical tetanus, large numbers of new spine-like processes began to emerge. These new postsynaptic structures more closely resembled dendritic filapodia than spines, persisted for approximately 40 minutes, and could be blocked by the NMDA receptor antagonist AP5. Interestingly, approximately 27% of these filapodia developed spine-like heads within one hour of the tetanus.

Engert and Bonhoeffer (1999) confirmed and extended these observations by combining the intracellular injection of a fluorescent dye into CA1 pyramidal cells and two-photon laser microscopy with a method of local superfusion to restrict the location of synapses activated along the parent dendrite to a sphere roughly 30 μm in diameter. LTP was found to coincide with the appearance of new dendritic spines, whereas either in control regions on the same dendrite or in slices where LTP was blocked, no significant spine growth occurred. New dendritic spines first appeared roughly 30 minutes after the induction of LTP and persisted for hours.

Both of these imaging studies indicated that the induction of LTP is temporally coupled with the formation of new dendritic spines. What remains to be addressed is whether or not these new postsynaptic spines in fact form functional synapses with presynaptic boutons, and whether the potentiation could be maintained by a stable increase in synapse number.

Some early answers to these questions have come from the electron microscopic studies by Toni et al. (1999). These investigators addressed the problem of specificity by employing a fixation protocol that produced an electron-dense precipitate where Ca^{++} had accumulated in dendritic spines following high-frequency stimulation—in principle, restricting their analysis to only potentiated synapses. Using this approach, they examined the ultrastructural correlates of LTP as a function of time after the induction of synaptic potentiation. Thirty minutes after the tetanus, the percentage of the precipitate-labeled spines that had a perforated postsynaptic density increased from 19% to 46%. However, this increase was only transient, reaching control levels after about 1 hour, but was followed by a threefold increase in the number of multiple-spine boutons—i.e., at least two dendritic spines contacting the same axonal terminal. Since this increase in multiple-spine boutons indicated possible synaptogenesis, Toni et al. (1999) used serial reconstructions to analyze the origin of the newly formed dendritic spines. They found that 89% of the multiple spine boutons that contacted unlabeled spines (i.e., those with no Ca^{++} precipitate) formed synapses with two different dendrites in control cultures. In contrast, 66% of the multiple-spine boutons that contacted labeled spines after LTP induction were formed from the two spines originating from the same dendrite. These data suggest that the late phase of LTP is associated with the formation of new, mature, and perhaps functional spine synapses contacting the same presynaptic terminals. Although the mechanisms by which these new synapses form is still in question (Fiala et al. 2002), one attractive feature of this model for synapse duplication is that the new spines contact a preexisting presynaptic bouton and thus maintain structural specificity in the processing of information.

In addition to the evidence for morphological changes in dendritic spines and the consequent postsynaptic remodeling of synapses during LTP, there is also evidence for structural plasticity at the presynaptic terminal. Using biophysical techniques and quantal analysis, Bolshakov et al. (1997) first examined this issue by analyzing the synaptic mechanisms that underlie the cAMP-dependent late phase of LTP.

LTP at hippocampal synapses where CA3 neurons contact CA1 neurons exhibits an early phase and a late phase, which can be distinguished by their underlying molecular mechanisms. Unlike the early phase, the late phase is dependent on both cAMP and protein synthesis. Quantal analysis of unitary synaptic transmission between a single presynaptic CA3 neuron and a single presynaptic CAl neuron suggests that, under certain conditions, the early phase of LTP involves an increase in the probability of release of a single quantum of transmitter from a single presynaptic release site, with no change either in the number of quanta that are released or in postsynaptic sensitivity to transmitter.

Bolshakov et al. (1997) found that the cAMP-induced late phase of LTP (L-LTP) also involves an increase in the number of quanta released in response to a single presynaptic action potential but that now the amplitude distribution of the synaptic response is no longer described by two Gaussians (with one peak at 0 failures and one at 4 pA unit response), but by a distribution that can best be fitted by three or more Gaussians each of which had a peak which was an integral multiple of the unit peak. These data suggested the interesting possibility that there may be an increase in the number of sites of synaptic transmission between a single CA3 and a single CA1 neuron following L-LTP. This multiquantal release during L-LTP could arise from one of three mechanisms. First, there could be an increase in the number of vesicles released from a single preexisting active zone. Although there is strong evidence that a single active zone releases only a single vesicle (Korn et al. 1994), this condition may not hold under all circumstances. However, to explain the fact that some EPSCs appear to be composed of four (or more) quanta during L-LTP, the postsynaptic receptors would have to be far from saturated in response to a single quantum of transmitter. A second possibility is that there is new growth of both presynaptic terminals and postsynaptic specializations, leading to an increase in the number of synapses. Finally, the finding that the size of the mean mEPSC increases during L-LTP suggests a third possible mechanism, the insertion into a preexisting presynaptic terminal of new active zones that can synchronously release transmitter. These new release sites would presumably contact new postsynaptic sites on the CA1 neuron. One attractive structural mechanism that could account for the change in synaptic function during L-LTP is the formation of perforated synapses. Geinisman (1993) had observed an increase in number of such synaptic subtypes associated with plasticity including hippocampal LTP. At these synapses, a projection or spinnule from the postsynaptic spine inserts into the presynaptic bouton, splitting the active zone into discrete regions. This mechanism could thus explain these results if each region can function as a separate release site with its own cluster of postsynaptic receptors. Moreover, because the active zones would be close to each other, this proximity could allow their synchronization during spontaneous release.

As a first step in examining the structural bases of the multiquantal release seen in cAMP-induced L-LTP, Ma et al. (1999) imaged synaptic vesicle turnover associated with evoked transmitter release in presynaptic terminals using the amphipathic membrane fluorescent dye FM 1–43

(Betz and Bewick 1992). This dye is taken up into synaptic vesicles during endocytosis following a period of evoked transmitter release. The dye can then be released from presynaptic terminals by a subsequent period of stimulated exocytosis. The difference in images of dye staining before and after the second period of stimulated release yields an activity-dependent level of presynaptic function and thus provides a profile of the population of active presynaptic terminals. By comparing such activity-dependent profiles before and after exposure to cAMP analogs, one can explore possible changes in the number of functional synapses during cAMP-dependent forms of long-lasting synaptic plasticity. Using this approach, Ma et al. found that the intensity of FM 1–43 staining is significantly increased 2 hours after application of the cAMP agonist Sp-cAMPs, and that this increase is blocked by protein synthesis inhibitors and by the PKA antagonist Rp-cAMPs. This Sp-cAMPs-induced increase in the number of functional synaptic contacts could result from the growth of new presynaptic boutons or by the activation of preexisting but functionally silent boutons. Comparison of the pattern of activity-independent staining with FM 1–43 before and after Sp-cAMPs suggests that the source for most of the new active boutons that appear after Sp-cAMPs treatment may be a population of preexisting but functionally silent boutons. Whether such silent boutons lack active zones which are then inserted to activate the bouton, or whether activation reflects a functional, rather than a structural change, remains to be determined.

Additional evidence for presynaptic modifications of the synapse during LTP comes from the study of Antonova et al. (2001) using immunoreactivity for both the postsynaptic receptor subunit GluR1 and the presynaptic vesicle-associated protein synaptophysin. They found that the long-lasting potentiation of synaptic transmission between cultured hippocampal neurons was accompanied by an NMDA-dependent increase in the number of clusters of postsynaptic glutamate receptors containing the subunit GluR1, as well as by a rapid (5–10 minutes) and long-lasting increase in the number of clusters of the presynaptic protein synaptophysin. Moreover, there was also an increase in the number of sites at which synaptophysin and GluR1 were colocalized (Fig. 3). The increase in synaptophysin and GluR1 puncta was not blocked by an inhibitor of protein synthesis but was blocked by an inhibitor of actin polymerization, suggesting that this increase may involve an actin-based redistribution of preexisting proteins. Imaging of GTP-synaptophysin in living neurons supports this idea that the new puncta arise from aggregation of more diffusely distributed material. Earlier studies had demonstrated that there is also an increase in the number of synaptophysin-GluR1 puncta during

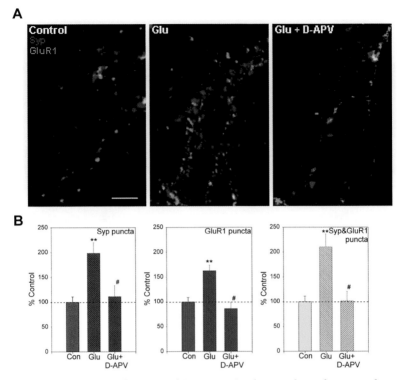

Figure 3. Glutamate produces rapid increases in the number of puncta that are immunoreactive (IR) for pre- and postsynaptic proteins. (*A*) Examples of synaptophysin (syp)-IR (*red*), GluR1-IR (*green*), and colocalization (either *yellow* or adjacent *red* and *green*) in a control dish (*left*), a dish fixed 5 minutes after brief application of 200 μM glutamate (*middle*), and a dish fixed after application of glutamate in the presence of 200 μM D-APV (*right*). Bar, 5 μm. (B) Average results from experiments like the one shown in *A* (*n* = 6 dishes in each group). In this and subsequent figures, ** = $p < 0.01$, * = $p < 0.05$ compared to control, and # = $p < 0.05$ compared to glutamate. The number of puncta in a representative field (94 μm by 142 μm) has been normalized to the number in comparable fields in control dishes from the same culture batch. The average control values (in number of puncta) were 47 ± 8 (syp), 75 ± 10 (GluR1), and 20 ± 3 (colocalized) (Modified from Antonova et al. 2001).

the protein synthesis-dependent late phase of LTP lasting more than 2 hours (Bozdagi et al. 2000), when there is actual growth of new synapses (Luscher et al. 2000). The results of Antonova et al. (2001) indicate that the number of synaptophysin puncta increases as early as 5–10 minutes after stimulation, before any synaptic growth is thought to occur. These new puncta may, therefore, represent an early stage in the assembly of new functional synapses, or they may be associated with the rapid conversion of "silent" presynaptic terminals to functional ones, as several previous studies have found an increase in functional terminals (as indicated by vesicle cycling) during early as well as late phase LTP. In either case, these findings support an emerging view that even the early stages of long-lasting plasticity involve at least some minimal change in microarchitecture and show that these changes can occur both pre- and postsynaptically in a coordinated fashion, as occurs during synaptic development.

Further support for coordinated changes in pre- and postsynaptic structure accompanying long-term plasticity have been provided by the real-time imaging of actin dynamics in cultured hippocampal cells (Colicos et al. 2001). In response to multiple trains of stimulation that induce LTP, this study has revealed simultaneous bilateral changes in actin dynamics that appear to coordinate an activity-dependent process of synaptic remodeling across the synaptic cleft. Some of these correlate well with the types of pre- and postsynaptic structural changes we have described above. For example, actin within already established puncta may progress through an intermediate torus shape prior to splitting off to form new presynaptic puncta that are contacted by newly formed filopodia-like protrusions from the postsynaptic neuron, which then become functional within hours. The time course of the appearance of the new presynaptic actin puncta is also similar to that which has been described during morphological changes associated with LTP in hippocampal slices. Both the actin budding described by Colicos et al. (2001) and the appearance of new synaptophysin puncta with LTP require glutamate receptor activation, cAMP signaling, and, at least in some cases, new protein synthesis (Bozdagi et al. 2000; Antonova et al. 2001).

MOLECULAR MECHANISMS UNDERLYING LEARNING-RELATED SYNAPTIC GROWTH IN *APLYSIA*

Insights into the molecules and mechanisms that may contribute to the storage of long-term memory first came from studies in *Aplysia*. Because the functional and structural changes that accompany long-term sensitization in *Aplysia* require new protein synthesis, Barzilai et al. (1989) utilized quantitative two-dimensional gels and [35S]methionine incorporation to examine changes in specific proteins in the sensory neurons in response to 5-HT. They found that 5-HT initiates a large increase in overall protein synthesis during training. Moreover, beyond these overall effects, 5-HT also produces three temporally discrete sets of changes in specific proteins that could be resolved on two-dimensional gels. First, 5-HT induces a rapid and transient increase at 30 minutes in the rate of synthesis of 10 proteins and a transient decrease in 5 proteins that subside within 1 hour and are in all cases dependent on transcription. These early changes are followed by at least two further rounds of changes in the expression of specific proteins, some of which are transient, and some of which persist for at least 24 hours. The 15 early proteins induced by repeated exposure to 5HT can also be induced by cAMP. These features—rapid induction, transcriptional dependence, and second-messenger mediation—suggest that this control might involve a gene cascade, whereby early regulatory proteins activate later effector genes. The early proteins induced by 5-HT and cAMP during the acquisition phase of long-term facilitation in *Aplysia* may therefore resemble the immediate-early gene products induced in vertebrate cells by growth factors.

INITIATION OF LONG-TERM FACILITATION REQUIRES THE ACTIVATION OF cAMP-RESPONSIVE GENES AND THE RECRUITMENT OF CREB-RELATED TRANSCRIPTION FACTORS

A number of studies have examined the regulation of immediate-early genes in learning and memory. For example, immediate-early genes are induced in the hippocampus and in neocortex by experimental treatments that lead to LTP. Despite these reports of gene induction by learning-related activity, it has so far proven difficult to assign function to any immediate-early genes in the brain.

In contrast to the mammalian brain, it has proven possible to interfere with the actions of genes in the sensory neurons of the gill-withdrawal reflex in *Aplysia* and to relate their activity to the induction of synaptic plasticity. Repeated presentations of 5-HT to sensory-motor cocultures produce a long-term enhancement in transmitter release that lasts for one or more days and requires for its induction both translation and transcription (Montarolo et al. 1986). Injection of cAMP into sensory neurons can induce long-term facilitation (Scholz and Byrne 1987; Schacher et al. 1988) and inhibitors of PKA block long-term facilitation (Ghirardi et al. 1992).

How PKA participates in the RNA and protein synthesis-dependent process was illustrated by an experiment carried out by Bacskai and associates (1993). They monitored the subcellular localization of the free PKA catalytic subunit and found that with one pulse of 5-HT, which produces short-term facilitation, the catalytic subunit was restricted to the cytoplasm. With repeated pulses of 5-HT, which induce long-term facilitation, the catalytic subunit translocated to the nucleus, where it presumably phosphorylates transcription factors and thereby regulates gene expres-

sion. Both cAMP and PKA are essential components of the signal transduction pathway for consolidating memories, not only in *Aplysia* but also
for certain types of memory in *Drosophila* and mammals. Several olfactory learning mutants in *Drosophila* map to the cAMP pathway (Drain et al.
1991; Davis et al. 1995; Davis 1996), indicating that blocking PKA function
blocks memory formation in flies. In parallel, the late but not the early
phase of LTP of the CA3-to-CAl synapse in the hippocampus is impaired
by pharmacological or genetic interference with PKA (Frey et al. 1993;
Huang et al. 1994; Abel et al. 1997). This interference blocks long-lasting
spatial memory in mice without interfering with short-term memory.

These studies suggest that the long-term enhancement of transmitter
release requires cAMP-related gene activation. Increases of cAMP concentration induce gene expression by activating transcription factors that
bind to the cAMP-responsive element (CRE) (Montminy et al. 1986). The
CRE-binding protein CREB (Hoeffler et al. 1988) is a major target of PKA
phosphorylation of a regulatory P-box, which in turn activates transcription by recruiting the CREB-binding protein CBP (Gonzalez and
Montminy 1989), a histone acetylase. During long-term facilitation in
Aplysia neurons, PKA appears to activate gene expression via an *Aplysia*
CREB (Dash et al. 1990). Dash et al. (1990) first tested the role of CREB
in long-term facilitation by microinjecting CRE oligonucleotides into sensory neurons. The CRE oligonucleotide binds to the CREB protein and
prevents it from binding to CRE sites in the regulatory regions of cAMP-
responsive genes. Although injection of the CRE oligonucleotide had no
effect on short-term facilitation, it selectively blocked long-term facilitation (Fig. 4A and 4B). Kaang et al. (1993) further explored the mechanisms of CREB activation in long-term facilitation by microinjecting two
constructs into *Aplysia* sensory neurons: an expression plasmid containing a chimeric *trans*-acting factor made by fusing the GAL4 DNA-binding
domain to the mammalian CRE-activation domain and a reporter plasmid containing the chloramphenicol acetyltransferase (CAT) gene under
the control of GAL4-binding sites. Following coinjection of these two
plasmids into sensory neurons, exposure to 5-HT produced a tenfold
stimulation of CAT expression (Fig. 4C and 4D). Finally, Bartsch et al.
(2000) directly demonstrated that CREB is essential for long-term facilitation. They cloned the *Aplysia* CREB gene and showed that blocking
expression of one of the CREB1 gene products (CREB1a protein) in sensory neurons by microinjection of antisense oligonucleotides or antibodies blocks long-term but not short-term facilitation. Not only is CREB1a
necessary for long-term facilitation, but it is also sufficient to induce long-

term facilitation, albeit in reduced form. Thus, sensory cell injection of recombinant CREB1a that was phosphorylated in vitro by PKA led to an increase in excitatory postsynaptic potential (EPSP) amplitude at 24 hours in the absence of any 5-HT stimulation.

The transcriptional switch in long-term facilitation is not only composed of the CREB1 regulatory unit. Bartsch and associates (1995) have found that another member of the CREB gene family, ApCREB2, a CRE-binding transcription factor constitutively expressed in sensory neurons, is a critical component of the genetic switch that converts short- to long-term facilitation. ApCREB2 resembles human CREB2 and mouse ATF4 (Hai et al. 1989; Karpinski et al. 1992), and, like CREB1b, ApCREB2 functions as a repressor of long-term facilitation. Thus, injection of anti-ApCREB2 antibodies into *Aplysia* sensory neurons causes a single pulse of 5-HT, which normally induces only short-term facilitation lasting minutes, to evoke facilitation that lasts more than 1 day. This response requires both transcription and translation and is accompanied by the growth of new synaptic connections (Fig. 5).

These studies reveal that long-term synaptic changes are governed by both positive and negative regulators, and that the transition from short-term facilitation to long-term facilitation requires the simultaneous removal of transcriptional repressors and activation of transcriptional activators. These transcriptional repressors and activators can interact with each other both physically and functionally. It is likely that the transition is a complex process involving temporally distinct phases of gene activation, repression, and regulation of signal transduction. The balance between CREB activator and repressor isoforms is also critically important in long-term behavioral memory, as first shown in *Drosophila*. Expression of an inhibitory form of CREB (dCREB-2b) blocks long-term olfactory memory but does not alter short-term memory (Yin et al. 1994). Overexpression of an activator form of CREB (dCREB-2a) increases the efficacy of massed training in long-term memory formation (Yin et al. 1995).

The CREB-mediated response to extracellular stimuli can be modulated by a number of kinases (PKA, CaMKII, CaMKIV, RSK2 MAPK, and PKC) and phosphatases (PP1 and Calcineurin). The CREB regulatory unit may therefore serve to integrate signals from various signal transduction pathways. This ability to integrate signaling, as well as to mediate activation or repression, may explain why CREB is so central to memory storage (Martin and Kandel 1996).

Recent studies by Guan et al. (2002) have examined directly the role of CREB-mediated responses in long-term synaptic integration by studying the long-term interactions of two opposing modulatory transmitters important for behavioral sensitization in *Aplysia*. Toward that end, they utilized a single bifurcated sensory neuron that contacts two spatially separated postsynaptic neurons (Martin et al. 1997b). They found that when a neuron receives input—at one set of synapses from the facilitatory transmitter 5-HT—and at the same time receives input from the inhibitory transmitter FMRFamide at another set of synapses, the synapse-specific long-term facilitation normally induced by 5-HT is suppressed, and synapse-specific long-term depression produced by FMRFamide dominates. These opposing inputs are integrated in the neuron's nucleus and are evident in the repression of the CAAT-box-enhanced-binding protein (C/EBP), a transcription regulator downstream from CREB that is critical for long-term facilitation. Whereas 5-HT induces C/EBP by activating CREB1 and recruiting the CREB-binding protein a histone acetylase, to

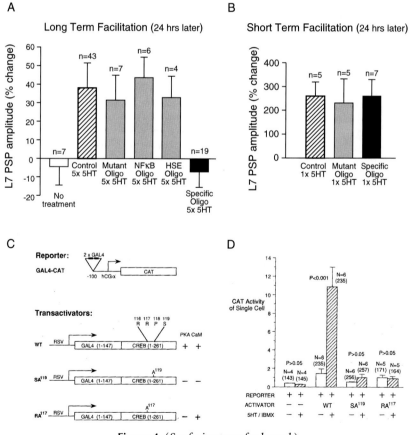

Figure 4. (*See facing page for legend.*)

acetylate histones, FMRFamide displaces CREB1 with CREB2 and recruits a histone deacetylase to deacetylate histones. When 5-HT and FMRFamide are given together, FMRFamide overrides 5-HT by recruiting CREB2 and the deacetylase to displace CREB1 and CBP, thereby inducing histone deacetylation and repression of C/EBP. Thus, the facilitatory and

---

Figure 4. (*A*, *B*) Injection of CRE oligonucleotides blocks 5-HT-induced long-term facilitation. (*A*) Summary of the blockade of the 5-HT-induced increase in EPSP amplitude by CRE injection. The height of each bar is the percentage change in the EPSP amplitude ± s.e.m. retested 24 hours after treatment. (A two-tailed t-test comparison of means indicated that the decrease in EPSP in cultures injected with CRE oligonucleolide is significantly different [$p< 0.05$] from the increase in the EPSP in the cells injected with either the mutant or NF-B oligonucleotides.) (*B*) Summary of the pooled data for short-term facilitation 24 hours after injection. In contrast to long-term facilitation, the 5-HT (5 μM) was applied after the EPSP was first depressed. Five stimuli were given with an interstimulus interval of 30 seconds, resulting in 70–80% depression in EPSP amplitude. Application of 5-HT after the fifth stimulus produced an increase in EPSP amplitude by the seventh stimulus. The increase in short-term facilitation was measured by calculating the percentage increase in the seventh EPSP amplitude as compared with the fifth EPSP amplitude. Because the facilitation here was of a depressed EPSP, the percentage facilitation is larger than the long-term, where only the nondepressed EPSP was examined. (*C*, *D*) 5-HT/IBMX leads to activation of CREB through the phosphorylation of PKA. (*C*) DNA constructs used for *trans*-activation experiments. DNA-binding domain of yeast GAL4 transcription factor (amino acids 1–147) fused to wild-type or mutated forms of mammalian CREB *trans*-activation domains (amino acids 1–261). Wild-type and mutated phosphorylation consensus sequences are indicated in single-letter amino acid code above constructs. The wild-type sequence (RRPS) is a phosphorylation substrate for both protein kinase A (PKA) and calcium/calmodulin-dependent kinase (CaM) at Ser-119. SA-119 contains a substitution of Ser-119 with Ala-119, preventing phosphorylation by either kinase. RA-117 contains a substitution of Arg-117 ~ with Ala-117, abolishing the PKA site, but leaving intact the CaMK consensus phosphorylation sequence. GAL4-CREB fusion proteins were constitutively expressed in pNEXd vector and injected with the reporter construct (GAL4-CAT) containing two copies of the GAL4 site upstream of human chorionic gonadotropin α-subunit gene driving the CAT gene. (*D*) Quantitative analysis of 5-HT-regulated CREB *trans*-activation. The wild-type (WT) GAL4-CREB *trans*-activator enhances 5-HT/IBMX-mediated expression of CAT, whereas mutant (SA-119 and RA-117) GAL4-CRED show no enhancement. Pairs of sensory clusters were injected and treated in parallel; each lane represents CAT activity of one sensory cluster. CAT activity is expressed as the percentage of substrate acetylated (X103); histograms show pooled area, N = number of animals, and value in parentheses is number of injected neurons. Mean, s.e.m., and p value from two-tailed paired test are shown. (*A*, *B* Reprinted, with permission from Dash et al. 1990; *C*, *D* reprinted, with permission, from Kaang et al. 1993.)

inhibitory modulatory transmitters that are important for long-term memory in *Aplysia* activate signal transduction pathways that alter nucleosome structure bidirectionally through acetylation and deacetylation of chromatin (Fig. 6).

MAPK ALSO CARRIES SIGNALS FROM THE SYNAPSE TO THE NUCLEUS

As discussed earlier, the transcriptional switch for the conversion of short- to long-term memory requires not only the activation of CREB1, but also the removal of the repressive action of CREB2, which lacks consensus sites for PKA phosphorylation (Bartsch et al. 1995). CREB2 does, however, have two conserved sites for MAPK phosphorylation. A number of features make MAPK an attractive candidate for playing a role in long-term facilitation. First, the MAPK system mediates graded responses to extracellular signals (Marshall 1995). For example, in PC12 cells, EGF causes transient activation of MAPK that results in cell proliferation, whereas NGF causes sustained activation of MAPK, which leads to its nuclear translocation and results in neuronal differentiation. This gradation of

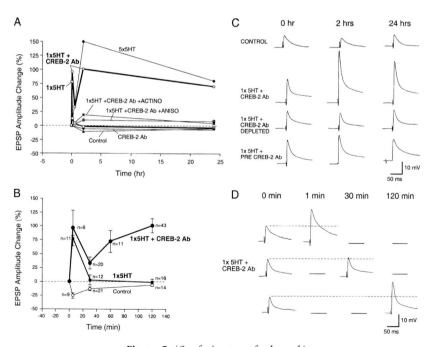

Figure 5. (*See facing page for legend.*)

response may be a molecular parallel to the gradation of synaptic strengthening in response to environmental stimuli, important for learning in both vertebrates and invertebrates. Similar to the PKA translocation, translocation of MAPK to the nucleus provides an attractive potential mechanism for the transition between the transcription-independent short-term forms and transcription-dependent long-term forms of memory. Furthermore, the MAPK pathway interacts with the cAMP pathway in PC12 cells and in neurons, indicating that MAPK might participate with PKA in the consolidation of long-term memory.

To test this possibility, Martin and colleagues (1997a) and Michael et al. (1998) examined the subcellular localization of an *Aplysia* ERK2 homolog in sensory-to-motor neuron cocultures during short- and long-term facilitation. Whereas MAPK immunoreactivity was predominantly

---

Figure 5. Time course of the effects of injection of ApCREB2 antiserum on short- and long-term facilitation. (A) Time course of EPSP amplitude changes recorded in motor neuron L7 in response to stimulation of the sensory neuron (expressed as percent change in the amplitude of the EPSP) after single and multiple applications of 5-HT to *Aplysia* sensorimotor neuron cocultures. Changes in EPSP amplitude after application of one 5-minute pulse of 5-MT (1 x 5-HT, short-term facilitation) and one 5-minute pulse of 5-HT paired with injection of anti-ApCREB2 antibodies (1 x 5-HT + CREB-2 Ab, both in boldface lines) are compared with changes in EPSP amplitude induced by five pulses of 5-HT (5 x 5-HT) at 2 and 24 hours. Whereas the EPSP facilitation decays rapidly after one pulse of 5-HT (with a return to baseline after 10 minutes), pairing one pulse with 5-HT with injection of aliti-APCREB2 antibodies induces a long-term facilitation paralleling that of 5 x 5-HT. This long-term facilitation is abolished by the application of the protein synthesis inhibitor anisomycin (1 x 5-HT + CREB-2 Ab + ANISO) or the RNA synthesis inhibitor actinomycin D (1 x 5-HT + CREB-2 Ab + ACTINO) during the training. The difference in EPSP amplitude at 2 hours between 5 x 5-HT and 1 x 5-HT + CREB-2 Ab may reflect the transient protein synthesis-dependent, but RNA synthesis-independent, component of long-term facilitation 2 hours after 5-HT stimulation. The controls are either untreated (*control*) or injected with ApCREB2 antiserum without 5-HT administration (CREB-2 Ab). (B) Comparison of the time course of the EPSP amplitude changes in the first 2 hours after application of a single 5-minute pulse of 5-HT with or without injection of CREB-2 antibody. The control cells were not exposed to 5-HT. (C) Example of EPSPs recorded in motoneuron L7 after stimulation of the sensory neuron before (*0 hr*) and 2 and 24 hours after 5-HT treatment. One pulse of 5-MT paired with the injection of an ApCREB2 antiserum induces a significant increase in EPSP amplitude at 2 and 24 hours, but injection with the preimmune serum (PRE-CREB-2 Ab) or depleted immune serum does not induce long-term facilitation. (D) Examples of EPSPs recorded at indicated times in cocultures injected with ApCREB2 antiserum paired with one 5-minute pulse of 5-HT. (Reprinted, with permission, from Bartsch et al. 1995.)

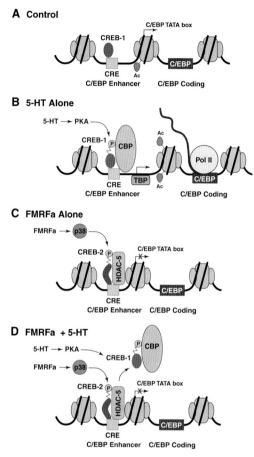

Figure 6. Diagram showing 5-HT and FMRFa bidirectionally regulate histone acetylation. (*A*) At the basal level, CREB1a resides on the C/EBP promoter and some lysine residues of histones are acetylated. (*B*) 5-HT, through PKA, phosphorylates CREB1 that binds to the C/EBP promoter. Phosphorylated CREB1 then forms a complex with CBP at the promoter. CBP then acetylates lysine residues of the histones (e.g., KS of H4). Acetylation modulates chromatin structure, enabling the transcription machinery to bind and induce gene expression. (*C*) FMRFa activates CREB2, which displaces CREB1 from the C/EBP promoter. HDAC5 is then recruited to deacetylate histones. As a result, the gene is repressed. (*D*) If the neuron is exposed to both FMRFa and 5-HT, CREBa is replaced by CREB2 at the promoter even though it might still be phosphorylated through the 5-HT-PKA pathway, and HDAC5 is then recruited to deacetylate histones, blocking gene induction. (Reprinted, with permission, from Guan et al. 2002.)

localized to the cytoplasm in both sensory and motor neurons during short-term facilitation, MAPK translocated into the nucleus of the presynaptic sensory neuron but not in the postsynaptic motor cell during

5-HT-induced long-term facilitation. Presynaptic but not postsynaptic nuclear translocation of MAPK was also triggered by elevations in intracellular cAMP, indicating that the cAMP pathway activates the MAPK pathway in a neuron-specific manner. Injection of either anti-MAPK antibodies or MAPK kinase inhibitors (PD98059) into the presynaptic sensory cell selectively blocked long-term facilitation without affecting short-term facilitation. Thus, MAPK appears to be necessary for the long-term form of facilitation. The involvement of MAPK in long-term plasticity may be quite general: Martin and associates (1997a) found that cAMP also activated MAPK in mouse hippocampal neurons, suggesting that MAPK may play a role in hippocampal long-term potentiation. The requirement for MAPK during hippocampal LTP has been shown by English and Sweatt (1996, 1997), who demonstrated that ERK1 is activated in CA1 pyramidal cells during LTP and that bath application of MAPK kinase inhibitors blocks LTP.

CONSOLIDATION OF LONG-TERM FACILITATION REQUIRES THE INDUCTION OF IMMEDIATE-EARLY GENES

The activation of adenylyl cyclase, the increase in cAMP concentration with the resulting dissociation of the catalytic subunit of PKA and its translocation to the nucleus, and the phosphorylation of CREB are all unaffected by inhibitors of protein synthesis. Where then does the protein synthesis-dependent step that characterizes the consolidation phase appear? Clearly, it must require an additional step—the activation of genes by CREB. To follow further the sequence of steps whereby CREB leads to the stable, self-perpetuating long-term process, Alberini and colleagues (1994) characterized the intermediary, immediate-early genes induced by cAMP and CREB. In a search for possible cAMP-dependent regulatory genes that might be interposed between constitutively expressed transcription factors and stable effector genes, Alberini and colleagues (1994) focused on the CCAAT-box-enhancer-binding protein (C/EBP) transcription factors. They cloned an *Aplysia* C/EBP homolog (ApC/EBP) and found that its expression was induced by exposure to 5-HT. Induction of ApC/EBP mRNA is rapid and transient, and occurs in the presence of protein synthesis inhibitors, characteristic for immediate-early genes (Fig. 7). To determine whether ApC/EBP is necessary for the induction and maintenance of long-term facilitation, Alberini and colleagues (1994) used three different approaches: First, they interfered with the binding of the ApC/EBP to its DNA-binding element (ERE) by injecting oligonucleotides that compete for the binding activity of the endoge-

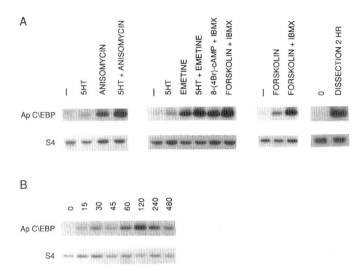

Figure 7. Induction of ApC/EBP mRNA. (*A*) ApC/EBP mRNA expression in CNS of untreated *Aplysia*, of *Aplysia* treated in vivo with the indicated drugs for 2 hours at 18°C, or dissected without treatment and kept at 18°C in culture medium. Four independent experiments are shown, in which 10 µg of total RNA extracted from CNSs of untreated (–) or treated *Aplysia*, as indicated, were electrophoresed, blotted, and hybridized with ^{32}P-labeled ApC/EBP (*top*) or S4 (*bottom*) probes. The latter encodes the *Aplysia* homolog of S4 ribosomal protein, which is constitutively expressed and used as a loading control. Zero indicates RNA extracted immediately after dissection of *Aplysia* CNS. Two-hour dissection represents RNA extracted from *Aplysia* CNS dissected and incubated in culture medium for 2 hours at 18°C. (*B*) Time course of ApC/EBP mRNA induction following 5-HT treatment. Times of treatment are indicated. Five mg of total RNA extracted form total CNS of in-vivo-treated *Aplysia* was analyzed as described in *B*. (Reprinted, with permission, from Alberini et al. 1994.)

nous ApC/EBP to its target sequence. Second, they blocked the synthesis of endogenous ApC/EBP by injecting ApC/EBP antisense RNA into the sensory cells. Third, they blocked ApC/EBP by injecting an anti-ApC/EBP antiserum into the sensory neurons. All of these blocked long-term facilitation but had no effect on short-term facilitation. Thus, the induction of ApC/EBP seems to serve as an intermediate component of a molecular switch activated during the consolidation period.

The existence of C/EBP, a cAMP- and CREB-regulated immediate-early gene that is itself a transcription factor and regulates other genes, leads to a model of sequential gene activation. CREB1a, CREB1b, CREB1c, and CREB2 represent the first level of control, since all are constitutively expressed. Stimuli that lead to long-term facilitation perturb the balance between CREB1-mediated activation and CREB2-mediated repression, through the action of PKA, MAPK, and possibly other kinases. This leads to the up-regulation of a family of immediate-early genes. Some of these immediate-early genes are transcription factors such as C/EBP; others are effectors, such as ubiquitin hydrolase, that contribute to consolidation by either extending the inducing signal or initiating the changes at the synapse that cause long-term facilitation.

SIGNALING EVENTS AND INITIATION OF SYNAPTIC GROWTH

How might this gene induction contribute to the growth of new sensory neuron synapses that accompanies long-term facilitation in *Aplysia*? Of the 15 early proteins Barzilai et al. (1989) observed to be specifically altered in expression during the acquisition of long-term facilitation, 6 have now been identified. Two proteins that increase (clathrin and tubulin) and 4 proteins that decrease their level of expression (NCAM-related cell adhesion molecules) seem to relate to structural changes.

Mayford et al. (1992) first focused on the four proteins, D1–D4, that decrease their expression in a transcriptionally dependent manner following the application of 5-HT or cAMP and found that they encoded different isoforms of an immunoglobulin-related cell adhesion molecule, designated apCAM, which shows greatest homology with NCAM in vertebrates and fasciclin II in *Drosophila*. Imaging of fluorescently labeled monoclonal antibodies to apCAM shows that not only is there a decrease in the level of expression, but that even preexisting protein is lost from the surface membrane of the sensory neurons within 1 hour after the addition of 5-HT. This transient modulation by 5-HT of cell adhesion molecules, therefore, may represent one of the early molecular steps required for initiating learning-related growth of synaptic connections. Indeed, blocking the expression of apCAM by a monoclonal antibody causes defasciculation, a step that appears to precede synapse formation in *Aplysia* (Keller and Schacher 1990).

To examine the mechanisms by which 5-HT modulates apCAM and the significance this modulation might have for the synaptic growth that is induced by 5-HT, Bailey et al. (1992a) combined thin-section electron microscopy with immunolabeling using a gold-conjugated monoclonal antibody specific to apCAM. They found that a 1-hour application of 5-HT led to a 50% decrease in the density of gold-labeled apCAM complexes at the surface of the sensory neuron. This down-regulation was particularly prominent at adherent processes of the sensory neurons and was achieved by a protein-synthesis-dependent activation of the endosomal pathway, leading to internalization and apparent degradation of apCAM. As is the case for the down-regulation at the level of expression, the 5-HT-induced internalization of apCAM can be simulated by cAMP. Concomitant with the down-regulation of apCAM, Hu et al. (1993) have demonstrated that, as part of this coordinated program for endocytosis, 5-HT and cAMP also induce an increase in the number of coated pits and coated vesicles in the sensory neurons and an increase in the expression of the light chain of clathrin (apClathrin). Since the apClathrin light chain contains the important functional domains of the two types of clathrin light chains found in vertebrates (LC_a and LC_b) which are thought to be essential for coated pit assembly and disassembly, the increase in clathrin may be an important component in the activation of the endocytic cycle required for the internalization of apCAM.

These initial effects of 5-HT on the surface and internal membrane systems of sensory neurons in *Aplysia* bear a striking similarity to the morphological changes induced in nonneuronal systems by protein growth factors, which suggests that modulatory transmitters important for learning, such as 5-HT, may serve a double function. In addition to producing a transient regulation of the excitability of neurons, with repeated or prolonged exposure they may also produce an action comparable to that of a growth factor, which results in a more persistent regulation of the architecture of the neuron.

To further define the mechanisms whereby 5-HT leads to apCAM down-regulation, Bailey and colleagues (1997) used epitope tags to examine the fate of the two apCAM isoforms (transmembrane and GPI-linked) and found that only the transmembrane form is internalized (Fig. 8). This internalization was blocked by overexpression of transmembrane apCAM with a point mutation in the two MAPK phosphorylation consensus sites, as well as by injection of a specific MAPK antagonist into sensory neurons. These data suggest that activation of the MAPK pathway is important for the internalization of apCAMs and may represent one of the initial and perhaps permissive stages of learning-related synaptic growth in *Aplysia*. Furthermore, the combined actions of MAPK both in the cytoplasm and in the nucleus suggest that MAPK plays multiple roles in long-lasting synaptic plasticity and appears to regulate each of the two distinctive processes that characterize the long-term process: activation of transcription and growth of new synaptic connections.

The differential down-regulation of the GPI-linked and transmembrane forms of apCAM raised the interesting possibility that learning-related synaptic growth in the adult may be initiated by an activity-dependent recruitment of specific isoforms of cell adhesion molecules. In

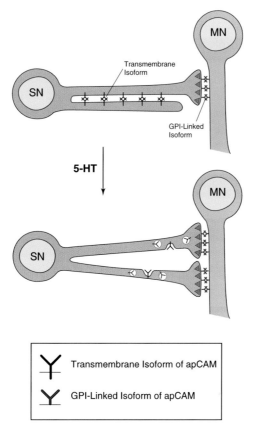

Figure 8. Regional specific down-regulation of the transmembrane isoform of apCAM. This model is based on the assumption that the relative concentration of the GPI-linked versus transmembrane isoforms of apCAM is highest at points of synaptic contact between the sensory neuron and motor neuron and reflects the results of studies done in dissociated cell culture. Thus, previously established connections might remain intact following exposure to 5-HT since they would be held in place by the adhesive, homophilic interactions of the GPI-linked isoforms, and the process of outgrowth from sensory neuron axons would be initiated by down-regulation of the transmembrane form at extrasynaptic sites of membrane apposition. In the intact ganglion, the axons of sensory neurons are likely to fasciculate not only with other sensory neurons, but also with the processes of other neurons and perhaps even glia. One of the attractive features of this model is that the mechanism for down-regulation is intrinsic to the sensory neurons. Thus, even if some of the sensory neuron axonal contacts in the intact ganglion were heterophilic in nature, i.e., with other neurons or glia, we would still expect the selective internalization of apCAM at the sensory neuron surface membrane at these sites of heterophilic apposition to destabilize adhesive contacts and to facilitate disassembly. (Reprinted, with permission, from Bailey et al. 1997.)

conjunction with the recruitment of the family of CREB isoforms described previously, this may provide a regulatory unit capable of acting sequentially at multiple nuclear, cytoplasmic, and plasma membrane sites during the early, inductive phases of the long-term process.

Is the down-regulation of the transmembrane isoform of apCAM required for the synaptic growth that accompanies 5-HT-induced long-term facilitation? To address this question, J.-H. Han et al. (in prep.) have recently overexpressed in *Aplysia* sensory neurons various HA-epitope-tagged recombinant apCAM molecules and investigated their effects on long-term facilitation. They found that overexpression of the transmembrane isoforms, but not the GPI-linked isoform of apCAM, blocked both the long-term synaptic facilitation and the long-term enhancement of synaptic growth. This inhibition of long-term facilitation by the overexpression of transmembrane apCAM was restored by interrupting the adhesive function of apCAM with the anti-HA antibody. In addition, long-term facilitation was completely blocked by the overexpression of the cytoplasmic tail portion of apCAM fused with GFP, designed to bind proteins such as MAP kinase p42 that might normally bind to the cytoplasm of the sensory neuron. These studies indicate that the extracellular domain of transmembrane apCAM has an inhibitory function that needs to be neutralized by internalization to induce long-term facilitation and that the cytoplasmic tail provides an interactive platform for both signal transduction and the internalization machinery.

Modulation of cell adhesion molecules appears to be a general mechanism in long-lasting forms of both developmental and learning-related synaptic plasticity (Martin and Kandel 1996; Murase and Schuman 1999; Benson et al. 2000). Thus, in forms of learning and memory such as passive avoidance learning in other animals in which chicks learn to suppress pecking behavior toward a bead that is coated with a bitter-tasting liquid, the synthesis of new cell adhesion molecules appears to be critical (Rose 1995). Relatedly, spatial learning and LTP are impaired in NCAM knock-out mice (Lüthl et al. 1994). Furthermore, increases in both the expression of polysialylated NCAM and the extracellular concentration of NCAM have been observed after hippocampal LTP, and LTP is inhibited by the application of either NCAM antibodies, or NCAM-blocking peptides, by the removal of polysialic acid by neuraminidase, and in NCAM-deficient mice (Becker et al. 1996; Muller et al. 1996; Cremer et al. 1998; Seki and Rutishauser 1998).

Addition of polysialic acid to NCAM may be functionally equivalent to the internalization of apCAM that occurs in *Aplysia* neurons in that both processes promote defasciculation, thereby allowing the growth of new synaptic connections (see also Landmesser et al. 1992; Tang et al. 1992). Other cell adhesion molecules, including the cadherins (Tang et al. 1998; Bozdagi et al. 2000; Tanaka et al. 2000) and integrins (Staubli et al. 1998), have also been implicated in regulating synaptic plasticity and learning and memory. Thus, in addition to their critical role in the development of the brain, cell adhesion molecules may also participate in the activity-dependent modulation of synaptic architecture during learning-related synaptic plasticity.

REGULATION OF LOCAL PROTEIN SYNTHESIS AND STABILIZATION OF SYNAPTIC GROWTH

Because most of the molecules present in the nervous system must be replaced at regular intervals, the persistent changes in neuronal function and structure thought to underlie the storage of long-term memory would seem to require some mechanism that can survive this molecular turnover (Crick 1984). One such mechanism is altered gene expression.

Studies in *Aplysia* have addressed this issue directly by examining in parallel the effects of inhibitors of protein and RNA synthesis on the structural and functional changes that accompany long-term facilitation. The long-term functional changes in synaptic strength evoked by 5-HT in dissociated cell cocultures of sensory neurons and motor neurons are dependent on new macromolecular synthesis during the period of transmitter application (Montarolo et al. 1986). Anisomycin and actinomycin-D block the long-lasting increases produced by 5-HT both in the synaptic potential and in varicosity number (Bailey et al. 1992b). These results suggest that macromolecular synthesis is required for the persistence of the long-lasting structural changes in the sensory cell and that this synthesis is coupled with the long-term functional modulation of sensory-to-motor synapses. A similar dependence on new protein synthesis has been reported for the in vivo structural changes in pleural sensory neurons, in which the cAMP-mediated increase in varicosity number can also be blocked by anisomycin (Nazif et al. 1991).

It is possible that the early and more transient aspects of synaptic structural plasticity, i.e., those that are less resistant to disruptive agents, may be limited to interactions involving preexisting macromolecules and may not require new synthesis. Some evidence for this notion comes from studies of long-term facilitation at the crayfish neuromuscular junction, where it has been shown that rapid (0.5–1 hour) morphological transformations involving modifications of preexisting presynaptic active zones, as well as the insertion of new release sites, can occur in the absence of the neuronal cell body and protein synthesis (Atwood et al. 1989; Wojtowicz et al. 1989). Even more rapid postsynaptic changes (within 15 minutes) have been described following the induction of long-term potentiation in the rat hippocampal slice (Lee et al. 1980; Chang and Greenough 1984). Such rapid presynaptic and postsynaptic remodeling might be triggered by a process of self-assembly from preexisting components. This would permit the local control of synaptic morphology and allow the initial structural changes to take place in minutes rather than hours or days. The stabilization of these initial structural changes might require an additional set of processes, including altered macromolecular synthesis, which would then facilitate the further elaboration of the newly formed synaptic connections, allowing them now to persist for the duration of the long-term memory.

Perhaps the best evidence for this idea has come from studies of synapse-specific long-term plasticity in *Aplysia* (Martin et al. 1997b; Casadio et al. 1999). In a culture system where a bifurcated *Aplysia* sensory neuron makes synapses with two spatially separated motor neurons, repeated application of 5-HT to one synapse produces a CREB-mediated, synapse-specific long-term facilitation that is accompanied by the growth of new synaptic connections and persists for at least 72 hours. This long-term facilitation, as well as the long-lasting synaptic growth, can be captured by a single pulse of 5-HT applied at the opposite sensory-to-motor neuron synapse. In contrast to the synapse-specific forms, cell-wide long-term facilitation generated by repeated pulses of 5-HT at the cell body is not associated with growth and does not persist beyond 48 hours. However, this cell-wide facilitation also can be captured and growth can be induced in a synapse-specific manner by a single pulse of 5-HT applied to one of the peripheral synapses.

CREB-mediated transcription thus appears to be necessary for the establishment of all four forms of long-term facilitation and for the initial maintenance of the synaptic plasticity at 24 hours. However, CREB-mediated transcription is not sufficient to maintain the changes beyond this time. To obtain persistent facilitation, and specifically to obtain the growth of new synaptic connections, one needs, in addition to CREB-mediated transcription, a marking signal produced by a single pulse of 5-HT applied to the synapse. This single pulse of 5-HT has at least two marking functions. First, it produces a PKA-mediated covalent modification that marks the captured synapse for growth. Second, it stimulates rapamycin-sensitive local protein synthesis, which is required for the long-term maintenance of the plasticity and stabilization of the growth beyond 24 hours.

Casadio et al. (1999) found that when the protein synthesis inhibitor emetine was applied to the synapse that received a single pulse of 5-HT, there was facilitation at the marked synapse at 24 hours, but this did not persist to 72 hours. Thus, whereas local protein synthesis is not required for synaptic capture at 24 hours, it is required at 72 hours. These findings suggested that synaptic growth required local protein synthesis and that the facilitation at 72 hours was blocked because the growth was blocked by inhibiting protein synthesis. To test this possibility, Casadio et al. (1999) imaged fluorescently labeled sensory neuron synapses and then treated one branch with 5 pulses of 5-HT and the other branch with a single pulse of 5-HT in the presence of inhibitors of protein synthesis. They found that although local protein synthesis inhibition blocked facilitation at 72 hours, it did not block synaptic growth at 24 hours. However, at the captured synapse the growth did not persist but returned to baseline by 72 hours. These results indicate that local protein synthesis is not required to initiate synaptic growth but is required for stabilization and persistence of that growth.

The finding of two distinct components for the marking signal first suggested that there is a mechanistic distinction between the initiation of long-term facilitation and synaptic growth (which require central transcription and translation but do not require local protein synthesis) and the stable maintenance of the long-term functional and structural changes which are dependent on local protein synthesis. What might constitute the mark necessary for stabilizing long-term facilitation? Since mRNAs are made in the cell body, the need for the local translation of some mRNAs suggests that these mRNAs may be dormant before they reach the marked synapse. If that is true, then the synaptic mark for stabilization might be a regulator of translation that is capable of activating mRNAs that are translationally dormant.

In search for components of the synaptic marking machinery required for stabilization of synapse-specific facilitation, K. Si et al. (in prep.) have recently identified the *Aplysia* homolog of CPEB (cytoplasmic polyadenylation element-binding protein), a protein capable of activating dormant mRNAs. This novel, neuron-specific isoform of CPEB is present in the processes of sensory neurons and is induced in the process by a single pulse of 5-HT. The induction of apCPEB is independent of transcription but requires new protein synthesis and is sensitive to rapamycin and to inhibitors of P13 kinase. Moreover, the induction of CPEB coincides with the polyadenylation of neuronal actin, and blocking apCPEB locally at the activated synapse blocks the long-term maintenance of synaptic facilitation but not its early expression at 24 hours. Thus, apCPEB has all the properties required of the local protein synthesis-dependent component of marking and supports the idea that there are separate mechanisms for initiation of the long-term process and its stabilization.

How might ApCPEB stabilize the late phase of long-term facilitation? As outlined above, the stability of long-term facilitation seems to result from the persistence of the structural changes in the synapses between sensory and motor neurons, the decay of which parallels the decay of the behavioral memory. These structural changes include the remodeling of preexisting facilitated synapses, as well as the growth and establishment of new synaptic connections. Reorganization and growth of new synapses has two broad requirements: (1) structural (changes in shape, size, and number) and (2) regulatory (where and when to grow). The genes involved in both of these aspects of synaptic growth might be potential targets of apCPEB. The structural aspects of the synapses are dynamically controlled by reorganization of the cytoskeleton, which can be achieved either by redistribution of preexisting cytoskeletal components or by their local synthesis. The observation that N-actin and Tα1 tubulin (K.C. Martin et al., unpubl.) are present in the peripheral population of mRNAs at the synapse and can be polyadenylated in response to 5-HT suggests that at least some of the structural components for synaptic growth can be controlled through apCPEB-mediated local synthesis (Kim and Lisman 1999). In addition, recently CPEB has been found to be involved in the regulation of local synthesis of EphA2 (Brittis et al. 2002), a member of the family of receptor tyrosine kinases, which have been implicated in axonal pathfinding and the formation of excitatory synapses in the mammalian brain. Thus, CPEB might contribute to the stabilization of learning-related synaptic growth by controlling the synthesis of both the structural molecules such as tubulin and N-actin and the regulatory molecules such as CAMK11 and members of the Ephrin family.

AN OVERALL VIEW

It is becoming increasingly clear that synaptic growth and synapse formation may represent one of the key signatures of the long-term memory process. The morphological correspondence between the studies of long-term sensitization in *Aplysia* and LTP in the mammalian hippocampus indicates that learning may resemble a process of neuronal growth and differentiation across a broad segment of the animal kingdom and suggests that new synapse formation may be a highly conserved feature for the storage of both implicit and explicit forms of long-term memory. Indeed, one of the unifying principles emerging from these studies is that despite the different ways by which each form of memory is induced, the subsequent steps required for conversion of their short-term memory to one of longer duration may be similar. This apparent similarity in some of the molecular steps may be because for both implicit and explicit memory storage, the synaptic growth may represent the final and self-sustaining change that stabilizes the long-term process. Despite this association, surprisingly little is known about the molecular mechanisms that underlie learning-related structural plasticity. Recent studies of the synaptic growth that accompanies long-term sensitization in *Aplysia* have begun to characterize the sequence of molecular events responsible for both the initiation and persistence of the structural change. This, in turn, has revealed that specific molecules and mechanisms important for cell growth and cytoplasm expansion during the development of the nervous system can be reutilized in the adult for the purposes of synaptic plasticity and memory storage.

For example, these studies indicate that long-term memory involves the flow of information from receptors on the cell surface to the genome, as seen in other processes of cell differentiation and growth (Fig. 9). Such changes may reflect a recruitment by environmental stimuli of developmental processes that are latent or inhibited in the fully differentiated neuron. Indeed, an increasing body of evidence suggests that the cell and molecular changes accompanying long-term memory storage share several features in common with the cascade of events that underlie neuronal differentiation and development. In both cases, there is a requirement for new protein and mRNA synthesis. These alterations can be initiated in the long-term process by modulatory transmitters that, in this respect, appear to mimic the effects of growth factors and hormones during the cell cycle and differentiation. Thus, modulatory transmitters important for learning not only activate a cascade of cytoplasmic events required for the short-term process, but also induce a genomic cascade by which the transmitter can exert a long-term regulation over both the excitability and structure of the neuron through changes in gene expression.

Studies in *Aplysia* have further demonstrated that the earliest stages of long-term memory formation are associated with modulation of an immunoglobulin-related cell adhesion molecule homologous to NCAM. With the emergence of the nervous system, the *Aplysia* NCAM becomes expressed exclusively in neurons and is specifically enriched at synapses. These cell adhesion molecules are maintained into adulthood, at which point they can be down-regulated by 5-HT, a modulatory transmitter important for both sensitization and classic conditioning in *Aplysia,* and by cAMP, a second messenger activated by 5-HT. This down-regulation appears to serve as a preliminary and permissive step for the growth of synaptic connections that accompanies the long-term process. Thus, a molecule used during development for cell adhesion and axon outgrowth

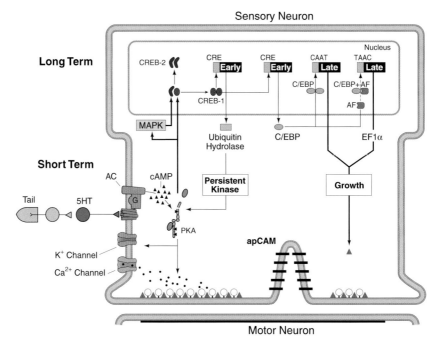

Figure 9. Effects of short- and long-term sensitization on the monosynaptic component of the gill-withdrawal reflex of *Aplysia*. In short-term sensitization (lasting minutes to hours) a single tail shock causes a transient release of serotonin that leads to covalent modification of preexisting proteins. The serotonin acts on a transmembrane serotonin receptor to activate the enzyme adenylyl cyclase (AC), which converts ATP to the second-messenger cAMP. In turn, cAMP recruits the cAMP-dependent protein kinase A (PKA) by binding to the regulatory subunits (*spindles*), causing them to dissociate from and free the catalytic subunits (*ovals*). These subunits can then phosphorylate substrates (channels and exocytosis machinery) in the presynaptic terminals, leading to enhanced transmitter availability and release. In long-term sensitization, repeated stimulation causes the level of cAMP to rise and persist for several minutes. The catalytic subunits can then translocate to the nucleus, and recruit the mitogen-activated protein kinase (MAPK). In the nucleus, PKA and MAPK phosphorylate and activate the cAMP response element-binding (CREB) protein and remove the repressive action of CREB-2, an inhibitor of CREB-1. CREB-1 in turn activates several immediate-response genes, including a ubiquitin hydrolase necessary for regulated proteolysis of the regulatory subunit of PKA. Cleavage of the (inhibitory) regulatory subunit results in persistent activity of PKA, leading to persistent phosphorylation of the substrate proteins of PKA. A second immediate-response gene activated by CREB-1 is C/EBP, which acts both as a homodimer and as a heterodimer with activating factor (AF) to activate downstream genes (including elongation factor 1a [EF1α]) that lead to the growth of new synaptic connections. (Reprinted, with permission, from Kandel 2001.)

is retained into adulthood, at which point it seems to restrain or inhibit growth until the molecule is rapidly and transiently decreased at the cell surface by a modulatory transmitter important for learning. The finding that 5-HT leads to the rapid down-regulation of only one isoform of apCAM (the transmembrane isoform) and not the others (the GPI-linked isoforms) raises the interesting possibility that learning-related synaptic growth in the adult may be initiated by an activity-dependent recruitment of specific isoforms of adhesion molecules, similar to the modulation of cell-surface receptors during the fine-tuning of synaptic connections in the developing nervous system. One consequence of isoform recruitment is that it would allow neuronal activity to regulate the surface expression of each isoform, a process that might take on additional functional significance if these surface molecules were distributed differentially along the three-dimensional extent of the neuron. These studies also suggest that processing and storage of information in the nervous system may rely on the same mechanisms utilized by other cells in the body to organize and regulate membrane trafficking important for growth.

Finally, insights from the molecular studies of learning and memory in *Aplysia* suggest that the critical time window for macromolecular synthesis that is a ubiquitous feature of long-term memory storage may be explained by a cascade of gene activation whereby one or more immediate-early genes control the transcription of late effector genes. The biological significance of an immediate-early gene-dependent response in long-term plasticity may reside in the necessity of a convergent checkpoint that turns on a genetic program, similar to the cascade of gene activation during cell differentiation. As is the case for development, in long-term memory a convergent checkpoint and cascade of gene activation may be critical to preserve important functions that ultimately rely on a small number of cells. Critical time windows have been previously described in other contexts, especially as part of developmental processes. For example, establishment of the differentiated state in DNA viruses often requires a sequence of gene activation whereby early regulatory genes lead to the maintained expression of later effector genes. A similar time window is evident in the later stages of *Drosophila* development where the steroid hormone ecdysone induces growth and molting by altering the expression of early genes that turn on the expression of later genes (Ashburner 1990).

The similarity between these critical periods and the one found in long-term memory suggests that aspects of the regulatory mechanisms underlying learning-related synaptic plasticity in the adult may eventually be understood in the context of the basic molecular logic used to refine synaptic connections during the later stages of neuronal development. Both processes appear to share a cascade of gene activation, with a critical time window during which the differentiated state is still labile and can be modified. That this feature is particularly well-developed in neurons, which can both grow and retract their synaptic connections on appropriate target cells in an activity-dependent fashion, may underlie their unique ability to respond to environmental stimuli which is essential for learning and memory storage.

ACKNOWLEDGEMENTS

This work is supported in part by the National Institutes of Health grant MH37134 to C.H.B., MH26212 to R.D.H. and by a grant from the Howard Hughes Medical Institute to E.R.K.

REFERENCES

Abel T.P.V., Nguyen V., Barad M., Deuel T.A., Kandel E.R., and Bourtchouladze R. 1997. Genetic demonstration of a role for PKA in the late phase of LTP and in hippocampus-based long-term memory. *Cell* **88:** 615–626.

Alberini C.M., Ghirardi M., Metz R., and Kandel E.R. 1994. C/EBP is an immediate-early gene required for the consolidation of long-term facilitation in *Aplysia*. *Cell* **76:** 1099–1114.

Antonova I., Arancio O., Trillat A.-C., Wang H.G., Zablow L., Udo H., Kandel E.R., and Hawkins R.D. 2001. Rapid increase in clusters of presynaptic proteins at onset of long-lasting potentiation. *Science* **294:** 1547–1550.

Ashburner M. 1990. Puff, genes and hormones revisited. *Cell* **61:** 1–3.

Atwood H.L., Dixon D., and Wojtowicz J.M. 1989. Rapid introduction of long-lasting synaptic changes at crustacean neuromuscular junctions. *J. Neurobiol.* **20:** 373–385.

Bacskai B.J., Hochner B., Mahaut-Smith M., Adams S.R., Kaang B.K., Kandel E.R., and Tsien R.Y. 1993. Spatially resolved dynamics of cAMP and protein kinase A subunits in *Aplysia* sensory neurons. *Science* **260:** 222–226.

Bailey C.H. and Chen M. 1983. Morphological basis of long-term habituation and sensitization in *Aplysia*. *Science* **220:** 91–93.

———. 1988a. Long-term memory in *Aplysia* modulates the total number of varicosities of single identified sensory neurons. *Proc. Natl. Acad. Sci.* **85:** 2373–2377.

———. 1988b. Long-term sensitization in *Aplysia* increases the number of presynaptic contacts onto the identified gill motor neuron L7. *Proc. Natl. Acad. Sci.* **85:** 9356–9359.

———. 1989. Time course of structural changes at identified sensory neuron synapses during long-term in *Aplysia*. *J. Neurosci.* **9:** 1774–1780.

Bailey C.H. and Kandel E.R. 1993. Structural changes accompanying memory storage. *Annu. Rev. Physiol.* **55:** 397–426.

Bailey C.H., Bartsch D., and Kandel E.R. 1996. Toward a molecular definition of long-term memory storage. *Proc. Natl. Acad. Sci.* **93:** 13445–13452.

Bailey C.H., Chen M., Keller F., and Kandel E.R. 1992a. Serotonin-mediated endocytosis of apCAM: An early step of learning-related synaptic growth in *Aplysia*. *Science* **256:** 645–649.

Bailey C.H., Montarolo P.G., Chen M., Kandel E.R., and Schacher S. 1992b. Inhibitors of protein and RNA synthesis block the structural changes that accompany long-term heterosynaptic plasticity in the sensory neurons of *Aplysia*. *Neuron* **9:** 749–758.

Bailey C.H., Kanng B.K., Chen M., Marin C., Lim C.S., Casadio A., and Kandel E.R. 1997. Mutation in the phosphorylation sites of MAP kinase blocks learning-related internalization of apCAM in *Aplysia* sensory neurons. *Neuron* **18:** 913–924.

Bartsch D., Ghirardi M., Casadio A., Giustetto M., Karl K.A., Zhu H., and Kandel E.R. 2000. Enhancement of memory-related long-term facilitation by ApAF, a novel transcription factor that acts downstream from both CREB1 and CREB2. *Cell* **103:** 595–608.

Bartsch D., Ghirardi M., Skehel P.A., Karl K.A., Herder S.P., Chen A., Bailey C.H., and Kandel E.R. 1995. *Aplysia* CREB2 represses long-term facilitation: Relief of repression converts transient facilitation into long-term functional and structural changes. *Cell* **83:** 979–992.

Barzilai A., Kennedy T.E., Sweatt J.D., and Kandel E.R. 1989. 5-HT modulates protein synthesis and the expression of specific proteins during long-term facilitation in *Aplysia* sensory neurons. *Neuron* **2:** 1577–1586.

Baudry M. and Davis J.L., Eds. 1991. *Long-term potentiation: A debate of current issues*. MIT Press, Cambridge, Massachusetts.

Becker C.G., Artola A., Gerardy-Schahn R., Becker T., Welzl H., and Schacher M. 1996. The polysialic acid modification of the neural cell adhesion molecule is involved in spatial learning and hippocampal long-term potentiation. *J. Neurosci. Res.* **45:** 143–152.

Benson D.L., Schnapp L.M., Shapiro L., and Huntley G.W. 2000. Making memories stick: Cell adhesive molecules in synaptic plasticity. *Trends Cell Biol.* **10:** 473–480.

Betz W.J. and Bewick G.S. 1992. Optical analysis of synaptic vesicle recycling at the frog neuromuscular junction. *Science* **255:** 200–203.

Bolshakov V.Y., Golan H., Kandel E.R., and Siegelbaum S.A. 1997. Recruitment of new sites of synaptic transmission during the cAMP-dependent late phase of LTP at CA3-CA1 synapses in the hippocampus. *Neuron* **19:** 635–375.

Bozdagi O., Shau W., Tanaka H., Benson D.L., and Huntley G.W. 2000. Increasing numbers of synaptic puncta during late-phase LTP: N-cadherin is synthesized, recruited to synaptic sites and required for potentiation. *Neuron* **28:** 245–259.

Brittis P.A., Lu, Q., and Flanagan J.G. 2002. Axonal protein synthesis provides a mechanism for localized regulation at an intermediate target. *Cell* **110:** 223–235.

Buchs P.A. and Muller D. 1996. Induction of long-term potentiation is associated with major ultrastructural changes of activated synapses. *Proc. Natl. Acad. Sci.* **93:** 8040–8045.

Casadio A., Martin K.C., Giustetto M., Zhu H., Chen M., Bartsch D., Bailey C.H., and Kandel E.R. 1999. A transient, neuron-wide form of CREB-mediated long-term facilitation can be stabilized at specific synapses by local protein synthesis. *Cell* **99:** 221–237.

Castellucci V.F., Blumenfeld H., Goelet P., and Kandel E.R. 1989. Inhibitor of protein synthesis blocks long-term behavioral sensitization in the isolated gill-withdrawal reflex of *Aplysia*. *Science* **220:** 91–93.

Chang F.-L.F. and Greenough W.T. 1984. Transient and enduring morphological correlates of synaptic activity and efficacy change in the rat hippocampal slice. *Brain Res.* **309:** 35–46.

Chang F.L., Isaac K.R., and Greenough W.T. 1991. Synapse formation occurs in asssocia-tion with the induction of long-term potentiation in two-year old rat hippocampus in vitro. *Neurobiol. Aging* **12:** 517–522.

Chavis P., Mollard P., Bockaert J., and Manzoni O. 1998. Visualization of cyclic AMP-reg-ulated presynaptic activity of cerebellar granule cells. *Neuron* **20:** 773–781

Colicos M.A., Collins B.E., Sailor M.J., and Goda Y. 2001. Remodeling of synaptic actin induced by photoconductive stimulation. *Cell* **107:** 605–616.

Crick F. 1984. Memory and molecular turnover. *Nature* **312:** 101–102.

Cremer H., Chazai G., Carleton A., Gordis C., Vincent J.D., and Lledo P.M. 1998. Long-term but not short-term plasticity at mossy fiber synapses is impaired in neural cell adhesion molecule deficient mice. *Proc. Natl. Acad. Sci.* **95:** 13242–13247.

Dale N., Kandel E.R., and Schacher S. 1987. Serotonin produces long-term changes in the excitability of *Aplysia* sensory neurons in culture that depend on new protein synthe-sis. *J. Neurosci.* **7:** 2232–2238.

Dash P.K., Hochner B., and Kandel E.R. 1990. Injection of the cAMP-responsive element into the nucleus of *Aplysia* sensory neurons blocks long-term facilitation. *Nature* **345:** 718–721.

Davis R.L. 1996. Physiology and biochemistry of *Drosophila* learning mutants. *Physiol. Rev.* **76:** 299–317.

Davis R.L, Cherry J., Dauwalder B., Han P.L., and Skoulakis E. 1995. The cyclic AMP sys-tem and *Drosophila* learning. *Mol. Cell. Biochem.* **149–150:** 271–278.

Drain P., Folkers E., and Quinn W.G. 1991. cAMP-dependent protein kinase and the dis-ruption of learning in transgenic flies. *Neuron* **6:** 71–82.

Durand G.M., Kovalchuk Y., and Konnerth A. 1996. Long-term potentiation and func-tional synapses induction in developing hippocampus. *Nature* **381:** 71–75.

Engert F. and Bonhoeffer T. 1999. Dendritic spine changes associated with hippocampal long-term synaptic plasticity. *Nature* **399:** 66–70.

English J.D. and Sweatt J.D. 1996. Activation of p42 mitogen-activated protein kinase in hippocampal long-term potentiation. *J. Biol. Chem.* **271:** 24329–24332.

———. 1997. A requirement for the mitogen-activated protein kinase cascade in hip-pocampal long-term potentiation. *J. Biol. Chem.* **272:** 19103–19106.

Fiala J.C., Allwardt B., and Harris K.M. 2002. Dendritic spines do not split during hip-pocampal LTP or maturation. *Nat. Neurosci.* **4:** 297–298.

Frey U., Huang Y.-Y., and Kandel E.R. 1993. Effects of cAMP simulate a late stage of LTP in hippocampal CA1 neurons. *Science* **260:** 1661–1664.

Frost W.N., Castellucci V.F., Hawkins R.D., and Kandel E.R. 1985. Monosynaptic connec-tions made by the sensory neurons of the gill- and siphon-withdrawal reflex in *Aplysia* participates in the storage of long-term memory for sensitization. *Proc. Natl. Acad. Sci.* **82:** 8266–8269.

Geinisman Y. 1993. Perforated axospinous synapses with multiple, completely, partitioned transmission zones: Probable structural intermediates in synaptic plasticity. *Hippocampus* **3:** 447–463.

———. 2000. Structural synaptic modifications associated with hippocampal LTP and behavioral learning. *Cereb. Cortex* **10:** 952–962.

Geinisman Y., de Toledo-Morrell L., and Morrell F. 1991. Induction of long-term potenti-ation is associated with an increase in the number of axospinous synapses with seg-mented postsynaptic densities. *Brain Res.* **566:** 77–88.

Geinisman Y., de Toledo-Morrell L., Morrell F., Persina I.S., and Beatty M.A. 1996. Synapse restructuring associated with the maintenance phase of hippocampal long-term potentiation. *J. Comp. Neurol.* **368:** 413–423.

Geinisman Y., de-Toledo-Morrell L., Morrell F., Heller R.E., Rossi M., and Parshall R.F. 1993. Structural synaptic correlates of long-term potentiation: Formation of axospinous synapses with multiple, completely partitioned transmission zones. *Hippocampus* **3:** 435–446.

Ghirardi M., Braha O., Hochner B., Montarolo P.G., Kandel E.R., and Dale N. 1992. Roles of PKA and PKC in facilitation of evoked and spontaneous transmitter release at depressed and nondepressed synapses in *Aplysia* sensory neurons. *Neuron* **9:** 479–489.

Glanzman D.L., Kandel E.R., and Schacher S. 1990. Target-dependent structural changes accompanying long-term synaptic facilitation in *Aplysia* neurons. *Science* **249:** 779–802.

Glanzman D.L., Mackey S.L., Hawkins R.D., Dyke A.M., Lloyd P.E., and Kandel E.R. 1989. Depletion of serotonin in the nervous system of *Aplysia* reduces the behavioral enhancement of gill withdrawal as well as the heterosynaptic facilitation produced by tail shock. *J. Neurosci.* **9:** 4200–4213.

Gonzalez G.A. and Montminy M.R. 1989. Cyclic AMP stimulates somatostatin gene transcription by phosphorylation of CREB at serine 133. *Cell* **59:** 675–680.

Guan Z., Giustetto M., Lomvardas S., Kim J.-H., Miniaci M.D., Schwartz J.H., Thanos D., and Kandel E.R. 2002. Integration of long-term memory related synaptic plasticity involves bidirectional regulation of gene expression and chromatin structure. *Cell* **111:** 483–493.

Hai T.W., Liu F., Coukos W.J., and Green M.K. 1989. Transcription factor ATF cDNA clones: An extensive family of leucine zipper proteins able to selectively form DNA-binding heterodimers (erratum in *Genes Dev.* [1990] **4:** 682). *Genes Dev.* **3:** 2083–2090.

Hochner B., Klein M., Schacher S., and Kandel E.R. 1986. Additional components in the cellular mechanisms of presynaptic facilitation contributes to behavioral dishabituation in *Aplysia*. *Proc. Natl. Acad. Sci.* **83:** 8794–8798.

Hoeffler J.P., Meyer T.E., Yun Y., Jameson J.L., and Habener J.F. 1988. Cyclic AMP-responsive DNA-binding protein: Structure based on a cloned placental cDNA. *Science* **242:** 1430–1433.

Hu Y., Barzilai A., Chen M., Bailey C.H., and Kandel E.R. 1993. 5-HT and cAMP induce the formation of coated pits and vesicles and increase the expression of clathrin light chain in sensory neurons of *Aplysia*. *Neuron* **10:** 921–929.

Huang Y.-Y., Li X.C., and Kandel E.R. 1994. cAMP contributes to mossy fiber LTP by initiating both a covalently mediated early phase and macromolecular synthesis-dependent late phase. *Cell* **79:** 69–79.

Isaac J.T.R., Crair M.C., Nicoll R.A., and Malenka R.C. 1997. Silent synapses during development of thalamocortical inputs. *Neuron* **18:** 1–20.

Kandel E.R. 2001. The molecular biology of memory storage: A dialogue between genes and synapses. *Science* **294:** 1030–1038.

Kaang B.K, Kandel E.R., and Grant S.G. 1993. Activation of cAMP-responsive genes by stimuli that produce long-term facilitation in *Aplysia* sensory neurons. *Neuron* **10:** 427–435.

Karpinski B.A., Morle G.D., Huggenvik J., Uhler M.D., and Leiden J.M. 1992. Molecular cloning of human CREB-2: An ATF/CREB transcription factor that can negatively regulate transcription from the cAMP response element. *Proc. Natl. Acad. Sci.* **89:** 4820–4824.

Keller Y. and Schacher S. 1990. Neuron-specific membrane glycoproteins promoting neurite fasciculation in *Aplysia californica*. *J. Cell Biol.* **111:** 2637–2650.

Klein M. and Kandel E.R. 1980. Mechanism of calcium current modulation underlying presynaptic facilitation and behavioral sensitization in *Aplysia*. *Proc. Natl. Acad. Sci.* **77:** 6912–6916.

Kim C.H. and Lisman J.E. 1999. A role of actin filaments in synaptic transmission and long-term potentiation. *J. Neurosci.* **19:** 4314–4321.

Korn H., Sur C., Charpier S., Legendre P., and Faber D.S. 1994. The one-vesicle hypothesis and multivesicular release. In *Molecular and cellular mechanisms of neurotransmitter release* (ed. L. Stjarne et al.), pp. 301–322. Raven Press, New York.

Kullman D.M. and Siegelbaum S.A. 1995. The site of expression of NMDA receptor-dependent long-term potentiation: New fuel for an old fire. *Neuron* **15:** 997–1002.

Landmesser I., Dahm I., Tang J., and Rutishauser U. 1992. Polysialic acid as a regulator of intramuscular nerve branching during embryonic development. *Neuron* **4:** 655–667.

Lee K., Schottler F., Oliver M., and Lynch G. 1980. Brief bursts of high-frequency stimulation produce two types of structural change in rat hippocampus. *J. Neurophysiol.* **4:** 247–258.

Liao D., Hessler N.A., and Malinow R. 1995. Activation of postsynaptically silent synapses during pairing-induced LTP in CA1 region of hippocampal slice. *Nature* **375:** 400–404.

Luscher C., Nicoll R.A., Malenka R.C., and Muller D. 2000. Synaptic plasticity and dynamic modulation of the postsynaptic membrane. *Nat. Neurosci.* **3:** 47–58.

Lüthl A., Laurent J.-P., Figurov A., Muller D., and Schachner D. 1994. Hippocampal long-term potentiation and neural cell adhesion molecules L1 and NCAM. *Nature* **372:** 777–779.

Ma L., Zablow L., Kandel E.R., and Siegelbaum S.A. 1999. Cyclic AMP induces functional presynaptic boutons in hippocampal CA3-CA1 neuronal cultures. *Nat. Neurosci.* **2:** 24–30.

Malenka R.C. and Nicoll R.A. 1997. Silent synapses speak up. *Neuron* **19:** 473–476.

———. 1999. Long-term potentiation: A decade of progress. *Science* **285:** 1870–1874.

Maletic-Savatic M., Malinow R., and Svoboda K. 1999. Rapid dendritic morphogenesis in CA1 hippocampal dendrites induced by synaptic activity. *Science* **283:** 1923–1927.

Marshall C.J. 1995. Specificity of receptor tyrosine kinase signaling: Transient versus sustained extracellular signal-regulated kinase activation. *Cell* **80:** 179–185.

Martin K.C. and Kandel E.R. 1996. Cell adhesion molecules, CREB and the formation of new synaptic connections during development and learning. *Neuron* **17:** 567–570.

Martin K.C., Michael D., Rose J.C., Barad M., Casadio A., Zhu H., and Kandel E.R. 1997a. MAP kinase translocates into the nucleus of the presynaptic cell and is required for long-term facilitation in *Aplysia*. *Neuron* **18:** 899–912.

Martin K.C., Casadio A., Zhu H., Yaping E., Rose J., Chen M., Bailey C.H., and Kandel E.R. 1997b. Synapse-specific long-term facilitation of *Aplysia* sensory somatic synapses: A function for local protein synthesis memory storage. *Cell* **91:** 927–938.

Mayford M., Barzilai A., Keller F., Schacher S., and Kandel E.R. 1992. Modulation of an NCAM-related adhesion molecule with long-term synaptic plasticity in *Aplysia*. *Science* **256:** 638–644.

Michael D., Martin K.C., Seger R., Ning M.M., Baston R., and Kandel E.R. 1998. Repeated pulses of serotonin required for long-term facilitation activate mitogen-activated protein kinase in sensory neurons of *Aplysia*. *Proc. Natl. Acad. Sci.* **95:** 1864–1869.

Montarolo P.G., Goelet P., Castellucci V.F., Morgan J., Kandel E.R., and Schacher S. 1986.

A critical period for macromolecular synthesis in long-term heterosynaptic facilitation in *Aplysia. Science* **234:** 1249–1254.

Montminy M.R., Sevarino K.A., Wagner J.A., Mandel G., and Goodman R.H. 1986. Identification of a cyclic-AMP-responsive element within the rat somatostatin gene. *Proc. Natl. Acad. Sci.* **83:** 6682–6686.

Muller D., Wang C., Skibo G., Toni N., Cremer H., Calaora V., Rougan G., and Kiss J.Z. 1996. PSA-NCAM is required for activity-induced synaptic plasticity. *Neuron* **17:** 413–422.

Murase S. and Schuman E.M. 1999. The role of cell adhesion molecules in synaptic plasticity and memory. *Curr. Opin. Cell Biol.* **11:** 549–553.

Nazif F.A., Byrne J.H., and Cleary L.J. 1991. cAMP induces long-term morphological changes in sensory neurons of *Aplysia. Brain Res.* **539:** 324–327.

Rose S.P. 1995. Cell-adhesion molecules, glucocrticoids and long-term memory formation. *Trends Neurosci.* **18:** 502–506.

Rusakov D.A., Davies H.A., Harrison E., Diana G., Richter-Levin G., Bliss T.V., and Stewart M.G. 1997. Ultrastructural synaptic correlates of spatial learning in rat hippocampus. *Neuroscience* **80:** 69–77.

Schacher S., Castellucci V.F., and Kandel E.R. 1988. cAMP evokes long-term facilitation *Aplysia* sensory neurons that requires new protein synthesis. *Science* **240:** 1667–1669.

Schwartz H., Castellucci V.F., and Kandel E.R. 1971. Functions of identified neurons and synapses in abdominal ganglion of *Aplysia* in absence of protein synthesis. *J. Neurophysiol.* **34:** 9639–9653.

Scholz K.P. and Byrne J.H. 1987. Long-term sensitization in *Aplysia:* Biophysical correlates in tail sensory neurons. *Science* **235:** 685–687.

Seki T. and Rutishauser U. 1998. Removal of polysialic acid-neural adhesion molecule induces aberrant mossy fiber innervation and ectopic synaptogeneiss in the hippocampus. *J. Neurosci.* **18:** 3757–3766.

Sorra K.E. and Harris K.M. 1998. Stability in synapse number and size at 2 hr after long-term potentiation in hippocampal area CA1. *J. Neurosci.* **18:** 658–671.

Staubli U., Chun D., and Lynch G. 1993. Time-dependent reversal of long-term potentiation by an integrin antagonist. *J. Neurosci.* **18:** 3460–3469.

Tanaka H., Shan W., Philips G.R, Arndt K., Bozdagi O., Shapiro L., Huntley G.W., Benson D.L., and Colman D.R. 2000. Molecular modification of N-cadherin in response to synaptic activity. *Neuron* **25:** 93–107.

Tang J., Landmesser L., and Rutishauser U. 1992. Polysialic acid influences specific pathfinding by avian motoneurons. *Neuron* **8:** 1031–1044.

Tang L., Hung C.P., and Schuman E.M. 1998. A role for the cadherin family of cell adhesion molecules in hippocampal long-term potentiation. *Neuron* **20:** 1165–1175.

Toni N., Buchs P.A., Nikonenko I., Bron C.R., and Muller D. 1999. LTP promotes formation of multiple spine synapses between a single axon terminal and a dendrite. *Nature* **402:** 421–425.

Trommald M., Hulleberg G., and Andersen P. 1996. Long-term potentiation is associated with new excitatory spine synapses on rat dentate granule cells. *Learn. Mem.* **3:** 218–228.

Wojtowicz J.M., Marin L., and Atwood H.L. 1989. Synaptic restructuring during long-term facilitation at the crayfish neuromuscular junction. *Can. J. Physiol. Pharmacol.* **67:** 167–171.

Wu G.Y., Malinow R., and Cline H.T. 1996. Maturation of a central glutamatergic synapse.

Science **274:** 972–976.

Yin J.C., Del Vecchio M., Zhou H., and Tully T. 1995. CREB as a memory modulator: Induced expression of a dCREB2 activator isoform enhances long-term memory in *Drosophila. Cell* **81:** 107–115.

Yin J.C., Wallach J.S., Del Vecchio M., Wilder E.L., Zhou H., Quinn W.G., and Tully T. 1994. Induction of a dominant negative CREB transgene specifically blocks long-term memory in *Drosophila. Cell* **79:** 49–58.

Yuste R. and Bonhoeffer T. 2001. Morphological changes in dendritic spines associated with long-term synaptic plasticity. *Annu. Rev. Neurosci.* **24:** 1071–1089.

15

Control of Synaptic Function by Local Protein Synthesis and Degradation

W. Bryan Smith, Baris Bingol,
Gentry N. Patrick, and Erin M. Schuman
Division of Biology 114-96
Caltech/Howard Hughes Medical Institute
Pasadena, California 91125

SYNAPTIC PLASTICITY—THE DYNAMIC MODIFICATION of functional synaptic strength between neurons of the central nervous system—is generally believed to be a physical mechanism underlying learning and memory. An individual neuron in the mammalian CNS may contain up to 10,000 synaptic connections. Furthermore, small groups of synapses can be independently regulated: Synaptic enhancement induced at one location on the dendritic arbor does not spread throughout the entire arbor (Andersen et al. 1977; Bonhoeffer et al. 1989; Schuman and Madison 1994; Engert and Bonhoeffer 1999). The precision with which a neuron is able to fine-tune its synaptic inputs is likely related to the information-processing capacity of that individual cell. Without the ability to independently regulate its inputs, the neuron probably could not distinguish among the myriad inputs it receives. Although dendritic spines are highly dynamic postsynaptic structures, sometimes growing and shrinking in response to neuronal activity, the overall size and structure of a neuron are relatively fixed in the adult animal (Fiala et al. 2002; Harris et al. 2003). Rather than constantly growing new synapses, changes in synaptic strength are achieved through changing the complement and concentration of synaptic proteins, as well as modifying protein function by covalent modifications. The local concentration of synaptic proteins is influenced by many processes, including protein trafficking, buffering, and sequestration, as well as protein synthesis and degradation. For example, in the presynaptic compartment, rapid axonal transport can provide many of the proteins and vesicles required to maintain and modulate synaptic function. In addition, the calcium-binding protein calmodulin can be buffered by the synaptic protein GAP43 (Baudier et al. 1991; Frey et al. 2000); the activity of some kinases is also effectively buffered by the binding of regulatory subunits. The most direct determinants of protein concentration, however, are protein synthesis and degradation. In this review, we explore the regulation of synaptic function by protein translation and ubiquitin-mediated degradation in dendrites. Both processes, synthesis and degradation, likely occur at or near synaptic sites involved in a given plasticity event, providing local control over synaptic strength.

LOCAL PROTEIN SYNTHESIS IN NEURONS

Although there are forms of short-lasting synaptic plasticity that rely only on transient enzymatic cascades, long-lasting changes require protein synthesis (Stanton and Sarvey 1984; Frey et al. 1988; Otani et al. 1989). This requirement for protein synthesis, when considered in the context of the complex geometry of the dendritic tree and the input specificity discussed above, poses an interesting question: How are newly synthesized proteins made available only at the synapses where they are needed?

At least three mechanisms may explain how neurons deliver the appropriate subset of proteins specifically to the modified synapses. In two of these, the proteins are synthesized in the cell body and then either shipped out to the correct synapses or specifically sequestered at the enhanced synapses via an activity-dependent synaptic tag (Schuman 1999; Martin and Kosik 2002). A more elegant solution to this problem, however, is provided by the local synthesis hypothesis. According to this idea, protein synthesis occurs specifically at synaptic sites where the new proteins are needed, thereby reducing the metabolic cost of activity-driven protein synthesis while simultaneously achieving region-specific protein delivery. Here, we present the various studies that demonstrate the occurrence of local protein synthesis (LPS) and the possible roles for this process in neural plasticity.

TRANSLATION MACHINERY

One of the earliest hints of the existence of local protein synthesis in neuronal processes was the electron microscopic detection of polyribosomes in the dendrites of dentate granule cells in the hippocampus (Steward and Levy 1982). In addition to observing these structures in the distal den-

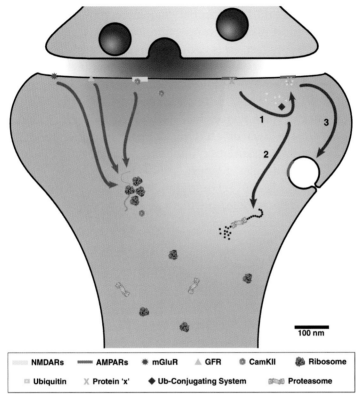

Figure 1. Synaptic protein synthesis and degradation. An individual synapse is shown with protein and membrane components roughly to scale. As illustrated by the blue gradient in the synaptic cleft, the presynaptic terminal (*top*) is releasing a single quantum of neurotransmitter onto an individual dendritic spine. Three pathways leading to increased protein synthesis at the synapse are shown (*gray arrows*): activation of metabotropic glutamate receptors (mGluR), growth factor receptors (GFR), and NMDA receptors (NMDARs). The green gradient near the top of the spine is an NMDAR-mediated Ca^{++} transient. Putative roles for the ubiquitin–proteasome system are also illustrated (1–3): activity-induced ubiquitination of synaptic proteins (1), ubiquitin-dependent protein degradation (2), lateral movement of AMPAR, followed by clathrin-dependent endocytosis (3). Protein x is a hypothetical target of the ubiquitin-conjugating system, the degradation of which is required for AMPAR endocytosis (see text). For the sake of clarity, only a small fraction of synaptic proteins are shown in the figure.

dritic compartment, Steward and Levy noted the preferential localization of ribosomal clusters near dendritic spines. These synapse-associated polyribosome clusters (SPRCs), situated in close proximity to the fundamental input units on CNS neurons, are positioned to respond rapidly to activity at nearby synapses (see Figure 1).

Since Steward and Levy's initial characterization of SPRCs 20 years ago, the anatomical evidence supporting local synthesis in neurons has steadily increased. In fact, an entire complement of translation machinery components has been detected in mature neurites, including messenger RNAs, endoplasmic reticulum, and markers for Golgi membranes (Steward and Reeves 1988; Torre and Steward 1996; Gardiol et al. 1999; Kacharmina et al. 2000; Pierce et al. 2001). Polyribosomes found at great distances from the nucleus are typically packaged into RNA granules: aggregates of mRNA, translation initiation and elongation factors, and a host of other molecules (Ainger et al. 1993; Knowles et al. 1996; Krichevsky and Kosik 2001). These granules, which are actively transported from the nucleus to the dendritic and axonal compartments, contain many of the components necessary to carry out regulated protein synthesis in these compartments.

Of critical importance in understanding the role of LPS in synaptic plasticity has been the identification of mRNAs present in the dendrites and axons. Using standard in situ hybridization techniques, a number of messages have been shown to exhibit somatodendritic localization (for review, see Steward and Schuman 2001). These include mRNAs that encode structural proteins such as MAP2 and β-actin, as well as those that encode plasticity-related proteins like the α subunit of calcium/calmodulin-dependent kinase (CamKIIα) and the NR1 subunit of the NMDA receptor. Not all of these messages, however, show identical subcellular distributions within the dendritic compartment. For example, the NR1 mRNA appears to be limited to the proximal domain of dendrites in the hippocampus (Miyashiro et al. 1994; Gazzaley et al. 1997), whereas the mRNA for CamKIIα is seen throughout the dendritic arbor (Burgin et al. 1990; Mayford et al. 1996).

mRNA TRAFFICKING

A number of studies have shown that the subcellular distribution of some mRNA transcripts is dynamically regulated in response to neuronal activity. Using high-resolution in situ hybridization and immunofluorescence microscopy, Bassell and colleagues investigated β-actin mRNA dynamics in developing cortical neurons in culture. In these experiments, the mRNA appeared in distinct granules that colocalized with components of the translation machinery. Furthermore, these granules were rapidly translocated into the dendritic and axonal growth cones in response to cAMP stimulation (Bassell et al. 1998). In more mature cultured neurons, the mRNAs for brain-derived neurotrophic factor (BDNF) and its TrkB receptor exhibited similar behavior: High-K$^+$-induced depolarization of cultured hippocampal neurons resulted in a redistribution of these mRNA species from a proximal to a more distal dendritic localization (Tongiorgi et al. 1997).

Perhaps the most thoroughly studied example of mRNA redistribution in neurons is the dynamic regulation of the mRNA for Arc, the activity-regulated cytoskeleton-associated protein (Lyford et al. 1995; Steward et al. 1998; Wallace et al. 1998). This series of studies took advantage of an ideal hippocampal anatomy: Axonal fibers innervating the dentate gyrus of the rat hippocampus are topographically distributed such that axons from a specific region of entorhinal cortex terminate in a specific layer of the dentate granule cell dendrites. This laminar organization of the hippocampal formation allowed the authors to specifically stimulate one region of the entorhinal cortex and show that Arc mRNA and protein accumulated precisely in the corresponding synaptic layer of the granule cell dendrites. Because these experiments were performed in mature animals in vivo, they provide very strong evidence for activity-dependent mRNA targeting to the dendrites. A potentially serious caveat to this study, however, is that maximum electroconvulsive shock (MECS) was used to induce the observed Arc redistribution. Although this provided the authors with sufficient Arc signal to be detectable, the stimulation patterns associated with MECS are quite different from the subtle stimuli that result in synaptic plasticity and may result in a pathological brain.

cis-Acting RNA Elements

A considerable amount of research effort has been devoted to characterizing the cis-regulatory regions involved in dendritic transport of specific mRNA molecules. In 1996, it was reported that sequences in the 3′ untranslated region (UTR) of the CamKIIα mRNA are necessary and sufficient for dendritic RNA trafficking (Mayford et al. 1996). Transgenic mice were constructed in which the lacZ coding sequence was placed upstream of either the 3′UTR of CamKIIα or, in a control construct, the polyadenylation sequence for bovine growth hormone (BGH). In mice expressing the lacZ-CamK3′UTR, both the lacZ mRNA and protein exhibited punctate distributions in the distal dendrites, while the control construct showed little or no dendritic localization. More recent data from the Mayford lab indicate that the 3′UTR of CamKIIα is necessary for sustained long-term potentiation (LTP) and memory consolidation (Miller et al. 2002). Using a targeted transgene approach, the authors replaced the 3´UTR of CamKIIα in mice with the BGH 3´UTR. The mutant mice showed dramatically altered CamKIIα mRNA distribution, with a majority of the transcripts being confined to the cell body layers in hippocampus and cortex. Consequently, the dendritic localization of the CamKIIα protein was reduced by over 75%. Hippocampal slices prepared from these mutant mice exhibited decreased late-phase LTP. In addition, in an important step toward implicating LPS in behavior, the mutants in this study were impaired in spatial as well as nonspatial memory tasks.

The trafficking of mRNAs in dendrites and axons is likely as dynamic as the trafficking of proteins. In an elegant set of experiments, Kosik and colleagues labeled the 3´UTR of the CamKIIα mRNA using a GFP/MS2 bacteriophage tagging system (Bertrand et al. 1998; Rook et al. 2000). In cultured hippocampal neurons, the authors described three distinct types of motion of the GFP-tagged 3´UTR: oscillatory motion, anterograde transport, and retrograde transport, with motion being skewed toward anterograde transport upon depolarization. Of particular interest in this study was that the GFP-tagged CamKIIα 3´UTR was shown to colocalize with synaptic markers, as determined by immunofluorescence microscopy. Taken together, these results demonstrate that synaptic activity is capable of driving CamKIIα mRNA into synaptic sites.

RNA-binding and Transport Proteins

Given that all mRNA synthesis occurs in the cell nucleus, an important question regarding mRNA transport concerns the identity of the molecular machinery involved in exporting RNA granules from the soma out to the dendrites. A tremendous amount of information from the development literature has been particularly instructive in this regard. For example, in the early stages of *Drosophila* development, a number of RNA-binding proteins are involved in the asymmetric distribution of mRNA and protein in the fertilized oocyte, which ultimately results in establishment of the anterior–posterior axis of the organism. A critical protein in this developmental process is the RNA-binding protein Staufen, which binds the 3´UTR of maternal mRNAs and transports them throughout the oocyte and developing embryo in a microtubule-dependent manner (Ferrandon et al. 1994; St. Johnston et al. 1991).

The essential role of Staufen in *Drosophila* development prompted the cloning of mammalian Staufen homologs by three independent groups (Kiebler et al. 1999; Wickham et al. 1999; DesGroseillers et al. 2001; Tang et al. 2001). As three of these studies illustrate, the mammalian Staufen proteins are also intimately involved with RNA transport in neurons. For example, overexpression of full-length Staufen results in an increase in total RNA in dendrites (Tang et al. 2001). Conversely, overexpressing the Staufen RNA-binding domain in the absence of the putative microtubule-binding region reduces the overall amount of RNA detectable in neuronal processes (Tang et al. 2001). This important observation indicates that Staufen is not only sufficient for RNA transport into dendrites, but also that interfering with the function of Staufen can affect overall RNA transport in the cell. Although the necessity of Staufen for RNA transport during synaptic enhancement remains to be demonstrated, it is clear that the RNA-binding and transport functions of this protein serve an important function in the maintenance of RNA distributions in neurons.

TRANSLATION REGULATION

If dendritic protein synthesis plays a meaningful role in input-specific synaptic enhancement, there must be translation regulation exerted on mRNAs that are shipped out to the neuronal processes. Without such control, the mRNA may be translated anywhere within the cell and at any point in time after it has been transcribed. The posttranscriptional modification and regulation of mRNA destined for export from the soma is therefore a critical component of the LPS hypothesis.

Significant progress has recently been made in understanding the regulation of mRNA translation. Although the molecular machinery underlying rapamycin-sensitive translation is discussed elsewhere in this volume, brief consideration of this signaling pathway is appropriate here, in the context of the specificity of translation regulation. Rapamycin, a protein synthesis inhibitor with immunosuppressive activity, has been a useful pharmacological agent for investigating regulated protein synthesis. Perhaps hinting at the existence of multiple translation regulatory systems, it is clear from the literature that only a subset of mRNAs inside a cell at a given time are sensitive to translation inhibition by rapamycin (Jefferies et al. 1994; Terada et al. 1994). At the *Aplysia* sensory-motor neuron synapse, serotonin-induced facilitation is completely blocked by the general protein synthesis inhibitor emetine but only partially inhibited by rapamycin (Casadio et al. 1999). In adult hippocampal neurons, the mammalian target of rapamycin (mTOR) and other proteins involved in cap-dependent translation initiation are localized to synaptic sites (Tang et al. 2002). Furthermore, Tang and colleagues showed that rapamycin treatment of hippocampal slices selectively attenuates late-phase LTP, while completely blocking BDNF-induced synaptic enhancement. In this case, the inhibition of synaptic plasticity observed in the presence of rapamycin was remarkably similar to that observed in the presence of a general protein synthesis inhibitor (Kang and Schuman 1996; Kang et al. 1997). Taken together, these findings suggest that rapamycin-sensitive translation plays an important role in synaptic plasticity.

CPE and IRES Sequences

Another potentially crucial element of translation control is the cytoplasmic polyadenylation element (CPE) and its cognate binding protein CPEB (Wells et al. 2000). Polyadenylation of mRNA transcripts is generally required for efficient translation, making temporal control over this process a candidate regulatory point in activity-driven protein synthesis. A detailed molecular analysis of CPEB signaling has revealed a putative role for this biochemical pathway in activity-driven polyadenylation and translation of the CPE-containing CamKIIα mRNA (Wu et al. 1998). In this model, an mRNA containing CPE sequences in its 3′UTR is rendered translationally dormant through an intricate series of protein–protein interactions involving CPEB, maskin, and the rate-limiting translation initiation factor eIF4E (Stebbins-Boaz et al. 1999). These messages become translationally competent as a result of aurora-kinase-catalyzed polyadenylation (Huang et al. 2002) and subsequent dissolution of the maskin–eIF4E interaction via the polyadenylation binding protein (Cao and Richter 2002). Although much remains to be learned in terms of understanding precisely how this process is involved in synaptic plasticity, the evidence that cytoplasmic polyadenylation is a potentially important checkpoint has been clearly established. Further investigations of the interactions between mTOR and CPEB-mediated cascades may provide the basis for an understanding of the complete series of reactions required to initiate translation of mRNAs in an activity-dependent fashion.

The stereotypical cap-dependent translation initiation through the CPEB pathway may only be part of the story, however. Another interesting component of regulated protein synthesis is the cap-independent translation mediated by internal ribosome entry sites (IRESs) present in some mRNAs. Although traditionally associated with bicistronic messages found in viral genomes, IRES sequences may also serve as regulators of protein synthesis in eukaryotic mRNAs. Importantly, IRESs have been identified in a number of dendritically localized messages, including those encoding Arc, CamKIIα, and neurogranin (Pinkstaff et al. 2001). Of particular interest is the observation that the neurogranin IRES confers preferential cap-independent translation initiation in the dendrites, whereas protein synthesis regulation in the soma is mediated by a cap-dependent mechanism. This finding raises the possibility that distinct mechanisms may differentially regulate somatic versus dendritic translation of a single mRNA species.

DEMONSTRATIONS OF LOCAL PROTEIN SYNTHESIS

With all of the components in place, and armed with the knowledge that mRNA granules containing translation machinery can be exported to the dendrites of living cells in an activity-dependent manner, it may seem a trivial leap to conclude that LPS is taking place in the dendritic compartment. However, the presence of mRNA and other components of the translation machinery does not necessarily indicate that all of these components are competent to produce functional proteins.

Radiolabeled Amino Acid Uptake

One of the earliest studies providing compelling evidence that protein synthesis may take place in neuronal processes employed ventricular injection of [³H]leucine into adult rats and subsequent autoradiographic and electron microscopic analysis (Kiss 1977). From these early experiments, Kiss observed that a majority of protein synthesis in dendrites was restricted to the proximal domain, although a small amount of [³H]leucine incorporation could be seen throughout the dendrites. It was not clear from this study, however, that the signal detected in distal processes was significantly higher than background levels. Furthermore, given the nature of these experiments, and the considerable delay between injection of the radioisotope and fixation of the sample for analysis, a somatic source of the radiolabeled proteins detected in the dendrites could not be ruled out.

A similar approach with analogous results was described in fresh hippocampal slices (Feig and Lipton 1993). Pairing electrical stimulation of Schaffer collateral axons with the cholinergic agonist carbachol resulted in increased [³H]leucine incorporation in the dendritic layers of the slices. Although this further suggested the involvement of LPS in plasticity, it was unfortunate that the stimulation protocol used failed to elicit any type of functional synaptic modification. In addition, as with the study by Kiss (1977), somatic protein synthesis as a source for the radiolabeled dendritic proteins could not be conclusively ruled out. Indeed, the fact that the cell body is such a large source of protein synthesis, with the neuronal processes being a relative sink, has plagued researchers interested in the local synthesis hypothesis from the beginning. To conclusively determine that the source of newly made protein in the dendrites cannot be attributed to proteins that were initially synthesized in the cell body, it is essential that the cell bodies and dendrites be somehow disconnected.

Synaptosome Preparations

Initial attempts at accomplishing this difficult task relied on subcellular fractionation techniques. Using a combination of biochemical tissue dissociation, filtration, and density gradient centrifugation of brain homogenates, it is possible to isolate small membranous particles termed synaptosomes. These synaptosomes are pinched-off membranous structures, which, when viewed through the electron microscope, appear to be resealed presynaptic terminals and their associated postsynaptic structures. A number of groups have employed this method of synaptosome preparation as a means of addressing the local synthesis hypothesis without the problem of potential somatic contributions.

Perhaps not surprisingly, CamKIIα mRNA is enriched in synaptosomes. Upon NMDA receptor activation, CamKIIα protein is specifically up-regulated in this biochemically purified preparation, and overall translation is reduced (Scheetz et al. 2000). From this work, it is clear that synaptic stimulation is not simply activating translation of all mRNAs at random, but rather that a subset of proteins are preferentially increased. Other studies in synaptosomes have shown that activation of metabotropic glutamate receptors increases synthesis of the fragile-X mental retardation protein (FMRP) (Weiler et al. 1997). The RNA-binding properties of this protein are critically involved in neuronal development (Darnell et al. 2001), and deletion of the FMRP gene results in a dramatic loss of protein synthesis detected in synaptosomes (Greenough et al. 2001).

With sufficiently pure synaptosome fractions, it should be possible to evaluate the protein synthesis competence of these structures in the absence of cell bodies. There are, however, a number of complications involved in interpreting the synaptosome data (for review, see Steward and Schuman 2001). In particular, a number of mRNAs exist in synaptosomes that are entirely absent from the dendrites of neurons as determined by in situ hybridization. These contaminants include the mRNA encoding glial fibrillary acidic protein, a protein that is found exclusively in glial cells. The presence of such an impurity in synaptosomes indicates that caution is needed in interpreting data acquired with this technique.

Acute Hippocampal Slices

An important study providing very strong evidence for a direct link between LPS and synaptic enhancement used a brute-force method of removing cell bodies as a potential protein source: Cell bodies in area CA1 of the hippocampus were disconnected from the synaptic neuropil by way of a microlesion with a dissecting knife (Kang and Schuman 1996). Having previously demonstrated protein-synthesis-dependent synaptic enhancement in hippocampal slices induced by neurotrophin treatment (Kang and Schuman 1995), Kang and Schuman used the microlesion technique to demonstrate further that this synaptic enhancement persisted even when the cell bodies were physically separated from the dendrites. Although this was the most convincing evidence to date, the argument could be made that the protein synthesis required for neurotrophin-induced enhancement was taking place in glial cells, interneurons, or axons—all of which are present in the dendritic layer of the hippocampus.

In addition to long-term potentiation, neurons in hippocampal slices exhibit long-term depression, or LTD. This form of synaptic modification can be induced in slices by bath application of the group 1 metabotropic glutamate receptor agonist DHPG ([RS]-3,5-dihydroxyphenylglycine) (Huber et al. 2000). Using a microlesion approach similar to that employed by Kang and Schuman, Huber and colleagues convincingly demonstrated that DHPG-induced LTD in area CA1 is blocked by inhibitors of protein synthesis. Because the net electrophysiological consequences of LTP and LTD are effectively opposite, it will be interesting to learn what proteins are differentially up-regulated and down-regulated in response to these two forms of synaptic plasticity.

Dissociated Culture Systems

The dissociated cell culture system has provided the most definitive data illustrating the ability of neurites to synthesize proteins. In a beautiful series of experiments, Martin and colleagues used cultured *Aplysia* neurons to examine the possibility that LPS contributes to long-term facilitation (LTF) in these neurons (Martin et al. 1997). Given the large size and relatively robust nature of *Aplysia* neurons, the authors were able to culture a single bifurcating sensory neuron synapsing onto two spatially separated motor neurons. Taking great care to avoid the cell body, they perfused serotonin locally onto one of the connections, leaving the other sensory-motor connection unperturbed. The result of this precise serotonin application was input-specific LTF: Only the synapses treated with serotonin exhibited synaptic enhancement. Importantly, the LTF was blocked by injection of a protein synthesis inhibitor into the presynaptic sensory neuron. Finally, to assess the ability of isolated neurites to synthesize proteins, 30–40 sensory neurons were cultured, and their cell bodies were removed and then tested for their ability to incorporate [^{35}S]methionine. Even in the absence of cell bodies, these isolated sensory neuron processes incorporated the radiolabeled amino acid in a protein-synthesis-dependent manner.

These exciting results provided the first definitive proof that neurites, independent of the cell body, are capable of producing proteins in response to synaptic enhancement-inducing stimuli. Although intriguing, the implications of the results of Martin and colleagues (1997) may be limited by the differences between vertebrate and invertebrate neurons. A presynaptic locus of protein-synthesis-dependent synaptic enhancement is also inconsistent with the ultrastructural data in vertebrate neurons: SPRCs are present in dendrites but have not been detected in axons of mature vertebrate neurons in the CNS. Nonetheless, it appears that there are both conceptual and molecular similarities in the invertebrate and vertebrate work that permit abundant cross-fertilization of ideas.

The dynamic visualization of dendritic protein synthesis was clearly demonstrated in a study carried out in cultured hippocampal neurons (Aakalu et al. 2001). Taking advantage of the visible fluorescence signal provided by green fluorescent protein (GFP), Aakalu and colleagues flanked a membrane-tethered GFP coding sequence with the 3′ and 5′ UTRs of CamKIIα, both to target the mRNA to the dendrites and to endow it with translation control elements. Neurons expressing this construct showed increased GFP production in response to BDNF. Increases in the distal dendrites were seen within 30 minutes after BDNF application, suggesting a local origin of the GFP signal. In an effort to rule out the cell body as a source of the GFP signal detected in the distal dendrites, the authors isolated dendrites using mechanical and optical techniques. In both cases, isolated dendrites continued to show increased GFP signal, which was completely blocked in the presence of a protein synthesis inhibitor. Lending additional support to the putative synaptic localization of the increased GFP, the authors quantitatively demonstrated that GFP hot spots in the dendrites were spatially correlated with synaptic markers. The findings by Aakalu and colleagues have been independently confirmed in a series of similar experiments (Job and Eberwine 2001). An important additional finding in this work was the observation of two distinct time courses of protein synthesis induction in the dendrites: a rapid-onset synthesis that is spatially correlated with sites of concentrated ribosomal immunoreactivity, and a slower, constitutive translation detected throughout the cell.

Axonal LPS

Whereas the majority of research in the field to date has focused on regulated translation in dendrites, the possible existence of axonal protein synthesis machinery has been described previously (Koenig and Giuditta 1999). It has also been shown that protein synthesis in the presynaptic compartment of dorsal root ganglion neurons regulates axon regeneration in response to injury (Twiss et al. 2000; Zheng et al. 2001). Furthermore, several recent studies have clearly implicated presynaptic LPS in growth cone responsiveness to local guidance cues during axon elongation (Campbell and Holt 2001; Brittis et al. 2002; Ming et al. 2002; Zhang and Poo 2002). Brittis and colleagues, using a high-density retinal explant culture system, determined that mechanically isolated axons continue to synthesize a GFP reporter. An important issue convincingly addressed in this study is the ability of distal processes to produce transmembrane proteins and deliver them to the cell surface. Although it remains unclear whether protein synthesis occurs in undamaged adult axons, these data indicate that LPS plays a critical role in the presynaptic compartment during axonal development and repair.

PROTEIN DEGRADATION AT SYNAPSES

In the preceding section, we described experiments illustrating that the availability of proteins, controlled by local synthesis, is important for synaptic function and plasticity. It follows, then, that any cellular process that regulates protein availability could be of importance in regulating synaptic function. Perhaps not surprisingly, several recent studies have implicated the ubiquitin–proteasome pathway in the control of synaptic development and plasticity. The ubiquitin-dependent degradation of regulatory proteins is important for many cellular processes, including cell-cycle progression, signal transduction, transcriptional regulation, receptor down-regulation, and endocytosis (Hershko and Ciechanover 1998). In recent years, it has become clear that the ubiquitin system is also utilized to control neuronal and synaptic function. The presence of ubiquitinated proteins in synaptic fractions from adult rat brains has been shown by immunohistochemical techniques (Chapman et al. 1994). Furthermore, immunostaining experiments with antibodies against ubiquitin and the α or β subunits of the proteasome show the presence of these proteins in the dendrites and spines of hippocampal neurons (Patrick et al., in prep.) and retinal growth cones (Campbell and Holt 2001). Spines and growth cones are the sites of excitatory synaptic contacts in the brain, suggesting that machinery responsible for ubiquitin-dependent degradation is present at mature and developing synapses. In this section, we describe the ubiquitin–proteasome pathway and discuss studies that indicate proteasomal degradation of proteins plays a role in synaptic development, plasticity, and neurodegenerative disease.

THE UBIQUITIN–PROTEASOME PATHWAY

The selective degradation of proteins via the ubiquitin–proteasome system involves three steps: recognition of the target protein via specific signals, marking of the target protein with a ubiquitin chain, and delivery of the ubiquitinated target protein to the 26S proteasome, a protein holocomplex that degrades the ubiquitinated proteins. The ubiquitination of the target protein is a highly regulated process. The protein degradation signal that is recognized by the ubiquitin-conjugating system may be a sequence motif within the protein itself, the presence or absence of a phosphate group, or damage within the protein. These signals usually result in recruitment of the ubiquitination machinery.

Ubiquitination Enzymes

Ubiquitination is a multistep enzymatic process, using three classes of enzymes (E1s, E2s, and E3s) and involving the sequential transfer of ubiquitin from these enzymes to the target protein (for review, see Hershko and Ciechanover 1998). First, ubiquitin needs to become activated. This activation is catalyzed by the ubiquitin-activating enzyme (E1) in an ATP-dependent reaction in which the carboxy-terminal glycine residue of ubiquitin binds to the active-site cysteine of an E1 in a thioester linkage (Haas et al. 1982).

It is important to note that the eventual specificity of protein ubiquitination is not dependent on the activity of E1s because there is usually a single E1 enzyme that catalyzes the activation of ubiquitin for all of the cellular ubiquitination reactions. Rather, the specificity of the ubiquitination reaction depends on the later steps of the ubiquitination process. There are a significant but limited number of ubiquitin-carrier enzymes (E2s) and a much larger number of ubiquitin ligases (E3s). Thus, the ubiquitination enzymes form a hierarchical cascade, where the substrate specificity of the overall ubiquitination reaction depends on the specific E2s and E3s that combine to ubiquitinate the substrate. Each E3 recognizes a set of substrates that share one or more signals for ubiquitination, and they pair up with one or a few E2s (for review, see Pickart 2001).

E2s

There are 13 genes coding for different E2-like proteins in the *Saccharomyces cerevisiae* genome, with an even greater number in higher organisms (Hochstrasser 1996). It is estimated that mammals express at least 20–30 different E2s. This increase in number reflects both multiple isoforms of the E2s (Jensen et al. 1995; Rajapurohitam et al. 1999) and the evolution of novel E2s (Hauser et al. 1998). E2s may have overlapping or specific functions. The specificity of the specific E2 functions is typically mediated by the E3 ligase. For example, there are cases in yeast where loss of function of an E2 results in accumulation of more than one substrate, where each of these substrates is ubiquitinated by distinct E3s (Tyers and Jorgensen 2000).

When the compartmentalized structure of synapses is considered, it is quite possible that there is a similar hierarchy in ubiquitination enzymes. There are likely only a handful of E1s at a synapse. Because the ubiquitination of synaptic proteins may be regulated by synaptic events, future research should identify a subset of synaptic E2s and E3s. Of course, it is also possible for target proteins to be ubiquitinated elsewhere and shipped to synapses, but when the rapidly changing composition of synapses is considered, it makes more sense for the machinery that regulates this process to be located at or near synaptic sites.

Functionally, E2s work as carriers of ubiquitin from E1 to E3s or to the substrate. Activated ubiquitin is transferred from E1 to an active-site cysteine of a ubiquitin-carrier enzyme (E2) in a *trans*-thiolation reaction, again involving the carboxy-terminal glycine of ubiquitin. There is a 14- to 16-kD core domain in E2s, which is ~35% conserved among family members. The carboxy-terminal and amino-terminal extensions of E2s are more variable and are involved in interactions with specific E3s (Mathias et al. 1998). These extensions may also serve as membrane anchors, bringing the E2s near substrates and E3s (Sommer and Jentsch 1993; Xie and Varshavsky 1999).

E3s

Ubiquitin is transferred from an E2 to the target protein, either directly or indirectly, with the aid of a ubiquitin ligase (E3). Ubiquitin is linked by its carboxyl terminus in an amide isopeptide linkage to an ε-amino group of the target protein's lysine residue. There are two classes of E3 enzymes: HECT domain E3s and RING finger E3s. HECT domain E3s accept and form thiolester intermediates with ubiquitin. Thus, in this case, the transfer of ubiquitin to the target protein occurs from the E3 (Huibregtse et al. 1995). Members of the other class, the so-called RING finger E3s, catalyze the direct transfer of ubiquitin from an E2 without the formation of a ubiquitin-E3 intermediate (Joazeiro and Weissman 2000). In this case, the E3 functions as an adapter, bringing the E2 and the target protein in proximity to one another so that the ubiquitination of the target protein can occur.

Ubiquitin Chain Formation

The signal that targets proteins to the proteasome is a polyubiquitin chain, which is formed by the addition of ubiquitin to lysine (K) 48 of the previous ubiquitin in the chain. Only substrates that contain many molecules of ubiquitin (e.g., ≥ 4) are efficiently recognized by the proteasome. When wild-type ubiquitin is replaced with a mutant ubiquitin that cannot form chains on K48, only a single ubiquitin can be attached to the substrate, accompanied by a large decrease in the degradation rate (Chau et al. 1989). Inability to form K48-linked chains in yeast is lethal, suggesting the importance of polyUb chain formation that is recognized by the proteasome (Finley et al. 1994). Ubiquitin chains may have different branching patterns, and lysine residues other than K48 may be used. In vivo, K11, K29, K48, and K63 of ubiquitin can all be used to form polyubiquitin chains, and these chains seem to function in distinct biological processes other than proteasome-dependent degradation (for review, see Pickart 2000). Furthermore, the number of ubiquitin molecules attached determines the outcome of ubiquitination. A single ubiquitin can efficiently signal for endocytosis instead of proteasomal degradation, as discussed below (Terrell et al. 1998).

The Proteasome

The 26S proteasome complex is composed of a catalytic 20S proteasome core and a regulatory 19S cap. The 20S core is a self-compartmentalizing assembly of 14 peptide subunits forming a barrel shape. The subunits of the 19S contain several ATPases that are involved in opening the pore of the 20S proteasome, unfolding the target proteins as they enter the proteasome, and translocating the proteins into the catalytic lumen of the core complex. Initial targeting of the substrates to the proteasome is probably accomplished through the recognition of the polyubiquitin chain by the non-ATPase subunits of the 19S cap (for review, see Voges et al. 1999).

De-ubiquitinating Enzymes

Similar to the presence of phosphatases that make phosphorylation reactions reversible, there are de-ubiquitinating enzymes (DUBs) that make ubiquitination reactions reversible. DUBs are cysteine proteases that generate free usable ubiquitin from a number of sources, including ubiquitin–protein conjugates, ubiquitin adducts, and ubiquitin precursors. More than 90 DUBs have been identified from different organisms (Chung and Baek 1999), making them the largest family of proteins in the ubiquitin system. This broad diversity illustrates the importance of ubiquitination reversibility.

There are two classes of DUBs: ubiquitin carboxy-terminal hydrolases (UCHs) and the ubiquitin-specific processing proteases (UBPs). UCHs usually cleave ubiquitin from the carboxyl terminus of small leaving groups or extended peptide chains by hydrolyzing the carboxy-terminal amides and esters of ubiquitin. UBP enzymes are responsible for removing ubiquitin from larger proteins and disassembling polyubiquitin chains. UCH enzymes are well conserved across species and have no apparent similarity to UBP enzymes. At least 12 UCH sequences have been identified from different organisms. There is one UCH identified in yeast, two in *Caenorhabditis elegans,* and three in humans. UBP enzymes have a 350-amino acid core catalytic domain and varying lengths of amino- and carboxy-terminal extensions, as well as some catalytic domain insertions. These extensions are thought to contribute to the substrate specificity and localization of different UBP enzymes. More than 80 full-length UBP sequences have been identified from different organisms (Chung and Baek 1999; Wilkinson 2000).

SYNAPTIC ROLES FOR UBIQUITIN AND PROTEASOMAL PROTEIN DEGRADATION

Synaptic Development

Ubiquitin-dependent processes are clearly important in synapse formation and regulation. For example, the *fat facets* (*faf*) gene in *Drosophila*, which codes for a DUB, is involved in the development of photoreceptors through modulation of the Ras signaling pathway (Wu et al. 1999). The *faf* phenotype can be suppressed by mutations in the proteasome, suggesting a critical role for faf in the de-ubiquitination of proteins that are normally degraded by the proteasome and involved at a key step during development (Huang and Fischer-Vize 1996). *Fam,* a *faf* homolog found in mouse, is bound to cell membranes at cell-to-cell contacts and also binds to AF-6, a downstream target of Ras (Wood et al. 1997; Taya et al. 1998).

Mutations in the *liquid facets* gene (*lqf*) were identified as dominant enhancers of the *faf* mutant eye phenotype in *Drosophila* (Cadavid et al. 2000). The *lgf* locus encodes a homolog of vertebrate epsin, a protein that initiates clathrin-dependent endocytosis. Both faf and lqf facilitate endocytosis. The function of lqf and faf in *Drosophila* eye is to prevent ectopic photoreceptor cell fate. The *faf* phenotype in *Drosophila* is extraordinarily sensitive to the level of lqf expression, suggesting that lqf may be a target of faf in the *Drosophila* eye.

Recently, *faf* has also been shown to be important for synapse development in *Drosophila* (DiAntonio et al. 2001). The neuromuscular junction (NMJ) of the body wall muscles of *Drosophila* is an easily accessible glutamatergic synapse. Similar to plasticity seen in CNS synapses, the *Drosophila* NMJ can undergo plasticity during development and in adult life. To study the mechanisms that regulate synaptic development in *Drosophila*, DiAntonio et al. screened for genes whose overexpression leads to synaptic growth abnormalities: They identified two mutant lines in which *faf* is overexpressed.

The endogenous *faf* transcript is widely and strongly expressed in the developing *Drosophila* CNS. Targeted overexpression of *faf* in *Drosophila* had both morphological and physiological consequences. Anatomical analysis revealed that *faf* overexpression leads to an increase in synaptic size, synaptic span (the extent of the muscle covered by the synapse), and the number of synaptic branches. These increases were not seen in flies that overexpress a nonfunctional *faf*. Furthermore, *faf* overexpression had a physiological phenotype: The evoked excitatory junctional potentials (EJPs) were markedly decreased despite the increased size of the synapse. This was also accompanied by a small decrease in both the amplitude and frequency of miniature EJPs. A large decrease in EJP with a small decrease in mEJP points to a decrease in the quantal content, which is a measure of the number of vesicles released by the excited nerve. The reduction in both quantal content and mEJP frequency suggests that *faf* overexpression leads to a defect in neurotransmitter release.

To test whether the *faf* overexpression phenotype was due to a disruption of ubiquitin-dependent protein degradation at the synapse, DiAntonio et al. overexpressed a yeast DUB in the nervous system of *Drosophila*. This yeast DUB has overlapping substrate specificity with faf (Wu et al. 1999). Its overexpression also led to a marked synaptic overgrowth and a severe reduction in presynaptic transmitter release, similar to *faf* overexpression. This indicates that antagonizing ubiquitin-dependent protein degradation via DUB overexpression leads to defects in synaptic development.

A screen for viable mutations that were lethal in combination with neuronal *faf* overexpression identified *highwire* (*hiw*). *hiw* codes for a RING finger type E3, and the *hiw* loss-of-function phenotype is very similar to *faf* overexpression phenotype (Wan et al. 2000). *faf* loss-of-function mutants have no defects in the morphology of their synapses, which may be due to redundant or compensatory mechanisms, but *faf* is required to suppress the electrophysiological phenotype of *hiw* loss-of-function mutants (DiAntonio et al. 2001). This suggests that *faf* acts to inhibit neurotransmitter release in a *hiw* loss-of-function background. The fact that the anatomical phenotype of *hiw* loss of function cannot be suppressed by the loss of function of *faf* indicates that these two phenotypes are mediated by different substrates of *hiw*. These data suggest that both hiw and faf control synaptic development through ubiquitination and de-ubiquitination of substrates critical for synaptic function. These substrates remain to be identified.

AXON GUIDANCE

Axons must navigate long distances to reach their target destinations. This is accomplished by sets of attractive and repellent molecular cues that steer the growth cone of a growing axon until the proper destination is reached. Netrin-1 and Semaphorin 3A (sema3a) are two such molecular cues in the CNS (Colamarino and Tessier-Lavigne 1995; Kobayashi et al. 1997). Netrin- and sema3a-induced guidance requires both protein synthesis and protein degradation (Campbell and Holt 2001; Ming et al. 2002). In particular, growth cones require translation in order to respond to chemoattractive gradients of netrin-1 or sema3a. Inhibitors of the proteasome blocked chemotropic responses to netrin-1 and growth cone collapse in response to lysophosphatidic acid (LPA). Netrin-1 and sema3a rapidly increased protein translation in growth cones. In addition, netrin-1 and LPA induced rapid accumulation of ubiquitin–protein conjugates in growth cones. These findings suggest that axon guidance involves an intricate balance between protein translation and ubiquitin-mediated proteolysis. Identifying the protein targets for both local synthesis and degradation is the obvious next step.

Synaptic Facilitation in *Aplysia*

One of the first observations that the ubiquitin–proteasome pathway is important for synaptic plasticity came from studies of facilitation of sensory-to-motor neuron synapses in *Aplysia*. This facilitation is presynaptic in nature and requires the action of a cAMP-dependent protein kinase (PKA) for both short-term (Klein and Kandel 1980; Siegelbaum et al. 1982) and long-term facilitation (Kandel and Schwartz 1982; Greenberg et al. 1987). Long-term facilitation (LTF), discussed earlier, requires the persistent activation of PKA, achieved by altering the levels of catalytic (C) and regulatory (R) subunits. The C subunits remain constant, but the R subunits are decreased during LTF (Greenberg et al. 1987; Bergold et al. 1990), leading to an increase in PKA activity. Hegde et al. determined that the loss of R subunits was achieved by the degradation of the R subunits by the ubiquitin–proteasome pathway (Hegde et al. 1993). They showed that degradation of the R subunits in reticulocyte lysates and in nerve tissue was blocked by the removal of proteasomes through ultracentrifugation. In addition, a series of higher-molecular-weight and putative ubiquitin conjugates appeared when recombinant R subunits were added to reticulocyte lysates in the presence of ubiquitin, ATP, and hemin (an inhibitor of proteasome activity). Furthermore, LTF produced by trains of serotonin application was completely blocked by inhibitors of the proteasome, further supporting the importance of ubiquitin-mediated proteolysis (Chain et al. 1999). In addition, there was a specific time window during which LTF was blocked by proteasome inhibitors: The application of lactacystin (an irreversible inhibitor of the proteasome) immediately after serotonin treatment blocked LTF, but application 3–6 hours later had no effect. Taken together, these data demonstrate that the activity of the proteasome is required early in LTF. The switch from short- to long-term facilitation requires CREB-mediated transcription and protein synthesis (Dash et al. 1990). Hegde et al. went on to identify an immediate-early gene product essential for LTF in *Aplysia*: a neuron-specific ubiquitin carboxy-terminal hydrolase (Hegde et al. 1997). This protein associates with the proteasome to increase proteasome activity, presumably by de-ubiquitinating proteins prior to their delivery to the proteasome barrel. Injection of antibodies or antisense oligonucleotides against the hydrolase blocked LTF induced by serotonin (Hegde et al. 1997). The de-ubiquitinating activity of the hydrolase may increase the activity of the proteasome in order to ensure the degradation of protein substrates that block the formation of long-term memory storage.

Synaptic Plasticity and Learning

Studies of E6-AP, the protein that is defective in the human disease Angelman's syndrome (AS), have also supported the idea that proteasomal protein degradation is important for both synaptic and behavioral plasticity. E6-AP was the first member of the HECT domain E3 family to be identified; it contains a conserved 350-amino acid region that defines the HECT domain (Huibregtse et al. 1995). E6-AP is required, together with the papillomavirus E6 oncoprotein, for the ubiquitination and degradation of the tumor suppressor p53 (Huibregtse et al. 1993; Scheffner et al. 1993, 1995). E6 serves as an adapter between E6-AP and p53, allowing E6-AP to catalyze the ubiquitination of p53. However, p53 is not the only target of E6-AP, and for its other substrates E6-AP does not need an adapter protein such as E6 to transfer the ubiquitin to the substrate. Mutations in the E6-AP gene (*Ube3a*) that cause AS result in mental retardation, seizures, an abnormal gut, tremor, and ataxia (Kishino et al. 1997). This disorder is associated with a maternally expressed, imprinted locus, mapping to chromosome 15q11-13. The molecular defects, due to point mutations, large deletions, complete absence of the gene, or imprinting mutations (Sutcliffe et al. 1994; Buiting et al. 1995; Horsthemke et al. 1997), lead to loss of E6-AP in those cell types where the paternal allele is silenced.

The mouse model that possesses a maternal *Ube3a* null mutation has demonstrated the importance of E6-AP for learning and memory (Jiang et al. 1998). The phenotype of these mice is similar to the phenotype of AS patients. For example, the *Ube3a* null mice exhibit impairments in bar-crossing and rotating-rod performance, which correlates with the ataxia and motor incoordination in human AS patients. Furthermore, the presence of inducible seizures and defects in context-dependent learning in the mouse model mirrors the high incidence of epilepsy and cognitive impairment in AS patients, respectively. Context-dependent learning in the mutant mice was examined using the conditioned fear paradigm in which animals are exposed to an electric shock paired with either the context (cage) or a cue (a tone). Following pairings, control animals typically exhibit conditioned freezing responses to either the cage or the tone. Immediately following pairing, the maternal *Ube3a* mutant mice exhibited freezing, indicating that they have normal sensory responses to the shock and normal short-term learning. After 24 hours, however, the *Ube3a*-deficient mice exhibited significantly less conditioned freezing to the context but normal conditioning to the auditory tone. This context-dependent learning deficit in the mutant mice may be due to a deficiency of Ube3a expression in the hippocampus, as the hippocampus has been shown to be important for contextual conditioning (Kim and Fanselow 1992).

The properties of synaptic transmission and plasticity were also examined in the *Ube3a*-deficient mice. LTP is considered a cellular mechanism of learning and memory (for review, see Malinow and Malenka 2002). The mutant mice exhibited normal basal synaptic transmission but reduced levels of LTP measured within an hour after the inducing stimulus of high-frequency stimulation. The authors of this study suggest a link between the reduced LTP in the mutant mice and the learning deficits observed in AS patients (Jiang et al. 1998).

Receptor Endocytosis

The ubiquitination of proteins does not always target them for degradation. As indicated above, the outcome of the ubiquitination reaction depends on both the number of ubiquitin moieties attached and the type of linkage between the individual ubiquitin units. If only one ubiquitin is attached to the target protein, it is called monoubiquitination. Monoubiquitination is involved in diverse cellular functions, including histone regulation, budding of retroviruses, and endocytosis of plasma membrane proteins (for review, see Hicke 2001).

The endocytosis of surface receptors is a common mechanism used to down-regulate receptor signaling (see Fig. 1). There are different endogenous internalization signals in the cytoplasmic domain of plasma membrane receptors. In addition to these endogenous signals, ubiquitin can be attached to the receptor to act as an internalization signal. Ubiquitin appears to be the most common internalization signal employed in yeast and higher eukaryotes. In yeast, there are several membrane proteins that must be ubiquitinated on their carboxy-terminal cytoplasmic domains in order for ligand-induced internalization to occur (for review, see Shaw et al. 2001). In mammalian cells, many receptors, including the epidermal growth factor receptor, platelet-derived growth factor receptor, and growth hormone receptor (GHR), are ubiquitinated in response to ligand binding (for review, see Bonifacino and Weissman 1998).

In addition to the receptors, in some cases, the proteins that compose the endocytic machinery are ubiquitinated. For example, the GHR requires the activity of ubiquitinating enzymes in order to be efficiently internalized, even though the ubiquitination of the receptor itself is not required (van Kerkhof et al. 2000). This suggests that an accessory protein(s), rather than the receptor itself, is the required target of ubiquitination.

One candidate ubiquitinated target of the endocytosis machinery is Eps15. By mapping the regions required for monoubiquitination on Eps15 and EpsR15, Polo et al. (2002) identified a region called the ubiquitin-interacting-motif (UIM) that is required for the monoubiquitination of these proteins. Polo et al. also showed that Nedd4, an E3 that also poly-ubiquitinates the epithelial sodium channel, catalyzes the monoubiquiti-nation of Eps15. As Eps15 interacts with clathrin-coated pits, it may form the connection between the monoubiquitinated receptors and the clathrin machinery through its UIM domain (Riezman 2002).

Recently, a role for the ubiquitin-dependent protein degradation machinery has been demonstrated in the down-regulation of G-protein-coupled receptors (GPCRs) (Shenoy et al. 2001). Ligand binding to GPCRs induces a conformational change in the receptor, leading to activation of cellular signaling events. This conformational change also induces phos-phorylation of the receptor by the G-protein-receptor kinases. The phos-phorylated receptor is recognized by the adapter protein β-arrestin, which uncouples the receptor from the downstream signaling events, leading to receptor desensitization. Furthermore, phosphorylation links the receptor to clathrin and adapter protein 2 (AP2), two key components of the endo-cytic machinery. Once internalized, receptors are either recycled back to the plasma membrane or degraded (for review, see Miller and Lefkowitz 2001; Pierce and Lefkowitz 2001; Luttrell and Lefkowitz 2002).

The regulation of GPCR endocytosis by ubiquitin-dependent protein modifications has been demonstrated previously (Shenoy et al. 2001). Shenoy et al. showed that the β2-adrenergic receptor (β2-AR) undergoes ubiquitination in response to ligand binding. Inhibition of proteasome activity does not lead to accumulation of the ubiquitinated receptor, indi-cating that the receptor itself is not the immediate target of the proteasome. β-arrestin2 also undergoes ubiquitination in response to ligand binding to β2-AR, but β-arrestin2 ubiquitination is more transient than the ubiquiti-nation of the receptor. This is apparently due to rapid de-ubiquitination, since ubiquitination of β-arrestin2 is only observed in the presence of DUB inhibitors. Ligand-induced ubiquitination of the β2-AR is dependent on its interaction with β-arrestin2 because β2-AR mutants that cannot bind to β-arrestin2 are not ubiquitinated in response to ligand binding. Furthermore, in cell lines lacking β-arrestin2, β2-AR cannot be ubiquitinated.

The requirement of β-arrestin2 binding for β2-AR ubiquitination suggests that β-arrestin2 acts to recruit the ubiquitination machinery to β2-AR. A yeast two-hybrid screen for E3s that interact with β-arrestin2 identified Mdm2. Mdm2 is an E3 and oncoprotein that acts as a negative regulator of p53 (Honda et al. 1997; Fang et al. 2000). Mdm2 can ubiqui-tinate both β2-AR and β-arrestin2 in vitro, but it is not required for β2-AR ubiquitination in vivo. In contrast, β-arrestin2 cannot be ubiquitinat-ed in cells that lack Mdm2. In cells lacking Mdm2, β2-AR internalization in response to ligand binding is markedly reduced, whereas receptor degradation occurs normally. This suggests that β-arrestin ubiquitination is required for receptor internalization, and β2-AR can be degraded even though it cannot be internalized (Shenoy et al. 2001).

Although the E3 that ubiquitinates the β2-AR has not been identified, it has been shown that blocking ubiquitination of β2-AR with lysine mutations blocks the degradation of the receptor. This suggests that ubiquitin-dependent mechanisms are involved in the receptor degradation. Furthermore, blocking the activity of the proteasome also blocks receptor degradation, although it is possible that proteasome is involved in the trafficking of the receptor to the lysosomes (Shenoy et al. 2001). Overall, these data suggest that ubiquitination machinery acts to regulate β2-AR internalization. This is accomplished by the regulation of ubiquitination of β-arrestin2 and possibly through other interactions of the receptor. Receptor ubiquitination is not required for β2-AR internalization but is required for its degradation. Rather, the ubiquitination of β-arrestin2 is critical for the internalization of the receptor.

Regulation of Glutamate Receptors

Glutamate is the major excitatory neurotransmitter in the mammalian brain. In the postsynaptic membrane of glutamatergic synapses there are two types of glutamate receptors: NMDA receptors (NMDARs) and AMPA receptors (AMPARs). NMDARs exhibit voltage-dependent calcium permeability, which allows NMDARs to control synaptic plasticity at many brain synapses (for review, see Izquierdo 1991). AMPARs mediate most of the basal responses to glutamate at excitatory synapse, and are formed from a combination of subunits (GluR1–4). In the hippocampus, AMPARs are composed of GluR1/2 and GluR2/3 subunits (Wenthold et al. 1996). It has been proposed that the number of functional AMPARs in the postsynaptic membrane can control synaptic strength (Luscher et al. 1999; Man et al. 2000; Wang and Linden 2000). As a consequence, the study of how AMPARs are delivered to the postsynaptic membrane, endocytosed, and recycled has been a major area of interest in the field of synaptic plasticity.

AMPAR trafficking in *C. elegans* is regulated by the clathrin endocytic machinery and ubiquitination (Burbea et al. 2002). In *C. elegans*, one of the non-NMDA ionotropic receptor subunits is encoded by the *glr-1* gene (Hart et al. 1995; Maricq et al. 1995). The GLR-1 glutamate receptors are localized to the sensory–interneuron synapse and are required for the mechanosensory nose-touch-avoidance behavior in *C. elegans* (Hart et al. 1995; Maricq et al. 1995; Rongo et al. 1998). Burbea et al. showed that GLR-1 is ubiquitinated. Mutations of lysine residues in GLR-1 that reduced ubiquitination increased the abundance of GLR-1 at synapses and altered locomotion behavior in a manner consistent with increased synaptic activity. In contrast, overexpression of ubiquitin decreased the amount of GLR-1 at synapses, an effect that was suppressed by mutations in unc-11, a clathrin adaptin protein (Burbea et al. 2002). The overexpression of Myc-ubiquitin decreased the density of synapses, as monitored by a GLR-1/GFP fusion, which cannot be simply explained by the ubiquitination of GLR-1 but likely involves the ubiquitination and/or degradation of other synaptic proteins. The ubiquitination of GLR-1 in *C. elegans,* leading to its down-regulation from synapses, likely represents a proteasome-independent use of ubiquitin, akin to the monoubiquitin-dependent internalization of receptors observed in yeast (for review, see Hicke 1999). Taken together, these data suggest that ubiquitination of GLR-1 receptors is important for regulating synaptic strength by regulating the abundance of the receptor at *C. elegans* synapses.

An involvement of ubiquitin and the proteasome in the regulation of mammalian glutamine receptors is suggested by studies of ligand-induced endocytosis of mammalian AMPAR subunits. Schuman and colleagues have observed that inhibition of proteasome activity blocked the agonist-mediated internalization of both GluR1 and GluR2 in hippocampal neurons. In addition, expression of a ubiquitin K48R mutant, in which polyubiquitin chain formation is inhibited, also blocked GluR1/2 endocytosis. These data suggest a role for the proteasome and polyubiquitination in the agonist-induced internalization of AMPARs. In light of the elegant work done in *C. elegans*, these data suggest that the mammalian receptors are internalized by different mechanisms. Interestingly, although AMPAR endocytosis is thought to underlie long-term depression (LTD) (Luscher et al. 1999; Man et al. 2000; Wang and Linden 2000), proteasome inhibitors did not prevent LTD. This suggests that a proteasome- and endocytosis-independent step is critical for synaptic depression. This step may involve the movement of AMPAR away from the synapse.

Indeed, recent work by Borgdorff and Choquet has provided evidence that AMPARs are highly mobile in the neuronal membrane. Using antibody-coupled beads, they observed regulated lateral diffusion of AMPARs, with periods of rapid diffusion alternating with periods of immobility (Borgdorff and Choquet 2002). Strong supporting evidence for this idea is also provided by Blanpied et al. (2002). Using clathrin-GFP to visualize endocytosis dynamically, they showed that the endocytic zone in mature hippocampal neurons is laterally juxtaposed to the synaptic zone. In addition, they showed that this endocytic zone is stable for long periods of time and is activity-independent. Taken together, these findings suggest that altering the efficacy of synaptic transmission to produce depression is not only mediated via a simple endocytosis of receptors—which may be regulated by the ubiquitin–proteasome pathway—but may also include the lateral movement of receptors in and out of the synaptic zone.

Neurodegeneration

Many of our clues for the way in which the ubiquitin–proteasome pathway may function in development and synaptic plasticity have been inspired by the study of neurodegenerative disorders. The identification of genes responsible for some neurodegenerative diseases has implicated the ubiquitin–proteasome pathway. In humans, many neurodegenerative diseases are associated with characteristic pathological hallmarks visible by standard light microscopy. Alzheimer's and Lewy body disease are the most common causes of dementia in elderly populations. These diseases are characterized by neurofibrillary tangles (NFT), neuritic plaques, and Lewy bodies (large dense inclusions). Familial Parkinson's disease is also characterized by Lewy bodies. In addition, large dense inclusions are characteristic of many of the polyglutamine repeat diseases such as Huntington's and motor neuron disease. Biochemical and immunohistochemical characterization of these inclusions, tangles, and neuritic plaques has indicated that they contain high concentrations of ubiquitin and ubiquitinated proteins. This has raised the interesting possibility that the ubiquitin–proteasome pathway contributes to these diseases.

Autosomal recessive juvenile parkinsonism (AR-JP) is one of the most common forms of familial Parkinson's disease. It is characterized by the selective and massive loss of dopaminergic neurons in the substantia nigra. Two of the genes associated with familial Parkinson's disease encode Parkin, an E3 ligase (Kitada et al. 1998), and UCH-L1, a ubiquitin carboxy-terminal hydrolase (Leroy et al. 1998). Both proteins are components of the ubiquitin–proteasome pathway. A third gene encodes α-synuclein, which is ubiquitinated and concentrated in Lewy bodies (Spillantini et al. 1997). Parkin has been shown to associate with the postsynaptic density, suggesting there may be a functional role for Parkin at postsynaptic sites (Fallon et al. 2002). In addition, the ubiquitin ligase activity of Parkin was shown to be responsible for the degradation of CDCrel-1, a synaptic-vesicle-associated protein (Zhang et al. 2000). Future studies of the ubiquitin–proteasome pathway in Parkinson's disease may shed light on the normal function of this pathway in neurons.

CONCLUDING REMARKS

As we learn more about the fundamental cellular processes responsible for neural plasticity, it is clear that the complex biochemical compartmentalization of synapses is critically involved in many normal functions of the neuron. As seen in *Aplysia*, local protein synthesis and degradation are essential for long-term facilitation of the siphon-withdrawal reflex. The ability to regulate synaptic protein concentrations by local synthesis and degradation in axons and dendrites of mammalian CNS neurons is also likely to be a key component of information processing in these cells. Evidence from experiments in the hippocampal system is particularly compelling in this regard, with critical roles for LPS and proteasome-mediated degradation having been clearly demonstrated in plasticity and disease.

Rather than relying on the cell body to accomplish the task of controlling protein levels at the thousands of synapses on the neuron, a more efficient system has evolved that permits much greater spatial specificity and temporal control over this critical task (see Fig. 1). As a neuron performs the complicated process of sorting through billions of incoming signals throughout its lifetime, spatio-temporal precision is of critical importance. Although the roles of local protein synthesis and degradation in this process remain to be determined, the specificity provided by these mechanisms lends itself well to the precision required in neuronal communication.

REFERENCES

Aakalu G., Smith W.B., Jiang C., Nguyen N., and Schuman E.M. 2001. Dynamic visualization of dendritic protein synthesis in hippocampal neurons. *Neuron* **30:** 489–502.

Ainger K., Avossa D., Morgan F., Hill S.J., Barry C., Barbarese E., and Carson J.H. 1993. Transport and localization of exogenous myelin basic protein mRNA microinjected into oligodendrocytes. *J. Cell Biol.* **123:** 431–441.

Andersen P., Sundberg S.H., Sveen O., and Wigstrom H. 1977. Specific long-lasting potentiation of synaptic transmission in hippocampal slices. *Nature* **266:** 736–737.

Bassell G.J., Zhang H., Byrd A.L., Femino A.M., Singer R.H., Taneja K.L., Lifshitz L.M., Herman I.M., and Kosik K.S. 1998. Sorting of beta-actin mRNA and protein to neurites and growth cones in culture. *J. Neurosci.* **18:** 251–265.

Baudier J., Deloulme J.C., Van Dorsselaer A., Black D., and Matthes H.W. 1991. Purification and characterization of a brain-specific protein kinase C substrate, neurogranin (p17). Identification of a consensus amino acid sequence between neurogranin and neuromodulin (GAP43) that corresponds to the protein kinase C phosphorylation site and the calmodulin-binding domain. *J. Biol. Chem.* **266:** 229–237.

Bergold P.J., Sweatt J.D., Winicov I., Weiss K.R., Kandel E.R., and Schwartz J.H. 1990. Protein synthesis during acquisition of long-term facilitation is needed for the persistent loss of regulatory subunits of the *Aplysia* cAMP-dependent protein kinase. *Proc. Natl. Acad. Sci.* **87:** 3788–3791.

Bertrand E., Chartrand P., Schaefer M., Shenoy S.M., Singer R.H., and Long R.M. 1998. Localization of ASH1 mRNA particles in living yeast. *Mol. Cell* **2:** 437–445.

Blanpied T.A., Scott D.B., and Ehlers M.D. 2002. Dynamics and regulation of clathrin coats at specialized endocytic zones of dendrites and spines. *Neuron* **36:** 435–449.

Bonhoeffer T., Staiger V., and Aertsen A. 1989. Synaptic plasticity in rat hippocampal slice cultures: Local "Hebbian" conjuction of pre- and postsynaptic stimulation leads to distributed synaptic enhancement. *Proc. Natl. Acad. Sci.* **86:** 8113–8117.

Bonifacino J.S. and Weissman A.M. 1998. Ubiquitin and the control of protein fate in the secretory and endocytic pathways. *Annu. Rev. Cell Dev. Biol.* **14:** 19–57.

Borgdorff A.J. and Choquet D. 2002. Regulation of AMPA receptor lateral movements. *Nature* **417:** 649–653.

Brittis P.A., Lu Q., and Flanagan J.G. 2002. Axonal protein synthesis provides a mechanism for localized regulation at an intermediate target. *Cell* **110:** 223–235.

Buiting K., Saitoh S., Gross S., Dittrich B., Schwartz S., Nicholls R.D., and Horsthemke B. 1995. Inherited microdeletions in the Angelman and Prader-Willi syndromes define an imprinting centre on human chromosome 15. *Nat. Genet.* **9:** 395–400.

Burbea M., Dreier L., Dittman J.S., Grunwald M.E., and Kaplan J.M. 2002. Ubiquitin and AP180 regulate the abundance of GLR-1 glutamate receptors at postsynaptic elements in *C. elegans. Neuron* **35:** 107–120.

Burgin K.E., Waxham M.N., Rickling S., Westgate S.A., Mobley W.C., and Kelly P.T. 1990. In situ hybridization histochemisty of Ca/calmodulin dependent protein kinase in developing rat brain. *J. Neurosci.* **10:** 1788–1789.

Cadavid A.L., Ginzel A., and Fischer J.A. 2000. The function of the *Drosophila* fat facets deubiquitinating enzyme in limiting photoreceptor cell number is intimately associated with endocytosis. *Development* **127:** 1727–1736.

Campbell D.S. and Holt C.E. 2001. Chemotropic responses of retinal growth cones mediated by rapid local protein synthesis and degradation. *Neuron* **32:** 1013–1026.

Cao Q. and Richter J.D. 2002. Dissolution of the maskin-eIF4E complex by cytoplasmic polyadenylation and poly(A)-binding protein controls cyclin B1 mRNA translation and oocyte maturation. *EMBO J.* **21:** 3852–3862.

Casadio A., Martin K.C., Giustetto M., Zhu H., Chen M., Bartsch D., Bailey C.H., and Kandel E.R. 1999. A transient, neuron-wide form of CREB-mediated long-term faciliation can be stabilized at specific synapses by local protein synthesis. *Cell* **99:** 221–237.

Chain D.G., Casadio A., Schacher S., Hegde A.N., Valbrun M., Yamamoto N., Goldberg A.L., Bartsch D., Kandel E.R., and Schwartz J.H. 1999. Mechanisms for generating the autonomous cAMP-dependent protein kinase required for long-term facilitation in *Aplysia. Neuron* **22:** 147–156.

Chapman A.P., Smith S.J., Rider C.C., and Beesley P.W. 1994. Multiple ubiquitin conjugates are present in rat brain synaptic membranes and postsynaptic densities. *Neurosci. Lett.* **168:** 238–242.

Chau V., Tobias J.W., Bachmair A., Marriott D., Ecker D.J., Gonda D.K., and Varshavsky A. 1989. A multiubiquitin chain is confined to specific lysine in a targeted short-lived protein. *Science* **243:** 1576–1583.

Chung C.H. and Baek S.H. 1999. Deubiquitinating enzymes: Their diversity and emerging roles. *Biochem. Biophys. Res. Commun.* **266:** 633–640.

Colamarino S.A. and Tessier-Lavigne M. 1995. The role of the floor plate in axon guidance. *Annu. Rev. Neurosci.* **18:** 497–529.

Darnell J.C., Jensen K.B., Jin P., Brown V., Warren S.T., and Darnell R.B. 2001. Fragile X mental retardation protein targets G quartet mRNAs important for neuronal function. *Cell* **107:** 489–499.

Dash P.K., Hochner B., and Kandel E.R. 1990. Injection of the cAMP-responsive element into the nucleus of *Aplysia* sensory neurons blocks long-term facilitation. *Nature* **345:** 718–721.

DesGroseillers L., Kuhl D., Richter D., and Kindler S. 2001. Two rat brain staufen isoforms differentially bind RNA. *J. Neurochem.* **76:** 155–165.

DiAntonio A., Haghighi A.P., Portman S.L., Lee J.D., Amaranto A.M., and Goodman C.S. 2001. Ubiquitination-dependent mechanisms regulate synaptic growth and function. *Nature* **412:** 449–452.

Engert F. and Bonhoeffer T. 1999. Dendritic spine changes associated with hippocampal long-term synaptic plasticity. *Nature* **399:** 66–70.

Fallon L., Moreau F., Croft B.G., Labib N., Gu W.J., and Fon E. A. 2002. Parkin and CASK/LIN-2 associate via a PDZ-mediated interaction and are co-localized in lipid rafts and postsynaptic densities in brain. *J. Biol. Chem.* **277:** 486–491.

Fang S., Jensen J.P., Ludwig R.L., Vousden K.H., and Weissman A.M. 2000. Mdm2 is a RING finger-dependent ubiquitin protein ligase for itself and p53. *J. Biol. Chem.* **275:** 8945–8951.

Feig S. and Lipton P. 1993. Pairing the cholinergic agonist carbachol with patterned Schaffer collateral stimulation initiates protein synthesis in hippocampal CA1 pyramidal cell dendrites via a muscarinic, NMDA-dependent mechanism. *J. Neurosci.* **13:** 1010–1021.

Ferrandon D., Elphick L., Nüsslein-Volhard C., and St Johnston D. 1994. Staufen protein associates with the 3′UTR of bicoid mRNA to form particles that move in a microtubule-dependent manner. *Cell* **79:** 1221–1232.

Fiala J.C., Allwardt B., and Harris K.M. 2002. Dendritic spines do not split during hippocampal LTP or maturation. *Nat. Neurosci.* **5:** 297–298.

Finley D., Sadis S., Monia B.P., Boucher P., Ecker D.J., Crooke S.T., and Chau V. 1994. Inhibition of proteolysis and cell cycle progression in a multiubiquitination-deficient yeast mutant. *Mol. Cell. Biol.* **14:** 5501–5509.

Frey D., Laux T., Xu L., Schneider C., and Caroni P. 2000. Shared and unique roles of CAP23 and GAP43 in actin regulation, neurite outgrowth, and anatomical plasticity. *J. Cell Biol.* **149:** 1443–1454.

Frey U., Krug M., Reymann K.G., and Matthies H. 1988. Anisomycin, an inhibitor of protein synthesis, blocks late phases of LTP phenomena in the hippocampal CA region in vitro. *Brain Res.* **452:** 57–65.

Gardiol A., Racca C., and Triller A. 1999. Dendritic and postsynaptic protein synthetic machinery. *J. Neurosci.* **19:** 168–179.

Gazzaley A.H., Benson D.L., Huntley G.W., and Morrison J.H. 1997. Differential subcellular regulation of NMDAR1 protein and mRNA in dendrites of dentate gyrus granule cells after perforant path transection. *J. Neurosci.* **17:** 2006–2017.

Greenberg S.M., Castellucci V.F., Bayley H., and Schwartz J.H. 1987. A molecular mechanism for long-term sensitization in *Aplysia*. *Nature* **329:** 62–65.

Greenough W.T., Klintsova A.Y., Irwin S.A., Galvez R., Bates K.E., and Weiler I.J. 2001. Synaptic regulation of protein synthesis and the fragile X protein. *Proc. Natl. Acad. Sci.* **98:** 7101–7106.

Haas A.L., Warms J.V., Hershko A., and Rose I.A. 1982. Ubiquitin-activating enzyme. Mechanism and role in protein-ubiquitin conjugation. *J. Biol. Chem.* **257:** 2543–2548.

Harris K.M., Fiala J.C., and Ostroff L. 2003. Structural changes at dendritic spine synapses during long-term potentiation. *Philos. Trans. R. Soc. Lond. B Biol. Sci.* **358:** 745–748.

Hart A.C., Sims S., and Kaplan J.M. 1995. Synaptic code for sensory modalities revealed by *C. elegans* GLR-1 glutamate receptor. *Nature* **378:** 82–85.

Hauser H.P., Bardroff M., Pyrowolakis G., and Jentsch S. 1998. A giant ubiquitin-conjugating enzyme related to IAP apoptosis inhibitors. *J. Cell Biol.* **141:** 1415–1422.

Hegde A.N., Goldberg A.L., and Schwartz J.H. 1993. Regulatory subunits of cAMP-dependent protein kinases are degraded after conjugation to ubiquitin: A molecular mechanism underlying long-term synaptic plasticity. *Proc. Natl. Acad. Sci.* **90:** 7436–7440.

Hegde A.N., Inokuchi K., Pei P., Casadio A., Ghirardi M., Chain D.G., Martin K.C., Kandel E.R., and Schwartz J.H. 1997. Ubiquitin C-terminal hydrolase is an immediate-early gene essential for long-term facilitation in *Aplysia. Cell* **89:** 115–126.

Hershko A. and Ciechanover A. 1998. The ubiquitin system. *Annu. Rev. Biochem.* **67:** 425–479.

Hicke L. 1999. Gettin' down with ubiquitin: Turning off cell-surface receptors, transporters and channels. *Trends Cell Biol.* **9:** 107–112.

——. 2001. Protein regulation by monoubiquitin. *Nat. Rev. Mol. Cell Biol.* **2:** 195–201.

Hochstrasser M. 1996. Ubiquitin-dependent protein degradation. *Annu. Rev. Genet.* **30:** 405–439.

Honda R., Tanaka H., and Yasuda H. 1997. Oncoprotein MDM2 is a ubiquitin ligase E3 for tumor suppressor p53. *FEBS Lett.* **420:** 25–27.

Horsthemke B., Dittrich B., and Buiting K. 1997. Imprinting mutations on human chromosome 15. *Hum. Mutat.* **10:** 329–337.

Huang Y. and Fischer-Vize J.A. 1996. Undifferentiated cells in the developing *Drosophila* eye influence facet assembly and require the Fat facets ubiquitin-specific protease. *Development* **122:** 3207–16.

Huang Y.S., Jung M.Y., Sarkissian M., and Richter J.D. 2002. N-methyl-D-aspartate receptor signaling results in Aurora kinase-catalyzed CPEB phosphorylation and alpha CaMKII mRNA polyadenylation at synapses. *EMBO J.* **21:** 2139–2148.

Huber K.M., Kayser M.S., and Bear M.F. 2000. Role for rapid dendritic protein synthesis in hippocampal mGluR-dependent long-term depression. *Science* **288:** 1254–1256.

Huibregtse J.M., Scheffner M., and Howley P.M. 1993. Cloning and expression of the cDNA for E6-AP, a protein that mediates the interaction of the human papillomavirus E6 oncoprotein with p53. *Mol. Cell. Biol.* **13:** 775–784.

Huibregtse J.M., Scheffner M., Beaudenon S., and Howley P.M. 1995. A family of proteins structurally and functionally related to the E6-AP ubiquitin-protein ligase. *Proc. Natl. Acad. Sci.* **92:** 2563–2567.

Izquierdo I. 1991. Role of NMDA receptors in memory. *Trends Pharmacol. Sci.* **12:** 128–129.

Jefferies H.B., Reinhard C., Kozma S.C., and Thomas G. 1994. Rapamycin selectively represses translation of the "polypyrimidine tract" mRNA family. *Proc. Natl. Acad. Sci.* **91:** 4441–4445.

Jensen J.P., Bates P.W., Yang M., Vierstra R.D., and Weissman A.M. 1995. Identification of a family of closely related human ubiquitin conjugating enzymes. *J. Biol. Chem.* **270:** 30408–30414.

Jiang Y.H., Armstrong D., Albrecht U., Atkins C.M., Noebels J.L., Eichele G., Sweatt J.D., and Beaudet A.L. 1998. Mutation of the Angelman ubiquitin ligase in mice causes increased cytoplasmic p53 and deficits of contextual learning and long-term potentiation. *Neuron* **21:** 799–811.

Joazeiro C.A. and Weissman A.M. 2000. RING finger proteins: Mediators of ubiquitin ligase activity. *Cell* **102:** 549–552.

Job C. and Eberwine J. 2001. Identification of sites for exponential translation in living dendrites. *Proc. Natl. Acad. Sci.* **98:** 13037–13042.

Kacharmina J.E., Job C., Crino P., and Eberwine J. 2000. Stimulation of glutamate receptor protein synthesis and membrane insertion within isolated neuronal dendrites. *Proc. Natl. Acad. Sci.* **97:** 11545–11550.

Kandel E.R. and Schwartz J.H. 1982. Molecular biology of learning: Modulation of transmitter release. *Science* **218:** 433–443.

Kang H. and Schuman E.M. 1995. Long-lasting neurotrophin-induced enhancment of synaptic transmission in the adult hippocampus. *Science* **267:** 1658–1662.

———. 1996. A requirement for local protein synthesis in neurotrophin-induced synaptic plasticity. *Science* **273:** 1402–1406.

Kang H., Welcher A.A., Shelton D., and Schuman E.M. 1997. Neurotrophins and time: Different roles for TrkB signaling in hippocampal long-term potentiation. *Neuron* **19:** 653–664.

Kiebler M.A., Hemraj I., Verkade P., Kohrmann M., Fortes P., Marion R.M., Ortin J., and Dotti C.G. 1999. The mammalian staufen protein localizes to the somatodendritic domain of cultured hippocampal neurons: Implications for its involvement in mRNA transport. *J. Neurosci.* **19:** 288–297.

Kim J.J. and Fanselow M.S. 1992. Modality-specific retrograde amnesia of fear. *Science* **256:** 675–677.

Kishino T., Lalande M., and Wagstaff J. 1997. UBE3A/E6-AP mutations cause Angelman syndrome. *Nat. Genet.* **15:** 70–73.

Kiss J. 1977. Synthesis and transport of newly formed proteins in dendrites of rat hippocampal pyramidal cells. An electron microscope autoradiographic study. *Brain Res.* **124:** 237–250.

Kitada T., Asakawa S., Hattori N., Matsumine H., Yamamura Y., Minoshima S., Yokochi M., Mizuno Y., and Shimizu N. 1998. Mutations in the parkin gene cause autosomal recessive juvenile parkinsonism. *Nature* **392:** 605–608.

Klein M. and Kandel E.R. 1980. Mechanism of calcium current modulation underlying presynaptic facilitation and behavioral sensitization in *Aplysia. Proc. Natl. Acad. Sci.* **77:** 6912–6916.

Knowles R.B., Sabry J.H., Martone M.E., Deerinck T.J., Ellisman M.H., Bassell G.J., and Kosik K.S. 1996. Translocation of RNA granules in living neurons. *J. Neurosci.* **16:** 7812–7820.

Kobayashi H., Koppel A.M., Luo Y., and Raper J.A. 1997. A role for collapsin-1 in olfactory and cranial sensory axon guidance. *J. Neurosci.* **17:** 8339–8352.

Koenig E. and Giuditta A. 1999. Protein-synthesizing machinery in the axon compartment. *Neuroscience* **89:** 5–15.

Krichevsky A.M. and Kosik K.S. 2001. Neuronal RNA granules: A link between RNA localization and stimulation-dependent translation. *Neuron* **32:** 683–696.

Leroy E., Anastasopoulos D., Konitsiotis S., Lavedan C., and Polymeropoulos M.H. 1998. Deletions in the Parkin gene and genetic heterogeneity in a Greek family with early onset Parkinson's disease. *Hum. Genet.* **103:** 424–427.

Luscher C., Xia H., Beattie E.C., Carroll R.C., von Zastrow M., Malenka R.C., and Nicoll R.A. 1999. Role of AMPA receptor cycling in synaptic transmission and plasticity. *Neuron* **24:** 649–658.

Luttrell L.M. and Lefkowitz R.J. 2002. The role of beta-arrestins in the termination and transduction of G-protein-coupled receptor signals. *J. Cell Sci.* **115:** 455–465.

Lyford G.L., Yamagata K., Kaufmann W.E., Barnes C.A., Sanders L.K., Copeland N.G., Gilbert D.J., Jenkins N.A., Lanahan A.A., and Worley P.F. 1995. Arc, a growth factor and activity-regulated gene, encodes a novel cytoskeleton-associated protein that is enriched in neuronal dendrites. *Neuron* **14:** 433–445.

Malinow R. and Malenka R.C. 2002. AMPA receptor trafficking and synaptic plasticity. *Annu. Rev. Neurosci.* **25:** 103–126.

Man H.Y., Lin J.W., Ju W.H., Ahmadian G., Liu L., Becker L.E., Sheng M., and Wang Y.T. 2000. Regulation of AMPA receptor-mediated synaptic transmission by clathrin-dependent receptor internalization. *Neuron* **25:** 649–662.

Maricq A.V., Peckol E., Driscoll M., and Bargmann C.I. 1995. Mechanosensory signalling in *C. elegans* mediated by the GLR-1 glutamate receptor. *Nature* **378:** 78–81.

Martin K.C. and Kosik K.S. 2002. Synaptic tagging—Who's it? *Nat. Rev. Neurosci.* **3:** 813–820.

Martin K.C., Casadio A., Zhu H., Yaping E., Rose J.C., Chen M., Bailey C.H., and Kandel E.R. 1997. Synapse-specific, long-term facilitation of *Aplysia* sensory to motor synapses: A function for local protein synthesis in memory storage. *Cell* **91:** 927–938.

Mathias N., Steussy C.N., and Goebl M.G. 1998. An essential domain within Cdc34p is required for binding to a complex containing Cdc4p and Cdc53p in *Saccharomyces cerevisiae. J. Biol. Chem.* **273:** 4040–4045.

Mayford M., Baranes D., Podsypanina K., and Kandel E.R. 1996. The 3′-untranslated region of CAMKII-alpha is a *cis*-acting signal for the localization and translation of mRNA in dendrites. *Proc. Natl. Acad. Sci.* **93:** 13250–13255.

Miller S., Yasuda M., Coats J.M., Jones Y., Martone M.E., and Mayford M. 2002. Disruption of dendritic translation of CamKIIalpha impairs stabilization of synaptic plasticity and memory consolidation. *Neuron* **36:** 507–519.

Miller W.E. and Lefkowitz R.J. 2001. Expanding roles for beta-arrestins as scaffolds and adapters in GPCR signaling and trafficking. *Curr. Opin. Cell Biol.* **13:** 139–145.

Ming G.-I., Wong S.T., Henley J., Yuan X.-B., Song H.-S., Spitzer N.C., and Poo M.-M. 2002. Adaptation in the chemotactic guidance of nerve growth cones. *Nature* **417:** 411–418.

Miyashiro K., Dichter M., and Eberwine J. 1994. On the nature and differential distribution of mRNAs in hippocampal neurites: Implications for neuronal functioning. *Proc. Natl. Acad. Sci.* **91:** 10800–10804.

Otani S., Marshall C.J., Tate W.P., Goddard G.V., and Abraham W.C. 1989. Maintenance of long-term potentiation in rat dentate gyrus requires protein synthesis but not messenger RNA synthesis immediately post-tetanization. *Neuroscience* **28:** 519–526.

Pickart C.M. 2000. Ubiquitin in chains. *Trends Biochem. Sci.* **25:** 544–548.

———. 2001. Mechanisms underlying ubiquitination. *Annu. Rev. Biochem.* **70:** 503–533.

Pierce J.P., Mayer T., and McCarthy J.B. 2001. Evidence for a satellite secretory pathway in neuronal dendritic spines. *Curr. Biol.* **11:** 351–355.

Pierce K.L. and Lefkowitz R.J. 2001. Classical and new roles of beta-arrestins in the regulation of G-protein-coupled receptors. *Nat. Rev. Neurosci.* **2:** 727–733.

Pinkstaff J.K., Chappell S.A., Mauro V.P., Edelman G.M., and Krushel L.A. 2001. Internal initiation of translation of five dendritically localized neuronal mRNAs. *Proc. Natl. Acad. Sci.* **98:** 2770–2775.

Polo S., Sigismund S., Faretta M., Guidi M., Capua M.R., Bossi G., Chen H., De Camilli P., and Di Fiore P.P. 2002. A single motif responsible for ubiquitin recognition and monoubiquitination in endocytic proteins. *Nature* **416:** 451–455.

Rajapurohitam V., Morales C.R., El-Alfy M., Lefrancois S., Bedard N., and Wing S.S. 1999. Activation of a UBC4-dependent pathway of ubiquitin conjugation during postnatal development of the rat testis. *Dev. Biol.* **212:** 217–228.

Riezman H. 2002. Cell biology: The ubiquitin connection. *Nature* **416:** 381–383.

Rongo C., Whitfield C.W., Rodal A., Kim S.K., and Kaplan J.M. 1998. LIN-10 is a shared component of the polarized protein localization pathways in neurons and epithelia. *Cell* **94:** 751–759.

Rook M.S., Lu M., and Kosik K.S. 2000. CaMKIIalpha 3′ untranslated region-directed mRNA translocation in living neurons: Visualization by GFP linkage. *J. Neurosci.* **20:** 6385–6393.

Scheetz A.J., Nairn A.C., and Constantine-Paton M. 2000. NMDA receptor-mediated control of protein synthesis at developing synapses. *Nat. Neurosci.* **3:** 211–216.

Scheffner M., Nuber U., and Huibregtse J.M. 1995. Protein ubiquitination involving an E1-E2-E3 enzyme ubiquitin thioester cascade. *Nature* **373:** 81–83.

Scheffner M., Huibregtse J.M., Vierstra R.D., and Howley P.M. 1993. The HPV-16 E6 and E6-AP complex functions as a ubiquitin-protein ligase in the ubiquitination of p53. *Cell* **75:** 495–505.

Schuman E.M. 1999. mRNA trafficking and local protein synthesis at the synapse. *Neuron* **23:** 645–648.

Schuman E.M. and Madison D.V. 1994. Locally distributed synaptic potentiation in the hippocampus. *Science* **263:** 532–536.

Shaw J.D., Cummings K.B., Huyer G., Michaelis S., and Wendland B. 2001. Yeast as a model system for studying endocytosis. *Exp. Cell Res.* **271:** 1–9.

Shenoy S.K., McDonald P.H., Kohout T.A., and Lefkowitz R. J. 2001. Regulation of receptor fate by ubiquitination of activated beta 2-adrenergic receptor and beta-arrestin. *Science* **294:** 1307–1313.

Siegelbaum S.A., Camardo J.S., and Kandel E.R. 1982. Serotonin and cyclic AMP close single K+ channels in *Aplysia* sensory neurones. *Nature* **299:** 413–417.

Sommer T. and Jentsch S. 1993. A protein translocation defect linked to ubiquitin conjugation at the endoplasmic reticulum. *Nature* **365:** 176–179.

Spillantini M.G., Schmidt M.L., Lee V.M., Trojanowski J.Q., Jakes R., and Goedert M. 1997. Alpha-synuclein in Lewy bodies. *Nature* **388:** 839–840.

Stanton P.K. and Sarvey J.M. 1984. Blockade of long-term potentiation in rat hippocampal CA1 region by inhibitors of protein synthesis. *J. Neurosci.* **4:** 3080–3084.

Stebbins-Boaz B., Cao Q., de Moor C.H., Mendez R., and Richter J.D. 1999. Maskin is a CPEB-associated factor that transiently interacts with eIF-4E. *Mol. Cell* **4:** 1017–1027.

Steward O. and Levy W.B. 1982. Preferential localization of polyribosomes under the base of dendritic spines in granule cells of the dentate gyrus. *J. Neurosci.* **2:** 284–291.

Steward O. and Reeves T.M. 1988. Protein-synthesis machinery beneath postsynaptic sites on CNS neurons: Association between polyribosomes and other organelles at the synaptic site. *J. Neurosci.* **8:** 176–184.

Steward O. and Schuman E.M. 2001. Protein synthesis at synaptic sites on dendrites. *Annu. Rev. Neurosci.* **24:** 299–325.

Steward O., Wallace C.S., Lyford G.L., and Worley P.F. 1998. Synaptic activation causes the mRNA for the IEG Arc to localize selectively near activated postsynaptic sites on dendrites. *Neuron* **21:** 741–751.

St Johnston D., Beuchle D., and Nüsslein-Volhard C. 1991. Staufen, a gene required to localize maternal RNAs in the *Drosophila* egg. *Cell* **66:** 51–63.

Sutcliffe J.S., Nakao M., Christian S., Orstavik K.H., Tommerup N., Ledbetter D.H., and Beaudet A.L. 1994. Deletions of a differentially methylated CpG island at the SNRPN gene define a putative imprinting control region. *Nat. Genet.* **8:** 52–58.

Tang S.J., Meulemans D., Vasquez L., Colaco N., and Schuman E.M. 2001. A role for staufen in the delivery of RNA to neuronal dendrites. *Neuron* **8:** 463–475.

Tang S.J., Reis G., Kang H., Gingras A.C., Sonenberg N., and Schuman E.M. 2002. A rapamycin-sensitive signaling pathway contributes to long-term synaptic plasticity. *Proc. Natl. Acad. Sci.* **99:** 467–472.

Taya S., Yamamoto T., Kano K., Kawano Y., Iwamatsu A., Tsuchiya T., Tanaka K., Kanai-Azuma M., Wood S.A., Mattick J.S., and Kaibuchi K. 1998. The Ras target AF-6 is a substrate of the fam deubiquitinating enzyme. *J. Cell Biol.* **142:** 1053–1062.

Terada N., Patel H.R., Takase K., Kohno K., Nairn A.C., and Gelfand E.W. 1994. Rapamycin selectively inhibits translation of mRNAs encoding elongation factors and ribosomal proteins. *Proc. Natl. Acad. Sci.* **91:** 11477–11481.

Terrell J., Shih S., Dunn R., and Hicke L. 1998. A function for monoubiquitination in the internalization of a G protein-coupled receptor. *Mol. Cell* **1:** 193–202.

Tongiorgi E., Righi M., and Cattaneo A. 1997. Activity-dependent dendritic targeting of BDNF and TrkB mRNAs in hippocampal neurons. *J. Neurosci.* **17:** 9492–9505.

Torre E.R. and Steward O. 1996. Protein synthesis within dendrites: Glycosylation of newly synthesized proteins in dendrites of hippocampal neurons in culture. *J. Neurosci.* **16:** 5967–5978.

Twiss J.L., Smith D.S., Chang B., and Shooter E.M. 2000. Translational control of ribosomal protein L4 mRNA is required for rapid neurite regeneration. *Neurobiol. Dis.* **7:** 416–428.

Tyers M. and Jorgensen P. 2000. Proteolysis and the cell cycle: With this RING I do thee destroy. *Curr. Opin. Genet. Dev.* **10:** 54–64.

van Kerkhof P., Govers R., Alves dos Santos C.M., and Strous G.J. 2000. Endocytosis and degradation of the growth hormone receptor are proteasome-dependent. *J. Biol. Chem.* **275:** 1575–1580.

Voges D., Zwickl P., and Baumeister W. 1999. The 26S proteasome: A molecular machine designed for controlled proteolysis. *Annu. Rev. Biochem.* **68:** 1015–1068.

Wallace C.S., Lyford G.L., Worley P.F., and Steward O. 1998. Differential intracellular sorting of immediate early gene mRNAs depends on signals in the mRNA sequence. *J. Neurosci.* **18:** 26–35.

Wan H.I., DiAntonio A., Fetter R.D., Bergstrom K., Strauss R., and Goodman C.S. 2000. Highwire regulates synaptic growth in *Drosophila*. *Neuron* **26:** 313–329.

Wang Y.T. and Linden D.J. 2000. Expression of cerebellar long-term depression requires postsynaptic clathrin-mediated endocytosis. *Neuron* **25:** 635–647.

Weiler I.J., Irwin S.A., Klintsova A.Y., Spencer C.M., Brazelton A.D., Miyashiro K., Comery T.A., Patel B., Eberwine J., and Greenough W.T. 1997. Fragile X mental retardation protein is translated near synapses in response to neurotransmitter activation. *Proc. Natl. Acad. Sci.* **94:** 5395–9400.

Wells D.G., Richter J.D., and Fallon J.R. 2000. Molecular mechanisms for activity-regulated protein synthesis in the synapto-dendritic compartment. *Curr. Opin. Neurobiol.* **10:** 132–137.

Wenthold R.J., Petralia R.S., Blahos J., II, and Niedzielski A.S. 1996. Evidence for multiple AMPA receptor complexes in hippocampal CA1/CA2 neurons. *J. Neurosci.* **16:** 1982–1989.

Wickham L., Duchaine T., Luo M., Nabi I.R., and DesGroseillers L. 1999. Mammalian staufen is a double-stranded RNA- and tubulin-binding protein which localizes to the rough endoplasmic reticulum. *Mol. Cell. Biol.* **19:** 2220–2230.

Wilkinson K.D. 2000. Ubiquitination and deubiquitination: Targeting of proteins for degradation by the proteasome. *Semin. Cell Dev. Biol.* **11:** 141–148.

Wood S.A., Pascoe W.S., Ru K., Yamada T., Hirchenhain J., Kemler R., and Mattick J.S. 1997. Cloning and expression analysis of a novel mouse gene with sequence similarity to the *Drosophila* fat facets gene. *Mech. Dev.* **63:** 29–38.

Wu L., Wells D., Tay J., Mendis D., Abbott M.-A., Barnitt A., Quinlan E., Heynen A., Fallon J.R., and Richter J.D. 1998. CPEB-mediated cytoplasmic polyadenylation and the regulation of experience-dependent translation of α-CAMKII mRNA at synapses. *Neuron* **21:** 1129–1139.

Wu Z., Li Q., Fortini M.E., and Fischer J.A. 1999. Genetic analysis of the role of the *Drosophila* fat facets gene in the ubiquitin pathway. *Dev. Genet.* **25:** 312–320.

Xie Y. and Varshavsky A. 1999. The E2-E3 interaction in the N-end rule pathway: The RING-H2 finger of E3 is required for the synthesis of multiubiquitin chain. *EMBO J.* **18:** 6832–6844.

Zhang X. and Poo M.M. 2002. Localized synaptic potentiation by BDNF requires local protein synthesis in the developing axon. *Neuron* **36:** 675–688.

Zhang Y., Gao J., Chung K.K., Huang H., Dawson V.L., and Dawson T.M. 2000. Parkin functions as an E2-dependent ubiquitin-protein ligase and promotes the degradation of the synaptic vesicle-associated protein, CDCrel-1. *Proc. Natl. Acad. Sci.* **97:** 13354–13359.

Zheng J.Q., Kelly T.K., Chang B., Ryazantsev S., Rajasekaran A.K., Martin K.C., and Twiss J.L. 2001. A functional role for intra-axonal protein synthesis during axonal regeneration from adult sensory neurons. *J. Neurosci.* **21:** 9291–9303.

16

Lymphocyte Growth

Daniel E. Bauer and Craig B. Thompson
Abramson Family Cancer Research Institute, University of
Pennsylvania, Philadelphia, Pennsylvania 19104

Size is one of the most obvious manifestations of natural variation. Differences in size abound between organisms—even within a single species, such as dogs, there can be over tenfold differences. Since the discovery of the cell as the basic unit of biology, the conundrum has been how such remarkable diversity of size can be achieved. For a long time, size differences were thought to be accounted for exclusively by differences in cell numbers. However, the sizes of cells within an individual organism may vary by orders of magnitude. For example, in most vertebrates a lymphocyte is several hundred times smaller than the largest neuron. Elegant studies have recently shown that the size of biological structures is determined not only by the number of cells but also by the size of those cells (Neufeld et al. 1998; see also Chapter 2).

Cell size is not infinitely variable. Cell growth and division are linked such that most dividing cells produce offspring of approximate parental size, a process referred to as *replicative cell division*. The outstanding progress in the understanding of the molecular control of the cell cycle over the past quarter century has perhaps overshadowed the importance of cell growth as an independent phenomenon. Specific circumstances that uncouple cell growth from cell proliferation demonstrate that understanding the mechanisms which regulate growth—the net acquisition of biomass—is one of the fundamental problems of cell biology, along with cell division, death, and differentiation. For example, the initial divisions of a fertilized egg do not involve cell growth, as each generation of blastomeres is smaller than the previous. The initial response of a lymphocyte to stimulation is blastogenesis, the transformation of a small resting cell into a large active one, independent of cell proliferation. In this chapter, we explore the regulation of cell growth and its relationship to nutrient metabolism, emphasizing the lymphocyte as a model cell whose size is governed by both external signals and internal constraints.

Biomass accumulation may be most spectacular during mammalian development, when a single cell gives rise to a full-sized organism. However, tissue homeostasis also absolutely depends on ongoing cell growth. The lymphoid compartment highlights the importance of regulated growth. During embryogenesis and throughout adulthood, hematopoietic precursors differentiate into billions of mature lymphocytes, which occupy lymphoid organs such as lymph nodes and spleen and circulate via blood and lymph. Mature lymphocytes are small dormant cells. When a single antigen-specific lymphocyte recognizes its cognate antigen, the cell first grows larger and then undergoes multiple rounds of replicative cell division to produce clonally derived effectors. For instance, in response to upper respiratory infection, this growth may be observed as intense swelling of cervical lymph nodes. At the end of an effective immune response, a wave of death occurs in the expanded clone of antigen-specific cells. The remaining lymphocytes return to the small size characteristic of quiescent cells. Therefore, what appears to be a static lymphocyte pool actually encompasses active cell growth and proliferation balanced by death. Furthermore, abnormal cell growth characterizes many diseases of the lymphoid compartment, including lymphoproliferative syndromes and autoimmunity. Malignancy is the ultimate consequence of dysregulated lymphocyte growth. In fact, most major classification schemes grade lymphomas on the basis of cell size.

What is the typical relationship between cell growth and cell proliferation? Killander and Zetterberg proposed in 1965 that a "critical mass" was required for a cell to initiate DNA synthesis. They demonstrated that, within a population of mouse fibroblasts, smaller cells had a longer G_1 phase, arguing that cells must reach a size threshold before entering S phase (Killander and Zetterberg 1965). Hartwell and colleagues extended these observations to budding yeast cells. These investigators showed that cell growth is dominant to proliferation; that is, inhibitors of growth block cell-cycle progression, whereas cell-cycle inhibitors do not block growth (Johnston et al. 1977). They proposed that yeast monitor cell size in G_1 at a phase of the cell cycle called Start (Hartwell and Unger 1977). Yen et al. (1975a) demonstrated that small lymphocytes spend more time than large lymphocytes growing before entering S phase. Yen and coworkers also showed that daughter lymphocytes produced by replicative division were larger in size than the starting population (Yen et al. 1975b).

The hypothesis that an absolute size threshold controls cell division is difficult to reconcile with actual cell behavior, however. For example, a strict size checkpoint would not allow dividing cells to have different volumes. Additionally, no mechanism by which a cell could directly measure its total size has been uncovered. A more refined version of the size-control hypothesis holds that cells do not measure absolute size but rather a more malleable surrogate. Some have proposed that cells actually measure protein synthesis capacity. For example, synthesis of the yeast G_1 cyclin Cln3 is inefficient, such that only at high levels of ribosomal capacity can enough Cln3 be generated to promote cell-cycle progression through Start (Polymenis and Schmidt 1997). Additional layers of control of Cln3 function may modulate the cell's response to a translation-dependent checkpoint (also see Chapter 4).

EXTRINSIC CELL GROWTH CONTROL

Unlike yeast, in which the Start size control depends on nutritional conditions, metazoan cells are not limited for growth and proliferation by the availability of nutrients in their environment. Instead, metazoan cells are regulated by external signals. These extracellular cues determine cell growth and proliferation, as well as survival (Conlon and Raff 1999; also see Chapter 3). Recently, Conlon and Raff showed that certain extracellular signal proteins could disentangle proliferation from growth. In experiments using rat Schwann cells, insulin-like growth factor-I (IGF-I) on its own promoted mainly cell growth, whereas glial growth factor (GGF) stimulated only cell division (Conlon et al. 2001). In constant IGF-I, and therefore constant cell growth, increasing concentrations of GGF led to decreasing cell-cycle times and decreasing cell size at division. Therefore, if a minimal size threshold exists for cell division, the starting Schwann cells in these experiments must have exceeded this threshold by a considerable margin. Alternatively, extracellular signals may be sufficient in metazoan cells to regulate division in the absence of a size checkpoint.

Recent experiments have shown that lymphocytes are subject to external constraints regulating cell growth. FL5.12 cells are a murine pro-B-cell line that requires interleukin-3 (IL-3) for its proliferation and survival (Nunez et al. 1990). When withdrawn from IL-3, these cells decrease progressively in size. However, these IL-3-deprived cells also rapidly initiate apoptosis, making it difficult to exclude the possibility that the atrophy is merely a by-product of commitment to death. Two strategies allowed investigators to conclude that IL-3 mediates an effect on cell growth independent of its inhibition of apoptosis. The first was to overexpress the anti-apoptotic protein Bcl-x$_L$ such that the cells remained viable even in the absence of IL-3. These apoptosis-resistant cells still undergo progressive atrophy, shrinking to approximately 50% of original volume within six days (Rathmell et al. 2000). Furthermore, the longer the cells are deprived of IL-3, the longer the atrophied cells take to reenter the cell cycle upon readdition of IL-3. The length of time required to reinitiate S phase is proportional to the degree of atrophy (Rathmell et al. 2000). This delay is consistent with the hypothesis that IL-3, a growth factor which couples cell growth and proliferation, depends on the attainment of a minimal size threshold before promoting S-phase entry.

The above experimental interpretation has been criticized because it involves overexpression of a gene that some groups have reported to be antiproliferative. More recently, this work has been confirmed using IL-3-dependent cells obtained from Bax$^{-/-}$ Bak$^{-/-}$ deficient animals (M. Harris and C. Thompson, unpubl.). Although resistant to apoptosis, these cells profoundly atrophy in response to four weeks of IL-3 withdrawal. Readdition of IL-3 induces an immediate and progressive increase in cell size. However, these cells do not reenter the cell cycle until two weeks after the replacement of growth factor, a time at which they have reached their pre-withdrawal size.

The second strategy that allowed the disentanglement of extrinsic growth control from protection from apoptosis was to assay the dose-dependence of the two effects. When IL-3-dependent cells were conditioned to grow in varying levels of IL-3 (up to 35-fold differences), cells in lower levels of IL-3 were always proportionally smaller (Vander Heiden et al. 2000; M. Harris and C. Thompson, unpubl.). All doses of IL-3 tested maintained cell viability. This change in cell size was not due merely to a lesser proportion of the population in G$_2$/M, since cells were smaller at each phase of the cell cycle. Thus, although there may be a minimal size below which cells cannot enter S phase without further growth, S-phase entry size is not absolute, but rather may be modulated depending on growth factor availability (Fig. 1).

Cell lines might have alterations in cell size control that do not reflect lymphocytes in vivo. However, many similarities have been found between factor-dependent lymphocyte lines and primary lymphocytes. T cells isolated from mice and cultured in the absence of growth factors atrophy, similar to FL5.12 cells deprived of IL-3 (Rathmell et al. 2000). Lymphocytes expressing a Bcl-x$_L$ transgene accumulate to high numbers

A

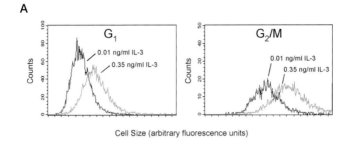

B

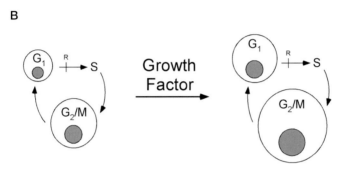

Figure 1. Extrinsic signals regulate cell size in proliferating lymphocytes. (*A*) Lymphocyte cell size is regulated by growth factors in a dose-dependent fashion. When IL-3-dependent lymphocyte cell lines are maintained in a relatively low concentration of IL-3 (0.01 ng/ml), they are significantly smaller than cells maintained in a higher but not saturating concentration of IL-3 (0.35 ng/ml). Living cells were stained with the cell-permeant DNA dye Hoechst 33342 and analyzed in a flow cytometer based on DNA content and on forward angle light scatter, which is proportional to cell size. Cells in higher cytokine levels are bigger at each phase of the cell cycle. (*B*) Schematic representation of the regulation of cell size in proliferating cells. Growth factors promote cell growth before a threshold size is reached at the restriction point (R) in G_1, where cell-cycle commitment occurs, as well as at later stages of the cell cycle.

in vivo, presumably because excess cells cannot be eliminated via apoptosis (Grillot et al. 1996). Interestingly, these Bcl-x_L transgenic lymphocytes are smaller than wild-type cells (Rathmell et al. 2000). These small cells have less protein per cell and require more time to enter S phase when stimulated in vitro, consistent with the idea that they have less critical mass than wild-type cells. The Bcl-x_L transgenic lymphocytes may be small because they receive fewer growth signals per cell than do wild-type lymphocytes, due to the excess number of cells competing for limiting trophic signals. This hypothesis has been tested directly by adoptive transfer experiments. When Bcl-x$_L$ transgenic T cells are introduced to a Bcl-x$_L$ transgenic recipient, the cells maintain pre-transfer cell size. However, when transferred to a nontransgenic recipient that has fewer T cells, the transferred cells significantly increase in size. Conversely, when non-transgenic T cells are transferred to a recipient with Bcl-x$_L$ transgenic T cells, they significantly decrease in size as compared to pre-transfer or following transfer to a nontransgenic recipient (Rathmell et al. 2000). Thus, cell size in vivo appears to be determined on a competitive basis.

Experiments to determine the components of the extrinsic growth signal for T cells have suggested that self-peptide/MHC engagement of the T-cell receptor plays an important role. Bcl-x$_L$ transgenic CD8 T cells transferred to hosts without endogenous lymphocytes (RAG1 deficient) grow larger. However, this enlargement is abrogated if the same cells are transferred to hosts that also lack self-peptide/MHC class I, the ligand that maintains CD8 T-cell survival in the absence of antigenic encounters (Rathmell et al. 2000).

Another signal involved in lymphocyte size control is cytokine stimulation. IL-7 addition prevents the atrophy of cultured T cells (Rathmell et al. 2001). IL-7 had dose-dependent effects on cell size, although no dose of IL-7 tested was sufficient to induce proliferation of a resting lymphocyte. Thus, cytokine signaling can be sufficient for lymphocyte growth but not for proliferation (Fig. 2). Other cytokines such as IL-2, IL-4, and IL-15 have been shown to have similar effects on the maintenance of T-cell size in vitro. Recent experiments have also demonstrated that appropriate lymphocyte trafficking in vivo is required for the ability of cells to access trophic support. T cells utilize chemokine signaling to migrate to T-cell zones in lymph nodes or the white pulp of the spleen. Isolated T cells in which chemokine signaling has been inhibited with pertussis toxin cannot migrate to T-cell zones when reintroduced to the mouse. These T cells are small as compared to untreated counterparts, which can receive appropriate in vivo growth signals. The inhibition of chemokine signaling does not have a direct effect on lymphocyte growth, since pertussis-toxin-treated cells are the same size as untreated cells in culture (J. Rathmell and C. Thompson, unpubl.).

In summary, lymphocytes require extracellular signals to regulate their basal size. This extrinsic control may take the form of antigen receptor engagement, cytokine stimulation, or both. These signals originate from specialized lymphoid compartments in vivo. Growth signals are limiting in the host, such that individual lymphocytes must compete for trophic support. Competition for antigen receptor engagement and cytokine signaling is thought to underlie lymphocyte homeostasis, the process by which the

A

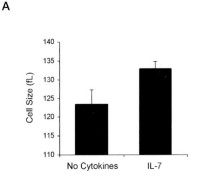

B

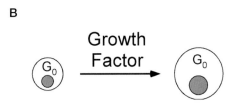

Figure 2. Extrinsic signals regulate cell size in nonproliferating lymphocytes. (*A*) When isolated T cells are cultured in the absence of cytokines, they progressively atrophy, as measured by Coulter counter. The Coulter counter calculates the volume displacement of individual particles. Addition of IL-7 to the culture is sufficient to maintain the lymphocyte size, although the IL-7 signal is insufficient to promote cell proliferation. (*B*) Schematic representation of the regulation of cell size by growth factors at a stage preceding the restriction point (R). Exogenous growth factors regulate the size of nonproliferating (G_0) lymphocytes.

number and distribution of lymphocytes are sustained despite the dynamic processes of lymphocyte development, activation, and death (Kedl et al. 2000; Ku et al. 2000). It appears that the same mechanisms which determine lymphocyte homeostasis also control cell size.

INTRINSIC CELL GROWTH CONTROL

External signals must operate on a cell-intrinsic apparatus. How signals effect control over size remains controversial. Clues as to the nature of the internal messages that promote cell growth have largely come from *Drosophila* genetics (see Chapters 2, 6, and 9). The Ras, Myc, PI3K, and Tor signaling cascades have each been positively linked to the regulation of cell-autonomous growth. Gain-of-function mutants in these pathways (by expression of positive components or disruption of negative regulators) display increased cell size, whereas loss-of-function mutants show decreased cell size. Transcription, translation, and metabolism have all been proposed to be fundamental processes involved in regulating cell size.

Transcriptional Growth Control

Perhaps the best-studied gene in lymphocyte growth control is the proto-oncogene c-*Myc*. It encodes a transcription factor proposed to regulate genes whose function is to effect cell growth that supports replicative cell division. The ability of Myc to promote cell mass accumulation was highlighted by experiments studying the effects of deregulated Myc expression on *Drosophila* development (Johnston et al. 1999; also see Chapter 2).

Several lines of evidence indicate that c-Myc regulates lymphocyte size. Expression of a dominant-negative c-Myc construct causes T-cell lines to decrease in size (Buckley et al. 2001). Conversely, c-Myc induction in a B-cell line leads to increased cell size, even in the presence of cell-cycle inhibition (Schuhmacher et al. 1999). c-Myc transgenic B cells are abnormally large (Langdon et al. 1986). This increased bulk occurs at all developmental stages, is independent of cell-cycle phase, and correlates with enhanced rates of protein synthesis (Iritani and Eisenman 1999).

Loss-of-function experiments in mice support the idea that endogenous c-Myc encourages lymphoid growth. One group generated T cells deficient for c-Myc by RAG1$^{-/-}$ chimerism (c-Myc$^{-/-}$ mice are embryonic lethal) (Douglas et al. 2001). These c-Myc-deficient thymocytes accumulate as small, double-negative (CD4$^-$CD8$^-$) cells unable to mature into larger thymocytes. Myc function can be antagonized by Mad, which represses transcription at Myc-responsive promoters (Luscher 2001). T-cell-specific expression of Mad1 also results in the accumulation of small, immature thymocytes (Iritani et al. 2002). Mad1-expressing cells are small, independent of cell-cycle phase, at each stage of thymic development.

Unexpectedly, mice with low levels of c-Myc expression do not have small cells. Rather, these mice, carrying combinations of hypomorphic and null c-Myc alleles, have small organs with fewer but normal-sized cells (Trumpp et al. 2001). Importantly, however, c-Myc-insufficient cells are not larger than normal, which would have been expected if biomass accumulation had remained intact with impaired cell division capacity. Therefore, the global function of endogenous c-Myc appears to be to promote cell growth coupled to cell proliferation.

Although the lymphocytes from these c-Myc-insufficient animals are of normal size, they inhabit c-Myc-insufficient hosts, so the cells may not receive trophic signals comparable to wild-type cells. To rigorously determine the cell-autonomous contribution of the c-Myc deficiency to lymphocyte size in these animals, their lymphocytes must be transferred to hosts expressing normal c-Myc. Only under such circumstances can growth in the context of equivalent external signals be compared. In vitro studies have shown that the c-Myc-insufficient lymphocytes have impaired proliferation but seemingly normal cell size in response to stimulation (Trumpp et al. 2001). These results contrast with the findings of c-Myc-deficient B cells generated via conditional knock-out. Although these cells demonstrate impaired proliferation in vivo, they fail to enlarge as do c-Myc-expressing cells when stimulated in vitro (de Alboran et al. 2001). Furthermore, Mad1 transgenic T cells do not enlarge in response to in vitro stimulation as do wild-type cells (Iritani et al. 2002). Can these experimental findings be reconciled? The absence or inhibition of c-Myc appears to result in the blockade of cell growth. Low levels of c-Myc may be enough to couple cell growth to proliferation. However, higher levels of c-Myc may saturate proliferation targets such that effects on cell growth predominate.

Mature lymphocytes find themselves in an impermanent respite between spurts of rapid growth. This process of quiescence was recently found to be actively maintained by the transcription factor lung Kruppel-like factor (LKLF) (Kuo et al. 1997). In a T-cell line containing an inducible LKLF construct, LKLF induction causes decreased protein synthesis and smaller cell size (Buckley et al. 2001). Evidence of LKLF's cell-autonomous role in lymphocyte size control comes from the finding that LKLF-deficient T cells are larger than wild-type T cells in chimeric animals (i.e., in the presence of equivalent extrinsic signals). Interestingly, LKLF appears to inhibit lymphocyte growth in part by antagonizing c-Myc, since c-*Myc* expression was repressed by LKLF induction. Transient transfection of T cell lines with c-*Myc* blocked the ability of LKLF to diminish cell size.

How does Myc mediate its effects on cell growth? Although cell-cycle regulatory factors are among the myriad c-Myc targets, the majority of c-Myc target genes appear to be involved in biosynthesis and bioenergetics (Eisenman 2001; Levens 2002). For example, Myc augments protein synthesis by activating the transcription of rRNA, ribosomal proteins, and translation initiation factors (Schmidt 1999). Furthermore, c-Myc directly activates glucose metabolism by promoting the transcription of glycolytic factors such as enolase, pyruvate kinase, and lactate dehydrogenase A

(Dang 1999; Osthus et al. 2000). Likewise, a DNA microarray profile showed that the majority of Mad1 repressed genes are involved in protein synthesis and metabolism (Iritani et al. 2002). Consistent with its ability to antagonize Myc function, LKLF induction depresses cellular metabolism as measured by ATP content or medium acidification (Buckley et al. 2001).

Translational Growth Control

The cytoplasmic serine/threonine kinase Tor has also been proposed to be a central regulator of cell size. Tor was first discovered because it is the target of the drug rapamycin, one of the most clinically effective drugs at inhibiting lymphocyte growth and proliferation. Rapamycin arrests proliferating cells in mid-G_1 phase at a position close or identical to the restriction point. The restriction point is the moment in the mammalian cell cycle after which the cell cycle can proceed without further input from extracellular signals. In budding yeast, Tor deficiency, inhibition via rapamycin treatment, or direct nutrient starvation each causes cells to switch from anabolic growth to a catabolic form of metabolism, characterized by autophagy, cell-cycle arrest, and the cessation of translation (Schmelzle and Hall 2000). Genetic experiments in fruit flies demonstrated that Tor is required for cell growth (Oldham et al. 2000; Zhang et al. 2000). Recent studies have also shown that mTOR regulates mammalian cell size at least partly via its downstream translational effectors S6K and 4EBP (Fingar et al. 2002). In lymphocytes, the ability of extrinsic factors to regulate cell size appears to depend on mTOR activity. For example, the effects of IL-7 on primary T-cell size are sensitive to rapamycin (Rathmell et al. 2001).

The regulation of mTOR activation remains controversial. Based on genetic epistasis in *Drosophila*, the PI3K pathway requires Tor activity to control cell size (Oldham et al. 2000; Zhang et al. 2000). Furthermore, biochemical hallmarks of mTOR activation such as S6K and 4EBP phosphorylation are sensitive to PI3K inhibitors (von Manteuffel et al. 1997). However, S6K alleles that are rapamycin-resistant yet PI3K inhibitor-sensitive, as well as genetic evidence that the effects of mTOR deficiency are more severe than those caused by the loss of upstream components of the PI3K pathway, suggest that mTOR and PI3K may be separable pathways (Weng et al. 1995; Oldham et al. 2000; Zhang et al. 2000). Some have proposed that mTOR may respond to amino acid availability, similar to yeast Tor. One report argued that mTOR is directly responsive to the lipid second-messenger phosphatidic acid (Fang et al. 2001). Another group provocatively suggested, based on mTOR's relatively weak affinity for ATP and exquisite sensitivity to inhibitors of cellular metabolism, that mTOR may be a direct ATP sensor (Dennis et al. 2001). Whether mTOR is downstream of growth factors in a traditional signal transduction sense or is merely permissive for growth factor signaling remains to be determined.

The PI3K signaling pathway appears integral to the control of mammalian cell size. Genetic experiments in mice have demonstrated that individual components of the PI3K pathway, such as the lipid kinase PI3K itself, the serine/threonine kinase Akt, or the PIP3 phosphatase, PTEN, regulate mammalian cell size (Shioi et al. 2000; Backman et al. 2001; Kwon et al. 20001; Tuttle et al. 2001; also see Chapter 18). Not unexpectedly, the PI3K pathway also impacts on lymphocyte size. In IL-3-dependent cells, LY294002 (a PI3K inhibitor) prevents cell growth after IL-3 reintroduction (Plas et al. 2001). In addition, Akt expression can increase cell size in the presence of IL-3 as well as mitigate IL-3-withdrawal-induced atrophy (Plas et al. 2001; Edinger and Thompson 2002). In the Jurkat T-cell line, PTEN expression or LY294002 treatment decreases cell size (Xu et al. 2002). LY294002 also prevents IL-7 maintenance of primary T-cell size in vitro (Rathmell et al. 2001). Finally, Akt transgenic T cells are larger than wild-type cells (Malstrom et al. 2001; J. Rathmell and C. Thompson, unpubl.). PI3K pathway effects on cell size appear to rely in part on mTOR activity, since the Akt-inducible increase in cell size can be blocked by rapamycin (Edinger and Thompson 2002).

Metabolic Growth Control

Strong data link protein synthesis with growth control. The role of other anabolic events in the regulation of biomass accumulation has been less investigated. Moreover, cellular metabolic changes are required to support the energy-intensive processes required for growth. Activation of nutrient metabolism is not limited to Myc, but appears to be a general outcome of growth factor signaling, on a par with cell growth, proliferation, and survival (Fig. 3) (Krauss et al. 2001). A number of laboratories have found that growth-factor-stimulated cells have increased ATP stores (Whetton and Dexter 1983; Vander Heiden et al. 1999). For example, Bcl-x_L-expressing cells deprived of IL-3 for 24 hours demonstrate a 50% reduction in the intracellular ATP/ADP ratio, concurrent with their cellular atrophy (Rathmell et al. 2000).

Growth factors appear to maintain ATP via direct effects on glucose metabolism. IL-3 stimulation of FL5.12 cells leads to a dose-dependent

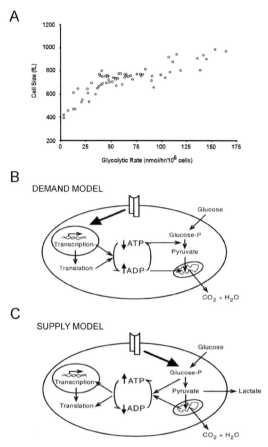

Figure 3. Relationship between cell size control and metabolism. (*A*) Larger cells have higher rates of glycolysis. Cell size was measured by Coulter counter, and glycolytic rate was measured by the conversion of radiolabeled glucose to water at the enolase step of glycolysis. (*B*) One possible explanation for this relationship is that extrinsic signals may increase cell size through direct regulation of gene expression and protein synthesis. The resulting depletion of energy equivalents such as ATP might induce a secondary increase in glycolysis and oxidative phosphorylation, as the cell attempts to maintain energy homeostasis. (*C*) Another possibility is that extrinsic signals may increase cell size by directly stimulating glycolysis. Increased ATP and macromolecular substrate levels might then fuel enhanced macromolecular synthesis at existing levels of transcription. Recent experiments support the hypothesis that lymphocyte metabolism is directly regulated by extrinsic signals.

increase in the rate of glycolysis (Vander Heiden et al. 2001). Dose-dependent IL-3 effects were detected on the expression level of Glut1, the principal lymphoid glucose transporter, as well as on hexokinase and phosphofructokinase (PFK) activities, the rate-limiting steps in glucose capture and glycolytic commitment, respectively (M. Harris and C. Thompson, unpubl.). Moreover, cells growing in higher concentrations of IL-3 secrete more lactate into the culture medium per glucose consumed. In other words, IL-3 promotes glucose utilization in excess of cellular demand. This result suggests that glycolytic stimulation is a direct response to growth factor signaling, and not merely a compensatory homeostatic response to increased energy demand (Fig. 3). Although both glycolysis and cell growth are stimulated by IL-3, the effects on glucose metabolism do not require cell growth. Treatment of cells with the protein synthesis inhibitor cycloheximide blocks growth-factor-induced increases in cell size, yet has no effect on growth factor induction of PFK activity (D. Bauer and C. Thompson, unpubl.).

Primary lymphocytes are subject to similar extracellular control of glucose metabolism. Bcl-x$_L$ transgenic lymphocytes, which receive reduced extrinsic signals per cell due to excess cell numbers, have decreased ATP content as compared to wild-type cells (Rathmell et al. 2000). T-cell-receptor engagement alone is insufficient to induce the increases in Glut1 expression required to support lymphocyte proliferation (Rathmell et al. 2000). Instead, engagement of the CD28 costimulatory receptor is required to up-regulate glucose uptake and glycolysis (Frauwirth et al. 2002). Cytokines like IL-2, IL-4, and IL-7 are sufficient to maintain glycolysis in activated T cells (Rathmell et al. 2001).

The effects of CD28 on glucose metabolism appear to be due to activation of the PI3K pathway, the same signaling cascade implicated in cell growth. By analogy, the PI3K pathway has previously been described as the major effector of insulin-induced glucose uptake (Kohn et al. 1996). CD28 costimulation induces Akt phosphorylation, and the inhibitory coreceptor CTLA4, which ablates the effects of CD28 on glycolysis, prevents Akt phosphorylation (Frauwirth et al. 2002). In lymphocyte cell lines, Akt expression is sufficient to increase ATP accumulation and glycolytic rate (Plas et al. 2001; Frauwirth et al. 2002).

Extrinsic signaling via the PI3K pathway has been shown to promote the transport of a multitude of nutrients in addition to glucose. Either cytokine addition or Akt expression can maintain the cell-surface localization of amino acid, lipoprotein, and iron transporters. This diverse nutrient uptake requires mTOR function, as rapamycin can block the maintenance of this transport (Edinger and Thompson 2002). Regulated glucose metabolism also requires mTOR activity, since IL-7 maintenance of glycolysis in primary T cells is also inhibited by rapamycin (Rathmell et al. 2001).

Unlike the direct role for extrinsic signals in regulating nutrient metabolism in cells, growth factors have indirect effects at the level of mitochondrial function. Growth-factor-induced PI3K signaling is sufficient to maintain mitochondrial potential in FL5.12 cells and primary T cells (Vander Heiden et al. 1997; Plas et al. 2001; Edinger and Thompson 2002). However, Akt induction, although it causes robust increases in glycolytic rate, has no effect on the rate of oxygen consumption (D. Bauer and C. Thompson, unpubl.). In fact, growth-factor-dependent glucose metabolism may suppress mitochondrial oxidative phosphorylation. For example, T cells fully stimulated with anti-CD3 and anti-CD28 increase their rate of oxygen consumption by nearly 4-fold when cultured in the presence of 25-fold reduced levels of glucose (the Crabtree effect) (Frauwirth et al. 2002). Furthermore, NADH measurements indicate that growth factor stimulation induces a large increase in NADH derived from glycolysis, while inhibiting mitochondrial TCA cycle generation of NADH (M. Harris and C. Thompson, unpubl.). To allow continued glycolytic flux, the end products pyruvate and NADH must be rapidly eliminated from the cytosol. Cytokine-stimulated cells increase both direct oxidization of NADH by lactate dehydrogenase and shuttling of NADH to the mitochondrial electron transport chain (D. Plas and C. Thompson, unpubl.). These results suggest a model in which growth factors permit the overproduction of substrates for mitochondrial electron transport. Strangely enough, activated cells are then less reliant on complete mitochondrial oxidation of nutrients via the TCA cycle to provide ATP. What then is the utility of these fully charged mitochondria?

Recently, Pfeiffer and coworkers have suggested that glycolysis, by providing a higher rate of ATP production than oxidative phosphorylation, may be advantageous for rapidly growing cells (Pfeiffer et al. 2001). Additionally, we propose that growth-factor-dependent provision of excess reducing substrates may allow the shift of mitochondria from a bioenergetic to a biosynthetic role (Fig. 4). This transition may allow cells to convert from quiescence to growth and proliferation. Protein synthesis has been the primary biosynthetic process considered in the regulation of cell growth. Clearly, other forms of macromolecular production are also essential for biomass accumulation. The pentose phosphate shunt is an alternative route of glucose metabolism that glucose-avid cells use to provide NADPH and ribose for lipid and nucleotide synthesis. The ongoing production of acetyl CoA is also required for synthesis of fatty acids and sterols, as well as for lipid modification and acetylation of proteins. Mitochondrial acetyl CoA formed from pyruvate can only contribute to these anabolic processes when shuttled back to the cytosol in the form of

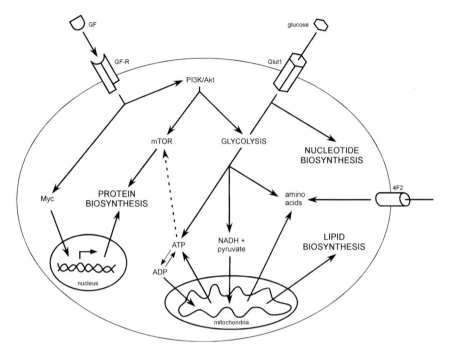

Figure 4. Lymphocyte growth requires the coordination of biosynthetic processes. This schematic represents a simplified circuit whereby extrinsic signals may inform intracellular pathways promoting growth. Growth factor (GF) engagement of growth factor receptors (GF-R) leads to changes in gene expression, in part due to the activation of Myc. Myc targets include genes that encode protein synthetic structural and regulatory factors, as well as glycolytic enzymes. Another major growth factor effector is the PI3K/Akt pathway. This pathway works on a posttranscriptional level to promote both protein synthesis and glycolysis. The induction of glucose uptake and glycolysis provides substrates for nucleotide synthesis, amino acid production, and mitochondrial electron transport, as well as ATP for energy-intensive growth. Glycolytic charging may allow a shift of mitochondria to a more synthetic function. Mitochondrial supply of citrate and amino acids supports lipid and protein biosyntheses, respectively. mTOR activation via both PI3K signaling and direct metabolic sensing likely plays a crucial role in promoting such an anabolic state.

citrate. Therefore, lipid biosynthesis demands that citrate does not transit through the TCA cycle. ATP citrate lyase is the enzyme that produces cytosolic acetyl CoA. Recent experiments in *Arabidopsis* have underscored the importance of ATP citrate lyase for cell growth. Transgenic plants expressing an antisense copy of ATP citrate lyase demonstrated small cells and slowed rates of development, with the phenotype severity proportional to the reduction in ATP citrate lyase activity (Fatland et al. 2000).

The importance of biosynthetic processes beyond translation is also underscored by recent experiments in which lymphocytes were maintained in defined nutrient conditions (L. Wang and C. Thompson, unpubl.). When primary T cells were cultured in decreasing concentrations of glucose in the presence of full anti-CD3 and anti-CD28 stimulation, cell proliferation and total biomass accumulation were drastically inhibited. However, the cells grown in low glucose were actually larger than cells grown in normal levels of glucose. These cells contained more protein and were bigger at each phase of the cell cycle. This apparent increase in growth may in fact be due to increased extrinsic growth signals per cell, since there were fewer cells in the low-glucose condition. This result is reminiscent of ribosomal subunit S6-deficient hepatocytes or E2F-deficient imaginal disc cells in which cell proliferation is disrupted to a greater degree than cell growth (Neufeld et al. 1998; Volarevic et al. 2000). The relative dependence of the cell cycle on glucose may be due to a requirement for nucleotide biosynthesis or some other unrecognized glucose checkpoint. This finding emphasizes that basic metabolic parameters, such as extracellular glucose availability, are absolutely required for cells to couple growth and proliferation. Metabolic derangements, in addition to impairment of growth factor signaling cascades, may result in the dysregulation of growth, proliferation, or both.

SUMMARY

Lymphocytes are a useful model for the study of metazoan cell size control. Their growth absolutely depends on extrinsic signals. These factors cause changes in gene expression, translational regulation, and cellular metabolism that promote net biosynthesis. Recent experiments have identified crucial pathways that impinge on these processes, although more exploration will be required to fully understand the precise mechanisms of growth control.

ACKNOWLEDGMENTS

The authors thank members of the Thompson laboratory for fruitful discussions, especially Aimee Edinger, Kenneth Frauwirth, Julian Lum, David Plas, and Jeffrey Rathmell for critical reading of the manuscript and Marian Harris, David Plas, Jeffrey Rathmell, and Lynn Wang for sharing unpublished results. Sean Armour, Ryan Cinalli, and Jill Desimini provided invaluable graphic assistance.

REFERENCES

Backman S.A., Stambolic V., Suzuki A., Haight J., Elia A., Pretorius J., Tsao M.S., Shannon P., Bolon B., Ivy G.O., and Mak T.W. 2001. Deletion of Pten in mouse brain causes seizures, ataxia and defects in soma size resembling Lhermitte-Duclos disease. *Nat. Genet.* **29**: 396–403.

Buckley A.F., Kuo C.T., and Leiden J.M. 2001. Transcription factor LKLF is sufficient to program T cell quiescence via a c-Myc-dependent pathway. *Nat. Immunol.* **2**: 698–704.

Conlon I. and Raff M. 1999. Size control in animal development. *Cell* **96**: 235–244.

Conlon I.J., Dunn G.A., Mudge A.W., and Raff M.C. 2001. Extracellular control of cell size. *Nat. Cell Biol.* **3**: 918–921.

Dang C.V. 1999. c-Myc target genes involved in cell growth, apoptosis, and metabolism. *Mol. Cell. Biol.* **19**: 1–11.

de Alboran I.M., O'Hagan R.C., Gartner F., Malynn B., Davidson L., Rickert R., Rajewsky K., DePinho R.A., and Alt F.W. 2001. Analysis of C-MYC function in normal cells via conditional gene-targeted mutation. *Immunity* **14**: 45–55.

Dennis P.B., Jaeschke A., Saitoh M., Fowler B., Kozma S.C., and Thomas G. 2001. Mammalian TOR: A homeostatic ATP sensor. *Science* **294**: 1102–1105.

Douglas N.C., Jacobs H., Bothwell A.L., and Hayday A.C. 2001. Defining the specific physiological requirements for c-Myc in T cell development. *Nat. Immunol.* **2**: 307–315.

Edinger A.L. and Thompson C.B. 2002. Akt maintains cell size and survival by increasing mTOR-dependent nutrient uptake. *Mol. Biol. Cell* **13**: 2276–2288.

Eisenman R.N. 2001. Deconstructing myc. *Genes Dev.* **15**: 2023–2030.

Fang Y., Vilella-Bach M., Bachmann R., Flanigan A., and Chen J. 2001. Phosphatidic acid-mediated mitogenic activation of mTOR signaling. *Science* **294**: 1942–1945.

Fatland B., Anderson M., Nikolau B.J., and Wurtele E.S. 2000. Molecular biology of cytosolic acetyl-CoA generation. *Biochem. Soc. Trans.* **28**: 593–595.

Fingar D.C., Salama S., Tsou C., Harlow E., and Blenis J. 2002. Mammalian cell size is controlled by mTOR and its downstream targets S6K1 and 4EBP1/eIF4E. *Genes Dev.* **16**: 1472–1487.

Frauwirth K.A., Riley J.L., Harris M.H., Parry R.V., Rathmell J.C., Plas D.R., Elstrom R.L., June C.H., and Thompson C.B. 2002. The CD28 signaling pathway regulates glucose metabolism. *Immunity* **16**: 769–777.

Grillot D.A., Merino R., Pena J.C., Fanslow W.C., Finkelman F.D., Thompson C.B., and Nunez G. 1996. bcl-x exhibits regulated expression during B cell development and activation and modulates lymphocyte survival in transgenic mice. *J. Exp. Med.* **183**: 381–391.

Hartwell L.H. and Unger M.W. 1977. Unequal division in *Saccharomyces cerevisiae* and its implications for the control of cell division. *J. Cell Biol.* **75**: 422–435.

Iritani B.M. and Eisenman R.N. 1999. c-Myc enhances protein synthesis and cell size during B lymphocyte development. *Proc. Natl. Acad. Sci.* **96**: 13180–13185.

Iritani B.M., Delrow J., Grandori C., Gomez I., Klacking M., Carlos L.S., and Eisenman R.N. 2002. Modulation of T-lymphocyte development, growth and cell size by the Myc antagonist and transcriptional repressor Mad1. *EMBO J.* **21**: 4820–4830.

Johnston G.C., Pringle J.R., and Hartwell L.H. 1977. Coordination of growth with cell division in the yeast *Saccharomyces cerevisiae*. *Exp. Cell Res.* **105**: 79–98.

Johnston L.A., Prober D.A., Edgar B.A., Eisenman R.N., and Gallant P. 1999. *Drosophila* myc regulates cellular growth during development. *Cell* **98**: 779–790.

Kedl R.M., Rees W.A., Hildeman D.A., Schaefer B., Mitchell T., Kappler J., and Marrack P. 2000. T cells compete for access to antigen-bearing antigen-presenting cells. *J. Exp. Med.* **192:** 1105–1113.

Killander D. and Zetterberg A. 1965. A quantitative cytochemical investigation of the relationship between cell mass and initiation of DNA synthesis in mouse fibroblasts in vitro. *Exp. Cell Res.* **40:** 12–20.

Kohn A.D., Summers S.A., Birnbaum M.J., and Roth R.A. 1996. Expression of a constitutively active Akt Ser/Thr kinase in 3T3-L1 adipocytes stimulates glucose uptake and glucose transporter 4 translocation. *J. Biol. Chem.* **271:** 31372–31378.

Krauss S., Brand M.D., and Buttgereit F. 2001. Signaling takes a breath: New quantitative perspectives on bioenergetics and signal transduction. *Immunity* **15:** 497–502.

Ku C.C., Murakami M., Sakamoto A., Kappler J., and Marrack P. 2000. Control of homeostasis of CD8+ memory T cells by opposing cytokines. *Science* **288:** 675–678.

Kuo C.T., Veselits M.L., and Leiden J.M. 1997. LKLF: A transcriptional regulator of single-positive T cell quiescence and survival. *Science* **277:** 1986–1990.

Kwon C.H., Zhu X., Zhang J., Knoop L.L., Tharp R., Smeyne R.J., Eberhart C.G., Burger P.C., and Baker S.J. 2001. Pten regulates neuronal soma size: A mouse model of Lhermitte-Duclos disease. *Nat. Genet.* **29:** 404–411.

Langdon W.Y., Harris A.W., Cory S., and Adams J.M. 1986. The c-myc oncogene perturbs B lymphocyte development in E-mu-myc transgenic mice. *Cell* **47:** 11–18.

Levens D. 2002. Disentangling the MYC web. *Proc. Natl. Acad. Sci.* **99:** 5757–5759.

Luscher B. 2001. Function and regulation of the transcription factors of the Myc/Max/Mad network. *Gene* **277:** 1–14.

Malstrom S., Tili E., Kappes D., Ceci J.D., and Tsichlis P.N. 2001. Tumor induction by an Lck-MyrAkt transgene is delayed by mechanisms controlling the size of the thymus. *Proc. Natl. Acad. Sci.* **98:** 14967–14972.

Neufeld T.P., de la Cruz A.F., Johnston L.A., and Edgar B.A. 1998. Coordination of growth and cell division in the *Drosophila* wing. *Cell* **93:** 1183–1193.

Nunez G., London L., Hockenbery D., Alexander M., McKearn J.P., and Korsmeyer S.J. 1990. Deregulated Bcl-2 gene expression selectively prolongs survival of growth factor-deprived hemopoietic cell lines. *J. Immunol.* **144:** 3602–3610.

Oldham S., Montagne J., Radimerski T., Thomas G., and Hafen E. 2000. Genetic and biochemical characterization of dTOR, the *Drosophila* homolog of the target of rapamycin. *Genes Dev.* **14:** 2689–2694.

Osthus R.C., Shim H., Kim S., Li Q., Reddy R., Mukherjee M., Xu Y., Wonsey D., Lee L.A., and Dang C.V. 2000. Deregulation of glucose transporter 1 and glycolytic gene expression by c-Myc. *J. Biol. Chem.* **275:** 21797–21800.

Pfeiffer T., Schuster S., and Bonhoeffer S. 2001. Cooperation and competition in the evolution of ATP-producing pathways. *Science* **292:** 504–507.

Plas D.R., Talapatra S., Edinger A.L., Rathmell J.C., and Thompson C.B. 2001. Akt and Bcl-xL promote growth factor-independent survival through distinct effects on mitochondrial physiology. *J. Biol. Chem.* **276:** 12041–12048.

Polymenis M. and Schmidt E.V. 1997. Coupling of cell division to cell growth by translational control of the G1 cyclin CLN3 in yeast. *Genes Dev.* **11:** 2522–2531.

Rathmell J.C., Farkash E.A., Gao W., and Thompson C.B. 2001. IL-7 enhances the survival and maintains the size of naive T cells. *J. Immunol.* **167:** 6869–6876.

Rathmell J.C., Vander Heiden M.G., Harris M.H., Frauwirth K.A., and Thompson C.B. 2000. In the absence of extrinsic signals, nutrient utilization by lymphocytes is insufficient to maintain either cell size or viability. *Mol. Cell* **6:** 683–692.

Schmelzle T. and Hall M.N. 2000. TOR, a central controller of cell growth. *Cell* **103:** 253–262.

Schmidt E.V. 1999. The role of c-myc in cellular growth control. *Oncogene* **18:** 2988–2996.

Schuhmacher M., Staege M.S., Pajic A., Polack A., Weidle U.H., Bornkamm G.W., Eick D., and Kohlhuber F. 1999. Control of cell growth by c-Myc in the absence of cell division. *Curr. Biol.* **9:** 1255–1258.

Shioi T., Kang P.M., Douglas P.S., Hampe J., Yballe C.M., Lawitts J., Cantley L.C., and Izumo S. 2000. The conserved phosphoinositide 3-kinase pathway determines heart size in mice. *EMBO J.* **19:** 2537–2548.

Trumpp A., Refaeli Y., Oskarsson T., Gasser S., Murphy M., Martin G.R., and Bishop J.M. 2001. c-Myc regulates mammalian body size by controlling cell number but not cell size. *Nature* **414:** 768–773.

Tuttle R.L., Gill N.S., Pugh W., Lee J.P., Koeberlein B., Furth E.E., Polonsky K.S., Naji A., and Birnbaum M.J. 2001. Regulation of pancreatic beta-cell growth and survival by the serine/threonine protein kinase Akt1/PKBalpha. *Nat. Med.* **7:** 1133–1137.

Vander Heiden M.G., Chandel N.S., Schumacker P.T., and Thompson C.B. 1999. Bcl-xL prevents cell death following growth factor withdrawal by facilitating mitochondrial ATP/ADP exchange. *Mol. Cell* **3:** 159–167.

Vander Heiden M.G., Chandel N.S., Williamson E.K., Schumacker P.T., and Thompson C.B. 1997. Bcl-xL regulates the membrane potential and volume homeostasis of mitochondria. *Cell* **91:** 627–637.

Vander Heiden M.G., Plas D.R., Rathmell J.C., Fox C.J., Harris M.H., and Thompson C.B. 2001. Growth factors can influence cell growth and survival through effects on glucose metabolism. *Mol. Cell. Biol.* **21:** 5899–5912.

Volarevic S., Stewart M.J., Ledermann B., Zilberman F., Terracciano L., Montini E., Grompe M., Kozma S.C., and Thomas G. 2000. Proliferation, but not growth, blocked by conditional deletion of 40S ribosomal protein S6. *Science* **288:** 2045–2047.

von Manteuffel S.R., Dennis P.B., Pullen N., Gingras A.C., Sonenberg N., and Thomas G. 1997. The insulin-induced signalling pathway leading to S6 and initiation factor 4E binding protein 1 phosphorylation bifurcates at a rapamycin-sensitive point immediately upstream of p70s6k. *Mol. Cell. Biol.* **17:** 5426–5436.

Weng Q.P., Andrabi K., Kozlowski M.T., Grove J.R., and Avruch J. 1995. Multiple independent inputs are required for activation of the p70 S6 kinase. *Mol. Cell. Biol.* **15:** 2333–2340.

Whetton A.D. and Dexter T.M. 1983. Effect of haematopoietic cell growth factor on intracellular ATP levels. *Nature* **303:** 629–631.

Xu Z., Stokoe D., Kane L.P., and Weiss A. 2002. The inducible expression of the tumor suppressor gene PTEN promotes apoptosis and decreases cell size by inhibiting the PI3K/Akt pathway in Jurkat T cells. *Cell Growth Differ.* **13:** 285–296.

Yen A., Fried J., Kitahara T., Strife A., and Clarkson B.D. 1975a. The kinetic significance of cell size. I. Variation of cell cycle parameters with size measured at mitosis. *Exp. Cell Res.* **95:** 295–302.

Yen A., Fried J., Kitahara T., Strife A., and Clarkson B.D. 1975b. The kinetic significance of cell size. II. Size distributions of resting and proliferating cells during interphase. *Exp. Cell Res.* **95:** 303–310.

Zhang H., Stallock J.P., Ng J.C., Reinhard C., and Neufeld T.P. 2000. Regulation of cellular growth by the *Drosophila* target of rapamycin dTOR. *Genes Dev.* **14:** 2712–2724.

17

Modulating Skeletal Muscle Hypertrophy and Atrophy: Signaling Pathways and Therapeutic Targets

David J. Glass and George D. Yancopoulos
Regeneron Pharmaceuticals
Tarrytown, New York 10591-6707

T HE SKELETAL MUSCLES OF ADULT ANIMALS are continuously shaped by forces and conditions that regulate size, composition, and strength. Weight-bearing exercise, as well as agents such as insulin-like growth factor-1 (IGF-1) and anabolic steroids, can promote muscle growth. On the other hand, the lack of exercise, nerve injuries, or glucocorticoids can cause muscle wasting, which is also associated with diseases such as cancer, AIDS, and sepsis. Although the growth of other biologic structures can involve increases in the number of cells composing that structure, as well as increases in the size of individual cells, the size of adult skeletal muscles is primarily regulated by modulating cell size as opposed to cell number. Muscles are composed of longitudinal groupings of many individual muscle fibers, with each myofiber corresponding to an elongated and multinucleated single cell, and it is the modulation of the diameters of these individual myofibers that results in changes in the size of adult muscles. Hypertrophy refers to increases in the diameter of individual myofibers, as well as to the corresponding increases in the girth of entire muscles, and atrophy refers to reciprocal decreases in size. Muscle hypertrophy is a natural adaptive response to load-bearing exercise and is associated with an enhanced rate of protein synthesis (Goldspink et al. 1983). This increase in protein synthesis allows new contractile filaments to be added to the preexisting muscle fiber, which in turn results in increased size and also enables the muscle to generate greater force.

It has recently been shown that increased weight-bearing, as well as agents that induce hypertrophy such as IGF-1, induce hypertrophy by stimulating the phosphatidylinositol 3-kinase (PI3K)/Akt pathway, resulting in the downstream activation of targets that are required for protein synthesis (Bodine et al. 2001a; Rommel et al. 2001). Although there are multiple ways of activating the PI3K/Akt pathway, weight-bearing exercise may lead to activation of the critical PI3K/Akt pathway by directly inducing muscle expression of IGF-1 (DeVol et al. 1990), which is able to induce hypertrophy of skeletal muscle (Vandenburgh et al. 1991), as most notably demonstrated in transgenic mice in which IGF-1 is overexpressed in skeletal muscle (Coleman et al. 1995; Musaro et al. 2001).

Lack of weight-bearing exercise, as might result from a sedentary lifestyle or forced bed rest, as well as nerve injuries and glucocorticoid treatment, can cause profound muscle atrophy. In addition, cancer and AIDS are disease conditions associated with skeletal muscle atrophy (Jagoe and Goldberg 2001). A variety of circulating proteins have been shown to induce atrophy, including the cachectic cytokine interleukin-1 (IL-1) and tumor necrosis factor (TNF) (Jagoe and Goldberg 2001); more recently it has been shown that the TGFβ family member myostatin can cause atrophy when administered to an adult animal (Zimmers et al. 2002).

Conditions that lead to skeletal muscle atrophy cause a decrease in the size of preexisting muscle fibers, due to increases in the rate of ATP-dependent ubiquitin-mediated proteolysis (Mitch and Goldberg 1996). Recent studies have shown that the expression of two muscle-specific ubiquitin ligase genes, *MuRF1* (Bodine et al. 2001b) and *MAFbx /Atrogin-1* (Bodine et al. 2001b; Gomes et al. 2001), is up-regulated in multiple settings of atrophy. Mice in which either the *MuRF1* or the *MAFbx /Atrogin-1* genes were deleted demonstrated sparing of skeletal muscle during atrophy conditions, as assessed by maintenance of skeletal muscle mass (Bodine et al. 2001b).

In this review, the aforementioned signaling pathways implicated in hypertrophy and atrophy are discussed in greater detail, with an eye toward understanding how to manipulate these pathways for therapeutic benefit in settings of muscle atrophy.

CHALLENGES IN COMBATING ATROPHY BY PROMOTING HYPERTROPHY VIA IGF-1

In theory, muscle atrophy can be combated either by promoting hypertrophic processes or by blocking atrophy pathways. Regarding the former possibility, one of the best-validated hypertrophy-promoting agents is IGF-1. As noted above, IGF-1 expression markedly increases during work-induced hypertrophy (DeVol et al. 1990). More importantly, IGF-1 treatment of muscle cells in vitro induces hypertrophy (Vandenburgh et al. 1991), and transgenic mice in which IGF-1 expression is increased using a muscle-specific promoter have muscles that are at least twofold greater in mass than those of wild-type mice (Coleman et al. 1995; Musaro et al. 2001). These studies indicate that IGF-1 causes muscle hypertrophy and may be normally involved in exercise-induced hypertrophic responses. In addition, muscle-specific overexpression of IGF-1 has been shown to ameliorate the dystrophic phenotype in the *MDX* mouse (Barton et al. 2002) and to accelerate regeneration (Rabinovsky et al. 2003). These observations further establish the potential benefit of IGF-1 to skeletal muscle, although these regenerative effects are more likely mediated by IGF actions on muscle satellite precursor cell proliferation and differentiation as opposed to those on myofiber hypertrophy.

The issue of directly exploiting IGF-1 as a treatment for skeletal muscle atrophy is complicated by several factors. First, the activity of administered IGF-1 seems to be blunted by circulating IGF-1-binding proteins (IGFBP1-6) that can bind and block IGF-1 activity (Firth and Baxter 2002). Furthermore, since IGF-1 can induce proliferation and a plethora of other responses in a wide variety of cell types, the broad application of IGF-1 may negatively affect other tissues, e.g., by promoting aberrant cellular proliferation. Although it still might be possible to administer IGF-1 directly, it is of interest to understand the signaling pathways activated by IGF-1, for the purpose of identifying downstream molecular targets that could be more easily modulated by pharmacologic intervention.

IGF-1 TRIGGERS THE PI3K/Akt SIGNALING PATHWAY DURING HYPERTROPHY

IGF-1 binding activates the IGF-1 receptor (IGFR), a receptor tyrosine kinase. The IGFR subsequently recruits the insulin receptor substrate (IRS-1), which results in the activation of at least two major signaling pathways: the Ras/Raf/Mek/ERK pathway (Rommel et al. 1999) and the PI3K/Akt pathway (Fig. 1A) (Rommel et al. 1999; Zimmerman and Moelling 1999). The Ras/Raf/Mek/ERK pathway is critical in mitosis-competent cells for cell proliferation and cell survival. However, in adult skeletal muscle, the function of the Ras/Raf/Mek/ERK pathway is less clear. Some studies have shown that the pathway downstream of Raf is actually inactivated during hypertrophy (Rommel et al. 1999) as a result of cross-talk from the PI3K/Akt pathway (Rommel et al. 1999; Zimmermann and Moelling 1999; Moelling et al. 2002). In contrast, genetic activation of the PI3K/Akt pathway has been shown to induce

A. Hypertrophy

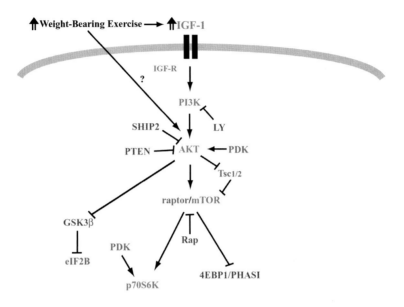

B. Atrophy

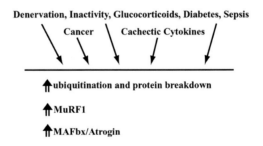

Figure 1. (*A*) Hypertrophy. Weight-bearing exercise leads to increased expression of insulin-like growth factor 1 (IGF-1). The IGF-1 signaling pathways relevant to hypertrophy are presented. Signaling molecules that have been shown to have a negative effect on hypertrophy are colored red. Proteins whose activation induces hypertrophy are shown in green. Proteins that have not been assayed for their role in hypertrophy are shown in blue. (GSK3β) Glycogen synthase kinase 3 beta; (LY - LY294002) a pharmacologic inhibitor of PI3K; (mTOR) mammalian target of rapamycin; (PI3K) phosphatidylinositol 3-kinase; (Rap) rapamycin, a pharmacologic inhibitor of mTOR; (SHIP2) SH2 domain containing inositol phosphatase. (*B*) Atrophy. Multiple different perturbations can induce skeletal muscle atrophy. Those which have been shown to cause the up-regulation of MuRF1 and MAFbx/Atrogin are illustrated. (MuRF1) Muscle RING finger 1; (MAFbx) muscle atrophy F-box protein.

skeletal muscle hypertrophy in vitro or in vivo (Bodine et al. 2001a; Rommel et al. 2001). Furthermore, muscle hypertrophy, induced by either increased load-bearing or by IGF-1, can be inhibited by pharmacologic inhibition or genetic blockade at various points along the PI3K/Akt pathway, in vitro or in vivo (Bodine et al. 2001a; Rommel et al. 2001). Therefore, the PI3K/Akt pathway appears necessary for at least some forms of hypertrophy, including that initiated by weight-bearing and IGF-1, and its activation is able to induce hypertrophy.

PI3K is a lipid kinase; it phosphorylates phosphatidylinositol-4,5-bis-phosphate, producing phosphatidylinositol-3,4,5-trisphosphate (Vivanco and Sawyers 2002; Matsui et al. 2003). Phosphatidylinositol-3,4,5-tris-phosphate is a membrane-binding site for two protein kinases: Akt1, a Ser/Thr-specific kinase also known as PKB (protein kinase B), and a second kinase called PDK1 (phosphoinositide-dependent protein kinase). Akt1 is phosphorylated by PDK1 and thereby activated upon transloca-tion to the membrane (Alessi et al. 1997; Andjelkovic et al. 1997). Once activated, Akt1 phosphorylates an ever-increasing number of identified substrates, including proteins that block apoptosis and induce protein synthesis, gene transcription, and cell proliferation (Vivanco and Sawyers 2002; Matsui et al. 2003).

Pharmacologic inhibition of PI3K blocks IGF1-mediated hypertro-phy in vitro (Bodine et al. 2001a; Rommel et al. 2001), and genetic activa-tion of PI3K causes muscle hypertrophy (Murgia et al. 2000). Knockout mice which are $Akt1^{-/-}$ are smaller than control littermates, demonstrating that Akt1 is required for normal organ growth (Chen et al. 2001). Transgenic mice that express a mutant, constitutively active form of Akt1 in cardiac muscle have hypertrophic hearts (Shioi et al. 2002). During skeletal muscle hypertrophy, endogenous Akt1 phosphorylation increases, as does the relative amount of Akt1 protein (Bodine et al. 2001a). Expression of a "dominant-negative" mutant of Akt1, which inhibits the endogenous activity of this protein, blocks IGF-1-mediated hypertrophy in vitro (Rommel et al. 2001). Expression of an activated form of Akt1 in skeletal muscle induces hypertrophy in vitro or in vivo (Rommel et al. 1999, 2001; Takahashi et al. 2002). These data indicate that Akt1 activity is required for IGF-1-mediated hypertrophy, and expression of activated Akt1 is able to induce muscle hypertrophy. Furthermore, the finding that Akt1 is activated subsequent to PI3K stimulation, and that Akt1 can reca-pitulate the hypertrophic effects seen with PI3K, suggests that PI3K and Akt1 are members of a linear pathway.

There are two other Akt-related genes, Akt2 and Akt3. Although sim-ilar in structure to Akt1, the proteins encoded by Akt2 and Akt3 apparently have distinct functions. For example, knockout mice that are *Akt2[−/−]* are normal in size but have a diabetic phenotype, due to insulin resistance (Cho et al. 2001). The PI3K/Akt2 pathway has been linked to glucose transport via the Glut4 transporter (Katome et al. 2003). Insulin sensitivity and energy metabolism as a result of glucose transport are not completely distinct issues from hypertrophy; during diabetes caused by insulin insensitivity, and a resultant down-regulation of the Akt pathways, skeletal muscle atrophy is also observed (Jagoe and Goldberg 2001). Furthermore, exercise and hypertrophy help to restore glucose homeostasis, probably by the coordinate regulation of Akt1 and Akt2. *Akt1[−/−] Akt2[−/−]* double knockout (DKO) mice were generated (Peng et al. 2003). These DKO mice died shortly after birth and demonstrated a significant decrease in individual muscle cell size in comparison to littermate controls (Peng et al. 2003).

Akt1 can be modulated either by directly controlling its phosphorylation state, or by altering the levels of the phosphorylated phosphoinositide which it binds at the cell membrane (Alessi et al. 1997). Akt1 activity depends on phosphorylation at two sites: Ser-473 and Thr-309 (Alessi et al. 1996). The phosphatase PP2A has been shown to dephosphorylate Akt1 (Andjelkovic et al. 1996; Resjo et al. 2002; Pankov et al. 2003). This phosphatase comprises three components, which are encoded by separate genes (Sontag 2001). PP2A has a wide array of substrates, and the gene that encodes the catalytic subunit of PP2A is ubiquitously expressed (Sontag 2001). However, there may be muscle-specific forms of the PP2A regulatory subunit (Tehrani et al. 1996) that could provide potential for the skeletal-muscle-specific regulation of PP2A.

The Akt1-binding lipid phosphatidylinositol-3,4,5-trisphosphate is dephosphorylated by two lipid phosphatases, SHIP (Damen et al. 1996) (for SH2 domain containing inositol phosphatase; there are two members of the SHIP family, SHIP1 and SHIP2) and PTEN (Fig. 1A) (Stambolic et al. 1998). Overexpression of SHIP2 blocks hypertrophy in muscle (Bodine et al. 2001a; Rommel et al. 2001). Expression of a dominant-negative mutant form of SHIP2 (d.n.SHIP2), in which the phosphatase has been rendered inactive, blocks the dephosphorylation of phosphatidylinositol-3,4,5-trisphosphate and allows Akt1 to remain active (Wada et al. 2001). Expression of d.n.SHIP2 in myotubes causes hypertrophy, coincident with an increase in the levels of phosphorylated Akt1 (Rommel et al. 2001). Similarly, overexpression of PTEN inactivates Akt1 (Stambolic et al. 1998) and causes a decrease in cell size (Goberdhan et al. 1999; Huang et al. 1999; Xu et al. 2002). Expression of a dominant-negative mutant form of PTEN in the heart induces cardiac hypertrophy (Crackower et al. 2002).

The inhibition of PTEN does not seem to be a promising clinical strategy for inducing muscle hypertrophy, since heterozygous *PTEN*[+/−] mice have an increased propensity for cancer (probably caused by activation of the Akt1 pathway in cells capable of proliferation [Waite and Eng 2002]). In the published report of a *SHIP2* knockout, SHIP2[−/−] animals die soon after birth due to hypoglycemia (Clement et al. 2001), perhaps due to an increase in Akt2 activity. However, these *SHIP2* mice are not genetically null for the entire *SHIP2* gene; in these animals, the first 18 exons were left intact (Clement et al. 2001). Given that the hypoglycemia phenotype was observed in the heterozygous *SHIP2*[+/−] animals, it seems possible that some portion of the protein is made and functions as a "dominant-negative" molecule, or as a deregulated protein fragment; however, northern and western data suggested that the SHIP2 mRNA and protein are absent in the knockouts (Clement et al. 2001). Whether SHIP2 might serve as a target for modulating hypertrophy depends on experiments that can discriminate between effects on glucose transport via Akt2 and stimulation of protein synthesis due to activation of Akt1.

In addition to the phosphatases that control Akt1 activity, Akt1 is modulated by proteins that bind it directly. One such protein is carboxy-terminal modulator protein (CTMP) (Maira et al. 2001), which binds specifically to the carboxy-terminal regulatory domain of Akt1. Binding of CTMP reduces the activity of Akt1 by inhibiting its phosphorylation. A second protein that modulates Akt1 activity is Trb3, the mammalian homolog of the *Drosophila* protein tribble (Du et al. 2003). Trb3 associates with Akt1 and decreases its activation by insulin. *Trb3* mRNA levels are regulated by fasting (Du et al. 2003). However, it is not known whether either CTMP or Trb3 plays a role in skeletal muscle.

HYPERTROPHY MEDIATORS DOWNSTREAM OF PI3K AND Akt: mTOR

Early experiments in yeast and *Drosophila* helped to define a particular pathway downstream of PI3K and Akt that can control cell size. For example, genetic loss or inhibition of IRS-1 (Bohni et al. 1999), PI3K (Leevers et al. 1996), TOR (also known as mTOR, FRAP, or RAFT-1 in mammals) (Zhang et al. 2000; Jacinto and Hall 2003), and p70 S6 kinase (Montagne et al. 1999) (p70S6K) all result in decreases in cell size in *Drosophila*. It is somewhat controversial whether these molecules represent a linear signaling pathway downstream of IGF-1 stimulation in mammalian cells. Whereas IGF-1 activates mTOR and p70S6K downstream of PI3K/Akt activation, amino acids can activate mTOR directly, causing a subsequent stimulation of p70S6K activity (Burnett et al. 1998; Hara et al. 1998). Thus, mTOR appears to have an important and central function in integrating a variety of growth signals, from simple nutritional stimulation to activation by protein growth factors, resulting in protein synthesis (Jacinto and Hall 2003). Akt phosphorylates mTOR (Nave et al. 1999), thereby possibly activating it (Scott et al. 1998; Nave et al. 1999), and both Akt phosphorylation (Bodine et al. 2001) and mTOR phosphorylation are increased during muscle hypertrophy (Fig. 1A) (Reynolds et al. 2002).

Rapamycin is a chemical that binds mTOR and inhibits its function. This reagent has been useful in elucidating the mTOR/p70S6K pathway. Rapamycin, when complexed with a protein called FK506-binding protein (FKBP12), binds and inhibits mTOR (Jacinto and Hall 2003). In vitro, when applied to myotube cultures, rapamycin blocks activation of p70S6K downstream of either activated Akt1 or IGF-1 stimulation (Rommel et al. 1999, 2001; Pallafacchina et al. 2002). However, rapamycin does not completely block IGF-mediated hypertrophy in vitro, which suggests that other pathways downstream of Akt1 but independent of mTOR play a role in some settings of hypertrophy. In contrast, in vivo treatment with rapamycin entirely blocks load-induced hypertrophy, and inhibits the activation of p70S6K normally observed in this hypertrophy model (Bodine et al. 2001a). Treatment with rapamycin during load-induced hypertrophy does not block activation of Akt1, again demonstrating that p70S6K activation requires the activation of mTOR (Bodine et al. 2001a). Experiments with rapamycin thus provide pharmacologic evidence for the activation of a linear mTOR/p70S6K pathway during hypertrophy (Fig. 1A).

Genetic support for a linear Akt1/mTOR/p70S6K pathway came recently from reports which demonstrated that the tuberous sclerosis complex-1 and -2 proteins (Tsc1 and Tsc2) inhibit mTOR. Akt1 phosphorylates Tsc2, thereby activating mTOR at least in part by disrupting the Tsc1–Tsc2 complex (Inoki et al. 2002). Furthermore, in insulin- or serum-stimulated cells, activation of p70S6K is inhibited by expression of the Tsc1–Tsc2 complex (Inoki et al. 2002; Tee et al. 2002). This finding demonstrates that introduction of a genetic inhibitor of mTOR, downstream of Akt1, inhibits the activation of p70S6K, adding genetic evidence for an Akt1/mTOR/p70S6K pathway. Since PDK1 has been shown to phosphorylate p70S6 kinase directly (Pullen et al. 1998), one might have assumed that activation of mTOR is dispensible in some settings of p70S6K activation. That may still be the case; alternatively, it may be that p70S6K must first be primed by another kinase, such as mTOR, before being activated by PDK1 (Fig. 1A) (Saitoh et al. 2002; Hannan et al. 2003).

In addition to stimulating p70S6K-mediated protein translation, activation of mTOR inhibits PHAS-1 (also known as 4E-BP), which is a negative regulator of the translation initiation factor eIF-4E (Hara et al. 1997). It has recently been shown that PHAS-1 can directly bind a protein called raptor, which also binds mTOR (Hara et al. 2002; Kim et al. 2002; Loewith et al. 2002). Mutations in PHAS-1 that inhibit interaction with raptor also inhibit mTOR-mediated phosphorylation of PHAS-1 (Choi et al. 2003). Finally, overexpression of raptor can enhance the phosphorylation of PHAS-1 by mTOR in vitro (Choi et al. 2003; Schalm et al. 2003). mTOR binds PHAS-1 via a TOR signaling (TOS) motif in PHAS-1; this same motif is also found in p70S6K (Schalm et al. 2003).

Raptor's binding to mTOR can be enhanced by a protein called GβL (G-protein β-subunit-like protein, pronounced "gable," also known as mLST8) (Loewith et al. 2002; Kim et al. 2003). Upon binding of GβL to mTOR, p70S6K activation is enhanced (Kim et al. 2003). GβL/mLST8 mRNA is expressed in rat (Rodgers et al. 2001) and human (Loewith et al. 2002) skeletal muscle.

In summary, mTOR can increase protein synthesis by modulating two distinct effectors, p70S6K and PHAS-1 (Fig. 1A). Blockade of PHAS-1 may be a potential route to increasing protein synthesis and, therefore, hypertrophy. Stimulation of p70S6K may be a second route. The same phosphatase that dephosphorylates Akt1, PP2A, has also been shown to dephosphorylate p70S6K (Peterson et al. 1999). PP2A therefore seems to be an important modulator of IGF-1 signaling.

ANOTHER HYPERTROPHY MEDIATOR DOWNSTREAM OF PI3K AND Akt: GSK3β

Glycogen synthase kinase 3 beta, GSK3β, is a distinct substrate of Akt1 that has been shown to modulate hypertrophy. GSK3β activity is inhibited by Akt1 phosphorylation (Cross et al. 1995). Expression of a dominant-negative, kinase-inactive form of GSK3β induces dramatic hypertrophy in skeletal myotubes (Rommel et al. 2001). In cardiac hypertrophy, GSK3β phosphorylation is also evident (Hardt and Sadoshima 2002), and expression of a dominant-negative form of GSK3β can induce cardiac hypertrophy (Hardt and Sadoshima 2002). Cardiac hypertrophy was also shown to proceed via a PI3K-dependent process, linking GSK3β to the PI3K/Akt pathway in the heart (Haq et al. 2000). GSK3β inhibits the translation initiation factor eIF2B and thereby blocks protein synthesis (Hardt and Sadoshima 2002). Therefore, GSK3β inhibition may induce hypertrophy by stimulating protein synthesis independently of the mTOR pathway.

SATELLITE CELLS: ROLE IN HYPERTROPHY?

Satellite cells are small mononuclear cells that reside next to skeletal muscle fibers. These cells are stimulated during injury, whereupon they proliferate and fuse to preexisting fibers, aiding in repair. Satellite cells are also stimulated and found to be required for load-induced skeletal muscle hypertrophy (Rosenblatt et al. 1994). The requirement for satellite cell proliferation and fusion in some instances of hypertrophy somewhat blurs the hypertrophy/proliferation distinction, suggesting that strategies to induce proliferation, survival, and fusion of satellite cells may be a distinct way to induce hypertrophy. However, it is not clear whether these approaches can be separated from activation of the Akt1 pathway, since Akt1 is also necessary for cell survival and proliferation in many mitosis-competent cells (Matsui et al. 2003).

CLENBUTEROL AND OTHER AGONISTS OF THE β2 ADRENERGIC RECEPTOR: DO THEY CONVERGE ON THE PI3K/Akt PATHWAY?

Agonists of the β-adrenergic receptor, such as clenbuterol, have been shown to induce skeletal muscle hypertrophy and to blunt atrophy (Hinkle et al. 2002). Most of the β2 adrenergic receptor signaling studies have focused on activation of cAMP pathways, mediated by the $G\alpha_s$ protein. However, it has also been reported that the Gβ/γ proteins activate the PI3K/Akt pathway (Crespo et al. 1994), raising the possibility that just as weight-bearing exercise and IGF-1 seem to converge on the PI3K/Akt pathway so as to mediate muscle hypertrophy, other hypertrophy-promoting agents may also converge on this critical pathway.

ATROPHY VIA INDUCTION OF UBIQUITIN LIGASE PATHWAYS

One might question whether skeletal muscle atrophy is simply the converse of skeletal muscle hypertrophy. However, during skeletal muscle atrophy, an entirely distinct process is stimulated: a dramatic increase in protein degradation and turnover. Furthermore, unique transcriptional pathways are activated, and these are not necessarily the converse of those seen during hypertrophy (Jagoe et al. 2002; Haddad et al. 2003). The stimulation of proteolysis was shown to occur at least in part due to an activation of the ubiquitin–proteasome pathway (Jagoe et al. 2002).

Ubiquitin is a short peptide that can be conjugated to specific protein substrates. A chain reaction may then ensue; a second ubiquitin peptide is ligated to the first, and a third to the second. In this way, a chain of poly-ubiquitin is built onto the substrate, and this ubiquitin chain targets the substrate to a structure called the proteasome, where the substrate is proteolyzed into small peptides (Jagoe et al. 2002).

The addition of ubiquitin to a protein substrate has come to be recognized as an exquisitely modulated process. This process requires three distinct enzymatic components: an E1 ubiquitin-activating enzyme, an E2 ubiquitin-conjugating enzyme, and an E3 ubiquitin-ligating enzyme. The E3 ubiquitin ligases are the components that confer substrate specificity. Several hundred distinct E3s have already been identified, and it is likely that each mediates the ubiquitination of a distinct set of substrates. Thus, ubiquitination may modulate signaling pathways, analogous to phosphorylation. Key signaling components may be activated by the enhanced proteolysis of an inhibitor protein, or signaling components may be directly inactivated via degradation.

The involvement of the ubiquitin–proteasome pathway in skeletal muscle atrophy is well established. Rates of protein breakdown increase during atrophy, and inhibition of the proteasome blocks these increases (Tawa et al. 1997). Furthermore, the amount of polyubiquitin conjugation per total protein increases during atrophy (Lecker et al. 1999), and mRNA levels of genes that encode distinct components of the ubiquitin pathway increase during atrophy (Fig. 1B) (Jagoe et al. 2002).

Differential expression screening studies, designed to identify markers of the atrophy process, identified two genes whose expression increases significantly in multiple models of skeletal muscle atrophy: *MuRF1* (Bodine et al. 2001b) (for *Muscle Ring Finger1*) and *MAFbx* (Bodine et al. 2001b) (for *Muscle Atrophy F-box*; also called *Atrogin-1* [Gomes et al. 2001]). Both of these genes were shown to encode E3 ubiquitin ligases (Bodine et al. 2001b). Expression of *MuRF1* and *MAFbx* is stimulated when the nerve innervating a muscle is cut, thus resulting in paralysis and severe atrophy. These genes are also up-regulated by simple immobilization of the muscle, or by treatment with a glucocorticoid, which causes muscle cachexia (Bodine et al. 2001b). More recently, a predictive experiment was performed, in which sepsis-induced atrophy was induced to determine whether this distinct model of atrophy might also increase expression of *MuRF1* and *MAFbx*. Both of these genes were up-regulated severalfold during sepsis (Wray et al. 2003), and this up-regulation could be blocked by a pharmacologic inhibitor of glucocorticoids (Fig. 1B) (Wray et al. 2003).

MuRF1 encodes a protein that contains three domains: a RING-finger domain (Borden and Freemont 1996) required for ubiquitin ligase activity (Kamura et al. 1999), a "B-box" whose function is unclear, and a "coiled-coil domain" that may be required for the formation of heterodimers between MuRF1 and the related protein MuRF2 (Centner et al. 2001). Proteins that have these three domains have been called RBCC proteins (for RING, B-box, coiled-coil) (Saurin et al. 1996) or TRIM proteins (for tripartite motif) (Reymond et al. 2001). MuRF1 has been demonstrated to have ubiquitin ligase activity that depends on the presence of the RING domain (Bodine et al. 2001b). Although particular substrates have not yet been demonstrated, MuRF1 has been shown to bind to the myofibrillar protein titin, at the M line (Centner et al. 2001; McElhinny et al. 2002; Pizon et al. 2002). Overexpression of MuRF1 results in the disruption of the subdomain of titin that binds MuRF1, suggesting that MuRF1 may play a role in titin turnover (McElhinny et al. 2002). MuRF1 has also been demonstrated to be in the nucleus, and indications that it interacts with transcription-regulating elements such as GMEB-1 suggest a potential role for MuRF1 in modulating transcription (McElhinny et al. 2002). This has not yet been demonstrated, however.

MAFbx/Atrogin-1 contains an F-box domain, a characteristic motif seen in a family of E3 ubiquitin ligases called SCFs (for Skp1, Cullin, F-box) (Jackson and Eldridge 2002). F-box-containing E3 ligases usually bind a substrate only after the substrate has first been posttranslationally modified, for example, by phosphorylation (Jackson and Eldridge 2002). This suggests the possibility of a signaling pathway in which a potential substrate is first phosphorylated as a response to an atrophy-induced stimulus and then degraded via MAFbx.

$MuRF1^{-/-}$ and $MAFbx^{-/-}$ mice appear phenotypically normal. However, under atrophy conditions, significantly less muscle mass is lost in either $MuRF1^{-/-}$ or $MAFbx^{-/-}$ animals in comparison to control littermates (Bodine et al. 2001b). This finding demonstrated for the first time that inhibition of discrete ubiquitin ligases can moderate the amount of muscle lost after an atrophy-inducing stimulus. Therefore, MuRF1 or MAFbx may be attractive targets for pharmacologic intervention. They may also serve as early markers of skeletal muscle atrophy, aiding in the diagnosis of muscle disease.

OTHER TRIGGERS OF ATROPHY:
THE TNFα/NF-κB ATROPHY PATHWAY

The secreted cytokine tumor necrosis factor alpha (TNFα) was purified as "cachectin," (Beutler et al. 1985) and shown to induce cachexia and, thus, skeletal muscle wasting (Fong et al. 1989). TNFα binding to its receptor induces the activation of the Rel/NF-κB (NF-κB) family of transcription factors (von Haehling et al. 2002). NF-κB activation, in turn, is required for cytokine-induced loss of skeletal muscle proteins (Ladner et al. 2003). Tumors that induce cachexia and, thus, muscle atrophy were shown to increase the transcription of *MAFbx/Atrogin-1* (Gomes et al. 2001). Thus, the TNFα/NF-κB pathway is a potential trigger of *MAFbx/Atrogin-1* expression and associated increases in ubiquitination. In addition to inducing atrophy directly, TNFα blocks myogenesis (Guttridge et al. 2000; Coletti et al. 2002) and thus may inhibit the ability of satellite cells to be recruited into the muscle, further enhancing the atrophic effect.

Activation of NF-κB is controlled by the I-κB kinase complex (IKK). Upon phosphorylation by IKK, I-κB is ubiquitinated and targeted to the proteasome for degradation. Inhibition of this process may provide a unique mechanism to inhibit atrophy induced by TNFα and related cytokines.

OTHER TRIGGERS OF ATROPHY: MYOSTATIN

The development of a strain of cows that yield excessive amounts of beef led to the isolation of myostatin, a TGFβ family member (Grobet et al. 1997; McPherron et al. 1997). The "double-muscled" cows were shown to have a mutation in *myostatin* (Grobet et al. 1997; Kambadur et al. 1997; McPherron and Lee 1997). Mice engineered to be *myostatin*-null similarly have a large increase in muscle mass relative to wild-type littermates (McPherron et al. 1997). However, when muscle obtained from *myostatin*$^{-/-}$ animals was analyzed, it was found to be larger as a result of an increase in the number of muscle fibers, and not as a result of hypertrophy. Therefore, it was thought that myostatin acts in a way that is distinct from the activation of atrophy or the inhibition of hypertrophy; it simply blocks the proliferation of muscle precursors, thus decreasing muscle mass. However, a more recent experiment has complicated matters. When adult animals are given an inhibitory antibody to myostatin, they undergo what appears to be muscle hypertrophy (Whittemore et al. 2003). In addition, mice given myostatin undergo muscle atrophy (Zimmers et al. 2002). Whether myostatin is acting directly on preexisting muscle fibers, or whether satellite-cell proliferation and fusion have a requisite role in the maintenance of normal muscle mass, remains to be seen. The possibility exists that blockade of the myostatin pathway may be a distinct route to inhibiting skeletal muscle atrophy. Inhibition of myostatin has already been shown to be beneficial for dystrophic muscle (Bogdanovich et al. 2002); however, this may be a unique circumstance due to the rapid turnover of muscle and the subsequent need for satellite-cell proliferation and differentiation in the dystrophic animal.

CONCLUSION

Recently, a considerable amount of progress has been made in understanding the signaling pathways that mediate skeletal muscle hypertrophy and atrophy. These findings give hope that novel drug targets may be found which block skeletal muscle atrophy and the gradual loss of strength seen even in normal aging. The lack of approved drugs for skeletal muscle disease highlights the need for continued research in this area.

ACKNOWLEDGMENTS

We thank Drs. P. R. Vagelos and L.S. Schleifer, as well as the rest of the Regeneron community, for their enthusiastic support and input. We thank Trevor Stitt for comments on this manuscript. We also thank colleagues at Procter & Gamble Pharmaceuticals for their collaborative input and support. Sincere apologies to scientific colleagues whose work was omitted from this review due to space constraints.

REFERENCES

Alessi D.R., Andjelkovic M., Caudwell B., Cron P., Morrice N., Cohen P., and Hemmings B.A. 1996. Mechanism of activation of protein kinase B by insulin and IGF-1. *EMBO J.* **15:** 6541–6551.

Alessi D.R., James S.R., Downes C.P., Holmes A.B., Gaffney P.R., Reese C.B., and Cohen P. 1997. Characterization of a 3-phosphoinositide-dependent protein kinase which phosphorylates and activates protein kinase Balpha. *Curr. Biol.* **7:** 261–269.

Andjelkovic M., Jakubowicz T., Cron P., Ming X.-F., Han J.-W., and Hemmings B.A. 1996. Activation and phosphorylation of a pleckstrin homology domain containing protein kinase (RAC-PK/PKB) promoted by serum and protein phosphatase inhibitors. *Proc. Natl. Acad. Sci.* **93:** 5699–5704.

Andjelkovic M., Alessi D.R., Meier R., Fernandez A., Lamb N.J., Frech M., Cron P., Cohen P., Lucocq J.M., and Hemmings B.A. 1997. Role of translocation in the activation and function of protein kinase B. *J. Biol. Chem.* **272:** 31515–31524.

Barton E.R., Morris L., Musaro A., Rosenthal N., and Sweeney H.L. 2002. Muscle-specific expression of insulin-like growth factor I counters muscle decline in mdx mice. *J. Cell Biol.* **157:** 137–148.

Beutler B., Mahoney J., Le Trang N., Pekala P., and Cerami A. 1985. Purification of cachectin, a lipoprotein lipase-suppressing hormone secreted by endotoxin-induced

RAW 264.7 cells. *J. Exp. Med.* **161:** 984–995.

Bodine S.C., Stitt T.N., Gonzalez M., Kline W.O., Stover G.L., Bauerlein R., Zlotchenko E., Scrimgeour A., Lawrence J.C., Glass D.J., and Yancopoulos G.D. 2001a. Akt/mTOR pathway is a crucial regulator of skeletal muscle hypertrophy and can prevent muscle atrophy in vivo. *Nat. Cell Biol.* **3:** 1014–1019.

Bodine S.C., Latres E., Baumhueter S., Lai V.K., Nunez L., Clarke B.A., Poueymirou W.T., Panaro F.J., Na E., Dharmarajan K., Pan Z.Q., Valenzuela D.M., DeChiara T.M., Stitt T.N., Yancopoulos G.D., and Glass D.J. 2001b. Identification of ubiquitin ligases required for skeletal muscle atrophy. *Science* **294:** 1704–1708.

Bogdanovich S., Krag T.O., Barton E.R., Morris L.D., Whittemore L.A., Ahima R.S., and Khurana T.S. 2002. Functional improvement of dystrophic muscle by myostatin blockade. *Nature* **420:** 418–421.

Bohni R., Riesgo-Escovar J., Oldham S., Brogiolo W., Stocker H., Andruss B.F., Beckingham K., and Hafen E. 1999. Autonomous control of cell and organ size by CHICO, a *Drosophila* homolog of vertebrate IRS1-4. *Cell* **97:** 865–875.

Borden K.L. and Freemont P.S. 1996. The RING finger domain: A recent example of a sequence-structure family. *Curr. Opin. Struct. Biol.* **6:** 396–401.

Burnett P.E., Barrow R.K., Cohen N.A., Snyder S.H., and Sabatini D.M. 1998. RAFT1 phosphorylation of the translational regulators p70 S6 kinase and 4E-BP1. *Proc. Natl. Acad. Sci.* **95:** 1432–1437.

Centner T., Yano J., Kimura E., McElhinny A.S., Pelin K., Witt C.C., Bang M.L., Trombitas K., Granzier H., Gregorio C.C., Sorimachi H., and Labeit S. 2001. Identification of muscle specific ring finger proteins as potential regulators of the titin kinase domain. *J. Mol. Biol.* **306:** 717–726.

Chen W.S., Xu P.Z., Gottlob K., Chen M.L., Sokol K., Shiyanova T., Roninson I., Weng W., Suzuki R., Tobe K., Kadowaki T., and Hay N. 2001. Growth retardation and increased apoptosis in mice with homozygous disruption of the Akt1 gene. *Genes Dev.* **15:** 2203–2208.

Cho H., Mu J., Kim J.K., Thorvaldsen J.L., Chu Q., Crenshaw E.B., III, Kaestner K.H., Bartolomei M.S., Shulman G.I., and Birnbaum M.J. 2001. Insulin resistance and a diabetes mellitus-like syndrome in mice lacking the protein kinase Akt2 (PKB beta). *Science* **292:** 1728–1731.

Choi K.M., McMahon L.P., and Lawrence J.C., Jr. 2003. Two motifs in the translational repressor PHAS-I required for efficient phosphorylation by mTOR and recognition by raptor. *J. Biol. Chem.* **278:** 19667–19673.

Clement S., Krause U., Desmedt F., Tanti J.-F., Behrends J., Pesesse X., Sasaki T., Penninger J., Doherty M., Malaisse W., Dumont J.E., Le Marchand-Brustel Y., Erneux C., Hue L., and Schurmans S. 2001. The lipid phosphatase SHIP2 controls insulin sensitivity. *Nature* **409:** 92–97.

Coleman M.E., DeMayo F., Yin K.C., Lee H.M., Geske R., Montgomery C., and Schwartz R.J. 1995. Myogenic vector expression of insulin-like growth factor I stimulates muscle cell differentiation and myofiber hypertrophy in transgenic mice. *J. Biol. Chem.* **270:** 12109–12116.

Coletti D., Yang E., Marazzi G., and Sassoon D. 2002. TNFalpha inhibits skeletal myogenesis through a PW1-dependent pathway by recruitment of caspase pathways. *EMBO J.* **21:** 631–642.

Crackower M.A., Oudit G.Y., Kozieradzki I., Sarao R., Sun H., Sasaki T., Hirsch E., Suzuki A., Shioi T., Irie-Sasaki J., Sah R., Cheng H.Y., Rybin V.O., Lembo G., Fratta L., Oviveirados-Santos A.J., Benovic J.L., Kahn C.R., Izumo S., Steinberg S.F., Wymann M.P., Backx P.H., and Penninger J.M. 2002. Regulation of myocardial contractility and cell size by distinct PI3K-PTEN signaling pathways. *Cell* **110:** 737–749.

Crespo P., Xu N., Simonds W.F., and Gutkind J.S. 1994. Ras-dependent activation of MAP kinase pathway mediated by G-protein beta gamma subunits. *Nature* **369:** 418–420.

Cross D.A., Alessi D.R., Cohen P., Andjelkovich M., and Hemmings B.A. 1995. Inhibition of glycogen synthase kinase-3 by insulin mediated by protein kinase B. *Nature* **378:** 785–789.

Damen J.E., Liu L., Rosten P., Humphries R.K., Jefferson A.B., Majerus P.W., and Krystal G. 1996. The 145-kDa protein induced to associate with Shc by multiple cytokines is an inositol tetraphosphate and phosphatidylinositol 3,4,5-triphosphate 5-phosphatase. *Proc. Natl. Acad. Sci.* **93:** 1689–1693.

DeVol D.L., Rotwein P., Sadow J.L., Novakofski J., and Bechtel P.J. 1990. Activation of insulin-like growth factor gene expression during work-induced skeletal muscle growth. *Am. J. Physiol.* **259:** E89–E95.

Du K., Herzig S., Kulkarni R.N., and Montminy M. 2003. TRB3: A tribbles homolog that inhibits Akt/PKB activation by insulin in liver. *Science* **300:** 1574–1577.

Firth S.M. and Baxter R.C. 2002. Cellular actions of the insulin-like growth factor binding proteins. *Endocr. Rev.* **23:** 824–854.

Fong Y., Moldawer L.L., Marano M., Wei H., Barber A., Manogue K., Tracey K.J., Kuo G., Fischman D.A., and Cerami A., et al. 1989. Cachectin/TNF or IL-1 alpha induces cachexia with redistribution of body proteins. *Am. J. Physiol.* **256:** R659–R665.

Goberdhan D.C., Paricio N., Goodman E.C., Mlodzik M., and Wilson C. 1999. *Drosophila* tumor suppressor PTEN controls cell size and number by antagonizing the Chico/PI3-kinase signaling pathway. *Genes Dev.* **13:** 3244–3258.

Goldspink D.F., Garlick P.J., and McNurlan M.A. 1983. Protein turnover measured in vivo and in vitro in muscles undergoing compensatory growth and subsequent denervation atrophy. *Biochem. J.* **210:** 89–98.

Gomes M.D., Lecker S.H., Jagoe R.T., Navon A., and Goldberg A.L. 2001. Atrogin-1, a muscle-specific F-box protein highly expressed during muscle atrophy. *Proc. Natl. Acad. Sci.* **98:** 14440–14445.

Grobet L., Martin L.J., Poncelet D., Pirottin D., Brouwers B., Riquet J., Schoeberlein A., Dunner S., Menissier F., Massabanda J., Fries R., Hanset R., and Georges M. 1997. A deletion in the bovine myostatin gene causes the double-muscled phenotype in cattle. *Nat. Genet.* **17:** 71–74.

Guttridge D.C., Mayo M.W., Madrid L.V., Wang C.Y., and Baldwin A.S., Jr. 2000. NF-kappaB-induced loss of MyoD messenger RNA: Possible role in muscle decay and cachexia. *Science* **289:** 2363–2366.

Haddad F., Roy R.R., Zhong H., Edgerton V.R., and Baldwin K.M. 2003. Atrophy responses to muscle inactivity. II. Molecular markers of protein deficits. *J. Appl. Physiol.* **95:** 791–802.

Hannan K.M., Thomas G., and Pearson R.B. 2003. Activation of S6K1 (p70 ribosomal protein S6 kinase 1) requires an initial calcium-dependent priming event involving formation of a high-molecular-mass signalling complex. *Biochem. J.* **370:** 469–477.

Haq S., Choukroun G., Kang Z.B., Ranu H., Matsui T., Rosenzweig A., Molkentin J.D., Alessandrini A., Woodgett J., Hajjar R., Michael A., and Force T. 2000. Glycogen synthase kinase-3beta is a negative regulator of cardiomyocyte hypertrophy. *J. Cell Biol.* **151:** 117–130.

Hara K., Maruki Y., Long X., Yoshino K., Oshiro N., Hidayat S., Tokunaga C., Avruch J., and

Yonezawa K. 2002. Raptor, a binding partner of target of rapamycin (TOR), mediates TOR action. *Cell* **110:** 177–189.

Hara K., Yonezawa K., Weng Q.P., Kozlowski M.T., Belham C., and Avruch J. 1998. Amino acid sufficiency and mTOR regulate p70 S6 kinase and eIF-4E BP1 through a common effector mechanism (erratum in *J. Biol. Chem.* [1998] **273:** 22160). *J. Biol. Chem.* **273:** 14484–14494.

Hara K., Yonezawa K., Kozlowski M.T., Sugimoto T., Andrabi K., Weng Q.P., Kasuga M., Nishimoto I., and Avruch J. 1997. Regulation of eIF-4E BP1 phosphorylation by mTOR. *J. Biol. Chem.* **272:** 26457–26463.

Hardt S.E. and Sadoshima J. 2002. Glycogen synthase kinase-3beta: A novel regulator of cardiac hypertrophy and development. *Circ. Res.* **90:** 1055–1063.

Hinkle R.T., Hodge K.M., Cody D.B., Sheldon R.J., Kobilka B.K., and Isfort R.J. 2002. Skeletal muscle hypertrophy and anti-atrophy effects of clenbuterol are mediated by the beta2-adrenergic receptor. *Muscle Nerve* **25:** 729–734.

Huang H., Potter C.J., Tao W., Li D.M., Brogiolo W., Hafen E., Sun H., and Xu T. 1999. PTEN affects cell size, cell proliferation and apoptosis during *Drosophila* eye development. *Development* **126:** 5365–5372.

Inoki K., Li Y., Zhu T., Wu J., and Guan K.L. 2002. TSC2 is phosphorylated and inhibited by Akt and suppresses mTOR signalling. *Nat. Cell Biol.* **4:** 648–657.

Jacinto E. and Hall M.N. 2003. TOR signalling in bugs, brain, and brawn. *Nat. Rev. Mol. Cell. Biol.* **4:** 117–126.

Jackson P.K. and Eldridge A.G. 2002. The SCF ubiquitin ligase: An extended look. *Mol. Cell* **9:** 923–925.

Jagoe R.T. and Goldberg A.L. 2001. What do we really know about the ubiquitin-proteasome pathway in muscle atrophy? *Curr. Opin. Clin. Nutr. Metab. Care* **4:** 183–190.

Jagoe R.T., Lecker S.H., Gomes M., and Goldberg A.L. 2002. Patterns of gene expression in atrophying skeletal muscles: Response to food deprivation. *FASEB J.* **16:** 1697–1712.

Kambadur R., Sharma M., Smith T.P., and Bass J.J. 1997. Mutations in myostatin (GDF8) in double-muscled Belgian Blue and Piedmontese cattle. *Genome Res.* **7:** 910–916.

Kamura T., Koepp D.M., Conrad M.N., Skowyra D., Moreland R.J., Iliopoulos O., Lane W.S., Kaelin W.G., Jr., Elledge S.J., Conaway R.C., Harper J.W., and Conaway J.W. 1999. Rbx1, a component of the VHL tumor suppressor complex and SCF ubiquitin ligase. *Science* **284:** 657–661.

Katome T., Obata T., Matsushima R., Masuyama N., Cantley L.C., Gotoh Y., Kishi K., Shiota H., and Ebina Y. 2003. Use of RNA-interference-mediated gene silencing and adenoviral overexpression to elucidate the roles of AKT/PKB-isoforms in insulin actions. *J. Biol. Chem.* **278:** 28312–28323.

Kim D.H., Sarbassov D.D., Ali S.M., King J.E., Latek R.R., Erdjument-Bromage H., Tempst P., and Sabatini D.M. 2002. mTOR interacts with raptor to form a nutrient-sensitive complex that signals to the cell growth machinery. *Cell* **110:** 163–175.

Kim D.H., Sarbassov D.D., Ali S.M., Latek R.R., Guntur K.V., Erdjument-Bromage H., Tempst P., and Sabatini D.M. 2003. GbetaL, a positive regulator of the rapamycin-sensitive pathway required for the nutrient-sensitive interaction between raptor and mTOR. *Mol. Cell* **11:** 895–904.

Ladner K.J., Caligiuri M.A., and Guttridge D.C. 2003. Tumor necrosis factor-regulated biphasic activation of NF-kappa B is required for cytokine-induced loss of skeletal muscle gene products. *J. Biol. Chem.* **278:** 2294–2303.

Lecker S.H., Solomon V., Price S.R., Kwon Y.T., Mitch W.E., and Goldberg A.L. 1999.

Ubiquitin conjugation by the N-end rule pathway and mRNAs for its components increase in muscles of diabetic rats. *J. Clin. Invest.* **104:** 1411–1420.

Leevers S.J., Weinkove D., MacDougall L.K., Hafen E., and Waterfield M.D. 1996. The *Drosophila* phosphoinositide 3-kinase Dp110 promotes cell growth. *EMBO J.* **15:** 6584–6594.

Loewith R. Jacinto E., Wullschleger S., Lorberg A., Crespo J.L., Bonenfant D., Oppliger W., Jenoe P., and Hall M.N. 2002. Two TOR complexes, only one of which is rapamycin sensitive, have distinct roles in cell growth control. *Mol. Cell* **10:** 457–468.

Maira S.M., Galetic I., Brazil D.P., Kaech S., Ingley E., Thelen M., and Hemmings B.A. 2001. Carboxyl-terminal modulator protein (CTMP), a negative regulator of PKB/Akt and v-Akt at the plasma membrane. *Science* **294:** 374–380.

Matsui T., Nagoshi T., and Rosenzweig A. 2003. Akt and PI 3-kinase signaling in cardiomyocyte hypertrophy and survival. *Cell Cycle* **2:** 220–223.

McElhinny A.S., Kakinuma K., Sorimachi H., Labeit S., and Gregorio C.C. 2002. Muscle-specific RING finger-1 interacts with titin to regulate sarcomeric M-line and thick filament structure and may have nuclear functions via its interaction with glucocorticoid modulatory element binding protein-1. *J. Cell Biol.* **157:** 125–136.

McPherron A.C. and Lee S.J. 1997. Double muscling in cattle due to mutations in the myostatin gene. *Proc. Natl. Acad. Sci.* **94:** 12457–12461.

McPherron A.C., Lawler A.M., and Lee S.J. 1997. Regulation of skeletal muscle mass in mice by a new TGF-beta superfamily member. *Nature* **387:** 83–90.

Mitch W.E. and Goldberg A.L. 1996. Mechanisms of muscle wasting. The role of the ubiquitin-proteasome pathway. *N. Engl. J. Med.* **335:** 1897–1905.

Montagne J., Stewart M.J., Stocker H., Hafen E., Kozma S.C., and Thomas G. 1999. *Drosophila* S6 kinase: A regulator of cell size. *Science* **285:** 2126–2129.

Murgia M., Serrano A.L., Calabria E., Pallafacchina G., Lomo T., and Schiaffino S. 2000. Ras is involved in nerve-activity-dependent regulation of muscle genes. *Nat. Cell Biol.* **2:** 142–147.

Musaro A., McCullagh K., Paul A., Houghton L., Dobrowolny G., Molinaro M., Barton E.R., Sweeney H.L., and Rosenthal N. 2001. Localized Igf-1 transgene expression sustains hypertrophy and regeneration in senescent skeletal muscle. *Nat. Genet.* **27:** 195–200.

Nave B.T., Ouwens M., Withers D.J., Alessi D.R., and Shepherd P.R. 1999. Mammalian target of rapamycin is a direct target for protein kinase B: Identification of a convergence point for opposing effects of insulin and amino-acid deficiency on protein translation. *Biochem. J.* **344:** 427–431.

Pallafacchina G., Calabria E., Serrano A.L., Kalhovde J.M., and Schiaffino S. 2002. A protein kinase B-dependent and rapamycin-sensitive pathway controls skeletal muscle growth but not fiber type specification. *Proc. Natl. Acad. Sci.* **99:** 9213–9218.

Pankov R., Cukierman E., Clark K., Matsumoto K., Hahn C., Poulin B., and Yamada K.M. 2003. Specific beta 1 integrin site selectively regulates Akt/protein kinase B signaling via local activation of protein phosphatase 2A. *J. Biol. Chem.* **278:** 18671–18681.

Peng X.D., Xu P.Z., Chen M.L., Hahn-Windgassen A., Skeen J., Jacobs J., Sundararajan D., Chen W.S., Crawford S.E., Coleman K.G., and Hay N. 2003. Dwarfism, impaired skin development, skeletal muscle atrophy, delayed bone development, and impeded adipogenesis in mice lacking Akt1 and Akt2. *Genes Dev.* **17:** 1352–1365.

Peterson R.T., Desai B.N., Hardwick J.S., and Schreiber S.L. 1999. Protein phosphatase 2A interacts with the 70-kDa S6 kinase and is activated by inhibition of FKBP12-rapamycinassociated protein. *Proc. Natl. Acad. Sci.* **96:** 4438–4442.

Pizon V., Iakovenko A., Van Der Ven P.F., Kelly R., Fatu C., Furst D.O., Karsenti E., and Gautel M. 2002. Transient association of titin and myosin with microtubules in nascent myofibrils directed by the MURF2 RING-finger protein. *J. Cell Sci.* **115:** 4469–4482.

Pullen N., Dennis P.B., Andjelkovic M., Dufner A., Kozma S.C., Hemmings B.A., and Thomas G. 1998. Phosphorylation and activation of p70s6k by PDK1. *Science* **279:** 707–710.

Rabinovsky E.D., Gelir E., Gelir S., Lui H., Kattash M., DeMayo F.J., Shenaq S.M., and Schwartz R.J. 2003. Targeted expression of IGF-1 transgene to skeletal muscle accelerates muscle and motor neuron regeneration. *FASEB J.* **17:** 53–55.

Resjo S., Goransson O., Harndahl L., Zolnierowicz S., Manganiello V., and Degerman E. 2002. Protein phosphatase 2A is the main phosphatase involved in the regulation of protein kinase B in rat adipocytes. *Cell. Signal.* **14:** 231–238.

Reymond A., Meroni G., Fantozzi A., Merla G., Cairo S., Luzi L., Riganelli D., Zanaria E., Messali S., Cainarca S., Guffanti A., Minucci S., Pelicci P.G., and Ballabio A. 2001. The tripartite motif family identifies cell compartments. *EMBO J.* **20:** 2140–2151.

Reynolds T.H., IV, Bodine S.C., and Lawrence J.C., Jr. 2002. Control of Ser2448 phosphorylation in the mammalian target of rapamycin by insulin and skeletal muscle load. *J. Biol. Chem.* **277:** 17657–17662.

Rodgers B.D., Levine M.A., Bernier M., and Montrose-Rafizadeh C. 2001. Insulin regulation of a novel WD-40 repeat protein in adipocytes. *J. Endocrinol.* **168:** 325–332.

Rommel C., Bodine S.C., Clarke B.A., Rossman R., Nunez L., Stitt T.N., Yancopoulos G.D., and Glass D.J. 2001. Mediation of IGF-1-induced skeletal myotube hypertrophy by PI(3)K/Akt/mTOR and PI(3)K/Akt/GSK3 pathways. *Nat. Cell Biol.* **3:** 1009–1013.

Rommel C., Clarke B.A., Zimmermann S., Nunez L., Rossman R., Reid K., Moelling K., Yancopoulos G.D., and Glass D.J. 1999. Differentiation stage-specific inhibition of the raf-MEK-ERK pathway by Akt. *Science* **286:** 1738–1741.

Rosenblatt J.D., Yong D., and Parry D.J. 1994. Satellite cell activity is required for hypertrophy of overloaded adult rat muscle. *Muscle Nerve* **17:** 608–613.

Saitoh M., Pullen N., Brennan P., Cantrell D., Dennis P.B., and Thomas G. 2002. Regulation of an activated S6 kinase 1 variant reveals a novel mammalian target of rapamycin phosphorylation site. *J. Biol. Chem.* **277:** 20104–20112.

Saurin A.J., Borden K.L., Boddy M.N., and Freemont P.S. 1996. Does this have a familiar RING? *Trends Biochem. Sci.* **21:** 208–214.

Schalm S.S., Fingar D.C., Sabatini D.M., and Blenis J. 2003. TOS motif-mediated raptor binding regulates 4E-BP1 multisite phosphorylation and function. *Curr. Biol.* **13:** 797–806.

Scott P.H., Brunn G.J., Kohn A.D., Roth R.A., and Lawrence J.C. 1998. Evidence of insulin-stimulated phosphorylation and activation of the mammalian target of rapamycin mediated by a protein kinase B signaling pathway. *Proc. Natl. Acad. Sci.* **95:** 7772–7777.

Shioi T., McMullen J.R., Kang P.M., Douglas P.S., Obata T., Franke T.F., Cantley L.C., and Izumo S. 2002. Akt/protein kinase B promotes organ growth in transgenic mice. *Mol. Cell. Biol.* **22:** 2799–2809.

Sontag, E. 2001. Protein phosphatase 2A: The Trojan Horse of cellular signaling. *Cell. Signal.* **13:** 7–16.

Stambolic V., Suzuki A., de la Pompa J.L., Brothers G.M., Mirtsos C., Sasaki T., Ruland J., Penninger J.M., Siderovski D.P., and Mak T.W. 1998. Negative regulation of PKB/Akt-dependent cell survival by the tumor suppressor PTEN. *Cell* **95:** 29–39.

Takahashi A., Kureishi Y., Yang J., Luo Z., Guo K., Mukhopadhyay D., Ivashchenko Y., Branellec D., and Walsh K. 2002. Myogenic Akt signaling regulates blood vessel recruitment during myofiber growth. *Mol. Cell. Biol.* **22:** 4803–4814.

Tawa N.E., Jr., Odessey R., and Goldberg A.L. 1997. Inhibitors of the proteasome reduce the accelerated proteolysis in atrophying rat skeletal muscles. *J. Clin. Invest.* **100:** 197–203.

Tee A.R., Fingar D.C., Manning B.D., Kwiatkowski D.J., Cantley L.C., and Blenis J. 2002. Tuberous sclerosis complex-1 and -2 gene products function together to inhibit mammalian target of rapamycin (mTOR)-mediated downstream signaling. *Proc. Natl. Acad. Sci.* **99:** 13571–13576.

Tehrani M.A., Mumby M.C., and Kamibayashi C. 1996. Identification of a novel protein phosphatase 2A regulatory subunit highly expressed in muscle. *J. Biol. Chem.* **271:** 5164–5170.

Vandenburgh H.H., Karlisch P., Shansky J., and Feldstein R. 1991. Insulin and IGF-I induce pronounced hypertrophy of skeletal myofibers in tissue culture. *Am. J. Physiol.* **260:** C475–C484.

Vivanco I. and Sawyers C.L. 2002. The phosphatidylinositol 3-kinase AKT pathway in human cancer. *Nat. Rev. Cancer* **2:** 489–501.

von Haehling S., Genth-Zotz S., Anker S.D., and Volk H.D. 2002. Cachexia: A therapeutic approach beyond cytokine antagonism. *Int. J. Cardiol.* **85:** 173–183.

Wada T., Sasaoka T., Funaki M., Hori H., Murakami S., Ishiki M., Haruta T., Asano T., Ogawa W., Ishihara H., and Kobayashi M. 2001. Overexpression of SH2-containing inositol phosphatase 2 results in negative regulation of insulin-induced metabolic actions in 3T3-L1 adipocytes via its 5′-phosphatase catalytic activity. *Mol. Cell. Biol.* **21:** 1633–1646.

Waite K.A. and Eng C. 2002. Protean PTEN: Form and function. *Am. J. Hum. Genet.* **70:** 829–844.

Whittemore L.A., Song K., Li X., Aghajanian J., Davies M., Girgenrath S., Hill J.J., Jalenak M., Kelley P., Knight A., Maylor R., O'Hara D., Pearson A., Quazi A., Ryerson S., Tan X.Y., Tomkinson K.N., Veldman G.M., Widom A., Wright J.F., Wudyka S., Zhao L., and Wolfman N.M. 2003. Inhibition of myostatin in adult mice increases skeletal muscle mass and strength. *Biochem. Biophys. Res. Commun.* **300:** 965–971.

Wray C.J., Mammen J.M., Hershko D.D., and Hasselgren P.O. 2003. Sepsis upregulates the gene expression of multiple ubiquitin ligases in skeletal muscle. *Int. J. Biochem. Cell Biol.* **35:** 698–705.

Xu Z., Stokoe D., Kane L.P., and Weiss A. 2002. The inducible expression of the tumor suppressor gene PTEN promotes apoptosis and decreases cell size by inhibiting the PI3K/Akt pathway in Jurkat T cells. *Cell Growth Differ.* **13:** 285–296.

Zhang H., Stallock J.P., Ng J.C., Reinhard C., and Neufeld T.P. 2000. Regulation of cellular growth by the *Drosophila* target of rapamycin dTOR. *Genes Dev.* **14:** 2712–2724.

Zimmermann S. and Moelling K. 1999. Phosphorylation and regulation of raf by akt (protein kinase B). *Science* **286:** 1741–1744.

Zimmers T.A., Davies M.V., Koniaris L.G., Haynes P., Esquela A.F., Tomkinson K.N., McPherron A.C., Wolfman N.M., and Lee S.J. 2002. Induction of cachexia in mice by systemically administered myostatin. *Science* **296:** 1486–1488.

18

Mechanisms Controlling Heart Growth in Mammals

Julie R. McMullen[1] and Seigo Izumo[1,2]
[1]Cardiovascular Division, Beth Israel Deaconess Medical Center
Harvard Medical School, Boston, Massachusetts 02215
[2]Novartis Institutes for Biomedical Research Cardiovascular Research
Cambridge, Massachusetts 02139

THE HEART IS A FOUR-CHAMBERED ORGAN (two ventricles and two atria) responsible for circulating blood throughout the body. Growth of the mammalian heart, for the most part, is regulated by functional demand. The heart is composed of a number of cell types, including cardiac muscle cells (cardiac myocytes), extracellular matrix, interstitial cells such as fibroblasts, and blood vessels. During postnatal life, growth of the heart occurs primarily as a result of an increase in the size of ventricular cardiac myocytes, known as hypertrophy. Even though ventricular cardiac myocytes make up only one-third of the total cell number, they account for ~70–80% of the heart's mass (Rakusan 1984). For this reason, the mechanisms controlling the growth of ventricular myocytes form the main focus of this review. Growth of cardiac myocytes requires the induction of a number of events, including changes at the level of gene expression, an increase in the overall rate of protein synthesis, and organization of contractile proteins into sarcomeric units.

CARDIAC MYOCYTES

Cardiac myocytes are specialized muscle cells composed of bundles of myofibrils that contain myofilaments. The myofibrils have distinct, repeating, micro-anatomical units, called sarcomeres (Fig. 1). The sar-

A

B

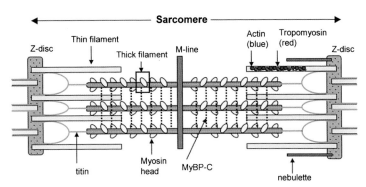

Figure 1. (*A*) A cardiac myocyte composed of bundles of myofibrils. (Reprinted, with permission, from Mann et al. 1991 [copyright Lippincott Williams & Wilkins].) (*B*) Myocyte sarcomeric proteins. The sarcomere lies between two Z discs and is composed of a number of major and minor proteins organized into thick, thin, titin, and nebulette filaments.

comere is the basic contractile unit of the heart. Approximately 50 sarcomeres end to end make up a myofibril, and a bundle of 50–100 myofibrils makes up a muscle cell. The sarcomere is defined as the region of the myofilament structures between two Z discs. The sarcomere consists of a number of major and minor proteins organized into thick filaments (including myosin heavy chain [MHC] and myosin light chain [MLC]), thin filaments (including actin, tropomyosin, troponin C, troponin T, troponin I), and titin and nebulette filaments (Gregorio and Antin 2000; Russell et al. 2000; Sanger et al. 2000; Sanger and Sanger 2001). Z discs are an anchor site for actin filaments, titin, and nebulette filaments, allowing the generation of force by contraction. The M line is the region where the myosin tails are linked and organized.

GROWTH OF THE HEART AT THE CELLULAR LEVEL

Cell growth and cell proliferation are two distinct processes. Proliferation can be defined as cell division, which leads to an increase in cell number, whereas cell growth is defined as macromolecular synthesis, which leads to an increase in cell mass or size (Schmelzle and Hall 2000). The growth of the heart occurs at the cellular level by an increase in the number of cells (hyperplasia, proliferation) or an increase in the size of cells (hypertrophy). In cardiac myocytes, the importance of these mechanisms depends largely on the stage of development and the type of growth stimulus (Zak 1984a,b).

CATEGORIES OF HEART GROWTH

Growth of the heart can be categorized into a number of stages and categories (Fig. 2). These include the embryonic and fetal stages of development occurring in utero, the rapidly growing phase during postnatal development, aging (or senescence), compensatory growth of the adult heart in response to stimuli such as exercise or pathological stimuli (e.g., overload), cardiomyopathy, decompensated growth, and heart failure (Fig. 2).

   Below, we briefly summarize the mechanisms responsible for the normal stages of heart growth; i.e., developmental growth, postnatal growth, and aging. We then describe in more detail the mechanisms that contribute to compensatory growth of the heart. Compensatory growth of

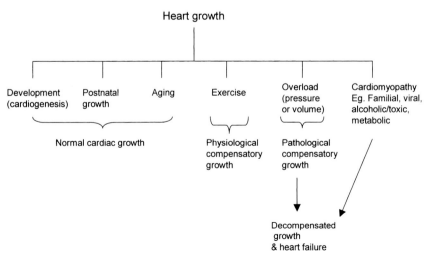

Figure 2. Stages and categories of heart growth.

the adult heart is the main focus of this review for two reasons. First, compensatory growth of the adult heart is associated largely with cell growth rather than cell proliferation. Second, much scientific research has focused on characterizing the signal transduction pathways that are associated with cardiac hypertrophy and failure because it continues to be one of the leading causes of morbidity and mortality in industrialized countries (Levy et al. 1990; Cohn et al. 1997).

Development and Postnatal Growth

The heart is one of the first organs that can be morphologically recognized. The heart develops from the mesodermal germ layer, initially forming a linear heart tube consisting of an inner endothelial tube and an outer myocardial tube. Normal cardiac morphogenesis requires hyperplasia and hypertrophy of cardiac myocytes, both of which are controlled by genetic factors as well as by load (Icardo 1984; Harvey 2002; Olson and Schneider 2003; Solloway and Harvey 2003). Growth signals from the epicardium (thin layer of cells surrounding the heart) and the endocardium (specialized endothelial lining of the heart) are critical for the proliferation and hypertrophic growth of myoctes (Olson and Schneider 2003; Solloway and Harvey 2003).

In mammals, at birth or within weeks thereafter, the majority of cardiac myocytes lose their ability to proliferate, and growth is restricted to hypertrophy of cardiac myocytes (Soonpaa et al. 1996). After a brief spurt in heart growth (relative to body weight) during the early postnatal period, the increase in heart weight is proportional to the increase in body weight (Rakusan 1984).

Aging Process

The term senescence is often used to refer to animals at an age associated with 50% mortality of the colony (Lakatta et al. 2001). Aging animals and humans free of heart disease develop mild left ventricular hypertrophy as a consequence of age-related decreases in the distensibility of the peripheral vasculature. The increase in heart mass is due to an increase in cardiac myocyte size (Rakusan 1984).

Compensatory Growth of the Adult Heart

In the adult, growth of the heart is closely matched to its functional load, and under normal circumstances is mainly constitutive in nature (Zak

1984a,b). Compensatory growth of the adult heart occurs in response to changes in functional load. Many mammalian cells are highly differentiated and reside in a quiescent state. Upon appropriate stimulation, they can reenter the cell cycle and undergo cell division. In contrast, it is generally believed that shortly after the postnatal period the majority of cardiac myocytes are unable to reenter the cell cycle. Thus, growth of the adult heart is largely due to an increase in myocyte size (Zak 1984a,b; Hudlicka and Brown 1996). The inability of postnatal cardiac myocytes to divide has come under some debate in the last decade (Anversa et al. 2002; Anversa and Nadal-Ginard 2002; Pasumarthi and Field 2002). However, estimates of DNA labeling indicate that DNA synthesis is taking place in a minute fraction of the total adult cardiac myocyte population (mouse ventricle: 0.04%; human: 0.052%) (Nakagawa et al. 1988; Soonpaa et al. 1996; MacLellan and Schneider 2000; Anversa et al. 2002; Pasumarthi and Field 2002). Furthermore, there is a critical distinction between DNA synthesis and cell division in the adult cardiac myocyte. DNA synthesis can result in either multinucleation or polyploidy in the absence of cell division.

The term cardiac hypertrophy is most commonly used to describe compensatory growth of the adult heart and has traditionally been defined as an increase in the mass of the heart, produced largely by an increase in the size of terminally differentiated cardiac ventricular myocytes (Zak 1984a,b; Hudlicka and Brown 1996; Schluter and Piper 1999). Because the heart also has two atria, and the ventricle also consists of extracellular matrix and connective tissue cells, this is somewhat of an oversimplification. Although it is not the focus of this review, it is worth noting that cardiac hypertrophy also involves an increase in atrial mass, and an increase in the number of various types of connective tissue cells, as well as deposition of increased amounts of connective tissue proteins in the interstitial spaces (Ferrans 1984). Cardiac hypertrophy is also classically considered to occur in the left ventricle of the adult. The right ventricle pumps blood to the lungs, and the left ventricle pumps blood to the rest of the body. Pulmonary resistance is lower than peripheral resistance; thus, the pressure load on the left ventricle is greater than that on the right. This consequently leads to the larger muscle mass characteristic of the adult left ventricle compared to the right (Anversa et al. 1992).

Compensated Growth: Pathological Versus Physiological

Compensated cardiac growth in the adult can broadly be categorized as either physiological or pathological (Fig. 2). Physiological hypertrophy is characterized by a normal organization of cardiac structure, normal or enhanced cardiac function, and a relatively normal pattern of cardiac gene expression (Schaible and Scheuer 1984; Kaplan et al. 1994; Fagard 1997), whereas pathological hypertrophy is associated with an altered pattern of cardiac gene expression, fibrosis, cardiac dysfunction, and increased morbidity and mortality (Fig. 3) (Ferrans 1984; Schaible and Scheuer 1984; Izumo et al. 1988; Levy et al. 1990; Cohn et al. 1997; Hunter and Chien 1999).

Pathological cardiac hypertrophy occurs in response to diverse cardiovascular disorders, including hypertension, atherosclerosis, valve disease, myocardial infarction, and cardiomyopathy. When disease causes pressure or volume overload of the heart, wall stress on the left ventricle increases. To counterbalance the chronic increase in wall stress, the heart triggers a hypertrophic response (Cooper 1987; Sugden and Clerk 1998; Hunter and Chien 1999). Initially, the enlargement of cardiac myocytes

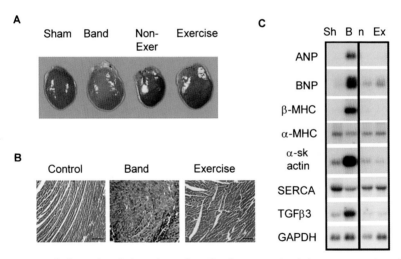

Figure 3. Pathological and physiological cardiac hypertrophy. (A) Representative pictures of hearts from mice subjected to a pathological stimulus (aortic banding [Band] for 1 week), a physiological stimulus (chronic exercise: swimming training for 4 weeks [Exercise]), or no stimulus (sham-operated [Sham] or non-exercise-trained mice [Non-Exer]). (B) Histological analysis of heart sections stained with Masson's trichrome. Representative sections from the LV wall of control mice (non-exercise trained), aortic banded mice (Band) and swimming mice (Exercise). Bars, 10 μM. Sections from sham-operated mice were similar to those from non-swim mice (control). (C) Cardiac gene expression in response to aortic banding or chronic exercise training. Representative northern blot showing total RNA from ventricles of sham (Sh), band (B), non-exercise trained (n), and exercise (Ex). Expression of GAPDH was determined to verify equal loading of RNA. See Table 1 for other abbreviations.

Table 1. Abbreviations used in this chapter

Abbreviation	Name	Abbreviation	Name
AC	Adenyl cyclase	JNK	c-Jun amino-terminal kinase
ACE	Angiotensin converting enzyme	*c-jun*	*jun* oncogene
Ang II	Angiotensin II	LIF	Leukemia inhibitory factor
ANP	Atrial natriuretic peptide	MAPK	Mitogen-activated protein kinase
AR	Adrenergic receptor	MCIP	Myocyte-enriched calcineurin-interacting protein
AT_1/AT_2	Ang II receptors 1 & 2	MEF2	Myocyte enhancer factor 2
BNP	Brain natriuretic peptide	MHC	Myosin heavy chain
ca	Constitutively active	MLC	Myosin light chain
CaMK	Ca^{++}/calmodulin-dependent protein kinase	MLCK	Myosin light-chain kinase
cAMP	Cyclic adenosine monophosphate	mTOR	Mammalian target of rapamycin
CREB	cAMP response element-binding protein	*c-myc*	*myc* oncogene
CT-1	Cardiotrophin-1	NE	Norepinephrine, noradrenaline
DG	Diacylglycerol	NFAT	Nuclear factor of activated T cells
dn	Dominant negative	PE	Phenylephrine
4E-BP1	4E binding protein 1	PGF2α	Prostaglandin F2α
ECM	Extracellular matrix	PI3K	Phosphoinositide 3-kinase
EGF	Epidermal growth factor	PIP_2	Phosphatidyl 4,5-biphosphate
EGFR	EGF receptor	PIP_3	Phosphatidylinositol 3,4,5-triphosphate
eIF4E	Eukaryotic initiation factor 4E	PKA	Protein kinase A
ErbB	EGF family tyrosine kinase receptors	PKC	Protein kinase C
ERK	Extracellular signal-regulated kinase	PLA_2	Phospholipase A_2
ET-1	Endothelin-1	PLC	Phospholipase C
ET_A/ET_B	Endothelin-1 receptors A & B	PLD	Phospholipase D

Table 1. (*continued*)

Abbreviation	Name	Abbreviation	Name
FAK	Focal adhesion kinase	PTEN	Phosphatase and tensin homolog
FAO	Fatty acid oxidation	Ras	Ras oncogene
FGF	Fibroblast growth factor	RAS	Renin angiotensin system
c-*fos*	c-*fos* oncogene	ROCK	RhoKinase
Gα, Gβγ	Subunits of heterotrimeric G proteins	rRNA	Ribosomal RNA
GATA4	GATA-binding protein 4	RTK	Receptor tyrosine kinase
GDP	Guanosine diphosphate	S6	40S ribosomal S6 protein
GH	Growth hormone	SAPK	Stress-activated protein kinase
Gp130	Glycoprotein 130	SERCA	Sarcoplasmic reticulum Ca^{++}-ATPase
GPCR	Heterotrimeric G-protein-coupled receptor	S6K1	70-kDa ribosomal S6 kinase
GSK3β	Glycogen synthase kinase 3β	STAT	Signal transducer and activator of transcription
GTP	Guanosine triphosphate	T_3	Triiodothyronine
HATs	Histone acetyltransferases	T_4	Thyroxine
HB-EGF	Heparin-binding EGF-like growth factor	TGFβ	Transforming growth factor β
HDAC	Histone deacetylase	TH	Thyroid hormone
IGF1	Insulin-like growth factor 1	TNFα	Tumor necrosis factor α
IGF1R	IGF1 receptor	5′TOP	5′Terminal oligopyrimidine
IL	Interleukin	TR	Thyroid hormone receptors
IP_3	Inositol 1,4,5-triphosphate	TSC	Tuberous sclerosis complex
JAK	Janus kinases	UBF	Upstream binding factor

and the formation of new sarcomeres serves to normalize wall stress and permit normal cardiovascular function at rest; i.e., compensated growth. However, function in the hypertrophied heart may eventually decompensate, leading to left ventricle dilation, increased interstitial fibrosis (resulting in increased myocardial stiffness), and heart failure (decompensated or

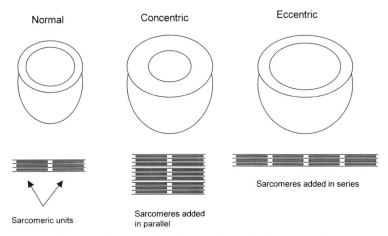

Figure 4. Concentric and eccentric cardiac hypertrophy.

maladaptive growth; Fig. 2). Thus, the increase in mass associated with pathological hypertrophy is due in large part to hypertrophy of cardiac myocytes; however, hyperplasia of fibroblasts, and the accumulation of extracellular matrix components, including collagens, also contributes (Weber and Brilla 1991; Weber et al. 1993). It should be noted that decompensated growth of the heart is commonly considered a progression occurring after compensated growth of the heart. However, some reports suggest that decompensated growth may also occur in the absence of compensated growth (Liao et al. 2001; Petrich et al. 2002, 2003; Yamamoto et al. 2003).

Physiological cardiac hypertrophy occurs in response to increased physical activity or chronic exercise training (Schaible and Scheuer 1984; Hudlicka and Brown 1996; Froelicher and Myers 2000). More active animals of the same or related species have heavier hearts: e.g., wild hare compared to domestic rabbit, wild rodents compared to laboratory rodents (Schaible and Scheuer 1984). In humans and animals, intense, prolonged exercise training results in an increase in cardiac mass (Schaible and Scheuer 1984; Kaplan et al. 1994; Fagard 1997; Froelicher and Myers 2000; McMullen et al. 2003). Unlike pathological hypertrophy, exercise-induced cardiac hypertrophy does not decompensate into dilated cardiomyopathy or heart failure.

Concentric and Eccentric Cardiac Hypertrophy

Depending on the nature of the initiating stimulus, the increase in heart mass during compensatory hypertrophy produces characteristic alterations in the volume of the cardiac cavities and in the thickness of its walls (Fig. 4). Morphologically distinct forms of pathological or physiological cardiac hypertrophy have been identified and classified as either concentric or eccentric (Ferrans 1984).

Pathological Hypertrophy

Pathological hypertrophy caused by chronic pressure overload (e.g., hypertension, left ventricular outflow obstruction, aortic coarction) produces an increase in systolic wall stress and results in concentric ventricular hypertrophy. Concentric hypertrophy is characterized by a parallel pattern of sarcomere addition, leading to an increase in myocyte cell width. Morphologically, this cellular adaptation results in hearts with thick walls and relatively small cavities (Fig. 4). In contrast, pathological hypertrophy caused by chronic volume overload (e.g., aortic regurgitation, arteriovenous fistulas) results in eccentric left ventricular hypertrophy. Eccentric hypertrophy is associated with a series pattern of sarcomere addition, leading to an increase in myocyte cell length. Morphologically, this cellular adaptation results in hearts with large dilated cavities and relatively thin walls (Fig. 4) (Ferrans 1984). Regional hypertrophy that occurs in viable myocardium adjacent to and remote from an area of infarction also has the characteristics of eccentric hypertrophy. There are exceptions to the principle that pathological hypertrophy occurs as a result of excessive increases in external work. For example, cardiomyopathy produced by inherited mutations in components of the sarcomere can result in massive asymmetric or concentric hypertrophy in the absence of augmented peripheral hemodynamic requirements. It is possible that the massive myofibrillar disarray that characterizes this genetic form of hypertrophy increases internal cardiac work, which in turn increases cardiac mass (Walsh 2001).

Physiological Hypertrophy

Isotonic exercise such as running, walking, cycling, and swimming involves movement of large muscle groups and produces eccentric hypertrophy by volume overload (Schaible and Scheuer 1984; Froelicher and Myers 2000). In contrast, isometric exercise (e.g., weight lifting) involves developing muscular tension against resistance without much movement. Such exercise causes a pressure load on the heart rather than a flow load and results in concentric hypertrophy. Flow does not increase much during isometric exercise because of greater pressure within the active muscle groups (Schaible and Scheuer 1984; Froelicher and Myers 2000).

OVERVIEW OF VENTRICULAR MYOCYTE HYPERTROPHY

Hypertrophy of ventricular myocytes is usually associated with stimulation of a hypertrophic program of gene expression (e.g., induction of immediate early genes such as c-*fos*, c-*jun*, c-*myc*; reactivation of fetal genes such as atrial natriuretic peptide [ANP], β-MHC, skeletal α-actin), an increase in the overall rate of protein synthesis, and organization of contractile proteins into sarcomeric units (Izumo et al. 1988; Chien et al. 1993; Sugden and Clerk 1998; Aoki and Izumo 2001). Factors including vasoactive substances, growth factors, hormones, cytokines (interleukin-1 [IL-1], IL-6, cardiotrophin-1 [CT-1]) and changes in energy metabolism can all act to trigger a series of intracellular signaling events (Fig. 5). These

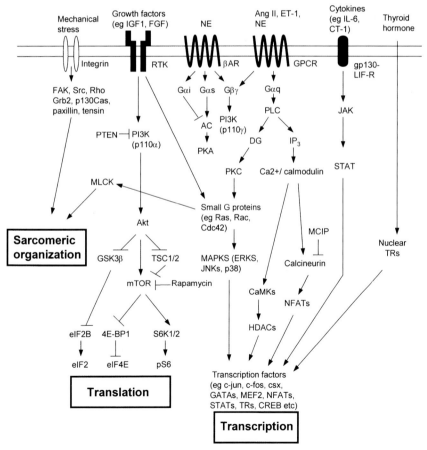

Figure 5. Signaling cascades that have been implicated in mediating cardiac hypertrophy. See Table 1 for abbreviations.

signals are transduced by signaling pathways including protein kinase C (PKC), protein kinase A (PKA), mitogen-activated protein kinases (MAPKs), tyrosine kinases, calcineurin, and phosphoinositide 3-kinase (PI3K). Many of these pathways activate transcription factors, which interact with DNA and transcriptional cofactors to either promote or suppress transcription. Ultimately, messenger RNA (mRNA) and ribosomal RNA (rRNA) are transported out of the nucleus to the endoplasmic reticulum so that translation may occur on ribosomes. Protein synthesis is the net result of increased ribosome biogenesis and protein translation.

Systems Used to Investigate Signaling Pathways Involved for the Induction of Cardiac Hypertrophy

A number of different systems have been used to study the control of growth in the heart. These include neonatal rat ventricular myocytes in primary culture, perfused rodent hearts, and transgenic and gene-targeted mice. The advantages and disadvantages of each model have been discussed in detail in a number of reviews (Francis and Carlyle 1993; Sugden and Clerk 1998; MacLellan and Schneider 2000; Izumo and Pu 2004). For instance, the isolated rat ventricular myocyte system has the advantage of studying the responses of cardiac myocytes without the complexity of interactions with non-myocytes, the extracellular matrix (ECM), and circulating growth factors. However, these interactions may play an important role in the in vivo control of cardiac myocyte growth. For this reason, the generation of cardiac-specific transgenics and gene-targeted mice is also critical.

For studying the role of signaling molecules in the adult heart, the α-MHC promoter has been commonly used to achieve cardiac myocyte–specific gene expression in transgenic mice (Izumo and Shioi 1998; Aoki and Izumo 2001). Two further developments were the introduction of the Cre-loxP system to generate cardiac-specific knockout mice and the use of the tetracycline-dependent *trans*-activator (tTA) for the conditional activation of a transgene in a cardiac muscle–specific manner (Izumo and Shioi 1998; Aoki and Izumo 2001). More recently, DNA microarrays have been used to examine the expression profiles of a vast number of genes in cardiac hypertrophy and failure (Cook and Rosenzweig 2002; Izumo and Pu 2004). Using human tissue and tissue from animal disease models, microarray studies have identified transcriptional changes associated with cardiac hypertrophy and heart failure (Friddle et al. 2000; Hwang et al. 2000; Barrans et al. 2001; McMullen et al. 2004).

Hypertrophic Triggers: Mechanical Stimuli, Vasoactive Substances, Growth Factors, Cytokines, Hormones, and Energy Metabolism

Hypertrophic stimuli can act directly to induce hypertrophy of cardiac myocytes, or act via paracrine or autocrine mechanisms (Fig. 5).

Mechanical Stress

Mechanical stress (e.g., stretch) of neonatal or adult cardiac myocytes and papillary muscle, and isolated hearts subjected to increased load, results in stimulation of protein synthesis (Cooper 1987; Sadoshima and Izumo 1997; Ruwhof and van der Laarse 2000). A mechanosensitive molecule is assumed to have some interaction with the plasma membrane and associated proteins in order to sense the tension of the membrane in response to mechanical stress (Sadoshima and Izumo 1997). There is evidence that mechanotransduction may be mediated by stretch-activated sarcolemmal ion channels, G-protein-coupled receptors (GPCR), Na^+/H^+ antiporters, tyrosine-kinase-containing receptors, glycoprotein 130 (gp130), and/or an extracellular matrix–integrin linked pathway (Nyui et al. 1998; Yamazaki et al. 1998; Ruwhof and van der Laarse 2000; Aoki and Izumo 2001). The cell-surface mechanotransducers then activate signal transduction pathways that initiate gene transcription and protein synthesis. A number of signaling cascades have been implicated in response to mechanical stress. These include PKC, MAPKs, the Rho family of small G proteins, stress-activated protein kinase (SAPK or c-Jun amino-terminal kinase [JNK]), and janus kinases (JAK)/signal transducer and activator of transcription (STAT) (Aikawa et al. 1999; Ruwhof and van der Laarse 2000; Aoki and Izumo 2001). Integrins bind to ECM proteins, and the cytoplasmic domains serve as docking sites for signaling molecules including nonreceptor tyrosine kinases (e.g., focal adhesion kinase [FAK]), adapter proteins (e.g., Grb2 and p130Cas), and structural proteins (e.g., paxillin and tensin) (Ross and Borg 2001; Kuppuswamy 2002). Autocrine/paracrine growth factors are thought to play an important role in the pathogenesis of stretch-induced cardiac hypertrophy. In response to mechanical stimuli, a number of vasoactive substances and growth factors (e.g., angiotensin II [Ang II], endothelin 1 [ET-1], insulin-like growth factor 1 [IGF1]) are expressed and/or secreted (Sadoshima et al. 1993; Yamazaki et al. 1995; Sadoshima and Izumo 1997; Ruwhof and van der Laarse 2000). These factors induce hypertrophic responses when applied exogenously to cardiac myocytes (see below).

Vasoactive Substances

Each of these vasoactive substances (Ang II, ET-1, prostaglandin F2α [PGF2α]) acts as a growth factor in cultured cardiac myocytes.

1. Ang II is the principal biologically active product of the renin–angiotensin system (RAS) and binds to GPCR. At least two pharmacologically distinct receptors have been cloned (AT_1 and AT_2). In rodents, two AT_1 receptor isoforms exist (AT_{1A} and AT_{1B}) (de Gasparo et al. 1995; Lorell 1999). All components of the RAS have been identified in the heart, suggesting the existence of a local RAS (Yamazaki and Yazaki 1997; Lijnen and Petrov 1999). The cardiac RAS is activated in cardiac hypertrophy induced by hemodynamic overload, and it is well established that angiotensin-converting enzyme (ACE) inhibitors (block Ang II formation) prevent the development of pressure-overload-induced cardiac hypertrophy in animal models and in humans (Sadoshima et al. 1996; Zhu et al. 1997; Lijnen and Petrov 1999; Yamazaki et al. 1999; Devereux 2000; Modesti et al. 2000).

   Ang II induces a hypertrophic response in cultured neonatal ventricular myocytes by activating a number of signaling pathways including tyrosine kinases, MAPKs, and intracellular calcium (Sadoshima and Izumo 1993; Miyata and Haneda 1994; Sadoshima et al. 1995). However, although Ang II is considered a hypertrophic factor, whether it is directly hypertrophic remains controversial. Its hypertrophic action may also involve autocrine and paracrine mechanisms (Yamazaki and Yazaki 1997; Hamawaki et al. 1998; Sugden and Clerk 1998).

   Despite Ang II receptors lacking a tyrosine kinase domain, Ang II can also activate tyrosine kinases (Sadoshima and Izumo 1996; Eguchi and Inagami 2000). Ang II was shown to activate tyrosine kinases via *trans*-activation of the epidermal growth factor receptor (EGFR; Eguchi et al. 1998), and it was recently reported that EGFR plays a critical role in cardiac hypertrophy induced by Ang II (Kagiyama et al. 2002; Thomas et al. 2002).

   Ang II receptor knockout mice (AT_{1A}/AT_{1B} or AT_2) did not display a cardiac phenotype under basal conditions (Hein et al. 1995; Ichiki et al. 1995; Oliverio et al. 1998). However, transgenic mice overexpressing AT_1 receptors specifically in cardiac myocytes developed cardiac hypertrophy (Paradis et al. 2000). Unexpectedly, pressure overload (a state known to activate the RAS) and stretch-induced hypertrophy still occurred in AT_{1A} knockout mice and their cardiac myocytes, respectively (Hamawaki et al. 1998; Harada et al. 1998; Kudoh et al. 1998), although AT_{1B} may also play a role in this response. Interestingly, cardiac-specific overexpression of the AT_2 receptor resulted in no cardiac phenotype (Masaki et al. 1998), but pressure-overload-induced hypertrophy was blocked in AT_2 knockout mice (Senbonmatsu et al. 2000).

2. Endothelin is one of the most potent vasoconstrictors known (Yanagisawa et al. 1988). Three isoforms of endothelin have been identified (ET-1, ET-2, and ET-3). ET-1 is the predominant cardiac endothelin and has been implicated in playing a role in cardiac hypertrophy and heart failure (Giannessi et al. 2001; Kedzierski and Yanagisawa 2001). In the heart, endothelial cells are the principal sites of generation of ET-1, but it is also produced by cardiac myocytes and cardiac fibroblasts. Plasma ET-1 levels and/or gene expression was elevated in animal models of pathological cardiac hypertrophy (Ueno et al. 1999; Modesti et al. 2000), plasma concentrations of endothelin were elevated in patients with heart failure (Miyauchi and Masaki 1999), and endothelin receptor blockers have been used successfully in animal models of heart failure to inhibit cardiac hypertrophy induced by pressure overload (Miyauchi and Goto 1999).

   ET-1 is a potent hypertrophic stimulus in neonatal cardiac myocytes (Shubeita et al. 1990). ET-1 binds to two GPCRs: ET_A and ET_B receptors. Both receptors are found on cardiac myocytes, but the ET_A receptor accounts for 90% of endothelin receptors on cardiac myocytes (Kedzierski and Yanagisawa 2001). ET-1 was shown to activate MAPKs in cardiac myocytes (Bogoyevitch et al. 1994; Miyauchi and Masaki 1999). Ang II and isoproterenol increased ET-1 levels in cultured rat cardiac myocytes, and an ET_A antagonist reduced Ang II-induced cardiac myocyte hypertrophy in vitro (Ito et al. 1993a; Morimoto et al. 2001). Unexpectedly, cardiac myocyte–specific ET_A knockout mice displayed an unaltered hypertrophic response to Ang II and isoproterenol (Kedzierski et al. 2003).

3. PGF2α is derived from arachidonic acid and is thought to mediate its actions via interaction with prostanoid FP receptors, which belong to the GPCR superfamily (Abramovitz et al. 1994). PGF2α stimulated hypertrophic growth in cultured neonatal and adult rat ventricular myocytes (Adams et al. 1996; Lai et al. 1996; Kunapuli et al. 1998). Chronic administration of an agonist analog of PGF2α

resulted in cardiac hypertrophy in vivo (Lai et al. 1996). Furthermore, PGF2α was elevated in animals with pathological hypertrophy induced by pressure overload or myocardial infarction (Chazov et al. 1979; Lai et al. 1996), and inhibitors of PG synthase blocked cardiac hypertrophy induced by hypertension (Kentera et al. 1979).

Cytokines

Cytokines, initially characterized by their effects on components of the immune system, have been implicated in normal and pathologic cardiac growth. Cytokines of the interleukin-6 and cardiotrophin family (e.g., IL-6, leukemia inhibitory factor [LIF], CT-1, ciliary neurotrophic factor, and oncostatin M) activate the gp130 transmembrane receptor and rapidly stimulate cytoplasmic JAKs (MacLellan and Schneider 2000). This receptor can also promote activation of the PI3K, extracellular signal-regulated kinase (ERK), and JNK pathways (Nyui et al. 1998; Kodama et al. 2000). In contrast, IL-1 and tumor necrosis factor alpha (TNFα) use a distinct pathway that involves activation of a phosphatidylcholine-specific phospholipase C (PLC) with generation of diacylglycerol (DG). LIF, CT-1, and TNFα induced cardiac myocyte hypertrophy in vitro (Pennica et al. 1995; Wollert et al. 1996; Kodama et al. 1997).

Mice with continuous activation of gp130 in their hearts (double transgenics for IL-6 and IL-6 receptors) displayed cardiac hypertrophy (Hirota et al. 1995). Transgenic mice expressing a dominant negative (dn) mutant of gp130 showed an attenuated response to cardiac hypertrophy induced by aortic banding (Uozumi et al. 2001). Cardiac-specific gp130 knockout mice did not display any obvious cardiac phenotype at baseline (Hirota et al. 1999). However, in response to pressure overload, the mice developed dilated cardiomyopathy associated with myocyte apoptosis (Hirota et al. 1999). Overexpression of secreted TNFα in the heart led to dilated cardiomyopathy in mice (Kubota et al. 1997; Bryant et al. 1998; Sivasubramanian et al. 2001).

Growth Factors

Locally acting polypeptide growth factors, synthesized by cardiac myocytes themselves, or by adjacent endocardial or microvessel endothelial cells, have been shown to be differentially expressed during heart development as well as during compensatory growth of the adult heart (Schneider and Parker 1990; MacLellan and Schneider 2000). Growth factors, which signal through receptor tyrosine kinases, include IGF1, transforming growth factor beta (TGFβ), fibroblast growth factors (FGFs), and neuregulins.

1. Mice deficient for IGF1 or the IGF1 receptor (IGF1R) are significantly growth retarded (Baker et al. 1993; Liu et al. 1993). IGF1 is produced by the heart and has been shown to promote cardiac myocyte hyperplasia and hypertrophy, improve contractility, and inhibit apoptosis (Ren et al. 1999). IGF1 caused hypertrophy of adult cultured cardiac myocytes (Delaughter et al. 1999) and increased protein synthesis in isolated cardiac myocytes (Fuller et al. 1992; Ito et al. 1993b; Decker et al. 1995; Lavandero et al. 1998). Transgenic mice overexpressing IGF1 specifically in the heart displayed an increase in heart size (Reiss et al. 1996). Interestingly, the increase in heart size was due to hyperplasia rather than hypertrophy of cardiac myocytes. In contrast, transgenic mice with cardiac-specific overexpression of IGF1R displayed cardiac hypertrophy that was associated with hypertrophy of cardiac myocytes (McMullen et al. 2004). The two major pathways of IGF1R signaling in cardiac myocytes are the PI3K pathway and the MAPK pathway (Ren et al. 1999; Liu et al. 2001). However, the PI3K(p110α isoform) pathway appears to account for the growth-promoting effects of IGF1R (McMullen et al. 2004).

2. Transforming growth factor beta 1 (TGFβ1) belongs to a superfamily that has been implicated in playing a role in heart growth during development and in the adult heart in response to pathological stimuli (Millan et al. 1991; Schneider et al. 1992; MacLellan and Schneider 2000). TGFβ1 is expressed at high levels in cardiac myocytes and fibroblasts during both these periods. TGFβ1 is produced locally, released into the immediate environment, and mediates its effects on growth through paracrine or autocrine mechanisms (Takahashi et al. 1994; MacLellan and Schneider 2000). It was demonstrated, using TGFβ1-deficient mice, that TGFβ1 was necessary for cardiac hypertrophy stimulated by subpressor doses of Ang II (Schultz Jel et al. 2002). Furthermore, overexpression of TGFβ1 induced pathological cardiac hypertrophy in mice (Rosenkranz et al. 2002).

3. FGFs are produced by the heart and have been shown to activate genes and signaling pathways associated with cardiac hypertrophy (Bogoyevitch et al. 1994; Kardami et al. 1995; Schultz et al. 1999). Both acidic and basic FGFs (aFGFs and bFGFs [also known as

FGF2]) can induce myocyte growth. Acidic FGF produces a hyper-plastic response, whereas bFGF stimulates an increase in protein synthesis with resultant hypertrophy (Parker et al. 1990; Cummins 1993). Ang II was shown to activate FGF2 expression and release from cardiac myocytes and fibroblasts (Fischer et al. 1997). Furthermore, mice lacking FGF2 display an impaired hypertrophic response to Ang II, suggesting that FGF2 is a critical mediator of cardiac hypertrophy via autocrine/paracrine actions (Pellieux et al. 2001). Using FGF2 knockout mice, it was shown that FGF2 is required for a full hypertrophic response to pressure overload (Schultz et al. 1999). Interestingly, transgenic mice with cardiac overexpression of FGF2 did not display cardiac hypertrophy (Sheikh et al. 2001).

4. Neuregulins are a group of membrane-bound or secreted peptides that promote growth, differentiation, and survival during develop-ment and oncogenesis (Marchionni 1995). Neuregulins are ago-nists for the epidermal growth factor (EGF) family of tyrosine kinase receptors, which include the EGF receptor (EGFR), ErbB2, ErbB3, and ErbB4 receptors. Targeted disruption of the gene for neuregulin-1 resulted in embryonic lethality due to defects in the developing heart (Lee et al. 1995; Meyer and Birchmeier 1995). The receptors Erb2 and Erb4 are expressed on neonatal and adult ven-tricular myocytes, and neuregulin-1 induced a hypertrophic response in neonatal and adult cardiac myocytes (Zhao et al. 1998; Baliga et al. 1999). Neuregulin-1 has been associated with increases in fetal genes including ANP and skeletal α-actin, as well as increased protein synthesis and the organization of actin filaments into myofibrils (Baliga et al. 1999). More recently, it was shown that cardiac-restricted deletion of ErbB2 receptors in adult cardiac myocytes led to dilated cardiomyopathy (Crone et al. 2002; Ozcelik et al. 2002).

Heparin-binding EGF-like growth factor (HB-EGF) is a mem-ber of the EGF family of growth factors that activates the EGFR and related receptor tyrosine kinase, ErbB4. HB-EGF null mice have dilated cardiac chambers and depressed cardiac function (Iwamoto et al. 2003).

Hormones

1. Thyroid hormone (TH) is often considered the classic hormonal mediator of normal postnatal cardiac hypertrophy. Developmental

growth of the heart is significantly reduced by inhibition of thyroid gland activity, and administration of excess TH to animals caused an increase in heart weight (Bedotto et al. 1989; Hudlicka and Brown 1996). Furthermore, hyperthyroidism has been reported to result in increased cardiac mass in humans (Ching et al. 1996). The thyroid gland secretes two biologically active hormones: thyroxine (T_4, prohormone) and triiodothyronine (T_3). T_3 and T_4 diffuse across the plasma membrane because of their lipid solubility, and T_4 is converted to T_3 (although the mouse heart appears uniquely dependent on serum T_3 since there is no appreciable conversion of T_4 to T_3) (Danzi and Klein 2002; Dillmann 2002). Circulating and the newly synthesized T_3 pass through the nuclear membrane to bind to specific TH receptors (TRs) (Lazar 1993; Mangelsdorf et al. 1995). TRs belong to a superfamily of nuclear hormone receptors (Lazar and Chin 1990; Harvey and Williams 2002). Mammalian TRs are encoded by two genes, TRα and TRβ; these generate multiple TR proteins by alternative splicing (α1, α2; β1, β2, β3). TRα1, TRα2, and TRβ1 are expressed in the heart (Lazar 1993; Harvey and Williams 2002). TH binds to the nuclear receptors, which then act as transcription factors to directly repress or activate cardiac genes (Lazar 1993; Danzi and Klein 2002; Dillmann 2002). TH has been reported to regulate a number of cardiac proteins, including α-MHC, β-MHC, cardiac troponin, sarcoplasmic reticulum Ca 2^{++} ATPase (SERCA), and voltage-gated potassium channels (Izumo et al. 1986; Rohrer and Dillmann 1988; Nishiyama et al. 1998; Danzi and Klein 2002).

The TH-induced increase in heart size was absent in TRβ knockout mice but unaffected in TRα knockout mice, suggesting that the β subtype is responsible for the growth phenotype (Weiss et al. 2002). TH-induced cardiac hypertrophy may be related to local activation of the cardiac RAS because the hypertrophic response to TH was reversed with ACE inhibitors (Asahi et al. 2001; Basset et al. 2001).

2. Growth hormone (GH) works as an endocrine promoter of hepatic production of IGF1 that ultimately mediates GH action on peripheral tissues. The heart contains GH and IGF1R (Colao et al. 2001). Studies suggest that the rate of GH secretion has a prominent role in postnatal cardiac development. Implantation of GH-secreting tumors in rats induced cardiac hypertrophy (Colao et al. 2001), and GH has been shown to stimulate cardiac growth in ani-

mal and human studies (Cittadini et al. 1996; Fazio et al. 1996; Colao et al. 2001; Khan et al. 2002).

IGF1 was considered responsible for mediating the trophic effect of GH in cardiac myocytes because the addition of IGF1, but not GH, to cardiac myocytes induced cell hypertrophy that was accompanied by expression of muscle-specific genes (Ito et al. 1993b; Donath et al. 1994). However, this was challenged by a study in which GH induced hypertrophy of neonatal cardiac myocytes directly, in the absence of any effect on IGF1 mRNA (Lu et al. 2001).

3. Adrenergic hormones. Cardiac hypertrophy is often associated with an increase in intra-cardiac sympathetic nerve activity and with elevated plasma catecholamines. Catecholamines activate adrenergic receptors (ARs), of which there are three major subfamilies: α_1-AR, α_2-AR, and β-AR. These receptors belong to the large GPCR superfamily (Scheuer 1999; Lomasney and Allen 2001; Rockman et al. 2002). AR agonists such as norepinephrine (NE, noradrenaline), phenylephrine (PE), and isoproterenol induce cardiac myocyte hypertrophy. Prolonged infusion of subpressor doses of NE increased the mass of the heart, suggesting that NE has direct hypertrophic effects on cardiac myocytes independent of effects on afterload (Yamazaki and Yazaki 2000). ARs are coupled to $G\alpha q$, $G\alpha s$, and $G\alpha i$, resulting in modulation of adenyl cyclase (AC), PLC, and ion channels (Rockman et al. 2002).

The mammalian heart expresses at least six types of ARs: three β-ARs (β_1, β_2, β_3) and three α_1-ARs (α_{1A}, α_{1B}, α_{1D}), with β_1-ARs predominating (Xiang and Kobilka 2003). However, adrenergic stimulation of neonatal myocyte protein synthesis and growth appears to occur mainly through the α_1-ARs and is transduced by Gq and pertussis toxin–sensitive G proteins, activating signaling pathways including PLC, phospholipase D (PLD), phospholipase A_2 (PLA$_2$), PKC, and Ca^{++} channels (Scheuer 1999; Koch et al. 2000; Xiang and Kobilka 2003). Transgenic mice with cardiac-specific expression of constitutively active α_{1B}-AR developed cardiac hypertrophy in the absence of changes in blood pressure, and the hypertrophy was associated with an increase in myocyte size and increased activation of PLC (Milano et al. 1994). Additional studies have supported the suggestion that the α_{1B}-AR is the main receptor that mediates catecholamine-induced cardiac hypertrophy independent of blood pressure effects (Zuscik et al. 2001; Vecchione et al. 2002). More recently, using $\alpha_{1A/C}$ and α_{1B} double knockout mice, it was shown that α_1-ARs in male mice are required for normal postnatal cardiac growth (O'Connell et al. 2003).

Energy Metabolism

In the normal adult mammalian heart, fatty acids serve as the chief energy substrate, providing the greatest yield of ATP per mole compared to other substrates such as glucose or lactate. In contrast, the fetal heart, which functions in a relatively hypoxic environment, relies predominantly on glucose and lactate as substrates for ATP production, because the glycolytic production of ATP does not require oxygen. During the postnatal period, the developing heart undergoes a marked switch to fatty acid as its preferred energy substrate. This coincides with postnatal mitochondrial proliferation and increased myocardial oxygen consumption, providing the adult heart with a large capacity for ATP production. During pathological hypertrophic growth, the myocardium shifts back toward a reliance on glycolysis as the primary pathway for energy production, with a reduction in cardiac fatty acid oxidation (FAO) rates. This is consistent with reinduction of the fetal energy metabolic program (Lehman and Kelly 2002a,b).

There is evidence to suggest that altered myocardial energy substrate utilization may serve as a hypertrophic trigger. Inborn errors of myocardial FAO are a cause of inherited childhood hypertrophic cardiomyopathy (Kelly and Strauss 1994), and animals given drugs that deprive the FAO pathway of substrate display cardiac hypertrophy (Greaves et al. 1984; Rupp and Jacob 1992). Furthermore, mice with cardiac-specific deficiency of the glucose transporter GLUT4 develop cardiac hypertrophy that is associated with increased expression of the glucose transporter GLUT1 and an increase in basal cardiac glucose transport (Abel et al. 1999). In addition, insulin, which has direct effects on glucose transport, glycolysis, glucose oxidation, glycogen synthesis, and protein synthesis in the heart has also been shown to regulate metabolism and cardiac size (Belke et al. 2002). Hearts of cardiac-myocyte-selective insulin receptor knockout mice were smaller and showed increased reliance on glycolysis with reduced FAO rates. Together, these data suggest that an alteration in energy substrate is sufficient for the induction of cardiac hypertrophy.

Intracellular Signaling Pathways

Several intracellular signaling pathways (Fig. 5) have been implicated for the induction of cardiac hypertrophy. These include G proteins (heterotrimeric and the small GTP-binding proteins), PLC/PKC, tyrosine kinase pathway, MAPK cascades, calcium signaling (calcineurin/nuclear factor of activated

T cells [NFAT]), JAK/STAT, integrin signaling, and the PI3K pathway (Aoki and Izumo 2001; Molkentin and Dorn 2001; Sugden 2001).

G Proteins

G proteins are a large family of proteins that are divided into two subgroups: heterotrimeric G proteins and small-molecular-weight monomeric G proteins (small G proteins).

1. Heterotrimeric G proteins are composed of three subunits (α, β, and γ) and couple to receptors known as G-protein-coupled receptors (GPCRs). Binding of an agonist to the GPCR causes dissociation of G proteins into $G\alpha$ and $G\beta\gamma$ subunits. These subunits transduce the receptor signals by activating downstream pathways (Gutkind 1998a,b; Rockman et al. 2002). Isoforms of the trimeric G proteins are primarily determined by the isoform of the α subunits, which fall into four large subfamilies based on amino acid sequence similarity: Gs, Gi, Gq (e.g., $G\alpha_q$, $G\alpha_{11}$), and G_{12} (Simon et al. 1991; Neer 1995).

   Ang II, ET-1, PE, and PGF2α all promote a hypertrophic response in cultured cardiac myocytes by binding to seven-transmembrane receptors, which are coupled to heterotrimeric G proteins of the Gq family (McKinsey and Olson 1999; Clerk and Sugden 2000). Ligand binding to these receptors activates Gq, which in turn activates a number of signaling pathways including PLC, PKC, MAPK, and PKA (Clerk and Sugden 1999; McKinsey and Olson 1999).

   Cardiac-specific overexpression of $G\alpha_q$ in transgenic mice induced cardiac hypertrophy that was associated with altered cardiac gene expression and cardiac dysfunction (D'Angelo et al. 1997). Additionally, mice lacking the G proteins $G\alpha_q$ and $G\alpha_{11}$ in cardiac myocytes did not develop cardiac hypertrophy in response to pressure overload, suggesting that the $G_{q/11}$ pathway is important for the induction of pathological cardiac hypertrophy (Wettschureck et al. 2001).

2. The small (~21 kD) GTP-binding proteins act as molecular switches that link cell membrane receptors to downstream signaling pathways. In the GDP-bound form they are inactive, and they are activated by the exchange of GDP for GTP. This family of proteins consists of five subfamilies: Ras, Rho, ADP ribosylation factors, Rab, and Ran. To date, Ras, Rho, and Rab family members have

been implicated in the development of cardiac hypertrophy (Clerk and Sugden 2000; Clerk et al. 2001; Wu et al. 2001). Ras and Rho isoforms are activated in myocytes in response to Ang II, ET-1, PE, or mechanical stress (Ramirez et al. 1997; Aoki et al. 1998; Aikawa et al. 1999; Chiloeches et al. 1999; Clerk and Sugden 2000; Clerk et al. 2001). Ras isoforms are believed to regulate cell growth via MAPKs and PI3Ks. The pathways activated by the Rho family (RhoA, Rac1, Cdc42) are currently less well understood, but appear to involve MAPKs, RhoKinase (ROCK), and myosin light-chain kinase (MLCK, induces sarcomeric organization) (Aoki et al. 2000; Clerk and Sugden 2000). Transgenic mice that express constitutively active (ca) Ras, or ca Rac1, or overexpress Rab1a, specifically in the heart, developed hypertrophy (Hunter et al. 1995; Sussman et al. 2000; Wu et al. 2001).

PLC/PKC Pathway

The "classic" signaling pathway of Gq is the activation of PLCβ, which hydrolyzes phosphatidyl 4,5-biphosphate (PIP_2) to produce inositol 1,4,5-triphosphate (IP_3) and DG. IP_3 mobilizes intracellular Ca^{++}, and DG activates PKC (Sugden and Clerk 1998; Adams and Brown 2001; Aoki and Izumo 2001). The increase in cytosolic Ca^{++} can mediate hypertrophic signals via Ca^{++}/calmodulin-dependent enzymes and phosphatases (see p. 574: Calcium Signaling).

PKC consists of a family of isozymes, which are divided into three major groups based on their structural and regulatory properties: the conventional Ca^{++}-dependent isoforms (cPKC, including PKCα, PKCβ, and PKCγ), the novel Ca^{++}-independent isoforms (nPKC, including PKCδ, PKCε, PKCη, and PKCθ) and the atypical isoforms (aPKC, including PKCζ and PKCλ). Cardiac myocytes are reported to express cPKCα, nPKCδ, nPKCε, and aPKCλ. The presence of cPKCβ remains controversial (Mackay and Mochly-Rosen 2001; Sabri and Steinberg 2003).

PKC has been postulated to be a main signal transducer in hypertrophic signaling. NE, Ang II, ET-1, PE, and mechanical stress have all been reported to increase the activation of PKC in cardiac myocytes, leading to activation of MAPKs (Clerk et al. 1994; Clerk and Sugden 1999). PKC appears to play a role in adaptive and maladaptive cardiac responses. Cardiac-specific transgenic mice overexpressing PKCβ or PKCδ developed pathological cardiac hypertrophy (Bowman et al. 1997; Wakasaki et al. 1997; Chen et al. 2001). However, PKCβ is not essential for hypertrophy, since PKCβ knockout mice responded normally to PE or aortic banding

(Roman et al. 2001). In contrast, mice overexpressing a cardiac-specific ca PKCε mutant displayed concentric cardiac hypertrophy with a physiological phenotype; i.e., normal function and no fibrosis (Takeishi et al. 2000).

Tyrosine Kinase Pathway

Receptor tyrosine kinases and non-receptor protein tyrosine kinases are important transducers for the induction of cardiac hypertrophy (Aoki and Izumo 2001). The tyrosine kinase pathway is typically activated by growth factors (e.g., IGF1 and FGF) and their receptor tyrosine kinases (RTKs). These receptors undergo ligand-induced dimerization, leading to enzymatic activation and autophosphorylation of tyrosine residues on the cytoplasmic domain. These phosphotyrosine domains recruit non-receptor tyrosine kinases such as Src family protein kinases and adapter molecules such as Grb2, ultimately resulting in activation of intracellular signaling pathways including the Ras-MAPK and the PI3K pathways (Schlessinger 2000; Aoki and Izumo 2001).

MAPK Pathway

MAPKs are ubiquitously expressed signaling molecules that play an important role in mediating cell growth by connecting signals from receptors to the nucleus (Widmann et al. 1999; Pearson et al. 2001). MAPKs are divided into three subfamilies based on the terminal kinase in the pathway: the ERKs, the JNKs, and the p38 MAPKs (Clerk and Sugden 1999; Widmann et al. 1999; Pearson et al. 2001).

1. To date, five ERK proteins have been identified (ERK1–ERK5). ERK1 and ERK2 are the most abundantly expressed and have been the best characterized in the heart (Bueno and Molkentin 2002). ERK1 and ERK2 are regulated by the MAPK kinases, MEK1 and MEK2. Hypertrophic stimuli such as ET-1, NE, and PE stimulated ERK1/2 in cardiac myocytes in culture (Clerk et al. 1994; Sugden and Clerk 1998), and ERK1/2 were activated by pressure overload (Rapacciuolo et al. 2001; Takeishi et al. 2001). Transgenic mice expressing ca MEK1 (specifically activates ERK1/2, but not JNKs or p38) in the heart displayed cardiac hypertrophy with a physiological phenotype (Bueno et al. 2000). Transgenic mice expressing a cardiac-specific ca MEK5 mutant (activates ERK5) displayed eccentric cardiac hypertrophy (Nicol et al. 2001).

2. JNK protein kinases are encoded by three genes: *Jnk1, Jnk2,* and *Jnk3.* All three are expressed in the heart and are regulated by MKK4/7 (Davis 2000; Pearson et al. 2001). JNKs are activated in response to mechanical stress or GPCR activation and are necessary for the development of cardiac myocyte hypertrophy in culture (Komuro et al. 1996; Choukroun et al. 1998). Activation of JNK has been reported in animal models of pressure-overload-induced cardiac hypertrophy and myocardial infarction, and in heart failure in humans (Li et al. 1998; Yano et al. 1998; Choukroun et al. 1999; Cook et al. 1999). Furthermore, in a study in which JNK activity was inhibited in the heart by a dominant negative (dn) MAPK kinase (MKK4), pressure-overload-induced hypertrophy was attenuated (Choukroun et al. 1999). These studies suggest that JNKs are necessary regulators of cardiac hypertrophy in response to a pathological stress. However, a more direct assessment of JNKs suggests that they antagonize cardiac growth. dn JNK1/2 transgenic mice and JNK1/2 gene-targeted mice displayed an enhanced hypertrophic response to pressure overload (Liang et al. 2003).

3. p38 is regulated by MKK3/6 and is activated in isolated cardiac myocytes in response to mechanical stress, GPCR ligands (Ang II, ET-1, and PE), and mitogens (Clerk et al. 1998; Pearson et al. 2001). p38 MAPK activity was elevated in a mouse model of pressure-overload-induced hypertrophy (Wang et al. 1998) and in failing human hearts (Cook et al. 1999). Overexpression of activated MKK3 or MKK6 in cultured neonatal cardiac myocytes induced hypertrophy in vitro (Zechner et al. 1997; Nemoto et al. 1998; Wang et al. 1998). However, results from transgenic mice have led to discrepancies. MKK3 and MKK6 transgenic mice developed heart failure, but this was not associated with hypertrophy of cardiac myocytes (Liao et al. 2001). In contrast, transgenic mice expressing a ca TAK1 (upstream of MKK3/6 and p38) mutant specifically in the heart developed cardiac hypertrophy (Zhang et al. 2000). However, TAK1 is considerably upstream of p38 and may regulate other signaling effectors such as IκB kinase (Liang and Molkentin 2003). Recent work suggests that p38 signaling has an anti-hypertrophic role in myocytes within the adult heart. Transgenic mice expressing dn mutants of MKK3, MKK6, and p38(α or β) displayed greater hypertrophic responses to pressure overload than nontransgenic controls (Braz et al. 2003; Zhang et al. 2003).

Calcium Signaling

In addition to its well-known involvement in regulating cardiac contractility, increases in intracellular calcium have also been associated with cardiac myocyte growth (Frey et al. 2000). A number of studies have shown that calcium/calmodulin acts as an important second messenger for signals including Ang II, ET-1, α-adrenergic agents, and mechanical stretch (Frey et al. 2000; Aoki and Izumo 2001; Sugden 2001). The best-described calcium-dependent signaling molecules include calcineurin and calcium/calmodulin-dependent protein kinases (CaMKs). Calcineurin has been implicated as a regulator of the hypertrophic response in conjunction with the NFAT family of transcription factors (Olson and Williams 2000). NFAT translocates to the nucleus, where it associates with other transcription factors such as GATA4 and myocyte enhancer factor 2 (MEF2) to regulate the expression of target genes (Frey et al. 2000; Wilkins and Molkentin 2002). Despite some initial controversy, it now appears that calcineurin plays a critical role in regulating heart growth (Wilkins and Molkentin 2002). Calcineurin signaling has been implicated for the induction of pathological and physiological cardiac hypertrophy. Cardiac-specific overexpression of the calcineurin inhibitory protein, myocyte-enriched calcineurin-interacting protein (MCIP) 1, in transgenic mice was shown to inhibit the hypertrophic response in mice with cardiac-restricted overexpression of activated calcineurin, to β-adrenergic receptor stimulation or exercise training (Rothermel et al. 2001).

JAK/STAT Pathway

JAKs are protein tyrosine kinases that associate with cytokine receptors (Pellegrini and Dusanter-Fourt 1997). LIF, CT-1, and other members of the IL-6 cytokine family activated the gp130 receptor associated with the LIF receptor. Once activated, this receptor interacts with JAK1, causing its activation, which in turn phosphorylates the STAT class of transcription factors (Kodama et al. 1997; Aoki and Izumo 2001; Molkentin and Dorn 2001). STAT3 induced genes involved in hypertrophic pathways (Kunisada et al. 1998), and transgenic mice with cardiac-specific overexpression of STAT3 displayed cardiac hypertrophy (Kunisada et al. 2000). In contrast, a dn STAT3 adenovirus attenuated LIF-induced protein synthesis in cardiac myocytes in vitro (Kunisada et al. 1998).

PI3K Pathway

PI3Ks are a family of lipid kinases that induce signals by phosphorylating the hydroxyl group at position 3 of membrane lipid phosphoinositides.

PI3Ks regulate a number of physiological functions, including membrane trafficking, adhesion, actin rearrangement, cell growth, and cell survival (Toker and Cantley 1997; Vanhaesebroeck et al. 1997). Activation of PI3Ks is coupled to both receptor tyrosine kinases (e.g., insulin and IGF1R) and GPCRs. There are multiple isoforms of PI3Ks which are divided into three classes (I, II, and III) and which have a number of subunits (Toker and Cantley 1997; Vanhaesebroeck et al. 1997). To date, one isoform of PI3K (p110α subunit from the class IA PI3Ks, coupled to receptor tyrosine kinases) has been implicated in playing a role in cardiac growth at baseline (Shioi et al. 2000). Cardiac expression of a ca mutant of PI3K(p110α) resulted in physiological cardiac hypertrophy due to an increase in myocyte size. In contrast, mice expressing a dn PI3K(p110α) mutant had significantly smaller hearts (Shioi et al. 2000). PI3K(p110γ) does not regulate heart size under basal conditions (Crackower et al. 2002), but appears to play a role for the induction of hypertrophy in response to β-adrenergic receptor stimulation. PI3K(p110γ)-deficient mice infused with isoproterenol displayed an attenuated hypertrophic response (Oudit et al. 2003).

The phosphatase and tensin homolog (PTEN), a tumor suppressor, has been implicated in the PI3K-signaling pathway, by lowering the levels of the PI3K product, PIP$_3$, and hence antagonizing PI3K signaling (Cantley and Neel 1999). It was shown that PTEN is an important negative regulator of cardiac hypertrophy. Mice with cardiac myocyte inactivation of PTEN displayed cardiac hypertrophy (Crackower et al. 2002).

Akt, a serine threonine kinase (also known as protein kinase B), is the best-characterized target of PI3K (Chan et al. 1999). Transgenic mice expressing a ca mutant of Akt specifically in the heart displayed cardiac hypertrophy (Matsui et al. 2002; Shioi et al. 2002). By crossing ca Akt mice with dn PI3K mice, it was shown that Akt is genetically downstream of PI3K (the heart size of double transgenics was similar to that of ca Akt mice alone). Furthermore, rapamycin attenuated the increase in heart size in ca Akt transgenics, demonstrating that Akt controlled heart size, at least in part, in a mammalian target of rapamycin (mTOR)-dependent manner (Shioi et al. 2002).

Glycogen synthase kinase 3 (GSK3), a cellular substrate of Akt, was initially identified for its role in glycogen metabolism. More recently, it has been shown that GSK3 is an important regulatory kinase with a number of cellular targets, including cytoskeletal proteins and transcription factors. GSK3 consists of two isoforms: GSK3α and GSK3β (Ferkey and Kimelman 2000; Harwood 2001; Hardt and Sadoshima 2002). Both isoforms are expressed in the heart; however, to date most studies have only examined the role of GSK3β (Hardt and Sadoshima 2002). Studies suggest that GSK3β negatively regulates heart growth and that inhibition of

GSK3β by hypertrophic stimuli is an important mechanism for the stimulation of cardiac growth (Haq et al. 2000; Morisco et al. 2000, 2001; Antos et al. 2002; Badorff et al. 2002).

Tuberous sclerosis (TSC) genes (*TSC1* and *TSC2*) have been implicated as negative regulators of cell growth (Gao et al. 2002; Potter et al. 2002). Recently, it was shown that Akt regulates growth, in part, by directly phosphorylating TSC2 (Inoki et al. 2002; Potter et al. 2002), and that TSC1/2 are critical for the control of S6K activation via TOR (Radimerski et al. 2002; Jacinto and Hall 2003). The TSC1 and TSC2 gene products are also known as hamartin and tuberin, respectively, and are present in cardiac myocytes (Johnson et al. 2001). Cardiac myocyte deletion of TSC1 in mice resulted in cardiac hypertrophy (L. Meikle and D. Kwiatkowski, unpubl.).

Other Hypertrophic Triggers, Signaling Pathways, and Cross-talk

The hypertrophic triggers and signaling pathways described above are relatively well characterized. However, it is by no means a complete list. Additional hypertrophic triggers and signaling pathways are continually being discovered and described. Furthermore, it is becoming clear that in many cases there is considerable cross-talk between intracellular signal transduction pathways (MacLellan and Schneider 2000; Aoki and Izumo 2001; Molkentin and Dorn 2001; Frey and Olson 2003).

Transcription Factors

To regulate the long-term alterations in gene expression that are associated with cardiac hypertrophy, intracellular signal transduction pathways must be coupled with transcription factors in the nucleus. Numerous transcription factors have been implicated for the activation of cardiac genes in response to hypertrophic stimuli. These include GATA4, GATA6, Csx/Nkx2.5, MEF2, c-jun, c-fos, c-myc, nuclear factor-κB and NFAT (Fig. 5) (Sadoshima and Izumo 1997; Aoki and Izumo 2001; Akazawa and Komuro 2003). Activation of a number of these transcription factors gives rise to changes in gene expression that have been referred to as reactivation of the "fetal gene program"(Izumo et al. 1988; Chien et al. 1991). This terminology has been used because several genes not normally expressed in the adult ventricle, but seen in the embryonic and neonatal heart, are reexpressed in the stressed heart. Examples include the reexpression of ANP, brain natriuretic peptide (BNP), and genes for fetal isoforms of contractile proteins, such as skeletal α-actin, atrial MLC-1, and β-MHC. This can be accompanied by down-regulation of genes normally expressed at higher levels in the adult than in the embryonic ventricle, such as α-MHC and SERCA2a (Izumo et al. 1988; Chien et al. 1991; MacLellan and Schneider 2000). The significance of these changes with regard to their direct effects on cardiac growth is not well understood. However, there is now reasonably convincing evidence to suggest that ANP/ BNP signaling represents an antihypertrophic regulatory circuit within cardiac myocytes that antagonizes the growth response (Horio et al. 2000; Kishimoto et al. 2001; Holtwick et al. 2003; Molkentin 2003).

Cardiac transcription factors are defined as essential transcriptional activators that are expressed predominantly in the heart and that regulate cardiac genes encoding structural or regulatory proteins characteristic of cardiac myocytes (Akazawa and Komuro 2003). Cardiac transcription factors include GATA family transcription factors, MEF2 transcription factors, and the homeobox transcription factor Csx/Nkx-2.5.

GATA4 has been implicated in the regulation of a number of cardiac-specific genes including ANP, BNP, α-MHC, β-MHC, MLC1/3, cardiac troponin C, cardiac troponin I, and cardiac sodium–calcium exchanger (Charron et al. 1999; Akazawa and Komuro 2003). Furthermore, GATA4 induced hypertrophy of cultured cardiac myocytes, and transgenic mice with mild overexpression of GATA4 displayed cardiac hypertrophy (Liang et al. 2001).

MEF2 binding A/T-rich DNA sequences have been identified within the promoter regions of many cardiac genes including α-MHC, MLC1/3, MLC2v, skeletal α-actin, SERCA, cardiac troponin T, cardiac troponin C, cardiac troponin I, desmin, and dystropin. MEF2 activity is enhanced in response to hypertrophic stimulation and is required for mediating hypertrophic features characteristic of many intracellular signaling pathways (Akazawa and Komuro 2003).

The role of Csx/Nkx-2.5 in the regulation of postnatal growth of the heart is less clear. However, its expression, along with some of its downstream targets (ANP, β-MHC, and cardiac α-actin), was up-regulated in hypertrophied hearts (Akazawa and Komuro 2003).

Control of Cardiac Gene Expression by Chromatin-modifying Enzymes

Histone-dependent packaging of genomic DNA into chromatin is a central mechanism for gene regulation. Nucleosomes, the basic unit of chromatin, interact to create a highly compact structure that limits access of genomic DNA to transcription factors, thus repressing gene expression (McKinsey et al. 2002). Residues within histone tails that protrude from the nucleosome are subject to diverse posttranslational modifications, including phosphorylation, acetylation, methylation, and ubiquitination. These modifications govern the higher-order structure of chromatin and, thus, gene expression. The most well characterized modification of histone tails is lysine-directed acetylation, which is catalyzed by histone acetyltransferases (HATs). Acetylation is thought to stimulate gene expression by destabilizing the histone–histone and histone–DNA interactions that limit access of transcription factors to DNA. The stimulatory effects of HATs on gene expression are antagonized by histone deacetylases (HDACs), which promote chromatin condensation and thus repress transcription. Chromatin-remodeling enzymes have been implicated in the reexpression of the fetal gene program, often associated with the hypertrophic growth of cardiac myocytes (McKinsey et al. 2002).

Members of the MEF2, GATA, and NFAT families of transcription factors are capable of associating with a HAT, p300/CBP (CREB-binding protein), which stimulates their transcriptional activity. Thus, it seems likely that p300/CBP positively regulates cardiac hypertrophy. Class II HDACs appear to be negative regulators of cardiac hypertrophy via their interactions with MEF2 (McKinsey et al. 2002; Zhang et al. 2002). In contrast, class I HDACs appear to be pro-hypertrophic (McKinsey et al. 2002). Recently, it was reported that overexpression of the atypical transcriptional corepressor homeodomain-only protein (Hop) resulted in cardiac hypertrophy via recruitment of a class I HDAC (Kook et al. 2003).

Translational Control

A defining character of cardiac growth is a net increase in protein synthesis above protein degradation. Under normal circumstances, these two processes are matched.

Protein Synthesis

Protein synthesis involves three steps: initiation, elongation, and termination (Rhoads 1999; Orphanides and Reinberg 2002). The overall rate of protein synthesis is determined by both its efficiency (the rate at which new peptide chains are synthesized per ribosome) and its capacity (the relative abundance of ribosomes, ribosome biogenesis).

1. Efficiency of protein synthesis is regulated principally at the level of initiation; i.e., formation of the methionyl-tRNA initiation complex (Rhoads 1999; Dever 2002). The molecular mechanisms regulating translational initiation and elongation have not been studied

extensively in the heart, but a number of signaling pathways have been shown to alter the activities of initiation and elongation factors in various cell types. These include PKC, cAMP, MAPK, calcium/calmodulin, and PI3K (Kleijn et al. 1998; Sonenberg and Gingras 1998; Rhoads 1999). To date, with respect to translational control in the heart, PI3K-dependent pathways have received the most attention. A downstream target of the PI3K pathway, ribosomal S6 kinase (S6K1, p70S6K), is thought to interact with the translational machinery of the ribosome by the phosphorylation of S6. S6 is a component of 40S ribosomal proteins, positioned at the interface between 40S and 60S ribosomal proteins, and localized to regions involved in mRNA and tRNA recognition (Brown and Schreiber 1996; Fumagalli and Thomas 2000). It has been suggested that S6 phosphorylation may be a rate-limiting event in protein synthesis because it is believed to modulate the translation of a family of mRNAs termed 5′TOPs (mRNAs characterized by a *ter*minal *o*ligo*p*yrimidine tract in the 5′ untranslated region; Chou and Blenis 1995; Thomas and Hall 1997). These mRNAs encode components of the protein synthetic apparatus such as ribosomal subunit proteins and many biosynthetic enzymes that are essential for cell growth (Brown and Schreiber 1996; Dufner and Thomas 1999). S6K1 activity and S6 phosphorylation were elevated in hearts of mice subjected to pressure overload (Shioi et al. 2003).

Another potential protein synthesis regulator is the translation initiation factor eIF-4E, whose phosphorylation is rate-limiting for translational initiation (Kleijn et al. 1998; Sonenberg and Gingras 1998; Rhoads 1999; Dever 2002). Phosphorylation of eIF-4E was elevated in a canine model of hypertrophy induced by acute pressure overload (Wada et al. 1996). Ras/ERK signaling was reported to activate protein synthesis via regulation of S6K1 and eIF-4E in adult cardiac myocytes in culture (Wang and Proud 2002).

The mammalian target of rapamycin (mTOR) is thought to control the translational machinery via the activation of S6K1 and via inhibition of 4E-BP1 (eIF4E inhibitor) (Schmelzle and Hall 2000). Rapamycin, a lipophilic macrolide, inhibits growth by forming a gain-of-function inhibitory complex with FKBP12 (FK506-binding protein, m.w. of 12 kD) (Thomas and Hall 1997; Schmelzle and Hall 2000). This complex binds to mTOR, preventing activation of mTOR targets including S6K1 and the 40S ribosomal S6 protein. Rapamycin inhibited Ang II and PE-induced increases in protein synthesis in cardiac myocytes in vitro (Sadoshima and Izumo 1995;

Boluyt et al. 1997). Recently, we reported that rapamycin attenuated and regressed pressure-overload-induced cardiac hypertrophy in mice (Shioi et al. 2003; McMullen et al. 2004).

2. In the long term, ribosome biogenesis is considered most likely to account for increased protein synthesis during cardiac hypertrophy. This is based on the fact that at least 80–90% of existing ribosomes in adult hearts (not undergoing hypertrophy) are in the form of polysomes (i.e., they are already engaged in synthesizing proteins) (Hannan and Rothblum 1995; Hannan et al. 2003). Ribosomes are complexes of ribosomal proteins and ribosomal RNA (rRNA), and their cytoplasmic content is tightly coupled to the rate of growth (Hannan and Rothblum 1995; Brandenburger et al. 2001; Hannan et al. 2003). In eukaryotes, ribosome biogenesis requires the coordination of the synthesis of four rRNA species transcribed by RNA polymerase I (18S, 28S, 5.8S) and RNA polymerase III (5S), and ~70–80 ribosomal proteins whose genes are transcribed by RNA polymerase II. The 18S, 5.8S, and 28S ribosomal RNAs result from processing of a single primary transcript, the 45S pre-ribosomal RNA (45S rRNA). In cardiac myocytes, it has been proposed that hypertrophic stimuli (by largely undefined pathways) increase the activity of the transcription factor upstream binding factor (UBF), resulting in accelerated rates of the 45S ribosomal gene (rDNA) transcription and rRNA synthesis. The increased pool of rRNA, together with an increased translation of ribosomal proteins, results in increased translational capacity (Brandenburger et al. 2001; Hannan et al. 2003).

Studies with neonatal cardiac myocytes in culture and intact hearts suggest that transcription of the ribosomal genes is a key regulatory event for protein synthesis. RNA polymerase I activity was elevated fivefold in nuclei isolated from rat hearts subjected to pressure overload compared to that of sham-operated animals. Furthermore, increased rates of rRNA synthesis were observed during beating-induced hypertrophy of cultured neonatal cardiac myocytes. Similar increases in rDNA transcription have been observed in the heart in response to hypertrophic stimuli such as NE, ET-1, adrenergic agonists, and contraction. In some cases, the increase in the rate of rDNA transcription was correlated with the activation of UBF (Hannan et al. 1995, 1996; Hannan and Rothblum 1995; Luyken et al. 1996). More recently, it was shown that cardiac hypertrophy induced by pressure overload in mice was associated with increased expression of UBF (Brandenburger et al. 2003).

Protein Degradation

Protein degradation appears to be modestly increased in cardiac hypertrophy and may play a role in the distinctive geometry of the ventricles in response to pressure or volume overloading, regression of hypertrophy, and cardiac atrophy. Posttranslational processes are increasingly being recognized as important factors in the production of the cardiac phenotype in cardiac hypertrophy and failure (Walsh 2001).

Signaling Mechanisms Responsible for Sarcomeric Organization

The sarcomere is the basic contractile unit of the heart. The formation of sarcomeres is a complex event that involves the synthesis of more than 50 proteins (Zak 1984a,b; Vigoreaux 1994). These proteins then become aggregated into filaments, which in turn are organized into specific tridimensional arrays and aligned with respect to other contractile elements already present in the cell (Zak 1984a,b; Sanger et al. 2000). The ultimate purpose of cardiac hypertrophy is to accommodate the increased workload (e.g., strain of pressure or volume overload) by increasing the contractile capacity. Therefore, organization of sarcomeres, and consequent increase in contractile units, is an essential part of cardiac hypertrophy (to maximize force generation).

The signaling mechanisms responsible for sarcomere organization are not well understood. Rac1 and RhoA have been implicated as potential signal intermediates for sarcomere organization (Aoki et al. 1998; Hoshijima et al. 1998; Pracyk et al. 1998). It has also been suggested that the calcium/calmodulin-dependent regulation of MLCK plays a role in sarcomeric organization. MLCK is necessary and sufficient for sarcomere organization, and hypertrophic stimuli (Ang II or PE) were shown to activate MLC-2v in cultured cardiac myocytes as well as in the adult heart in vivo (Aoki et al. 2000).

FAK and p130Cas (substrates for the nonreceptor tyrosine kinase Src) are localized to Z lines and were shown to be critical for the assembly of sarcomeric units in cardiac myocytes in culture (Kovacic-Milivojevic et al. 2001). Nebulette, a modular cardiac protein that exhibits a high degree of homology with skeletal muscle nebulin, has been implicated in myofibril organization and function. It was shown to play a role in the structure and stability of the cardiac Z line (Moncman and Wang 2002).

Cell Death

Cardiac growth is also regulated by cell death. Cell death can occur in an uncontrolled manner via necrosis, or by a highly regulated programmed mechanism termed apoptosis (MacLellan and Schneider 1997; Haunstetter and Izumo 1998; Kang et al. 2002; Kang and Izumo 2003). Apoptosis is controlled by a complex interaction of numerous pro-survival and pro-death signals and appears to play an important role in the determination of heart shape and chamber formation during cardiogenesis, as well as contributing to altered cardiac chamber geometry in response to pathological stimuli (Thompson 1995; Gill et al. 2002). Possible triggers for apoptosis include oxidative stress, calcium overload, mitochondrial defects, stimulation of pro-apoptotic factors, or loss of cardiac myocyte survival factors. A key phenomenon of apoptotic cell death is the activation of a unique class of aspartate-specific proteases termed caspases. Activation of caspases may take place either by a death-receptor pathway or by a mitochondrial pathway. The death-receptor-mediated pathway involves the binding of death ligand (e.g., Fas ligand, TNFα) to a membrane-bound death receptor (e.g., Fas, TNF receptor). In the mitochondrial pathway, an apoptotic insult causes the mitochondria to release cytochrome c. This pathway is tightly regulated by the Bcl-2 superfamily proteins (e.g., bcl-2, bcl-xL, bax, bad). Caspase inhibitors have also been identified in mammals (e.g., inhibitors of apoptosis proteins, IAPs). In the heart, reports suggest that in the setting of oxidative stress, the primary apoptotic pathway is mediated by the mitochondria (MacLellan and Schneider 1997; Haunstetter and Izumo 1998; Kang et al. 2002; Kang and Izumo 2003).

Growth factors (e.g., IGF1), and cytokines (e.g., CT-1) are reported to have anti-apoptotic effects, in part, via PI3K and/or ERK signaling (Parrizas et al. 1997; Sheng et al. 1997; Haunstetter and Izumo 1998; Kang et al. 2002). In contrast, signaling via Gαq or Gαs in transgenic mice was shown to promote cardiac myocyte apoptosis (Adams et al. 1998; Geng et al. 1999). Calcineurin has been found to have both pro- and anti-apoptotic effects, depending on the cellular context (De Windt et al. 2000; Saito et al. 2000; Pu et al. 2003). It was recently reported that the mitochondrial death protein Nix was induced by Gq-dependent cardiac hypertrophy and triggered apoptosis (Yussman et al. 2002). Gp130, JNK, and p38 MAPK have been associated with increased apoptosis in cardiac myocytes (Wang et al. 1998; Hirota et al. 1999; Kang et al. 2002).

Signaling Pathways Responsible for the Induction of Different Categories of Cardiac Growth

Pathological and Physiological Cardiac Hypertrophy

An unresolved question in cardiac biology has been whether distinct signaling pathways are responsible for the development of pathological and physiological cardiac hypertrophy in the adult. Physiological hypertrophy is characterized by a normal organization of cardiac structure, and normal or enhanced cardiac function; pathological hypertrophy is associated with an altered pattern of cardiac gene expression, fibrosis, and cardiac dysfunction. The elucidation of signaling cascades that play distinct roles in these two forms of hypertrophy will most likely be critical for the development of more effective strategies to treat heart failure. One potential therapeutic strategy would be to inhibit the pathological growth process while augmenting the physiological growth process.

A large number of transgenic overexpression and knockout studies have implicated a wide array of intracellular signaling pathways in the induction of pathological hypertrophy. These include AT_1 (Hein et al. 1997; Paradis et al. 2000), PKCβ (Bowman et al. 1997; Wakasaki et al. 1997), ras (Hunter et al. 1995), $G\alpha_{11}$ and/or Gαq (D'Angelo et al. 1997; Wettschureck et al. 2001). There have been fewer reports of signaling molecules implicated for the induction of physiological hypertrophy. These include PI3K(p110α) (Shioi et al. 2000), calcineurin (Rothermel et al. 2001), IGF1/IGF1R (Duerr et al. 1995; McMullen et al. 2004), and ERK1/2 (Bueno et al. 2000). We recently reported that PI3K (p110α) played a critical role for the induction of exercise-induced hypertrophy but not pressure-overload-induced hypertrophy (McMullen et al. 2003). On the basis of this work and other reports in the literature, we present a working model of signaling cascades that may be important for the induction of pathological and physiological cardiac hypertrophy (Fig. 6).

Other Categories

Currently, less is known regarding the signaling cascades responsible for the induction of eccentric versus concentric hypertrophy, and compensated versus decompensated growth of the heart. Of note, MEK5 has been implicated for the induction of eccentric hypertrophy (Nicol et al. 2001), and Mst1 (a prominent myelin basic protein kinase activated by pro-apoptotic stimuli in cardiac myocytes) and JNK may contribute to decompensated growth of the heart (Petrich et al. 2003; Yamamoto et al. 2003).

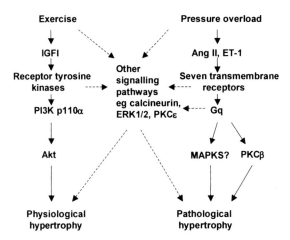

Figure 6. Model illustrating signaling pathways that may be involved in the development of pathological and physiological cardiac hypertrophy.

SUMMARY

Growth of the heart requires changes at the level of gene expression, an increase in the overall rate of protein synthesis, and sarcomere assembly. The hypertrophic response is initiated by a vast number of effectors, which activate multiple intracellular signaling pathways to serve this purpose. In the last decade, transgenic and knockout mice have greatly accelerated progress made in this field. An on-going challenge is to elucidate signaling cascades that directly influence cardiac growth, as opposed to those that affect growth as a result of impairment of cardiac pump function. Defining and dissecting signaling pathways that play distinct roles in the different categories of heart growth (i.e., physiological versus pathological growth, eccentric versus concentric growth, compensated versus decompensated growth) should result in better therapeutics to treat cardiac hypertrophy and failure.

ACKNOWLEDGMENTS

The authors thank David Kwiatkowski and Lynsey Meikle for sharing unpublished results, and William T. Pu for critical reading of the chapter. Due to the broad topic area and extensive volume of literature, in some instances we have cited reviews rather than the primary references. We apologize to authors of primary references for omitting these citations.

REFERENCES

Abel E.D., Kaulbach H.C., Tian R., Hopkins J.C., Duffy J., Doetschman T., Minnemann T., Boers M.E., Hadro E., Oberste-Berghaus C., Quist W., Lowell B.B., Ingwall J.S., and Kahn B.B. 1999. Cardiac hypertrophy with preserved contractile function after selective deletion of GLUT4 from the heart. *J. Clin. Invest.* **104:** 1703–1714.

Abramovitz M., Boie Y., Nguyen T., Rushmore T.H., Bayne M.A., Metters K.M., Slipetz D.M., and Grygorczyk R. 1994. Cloning and expression of a cDNA for the human prostanoid FP receptor. *J. Biol. Chem.* **269:** 2632–2636.

Adams J.W. and Brown J.H. 2001. G-proteins in growth and apoptosis: Lessons from the heart. *Oncogene* **20:** 1626–1634.

Adams J.W., Sakata Y., Davis M.G., Sah V.P., Wang Y., Liggett S.B., Chien K.R., Brown J.H., and Dorn G.W., II. 1998. Enhanced Galphaq signaling: A common pathway mediates cardiac hypertrophy and apoptotic heart failure. *Proc. Natl. Acad. Sci.* **95:** 10140–10145.

Adams J.W., Migita D.S., Yu M.K., Young R., Hellickson M.S., Castro-Vargas F.E., Domingo J.D., Lee P.H., Bui J.S., and Henderson S.A. 1996. Prostaglandin F2 alpha stimulates hypertrophic growth of cultured neonatal rat ventricular myocytes. *J. Biol. Chem.* **271:** 1179–1186.

Aikawa R., Komuro I., Yamazaki T., Zou Y., Kudoh S., Zhu W., Kadowaki T., and Yazaki Y. 1999. Rho family small G proteins play critical roles in mechanical stress-induced hypertrophic responses in cardiac myocytes. *Circ. Res.* **84:** 458–466.

Akazawa H. and Komuro I. 2003. Roles of cardiac transcription factors in cardiac hypertrophy. *Circ. Res.* **92:** 1079–1088.

Antos C.L., McKinsey T.A., Frey N., Kutschke W., McAnally J., Shelton J.M., Richardson J.A., Hill J.A., and Olson E.N. 2002. Activated glycogen synthase-3 beta suppresses cardiac hypertrophy in vivo. *Proc. Natl. Acad. Sci.* **99:** 907–912.

Anversa P. and Nadal-Ginard B. 2002. Myocyte renewal and ventricular remodelling. *Nature* **415:** 240–243.

Anversa P., Capasso J.M., Olivetti G., and Sonnenblick E.H. 1992. Cellular basis of ventricular remodeling in hypertensive cardiomyopathy. *Am. J. Hypertens.* **5:** 758–770.

Anversa P., Leri A., Kajstura J., and Nadal-Ginard B. 2002. Myocyte growth and cardiac repair. *J. Mol. Cell. Cardiol.* **34:** 91–105.

Aoki H. and Izumo S. 2001. Signal transduction of cardiac myocyte hypertrophy. In *Heart physiology and pathology* (ed. N. Sperelakis et al.), pp. 1065-1086. Academic Press, San Diego, California.

Aoki H., Izumo S., and Sadoshima J. 1998. Angiotensin II activates RhoA in cardiac myocytes: A critical role of RhoA in angiotensin II-induced premyofibril formation. *Circ. Res.* **82:** 666–676.

Aoki H., Sadoshima J., and Izumo S. 2000. Myosin light chain kinase mediates sarcomere organization during cardiac hypertrophy in vitro. *Nat. Med.* **6:** 183–188.

Asahi T., Shimabukuro M., Oshiro Y., Yoshida H., and Takasu N. 2001. Cilazapril prevents cardiac hypertrophy and postischemic myocardial dysfunction in hyperthyroid rats. *Thyroid* **11:** 1009–1015.

Badorff C., Ruetten H., Mueller S., Stahmer M., Gehring D., Jung F., Ihling C., Zeiher A.M., and Dimmeler S. 2002. Fas receptor signaling inhibits glycogen synthase kinase 3 beta and induces cardiac hypertrophy following pressure overload. *J. Clin. Invest.* **109:** 373–381.

Baker J., Liu J.P., Robertson E.J., and Efstratiadis A. 1993. Role of insulin-like growth factors in embryonic and postnatal growth. *Cell* **75:** 73–82.

Baliga R.R., Pimental D.R., Zhao Y.Y., Simmons W.W., Marchionni M.A., Sawyer D.B., and Kelly R.A. 1999. NRG-1-induced cardiomyocyte hypertrophy. Role of PI-3-kinase, p70(S6K), and MEK-MAPK-RSK. *Am. J. Physiol.* **277:** H2026–H2037.

Barrans J.D., Stamatiou D., and Liew C. 2001. Construction of a human cardiovascular cDNA microarray: Portrait of the failing heart. *Biochem. Biophys. Res. Commun.* **280:** 964–969.

Basset A., Blanc J., Messas E., Hagege A., and Elghozi J.L. 2001. Renin-angiotensin system contribution to cardiac hypertrophy in experimental hyperthyroidism: An echocardiographic study. *J. Cardiovasc. Pharmacol.* **37:** 163–172.

Bedotto J.B., Gay R.G., Graham S.D., Morkin E., and Goldman S. 1989. Cardiac hypertrophy induced by thyroid hormone is independent of loading conditions and beta adrenoceptor blockade. *J. Pharmacol. Exp. Ther.* **248:** 632–636.

Belke D.D., Betuing S., Tuttle M.J., Graveleau C., Young M.E., Pham M., Zhang D., Cooksey R.C., McClain D.A., Litwin S.E., Taegtmeyer H., Severson D., Kahn C.R., and Abel E.D. 2002. Insulin signaling coordinately regulates cardiac size, metabolism, and contractile protein isoform expression. *J. Clin. Invest.* **109:** 629–639.

Bogoyevitch M.A., Glennon P.E., Andersson M.B., Clerk A., Lazou A., Marshall C.J., Parker P.J., and Sugden P.H. 1994. Endothelin-1 and fibroblast growth factors stimulate the mitogen-activated protein kinase signaling cascade in cardiac myocytes. The potential role of the cascade in the integration of two signaling pathways leading to myocyte hypertrophy. *J. Biol. Chem.* **269:** 1110–1119.

Boluyt M.O., Zheng J.S., Younes A., Long X., O'Neill L., Silverman H., Lakatta E.G., and Crow M.T. 1997. Rapamycin inhibits alpha 1-adrenergic receptor-stimulated cardiac myocyte hypertrophy but not activation of hypertrophy-associated genes. Evidence for involvement of p70 S6 kinase. *Circ. Res.* **81:** 176–186.

Bowman J.C., Steinberg S.F., Jiang T., Geenen D.L., Fishman G.I., and Buttrick P.M. 1997. Expression of protein kinase C beta in the heart causes hypertrophy in adult mice and sudden death in neonates. *J. Clin. Invest.* **100:** 2189–2195.

Brandenburger Y., Jenkins A., Autelitano D.J., and Hannan R.D. 2001. Increased expression of UBF is a critical determinant for rRNA synthesis and hypertrophic growth of cardiac myocytes. *FASEB J.* **15:** 2051–2053.

Brandenburger Y., Arthur J.F., Woodcock E.A., Du X.J., Gao X.M., Autelitano D.J., Rothblum L.I., and Hannan R.D. 2003. Cardiac hypertrophy in vivo is associated with increased expression of the ribosomal gene transcription factor UBF. *FEBS Lett.* **548:** 79–84.

Braz J.C., Bueno O.F., Liang Q., Wilkins B.J., Dai Y.S., Parsons S., Braunwart J., Glascock B.J., Klevitsky R., Kimball T.F., Hewett T.E., and Molkentin J.D. 2003. Targeted inhibition of p38 MAPK promotes hypertrophic cardiomyopathy through upregulation of calcineurin-NFAT signaling. *J. Clin. Invest.* **111:** 1475–1486.

Brown E.J. and Schreiber S.L. 1996. A signaling pathway to translational control. *Cell* **86:** 517–520.

Bryant D., Becker L., Richardson J., Shelton J., Franco F., Peshock R., Thompson M., and Giroir B. 1998. Cardiac failure in transgenic mice with myocardial expression of tumor necrosis factor-alpha. *Circulation* **97:** 1375–1381.

Bueno O.F. and Molkentin J.D. 2002. Involvement of extracellular signal-regulated kinases 1/2 in cardiac hypertrophy and cell death. *Circ. Res.* **91:** 776–781.

Bueno O.F., De Windt L.J., Tymitz K.M., Witt S.A., Kimball T.R., Klevitsky R., Hewett T.E., Jones S.P., Lefer D.J., Peng C.F., Kitsis R.N., and Molkentin J.D. 2000. The MEK1-

ERK1/2 signaling pathway promotes compensated cardiac hypertrophy in transgenic mice. *EMBO J.* **19:** 6341–6350.

Cantley L.C. and Neel B.G. 1999. New insights into tumor suppression: PTEN suppresses tumor formation by restraining the phosphoinositide 3-kinase/AKT pathway. *Proc. Natl. Acad. Sci.* **96:** 4240–4245.

Chan T.O., Rittenhouse S.E., and Tsichlis P.N. 1999. AKT/PKB and other D3 phospho-inositide-regulated kinases: Kinase activation by phosphoinositide-dependent phosphorylation. *Annu. Rev. Biochem.* **68:** 965–1014.

Charron F., Paradis P., Bronchain O., Nemer G., and Nemer M. 1999. Cooperative interaction between GATA-4 and GATA-6 regulates myocardial gene expression. *Mol. Cell. Biol.* **19:** 4355–4365.

Chazov E.I., Pomoinetsky V.D., Geling N.G., Orlova T.R., Nekrasova A.A., and Smirnov V.N. 1979. Heart adaptation to acute pressure overload: An involvement of endogenous prostaglandins. *Circ. Res.* **45:** 205–211.

Chen L., Hahn H., Wu G., Chen C.H., Liron T., Schechtman D., Cavallaro G., Banci L., Guo Y., Bolli R., Dorn G.W., II, and Mochly-Rosen D. 2001. Opposing cardioprotective actions and parallel hypertrophic effects of delta PKC and epsilon PKC. *Proc. Natl. Acad. Sci.* **98:** 11114–11119.

Chien K.R., Knowlton K.U., Zhu H., and Chien S. 1991. Regulation of cardiac gene expression during myocardial growth and hypertrophy: Molecular studies of an adaptive physiologic response. *FASEB J.* **5:** 3037–3046.

Chien K.R., Zhu H., Knowlton K.U., Miller-Hance W., van-Bilsen M., O'Brien T.X., and Evans S.M. 1993. Transcriptional regulation during cardiac growth and development. *Annu. Rev. Physiol.* **55:** 77–95.

Chiloeches A., Paterson H.F., Marais R., Clerk A., Marshall C.J., and Sugden P.H. 1999. Regulation of Ras.GTP loading and Ras-Raf association in neonatal rat ventricular myocytes by G protein-coupled receptor agonists and phorbol ester. Activation of the extracellular signal-regulated kinase cascade by phorbol ester is mediated by Ras. *J. Biol. Chem.* **274:** 19762–19770.

Ching G.W., Franklyn J.A., Stallard T.J., Daykin J., Sheppard M.C., and Gammage M.D. 1996. Cardiac hypertrophy as a result of long-term thyroxine therapy and thyrotoxicosis. *Heart* **75:** 363–368.

Chou M.M. and Blenis J. 1995. The 70 kDa S6 kinase: Regulation of a kinase with multiple roles in mitogenic signalling. *Curr. Opin. Cell Biol.* **7:** 806–814.

Choukroun G., Hajjar R., Kyriakis J.M., Bonventre J.V., Rosenzweig A., and Force T. 1998. Role of the stress-activated protein kinases in endothelin-induced cardiomyocyte hypertrophy. *J. Clin. Invest.* **102:** 1311–1320.

Choukroun G., Hajjar R., Fry S., del Monte F., Haq S., Guerrero J.L., Picard M., Rosenzweig A., and Force T. 1999. Regulation of cardiac hypertrophy in vivo by the stress-activated protein kinases/c-Jun NH(2)-terminal kinases. *J. Clin. Invest.* **104:** 391–398.

Cittadini A., Stromer H., Katz S.E., Clark R., Moses A.C., Morgan J.P., and Douglas P.S. 1996. Differential cardiac effects of growth hormone and insulin-like growth factor-1 in the rat. A combined in vivo and in vitro evaluation. *Circulation* **93:** 800–809.

Clerk A. and Sugden P.H. 1999. Activation of protein kinase cascades in the heart by hypertrophic G protein-coupled receptor agonists. *Am. J. Cardiol.* **83:** 64H–69H.

―――. 2000. Small guanine nucleotide-binding proteins and myocardial hypertrophy. *Circ. Res.* **86:** 1019–1023.

Clerk A., Michael A., and Sugden P.H. 1998. Stimulation of the p38 mitogen-activated protein kinase pathway in neonatal rat ventricular myocytes by the G protein-coupled receptor agonists, endothelin-1 and phenylephrine: A role in cardiac myocyte hypertrophy? *J. Cell Biol.* **142:** 523–535.

Clerk A., Bogoyevitch M.A., Anderson M.B., and Sugden P.H. 1994. Differential activation of protein kinase C isoforms by endothelin-1 and phenylephrine and subsequent stimulation of p42 and p44 mitogen-activated protein kinases in ventricular myocytes cultured from neonatal rat hearts. *J. Biol. Chem.* **269:** 32848–32857.

Clerk A., Pham F.H., Fuller S.J., Sahai E., Aktories K., Marais R., Marshall C., and Sugden P.H. 2001. Regulation of mitogen-activated protein kinases in cardiac myocytes through the small G protein Rac1. *Mol. Cell. Biol.* **21:** 1173–1184.

Cohn J.N., Bristow M.R., Chien K.R., Colucci W.S., Frazier O.H., Leinwand L.A., Lorell B.H., Moss A.J., Sonnenblick E.H., Walsh R.A., Mockrin S.C., and Reinlib L. 1997. Report of the National Heart, Lung, and Blood Institute Special Emphasis Panel on Heart Failure Research. *Circulation* **95:** 766–770.

Colao A., Marzullo P., Di Somma C., and Lombardi G. 2001. Growth hormone and the heart. *Clin. Endocrinol.* **54:** 137–154.

Cook S.A. and Rosenzweig A. 2002. DNA microarrays: Implications for cardiovascular medicine. *Circ. Res.* **91:** 559–564.

Cook S.A., Sugden P.H., and Clerk A. 1999. Activation of c-Jun N-terminal kinases and p38-mitogen-activated protein kinases in human heart failure secondary to ischaemic heart disease. *J. Mol. Cell. Cardiol.* **31:** 1429–1434.

Cooper G.T. 1987. Cardiocyte adaptation to chronically altered load. *Annu. Rev. Physiol.* **49:** 501–518.

Crackower M.A., Oudit G.Y., Kozieradzki I., Sarao R., Sun H., Sasaki T., Hirsch E., Suzuki A., Shioi T., Irie-Sasaki J., Sah R., Cheng H.Y., Rybin V.O., Lembo G., Fratta L., Oliveira-dos-Santos A.J., Benovic J.L., Kahn C.R., Izumo S., Steinberg S.F, Wymann M.P., Backx P.H., and Penninger J.M. 2002. Regulation of myocardial contractility and cell size by distinct PI3K-PTEN signaling pathways. *Cell* **110:** 737–749.

Crone S.A., Zhao Y.Y., Fan L., Gu Y., Minamisawa S., Liu Y., Peterson K.L., Chen J., Kahn R., Condorelli G., Ross J., Jr., Chien K.R., and Lee K.F. 2002. ErbB2 is essential in the prevention of dilated cardiomyopathy. *Nat. Med.* **8:** 459–465.

Cummins P. 1993. Fibroblast and transforming growth factor expression in the cardiac myocyte. *Cardiovasc. Res.* **27:** 1150–1154.

D'Angelo D.D., Sakata Y., Lorenz J.N., Boivin G.P., Walsh R.A., Liggett S.B., and Dorn G.W., II. 1997. Transgenic Galphaq overexpression induces cardiac contractile failure in mice. *Proc. Natl. Acad. Sci.* **94:** 8121–8126.

Danzi S. and Klein I. 2002. Thyroid hormone-regulated cardiac gene expression and cardiovascular disease. *Thyroid* **12:** 467–472.

Davis R.J. 2000. Signal transduction by the JNK group of MAP kinases. *Cell* **103:** 239–252.

Decker R.S., Cook M.G., Behnke-Barclay M., and Decker M.L. 1995. Some growth factors stimulate cultured adult rabbit ventricular myocyte hypertrophy in the absence of mechanical loading. *Circ. Res.* **77:** 544–555.

de Gasparo M., Husain A., Alexander W., Catt K.J., Chiu A.T., Drew M., Goodfriend T., Harding J.W., Inagami T., and Timmermans P.B. 1995. Proposed update of angiotensin receptor nomenclature. *Hypertension* **25:** 924–927.

Delaughter M.C., Taffet G.E., Fiorotto M.L., Entman M.L., and Schwartz R.J. 1999. Local insulin-like growth factor I expression induces physiologic, then pathologic, cardiac hypertrophy in transgenic mice. *FASEB J.* **13:** 1923–1929.

Dever T.E. 2002. Gene-specific regulation by general translation factors. *Cell* **108:** 545–556.

Devereux R.B. 2000. Therapeutic options in minimizing left ventricular hypertrophy. *Am. Heart J.* **139:** S9–S14.

De Windt L.J., Lim H.W., Taigen T., Wencker D., Condorelli G., Dorn G.W., II, Kitsis R.N., and Molkentin J.D. 2000. Calcineurin-mediated hypertrophy protects cardiomyocytes from apoptosis in vitro and in vivo: An apoptosis-independent model of dilated heart failure. *Circ. Res.* **86:** 255–263.

Dillmann W.H. 2002. Cellular action of thyroid hormone on the heart. *Thyroid* **12:** 447–452.

Donath M.Y., Zapf J., Eppenberger-Eberhardt M., Froesch E.R., and Eppenberger H.M. 1994. Insulin-like growth factor I stimulates myofibril development and decreases smooth muscle alpha-actin of adult cardiomyocytes. *Proc. Natl. Acad. Sci.* **91:** 1686–1690.

Duerr R.L., Huang S., Miraliakbar H.R., Clark R., Chien K.R., and Ross, J., Jr. 1995. Insulin-like growth factor-1 enhances ventricular hypertrophy and function during the onset of experimental cardiac failure. *J. Clin. Invest.* **95:** 619–627.

Dufner A. and Thomas G. 1999. Ribosomal S6 kinase signaling and the control of translation. *Exp. Cell Res.* **253:** 100–109.

Eguchi S. and Inagami T. 2000. Signal transduction of angiotensin II type 1 receptor through receptor tyrosine kinase. *Regul. Pept.* **91:** 13–20.

Eguchi S., Numaguchi K., Iwasaki H., Matsumoto T., Yamakawa T., Utsunomiya H., Motley E.D., Kawakatsu H., Owada K.M., Hirata Y., Marumo F., and Inagami T. 1998. Calcium-dependent epidermal growth factor receptor transactivation mediates the angiotensin II-induced mitogen-activated protein kinase activation in vascular smooth muscle cells. *J. Biol. Chem.* **273:** 8890–8896.

Fagard R.H. 1997. Impact of different sports and training on cardiac structure and function. *Cardiol. Clin.* **15:** 397–412.

Fazio S., Sabatini D., Capaldo B., Vigorito C., Giordano A., Guida R., Pardo F., Biondi B., and Sacca L. 1996. A preliminary study of growth hormone in the treatment of dilated cardiomyopathy. *N. Engl. J. Med.* **334:** 809–814.

Ferkey D.M. and Kimelman D. 2000. GSK-3: New thoughts on an old enzyme. *Dev. Biol.* **225:** 471–479.

Ferrans V.J. 1984. Cardiac hypertrophy: Morphological aspects. In *Growth of the heart in health and disease* (ed. R. Zak), pp. 187-239. Raven Press, New York.

Fischer T.A., Ungureanu-Longrois D., Singh K., de Zengotita J., DeUgarte D., Alali A., Gadbut A.P., Lee M.A., Balligand J.L., Kifor I., Smith T.W., and Kelly R.A. 1997. Regulation of bFGF expression and ANG II secretion in cardiac myocytes and microvascular endothelial cells. *Am. J. Physiol.* **272:** H958–H968.

Francis G.S. and Carlyle W.C. 1993. Hypothetical pathways of cardiac myocyte hypertrophy: Response to myocardial injury. *Eur. Heart. J.* (suppl. J) **14:** 49–56.

Frey N. and Olson E.N. 2003. Cardiac hypertrophy: The good, the bad, and the ugly. *Annu. Rev. Physiol.* **65:** 45–79.

Frey N., McKinsey T.A., and Olson E.N. 2000. Decoding calcium signals involved in cardiac growth and function. *Nat. Med.* **6:** 1221–1227.

Friddle C.J., Koga T., Rubin E.M., and Bristow J. 2000. Expression profiling reveals distinct sets of genes altered during induction and regression of cardiac hypertrophy. *Proc. Natl. Acad. Sci.* **97:** 6745–6750.

Froelicher V.F. and Myers J.N., Eds. 2000. In *Exercise and the heart*. W.B. Saunders, Philadelphia, Pennsylvania.

Fuller S.J., Mynett J.R., and Sugden P.H. 1992. Stimulation of cardiac protein synthesis by insulin-like growth factors. *Biochem. J.* (Pt. 1) **282:** 85–90.

Fumagalli S. and Thomas G. 2000. S6 phosphorylation and signal transduction. In *Translational control of gene expression* (ed. N. Sonenberg et al.), pp. 695-717. Cold Spring Harbor Laboratory Press, Cold Spring Harbor, New York.

Gao X., Zhang Y., Arrazola P., Hino O., Kobayashi T., Yeung R.S., Ru B., and Pan D. 2002. Tsc tumour suppressor proteins antagonize amino-acid-TOR signalling. *Nat. Cell Biol.* **4:** 699–704.

Geng Y.J., Ishikawa Y., Vatner D.E., Wagner T.E., Bishop S.P., Vatner S.F., and Homcy C.J. 1999. Apoptosis of cardiac myocytes in Gsalpha transgenic mice. *Circ. Res.* **84:** 34–42.

Giannessi D., Del Ry S., and Vitale R.L. 2001. The role of endothelins and their receptors in heart failure. *Pharmacol. Res.* **43:** 111–126.

Gill C., Mestril R., and Samali A. 2002. Losing heart: The role of apoptosis in heart disease—A novel therapeutic target? *FASEB J.* **16:** 135–146.

Greaves P., Martin J., Michel M.C., and Mompon P. 1984. Cardiac hypertrophy in the dog and rat induced by oxfenicine, an agent which modifies muscle metabolism. *Arch. Toxicol. Suppl.* **7:** 488–493.

Gregorio C.C. and Antin P.B. 2000. To the heart of myofibril assembly. *Trends Cell Biol.* **10:** 355–362.

Gutkind J.S. 1998a. Cell growth control by G protein-coupled receptors: From signal transduction to signal integration. *Oncogene* **17:** 1331–1342.

———. 1998b. The pathways connecting G protein-coupled receptors to the nucleus through divergent mitogen-activated protein kinase cascades. *J. Biol. Chem.* **273:** 1839–1842.

Hamawaki M., Coffman T.M., Lashus A., Koide M., Zile M.R., Oliverio M.I., DeFreyte G., Cooper G.T., and Carabello B.A. 1998. Pressure-overload hypertrophy is unabated in mice devoid of AT1A receptors. *Am. J. Physiol.* **274:** H868–H873.

Hannan R.D. and Rothblum L.I. 1995. Regulation of ribosomal DNA transcription during neonatal cardiomyocyte hypertrophy. *Cardiovasc. Res.* **30:** 501–510.

Hannan R.D., Jenkins A., Jenkins A.K., and Brandenburger Y. 2003. Cardiac hypertrophy: A matter of translation. *Clin. Exp. Pharmacol. Physiol.* **30:** 517–527.

Hannan R.D., Luyken J., and Rothblum L.I. 1995. Regulation of rDNA transcription factors during cardiomyocyte hypertrophy induced by adrenergic agents. *J. Biol. Chem.* **270:** 8290–8297.

———. 1996. Regulation of ribosomal DNA transcription during contraction-induced hypertrophy of neonatal cardiomyocytes. *J. Biol. Chem.* **271:** 3213–3220.

Haq S., Choukroun G., Kang Z.B., Ranu H., Matsui T., Rosenzweig A., Molkentin J.D., Alessandrini A., Woodgett J., Hajjar R., Michael A., and Force T. 2000. Glycogen synthase kinase-3beta is a negative regulator of cardiomyocyte hypertrophy. *J. Cell Biol.* **151:** 117–130.

Harada K., Komuro I., Zou Y., Kudoh S., Kijima K., Matsubara H., Sugaya T., Murakami K., and Yazaki Y. 1998. Acute pressure overload could induce hypertrophic responses in the heart of angiotensin II type 1a knockout mice. *Circ. Res.* **82:** 779–785.

Hardt S.E. and Sadoshima J. 2002. Glycogen synthase kinase-3beta: A novel regulator of cardiac hypertrophy and development. *Circ. Res.* **90:** 1055–1063.

Harvey C.B. and Williams G.R. 2002. Mechanism of thyroid hormone action. *Thyroid* **12:** 441–446.

Harvey R.P. 2002. Patterning the vertebrate heart. *Nat. Rev. Genet.* **3:** 544–556.

Harwood A.J. 2001. Regulation of GSK-3: A cellular multiprocessor. *Cell* **105:** 821–824.

Haunstetter A. and Izumo S. 1998. Apoptosis: Basic mechanisms and implications for cardiovascular disease. *Circ. Res.* **82:** 1111–1129.

Hein L., Barsh G.S., Pratt R.E., Dzau V.J., and Kobilka B.K. 1995. Behavioural and cardiovascular effects of disrupting the angiotensin II type-2 receptor in mice. *Nature* **377:** 744–747.

Hein L., Stevens M.E., Barsh G.S., Pratt R.E., Kobilka B.K., and Dzau V.J. 1997. Overexpression of angiotensin AT1 receptor transgene in the mouse myocardium produces a lethal phenotype associated with myocyte hyperplasia and heart block. *Proc. Natl. Acad. Sci.* **94:** 6391–6396.

Hirota H., Yoshida K., Kishimoto T., and Taga T. 1995. Continuous activation of gp130, a signal-transducing receptor component for interleukin 6-related cytokines, causes myocardial hypertrophy in mice. *Proc. Natl. Acad. Sci.* **92:** 4862–4866.

Hirota H., Chen J., Betz U.A., Rajewsky K., Gu Y., Ross, J., Jr., Muller W., and Chien K.R. 1999. Loss of a gp130 cardiac muscle cell survival pathway is a critical event in the onset of heart failure during biomechanical stress. *Cell* **97:** 189–198.

Holtwick R., van Eickels M., Skryabin B.V., Baba H.A., Bubikat A., Begrow F., Schneider M.D., Garbers D.L., and Kuhn M. 2003. Pressure-independent cardiac hypertrophy in mice with cardiomyocyte-restricted inactivation of the atrial natriuretic peptide receptor guanylyl cyclase-A. *J. Clin. Invest.* **111:** 1399–1407.

Horio T., Nishikimi T., Yoshihara F., Matsuo H., Takishita S., and Kangawa K. 2000. Inhibitory regulation of hypertrophy by endogenous atrial natriuretic peptide in cultured cardiac myocytes. *Hypertension* **35:** 19–24.

Hoshijima M., Sah V.P., Wang Y., Chien K.R., and Brown J.H. 1998. The low molecular weight GTPase Rho regulates myofibril formation and organization in neonatal rat ventricular myocytes. Involvement of Rho kinase. *J. Biol. Chem.* **273:** 7725–7730.

Hudlicka O. and Brown M.D. 1996. Postnatal growth of the heart and its blood vessels. *J. Vasc. Res.* **33:** 266–287.

Hunter J.J. and Chien K.R. 1999. Signaling pathways for cardiac hypertrophy and failure. *N. Engl. J. Med.* **341:** 1276–1283.

Hunter J.J., Tanaka N., Rockman H.A., Ross J., Jr., and Chien K.R. 1995. Ventricular expression of a MLC-2v-ras fusion gene induces cardiac hypertrophy and selective diastolic dysfunction in transgenic mice. *J. Biol. Chem.* **270:** 23173–23178.

Hwang D.M., Dempsey A.A., Lee C.Y., and Liew C.C. 2000. Identification of differentially expressed genes in cardiac hypertrophy by analysis of expressed sequence tags. *Genomics* **66:** 1–14.

Icardo J.M. 1984. The growing heart: An anatomical perspective. In *Growth of the heart in health and disease* (ed. R. Zak), pp. 41–79. Raven Press, New York.

Ichiki T., Labosky P.A., Shiota C., Okuyama S., Imagawa Y., Fogo A., Niimura F., Ichikawa I., Hogan B.L., and Inagami T. 1995. Effects on blood pressure and exploratory behaviour of mice lacking angiotensin II type-2 receptor. *Nature* **377:** 748–750.

Inoki K., Li Y., Zhu T., Wu J., and Guan K.L. 2002. TSC2 is phosphorylated and inhibited by Akt and suppresses mTOR signalling. *Nat. Cell Biol.* **4:** 648–657.

Ito H., Hirata Y., Adachi S., Tanaka M., Tsujino M., Koike A., Nogami A., Murumo F., and Hiroe M. 1993a. Endothelin-1 is an autocrine/paracrine factor in the mechanism of angiotensin II-induced hypertrophy in cultured rat cardiomyocytes. *J. Clin. Invest.* **92:** 398–403.

Ito H., Hiroe M., Hirata Y., Tsujino M., Adachi S., Shichiri M., Koike A., Nogami A., and Marumo F. 1993b. Insulin-like growth factor-I induces hypertrophy with enhanced expression of muscle specific genes in cultured rat cardiomyocytes. *Circulation* **87:** 1715–1721.

Iwamoto R., Yamazaki S., Asakura M., Takashima S., Hasuwa H., Miyado K., Adachi S., Kitakaze M., Hashimoto K., Raab G., Nanba D., Higashiyama S., Hori M., Klagsbrun M., and Mekada E. 2003. Heparin-binding EGF-like growth factor and ErbB signaling is essential for heart function. *Proc. Natl. Acad. Sci.* **100:** 3221–3226.

Izumo S. and Pu T. 2004. The molecular basis of heart failure. In *Heart failure* (ed. D.L. Mann), pp. 10–40. W.B. Saunders, Philadelphia, Pennsylvania.

Izumo S. and Shioi T. 1998. Cardiac transgenic and gene-targeted mice as models of cardiac hypertrophy and failure: A problem of (new) riches. *J. Card. Fail.* **4:** 263–270.

Izumo S., Nadal-Ginard B., and Mahdavi V. 1986. All members of the MHC multigene family respond to thyroid hormone in a highly tissue-specific manner. *Science* **231:** 597–600.

———. 1988. Protooncogene induction and reprogramming of cardiac gene expression produced by pressure overload. *Proc. Natl. Acad. Sci.* **85:** 339–343.

Jacinto E. and Hall M.N. 2003. Tor signalling in bugs, brain and brawn. *Nat. Rev. Mol. Cell. Biol.* **4:** 117–126.

Johnson M.W., Kerfoot C., Bushnell T., Li M., and Vinters H.V. 2001. Hamartin and tuberin expression in human tissues. *Mod. Pathol.* **14:** 202–210.

Kagiyama S., Eguchi S., Frank G.D., Inagami T., Zhang Y.C., and Phillips M.I. 2002. Angiotensin II-induced cardiac hypertrophy and hypertension are attenuated by epidermal growth factor receptor antisense. *Circulation* **106:** 909–912.

Kang P.M. and Izumo S. 2003. Apoptosis in heart: Basic mechanisms and implications in cardiovascular diseases. *Trends Mol. Med.* **9:** 177–182.

Kang P.M., Yue P., and Izumo S. 2002. New insights into the role of apoptosis in cardiovascular disease. *Circ. J.* **66:** 1–9.

Kaplan M.L., Cheslow Y., Vikstrom K., Malhotra A., Geenen D.L., Nakouzi A., Leinwand L.A., and Buttrick P.M. 1994. Cardiac adaptations to chronic exercise in mice. *Am. J. Physiol.* **267:** H1167–H1173.

Kardami E., Liu L., Kishore S., Pasumarthi B., Doble B.W., and Cattini P.A. 1995. Regulation of basic fibroblast growth factor (bFGF) and FGF receptors in the heart. *Ann. N.Y. Acad. Sci.* **752:** 353–369.

Kedzierski R.M. and Yanagisawa M. 2001. Endothelin system: The double-edged sword in health and disease. *Annu. Rev. Pharmacol. Toxicol.* **41:** 851–876.

Kedzierski R.M., Grayburn P.A., Kisanuki Y.Y., Williams C.S., Hammer R.E., Richardson J.A., Schneider M.D., and Yanagisawa M. 2003. Cardiomyocyte-specific endothelin A receptor knockout mice have normal cardiac function and an unaltered hypertrophic response to angiotensin II and isoproterenol. *Mol. Cell. Biol.* **23:** 8226–8232.

Kelly D.P. and Strauss A.W. 1994. Inherited cardiomyopathies. *N. Engl. J. Med.* **330:** 913–919.

Kentera D., Susic D., and Zdravkovic M. 1979. Effects of verapamil and aspirin on experimental chronic hypoxic pulmonary hypertension and right ventricular hypertrophy in rats. *Respiration* **37:** 192–196.

Khan A.S., Sane D.C., Wannenburg T., and Sonntag W.E. 2002. Growth hormone, insulin-like growth factor-1 and the aging cardiovascular system. *Cardiovasc. Res.* **54:** 25–35.

Kishimoto I., Rossi K., and Garbers D.L. 2001. A genetic model provides evidence that the receptor for atrial natriuretic peptide (guanylyl cyclase-A) inhibits cardiac ventricular myocyte hypertrophy. *Proc. Natl. Acad. Sci.* **98:** 2703–2706.

Kleijn M., Scheper G.C., Voorma H.O., and Thomas A.A. 1998. Regulation of translation initiation factors by signal transduction. *Eur. J. Biochem.* **253:** 531–544.

Koch W.J., Lefkowitz R.J., and Rockman H.A. 2000. Functional consequences of altering myocardial adrenergic receptor signaling. *Annu. Rev. Physiol.* **62:** 237–260.

Kodama H., Fukuda K., Pan J., Makino S., Baba A., Hori S., and Ogawa S. 1997. Leukemia inhibitory factor, a potent cardiac hypertrophic cytokine, activates the JAK/STAT pathway in rat cardiomyocytes. *Circ. Res.* **81:** 656–663.

Kodama H., Fukuda K., Pan J., Sano M., Takahashi T., Kato T., Makino S., Manabe T., Murata M., and Ogawa S. 2000. Significance of ERK cascade compared with JAK/STAT and PI3-K pathway in gp130-mediated cardiac hypertrophy. *Am. J. Physiol. Heart Circ. Physiol.* **279:** H1635–H1644.

Komuro I., Kudo S., Yamazaki T., Zou Y., Shiojima I., and Yazaki Y. 1996. Mechanical stretch activates the stress-activated protein kinases in cardiac myocytes. *FASEB J.* **10:** 631–636.

Kook H., Lepore J.J., Gitler A.D., Lu M.M., Wing-Man Yung W., Mackay J., Zhou R., Ferrari V., Gruber P., and Epstein J.A. 2003. Cardiac hypertrophy and histone deacetylase-dependent transcriptional repression mediated by the atypical homeodomain protein Hop. *J. Clin. Invest.* **112:** 863–871.

Kovacic-Milivojevic B., Roediger F., Almeida E.A., Damsky C.H., Gardner D.G., and Ilic D. 2001. Focal adhesion kinase and p130Cas mediate both sarcomeric organization and activation of genes associated with cardiac myocyte hypertrophy. *Mol. Biol. Cell* **12:** 2290–2307.

Kubota T., McTiernan C.F., Frye C.S., Slawson S.E., Lemster B.H., Koretsky A.P., Demetris A.J., and Feldman A.M. 1997. Dilated cardiomyopathy in transgenic mice with cardiac-specific overexpression of tumor necrosis factor-alpha. *Circ. Res.* **81:** 627–635.

Kudoh S., Komuro I., Hiroi Y., Zou Y., Harada K., Sugaya T., Takekoshi N., Murakami K., Kadowaki T., and Yazaki Y. 1998. Mechanical stretch induces hypertrophic responses in cardiac myocytes of angiotensin II type 1a receptor knockout mice. *J. Biol. Chem.* **273:** 24037–24043.

Kunapuli P., Lawson J.A., Rokach J.A., Meinkoth J.L., and FitzGerald G.A. 1998. Prostaglandin F2alpha (PGF2alpha) and the isoprostane, 8, 12-iso-isoprostane F2alpha-III, induce cardiomyocyte hypertrophy. Differential activation of downstream signaling pathways. *J. Biol. Chem.* **273:** 22442–22452.

Kunisada K., Tone E., Fujio Y., Matsui H., Yamauchi-Takihara K., and Kishimoto T. 1998. Activation of gp130 transduces hypertrophic signals via STAT3 in cardiac myocytes. *Circulation* **98:** 346–352.

Kunisada K., Negoro S., Tone E., Funamoto M., Osugi T., Yamada S., Okabe M., Kishimoto T., and Yamauchi-Takihara K. 2000. Signal transducer and activator of transcription 3 in the heart transduces not only a hypertrophic signal but a protective signal against doxorubicin-induced cardiomyopathy. *Proc. Natl. Acad. Sci.* **97:** 315–319.

Kuppuswamy D. 2002. Importance of integrin signaling in myocyte growth and survival. *Circ. Res.* **90:** 1240–1242.

Lai J., Jin H., Yang R., Winer J., Li W., Yen R., King K.L., Zeigler F., Ko A., Cheng J., Bunting S., and Paoni N.F. 1996. Prostaglandin F2 alpha induces cardiac myocyte hypertrophy in vitro and cardiac growth in vivo. *Am. J. Physiol.* **271:** H2197–H2208.

Lakatta E.G., Zhou Y., and Xiao R. 2001. Aging of the cardiovascular system. In *Heart physiology and pathophysiology* (ed. N. Sperelakis et al.), pp. 737–760. Academic Press, San Diego, California.

Lavandero S., Foncea R., Perez V., and Sapag-Hagar M. 1998. Effect of inhibitors of signal transduction on IGF-1-induced protein synthesis associated with hypertrophy in cultured neonatal rat ventricular myocytes. *FEBS Lett.* **422:** 193–196.

Lazar M.A. 1993. Thyroid hormone receptors: Multiple forms, multiple possibilities. *Endocr. Rev.* **14:** 184–193.

Lazar M.A. and Chin W.W. 1990. Nuclear thyroid hormone receptors. *J. Clin. Invest.* **86:** 1777–1782.

Lee K.F., Simon H., Chen H., Bates B., Hung M.C., and Hauser C. 1995. Requirement for neuregulin receptor erbB2 in neural and cardiac development. *Nature* **378:** 394–398.

Lehman J.J. and Kelly D.P. 2002a. Gene regulatory mechanisms governing energy metabolism during cardiac hypertrophic growth. *Heart Fail Rev.* **7:** 175–185.

———. 2002b. Transcriptional activation of energy metabolic switches in the developing and hypertrophied heart. *Clin. Exp. Pharmacol. Physiol.* **29:** 339–345.

Levy D., Garrison R.J., Savage D.D., Kannel W.B., and Castelli W.P. 1990. Prognostic implications of echocardiographically determined left ventricular mass in the Framingham Heart Study. *N. Engl. J. Med.* **322:** 1561–1566.

Li W.G., Zaheer A., Coppey L., and Oskarsson H.J. 1998. Activation of JNK in the remote myocardium after large myocardial infarction in rats. *Biochem. Biophys. Res. Commun.* **246:** 816–820.

Liang Q. and Molkentin J.D. 2003. Redefining the roles of p38 and JNK signaling in cardiac hypertrophy: Dichotomy between cultured myocytes and animal models. *J. Mol. Cell. Cardiol.* **35:** 1385–1394.

Liang Q., Bueno O.F., Wilkins B.J., Kuan C.Y., Xia Y., and Molkentin J.D. 2003. c-Jun N-terminal kinases (JNK) antagonize cardiac growth through cross-talk with calcineurin-NFAT signaling. *EMBO J.* **22:** 5079–5089.

Liang Q., De Windt L.J., Witt S.A., Kimball T.R., Markham B.E., and Molkentin J.D. 2001. The transcription factors GATA4 and GATA6 regulate cardiomyocyte hypertrophy in vitro and in vivo. *J. Biol. Chem.* **276:** 30245–30253.

Liao P., Georgakopoulos D., Kovacs A., Zheng M., Lerner D., Pu H., Saffitz J., Chien K., Xiao R.P., Kass D.A., and Wang Y. 2001. The in vivo role of p38 MAP kinases in cardiac remodeling and restrictive cardiomyopathy. *Proc. Natl. Acad. Sci.* **98:** 12283–12288.

Lijnen P. and Petrov V. 1999. Renin-angiotensin system, hypertrophy and gene expression in cardiac myocytes. *J. Mol. Cell. Cardiol.* **31:** 949–970.

Liu J.P., Baker J., Perkins A.S., Robertson E.J., and Efstratiadis A. 1993. Mice carrying null mutations of the genes encoding insulin-like growth factor I (Igf-1) and type 1 IGF receptor (Igf1r). *Cell* **75:** 59–72.

Liu T., Lai H., Wu W., Chinn S., and Wang P.H. 2001. Developing a strategy to define the effects of insulin-like growth factor-1 on gene expression profile in cardiomyocytes. *Circ. Res.* **88:** 1231–1238.

Lomasney J.W. and Allen L.F. 2001. Adrenergic receptors in the cardiovascular system. In *Heart physiology and pathophysiology* (ed. N. Sperelakis et al.), pp. 599–608. Academic Press, San Diego, California.

Lorell B.H. 1999. Role of angiotensin AT1, and AT2 receptors in cardiac hypertrophy and disease. *Am. J. Cardiol.* **83:** 48H–52H.

Lu C., Schwartzbauer G., Sperling M.A., Devaskar S.U., Thamotharan S., Robbins P.D., McTiernan C.F., Liu J.L., Jiang J., Frank S.J., and Menon R.K. 2001. Demonstration of direct effects of growth hormone on neonatal cardiomyocytes. *J. Biol. Chem.* **276:** 22892–22900.

Luyken J., Hannan R.D., Cheung J.Y., and Rothblum L.I. 1996. Regulation of rDNA transcription during endothelin-1-induced hypertrophy of neonatal cardiomyocytes.

Hyperphosphorylation of upstream binding factor, an rDNA transcription factor. *Circ. Res.* **78:** 354–361.

Mackay K. and Mochly-Rosen D. 2001. Localization, anchoring, and functions of protein kinase C isozymes in the heart. *J. Mol. Cell. Cardiol.* **33:** 1301–1307.

MacLellan W.R. and Schneider M.D. 1997. Death by design. Programmed cell death in cardiovascular biology and disease. *Circ. Res.* **81:** 137–144.

———. 2000. Genetic dissection of cardiac growth control pathways. *Annu. Rev. Physiol.* **62:** 289–319.

Mangelsdorf D.J., Thummel C., Beato M., Herrlich P., Schutz G., Umesono K., Blumberg B., Kastner P., Mark M., and Chambon P., et al. 1995. The nuclear receptor superfamily: The second decade. *Cell* **83:** 835–839.

Mann D.L., Urabe Y., Kent R.L., Vinciguerra S., and Cooper G., IV. 1991. Cellular versus myocardial basis for the contractile dysfunction of hypertrophied mycocardium. *Cir. Res.* **68:** 402–415.

Marchionni M.A. 1995. Cell-cell signalling. neu tack on neuregulin. *Nature* **378:** 334–335.

Masaki H., Kurihara T., Yamaki A., Inomata N., Nozawa Y., Mori Y., Murasawa S., Kizima K., Maruyama K., Horiuchi M., Dzau V.J., Takahashi H., Iwasaka T., Inada M., and Matsubara H. 1998. Cardiac-specific overexpression of angiotensin II AT2 receptor causes attenuated response to AT1 receptor-mediated pressor and chronotropic effects. *J. Clin. Invest.* **101:** 527–535.

Matsui T., Li L., Wu J.C., Cook S.A., Nagoshi T., Picard M.H., Liao R., and Rosenzweig A. 2002. Phenotypic spectrum caused by transgenic overexpression of activated Akt in the heart. *J. Biol. Chem.* **277:** 22896–22901.

McKinsey T.A. and Olson E.N. 1999. Cardiac hypertrophy: Sorting out the circuitry. *Curr. Opin. Genet. Dev.* **9:** 267–274.

McKinsey T.A., Zhang C.L., and Olson E.N. 2002. Signaling chromatin to make muscle. *Curr. Opin. Cell Biol.* **14:** 763–772.

McMullen J.R., Shioi T., Zhang L., Tarnavski O., Sherwood M.C., Kang P.M., and Izumo S. 2003. Phosphoinositide 3-kinase(p110 alpha) plays a critical role for the induction of physiological, but not pathological, cardiac hypertrophy. *Proc. Natl. Acad. Sci.* **100:** 12355–12360.

McMullen J.R., Shioi T., Huang W.Y., Zhang L., Tarnavski O., Bisping E., Schinke M., Kong S., Sherwood M.C., Brown J., Riggi L., Kang P.M., and Izumo S. 2004. The insulin-like growth factor 1 receptor induces physiological heart growth via the phosphoinositide 3-kinase(p110alpha) pathway. *J. Biol. Chem.* **279:** 4782–4793.

Meyer D. and Birchmeier C. 1995. Multiple essential functions of neuregulin in development. *Nature* **378:** 386–390.

Milano C.A., Dolber P.C., Rockman H.A., Bond R.A., Venable M.E., Allen L.F., and Lefkowitz R.J. 1994. Myocardial expression of a constitutively active alpha 1B-adrenergic receptor in transgenic mice induces cardiac hypertrophy. *Proc. Natl. Acad. Sci.* **91:** 10109–10113.

Millan F.A., Denhez F., Kondaiah P., and Akhurst R.J. 1991. Embryonic gene expression patterns of TGF beta 1, beta 2 and beta 3 suggest different developmental functions in vivo. *Development* **111:** 131–143.

Miyata S. and Haneda T. 1994. Hypertrophic growth of cultured neonatal rat heart cells mediated by type 1 angiotensin II receptor. *Am. J. Physiol.* **266:** H2443–H2451.

Miyauchi T. and Goto K. 1999. Heart failure and endothelin receptor antagonists. *Trends Pharmacol. Sci.* **20:** 210–217.

Miyauchi T. and Masaki T. 1999. Pathophysiology of endothelin in the cardiovascular system. *Annu. Rev. Physiol.* **61:** 391–415.

Modesti P.A., Vanni S., Bertolozzi I., Cecioni I., Polidori G., Paniccia R., Bandinelli B., Perna A., Liguori P., Boddi M., Galanti G., and Serneri G.G. 2000. Early sequence of cardiac adaptations and growth factor formation in pressure- and volume-overload hypertrophy. *Am. J. Physiol. Heart. Circ. Physiol.* **279:** H976–H985.

Molkentin J.D. 2003. A friend within the heart: Natriuretic peptide receptor signaling. *J. Clin. Invest.* **111:** 1275–1277.

Molkentin J.D. and Dorn I.G., II. 2001. Cytoplasmic signaling pathways that regulate cardiac hypertrophy. *Annu. Rev. Physiol.* **63:** 391–426.

Moncman C.L. and Wang K. 2002. Targeted disruption of nebulette protein expression alters cardiac myofibril assembly and function. *Exp. Cell Res.* **273:** 204–218.

Morimoto T., Hasegawa K., Wada H., Kakita T., Kaburagi S., Yanazume T., and Sasayama S. 2001. Calcineurin-GATA4 pathway is involved in beta-adrenergic agonist-responsive endothelin-1 transcription in cardiac myocytes. *J. Biol. Chem.* **276:** 34983–34989.

Morisco C., Seta K., Hardt S.E., Lee Y., Vatner S.F., and Sadoshima J. 2001. Glycogen synthase kinase 3beta regulates GATA4 in cardiac myocytes. *J. Biol. Chem.* **276:** 28586–28597.

Morisco C., Zebrowski D., Condorelli G., Tsichlis P., Vatner S.F., and Sadoshima J. 2000. The Akt-glycogen synthase kinase 3beta pathway regulates transcription of atrial natriuretic factor induced by beta-adrenergic receptor stimulation in cardiac myocytes. *J. Biol. Chem.* **275:** 14466–14475.

Nakagawa M., Hamaoka K., Hattori T., and Sawada T. 1988. Postnatal DNA synthesis in hearts of mice: Autoradiographic and cytofluorometric investigations. *Cardiovasc. Res.* **22:** 575–583.

Neer E.J. 1995. Heterotrimeric G proteins: Organizers of transmembrane signals. *Cell* **80:** 249–257.

Nemoto S., Sheng Z., and Lin A. 1998. Opposing effects of Jun kinase and p38 mitogen-activated protein kinases on cardiomyocyte hypertrophy. *Mol. Cell. Biol.* **18:** 3518–3526.

Nicol R.L., Frey N., Pearson G., Cobb M., Richardson J., and Olson E.N. 2001. Activated MEK5 induces serial assembly of sarcomeres and eccentric cardiac hypertrophy. *EMBO J.* **20:** 2757–2767.

Nishiyama A., Kambe F., Kamiya K., Seo H., and Toyama J. 1998. Effects of thyroid status on expression of voltage-gated potassium channels in rat left ventricle. *Cardiovasc. Res.* **40:** 343–351.

Nyui N., Tamura K., Mizuno K., Ishigami T., Kihara M., Ochiai H., Kimura K., Umemura S., Ohno S., Taga T., and Ishii M. 1998. gp130 is involved in stretch-induced MAP kinase activation in cardiac myocytes. *Biochem. Biophys. Res. Commun.* **245:** 928–932.

O'Connell T.D., Ishizaka S., Nakamura A., Swigart P.M., Rodrigo M.C., Simpson G.L., Cotecchia S., Rokosh D.G., Grossman W., Foster E., and Simpson P.C. 2003. The alpha(1A/C)- and alpha(1B)-adrenergic receptors are required for physiological cardiac hypertrophy in the double-knockout mouse. *J. Clin. Invest.* **111:** 1783–1791.

Oliverio M.I., Kim H.S., Ito M., Le T., Audoly L., Best C.F., Hiller S., Kluckman K., Maeda N., Smithies O., and Coffman T.M. 1998. Reduced growth, abnormal kidney structure, and type 2 (AT2) angiotensin receptor-mediated blood pressure regulation in mice lacking both AT1A and AT1B receptors for angiotensin II. *Proc. Natl. Acad. Sci.* **95:** 15496–15501.

Olson E.N. and Schneider M.D. 2003. Sizing up the heart: Development redux in disease. *Genes Dev.* **17:** 1937–1956.

Olson E.N. and Williams R.S. 2000. Calcineurin signaling and muscle remodeling. *Cell* **101:** 689–692.

Orphanides G. and Reinberg D. 2002. A unified theory of gene expression. *Cell* **108:** 439–451.

Oudit G.Y., Crackower M.A., Eriksson U., Sarao R., Kozieradzki I., Sasaki T., Irie-Sasaki J., Gidrewicz D., Rybin V.O., Wada T., Steinberg S.F., Backx P.H., and Penninger J.M. 2003. Phosphoinositide 3-kinase gamma-deficient mice are protected from isoproterenol-induced heart failure. *Circulation* **108:** 2147–2152.

Ozcelik C., Erdmann B., Pilz B., Wettschureck N., Britsch S., Hubner N., Chien K.R., Birchmeier C., and Garratt A.N. 2002. Conditional mutation of the ErbB2 (HER2) receptor in cardiomyocytes leads to dilated cardiomyopathy. *Proc. Natl. Acad. Sci.* **99:** 8880–8885.

Paradis P., Dali-Youcef N., Paradis F.W., Thibault G., and Nemer M. 2000. Overexpression of angiotensin II type I receptor in cardiomyocytes induces cardiac hypertrophy and remodeling. *Proc. Natl. Acad. Sci.* **97:** 931–936.

Parker T.G., Packer S.E., and Schneider M.D. 1990. Peptide growth factors can provoke "fetal" contractile protein gene expression in rat cardiac myocytes. *J. Clin. Invest.* **85:** 507–514.

Parrizas M., Saltiel A.R., and LeRoith D. 1997. Insulin-like growth factor 1 inhibits apoptosis using the phosphatidylinositol 3′-kinase and mitogen-activated protein kinase pathways. *J. Biol. Chem.* **272:** 154–161.

Pasumarthi K.B. and Field L.J. 2002. Cardiomyocyte cell cycle regulation. *Circ. Res.* **90:** 1044–1054.

Pearson G., Robinson F., Beers, Gibson T., Xu B.E., Karandikar M., Berman K., and Cobb M.H. 2001. Mitogen-activated protein (MAP) kinase pathways: Regulation and physiological functions. *Endocr. Rev.* **22:** 153–183.

Pellegrini S. and Dusanter-Fourt I. 1997. The structure, regulation and function of the Janus kinases (JAKs) and the signal transducers and activators of transcription (STATs). *Eur. J. Biochem.* **248:** 615–633.

Pellieux C., Foletti A., Peduto G., Aubert J.F., Nussberger J., Beermann F., Brunner H.R., and Pedrazzini T. 2001. Dilated cardiomyopathy and impaired cardiac hypertrophic response to angiotensin II in mice lacking FGF-2. *J. Clin. Invest.* **108:** 1843–1851.

Pennica D., King K.L., Shaw K.J., Luis E., Rullamas J., Luoh S.M., Darbonne W.C., Knutzon D.S., Yen R., and Chien K.R., et al. 1995. Expression cloning of cardiotrophin 1, a cytokine that induces cardiac myocyte hypertrophy. *Proc. Natl. Acad. Sci.* **92:** 1142–1146.

Petrich B.G., Molkentin J.D., and Wang Y. 2003. Temporal activation of c-Jun N-terminal kinase in adult transgenic heart via cre-loxP-mediated DNA recombination. *FASEB J.* **17:** 749–751.

Petrich B.G., Gong X., Lerner D.L., Wang X., Brown J.H., Saffitz J.E., and Wang Y. 2002. c-Jun N-terminal kinase activation mediates downregulation of connexin43 in cardiomyocytes. *Circ. Res.* **91:** 640–647.

Potter C.J., Pedraza L.G., and Xu T. 2002. Akt regulates growth by directly phosphorylating Tsc2. *Nat. Cell Biol.* **4:** 658–665.

Pracyk J.B., Tanaka K., Hegland D.D., Kim K.S., Sethi R., Rovira II, Blazina D.R., Lee L., Bruder J.T., Kovesdi I., Goldshmidt-Clermont P.J., Irani K., and Finkel T. 1998. A requirement for the rac1 GTPase in the signal transduction pathway leading to cardiac myocyte hypertrophy. *J. Clin. Invest.* **102:** 929–937.

Pu W.T., Ma Q., and Izumo S. 2003. NFAT transcription factors are critical survival factors that inhibit cardiomyocyte apoptosis during phenylephrine stimulation in vitro. *Circ. Res.* **92:** 725–731.

Radimerski T., Montagne J., Hemmings-Mieszczak M., and Thomas G. 2002. Lethality of *Drosophila* lacking TSC tumor suppressor function rescued by reducing dS6K signaling. *Genes Dev.* **16:** 2627–2632.

Rakusan K. 1984. Cardiac growth, maturation, and aging. In *Growth of the heart in health and disease* (ed. R. Zak), pp. 131–164. Raven Press, New York.

Ramirez M.T., Sah V.P., Zhao X.L., Hunter J.J., Chien K.R., and Brown J.H. 1997. The MEKK-JNK pathway is stimulated by alpha1-adrenergic receptor and ras activation and is associated with in vitro and in vivo cardiac hypertrophy. *J. Biol. Chem.* **272:** 14057–14061.

Rapacciuolo A., Esposito G., Caron K., Mao L., Thomas S.A., and Rockman H.A. 2001. Important role of endogenous norepinephrine and epinephrine in the development of in vivo pressure-overload cardiac hypertrophy. *J. Am. Coll. Cardiol.* **38:** 876–882.

Reiss K., Cheng W., Ferber A., Kajstura J., Li P., Li B., Olivetti G., Homcy C.J., Baserga R., and Anversa P. 1996. Overexpression of insulin-like growth factor-1 in the heart is coupled with myocyte proliferation in transgenic mice. *Proc. Natl. Acad. Sci.* **93:** 8630–8635.

Ren J., Samson W.K., and Sowers J.R. 1999. Insulin-like growth factor I as a cardiac hormone: Physiological and pathophysiological implications in heart disease. *J. Mol. Cell. Cardiol.* **31:** 2049–2061.

Rhoads R.E. 1999. Signal transduction pathways that regulate eukaryotic protein synthesis. *J. Biol. Chem.* **274:** 30337–30340.

Rockman H.A., Koch W.J., and Lefkowitz R.J. 2002. Seven-transmembrane-spanning receptors and heart function. *Nature* **415:** 206–212.

Rohrer D. and Dillmann W.H. 1988. Thyroid hormone markedly increases the mRNA coding for sarcoplasmic reticulum Ca2+-ATPase in the rat heart. *J. Biol. Chem.* **263:** 6941–6944.

Roman B.B., Geenen D.L., Leitges M., and Buttrick P.M. 2001. PKC-beta is not necessary for cardiac hypertrophy. *Am. J. Physiol. Heart. Circ. Physiol.* **280:** H2264–H2270.

Rosenkranz S., Flesch M., Amann K., Haeuseler C., Kilter H., Seeland U., Schluter K.D., and Bohm M. 2002. Alterations of beta-adrenergic signaling and cardiac hypertrophy in transgenic mice overexpressing TGF-beta(1). *Am. J. Physiol. Heart Circ. Physiol.* **283:** H1253–H1262.

Ross R.S. and Borg T.K. 2001. Integrins and the myocardium. *Circ. Res.* **88:** 1112–1119.

Rothermel B.A., McKinsey T.A., Vega R.B., Nicol R.L., Mammen P., Yang J., Antos C.L., Shelton J.M., Bassel-Duby R., Olson E.N., and Williams R.S. 2001. Myocyte-enriched calcineurin-interacting protein, MCIP1, inhibits cardiac hypertrophy in vivo. *Proc. Natl. Acad. Sci.* **98:** 3328–3333.

Rupp H. and Jacob R. 1992. Metabolically-modulated growth and phenotype of the rat heart. *Eur. Heart. J.* (suppl. D) **13:** 56–61.

Russell B., Motlagh D., and Ashley W.W. 2000. Form follows function: How muscle shape is regulated by work. *J. Appl. Physiol.* **88:** 1127–1132.

Ruwhof C. and van der Laarse A. 2000. Mechanical stress-induced cardiac hypertrophy: Mechanisms and signal transduction pathways. *Cardiovasc. Res.* **47:** 23–37.

Sabri A. and Steinberg S.F. 2003. Protein kinase C isoform-selective signals that lead to cardiac hypertrophy and the progression of heart failure. *Mol. Cell. Biochem.* **251:** 97–101.

Sadoshima J. and Izumo S. 1993. Molecular characterization of angiotensin II-induced hypertrophy of cardiac myocytes and hyperplasia of cardiac fibroblasts. Critical role of the AT1 receptor subtype. *Circ. Res.* **73:** 413–423.

———. 1995. Rapamycin selectively inhibits angiotensin II-induced increase in protein synthesis in cardiac myocytes in vitro. Potential role of 70-kD S6 kinase in angiotensin II-induced cardiac hypertrophy. *Circ. Res.* **77:** 1040–1052.

————. 1996. The heterotrimeric G q protein-coupled angiotensin II receptor activates p21 ras via the tyrosine kinase-Shc-Grb2-Sos pathway in cardiac myocytes. *EMBO J.* **15:** 775–787.

————. 1997. The cellular and molecular response of cardiac myocytes to mechanical stress. *Annu. Rev. Physiol.* **59:** 551–571.

Sadoshima J., Malhotra R., and Izumo S. 1996. The role of the cardiac renin-angiotensin system in load-induced cardiac hypertrophy. *J. Card. Fail.* **2:** S1–S6.

Sadoshima J., Qiu Z., Morgan J.P., and Izumo S. 1995. Angiotensin II and other hypertrophic stimuli mediated by G protein-coupled receptors activate tyrosine kinase, mitogen-activated protein kinase, and 90-kD S6 kinase in cardiac myocytes. The critical role of Ca(2+)-dependent signaling. *Circ. Res.* **76:** 1–15.

Sadoshima J., Xu Y., Slayter H.S., and Izumo S. 1993. Autocrine release of angiotensin II mediates stretch-induced hypertrophy of cardiac myocytes in vitro. *Cell* **75:** 977–984.

Saito S., Hiroi Y., Zou Y., Aikawa R., Toko H., Shibasaki F., Yazaki Y., Nagai R., and Komuro I. 2000. beta-Adrenergic pathway induces apoptosis through calcineurin activation in cardiac myocytes. *J. Biol. Chem.* **275:** 34528–34533.

Sanger J.W. and Sanger J.M. 2001. Fishing out proteins that bind to titin. *J. Cell Biol.* **154:** 21–24.

Sanger J.W., Ayoob J.C., Chowrashi P., Zurawski D., and Sanger J.M. 2000. Assembly of myofibrils in cardiac muscle cells (see discussion). *Adv. Exp. Med. Biol.* **481:** 89–105.

Schaible T.F. and Scheuer J. 1984. Response of the heart to exercise training. In *Growth of the heart in health and disease* (ed. R. Zak), pp. 381–420. Raven Press, New York.

Scheuer J. 1999. Catecholamines in cardiac hypertrophy. *Am. J. Cardiol.* **83:** 70H–74H.

Schlessinger J. 2000. Cell signaling by receptor tyrosine kinases. *Cell* **103:** 211–225.

Schluter K.D. and Piper H.M. 1999. Regulation of growth in the adult cardiomyocytes. *FASEB J.* (suppl.) **13:** S17–S22.

Schmelzle T. and Hall M.N. 2000. TOR, a central controller of cell growth. *Cell* **103:** 253–262.

Schneider M.D. and Parker T.G. 1990. Cardiac myocytes as targets for the action of peptide growth factors. *Circulation* **81:** 1443–1456.

Schneider M.D., McLellan W.R., Black F.M., and Parker T.G. 1992. Growth factors, growth factor response elements, and the cardiac phenotype. *Basic Res. Cardiol.* (suppl. 2) **87:** 33–48.

Schultz J.E., Witt S.A., Nieman M.L., Reiser P.J., Engle S.J., Zhou M., Pawlowski S.A., Lorenz J.N., Kimball T.R., and Doetschman T. 1999. Fibroblast growth factor-2 mediates pressure-induced hypertrophic response. *J. Clin. Invest.* **104:** 709–719.

Schultz Jel J., Witt S.A., Glascock B.J., Nieman M.L., Reiser P.J., Nix S.L., Kimball T.R., and Doetschman T. 2002. TGF-beta1 mediates the hypertrophic cardiomyocyte growth induced by angiotensin II. *J. Clin. Invest.* **109:** 787–796.

Senbonmatsu T., Ichihara S., Price E., Jr., Gaffney F.A., and Inagami T. 2000. Evidence for angiotensin II type 2 receptor-mediated cardiac myocyte enlargement during in vivo pressure overload. *J. Clin. Invest.* **106:** R25–R29.

Sheikh F., Sontag D.P., Fandrich R.R., Kardami E., and Cattini P.A. 2001. Overexpression of FGF-2 increases cardiac myocyte viability after injury in isolated mouse hearts. *Am. J. Physiol. Heart Circ. Physiol.* **280:** H1039–H1050.

Sheng Z., Knowlton K., Chen J., Hoshijima M., Brown J.H., and Chien K.R. 1997. Cardiotrophin 1 (CT-1) inhibition of cardiac myocyte apoptosis via a mitogen-activated protein kinase-dependent pathway. Divergence from downstream CT-1 signals for myocardial cell hypertrophy. *J. Biol. Chem.* **272:** 5783–5791.

Shioi T., McMullen J.R., Tarnavski O., Converso K., Sherwood M.C., Manning W.J., and Izumo S. 2003. Rapamycin attenuates load-induced cardiac hypertrophy in mice. *Circulation* **107:** 1664–1670.

Shioi T., Kang P.M., Douglas P.S., Hampe J., Yballe C.M., Lawitts J., Cantley L.C., and Izumo S. 2000. The conserved phosphoinositide 3-kinase pathway determines heart size in mice. *EMBO J.* **19:** 2537–2548.

Shioi T., McMullen J.R., Kang P.M., Douglas P.S., Obata T., Franke T.F., Cantley L.C., and Izumo S. 2002. Akt/protein kinase B promotes organ growth in transgenic mice. *Mol. Cell. Biol.* **22:** 2799–2809.

Shubeita H.E., McDonough P.M., Harris A.N., Knowlton K.U., Glembotski C.C., Brown J.H., and Chien K.R. 1990. Endothelin induction of inositol phospholipid hydrolysis, sarcomere assembly, and cardiac gene expression in ventricular myocytes. A paracrine mechanism for myocardial cell hypertrophy. *J. Biol. Chem.* **265:** 20555–20562.

Simon M.I., Strathmann M.P., and Gautam N. 1991. Diversity of G proteins in signal transduction. *Science* **252:** 802–808.

Sivasubramanian N., Coker M.L., Kurrelmeyer K.M., MacLellan W.R., DeMayo F.J., Spinale F.G., and Mann D.L. 2001. Left ventricular remodeling in transgenic mice with cardiac restricted overexpression of tumor necrosis factor. *Circulation* **104:** 826–831.

Solloway M.J. and Harvey R.P. 2003. Molecular pathways in myocardial development: A stem cell perspective. *Cardiovasc. Res.* **58:** 264–277.

Sonenberg N. and Gingras A.C. 1998. The mRNA 5′ cap-binding protein eIF4E and control of cell growth. *Curr. Opin. Cell Biol.* **10:** 268–275.

Soonpaa M.H., Kim K.K., Pajak L., Franklin M., and Field L.J. 1996. Cardiomyocyte DNA synthesis and binucleation during murine development. *Am. J. Physiol.* **271:** H2183–H2189.

Sugden P.H. 2001. Signalling pathways in cardiac myocyte hypertrophy. *Ann. Med.* **33:** 611–622.

Sugden P.H. and Clerk A. 1998. Cellular mechanisms of cardiac hypertrophy. *J. Mol. Med.* **76:** 725–746.

Sussman M.A., Welch S., Walker A., Klevitsky R., Hewett T.E., Price R.L., Schaefer E., and Yager K. 2000. Altered focal adhesion regulation correlates with cardiomyopathy in mice expressing constitutively active rac1. *J. Clin. Invest.* **105:** 875–886.

Takahashi N., Calderone A., Izzo N.J., Jr., Maki T.M., Marsh J.D., and Colucci W.S. 1994. Hypertrophic stimuli induce transforming growth factor-beta 1 expression in rat ventricular myocytes. *J. Clin. Invest.* **94:** 1470–1476.

Takeishi Y., Ping P., Bolli R., Kirkpatrick D.L., Hoit B.D., and Walsh R.A. 2000. Transgenic overexpression of constitutively active protein kinase C epsilon causes concentric cardiac hypertrophy. *Circ. Res.* **86:** 1218–1223.

Takeishi Y., Huang Q., Abe J., Glassman M., Che W., Lee J.D., Kawakatsu H., Lawrence E.G., Hoit B.D., Berk B.C., and Walsh R.A. 2001. Src and multiple MAP kinase activation in cardiac hypertrophy and congestive heart failure under chronic pressure-overload: Comparison with acute mechanical stretch. *J. Mol. Cell. Cardiol.* **33:** 1637–1648.

Thomas G. and Hall M.N. 1997. TOR signalling and control of cell growth. *Curr. Opin. Cell Biol.* **9:** 782–787.

Thomas W.G., Brandenburger Y., Autelitano D.J., Pham T., Qian H., and Hannan R.D. 2002. Adenoviral-directed expression of the type 1A angiotensin receptor promotes cardiomyocyte hypertrophy via transactivation of the epidermal growth factor receptor. *Circ. Res.* **90:** 135–142.

Thompson C.B. 1995. Apoptosis in the pathogenesis and treatment of disease. *Science* **267:** 1456–1462.

Toker A. and Cantley L.C. 1997. Signalling through the lipid products of phosphoinositide-3-OH kinase. *Nature* **387:** 673–676.

Ueno M., Miyauchi T., Sakai S., Kobayashi T., Goto K., and Yamaguchi I. 1999. Effects of physiological or pathological pressure load in vivo on myocardial expression of ET-1 and receptors. *Am. J. Physiol.* **277:** R1321–R1330.

Uozumi H., Hiroi Y., Zou Y., Takimoto E., Toko H., Niu P., Shimoyama M., Yazaki Y., Nagai R., and Komuro I. 2001. gp130 plays a critical role in pressure overload-induced cardiac hypertrophy. *J. Biol. Chem.* **276:** 23115–23119.

Vanhaesebroeck B., Leevers S.J., Panayotou G., and Waterfield M.D. 1997. Phosphoinositide 3-kinases: A conserved family of signal transducers. *Trends Biochem. Sci.* **22:** 267–272.

Vecchione C., Fratta L., Rizzoni D., Notte A., Poulet R., Porteri E., Frati G., Guelfi D., Trimarco V., Mulvany M.J., Agabiti-Rosei E., Trimarco B., Cotecchia S., and Lembo G. 2002. Cardiovascular influences of alpha1b-adrenergic receptor defect in mice. *Circulation* **105:** 1700–1707.

Vigoreaux J.O. 1994. The muscle Z band: Lessons in stress management. *J. Muscle Res. Cell Motil.* **15:** 237–255.

Wada H., Ivester C.T., Carabello B.A., Cooper G.T., IV, and McDermott P.J. 1996. Translational initiation factor eIF-4E. A link between cardiac load and protein synthesis. *J. Biol. Chem.* **271:** 8359–8364.

Wakasaki H., Koya D., Schoen F.J., Jirousek M.R., Ways D.K., Hoit B.D., Walsh R.A., and King G.L. 1997. Targeted overexpression of protein kinase C beta2 isoform in myocardium causes cardiomyopathy. *Proc. Natl. Acad. Sci.* **94:** 9320–9325.

Walsh R.A. 2001. Molecular and cellular biology of the normal, hypertrophied, and failing heart. In *Hurst's the heart* (ed. V. Fuster et al.), pp. 115–125. McGraw-Hill, New York.

Wang L. and Proud C.G. 2002. Ras/Erk signaling is essential for activation of protein synthesis by Gq protein-coupled receptor agonists in adult cardiomyocytes. *Circ. Res.* **91:** 821–829.

Wang Y., Huang S., Sah V.P., Ross J., Jr., Brown J.H., Han J., and Chien K.R. 1998. Cardiac muscle cell hypertrophy and apoptosis induced by distinct members of the p38 mitogen-activated protein kinase family. *J. Biol. Chem.* **273:** 2161–2168.

Weber K.T. and Brilla C.G. 1991. Pathological hypertrophy and cardiac interstitium. Fibrosis and renin-angiotensin-aldosterone system. *Circulation* **83:** 1849–1865.

Weber K.T., Brilla C.G., and Janicki J.S. 1993. Myocardial fibrosis: Functional significance and regulatory factors. *Cardiovasc. Res.* **27:** 341–348.

Weiss R.E., Korcarz C., Chassande O., Cua K., Sadow P.M., Koo E., Samarut J., and Lang R. 2002. Thyroid hormone and cardiac function in mice deficient in thyroid hormone receptor-alpha or -beta: An echocardiograph study. *Am. J. Physiol. Endocrinol. Metab.* **283:** E428–E435.

Wettschureck N., Rutten H., Zywietz A., Gehring D., Wilkie T.M., Chen J., Chien K.R., and Offermanns S. 2001. Absence of pressure overload induced myocardial hypertrophy after conditional inactivation of Galphaq/Galpha11 in cardiomyocytes. *Nat. Med.* **7:** 1236–1240.

Widmann C., Gibson S., Jarpe M.B., and Johnson G.L. 1999. Mitogen-activated protein kinase: Conservation of a three-kinase module from yeast to human. *Physiol. Rev.* **79:** 143–180.

Wilkins B.J. and Molkentin J.D. 2002. Calcineurin and cardiac hypertrophy: Where have we been? Where are we going? *J. Physiol.* **541:** 1–8.

Wollert K.C., Taga T., Saito M., Narazaki M., Kishimoto T., Glembotski C.C., Vernallis A.B., Heath J.K., Pennica D., Wood W.I., and Chien K.R. 1996. Cardiotrophin-1 activates a distinct form of cardiac muscle cell hypertrophy. Assembly of sarcomeric units in series VIA gp130/leukemia inhibitory factor receptor-dependent pathways. *J. Biol. Chem.* **271:** 9535–9545.

Wu G., Yussman M.G., Barrett T.J., Hahn H.S., Osinska H., Hilliard G.M., Wang X., Toyokawa T., Yatani A., Lynch R.A., Robbins J., and Dorn G.W., II. 2001. Increased myocardial Rab GTPase expression: A consequence and cause of cardiomyopathy. *Circ. Res.* **89:** 1130–1137.

Xiang Y. and Kobilka B.K. 2003. Myocyte adrenoceptor signaling pathways. *Science* **300:** 1530–1532.

Yamamoto S., Yang G., Zablocki D., Liu J., Hong C., Kim S.J., Soler S., Odashima M., Thaisz J., Yehia G., Molina C.A., Yatani A., Vatner D.E., Vatner S.F., and Sadoshima J. 2003. Activation of Mst1 causes dilated cardiomyopathy by stimulating apoptosis without compensatory ventricular myocyte hypertrophy. *J. Clin. Invest.* **111:** 1463–1474.

Yamazaki T. and Yazaki Y. 1997. Is there major involvement of the renin-angiotensin system in cardiac hypertrophy? *Circ. Res.* **81:** 639–642.

———. 2000. Molecular basis of cardiac hypertrophy. *Z. Kardiol.* **89:** 1–6.

Yamazaki T., Komuro I., and Yazaki Y. 1999. Role of the renin-angiotensin system in cardiac hypertrophy. *Am. J. Cardiol.* **83:** 53H–57H.

Yamazaki T., Komuro I., Kudoh S., Zou Y., Nagai R., Aikawa R., Uozumi H., and Yazaki Y. 1998. Role of ion channels and exchangers in mechanical stretch-induced cardiomyocyte hypertrophy. *Circ. Res.* **82:** 430–437.

Yamazaki T., Komuro I., Kudoh S., Zou Y., Shiojima I., Mizuno T., Takano H., Hiroi Y., Ueki K., and Tobe K., et al. 1995. Angiotensin II partly mediates mechanical stress-induced cardiac hypertrophy. *Circ. Res.* **77:** 258–265.

Yanagisawa M., Kurihara H., Kimura S., Goto K., and Masaki T. 1988. A novel peptide vasoconstrictor, endothelin, is produced by vascular endothelium and modulates smooth muscle Ca2+ channels. *J. Hypertens.* (suppl.) **6:** S188–S191.

Yano M., Kim S., Izumi Y., Yamanaka S., and Iwao H. 1998. Differential activation of cardiac c-jun amino-terminal kinase and extracellular signal-regulated kinase in angiotensin II-mediated hypertension. *Circ. Res.* **83:** 752–760.

Yussman M.G., Toyokawa T., Odley A., Lynch R.A., Wu G., Colbert M.C., Aronow B.J., Lorenz J.N., and Dorn G.W., II. 2002. Mitochondrial death protein Nix is induced in cardiac hypertrophy and triggers apoptotic cardiomyopathy. *Nat. Med.* **8:** 725–730.

Zak R., Ed. 1984a. Overview of the growth process. In *Growth of the heart in health and disease*, pp. 1–24. Raven Press, New York.

———. 1984b. Factors controlling cardiac growth. In *Growth of the heart in health and disease*, pp. 165–185. Raven Press, New York.

Zechner D., Thuerauf D.J., Hanford D.S., McDonough P.M., and Glembotski C.C. 1997. A role for the p38 mitogen-activated protein kinase pathway in myocardial cell growth, sarcomeric organization, and cardiac-specific gene expression. *J. Cell Biol.* **139:** 115–127.

Zhang C.L., McKinsey T.A., Chang S., Antos C.L., Hill J.A., and Olson E.N. 2002. Class II histone deacetylases act as signal-responsive repressors of cardiac hypertrophy. *Cell* **110:** 479–488.

Zhang D., Gaussin V., Taffet G.E., Belaguli N.S., Yamada M., Schwartz R.J., Michael L.H., Overbeek P.A., and Schneider M.D. 2000. TAK1 is activated in the myocardium after pressure overload and is sufficient to provoke heart failure in transgenic mice. *Nat. Med.* **6:** 556–563.

Zhang S., Weinheimer C., Courtois M., Kovacs A., Zhang C.E., Cheng A.M., Wang Y., and Muslin A.J. 2003. The role of the Grb2-p38 MAPK signaling pathway in cardiac hypertrophy and fibrosis. *J. Clin. Invest.* **111:** 833–841.

Zhao Y.Y., Sawyer D.R., Baliga R.R., Opel D.J., Han X., Marchionni M.A., and Kelly R.A. 1998. Neuregulins promote survival and growth of cardiac myocytes. Persistence of ErbB2 and ErbB4 expression in neonatal and adult ventricular myocytes. *J. Biol. Chem.* **273:** 10261–10269.

Zhu Y.C., Zhu Y.Z., Gohlke P., Stauss H.M., and Unger T. 1997. Effects of angiotensin-converting enzyme inhibition and angiotensin II AT1 receptor antagonism on cardiac parameters in left ventricular hypertrophy. *Am. J. Cardiol.* **80:** 110A–117A.

Zuscik M.J., Chalothorn D., Hellard D., Deighan C., McGee A., Daly C.J., Waugh D.J., Ross S.A., Gaivin R.J., Morehead A.J., Thomas J.D., Plow E.F., McGrath J.C., Piascik M.T., and Perez D.M. 2001. Hypotension, autonomic failure, and cardiac hypertrophy in transgenic mice overexpressing the alpha 1B-adrenergic receptor. *J. Biol. Chem.* **276:** 13738–13743.

19

Regulation of Cell Growth in the Endocrine Pancreas

Kristin Roovers and Morris J. Birnbaum
Department of Medicine and Howard Hughes Medical Institute
University of Pennsylvania School of Medicine
Philadelphia, Pennsylvania 19104

ORGANS OF METAZOANS ARE SUBJECT to two distinct challenges in terms of growth. First, they have to maintain appropriate size during development, only achieving at maturity a final mass commensurate with the size and requirements of the organism. There is a second need for growth in both the adult and developing organism. To meet specific physiological responsibilities, many organs exhibit compensatory growth. For example, following partial hepatectomy, the liver has the ability to regenerate itself. On the other hand, the kidney, after unilateral nephrectomy, compensates by hypertrophy of the remaining organ. Several endocrine organs, including adrenal, parathyroid, and thyroid, show compensatory growth by both hyperplasia (an increase in cell number) and hypertrophy (an increase in cell volume). In this chapter, we discuss the contribution of islet/β-cell growth to the change in endocrine pancreatic mass that accompanies normal development or compensation for an increased requirement for insulin. In addition, we examine the role of the insulin/insulin-like growth factor (IGF) signaling pathway in the determination of β-cell growth and proliferation.

A few comments are required before we discuss β-cell growth in more detail. First, only recently have investigators become conscious of independent regulation of growth and proliferation, and therefore, studies in which these parameters are carefully examined are few. Second, considerable attention has been given to the concept of β-cell mass as an indication of fundamental functional capacity. As is true for most tissues, mass has generally been regarded as determined by the rate of proliferation of β-cells minus the rate of cell death, both by apoptosis and necrosis. Although readily performable assays exist for both these physiological events (e.g., BrdU incorporation and TUNEL staining for proliferation and apoptosis, respectively), in a relatively stable organ such as the endocrine pancreas, under typical, nonstressed conditions, these measures generally fall at or below the limits of detection. Thus, it has been difficult or impossible to obtain direct measures of β-cell turnover, except under manipulated or pathological conditions. In addition, neogenesis, the formation of new β-cells, appears to contribute to pancreatic endocrine mass throughout life, but it has not yet been feasible to measure this directly.

Another difficulty presents itself when approaching the issue of cell growth. In virtually all mammalian in vivo experiments, one measures cell size, which is, at best, a poor surrogate for growth. Perhaps this is best illustrated by experiments in which selective alteration of rates of *Drosophila* cell division produces inverse changes in cell size with no modification of growth rate (Neufeld et al. 1998). Similarly, enforcing arrest in G_1 in mammalian fibroblasts exposed to serum leads to a sustained increase in the size of the cell, whereas in the absence of a cell cycle inhibitor there is coordination of growth and proliferation and no change in cell size (Conlon et al. 2001; Fingar et al. 2002). Thus, one must exercise caution in interpreting increases in cell size as indicative of acceleration in growth, as it might instead reflect sustained or even decreased growth in the face of substantially attenuated proliferation.

THE ENDOCRINE PANCREAS AND β-CELLS

The pancreas is a complex endocrine and exocrine gland, essential for glucose homeostasis and digestion. The exocrine pancreas, which comprises 95–98% of the mature organ, is a highly branched structure composed of the acinar secretory cells producing the digestive enzymes, and the secreting ducts, continuous with the digestive tract. The endocrine pancreas, representing the remaining volume of the pancreas, consists of the islets of Langerhans, which are ovoid collections of cells embedded in the exocrine tissue. These islets, which in humans number 1 to 2 million, are highly vascularized and innervated. They are composed of four types of endocrine cells (α, β, δ, and PP cells), each secreting a specific hormone into the bloodstream. The α-cells produce glucagon, the β-cells insulin, the δ-cells somatostatin, and the PP cells secrete pancreatic polypeptide.

The β-cells, which are the most common and account for 60–75% of the cells in the islets, are generally located in the center of each islet. They tend to be surrounded by the α-cells, which make up 20% of the total, and the less abundant δ and PP cells.

Pancreatic β-cells have a remarkable ability to maintain blood glucose levels within a very narrow range for the lifetime of most individuals. Failure of this capacity is a fundamental part of the pathogenesis of all forms of diabetes mellitus. In type 1 diabetes (insulin-dependent diabetes mellitus, IDDM), the β-cell mass is reduced by autoimmune destruction of the endocrine pancreas. In type 2 diabetes (non-insulin-dependent diabetes mellitus, NIDDM), insulin resistance in peripheral tissues is a major contributor. However, diabetes only develops when there is an inadequate functional mass of β-cells, that is, when β-cells fail to compensate for the increased demand for insulin.

As discussed below, there is compelling evidence that β-cells are dynamic and can compensate for added metabolic demand in order to maintain euglycemia. Two types of compensation can occur: a functional one in which each β-cell secretes more insulin in response to a given challenge by glucose and a second one in which there is a change in β-cell mass. Functional adaptations, such as the changes in threshold and maximal capacity for glucose-induced insulin secretion, are also involved in the maintenance of glucose homeostasis. Nonetheless, the β-cell mass, as determined by both cell number and size, is the major factor in the amount of insulin that can be secreted. Although most recent attention has focused on the control of β-cell proliferation, neogenesis, and turnover, it is quite likely that the mass of the individual β-cell profoundly affects its secretory ability. β-cell size correlates in a strong positive manner with the cell's capability to synthesize and secrete insulin (Giordano et al. 1993).

PANCREATIC ISLET/β-CELL MASS CHANGES IN THE NORMAL INDIVIDUAL

The idea most generally accepted by β-cell researchers is that during the fetal period, endocrine cells arise from undifferentiated pancreatic ductal stem cells, which migrate into the exocrine pancreas to form the islets of Langerhans, whereas during adulthood, proliferation of mature β-cells predominates (for review, see Bonner-Weir 2000; Endlund 2002). However, since it is not possible to measure neogenesis directly, this model still has to be proven conclusively. For example, it is quite possible that cells capable of differentiating into β-cells reside in compartments other than the duct, and that neogenesis may persist past the neonatal period (Bernard-Kargar and Ktorza 2001). In addition, it is likely that the proliferative capacity of mature β-cells differs among species, with adult humans possessing a rather limited potential. In rodents, β-cell replication is high during late gestation and in the neonatal period, is reduced after weaning and in the initial months of life, but remains relatively stable late in life (Swenne 1983; Montanya et al. 2000). However, the data for older animals are difficult to rely on due to the scarcity of studies and the low absolute rates. There is a much better consensus for the absolute increase in β-cell mass throughout life, except for a brief period occurring at several weeks of age which is thought to be due to a burst in the rate of apoptosis (Finegood et al. 1995). Perhaps the most striking feature of β-cell mass is its near-constant ratio to the weight of the animal, at least in rodents. This is specific to β-cells, as both acinar tissue and non-insulin-producing β-cells remain stable in adulthood. There is a real paucity of studies on the size of individual β-cells, although the available data indicate that there is a doubling early in life (Montanya et al. 2000). In a detailed longitudinal study, β-cell area and pancreatic β-cell mass were determined throughout the entire postweaning life span of Lewis rats. Body weight increased progressively with a maximum value at 20 months, whereas cross-sectional area of individual β-cells increased progressively in the initial months, remained stable from months 7 to 10, and then increased again at month 20. The interesting conclusion from these studies is that expansion of β-cell mass early in life is due primarily to an increased number of cells, whereas in late adulthood hypertrophy accounts for the change (Montanya et al. 2000). The reason for the predominant role for hypertrophy in the older rat is not clear, although a limited proliferative potential for the β-cell was suggested (Remacle et al. 1980).

CHANGES IN RESPONSE TO PHYSIOLOGICAL CONDITIONS

In addition to the gradual increase in β-cell mass throughout the lifetime of the organism, the endocrine pancreas maintains euglycemia via compensatory changes that meet acute or more sustained increases in demand for insulin. These compensatory changes emphasize the dynamic nature of the β-cell and provide further evidence for their ability to proliferate and to grow (Bernard-Kargar and Ktorza 2001). Whereas expansion of β-cell mass in response to the insulin resistance of type 2 diabetes mellitus has received the most attention, alterations in growth also occur in more physiological settings.

The changes in β-cell mass during and after pregnancy and the mechanisms involved are a good illustration of the plasticity of the endocrine pancreas. In mammals, pregnancy results in profound changes in maternal metabolism and insulin secretion that allow optimal nutrient supply to the fetus. During the last third of pregnancy, marked insulin resistance develops. To accommodate this increased demand for insulin, it is essential that the islets undergo major structural and functional changes (Sorenson and Brelje 1997). The inability of the maternal islets to respond to the increased demand for insulin can lead to the development of gestational diabetes, a condition that, if left untreated, threatens the well-being of both mother and fetus. Adaptive changes that occur during normal pregnancy include augmented glucose-stimulated insulin secretion, decreased glucose stimulation threshold, β-cell hyperplasia and hypertrophy, and enhanced gap-junctional coupling among β-cells.

The adaptive changes of the islets to the increased demands of pregnancy have been described in rats and mice. In rats, islet cell proliferation was increased over age-matched controls by day 10, rose continuously until day 14, and then returned to control levels by day 18 (Parsons et al. 1992). Although islet numbers in pancreata of 16-day pregnant mice were unchanged, there was a 2-fold increase in islet area and a 2.4-fold increase in islet volume (Parsons et al. 1995). The expansion of the islet organ during pregnancy in the mouse occurs by enlargement of preexisting islets rather than by the development of new islets. However, much more striking is the change in cell volume that takes place in the maternal endocrine pancreas following birth. At the end of pregnancy and after delivery, there is an involution of β-cell mass (Scaglia et al. 1995). β-cell replication decreases significantly postpartum but returns to nonpregnant levels by 10 days postpartum. At day-4 postpartum, there was evidence for an increased number of apoptotic cells in the pancreas. Although in this study cell size was not different at the end of pregnancy as compared to nonpregnant controls, it was significantly decreased at 4 and 10 days postpartum as compared to the end of pregnancy, and at 10 days postpartum, cell size was significantly decreased as compared to control. Thus, the endocrine pancreas is capable of increasing and decreasing its mass by using the mechanisms of changes in rates of β-cell replication and of β-cell death, as well changes in β-cell size to achieve homeostasis.

There is a substantial amount of evidence that the hormonal signal driving the increase in β-cell mass during pregnancy in the rodent is provided by placental lactogen and possibly prolactin, in both cases via binding to the β-cell prolactin receptor. Changes in pancreatic endocrine capacity correlate well with levels of placental lactogen during pregnancy, and activation of the prolactin receptor on insulinoma cells provides a potent signal for improved secretory function and mitogenesis (Brelje et al. 1993; Sorenson et al. 1993; Parsons et al. 1995). The latter is due, at least in part, to an increase in cyclin D2 gene expression (Friedrichsen et al. 2003). This process was studied in transgenic mice by directing placental lactogen expression to β-cells with a rat insulin promoter (Vasavada et al. 2000). These mice displayed a marked increase in islet mass, which resulted from both hyperplasia and hypertrophy. Plasma insulin concentrations were inappropriately elevated, resulting in a 2-fold increase in pancreatic insulin content, and β-cell proliferation rates increased. This hyperplasia, together with a 20% increase in β-cell size, resulted in a 2-fold increase in islet mass and a 1.45-fold increase in islet number.

Clearly, placental lactogen is capable of providing both a mitogenic and a growth stimulus to β-cells, apparently with the latter moderately predominating. The signaling pathway that transmits the intracellular signal is a canonical JAK/STAT pathway. Binding of placental lactogen or prolactin to the prolactin receptor, which lacks intrinsic catalytic activity, leads to a conformational change in the receptor and an activation of the associated Janus kinase 2 (JAK2). This tyrosine kinase phosphorylates itself as well as the prolactin receptor, creating binding sites for the src homology 2 (SH2) domains of the signal transducer and activator of transcription 5 (STAT5). This protein is then itself phosphorylated by JAK2 on appropriate tyrosine residues, which instructs STAT5 to translocate to the nucleus, bind to DNA in a sequence-specific manner, and activate target genes. It is quite clear that STAT5 is required for the mitogenic actions of prolactin in β-cells, and that active STAT5 can mimic the effects of hormone with regard to proliferation (Friedrichsen et al. 2001, 2003). However, there are no data available regarding the role of this pathway in the control of β-cell size. Nonetheless, it is quite intriguing that the JAK/STAT pathway has been implicated in hypertrophy of other cell types, in particular, cardiac and smooth muscle (Kunisada et al. 1998; Kodama et al. 2000). In contrast, overexpression of JAK/STAT signaling in the *Drosophila* eye leads to a large eye, due only to an increase in ommatidia and cell number (Chen et al. 2002). This is in marked contrast to members of the insulin signaling pathway such as Akt/PKB, which augment organ size in fruit flies primarily via effects on cell growth (Verdu et al. 1999; Stocker and Hafen 2000).

An important, unresolved question regarding signaling by the prolactin receptor in β-cells is whether physiological responses are mediated exclusively by the nuclear targets of STAT5, or whether additional pathways are involved. In particular, significant controversy clouds the question of whether Janus kinases are capable of phosphorylating insulin receptor substrate (IRS) proteins, and thus mimicking insulin or insulin-like growth factor (IGF) signaling. A substantial number of studies have found phosphorylation of IRS-1 or IRS-2 upon cytokine-dependent engagement of JAK/STAT signaling in non-β cells (Yin et al. 1994; Platanias et al. 1996; Burfoot et al. 1997). However, Cousin et al. (1999) reported that JAK2/STAT5-induced β-cell proliferation is completely independent of IRS phosphorylation, although it still requires the permissive activity of PI 3′ kinase. Perhaps more important are the data indicating that JAK/STAT signaling is capable of activating PI-3′-kinase-dependent pathways, whether this occurs via phosphorylation of IRS or by some unexplained route. JAK/STAT stimulates not only PI 3′ kinase, but also the serine/threonine kinase Akt/PKB and phosphorylation of the mTOR-dependent targets p70 S6 kinase and 4E-BP1/PHAS (Cousin et al. 1999; Carvalho et al. 2003; Lekmine et al. 2003). As discussed below and in more detail in other chapters, these signaling molecules are strongly implicated as directly regulating cell growth.

CHANGES IN RESPONSE TO PATHOLOGICAL CONDITIONS AND EXPERIMENTAL MANIPULATIONS

Insulin Resistance

The most common pathological state that increases the demand for insulin, thus inciting a compensatory growth response in the endocrine pancreas, is insulin resistance; i.e., an inability of maximal concentrations of the hormone to elicit an appropriate metabolic response. Obesity is generally associated with insulin resistance, but this does not progress to full diabetes mellitus unless there is a relative or absolute failure of the insulin-secreting β-cells. In fact, glucose homeostasis remains close to normal in 75–80% of all obese subjects (Lingohr et al. 2002). Thus, the most common hypothesis for the development of diabetes mellitus is that early in the course of insulin resistance, there is a secondary increase in β-cell mass that allows adequate compensation, but in the genetically predisposed individual the β-cells ultimately fail. Although the mechanism underlying β-cell failure is not well understood, the deterioration of β-cells is believed to involve defects in insulin secretion as a result of protein and lipid toxicity and decreases in β-cell mass due to changes in rates of apoptosis and proliferation.

The contribution of cellular hypertrophy, as opposed to neogenesis, proliferation, and suppression of apoptosis, to the compensatory expansion of β-cell mass in insulin resistance has not been precisely defined. Interestingly, even the exact signal that informs the β-cell of insulin resistance and the need for increased secretion remains controversial. Although glucose itself is the most likely and appealing circulating factor (see below), examples of experimental situations in which compensatory β-cell changes can be ascertained in the absence of measurable increases in blood glucose have led to the suggestion that other hormones or neuronal signals may be crucial (Mittelman et al. 2000). The hormones proposed as stimulators of pancreatic endocrine growth under these conditions include IGF-1, the incretin hormone glucagon-like peptide 1 (GLP-1), and, most prominently, autocrine or paracrine signaling by insulin itself (Lingohr et al. 2002). In any case, although the exact mechanism and contribution of these hormones to alterations in β-cell mass are as yet unclear, there is little doubt that β-cell hypertrophy is a conspicuous feature of the response to the increased needs for insulin. For example, when rats are placed for 6 weeks on a high-fat diet, a well-established strategy for generating obese, insulin-resistant rodents, there is an ~50% increase in the total islet area expressed as a percentage of pancreatic area (Lingohr et al. 2002). This is accounted for in large part by a 35% increase in the mean β-cell diameter, and this change in cell size is limited to the β-cell, as α- and δ-cells remain unaltered. A similar phenomenon has been reported in other rodent models for insulin resistance. The nondiabetic Zucker fa/fa rats develop obesity due to a recessive mutation in the leptin receptor, and this leads to a compensatory 4-fold increase in β-cell mass (Milburn et al. 1995). Again, β-cells from these animals demonstrate a significant increase in cell size (Chan et al. 1999).

Glucose Infusion

Glucose represents a potent stimulus of pancreatic β-cell growth both in vitro and in vivo. In vitro, the proliferation rate of rodent β-cells goes up with increasing glucose concentrations (Schuppin et al. 1993). The potent effect of glucose on β-cell mass was clearly demonstrated in nondiabetic rats infused with glucose for 96 hours (Bonner-Weir et al. 1989). β-Cell mass was enhanced 50% in glucose-infused rats compared to saline-infused control rats. The mitotic index was increased 5-fold, providing evidence of augmented replication. The mean cross-sectional area of β-cells was increased 150%, as compared to control values. The effect of glucose on β-cell size was also obvious in rats made mildly diabetic from injections of the β-cell toxin streptozotocin (Bernard et al. 1999). Thus, β-cells have a large capacity to respond to glucose levels through inducing both hyperplasia and hypertrophy.

Several studies have attempted to distinguish between the direct effects of glucose and the actions of a secondary increase in insulin levels on β-cell growth. Hyperinsulinemia–euglycemia clamp experiments, in which insulin is increased but glucose is maintained at physiological levels by careful infusion of hexose, were performed in unrestrained rats over 48 hours (Bernard et al. 1998). Under these conditions, insulin infusion per se did not promote β-cell growth, arguing against a role for circulating insulin, at least for relatively short periods of time. In support of this conclusion, pancreatic islets transplanted from lean mice into obese hyperglycemic mice grew rapidly and intensively, whereas when they were implanted into hyperinsulinemic and euglycemic mice, islet growth was not observed (Montana et al. 1994b). Although short-term hyperinsulinemia per se is probably not a determinant of β-cell mass increase in adult rats, this does not exclude long-term effects of insulin on β-cell mass.

Pancreatectomy

The 85–95% partial pancreatectomy model has been used to study β-cell adaptation and changes in gene expression (Bonner-Weir et al. 1983; Jonas et al. 1999). An important advantage of this model is that alterations in β-cell function reflect exclusively reduced β-cell mass, without complicating genetic variables found in the inherited rodent obesity syndromes or the potentially confounding effects of streptozotocin. Although there is active regeneration in the first 10 days after surgery, by 4 weeks, the well-formed islets are exposed to hyperglycemia for a sufficient time to produce a stable reproducible model of diabetes. During the days after partial pancreatectomy, there is also an increase in β-cell mass caused by neogenesis and β-cell replication (Bonner-Weir et al. 1983). More recently, studies have revealed that an important component of the increase in β-cell mass is cell hypertrophy, with an 85% increase in β-cell size 4 weeks after partial pancreatectomy (Jonas et al. 1999). Although at this time β-cell hypertrophy is striking, turnover is surprisingly normal, with an unchanged replication rate and no obvious increase in the frequency of apoptotic bodies. Although β-cell mass cannot be increased sufficiently to prevent hyperglycemia, hypertrophy contributes to insulin production and helps prevent severe diabetic decompensation.

To examine the capacity of transplanted β-cells to modify their mass, partial pancreatectomy was performed either 14 days before or following transplantation of islets into otherwise normal rats (Montana et al. 1994a). β-Cells transplanted into normal rats increased their replication and size upon 95% pancreatectomy, accounting for a nearly 3-fold increase in total β-cell mass compared to normal control rats. Similar results were obtained when islets were transplanted into already pancreatecomized rats, although it took several days to achieve euglycemia. Thus, like endogenous β-cells, transplanted ones are able to respond to increased metabolic demand and to change their size and rates of replication.

SIGNALING PATHWAYS INVOLVED IN THE REGULATION OF β-CELL SIZE

As described above, there is considerable evidence that β-cell mass is dynamic, making use of changing rates of replication, neogenesis, and apoptosis, as well as β-cell size. Efforts are now under way to discern the role of specific intracellular cell signaling pathways in the regulation of these processes. The pathway most likely to be important to β-cell growth is insulin- or IGF-1-dependent alteration in signaling through PI-3′ kinase.

The insulin/insulin-like growth factor family of ligands and receptors controls growth, metabolism, and reproduction. Mice lacking the insulin receptor are born with about 10% growth retardation and with no apparent metabolic abnormalities, but soon after birth, metabolic control deteriorates, and death results as a result of severe diabetes (Kitamura et al. 2003). Similar phenotypes are observed for mice lacking expression of insulin-1 and insulin-2. The evidence for a growth-promoting role derives more from studies in mice with combined mutations of various elements of the IGF system. Ablation of the IGF-1 receptor results in embryos that are 45% of normal size, whereas combined deletion of the insulin and the IGF-1 receptors generates even smaller mice, about 30% of normal size. Because of lethality, tissue-specific deletion of these genes has proved more informative. Inactivation of the insulin receptor in β-cells results in impaired glucose-stimulated insulin secretion, leading to age-dependent glucose intolerance and occasional overt diabetes (Kulkarni et al. 1999). However, there is no obvious indication of impaired growth, at least early in life. β-Cell-specific deletion of the IGF-1 receptor in mice results in defective glucose-stimulated insulin secretion and impaired glucose tolerance. These mice display normal β-cell growth and development, with no alteration in β-cell mass (Kulkarni et al. 2002; Xuan et al. 2002). On the other hand, following 90% pancreatectomy, IGF-1 mRNA levels increased rapidly in the area of regenerating pancreas (Smith et al. 1991). Thus, IGF-1 still may play a role in the growth and differentiation of the pancreas tissue under some conditions.

At least one route of signaling by the insulin and IGF-1 receptors is via phosphorylation of a class of scaffolds typified by the insulin receptor substrate (IRS) proteins. The phenotypes of *IRS1* or *IRS2* null mice present a striking contrast (Tamemoto et al. 1994; Withers et al. 1998; for review, see Burks and White 2001; White 2002) between the functions of these molecules. Mice lacking IRS1 are 50% the size of control animals and display only mild insulin resistance. IRS2-deficient mice are normal in size but develop severe diabetes by age 10 weeks. Although there is a compensatory 2-fold expansion in β-cell mass in the IRS1 null mice, IRS2-deficient animals are born with a 50% reduction in β-cell mass and show no evidence of compensatory expansion even in the presence of insulin resistance. IRS1$^{-/-}$ IRS2$^{+/-}$ mice are even smaller in size than the IRS1$^{-/-}$ mice but remain glucose-tolerant because they maintain an adequate mass of functional β-cells. Also, IRS1$^{+/-}$ IRS2$^{-/-}$ mice are yet smaller than the IRS2$^{-/-}$ mice and have β-cell mass that is 10% of control mice. The targeted disruption of IRS1 and IRS2 genes in mice demonstrates that these insulin signaling molecules have distinct roles in the maintenance of glucose homeostasis, particularly at the level of the β-cell. Whereas IRS1 may be important for insulin secretion, IRS2 plays a key role in the control of β-cell mass. Consistent with this model, IRS-2 expression was enhanced in the regenerating pancreatic duct epithelium and in clusters of insulin-positive cells of 60% partial pancreatectomized rats (Jetton et al. 2001).

It is clear that mutations of insulin, IGF-1, IGF-2 or IGF receptor tyrosine kinases, or insulin receptor substrates (IRS-1 and IRS-2) lead to smaller organs in the mouse, but it is unclear whether this is due to a reduction in cell size, cell number, or both (Baserga 2000). Efstratiadis and colleagues proposed that the dwarfism observed in mice with null mutations in the IGF-1 and IGF-1 receptors was due to their indispensable participation in a proliferative pathway (Liu et al. 1993). In contrast, Bondy and associates demonstrated that IGF-1 promotes longitudinal bone size by increasing anabolic pathways in chondrocytes, resulting in larger cells without affecting cell number or proliferative rates (Wang et al. 1999). With regard to β-cells, no reports have appeared demonstrating reduced β-cell size after experimental manipulation of insulin or IGF-1 receptor function; it is unclear whether this is due to the lack of a role for these factors in β-cell growth, to the involvement of other compensating hormones, or to the absence of a careful assessment of cell size by the investigators.

Phosphorylation of IRS proteins on specific tyrosine residues leads to the recruitment of signaling molecules via their SH2 domains. The most important intermediate with regard to regulation of growth and metabolism appears to be the class 1A PI-3′ kinase. The role of PI-3′ kinase in β-cell growth has yet to be studied. In the heart, there is good evidence that PI-3′ kinase is involved in organ and cell size. Expression of an activated PI-3′ kinase in the mouse heart increases in heart size, whereas expression of a dominant inhibitory PI-3′ kinase in the heart leads to a decrease in cell and organ size (Shioi et al. 2000).

The serine/threonine protein kinase Akt/PKB is a major downstream target of PI-3′ kinase and is implicated in a plethora of cellular functions. The mammalian genome contains three closely related genes that encode the isoforms Akt1, Akt2, and Akt3, also known as PKBα, PKBβ, and PKBγ, all with high amino acid identity and broad tissue distribution. Overexpression of constitutively active Akt1 specifically in pancreatic β-cells of mice produces a significant increase in both β-cell size and total islet mass, as well as improved glucose tolerance and resistance to streptozotocin-induced diabetes mellitus (Bernal-Mizrachi et al. 2001; Tuttle et al. 2001). The 3- to 4-fold increase in β-cell size was not sufficient to account for the increase in β-cell mass, indicating an independent effect on cell number. Tuttle et al. (2001) found a paradoxical increase in apoptosis, indicating that the increased number of cells had to be due to either neogenesis or proliferation. One of the most striking features of the phenotype is that, despite the large increase in cell size, the β-cells retained a fair degree of normal function. Thus, they were capable of responding to an in vivo infusion of glucose with an appropriate secretory increase in insulin, although the dose-response curve appeared left-shifted. Electron microscopy of islets from the mice expressing active Akt showed remarkably normal ultrastructure, except for a modest decrease in the density of secretory granules (R.L. Tuttle and M.J. Birnbaum, unpubl.). It is unclear whether this is a direct result of increased Akt activity, or secondary to an elevation in basal insulin secretion.

Akt1-deficient mice display growth retardation during fetal and postnatal growth that persisted into adulthood (Chen et al. 2001; Cho et al. 2001a). There is no report on β-cell mass or cell size in the Akt1 null mouse, although these animals demonstrate normal glucose tolerance and β-cell mass. *Akt2* null mice do not have severe growth defects, but they do manifest a diabetes-like phenotype marked by insulin resistance (Cho et al. 2001b). Hyperglycemia in Akt2-deficient mice is accompanied by increased levels of serum insulin, suggesting that decreased responsiveness to the hormone in peripheral tissues may have resulted as a compensatory adjustment to maintain glucose homeostasis. Consistent with this idea, Akt2-deficient mice have a dramatic increase in the size and number of pancreatic islets, which would compensate for hyperglycemia by producing more insulin.

A molecule that integrates signals from growth factors and nutrients is the large, rapamycin-inhibitable protein, mammalian target of rapamycin (mTOR). The hormone-response regulation of mTOR targets involves activation of PI-3′ kinase and appears to go through both Akt-dependent and independent pathways. Amino acids, in particular leucine, exert a profound positive effect on mTOR activity. Unlike growth factors, amino acids are necessary and sufficient to activate the mTOR pathway. Although mTOR is involved in a number of biological responses including transcription and cell cycle regulation, the best-characterized is protein synthesis, and this is thought to be responsible for mTOR's strong influence on cell growth. mTOR has not been studied in vivo with regard to β-cell growth, but mice deficient in the ribosomal S6 kinase 1 (S6K1) demonstrate profound abnormalities in pancreatic endocrine function. S6K1-deficient mice are viable and fertile but display a reduction in body size during embryogenesis that is almost completely overcome by 12 weeks of age (Shima et al. 1998; Pende et al. 2000). The weak penetrance of the phenotype may arise from increased expression of S6K2 in these S6K1-deficient mice (Shima et al. 1998). *S6K1* null mice are hypoinsulinemic and glucose-intolerant (Pende et al. 2000). Insulin resistance was not observed in isolated muscle, but there was a sharp reduction in glucose-induced insulin secretion and in pancreatic insulin content. Analysis of transcription, synthesis, and degradation of insulin suggested no major lesions in insulin metabolism, raising the possibility that the effect was due to a reduction in β-cell mass. Morphometric analysis showed that the effect was due to a specific reduction in pancreatic endocrine mass accounted for by a selective decrease in β-cell size and not a decrease in β-cell number (Pende et al. 2000). One complication in the interpretation of these data is that, since the deficiency of S6K1 was not limited to β-cells, it is not clear that the defect is cell-autonomous. Since animals that are nutritionally deprived during fetal development also display abnormalities in β-cell growth, it is possible that a deficiency of S6K1 is mimicking early organismal starvation. Resolution of this question will require the generation of β-cell-specific *S6K1* knockout mice (see Chapter 9).

The activation of mTOR by glucose, amino acids, and insulin has been studied in some detail in both primary β-cells and β-cell lines

(McDaniel et al. 2002). As in most cell types, insulin stimulates mTOR-dependent phosphorylation of S6K and the inhibitor of translational initiation, PHAS-1/4E-BP1, in a leucine-dependent manner (see Chapters 7 and 10). Interestingly, glucose also stimulates phosphorylation of mTOR targets in β-cells, suggesting an intriguing potential link between hyperglycemia and growth. This possibility has not yet been studied. However, McDaniel and colleagues have suggested that the metabolic pathways by which leucine controls mTOR activity and insulin secretion are fundamentally the same. Again, this hypothesis needs further testing, particularly in the context of new mTOR-associated proteins suggested to be amino acid-sensing units.

A substrate of Akt originally identified through genetic studies in *C. elegans* is the Foxo subfamily of forkhead transcription factors. Phosphorylation of Foxo1 reduces its activity, at least in part, by excluding the factor from the nucleus (Biggs et al. 1999; Brunet et al. 1999). Foxo1 has been clearly implicated in β-cell growth and function, although, once again, how much control of cell growth contributes to its overall actions remains unknown. Targeting of a gain-of-function Foxo1 mutation to the liver and pancreatic β-cells results in diabetes arising from increased hepatic glucose production and impaired β-cell compensation (Nakae et al. 2002). One possible target of Foxo1 is the homeodomain transcription factor Pdx1, which plays an important role in islet development and β-cell function. In adult mice, Pdx1 promotes the expression of proinsulin, the glucose transporter, Glut2, and glucokinase, all of which are important for optimal glucose-sensitive insulin secretion. Pdx1 also positively regulates FGF and their receptors, which promote β-cell growth. Transgenic overexpression of Pdx1 restored β-cell mass and function in IRS2 null mice (Kushner et al. 2002). Similarly, haploinsufficiency for Foxo1 reverses β-cell failure in IRS2 null mice, presumably through increased expression of Pdx1 (Kitamura et al. 2002). Thus, insulin/IGF can potentially regulate β-cell mass by relieving Foxo1 inhibition of Pdx1 expression.

SUMMARY AND CONCLUDING THOUGHTS

The endocrine pancreas is undeniably a dynamic organ that can alter its mass depending on the physiological demands. Although poorly studied, β-cell hypertrophy has nonetheless established itself as an important mechanism by which mass and, quite probably, function are controlled. What remains unclear is the relative contribution of growth to the compensatory changes also involving neogenesis, proliferation, and increased survival.

Almost nothing is known concerning the cell-intrinsic mechanisms by which β-cells augment their own growth. The study of β-cell biology is critical because of its relevance to an epidemic human disease, diabetes mellitus. However, the highly differentiated state of β-cells makes them problematic to study. Insulinoma cell lines recapitulate only a small part of normal β-cell physiology, and, when islets are manipulated in vivo, it is challenging at best to distinguish the effects of circulating factors from the cell-autonomous control of growth. Nonetheless, a number of PI-3′ kinase and Akt effectors are implicated in the regulation of organ size, and make good candidates for regulators of β-cell growth. These include glycogen synthase kinase-3 (GSK-3), the tuberous sclerosis complex (TSC), and, as noted above, mTOR, S6K, and initiation factor 4E-binding protein (4E-BP) (Shima et al. 1998; Haq et al. 2000; Pende et al. 2000; Rommel et al. 2001; Antos et al. 2002; Vyas et al. 2002).

REFERENCES

Antos C.L., McKinsey T.A., Frey N., Kutschke W., McAnally J., Shelton J.M., Richardson J.A., Hill J.A., and Olson E.N. 2002. Activated glycogen synthase-3 beta suppresses cardiac hypertrophy in vivo. *Proc. Natl. Acad. Sci.* **99:** 907–912.

Baserga R. 2000. The contradictions of the insulin-like growth factor 1 receptor. *Oncogene* **19:** 5574–5581.

Bernal-Mizrachi E., Wen W., Stahlhut S., Welling C.M., and Permutt M.A. 2001. Islet β-cell expression of constitutively active Akt1/PKB alpha induces striking hypertrophy, hyperplasia, and hyperinsulinemia. *J. Clin. Invest.* **108:** 1631–1638.

Bernard C., Berthault M.F., Saulnier C., and Ktorza A. 1999. Neogenesis vs. apoptosis as main components of pancreatic β-cell mass changes in glucose-infused normal and mildly diabetic adult rats. *FASEB J.* **13:** 1195–1205.

Bernard C., Thibault C., Berthault M.F., Magnan C., Saulnier C., Portha B., Pralong W.F., Penicaud L., and Ktorza A. 1998. Pancreatic beta-cell regeneration after 48-h glucose infusion in mildly diabetic rats is not correlated with functional improvement. *Diabetes* **47:** 1058–1065.

Bernard-Kargar C. and Ktorza A. 2001. Endocrine pancreas plasticity under physiological and pathological conditions. *Diabetes* (suppl. 1) **50:** S30–S35.

Biggs W.H., Meisenhelder J., Hunter T., Cavenee W.K., and Arden K.C. 1999. Protein kinase B/Akt-mediated phosphorylation promotes nuclear exclusion of the winged helix transcription factor FKHR1. *Proc. Natl. Acad. Sci.* **96:** 7421–7426.

Bonner-Weir S. 2000. Islet growth and development in the adult. *J. Mol. Endocrinol.* **24:** 297–302.

Bonner-Weir S., Trent D.F., and Weir G.C. 1983. Partial pancreatectomy in the rat and subsequent defect in glucose-induced insulin release. *J. Clin. Invest.* **71:** 1544–1553.

Bonner-Weir S., Deery D., Leahy J.L., and Weir G.C. 1989. Compensatory growth of pancreatic beta-cells in adult rats after short-term glucose infusion. *Diabetes* **38:** 49–53.

Brelje T.C., Scharp D.W., Lacy P.E., Ogren L., Talamantes F., Robertson M., Friesen H.G., and Sorenson R.L. 1993. Effect of homologous placental lactogens, prolactins, and growth hormones on islet β-cell division and insulin secretion in rat, mouse, and human islets: Implication for placental lactogen regulation of islet function during pregnancy. *Endocrinology* **132:** 879–887.

Brunet A., Bonni A., Zigmond M.J., Lin M.Z., Juo P., Hu L.S., Anderson M.J., Arden K.C., Blenis J., and Greenberg M.E. 1999. Akt promotes cell survival by phosphorylating and inhibiting a forkhead transcription factor. *Cell* **96:** 857–868.

Burfoot M.S., Rogers N.C., Watling D., Smith J.M., Pons S., Paonessaw G., Pellegrini S., White M.F., and Kerr I.M. 1997. Janus kinase-dependent activation of insulin receptor substrate 1 in response to interleukin-4, oncostatin M, and the interferons. *J. Biol. Chem.* **272:** 24183–24190.

Burks D.J. and White M.F. 2001. IRS proteins and beta-cell function. *Diabetes* (suppl. 1) **50:** S140–S145.

Carvalho C.R., Carvalheira J.B., Lima M.H., Zimmerman S.F., Caperuto L.C., Amanso A., Gasparetti A.L., Meneghetti V., Zimmerman L.F., Velloso L.A., and Saad M.J. 2003. Novel signal transduction pathway for luteinizing hormone and its interaction with insulin: Activation of Janus kinase/signal transducer and activator of transcription and phosphoinositol 3-kinase/Akt pathways. *Endocrinology* **144:** 638–647.

Chan C.B., MacPhail R.M., Sheu L., Wheeler M.B., and Gaisano H.Y. 1999. Beta-cell hypertrophy in fa/fa rats is associated with basal glucose hypersensitivity and reduced SNARE protein expression. *Diabetes* **48:** 997–1005.

Chen H.W., Chen X., Oh S.W., Marinissen M.J., Gutkind J.S., and Hou S.X. 2002. mom identifies a receptor for the *Drosophila* JAK/STAT signal transduction pathway and encodes a protein distantly related to the mammalian cytokine receptor family. *Genes Dev.* **16:** 388–398.

Chen W.S., Xu P.Z., Gottlob K., Chen M.L., Sokol K., Shiyanova T., Roninson I., Weng W., Suzuki R., Tobe K., Kadowaki T., and Hay N. 2001. Growth retardation and increased apoptosis in mice with homozygous disruption of the Akt1 gene. *Genes Dev.* **15:** 2203–2208.

Cho H., Thorvaldsen J.L., Chu Q., Feng F., and Birnbaum M.J. 2001a. Akt1/PKBalpha is required for normal growth but dispensable for maintenance of glucose homeostasis in mice. *J. Biol. Chem.* **276:** 38349–38352.

Cho H., Mu J., Kim J.K., Thorvaldsen J.L., Chu Q., Crenshaw E.B., III, Kaestner K.H., Bartolomei M.S., Shulman G.I., and Birnbaum M.J. 2001b. Insulin resistance and a diabetes mellitus-like syndrome in mice lacking the protein kinase Akt2 (PKB beta). *Science* **292:** 1728–1731.

Conlon I.J., Dunn G.A., Mudge A.W., and Raff M.C. 2001. Extracellular control of cell size. *Nat. Cell Biol.* **3:** 918–921.

Cousin S.P., Hugl S.R., Myers M.G., Jr., White M.F., Reifel-Miller A., and Rhodes C.J. 1999. Stimulation of pancreatic beta-cell proliferation by growth hormone is glucose-dependent: Signal transduction via Janus kinase 2 (JAK2)/signal transducer and activator of transcription 5 (STAT5) with no crosstalk to insulin receptor substrate-mediated mitogenic signalling. *Biochem J.* **344:** 649–658.

Edlund H. 2002. Pancreatic organogenesis—Developmental mechanisms and implications for therapy. *Nat. Rev. Genet.* **3:** 524–532.

Finegood D.T., Scaglia L., and Bonner-Weir S. 1995. Dynamics of beta-cell mass in the growing rat pancreas. Estimation with a simple mathematical model. *Diabetes* **44:** 249–256.

Fingar D.C., Salama S., Tsou C., Harlow E., and Blenis J. 2002. Mammalian cell size is controlled by mTOR and its downstream targets S6K1 and 4EBP1/eIF4E. *Genes Dev.* **16:** 1472–1487.

Friedrichsen B.N., Galsgaard E.D., Nielsen J.H., and Moldrup A. 2001. Growth hormone- and prolactin-induced proliferation of insulinoma cells, INS-1, depends on activation of STAT5 (signal transducer and activator of transcription 5). *Mol. Endocrinol.* **15:** 136–148.

Friedrichsen B.N., Richter H.E., Hansen J.A., Rhodes C.J., Nielsen J.H., Billestrup N., and Moldrup A. 2003. Signal transducer and activator of transcription 5 activation is suffi- cient to drive transcriptional induction of cyclin D2 gene and proliferation of rat pan- creatic beta-cells. *Mol. Endocrinol.* **17:** 945–958.

Giordano E., Cirulli V., Bosco D., Rouiller D., Halban P., and Meda P. 1993. B-cell size influ- ences glucose-stimulated insulin secretion. *Am. J. Physiol.* **265:** C358–C364.

Haq S., Choukroun G., Kang Z.B., Ranu H., Matsui T., Rosenzweig A., Molkentin J.D., Alessandrini A., Woodgett J., Hajjar R., Michael A., and Force T. 2000. Glycogen syn- thase kinase-3beta is a negative regulator of cardiomyocyte hypertrophy. *J. Cell Biol.* **151:** 117–130.

Jetton T.L., Liu Y.Q., Trotman W.E., Nevin P.W., Sun X.J., and Leahy J.L. 2001. Enhanced expression of insulin receptor substrate-2 and activation of protein kinase B/Akt in regenerating pancreatic duct epithelium of 60%-partial pancreatectomy rats. *Diabetologia* **44:** 2056–2065.

Jonas J.C., Sharma A., Hasenkamp W., Ilkova H., Patane G., Laybutt R., Bonner-Weir S., and Weir G.C. 1999. Chronic hyperglycemia triggers loss of pancreatic β-cell differen- tiation in an animal model of diabetes. *J. Biol. Chem.* **274:** 14112–14121.

Kitamura T., Kahn C.R., and Accili D. 2003. Insulin receptor knockout mice. *Annu. Rev. Physiol.* **65:** 313–332.

Kitamura T., Nakae J., Kitamura Y., Kido Y., Biggs W.H., III, Wright C.V., White M.F., Arden K.C., and Accili D. 2002. The forkhead transcription factor Foxo1 links insulin signal- ing to Pdx1 regulation of pancreatic β-cell growth. *J. Clin. Invest.* **110:** 1839–1847.

Kodama H., Fukuda K., Pan J., Sano M., Takahashi T., Kato T., Makino S., Manabe T., Murata M., and Ogawa S. 2000. Significance of ERK cascade compared with JAK/STAT and PI3-K pathway in gp130-mediated cardiac hypertrophy. *Am. J. Physiol. Heart Circ. Physiol.* **279:** H1635–H1644.

Kulkarni R.N., Bruning J.C., Winnay J.N., Postic C., Magnuson M.A., and Kahn C.R. 1999. Tissue-specific knockout of the insulin receptor in pancreatic β-cells creates an insulin secretory defect similar to that in type 2 diabetes. *Cell* **96:** 329–339.

Kulkarni R.N., Holzenberger M., Shih D.Q., Ozcan U., Stoffel M., Magnuson M.A., and Kahn C.R. 2002. β-cell-specific deletion of the Igf1 receptor leads to hyperinsulinemia and glucose intolerance but does not alter beta-cell mass. *Nat. Genet.* **31:** 111–115.

Kunisada K., Tone E., Fujio Y., Matsui H., Yamauchi-Takihara K., and Kishimoto T. 1998. Activation of gp130 transduces hypertrophic signals via STAT3 in cardiac myocytes. *Circulation* **98:** 346–352.

Kushner J.A., Ye J., Schubert M., Burks D.J., Dow M.A., Flint C.L., Dutta S., Wright C.V., Montminy M.R., and White M.F. 2002. Pdx1 restores β-cell function in Irs2 knockout mice. *J. Clin. Invest.* **109:** 1193–1201.

Lekmine F., Uddin S., Sassano A., Parmar S., Brachmann S.M., Majchrzak B., Sonenberg N., Hay N., Fish E.N., and Platanias L.C. 2003. Activation of the p70 S6 kinase and phosphorylation of the 4E-BP1 repressor of mRNA translation by type I interferons. *J. Biol. Chem.* **278:** 27772–27780.

Lingohr M.K., Buettner R., and Rhodes C.J. 2002. Pancreatic beta-cell growth and survival—A role in obesity-linked type 2 diabetes? *Trends Mol. Med.* **8:** 375–384.

Liu J.P., Baker J., Perkins A.S., Robertson E.J., and Efstratiadis A. 1993. Mice carrying null mutations of the genes encoding insulin-like growth factor I (Igf-1) and type 1 IGF receptor (Igf1r). *Cell* **75:** 59–72.

McDaniel M.L., Marshall C.A., Pappan K.L., and Kwon G. 2002. Metabolic and autocrine regulation of the mammalian target of rapamycin by pancreatic beta-cells. *Diabetes* **51:** 2877–2885.

Milburn J.L., Jr., Hirose H., Lee Y.H., Nagasawa Y., Ogawa A., Ohneda M., BeltrandelRio H., Newgard C.B., Johnson J.H., and Unger R.H. 1995. Pancreatic beta-cells in obesity. Evidence for induction of functional, morphologic, and metabolic abnormalities by increased long chain fatty acids. *J. Biol. Chem.* **270:** 1295–1299.

Mittelman S.D., Van Citters G.W., Kim S.P., Davis D.A., Dea M.K., Hamilton-Wessler M., and Bergman R.N. 2000. Longitudinal compensation for fat-induced insulin resistance includes reduced insulin clearance and enhanced beta-cell response. *Diabetes* **49:** 2116–2125.

Montana E., Bonner-Weir S., and Weir G.C. 1994a. Transplanted β-cell response to increased metabolic demand. Changes in β-cell replication and mass. *J. Clin. Invest.* **93:** 1577–1582.

———. 1994b. Transplanted beta-cell replication and mass increase after 95% pancreatectomy. *Transplant Proc.* **26:** 657.

Montanya E., Nacher V., Biarnes M., and Soler J. 2000. Linear correlation between beta-cell mass and body weight throughout the lifespan in Lewis rats: Role of beta-cell hyperplasia and hypertrophy. *Diabetes* **49:** 1341–1346.

Nakae J., Biggs W.H., III, Kitamura T., Cavenee W.K., Wright C.V., Arden K.C., and Accili D. 2002. Regulation of insulin action and pancreatic beta-cell function by mutated alleles of the gene encoding forkhead transcription factor Foxo1. *Nat. Genet.* **32:** 245–253.

Neufeld T.P., de la Cruz A.F., Johnston L.A., and Edgar B.A. 1998. Coordination of growth and cell division in the *Drosophila* wing. *Cell* **93:** 1183–1193.

Parsons J.A., Bartke A., and Sorenson R.L. 1995. Number and size of islets of Langerhans in pregnant, human growth hormone-expressing transgenic, and pituitary dwarf mice: Effect of lactogenic hormones. *Endocrinology* **136:** 2013–2021.

Parsons J.A., Brelje T.C., and Sorenson R.L. 1992. Adaptation of islets of Langerhans to pregnancy: Increased islet cell proliferation and insulin secretion correlates with the onset of placental lactogen secretion. *Endocrinology* **130:** 1459–1466.

Pende M., Kozma S.C., Jaquet M., Oorschot V., Burcelin R., Le Marchand-Brustel Y., Klumperman J., Thorens B., and Thomas G. 2000. Hypoinsulinaemia, glucose intolerance and diminished beta-cell size in S6K1-deficient mice. *Nature* **408:** 994–997.

Platanias L.C., Uddin S., Yetter A., Sun X.J., and White M.F. 1996. The type I interferon receptor mediates tyrosine phosphorylation of insulin receptor substrate 2. *J. Biol. Chem.* **271:** 278–282.

Remacle C., De Clercq L., Delaere P., Many M.C., and Gommers A. 1980. Organ culture of the islets of Langerhans from young and senscent rats. *Cell Tissue Res.* **207:** 429–448.

Rommel C., Bodine S.C., Clarke B.A., Rossman R., Nunez L., Stitt T.N., Yancopoulos G.D., and Glass D.J. 2001. Mediation of IGF-1-induced skeletal myotube hypertrophy by PI(3)K/Akt/mTOR and PI(3)K/Akt/GSK3 pathways. *Nat. Cell Biol.* **3:** 1009–1013.

Scaglia L., Smith F.E., and Bonner-Weir S. 1995. Apoptosis contributes to the involution of β-cell mass in the post partum rat pancreas. *Endocrinology* **136:** 5461–5468.

Schuppin G.T., Bonner-Weir S., Montana E., Kaiser N., and Weir G.C. 1993. Replication of adult pancreatic-beta cells cultured on bovine corneal endothelial cell extracellular matrix. *In Vitro Cell. Dev. Biol. Anim.* **29A:** 339–344.

Shima H., Pende M., Chen Y., Fumagalli S., Thomas G., and Kozma S.C. 1998. Disruption of the p70(s6k)/p85(s6k) gene reveals a small mouse phenotype and a new functional S6 kinase. *EMBO J.* **17:** 6649–6659.

Shioi T., Kang P.M., Douglas P.S., Hampe J., Yballe C.M., Lawitts J., Cantley L.C., and Izumo S. 2000. The conserved phosphoinositide 3-kinase pathway determines heart size in mice. *EMBO J.* **19:** 2537–2548.

Smith F.E., Rosen K.M., Villa-Komaroff L., Weir G.C., and Bonner-Weir S. 1991. Enhanced insulin-like growth factor I gene expression in regenerating rat pancreas. *Proc. Natl. Acad. Sci.* **88:** 6152–6156.

Sorenson R.L. and Brelje T.C. 1997. Adaptation of islets of Langerhans to pregnancy: beta-cell growth, enhanced insulin secretion and the role of lactogenic hormones. *Horm. Metab. Res.* **29:** 301–307.

Sorenson R.L., Brelje T.C., and Roth C. 1993. Effects of steroid and lactogenic hormones on islets of Langerhans: A new hypothesis for the role of pregnancy steroids in the adaptation of islets to pregnancy. *Endocrinology* **133:** 2227–2234.

Stocker H. and Hafen E. 2000. Genetic control of cell size. *Curr. Opin. Genet. Dev.* **10:** 529–535.

Swenne I. 1983. Effects of aging on the regenerative capacity of the pancreatic B-cell of the rat. *Diabetes* **32:** 14–19.

Tamemoto H., Kadowaki T., Tobe K., Yagi T., Sakura H., Hayakawa T., Terauchi Y., Ueki K., Kaburagi Y., and Satoh S., et al. 1994. Insulin resistance and growth retardation in mice lacking insulin receptor substrate-1. *Nature* **372:** 182–186.

Tuttle R.L., Gill N.S., Pugh W., Lee J.P., Koeberlein B., Furth E.E., Polonsky K.S., Naji A., and Birnbaum M.J. 2001. Regulation of pancreatic beta-cell growth and survival by the serine/threonine protein kinase Akt1/PKBalpha. *Nat. Med.* **7:** 1133–1137.

Vasavada R.C., Garcia-Ocana A., Zawalich W.S., Sorenson R.L., Dann P., Syed M., Ogren L., Talamantes F., and Stewart A.F. 2000. Targeted expression of placental lactogen in the β-cells of transgenic mice results in β-cell proliferation, islet mass augmentation, and hypoglycemia. *J. Biol. Chem.* **275:** 15399–15406.

Verdu J., Buratovich M.A., Wilder E.L., and Birnbaum M.J. 1999. Cell-autonomous regulation of cell and organ growth in *Drosophila* by Akt/PKB. *Nat. Cell Biol.* **1:** 500–506.

Vyas D.R., Spangenburg E.E., Abraha T.W., Childs T.E., and Booth F.W. 2002. GSK-3beta negatively regulates skeletal myotube hypertrophy. *Am. J. Physiol. Cell Physiol.* **283:** C545–C551.

Wang J., Zhou J., and Bondy C.A. 1999. Igf1 promotes longitudinal bone growth by insulin-like actions augmenting chondrocyte hypertrophy. *FASEB J.* **13:** 1985–1990.

White M.F. 2002. IRS proteins and the common path to diabetes. *Am. J. Physiol. Endocrinol. Metab.* **283:** E413–E422.

Withers D.J., Gutierrez J.S., Towery H., Burks D.J., Ren J.M., Previs S., Zhang Y., Bernal D., Pons S., Shulman G.I., Bonner-Weir S., and White M.F. 1998. Disruption of IRS-2 causes type 2 diabetes in mice. *Nature* **391:** 900–904.

Xuan S., Kitamura T., Nakae J., Politi K., Kido Y., Fisher P.E., Morroni M., Cinti S., White M.F., Herrera P.L., Accili D., and Efstratiadis A. 2002. Defective insulin secretion in pancreatic β-cells lacking type 1 IGF receptor. *J. Clin. Invest.* **110:** 1011–1019.

Yin T., Tsang M.L., and Yang Y.C. 1994. JAK1 kinase forms complexes with interleukin-4 receptor and 4PS/insulin receptor substrate-1-like protein and is activated by interleukin-4 and interleukin-9 in T lymphocytes. *J. Biol. Chem.* **269:** 26614–26617.

20

Plant Cell Growth

Benoît Menand
Department of Cell and Developmental Biology
John Innes Centre, Norwich, NR47UH, United Kingdom

Christophe Robaglia
CEA Cadarache DSV DEVM
Laboratoire de Génétique et Biophysique des Plantes
UMR 163 CNRS CEA
Université Méditerranée UMR 163
F-13108 Saint-Paul-lez-Durance, France

T HE PRODUCTION OF PLANT CELLS is confined mainly to meristems, small specialized growth zones found in shoots, roots, and lateral areas of the plant. Meristems are highly organized structures where rapidly cycling cells are produced from a small central mass of cells that proliferate very slowly if at all, analogous to stem cells in animals. New progeny cells are displaced toward the periphery of a meristem where they progressively differentiate and organize into organs. The meristems are therefore dynamic structures continuously traversed by a flux of cells that divide and grow at different rates depending on their position in the meristem. The shoot and root meristems are defined during embryogenesis. Lateral meristems, mediating radial growth, and meristems of secondary roots are formed later during the development of the young plant. New meristems can also form spontaneously in culture from an undifferentiated mass of proliferating cells, thereby allowing the regeneration of a plant. Meristematic cells (Fig. 1) are small (~5 μm), densely packed with cytoplasm, and, in defined regions of the meristem, grow to a specific size before dividing. Growing and dividing meristematic cells display a high level of metabolic activity and macromolecular synthesis as required to generate new cell mass (Woodard et al. 1961). The nutrients needed to sustain the high growth rate of meristematic cells (sink tissue) are provided by the rest of the plant body.

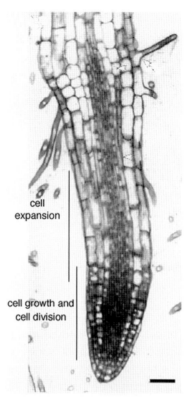

cell
expansion

cell growth and
cell division

Figure 1. Median longitudinal toluidine blue-stained section of an *Arabidopsis* root 2 days after germination. Bar, 50 μm. (Reprinted, with permission, from Scheres et al. 1994 [copyright The Company of Biologists].)

"SOLID" VERSUS "LIQUID" GROWTH

During the development of organs, cells lose their capacity to divide and can enlarge up to one thousand times (for review, see Cosgrove 2000). This process, called cell expansion, is caused by water accumulation in the vacuole and a concomitant increase in the surface area of the cell wall, and is thus mechanistically different from the growth that cells undergo during proliferation in meristems. These two types of growth can be tentatively referred to as "solid" growth, where cells increase in mass following the accumulation of macromolecules in the cytoplasm (and other compartments), and "liquid" growth, where cells enlarge by taking up water into the vacuole during expansion. However, the two processes may not be totally independent, since components of the solid growth process remain active during expansion. Cells continue to divide for a short while as they start to enlarge by expansion during the formation of an organ. Furthermore, although cell wall expansion is particularly spectacular during liquid growth, it also has to occur during solid growth. Meristematic cells are organized in defined rows and layers, and their growth is thus constrained in a two-dimensional space, suggesting that cell wall and cytoplasmic growth are coordinated, possibly via the cytoskeleton (Scheres et al. 1994; Vernoux et al. 2000). Expansion is also often accompanied by several cycles of endoreduplication of DNA, which is a simplified version of a mitotic cycle in which certain components of the cell division machinery remain active while others are progressively disconnected. Regulators of the transition to growth by expansion, such as the *ccs52* gene product from *Medicago sativa* and DNA topoisomerase VI, inhibit the mitotic cycle and promote endoreduplication (Cebolla et al. 1999; Sugimoto-Shirasu et al. 2002). Conversely, the overexpression of the *Arabidopsis* Cyclin D3;1, which is associated with cell proliferation, inhibits endoreduplication and promotes ectopic cell division (Dewitte et al. 2003).

Although solid cell growth is essential for plant growth, since it provides the cells that will be incorporated into organs, it represents a small contribution to the overall size of the plant. In contrast, liquid cell growth accounts for the bulk of the plant. Expansion also appears to be an economic means to increase in size to explore the environment for essential nutrients (Cosgrove 1997). In this chapter, we focus on the factors affecting solid cell growth during proliferation. Plant cell expansion has been the subject of many excellent reviews in recent years (Nicol and Hofte 1998; Cosgrove 2000).

THE LINK BETWEEN CELL GROWTH AND CELL DIVISION

The size reached by cells at the time of division and the rate of mitotic cell growth are poorly studied in plants. Thus, most of the current understanding of plant cell growth is derived indirectly from the analysis of related processes, such as cell division, the synthesis of macromolecules, and, more recently, the expression of genes affecting cell growth in other organisms. With the exceptions of postmeiotic gametophytic divisions and angiosperm embryogenesis, higher plants have no tissues where repeated cell divisions occur without cell growth (Mansfield and Briarty 1991). This illustrates that cell growth is an important aspect of cell division, but the two processes can be viewed separately. Many lower plants, like the alga families *Caulerpa* and *Vaucheria*, are noncellular and undergo growth and morphogenesis without the formation of new cells per se. This led to the concept dating back to 1878 that cell division does not drive plant growth and morphogenesis but is subordinate to a supracellular growth control that determines the size of organs or the overall size of the plant (Sachs 1878; Kaplan 1992). Even if cell division, i.e., the formation of new cell septa, does not occur in these species, the nuclei proliferate together with the associated cytoplasm, and therefore what was defined above as solid growth clearly occurs (Brown 1976). In a similar manner, after fecundation in angiosperms, a triploid tissue, the endosperm, is formed. This nutritive tissue, which fills a large part of the seed and is consumed during embryogenesis, is a syncytial structure where the number of nuclei increases rapidly without the production of new cells (Sorensen et al. 2002). The idea that a supracellular growth control exists in plants has received some experimental support. Plant cells in which cell division is inhibited by irradiation, expression of cell-cycle inhibitors, or deletion of *AINTEGUMENTA,* a transcription factor that maintains meristematic competence, can continue to increase in size (Haber 1962; Hemerley et al. 1995; Mizukami and Fischer 2000; De Veylder et al. 2001). It is not clear what the relative contributions of cytoplasmic growth and expansion are to this increase in size. Also, accelerated cell division due to the heterologous expression of the fission yeast cdc25 protein phosphatase in tobacco leads to a reduced size at division of the cells of lateral root meristems, but does not lead to an alteration in overall plant growth (McKibbin et al. 1998). Finally, recessive or semidominant mutations in *Arabidopsis* ribosomal protein genes lead to phenotypes equivalent to the *Drosophila* Minute syndrome, and inactivation of the *Arabidopsis* TOR kinase leads to an arrest of embryonic growth and division (Weijers et al. 2001; Menand et al. 2002). The effect of TOR inactivation in *Arabidopsis* is similar to the effect of inactivation of the TOR/S6K growth-controlling pathway in *Drosophila*, where a smaller body or developmental arrest results from reduced cell size (Montagne et al. 1999; Zhang et al. 2000). Altogether, these observations suggest that, although plants can compensate for an alteration of the cell division machinery by adjusting the number and size of their cells, they have an absolute requirement for the protein translation apparatus to sustain cell growth, as in other eukaryotes (Schmelzle and Hall 2000; Gingras et al. 2001). Interestingly, in both the *Drosophila* and *Arabidopsis* cases, cell growth arrest is not immediately lethal, since growth-arrested *Drosophila* larvae and *Arabidopsis tor* embryos can survive for some time while the cell division machinery is still active (Galloni and Edgar 1999; Menand et al 2002).

WHAT ARE THE GROWTH PHASES DURING
THE PLANT CELL CYCLE?

The accumulation of protein, RNA, and DNA during interphase of *Vicia faba* (broad bean) root meristems cells has been studied using both autoradiographic and microphotometric methods, after incubation of meristems with radioactive precursors (Woodard et al. 1961). These authors observed three periods that certainly correspond, at least in part, to G_1, S, and G_2, respectively: (1) a period of strong accumulation of RNA and proteins during telophase to post-telophase, (2) a period of histone and DNA accumulation during post-telophase to preprophase, and (3) another period of strong accumulation of RNA and proteins during pre-prophase to prophase. These findings indicate that protein and RNA synthesis occurs during both G_1 and G_2 in plant cells. The question of the regulation of different phases of the cell cycle by protein synthesis in plants has also been addressed, in onion root meristems. Inhibition of protein synthesis by anisomycin leads to a strong increase in the length of G_1, but also to a significant extension of G_2, confirming the need of protein synthesis for these two phases, most importantly for G_1 (Cuadrado et al. 1985). Complementary studies show that, in the presence of anisomycin, cell size as measured by protein content and surface area projection is less diminished at the end of S than at division, suggesting that the cell size requirement is more critical in G_1 than in G_2 (Cuadrado et al. 1986). On the basis of these findings, Cuadrado et al. (1986) proposed the hypothesis that the cell growth control during G_1 is related to the attainment of a critical size for DNA synthesis. Furthermore, they propose that there is no critical size requirement for division, and the length of G_2 negatively correlates with the size of the cell at termination of DNA synthesis, G_2 being shorter in a big cell than in a small cell. A negative correlation between G_2 time and cell size has been confirmed during physiological growth (without drug) where cell size shows little variation at termination of DNA synthesis and at mitosis initiation, whereas it is three times more variable during G_2 (Navarette et al. 1987).

GENETIC ANALYSIS OF PLANT CELL GROWTH

Proteins involved in cell growth control in animals, such as the major members of the TOR pathway, have homologs in *Arabidopsis* (Table 1). The only exception is 4E-BP, for which homologs are lacking in *Caenorhabditis elegans*, *Saccharomyces cerevisiae*, and *Arabidopsis*. These organisms may contain nonhomologous proteins performing the 4E-BP function (Cosentino et al. 2000; Freire et al. 2000; Long et al. 2002). In

Table 1. Compilation of data about plant (*Arabidopsis* if not specified) homologs of cell growth regulators in animals and yeast

Homolog of	Name(s)	Expression data	Activity and phenotype of mutants	Reference
mTOR TOR1 TOR2	AtTOR	embryo, endosperm, primary meristems, and primordia	*tor* null mutations lead to premature arrest of endosperm and embryo development	Menand et al. (2002)
P70 S6 kinase	AtS6k1 and AtS6k2		AtS6k2 phosphorylates specifically mammalian and plant S6 but is resistant to rapamycin in mammalian cells	Turck et al. (1998)
eIF4E	5 homologs		eIF(iso)4E knockout has no obvious phenotype	Duprat et al. (2002) A.G.I. (2000)
4E-BPs	no obvious homologs			
D cyclins	D cyclins	expression induced by sucrose and cytokinins	overexpression of cyclin D3 allows callus proliferation in the absence of cytokinins; overexpression of cyclin D2 in tobacco increases the plant growth rate by shortening G_1 without affecting cell size of meristematic and postmitotic cell	Riou-Khamlichi et al. (1999, 2000) Cockcroft et al. (2000)

p27^{Kip1}	7 homologs (AtKRP1 to AtKRP7)	overexpression of AtKRP2 leads to plants with fewer but larger cells than wild type (no examination of vacuole/cytosol ratio)	De Veylder et al. (2001)
pRB	MAT3 (*Chlamydomonas*)	small size phenotype due to initiation of division at a below-normal size and extra rounds of divisions	Umen and Goodenough (2001), de Jager and Murray (1999)
	RB pathway proteins in *Arabidopsis*		
α4 TAP42	TAP46	associated with a type 2A protein phosphatase in vivo	Harris et al. (1999)
PTEN	3 homologs AtPTEN1 to AtPTEN3 *AtPTEN1* is expressed exclusively in pollen grains	RNA interference of *AtPTEN1* causes pollen cell death after mitosis	Gupta et al. (2002)
PDK1	AtPDK1	able to bind human PKB in vitro	Deak et al. (1999)

(Arabidopsis Genome Initiative [A.G.I.] 2000).

addition, homologs of animal cell growth controllers might not have equivalent functions in plants. For example, AtPTEN1, an *Arabidopsis* homolog of the animal tumor suppressor PTEN, is involved in pollen development (Table 1) (Gupta et al. 2002).

The genetic analysis of TOR in a land plant has been undertaken in *Arabidopsis thaliana*. The *TOR* gene in this species (*AtTOR*) is unique, and its inactivation leads to the arrest of embryonic growth at a stage where cells begin to grow in wild-type plants. *AtTOR* is expressed in the plant's proliferation zones where cell growth occurs (Menand et al. 2002). However, the study of the plant TOR pathway is somewhat hampered by the natural resistance of plants to rapamycin, which specifically inactivates TOR in yeast and animals and has been an important pharmacological agent to genetically and biochemically dissect this pathway. Plants may be naturally resistant to rapamycin because this drug is a secondary metabolite produced by a soil bacterium, *Streptomyces hygroscopicus*. *C. elegans*, which also inhabits the soil, is also naturally resistant to rapamycin.

Although most cells grow 2-fold before dividing once, the unicellular green alga *Chlamydomonas reinhardtii* grows up to 32-fold during daylight, then rapidly divides several times to return to its original size. Cells containing a disruption of the *mat3* gene, encoding a Retinoblastoma (Rb) protein homolog, initiate division at a size 40% smaller than normal and do two or three extra rounds of division, leading to a small-cell-size phenotype (Umen and Goodenough 2001). Therefore, the algal Rb homolog is involved in two different cell-size-related decisions. The first is as a cell-size-dependent repressor of cell cycle progression linking commitment size to progression into the S phase, and the second where it limits the number of mitotic divisions. *mat3* mutant cells have a normal growth rate and do not show a shortened G_1 phase (in fact a little bit longer), probably because *mat3* daughter cells, which are born 4-fold smaller than wild-type cells, need to grow for a longer time to attain the commitment size, although their commitment size is smaller. Thus, the function of *C. reinhardtii* Rb is different from that of mammalian Rb, whose deletion leads to G_1 shortening and precocious entry into S phase (Cross and Roberts 2001). Genetic analysis of the land plant Rb homologs has not yet been reported, but the phylogenetic relatedness of land plants to green algae suggests that the function of their Rb genes in cell growth regulation may be different from that of animals.

HORMONES AND CELL GROWTH

Among the different phytohormones, auxins and cytokinins historically have been linked to the capacity of plant cells to proliferate. Exogenous application of these hormones is often sufficient to reactivate cell division of differentiated cells and tissues "in vitro" (Kende and Zeevaart 1997). However, auxins and cytokinins may have a role primarily in cell growth. The effect of hormones on ribosome biosynthesis, which is a determinant of cytoplasmic cell growth, has been tested in *Arabidopsis* plantlets by measuring the transcription of ribosomal RNAs (Gaudino and Pikaard 1997). Of five hormones tested, including gibberellic acid, absissic acid, ethylene, 2,4D (an auxin), and kinetin (a cytokinin), only cytokinin elicited a response. Cytokinin stimulated Pol I-dependent rRNA transcription within 1 hour of treatment. This stimulation is not a consequence of cell division since DNA content was not increased within 24 hours of treatment. This study is in agreement with previous work describing rapid stimulation of nuclear RNA Pol I activity by cytokinin in pumpkin cotyledons (Ananiev et al. 1987). Furthermore, cytokinins increase the expression of cyclin D3, overexpression of which is sufficient to induce calli (a group of undifferentiated cells undergoing cell growth and proliferation) formation in the absence of cytokinin (Riou-Khamlichi et al. 1999). Together, these results suggest that cytokinins stimulate cell growth, which leads to cyclin D3 activation and cell proliferation. AtTOR could be a mediator of this pathway, since RNA Pol I and D cyclins are under the control of TOR in both yeast and mammalian cells.

Plant cells are surrounded by a rigid cell wall composed of polymers of cellulose and lignin. Mechanisms must exist to remodel the cell wall during cell growth. Treatment of a narrow region of the meristem with extensin, a protein known to promote the loosening of the cell wall, causes the emergence of a new leaf at the site of application (Pien et al. 2001). Comparable results can be obtained when applying the phytohormone auxin, which is known to stimulate both cell expansion and cell division (Reinhardt et al. 2000). Down-regulation of the auxin-binding protein ABP1 alters the auxin-stimulated expansion of cultured dividing cells without modifying the cell size at division (Chen et al. 2001). This, together with the observation that *abp1* early embryonic cells fail to grow, suggests that one of the functions of auxin might be to mediate the remodeling and extension of the cell wall, whatever the mode of growth.

A simple model reconciling the seemingly double requirements for auxin and cytokinin in the separate processes of division and growth can be proposed. Cytokinin might stimulate division as a consequence of activating cell growth (protein and nucleic acid synthesis), whereas auxin mediates cell wall extension and is thus necessary to accommodate the cytokinin-stimulated increase in cell mass. A general role for auxin in cell wall extension would explain its presence during both cell expansion and cell growth coupled to cell division.

ROLE OF NUTRIENTS IN PLANT CELL GROWTH

In plants, as in other organisms, starvation affects anabolic processes, which are essential for cell growth and proliferation. Plants produce their own amino acids and are thus not subject to direct amino acid starvation, but they are dependent on the availability of nitrogen and other mineral elements. Furthermore, although plants are autotrophs for carbon, some of their organs and non-photosynthetic cells (notably roots and meristems) can be considered as heterotrophs (sink tissues). Studies carried out on root tips and cell cultures, and confirmed in maize plantlets deprived of light, have shown that plant cells respond to sugar starvation by activating the use of alternative carbon sources (lipid, starch, and proteins) and by repressing cellular functions that consume sugars, including protein synthesis and cell cycle progression (for review, see Brouquisse et al. 1998; Yu 1999). Cells deprived of a carbon source also display an increase in the size of the vacuole and an induction of autophagic activity (Aubert et al. 1996). Thus, plant cells respond to starvation by down-regulating growth in a manner similar to that observed for other organisms. The pathways that mediate these responses in plants are not known.

CONCLUSIONS

The study of the control of plant cell cytoplasmic (solid) growth is still in its infancy. The comparative analysis of growth functions already identified in model organisms such as yeast and *Drosophila* will now certainly progress rapidly in plants, given the recent sequencing of the *Arabidopsis thaliana* genome. An important area of future research will be the understanding of the integration of cell growth with cell differentiation and morphogenesis at the level of the organism. In this case, comparative genomics might be of limited use since plants and animals acquired multicellularity independently (Meyerowitz 2002). Plant cell behavior is essentially dependent on positional information rather than on cell lineage (van der Berg et al. 1995; Berger et al. 1998). Thus, the study of cell growth in plants might reveal new paradigms.

ACKNOWLEDGMENTS

We thank Mike Hall for suggesting the terms "solid" and "liquid" in the definition of plant cell growth modes. B.M. is supported by EMBO long-term fellowship ALTF 89-2002.

REFERENCES

Ananiev E.D., Karagyozov L.K., and Karanov E.N. 1987. Effect of cytokinins on ribosomal RNA gene expression in excised cotyledons of *Cucurbita pepo* L. *Planta* **170:** 370–378.

Arabidopsis Genome Initiative (A.G.I.). 2000. Analysis of the genome sequence of the flowering plant *Arabidopsis thaliana*. *Nature* **408:** 796–815.

Aubert S., Gout E., Bligny R., Marty-Mazars D., Barrieu F., Alabouvette J., Marty F., and Douce R. 1996. Ultrastructure and biochemical characterization of autophagy in higher plant cells subjected to carbon deprivation: Control by the supply of mitonchondria with respiratory substrates. *J. Cell Biol.* **133:** 1251–1263.

Berger F., Haseloff J., Schiefelbein J., and Dolan L. 1998. Positional information in root epidermis is defined during embryogenesis and acts in domains with strict boundaries. *Curr. Biol.* **8:** 421–430.

Brouquisse R., Gaudillère J.-P., and Raymond P. 1998. Induction of a carbon-starvation-related proteolysis in whole maize plants submitted to light/dark cycles and to extended darkness. *Plant Physiol.* **117:** 1281–1291.

Brown R. 1976. Significance of division in the higher plants. In *Cell division in higher plants* (ed. M.M. Yeoman), pp. 3–46. Academic Press, New York.

Cebolla A., Vinardell J.M., Kiss E., Olah B., Roudier F., Kondorosi A., and Kondorosi E. 1999. The mitotic inhibitor *ccs52* is required for endoreduplication and ploidy-dependent cell enlargement in plants. *EMBO J.* **18:** 4476–4484.

Chen J.G., Ullah H., Young J.C., Sussman M.R., and Jones A.M. 2001. ABP1 is required for organized cell elongation and division in *Arabidopsis* embryogenesis. *Genes Dev.* **15:** 902–911.

Cockcroft C.E., den Boer B.G.W., Healy J.M.S., and Murray J.A.H. 2000. Cyclin D control of growth rate in plants. *Nature* **405:** 575–579.

Cosentino G.P., Schmelzle T., Haghighat A., Helliwell S.B., Hall M.N., and Sonenberg N. 2000. Eap1p, a novel eukaryotic translation initiation factor 4E-associated protein in *Saccharomyces cerevisiae*. *Mol. Cell. Biol.* **20:** 4604–4613.

Cosgrove D.J. 1997. Relaxation on a high-stress environment: The molecular bases of extensible cell walls and cell enlargement. *Plant Cell* **9:** 1031–1041.

———. 2000. Loosening of plant cell walls by expansins. *Nature* **407:** 321–326.

Cross F.R. and Roberts J.M. 2001. Retinoblastoma protein: Combating algal bloom. *Curr. Biol.* **11:** 824–827.

Cuadrado A., Navarrete M.H., and Canovas J.L. 1985. The effect of partial protein synthesis inhibition on cell proliferation in higher plants. *J. Cell Sci.* **76:** 97–104.

———. 1986. Regulation of G_1 and G_2 by cell size in higher plants. *Cell Biol. Int. Rep.* **10:** 223–230.

Deak M., Casamayor A., Currie R.A., Downes C.P., and Alessi D.R. 1999. Characterisation of a plant 3-phosphoinositide-dependent protein kinase-1 homologue which contains a pleckstrin homology domain. *FEBS Lett.* **451:** 220–226.

de Jager S.M. and Murray J.A. 1999. Retinoblastoma proteins in plants. *Plant Mol. Biol.* **41:** 295–299.

De Veylder L., Beeckman T., Beemster G.T., Krols L., Terras-Landrieu I., Van Der Schueren E., Maes S., Naudts M., and Inzé D. 2001. Functional analysis of cyclin-dependent kinase inhibitors of *Arabidopsis*. *Plant Cell* **13:** 1653–1668.

Dewitte W., Riou-Khamlichi C., Scofield S., Healy J.M., Jacqmard A., Kilby N.J., and Murray J.A. 2003. Altered cell cycle distribution, hyperplasia, and inhibited differentiation in *Arabidopsis* caused by the D-type cyclin CYCD3. *Plant Cell* **15**: 79–92.

Duprat A., Caranta C., Revers F., Menand B., Browning K.S., and Robaglia C. 2002. The *Arabidopsis* eucaryotic initiation factor (iso)4E is dispensable for plant growth but required for susceptibility to potyviruses. *Plant J.* **32**: 927–934.

Freire M.A., Tourneur C., Granier F., Camonis J., El Amrani A., Browning K.S., and Robaglia C. 2000. Plant lipoxygenase 2 is a translation initiation factor-4E-binding protein. *Plant Mol. Biol.* **44**: 129–140.

Galloni M. and Edgar B.A. 1999. Cell-autonomous and non-autonomous growth-defective mutants of *Drosophila melanogaster*. *Development* **126**: 2365–2375.

Gaudino R.J. and Pikaard C.S. 1997. Cytokinin induction of RNA polymerase I transcription in *Arabidopsis thaliana*. *J. Biol. Chem.* **272**: 6799–6804.

Gingras A.C., Raught B., and Sonenberg N. 2001. Regulation of translation initiation by FRAP/mTOR. *Genes Dev.* **15**: 807–826.

Gupta R., Ting J.T.L., Sokolov L.N.S., Johnson S.A., and Luan S. 2002. A tumor suppressor homolog, AtPTEN1, is essential for pollen development in *Arabidopsis*. *Plant Cell* **14**: 2495–2507.

Haber A.H. 1962. Non essentiality of concurrent cell divisions for degree of polarization of leaf growth. I. Studies with radiation-induced mitotic inhibition. *Am. J. Bot.* **49**: 583–589.

Harris D.M., Myrick T.L., and Rundle S.J. 1999. The *Arabidopsis* homolog of yeast TAP42 and mammalian α4 binds to the catalytic subunit of protein phosphatase 2A and is induced by chilling. *Plant Physiol.* **121**: 606–617.

Hemerly A., de Almeida Engler J., Bergounioux C., Van Montagu M., Engler G., Inzé D., and Ferreira P. 1995. Dominant negative mutants of the cdc2 kinase uncouple cell division from iterative plant development. *EMBO J.* **14**: 3925–3936.

Kaplan D.R. 1992. The relationship of cells to organisms in plants: Problems and implications of an organismal perspective. *Int. J. Plant Sci.* **153**: 528–537.

Kende H. and Zeevaart J.A.D. 1997. The five "classical" plant hormones. *Plant Cell* **9**: 1197–1210.

Long X., Spycher C., Han Z., Rose A., Muller F., and Avruch J. 2002. TOR deficiency in *C. elegans* causes developmental arrest and intestinal atrophy by inhibition of mRNA translation. *Curr. Biol.* **12**: 1448–1461.

Mansfield S.G. and Briarty L.G. 1991. Early embryogenesis in *Arabidopsis*. II. The developing embryo. *Can. J. Bot.* **69**: 461–476.

McKibbin R.S., Halford N.G., and Francis D. 1998. Expression of fission yeast *cdc25* alters the frequency of lateral root formation in transgenic tobacco. *Plant Mol. Biol.* **36**: 601–612.

Menand B., Desnos T., Nussaume L., Berger F., Bouchez D., Meyer C., and Robaglia C. 2002. Expression and disruption of the *Arabidopsis TOR* (target of rapamycin) gene. *Proc. Natl. Acad. Sci.* **99**: 6422–6427.

Meyerowitz E.M. 2002. Plants compared to animals: The broadest comparative study of development. *Science* **295**: 1482–1485.

Mizukami Y. and Fischer R.L. 2000. Plant organ size control: AINTEGUMENTA regulates growth and cell number during organogenesis. *Proc. Natl. Acad. Sci.* **97**: 942–947.

Montagne J., Stewart M.J., Stocker H., Hafen E., Kozma S.C., and Thomas G. 1999. *Drosophila* S6 kinase: A regulator of cell size. *Science* **285**: 2126–2129.

Navarette M.H., Cuadrado A., Escalera M., and Canovas J. 1987. Regulation of G2 by cell size contributes to maintaining cell size variability within certain limits in higher plants. *J. Cell Sci.* **87:** 635–641.

Nicol F. and Hofte H. 1998. Plant cell expansion: Scaling the wall. *Curr. Opin. Plant Biol.* **1:** 12–17.

Pien S., Wyrzykowska J., McQueen-Mason S., Smart C., and Fleming A. 2001. Local expression of expansin induces the entire process of leaf development and modifies leaf shape. *Proc. Natl. Acad. Sci.* **98:** 11812–11817.

Reinhardt D., Mandel T., and Kuhlemeier C. 2000. Auxin regulates the initiation and radial position of plant lateral organs. *Plant Cell* **12:** 507–518.

Riou-Khamlichi C., Huntley R., Jacqmard A., and Murray J.A.H. 1999. Cytokinin activation of *Arabidopsis* cell division trough a D-type cyclin. *Science* **283:** 1541–1544.

Riou-Khamlichi C., Menges M., Healy J., and Murray J.A.H. 2000. Sugar control of the plant cell cycle: Differential regulation of *Arabidopsis* D-type cyclin gene expression. *Mol. Cell. Biol.* **20:** 4513–4521.

Sachs J. 1878. Über die anordnung der zellen in jüngsten pflanzenteilen. *Arb. Bot. Inst. Wurzburg* **2:** 46–104.

Scheres B., Wolkenfelt H., Willemsen V., Terlouw M., Lawson E., Dean C., and Weisbeek P. 1994. Embryonic origin of the *Arabidopsis* primary root and root meristem initials. *Development* **120:** 2475–2487.

Schmelzle T. and Hall M.N. 2000. TOR, a central controller of cell growth. *Cell* **103:** 253–262.

Sorensen M.B., Mayer U., Lukowitz W., Robert H., Chambrier P., Jurgens G., Somerville C., Lepiniec L., and Berger F. 2002. Cellularisation in the endosperm of *Arabidopsis thaliana* is coupled to mitosis and shares multiple components with cytokinesis. *Development* **129:** 5567–5576.

Sugimoto-Shirasu K., Stacey N.J., Corsar J., Roberts K., and McCann M.C. 2002. DNA topoisomerase VI is essential for endoreduplication in *Arabidopsis. Curr. Biol.* **12:** 1782–1786.

Turck F., Kozma S.C., Thomas G., and Nagy F. 1998. A heat-sensitive *Arabidopsis thaliana* kinase substitutes for human p70s6k function in vivo. *Mol. Cell. Biol.* **18:** 2038–2044.

Umen J.G. and Goodenough U.W. 2001. Control of cell division by a retinoblastoma protein homolog in *Chlamydomonas. Genes Dev.* **15:** 1652–1661.

van den Berg C., Willemsen V., Hage W., Weisbeek P., and Scheres B. 1995. Cell fate in the *Arabidopsis* root meristem determined by directional signalling. *Nature* **378:** 62–65.

Vernoux T., Autran D., and Traas J. 2000. Developmental control of cell division patterns in the shoot apex. *Plant Mol. Biol.* **43:** 569–581.

Weijers D., Franke-van Dijk M., Vencken R.J., Quint A., Hooykaas P., and Offringa R. 2001. An *Arabidopsis* minute-like phenotype caused by a semi-dominant mutation in a ribosomal protein S5 gene. *Development* **128:** 4289–4299.

Woodard J., Rasch E., and Swift H. 1961. Nucleic acid and protein metabolism during the mitotic cycle in *Vicia faba. J. Biophys. Biochem. Cytol.* **9:** 445–462.

Yu S.-M. 1999. Cellular and genetic responses of plants to sugar starvation. *Plant Physiol.* **121:** 687–693.

Zhang H., Stallock J.P., Ng J.C., Reinhard C., and Neufeld T.P. 2000. Regulation of cellular growth by the *Drosophila* target of rapamycin dTOR. *Genes Dev.* **14:** 2712–2724.

Subject Index

A

Acanthamoeba castellanii, RNA polymerase transcription, 383–387
Actin filaments, cardiac, 550
Activation-loop kinase, 273–275
Adenosine deaminase-related growth factors (ADGF), 40
Adhesion molecules. *See* Synaptic growth, and memory storage.
Adipocytes, TOR regulation of, 198
Adrenergic receptors
    in cardiac hypertrophy, 566–567
    in skeletal muscle hypertrophy, 538
Aging, cardiac, 552
Akt (protein kinase B) signaling
    in cardiac hypertrophy, 573
    cross-talk with TOR, 180–182
    insulin binding in *Drosophila,* 172–174, 204
    lymphocyte growth, 521
    in nematodes, 173
    and pancreatic cell size, 616–617
    in skeletal muscle hypertrophy and atrophy, 531–535
    in translation initiation, 312–313
Alzheimer's disease, ubiquitin–proteasome system, 498
Amino acid permeases, TOR effects, 146–148
Amino acids
    in local protein synthesis, 481–482
    in TOR signaling, 206–208
Ammonia permase, 147
Angelman's syndrome, 492
Aphidocolin, in Schwann cells, 88, 91–93

Aplysia californica synaptic growth
    local protein synthesis, 483–484, 491–492
    memory storage, 433–436, 444–445, 463
Apoptosis
    cardiac, 580–581
    in lymphocytes, 512
Arabidpopsis. See Plant cell growth.
Ascaris lumbricoides.
    See also Nematodes.
    growth patterns, 7, 9–10
ATP
    in cardiac hypertrophy, 567–568
    in lymphocyte growth, 522–524
Atrial natriuretic peptide, 557, 575
Autophagy, 413–429
    genetic factors, 417–418
    in higher eukaryotes, 425–426
    phenotypes of yeast mutants, 418–419
    physiological role, 425
    proteins
        Apg1 protein kinase complex, 421–422
        Apg8 lipidation system, 421
        Apg12 conjugation system, 420–421
        functions of, 423–424
        PI 3-kinase complex, 422–423
    regulators, 424–425
    TOR regulation of, 150–151, 198
    vacuolar protein degradation, 416
Auxins, 632–633
Axons, synaptic protein synthesis, 485, 491–498
    in *Aplysia,* 491–492
    glutamate receptors, 496–498